ADVANCED ENGINEERING AND TECHNOLOGY

PROCEEDINGS OF THE 2014 ANNUAL CONGRESS ON ADVANCED ENGINEERING AND TECHNOLOGY (CAET 2014), HONG KONG, 19–20 APRIL 2014

Advanced Engineering and Technology

Editors

Liquan Xie
Department of Hydraulic Engineering, Tongji University, Shanghai, China

Dianjian Huang
National Research Center of Work Safety Administration, Beijing, China

CRC Press is an imprint of the
Taylor & Francis Group, an **informa** business

A BALKEMA BOOK

CRC Press/Balkema is an imprint of the Taylor & Francis Group, an informa business

© 2014 Taylor & Francis Group, London, UK

Typeset by MPS Limited, Chennai, India

All rights reserved. No part of this publication or the information contained herein may be reproduced, stored in a retrieval system, or transmitted in any form or by any means, electronic, mechanical, by photocopying, recording or otherwise, without written prior permission from the publisher.

Although all care is taken to ensure integrity and the quality of this publication and the information herein, no responsibility is assumed by the publishers nor the author for any damage to the property or persons as a result of operation or use of this publication and/or the information contained herein.

Published by: CRC Press/Balkema
P.O. Box 11320, 2301 EH Leiden, The Netherlands
e-mail: Pub.NL@taylorandfrancis.com
www.crcpress.com – www.taylorandfrancis.com

ISBN: 978-1-138-02636-0 (Hardback)
ISBN: 978-1-315-76474-0 (eBook PDF)

Table of contents

Preface	XIII

Integration of hot-water cooling and evaporative cooling system for datacenter *M.H. Kim, S.W. Ham & J.W. Jeong*	1
Commissioning of desiccant and evaporative cooling-assisted 100% outdoor air system *M.H. Kim & J.W. Jeong*	7
Liquid desiccant system performance for cooling water temperature variation in the absorber *J.Y. Park & J.W. Jeong*	13
Mechanical analysis on the tower-girder fixed region of Chongqing Jiayue bridge *Y. Pan & Y. Deng*	19
The development of intelligent aided system for safety assessment of buildings in post-earthquake based on web *B. Sun, L. Zhang, H. Chen & X. Chen*	25
Impact of PV/T system on energy savings and CO_2 emissions reduction in a liquid desiccant-assisted evaporative cooling system *E.J. Lee & J.W. Jeong*	31
Study on multi-zone airflow model calibration process and validation *S.K. Han & J.W. Jeong*	37
Impact of demand controlled ventilation on energy saving and indoor air quality enhancement in underground parking facilities *H.J. Cho & J.W. Jeong*	43
Technique for detecting vertical prestress based on vibration principle *X. Shu, M. Shen, X. Zhong, T. Yang & H. Chen*	49
Influence analysis of ground asymmetric load on spandrel with different bases of existing open trench tunnels *S. Zeng, J. Cao, Q. Pu & H. Liu*	59
Sensitivity analysis of prestressing loss of long-curved tendons in tensioned phase *J. Tian, X. Fang, L. Qin, W. Xie & J. Tong*	65
Numerical analysis of the wind field on a low-rise building group *C. Zhang & W.B. Yuan*	71
Modeling of a tunnel form building using Ruaumoko 2D program *N.H. Hamid & S.A. Anuar*	77
A simplified design method for CFST frame-core wall structure with viscous dampers *F. Ren, C. Wu & J. Liang*	83
Experiment research on the bending behavior of stainless steel reinforced concrete slabs *Y. Zhang, G. Zhang, X. Zhou & Z. Chen*	89

Effect of geogrid-confinement on swelling deformation of expansive soil *J. Ding, R. Chen & J. Liu*	95
A micro-mechanics study of saturated asphalt concrete *Y. Yuan & Y.H. Zhao*	103
Numerical analysis of SRC columns with different cross-section of encased steel *H.J. Jiang, Y.H. Li & Z.Q. Yang*	109
Modeling hysteresis loops of corner beam-column joint using Ruaumoko program *A.G. Kay Dora & N.H. Hamid*	115
Prediction of ship resistance in muddy navigation area using RANSE *Z. Gao, H. Yang & Q. Pang*	121
Numerical analysis of depth effects on open flow over a horizontal cylinder *Y. Zhu, L. Xie & X. Liang*	127
Choice of azimuth axis and stability analysis of surrounding rock of Lianghekou underground powerhouse *Z.Z. Wang, J.H. Zhang, H.Y. Zeng, C.G. Liao & D.Q. Hou*	133
Experimental study on the chloride-permeation resistance of marine concrete mixed with slag and fly ash *S. Cheng, Su Cheng & C. Hu*	143
Experiments on coupling effect of upward seepage on slope soil detachment *X. Liang, L. Xie & Y. Zhu*	147
Seismic safety evaluation of gravity dam in aftershocks *S. Fan, J. Chen & S. Lv*	153
Research on the features of the wide range continuous rainstorms in Guangxi in the conditions of three types of trough *W. Liang, M. Huang, M. Qu & X. Li*	159
Precipitation analysis in Wuhan, China over the past 62 years *S. Deng, B.H. Lu, H.W. Zhang & E. Akwei*	165
Climatological anomalies of 500 hPa height fields for heavy rainfalls over the Huaihe River Basin *J. He, G.H. Lu, Z.Y. Wu, H. He, Y. Yang & J.Y. Huang*	171
A method on long-term forecasting of meteorological droughts in the southwest of China *G.H. Lu, L. Dong & Z.Y. Wu*	177
Comparison on river flow regime between pre- and post-dam *X.S. Ai, J.J. Tu, Z.Y. Gao, S. Sandoval-Solis & Q.J. Dong*	183
Spatial-temporal variation and trend of meteorological variables in Jiangsu, China *Z.Y. Wu, Y. Mao, G.H. Lu, Y. Yang & J.H. Zhang*	189
Trichloroethylene biodegradation under sulfate reduction conditions *Y. Guo & K. Cui*	195
Tidal range analysis of the central Jiangsu coast based on the measured tide level data *Y. Kang & X. Ding*	201
Research of Linyi Municipal groundwater resources management information system based on GIS *Z. Cui, J. Yu, W. Cui, X. Gao, Y. Li & G. Ren*	207

Evolution of water exchanging period in Lvdao lagoon induced by different water depth *X. Pan, D. Li, B. Liang & P. Du*	213
Mechanics analysis of arch braced aqueduct *X. Lu, C. Li, Y. Ye & Y. Zhao*	219
Spatial coupling relationship between settlement and land and water resources – based on irrigation scale – A case study of Zhangye Oasis *L.C. Wang, R.W. Wu & J. Gao*	225
Numerical simulation of tidal current near the artificial island sea area *Z. Yin, Z. Wang, T. Xu, N. Ma & W. Chi*	233
Using ANN and SFLA for the optimal selection of allocation and irrigation modes *J.X. Xu, Y.J. Yan, Y. Zhao, H.Y. Zhang & Y. Gao*	239
Influences of cement temperature on the performance of concrete mixed with superplasticizer *S.-Y. Wang, W.-K. Guo & Z.-Y. Gao*	243
Application of ceramic filter in seawater desalination pretreatment *Y. Lv, J. Yang, J. Shi & H. Liang*	251
Simulation of sedimentation in the channel of Lianyungang Harbor by using 3D sediment transport model of FVCOM *B. Yan, H. Yang, Q. Zhang & Y. Wu*	257
Assessment based on Monte Carlo for sample rotation under stratified cluster sampling *Y. Fu, G. Gao & S.X. Liu*	265
Reliability and validity evaluation based on Monte Carlo simulations by computer in three-stage cluster sampling on multinomial sensitive question survey *S. Yang, K. Chen, X. Chen, J. Shi, W. Li, Q. Du, Z. Jin & G. Gao*	271
Parameter estimation for sample rotation of successive survey in stratified sampling and its application *S. Liu, Y. Fan & G. Gao*	277
Earthquake influence field dynamic assessment system *B. Sun, X. Chen, H. Chen & Y. Zhong*	283
Analysis of safety-accident on residential building construction *H.-B. Zhou, X.-X. He, X. Lu, M.-S. Zhou & L. Feng*	289
Study on the harmony of emergency management to urban major accidents *D. Huang*	299
The prediction of razor clams shelf-life based on kinetics model *Q. He, X.-D. Lin, L. Wang, L.-N. Feng & M. Xiong*	305
DEA model evaluating the supply efficiency of rural productive public goods in Hubei Province, China *F. Luo, Y. Zhou, C. Sun & L. Chen*	311
The experimental study of the compressive mechanical behaviors of the H-shaped grouted masonry dry-stacked without mortar *N. Lou, G. Yi & L. Zhang*	317
Evaluation based on Monte Carlo simulation under stratified three-stage sampling *K.J. Chen, Y.B. Fan & G. Gao*	323

Nonlinear stability analysis for composite laminated shallow spherical shells in hydrothermal environment   329
S.-Y. Jiang, Y.-S. Deng & J.-C. Xu

Evidential uncertainty quantification in fatigue damage prognosis   335
H. Tang, J. Li & L. Deng

Side force proportional control with vortex generators fixed on tip of slender   341
J. Zhai, W.W. Zhang, C.Q. Gao, Y.H. Zhang & L. Liu

Characteristic of unsteady aerodynamic loads with a synthetic jet at airfoil trailing edge   347
C.Q. Gao, W.W. Zhang, J. Zhai & X.B. Liu

Simulation analysis on surface subsidence near tunnel influenced by lining partially flooded in seepage field   353
Z. Jiang, F. Ma, X. Dong & X. Zhang

Numerical simulation and dynamic response analysis of wedge water impact   359
Y.Q. Zhang, F. Xu & S.Y. Jin

Aeroelastic study on a flexible delta wing   365
J.G. Quan, Z.Y. Ye & W.W. Zhang

Efficient numerical simulation based on computational fluid dynamics for the dynamic stability analysis of flexible aircraft   373
W.G. Liu, R. Mei, W.W. Zhang & Z.Y. Ye

Hydro-mechanical analysis on the fluid circulation in a coaxial cylinder rheometer   379
J. Xing, S. Ding & J. Xu

Analysis of adhesively bonded pipe joints under torsion by finite element method   383
J.Y. Wu, H. Yuan & J. Han

Plastic deformation of Honeycomb sandwich plate under impact loading in ansys   387
Q. Huan & Y. Li

The responses of self-anchored suspension bridges to aerostatic wind   393
T.-F. Li, C.-H. Kou, C.-S. Kao, J. Jang & P.-Y. Lin

Distribution features and protection for heritage in the metropolitan area   399
Y. Zhang, X. Xi & X. Wang

Landscape planning for city parks with rainwater control and utilization   409
Y. Yang & X. Xi

The study on the evolution of wind-sand flow with non-spherical particle   419
S. Ren, Y. Luo & X. Hu

Comprehensive ecosystem assessment based on the integrated coastal zone management in Jiaozhou Bay area   425
C. Wang, J. Dai, Z. Guo & Q. Li

Spatial variation and source apportionment of coastal water pollution in Xiamen Bay, Southeast China   431
Q. Li, C. Wang, Y. Ouyang & J. Dai

Isolation and identification and characterization of one phthalate-degrading strain from the active sludge of sewage treatment plant   437
B. Li, J. Cao, C. Xing, Z. Wang & L. Cui

Seedling emergence and growth of Ricinus communis L. grown in soil contaminated by Lead/Zinc tailing 445
X. Yi, L. Jiang, Q. Liu, M. Luo & Y. Chen

The acute toxicity of the organic extracts from two effluents of different wastewater treatments 453
H. Zhao, Z. Huang & B. Li

Aerobic degradation of highly chlorinated biphenyl by *Pseudomonas putida* HP isolated from contaminated sediments 459
C. Wang, J. Cui, J. Wang, S. Cui & Y. Sun

Research on transport and diffusion of suspended sediment in reclamation project 465
X. Guo, C. Chen & J. Tang

Isolation and characteristics of two novel crude oil-degrading bacterial strains 473
Y. Gao, X. Gong, H. Li, M. Ma, J. Wang, S. Cui & C. Wang

The challenges of water resources and environmental impact of China mainland unconventional gas exploration and development 479
T. Yue, S. Hu, J. Peng, F. Chu & Y. Zhang

Optimization of 4-chlorophenol in the electrochemical system using a RSM 487
H. Wang & Z.Y. Bian

Integrated watershed management framework for abating nonpoint source pollution in the Lake Dianchi Basin, China 493
C. Wang, M. Peng, R. Zhou, S. Yang, Z. He, C. Duan, G. Zhi, Z. Li & L. He

Study and analysis on spatio-temporal variation of water quality of the middle and lower reaches of Hanjiang River 503
L. Song & T. Li

Different ratio of rice straw and vermiculite mixing with sewage sludge during composting 511
M. Yu, Y.X. Xu, H. Xiao, W.H. An, H. Xi, Y.L. Wang, A.L. Shen, J.C. Zhang & H.L. Huang

Heavy metal contamination in surface sediments of the Pearl River Estuary, Southern China 517
M. Song, P. Yin, L. Zhao, Z. Jiao & S. Duan

The design of water pollutant emission allowance control scheme in the city of Shenzhen, China 523
X. Xiao, X. Ma & P. Song

Enhanced waste activated sludge digestion by thermophilic consortium 529
Y.L. Wang, Z.W. Liang, S.Y. Yang & Y. Yang

The effects of reversible lane's application on vehicle emissions in Chaoyang Road in Beijing 535
X. Yao, L. Lei & J. Yang

Research on system evaluation framework of green elderly community for Chinese population change in next decade 545
Z. Qiu, Y. Ming & J. Wang

The survey of EEWH–Daylight evaluation system by LEED-Daylight standard – Taking Beitou Library as an example 553
Y.M. Su & Y.S. Wu

Slaughterhouse wastewater treatment by hydrolysis and Bardenpho process 559
A. Li, Y. Zhang, H. Guo & T. Pan

Effects of the construction of large offshore artificial island on hydrodynamic environment 567
N. Zhang, C. Chen & H. Yang

The thinking and exploration on building a characteristic apartment system for
the aged in city 573
J. Dai, Z. Wang & Z. Qiu

Effect of membrane characteristics of alginate-chitosan microcapsule on immobilized
recombinant *Pichia pastoris* cells 579
Q. Li, W. Deng, Q. Zhang, Y. Zhao & W. Xue

Working characteristics of natural circulating evaporator 585
F. Zhang, Y. Zhong, H. Du, X. Liu, L. Yu & H. Yang

Guaranteed appropriate hydrolysis of meat by adding bromelain and zinc sulfate 591
L. He, L. Xue, Q.J. Zhu, S. Song, M. Duan, L. Deng, J. Chen & X. Zhang

Isolation and identification of a pectinase high-yielding fungus and optimization of
its pectinase-producing condition 599
S.Y. Jiang, Y.Q. Guo, L.Z. Zhao, L.B. Yin & K. Xiao

Study on flavonoids content and antioxidant activity of *muskmelon* with different
pulp color 605
X. Sun, T. Li, L. Jiang & X. Luo

Flavonoids intake estimation of urban and rural residents in Liaoning Province 611
Z. Wu, L. Yang, S. Li, W. Chen, D. Jiang, Y. Wang, Q. Zhang & X. Zhang

Effect of 6-Benzyladenine, Gibberellin combined with fruit wax coating treatment
on the lignification of postharvest Betel nut 619
Y.G. Pan & W.W. Zhang

A new approach for recycling potash from the mother liquor B in beet sugar industry 625
S.-M. Zhu, W. Long, X. Fu, L. Zhu & S.-J. Yu

Review of *Brettanomyces* of grape wine and its inhibition methods 635
W. Lin, X. You & T. Feng

Extraction technology of crude polysaccharide from *Stropharia rugoso-annulata* 641
J.C. Chen, P.F. Lai, M.J. Weng & H.S. Shen

Study on dietary fiber extracted from *bean curd residue* by enzymic method 647
H. Lai, G. Su, W. Su, S. Zhang & S. Shou

Protective effects of *Portulaca oleracea* on alloxan-induced oxidative stress in
HIT-T15 pancreatic β cells 653
J.L. Song & G. Yang

Study of gas-oil diffusion coefficient in porous media under high temperature and pressure 659
P. Guo, H. Tu, A. Ye, Z. Wang, Y. Xu & Z. Ou

The significance of biomarkers in nipple discharge and serum in diagnosis of breast cancer 665
G.P. Wang, M.H. Qin, Y.A. Liang, H. Zhang, Z.F. Zhang & F.L. Xu

Experimental study on minimum miscible pressure of rich gas flooding in light oil reservoir 673
P. Guo, Z. Ou, A. Ye, J. Du & Z. Wang

The traumatic characteristics of inpatients injured in Lushan Earthquake 679
Y.M. Li, J.W. Gu, C. Zheng, B. Yang & N.Y. Sun

Study on immunomodulatory activity of a polysaccharide from *blueberry*   685
X. Sun & X. Meng

Effects of sodium additives on deacidification and weight gain of squid (*Dosidicus gigas*)   691
Y.N. Wang, S.W. Liu, G.B. Wang, Z.B. Li & Q.C. Zhao

Preservation effects of active packaging membrane from allyl isothiocyanate molecularly
imprinted polymers cooperating chitosan on chilled meat   697
K. Lu, Y. Huang, J. Wu & Q. Zhu

Optimization of acid hydrolysis and ethanol precipitation assisted extraction of pectin
from navel orange peel   707
L.B. Yin, K. Xiao, L.Z. Zhao, J.F. Li & Y.Q. Guo

Optimization of ultrasonic-assisted enzymatic extraction of phenolics from broccoli
inflorescences   713
H. Wu, J.X. Zhu, L. Yang, R. Wang & C.R. Wang

Author index   723

Advanced Engineering and Technology – Xie & Huang (Eds)
© 2014 Taylor & Francis Group, London, ISBN 978-1-138-02636-0

Preface

Some recent essential ideas and advanced techniques have been presented to overcome the current issues in engineering and technology. Annual Congress on Advanced Engineering and Technology (CAET) aims to promote technological progress and activities, technical transfer and cooperation, and opportunities for engineers and researchers to maintain and improve scientific and technical competence in engineering fields, which include civil engineering, hydraulic engineering, environmental engineering, modelling & data analysis, chemistry & biochemical engineering, and other related fields.

In 2014, CAET will be held in Hong Kong during 19-20 April 2014. The 4th Workshop on Applied Mechanics and Civil Engineering (AMCE 2014) is an important track of CAET 2014 and focuses on the frontier research of applied mechanics and civil engineering.

110 technical papers from CAET 2014 (including AMCE 2014) are published in the proceedings. Each of the papers has been peer reviewed by recognized specialists and revised prior to acceptance for publication. The topics of the contributed papers covered topics such as:

- building engineering, e.g. steel reinforced concrete (SRC) columns, CFST frame-core wall structure, stainless steel reinforced concrete slabs, tunnel form building, honeycomb sandwich plate, prestress detection based on vibration
- geotechnical engineering, e.g. open trench tunnels, underground parking facilities, large-dimension expansion model tests of expansive soil
- road and bridge engineering, e.g. chongqing jiayue bridge crossing jialing river, micro-mechanics of saturated asphalt concrete, reversible lanes
- hydraulic engineering, e.g. seismic safety of gravity dam, underground powerhouse in Lianghekou hydropower station, sedimentation in Lianyungang Harbor channel, pipeline on seabed, marine concrete, suspended sediment transport at the artificial island sea area
- environmental engineering, e.g. trichloroethylene biodegradation, effects of the offshore island construction on environment, transport and diffusion of suspended sediment in reclamation project
- pollution and control, e.g. aerobic degradation of highly chlorinated biphenyl, novel crude oil-degrading bacterial strains, heavy metal contamination in surface sediments of the Pearl River Estuary in Southern China, sludge digestion by thermophilic consortium, coastal water pollution at Xiamen Bay in Southeast China, water pollutant emission allowance control
- water resources and water treatment, e.g. water quality of the middle and lower reaches of Hanjiang River, rainwater control and utilization, rainstorms features in Guangxi China, municipal groundwater resources management information system based on GIS, irrigation modes, slaughterhouse wastewater treatment, ceramic filter in seawater desalination pretreatment
- mechanics in engineering, e.g. pipeline failure under torsion load, fatigue damage prognosis under epistemic uncertainty
- water and soil conservation, e.g. seepage effects on soil erosion
- climate change and environmental dynamics, e.g. meteorological variables in Jiangsu/China, meteorological droughts in southwestern China, heavy rainfalls over the Huaihe River Basin, precipitation analysis in Wuhan/China in recent 62 years
- intelligent safety system, e.g. safety assessment of buildings in post-earthquake based on web, earthquake influence field dynamic assessment system, wind field on a low-rise building group, safety accidents on residential building construction
- numerical software and applications
- Chemistry, biochemical and food engineering
- Modelling, computing and data analysis

Although these papers represent only modest advances toward overcoming major scientific problems in engineering, some of the technologies might be key factors in the success of future engineering advances. It is expected that this book will stimulate new ideas, methods and applications in ongoing engineering advances.

Last but not least, we would like to express our deep gratitude to all authors, reviewers for their excellent work, and Léon Bijnsdorp, Lukas Goosen and other editors from Taylor & Francis Group for their wonderful work.

Liquan Xie
College of Civil Engineering, Tongji University, Shanghai, China
E-mail: xie_liquan@tongji.edu.cn

Dianjian Huang
National Research Center of Work Safety Administration, Beijing, China
E-mail: huangboss2008@sina.com

Integration of hot-water cooling and evaporative cooling system for datacenter

M.H. Kim, S.W. Ham & J.W. Jeong
Division of Architectural Engineering, College of Engineering, Hanyang University, Seoul, Republic of Korea

ABSTRACT: The main purpose of this paper is to introduce a hot-water cooling system integrated with a desiccant and evaporative cooling system for a data center cooling. The applicability of the suggested system is evaluated using a heat transfer simulation. Then, the energy saving potential of the proposed system is determined in relation to a conventional air-cooling system. The proposed system is quantitatively simulated by using TRNSYS 16 and a commercial equation-solver program. The results show that the proposed system can save more than 80% of the operating power for a data center and reduce the power usage effectiveness value from 1.5 to 1.15. It is also found that the proposed system consumes approximately 80% less annual operating energy compared with the conventional air-cooling system.

1 INTRODUCTION

Because the energy consumption of data centers is rapidly increasing by following the bulk of rack energy consumption, the interest in the HVAC systems to extract the heat dissipated from the rack, which consume more than half of the data center energy, has also increased. A data center cooling system is commonly called computer room air conditioning (CRAC) or computer room air handler (CRAH).

There are two types of data center cooling systems: air-cooling systems and water-cooling systems. Conventionally, data centers have been cooled by air-cooling systems. An air-cooling system supplies conditioned air to extract heat from the server racks in the data center. In order to reduce the energy consumption of a conventional CRAC system, studies on air-side economizer systems have been conducted (Lui 2010).

On the other hand, many researchers have studied water-cooling systems. A water-cooling system uses water as the heat rejection medium. Recently, there has been increased interest in hot-water cooling systems that use an energy efficient rack cooling system (Meijer 2010). In a hot-water cooling system, all or a part of (i.e. the CPU) the rack is cooled by using hot water (45–70°C). This water-cooling system may cool the rack, but other parts of the cooling load such as the UPS, occupancy, and lighting should be treated by a conventional air-cooling system.

The hot water produced by the rack can also serve as a heat source for the building or other parts of the industry. The heat obtained from the rack by the hot-water cooling system could be used for regenerating the desiccant solution in a liquid-desiccant (LD) unit. A desiccant and evaporative cooling system can be used for the air-cooling part of the data center. It can also overcome the limitation of the evaporative cooling effect in a hot and humid climate (Gandhidasan 1990).

The main purpose of this paper is to evaluate the integration of a hot-water cooling system into a desiccant and evaporative cooling system (LD-IDECOAS). An operating scheme is developed to estimate the energy performance of the proposed system. By simulating the proposed system, the energy saving potential is estimated with respect to a conventional air-cooling system.

2 SYSTEM OVERVIEW

2.1 Desiccant and evaporative cooling system

The proposed system consists of an LD unit, an indirect evaporative cooler (IEC), and a direct evaporative cooler (DEC) in the process air stream. During the cooling season, hot and humid outdoor air is primarily dehumidified by the LD, and then sensibly cooled by the IEC. The isentropic cooling of the process air in the DEC makes it possible to meet a target cold deck supply air temperature. In intermediate seasons, when the outdoor air is dry enough to reach the target supply air temperature using only the IEC and DEC, the LD is not operated.

2.2 Hot-water cooling system

A hot-water cooling system can cover rack cooling loads ranging from a CPU to the entire number of chips in the rack. When hot-water cooling is used for a rack in the form of manifold micro-channel (MMC) cooling (Copeland et al. 1997; Zimmermann et al. 2012), it can remove the entire cooling load of the rack.

From the literature (Zimmermann et al. 2012), the feasible hot-water inlet temperature for a rack ranges from 45°C to 70°C. When the hot-water temperature is higher than 60°C after heat is rejected to the solution, the hot water is cooled to the target hot-water temperature using water-side free cooling with the operation of a cooling tower. When the hot-water temperature is lower than 60°C, the hot water is continuously re-circulated until reaching 60°C using a three-way valve.

2.3 Integration of hot-water cooling system and LD-IDECOAS

By integrating the hot-water cooling system and LD-IDECOAS, the dissipated heat from a rack is rejected, and then the heat is reused as a heat source for regenerating the solution. In order to regenerate the liquid desiccant solution, it should be heated to the target solution temperature. In this research, the target solution temperature is set to 55°C, and the solution mass flow rate is varied depending on the latent load of the outdoor air and internal environment. In the intermediate and winter seasons, because the operation of the liquid desiccant system is not necessary, the hot water is maintained at 60°C by using water-side free cooling.

3 ENERGY SIMULATION

3.1 Data center simulation model

The model data center for the simulation has a 7000 m^2 floor area, and 70% (4900 m^2) of this floor area is used for the rack. It is also assumed that the rack dissipates 1280 W/m^2. The U-values of the exterior wall and roof are 0.511 W/m^{2}K and 0.316 W/m^{2}K, respectively. The window to wall ratio is 0.25. The required ventilation rate is set at 2.5 L/s per person in the office space and 0.3 L/s per square meter floor area in the office space, computer room, and electric equipment rooms, in accordance with ASHRAE standard 62.1-2007 (2007). The entire system performance of the UPS, PDU, transformers, and power distribution has an effectiveness of 80%, and it contains 16% of the data center cooling load. The heat density of the lighting fixtures is 19 W/m^2, and the 50 occupants generate 100 W of sensible heat and 130 W of latent heat per person based on the ISO-7730.

The setpoint of the supply air on the cold aisle is set to a 27°C dry-bulb temperature and 15°C dew point temperature within the range of the ASHRAE TC 9.9 recommendation. It is also assumed that the exhaust air from the rack has a 37°C dry-bulb temperature, while maintaining a 10°C temperature difference between the cold aisle (i.e. supply air) and hot aisle (i.e. return air).

In the office area, the target indoor condition is set to a 24°C dry-bulb temperature and 55% relative humidity. To maintain this target indoor condition, the supply air temperature is set to a 15°C dry-bulb temperature. Each parameter of the data center and office cooling load is estimated by using the TRNSYS 16 program (Klein et al. 2004). Then, the system performance of the hot water and LD-IDECOAS is analyzed by using the Engineering Equation Solver (EES) program (Klein 2012). In order to compare the thermal performance of the proposed system, a conventional air-cooling system with an air-side economizer (i.e. CASE 1) is also evaluated.

(1) CASE 1: air-cooling system
As shown in Figure 1a, the conventional CRAC system (CASE 1) consists of a conventional cooling coil. The components of this system include the cooling coil, chiller, and cooling tower. In order to enhance the cooling performance of the CRAC system, an air-side economizer is considered (CASE 1).

(2) CASE 2: hot-water cooling and LD-IDECOAS
Because the hot-water cooling system cannot control the other cooling loads such as the occupancy, lighting, and UPS, the LD-IDECOAS should be used to remove the other cooling loads (Fig. 1b). The hot-water cooling system covers all the server racks (CASE 2).

4 SIMULATION RESULT

4.1 *Thermal performance of hot-water cooling system*

As shown in Figure 2, in order to use the hot water generated from the rack to regenerate the liquid desiccant solution and maintain the setpoint temperature of the hot water, two heat exchangers (i.e. HX1 and HX2) and two 3-way valves are required. The return hot water (state 1) from the

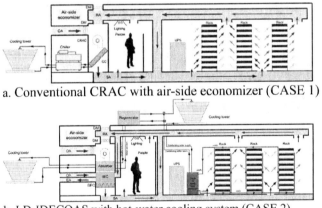

Figure 1. Data center cooling systems.

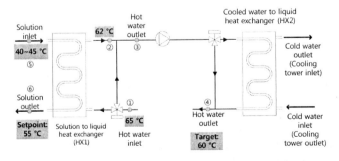

Figure 2. Heat exchanger between hot water and solution.

Table 1. Peak power consumption.

		CASE 1	CASE 2
Hot-water cooling tower (HWCT)	Pump [kW]	–	42.0
	Fan [kW]	–	55.1
Solution cooling tower (SCT)	Pump [kW]	–	15.7
	Fan [kW]	–	44.0
Cooling tower (CT)	Pump [kW]	75.0	–
	Fan [kW]	110.1	–
Chiller	Chiller [kW]	2450.2	–
	Pump [kW]	30.0	–
Hot-water system (HW)	Pump [kW]	–	15.7
Main Fan [kW]		1160.0	445.1
Total [kW]		3825.2	557.9
PUE [–]		1.8	1.3

rack is normally maintained at 65°C, whereas the target solution temperature (state 6) is met by controlling the 3-way valve for the return hot water in HX1. When the outlet hot-water temperature (state 3) is higher than the target supply hot-water temperature for the rack (i.e. 60°C), the hot water is cooled using cold water from the cooling tower in HX2. On the other hand, if the outlet hot-water temperature (state 3) is lower than the target temperature, the hot-water is re-circulated until it reaches the target temperature. The thermodynamic properties of LiCl solution and water are used inserted in the EES program (Klein 2012).

4.2 Peak power consumption

The power usage effectiveness (PUE) introduced by Green Grid in 2006 is the most widely used indicator for the power efficiency of a data center. As shown in Equation 1, PUE is the ratio of the total amount of power usage in a data center to the power usage of the IT equipment (i.e. rack).

$$PUE = \frac{Total\ Energy}{IT\ Energy} = \frac{Cooling + Power + Lighting + IT}{IT} \qquad (1)$$

Table 1 shows the cooling power for each system case. It shows that the proposed system (i.e. CASE 2) can save 85.4% of the peak power consumption compared with the air-cooling system (i.e. CASE 1). Based on Equation 1, it is also possible to see that the proposed system can improve the existing data center PUE from 1.8 to 1.3. These energy savings are mainly caused by the fact that the proposed system may not need the chiller, in contrast to the air-cooling system.

4.3 Monthly and annual energy consumption

In order to consider the part load reduction of the cooling system by using PUE terms, the monthly average PUE is suitable to indicate the cooling performance, depending on the climatic conditions. Figure 3 also presents the mean outdoor air dry-bulb temperature. As shown in Figure 3, in CASE 1, PUE varies depending on the outdoor air condition. During the heating (i.e. Jan, Feb, Mar, Nov, and Dec) and intermediate (Apr, May, and Oct) seasons, an air-side economizer is used to reduce the cooling coil load, which greatly reduces the chiller operation.

Meanwhile, CASE 2 presents a nearly continuous PUE regardless of the outdoor air condition. This is because the power consumption of the desiccant and evaporative cooling system may not have a significant impact on the total data center power consumption. This means that the desiccant and evaporative cooling system can efficiently alter the conventional chiller operation.

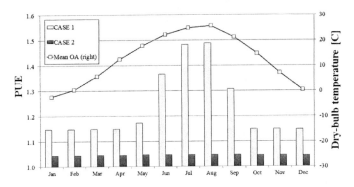

Figure 3. Monthly average PUE.

CASE 2 can save approximately 80% of the annual operating energy consumption compared to CASE 1. This is mainly because the fan flow rate is significantly reduced by the hot-water cooling system, and the supply air temperature is sufficiently met without operating the chiller.

5 CONCLUSIONS

In this research, data center cooling applications were categorized in two types: air-cooling systems and water-cooling systems. Then, in order to overcome the limitation of the system cooling efficiency of a conventional air-cooling system, the integration of a liquid-desiccant and evaporative cooling system with a hot-water cooling system was suggested.

An energy simulation model was used to quantitatively evaluate the energy saving potential of the proposed system compared to a conventional air-cooled system. The results showed that the proposed system could save more than 85% of the data center power consumption of an air-cooling system. Finally, the proposed system could save more than 80% of the annual operating cooling energy used by the air-cooling system.

ACKNOWLEDGEMENT

This work was supported by a National Research Foundation of Korea (NRF) grant (No. 2012001927) and a Korea Evaluation Institute of Industrial Technology grant funded by the Korean government (No. 10045236).

REFERENCES

ASHRAE Standard 62.1-2007. Ventilation for Acceptable Indoor Air Quality, Atlanta GA, American Society of Heating, Refrigerating and Air Conditioning Engineers, 2007.
Copeland, D., Behnia, M., Nakayama, W. 1997. Manifold Microchannel Heat Sinks: Isothermal Analysis. *IEEE Transactions on Components, Packaging, and Manufacturing Technology Part A* 20(2): 96–102.
Gandhidasan, P. 1990. Analysis of a solar space cooling system using liquid desiccants. *In: Proceedings of the energy conversion engineering conference*: 162–6.
Klein, S.A., et al. 2004. TRNSYS 16, a TRaNsientSYstem Simulationprogram. Solar Energy Laboratory, Univ. of Wisconsin-Madison, Wisconsin.
Klein, S.A. 2012. EES—Engineering Equation Solver, F-ChartSoftware, Available from: <www.fchart.com>.
Lui, Y.Y. 2010. Waterside and Airside Economizers Design Considerations for Data Center Facilities. *ASHRAE Transactions* 116(1): 98–108.
Meijer, G.I. 2010. Cooling energy-hungry data centers. *Science* 328: 318–9.
Zimmermann, S. et al. 2012. Hot water cooled electronics: Exergy analysis and waste heat reuse feasibility. *International Journal of Heat and Mass Transfer* 55(23): 6391–6399.

Commissioning of desiccant and evaporative cooling-assisted 100% outdoor air system

M.H. Kim & J.W. Jeong
Division of Architectural Engineering, College of Engineering, Hanyang University, Seoul, Republic of Korea

ABSTRACT: The main purpose of this study was to develop a desiccant and evaporative cooling-assisted 100% outdoor air system (LD-IDECOAS). The pilot system of the LD-IDECOAS was installed and an automatic control and monitoring system was set up for a test bed of a building. The draft test was conducted during the peak-cooling season, and the cooling performance was achieved. The commissioning processes included two options. Test 1 involved controlling the return air humidity level to enhance the cooling performance of an Indirect Evaporative Cooler (IEC). In test 2, the supply water temperature was controlled for the Liquid Desiccant (LD) unit to adjust the dehumidification rate. The test 1 result showed that the impact of the DEC operation on the humidity level of indoor was limited, but the DEC had an significant effect on the cooling performance of the proposed system. The supply cooling water temperature also affected the dehumidification rate of the LD unit. In test 2, an increase in the supply water temperature for the LD unit led to an enhancement of the cooling performance of IEC, but it is also led to an increase in the supply air temperature.

1 INTRODUCTION

Recently, research has been conducted on the use of indirect and direct evaporative cooling to assist 100% outdoor air system (IDECOAS) as an energy efficient alternative to a conventional variable air volume (VAV) system (Kim et al. 2011; Kim et al. 2012; Kim and Jeong 2013). Moreover, evaporative cooling systems have gained attention as environmentally friendly air conditioning systems because they use the latent heat of water evaporation. However, an evaporative cooling system still has a limited cooling performance in a hot and humid climate. To overcome this limitation, the integration of a liquid-desiccant (LD) system and IDECOAS has been suggested to enhance the thermal performance of the evaporative cooling system (Kim et al. 2013).

The proposed desiccant and evaporative cooling-assisted 100% outdoor air system (LD-IDECOAS) supplies 100% outdoor air to meet the required space cooling load. In a hot and humid season, the entering outdoor air to be delivered to the space first passes through the LD unit, which dehumidifies it. Then, the processed air passes through the indirect evaporative cooler (IEC) and is sensibly cooled. Finally, the direct evaporative cooler (DEC) further reduces the process air temperature using adiabatic cooling, and the setpoint of the supply air condition is maintained. The thermal performance of the LD unit, IEC, and DEC significantly affects the energy saving potential of the proposed system.

In this research, an LD-IDECOAS pilot system was designed to verify the developed concept of the proposed system. Then, experimental research was conducted using this pilot system. Based on observations of the thermal performance of the draft test during the cooling season, the commissioning processes were investigated to enhance the thermal performance and determine the characteristics of the proposed system.

2 DESCRIPTION OF PILOT SYSTEM

The LD-IDECOAS pilot system was installed in a 6 m × 7.2 m fourth floor office in a four-story building at Incheon in South Korea. The setpoints for the room included a 24°C dry-bulb temperature (DBT) and 60% relative humidity (RH). The DBT of the supply air was 15°C with a saturated condition. The supply air flow rate was set at 2000 m³/h.

The pilot system consisted of an LD unit, IEC, DEC pad, a cooling tower, an air-cooled chiller, two water storage tanks (cooling water and hot water), a gas-fired water boiler, SHE, and two fans (supply and return). The dehumidified outdoor air from the LD unit was conditioned by passing through the IEC and DEC. The SA flow rate was adjusted based on the space sensible cooling load. The IEC and DEC treated the sensible cooling load of supply air-conditioning load, and the LD eliminated the latent cooling load of that. The adiabatic cooling process of the DEC reduced the process air temperature, but increased the humidity.

An automatic control system was also developed to achieve the sequence of operation. The sequence of operation was defined by the outdoor air condition and room condition. The automatic control system utilized a programmable logic controller (PLC). The acquired data and operating states were monitored and recorded by a personal computer that was connected using Ethernet. The operating data were collected at 1-min intervals.

A total of 11 temperature and humidity sensors, 4 liquid temperature sensors, 2 liquid flow meters, and 4 air flow meters were installed in ducts or pipes to observe the thermal behavior of the proposed system components.

3 TEST RESULTS

3.1 Draft test

3.1.1 Test conditions

A draft test of the LD-IDECOAS pilot system was conducted on August 12, 2013, during the peak-cooling season. The supply water temperature for IEC and DEC was 28.0°C. No sensible heat and latent heat were generated during the test. The target supply cooling water temperature was set to 25°C. This temperature was used to enhance the usage potential of the water-side free cooling operation. In order to maintain the target cooling water temperature for the LD unit, an air-cooled chiller was operated during the test. The 51.9°C hot water produced by a gas-fired boiler was supplied to the LD unit for regeneration. Both the supply and return air flow rates were set to 2000 m³/h.

3.1.2 Thermal performance of components

Figure 1 shows the draft test result of the psychrometric process of the proposed system. As shown in Figure 1, the hot and humid outdoor air was dehumidified and sensibly cooled by the LD unit. This sensible cooling was due to the heat reclaiming between the process air and supplied cooling water by the air-water heat exchanger located on the upstream of the LD unit. The supply air was sensibly cooled at the IEC, and then adiabatically cooled and humidified at the DEC. The supply air reached 20.5°C DBT and 91.8% RH. The test room was maintained at 24.2°C DBT and 72.4% RH.

The sensible and latent cooling capacities of the LD system were 3.1 kW and 12.9 kW, respectively. From this result, one can see that the latent cooling capacity of the LD system was 55.5% less than its rated latent cooling capacity. One may expect that the supply cooling water temperature (i.e. 25.7°C) during the test was higher than the rated supply cooling water temperature (i.e. 6°C).

The sensible cooling capacity of IEC was 1.7 kW, which was 80.5% less than that under the rated condition (i.e. 8.7 kW). The observed effectiveness of IEC was 36.1%, but its rated effectiveness is 84% under the conditions of 31.6°C DBT and 45% RH on the primary side and 24.5°C DBT and 39% RH on the secondary side. This effectiveness drop might have been caused by the fact that the RH of the secondary side return air was much higher than the rated condition, although

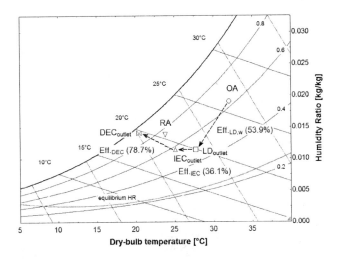

Figure 1. Psychrometric performance during draft test.

the DBT of the return air was close to the rated condition. The effectiveness of DEC was 78.7%, which showed that the supply air was cooled nearly to the saturated condition. These test results showed that commissioning was required to enhance the thermal performance of the LD system and IEC.

3.2 *Commissioning*

Because of the unexpected thermal performance of LD-IDECOAS in the first draft experimental results, it was recognized that a commissioning process was required to solve the deterioration of the cooling performance of both the LD unit and IEC. Following the pilot system test, two commissioning tests to enhance the thermal performance of LD and IEC were conducted. As a first test (test 1), the humidity level of the return air was adjusted to enhance the evaporative cooling performance of IEC. The humidity level control could be obtained using an on-off control of the DEC to reduce the humidity level of the room. In the second test (test 2), the supply cooling water temperature for the LD unit was adjusted to determine the influence on the dehumidification rate. This test also included the operation of the IEC and LD.

3.2.1 *Test 1*

During the draft test, the effectiveness of IEC did not reach the rated value (i.e. 83%), because the wet-bulb temperature (WBT) of the return air (i.e. 20.5°C) was much higher than the rated condition (i.e. 15.6°C). The first commissioning test was conducted by shutting off the DEC to reduce the humidity level of the space.

The outdoor air condition, supply and return air flow rates, and supply cooling water temperature were similar to the draft test conditions. As shown in Figure 1, the LD dehumidified and cooled the supply air at 27.8°C DBT and 43.1% RH. Then, the air passing through the IEC showed a 25.2°C DBT and 43.1% RH. This supply air conditioned the space at 27.6°C DBT and 46.1% RH, and the WBT of the return air was 19.3°C, which was similar to the draft test. One may conclude that the setpoint of the supply air temperature could not be met without the DEC operation, which led to a high DBT and WBT for the space. The IEC effectiveness and cooling capacity were 31.2% and 1.8 kW, respectively. From these results, the DEC should be operated to maintain the target space DBT, and the cooling contribution of DEC is not negligible.

3.2.2 *Test 2*

To observe the thermal performance variation of the proposed system, test 2 was conducted by varying the supply cooling water temperature. Figure 3 shows the psychrometric processes to

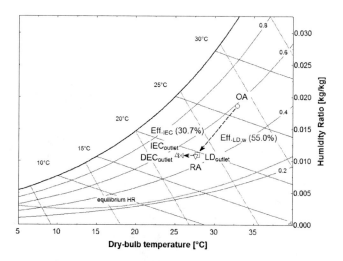

Figure 2. Psychrometric performance in test 1.

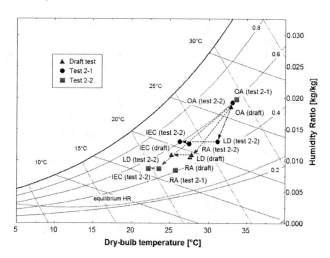

Figure 3. Psychrometric performance in test 2.

observe the variation in the thermal performance of the proposed system on the test 2. The flow rates of the supply and return air were identical to the draft test (i.e. 2000 m³/h). The supply cooling water temperature was varied from 20°C (test 2-1) to 30°C (test 2-2). The DEC was not activated to maintain the low WBT of the return air. Test 2-1 was conducted to observe the system performance when the cooling water temperature was reduced to 20°C. The outdoor air at 33.0°C DBT and 61.4% RH was both dehumidified and cooled at 23.6°C DBT and 42.3% RH by the LD system. The supply air was sensibly cooled at IEC to 22.2°C, and a lower IEC outlet temperature was shown compared to test 1. In test 2-2, the outdoor air was dehumidified and cooled at LD to 31.2°C and 42.9%, and the supply air was cooled at IEC to 26.3°C, with a 20.8°C WBT for the return air.

In test 2, one can see that the dehumidification effectiveness of the LD unit was enhanced from 43.7% to 70.8% when the supply cooling water temperature was decreased from 30°C to 20°C. The latent cooling capacity of the LD system was also increased from 9.7 kW to 19.4 kW.

The maximum cooling rate of IEC was shown by the temperature difference between the DBT of the primary channel inlet air and WBT of the secondary channel inlet air. An increase in the temperature difference from 6.6°C in test 2-1 to 10.4°C in test 2-2 led to an increase in the effectiveness of IEC from 21.0% in test 2-1 to 47.3% in test 2-2. This means that when the temperature difference was increased, a high cooling performance could be expected from IEC. This was primary because the supply cooling water temperature was varied in the LD unit.

The result shows that the lower supply cooling water temperature provided a higher sensible cooling capacity in the proposed system. For the energy efficient operation of the proposed system, water-side free cooling should be used. However, when the supply water temperature was maintained at less than 25°C or 30°C, additional chiller operation energy was consumed, and the water-side free cooling operation potential was reduced. This was because the cooling tower used for the water-side free cooling could not provide cooling water at less than the WBT of the outdoor air. Additionally, the cooling contribution of IEC was rapidly reduced by a reduction in the cooling water temperature.

4 DISCUSSIONS AND CONCLUSIONS

This paper demonstrated the design and installation of a pilot system, and then the first draft test was conducted. However, the thermal performance of the pilot system did not match the design condition, and two commissioning processes were developed and carried out. These commissioning processes were conducted to enhance and determine the performances of the system components.

(1) Eliminating the DEC operation reduced the return air humidity level in test 1. However, because of the high supply air temperature, the WBT was similar to that in the first draft test. One can also see that the cooling contribution of DEC was not negligible in the proposed system.

(2) The second commissioning (test-2) involved varying the supply cooling water temperature of the LD unit from 20 to 30°C. One can observe that the low supply cooling water temperature caused an increase in both the dehumidification performance of LD and the cooling performance of the system, along with a reduction in the cooling performance of the IEC. In contrast, a supply water temperature increase enhanced the cooling performance of the IEC, which is an energy efficient and environmentally friendly system. The cooling water temperature setpoint greater than 25°C was used to enhance the energy efficient performance using water-side free cooling. This low cooling water temperature provided a high cooling performance for the system, but additional chiller operation and energy consumption occurred.

No faults in the system components could be observed during the commissioning process. From the test results, the first draft test might be the most reliable and commonly operable condition.

ACKNOWLEDGEMENT

This work was supported by a National Research Foundation of Korea (NRF) grant (No. 2012001927) and a Korea Evaluation Institute of Industrial Technology grant funded by the Korean government (No. 10045236).

REFERENCES

Kim, M.H., Kim, J.H., Kwon, O.H., Choi, A.S. & Jeong, J.W. 2011. Energy conservation potential of an indirect and direct evaporative cooling assisted 100% outdoor air system. *Building Services Engineering Research and Technology* 32(4): 345–60.

Kim, M.H., Choi, A.S. & Jeong, J.W. 2012. Energy performance of an evaporative cooler assisted 100% outdoor air system in the heating season operation. *Energy and Buildings* 46: 402–409.

Kim, M.H. & Jeong, J.W. 2013. Cooling performance of a 100% outdoor air system integrated with indirect and direct evaporative coolers. *Energy* 52: 245–257.

Kim, M.H., Park, J.S. & Jeong, J.W. 2013. Energy saving potential of liquid desiccant in evaporative-cooling-assisted 100% outdoor air system. *Energy* 59: 726–736.

Liquid desiccant system performance for cooling water temperature variation in the absorber

J.Y. Park & J.W. Jeong
Division of Architectural Engineering, College of Engineering, Hanyang University, Seoul, Republic of Korea

ABSTRACT: This research is an experimental study for estimating the impact of the absorber cooling water temperature on the dehumidification performance of a liquid desiccant system. This paper addresses the dehumidification effectiveness variation with the cooling water temperature changes in the absorber based on the experiment data collected from the liquid desiccant system operation. The lithium-chloride (LiCl) solution is used as the liquid desiccant. The liquid desiccant system used in this research dehumidifies 2000 m^3/h process air and 10USRT air-cooled chiller may provide cooling water at any given temperature. The measurement of dehumidification effectiveness values are conducted during the summer in Seoul, South Korea when the outdoor air (i.e. process air) is hot and humid. Based on the experiment data showing the dehumidification performance variation of the liquid desiccant system for the cooling water temperature, a validation of the dehumidification effectiveness values of a liquid desiccant system is compared with existing model.

1 INTRODUCTION

In hot and humid summer season, air conditioning requires removal of latent heat from not only the building but also the ambient air over the course of air conditioning. According to this reason for controlling the latent load, the well-adapted air-conditioning system is necessary.

The conventional air-conditioning system consumes a lot of energy against latent heat load, as known by many literatures. Liquid desiccant air conditioning systems are being considered as a reliable alternative to the conventional HVAC system, which is passively controlling the latent load of the building.

This research is about the empirical analysis of the operational performance of Liquid Desiccant System. The experiment was performed assuming the hot and humid summer ambient air to measure the variation in dehumidification effectiveness of the system by varying the cooling water temperature being one of control parameters of the pilot system. The reliability of the experimental result was verified to compare with one of the prior researches suggesting the dehumidification effectiveness prediction model for liquid desiccant system (Chung (1994), Martin and Goswami (2000)).

2 LIQUID DESICCANT PILOT SYSTEM

The Liquid Desiccant System of this research was designed for the maximum air volume flow of 2000 m^3/h and using LiCl solution as a desiccant material. This System can be control the sensible heat and latent heat loads of the ambient air at the same time.

System is largely two parts, an absorber and a regenerator, each part equipped with a sensible heat exchanger for heat transfer between the ambient air and cooling water, as well as a wood

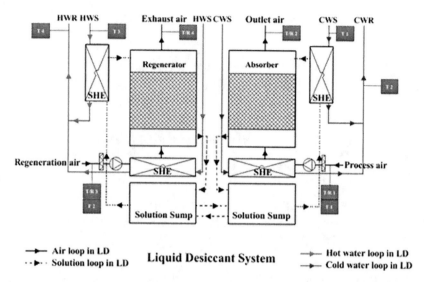

Figure 1. Schematic diagram of the liquid desiccant system.

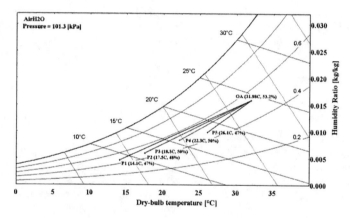

Figure 2. Experimental results for condition of outlet air in the absorber.

fiber honeycomb for moisture transfer between the ambient air and the desiccant solution (Fig 2.). Pilot system has been considered the dehumidification capacity of 12.3 g/kg, with the maximum sensible cooling capacity of 14.0 Kw.

The pilot system located at the highest floor of a 4-story building in Seoul, Korea, with the absorber and the regenerator equipped with the heat exchanger and a packed column respectively. Cooling water and Heating source supply in to progress pre-cooling and pre-heating for the ambient air and heat transfer involving the desiccant material. As for cooling water supply an air-cooled chiller of 10USRT-capacity was installed at the absorber for adjustment of the target water temperature, with a gas fired condensing boiler of 55.8 Kw-capacity further installed at the regenerator for controlling temperature of hot water being a heat source. Further research on the sources of energy consumption, cooling tower with 136.1 Kw-capacity has been installed for selective use of sources in supplying the absorber with cooling water.

In Fig. 1, monitoring for the state of the pilot system, 4 temperature and humidity sensors and 2 air volume flow meters and 4 temperature sensors installed at the absorber and regenerator.

Table 1. Value of the operating parameters of each point.

Point	T_{oa} [°C]	H_{oa} [%]	T_{cw} [°C]	Q_{ai} [m³/h]	Q_{si} [liter/m]	X_{si} [%]
Point 1.	32.5	53	10			
Point 2.	32.3	55	15			
Point 3.	32.3	52	16.5	1840	150	39
Point 4.	31.6	53	19.2			
Point 5.	31.2	53	24.4			

3 EXPERIMENT OF THE PILOT SYSTEM

3.1 Experiment conditions

Assuming the ambient air conditions for peak load, the experiment was performed for four hours 1 PM to 5 PM August 19 2013. The average dry-bulb temperature of ambient air was 31.98°C, and the average relative humidity measured at 53.2%. For accurate measurement of variation in dehumidification effectiveness of the pilot system by temperature variation in cooling water supplied, all other control parameters were fixed, and only changed cooling water temperature. The control parameters were set for a total of five measurements arranged with an hour-long interval.

As Table 1, a total of five experiment points were arranged by cooling water temperature, with all other control parameters fixed, only change the cooling water temperature. The cooling water temperature was varied by the air-cooled chiller from 10°C to 24.4°C. The air volume induced in to the absorber and the amount of the desiccant solution were maintained at 1840 m³/h and 150 liter/m, respectively. The ambient air dry-bulb temperature and the relative humidity measured around the respective 31.2–32.5°C and 52%–55%. The desiccant solution concentration was fixed at 39%.

3.2 Analysis of the experiment results

The experiment results were measured by the air volume flow meter and temperature and humidity sensors set out in the Fig. 1. The temperature and humidity of air were measured when induced in and exhausted via T/R1 and T/R2, respectively. As for cooling water, temperature and humidity was measured when induced in and exhausted at T1 and T2, respectively for configuration of the cooling water temperature, for each experiment point, when induced in. The real-time data of temperature and humidity upon induced-in and exhaustion of the ambient air temperature from the pilot system were kept in records for the eventual analysis of the experiment results.

As for system performance analysis, the temperature and humidity measured at each experiment point were incorporated in evaluation of the dehumidification effectiveness of pilot system by using the Eq. 1 and Eq. 2. For the reliability of the experimental result was verified by way of comparison with literature researches model to the dehumidification effectiveness prediction model for liquid desiccant system.

$$\varepsilon_y = \frac{\omega_{inlet} - \omega_{outlet}}{\omega_{inlet} - \omega_{equ}} \quad (1)$$

where ε_y = dehumidification effectiveness [%]; ω_{inlet} = inlet air humidity ratio [kg/kg]; ω_{outlet} = outlet air humidity ratio [kg/kg]; ω_{equ} = equilibrium humidity of desiccant solution [kg/kg]

$$\varepsilon_{equ} = 0.662 \times \frac{P_s}{B - P_s} \quad (2)$$

where P_s = vapor pressure of the liquid desiccant [kPa]; B = atmospheric pressure [kPa].

3.3 Experiment results

The experiment set up with the results assuming the randomly selected system stabilization time following the temperature variation of cooling water over four-hour running time of the pilot system. The experiment results were monitored the data upon lapse of 45 minutes from each experiment setting on Table 3. The result data comprises the temperature and humidity upon inlet and outlet of the ambient air at the absorber (Fig. 2).

As for point-to-point comparison, the data for point 1, the cooling water temperature at 10°C, represented the sensible cooling of 17.8°C via heat transfer to the ambient air with the dehumidification effect of 0.01122 kg/kg. At the point 2, the cooling water temperature at 15°C, represented the sensible cooling of 14.48°C via heat transfer to the ambient air with the dehumidification effect of 0.00997 kg/kg. The point 3, the cooling water temperature at 16.5°C, fairly hotter than what the air-cooled chiller setting due to the transiently hot ambient temperature and latent heat load, represented the sensible cooling of 13.68°C via heat transfer to the ambient air with the dehumidification effect of 0.0094 kg/kg. At point 4, the cooling water temperature at 19.2°C, represented the sensible cooling of 9.68°C via heat transfer to the ambient air with the dehumidification effect of 0.00754 kg/kg. The last for Point 5, the cooling water temperature at 24.4°C, represented the sensible cooling of 5.88°C via heat transfer to the ambient air with the dehumidification effect of 0.006 kg/kg.

The experimental results, it was confirmed that the value of the temperature and humidity of the ambient air passing through the absorber in response to the change in cooling water temperature changes. It was occurred simultaneously in the process of sensible cooling and dehumidification at the sensible heat exchanger and absorber, respectively. It is determined that the cooling water temperature was significant impact on these results.

3.4 Reliability of the experiment results

This research have been validated the reliability of the experiment results, based upon the experiment data in 3.3, by using the Eq. 1. and Eq. 2, estimating the pilot system effectiveness. The reliability of the experimental result was verified to compare with one of the prior researches suggesting the dehumidification effectiveness prediction model for liquid desiccant system (Chung (1994), Martin and Goswami (2000)). Before the comparison the case of value of the variables required for prediction models of each, it have applied the value of the variable that is planned in the experimental conditions. In addition, by comparing the change in the dehumidification effectiveness according to changes in the cooling water temperature supplied to the same experiment condition and analyzed the validity of the experimental results.

As shown Fig. 3, for variation in the dehumidification effectiveness by cooling water temperature variation as the pilot system represents the dehumidification effectiveness of 88.9% at point 1

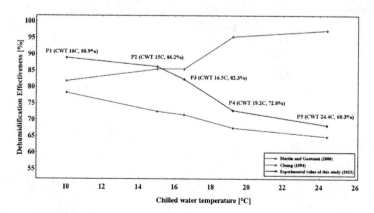

Figure 3. Validation of the reliability of the experiment results.

with the cooling water temperature at 10°C, 86.2% at point 2 with the cooling water temperature at 15°C, 82.3% at point 3 with the cooling water temperature at 16.5°C, 72.8% at point 4 with the cooling water temperature at 19.2°C and 68.3% at point 5 with the cooling water temperature at 24.4°C. While the cooling water temperature increases, the value of the dehumidification effectiveness decreases overall.

Findings similar to this result are described in previous studies Martin and Goswami (2000). According to this research, the difference of result between Martin Goswami (2000) and Chung (1994), Martin and Goswami (2000) have been considered to effect of cooling water temperature on the dehumidification effectiveness more than Chung (1994).

4 CONCLUSIONS

This research conducted the variation in the dehumidification effectiveness of the liquid desiccation system by changing in cooling water temperature varying from 10°C to 24.4°C for the changing the system effectiveness from 88.9% to 68.3%, with the sensible cooling effect varying from 17.8°C to 5.88°C. It estimated that influenced by the sensible cooling effect by cooling water supply and the desiccant solution.

As for performance of the pilot system upon variation in the cooling water temperature, the sensible cooling effect is considered directly influenced. However, the dehumidification effectiveness requires further experiment to verify the cooling water temperature influence. As for the comparison between the experiment results and literature models, the similarity with the predicted result from Martin and Goswami (2000) model was demonstrated, but the comparison with the Chung (1996) showed some disagreements. For this result, further research should be conducted based on analytical and/or empirical research methods.

Ongoing research will address the variation of control parameters affecting on dehumidification effectiveness, and the energy consumption of pilot system for the adaptation renewable energy sources for the heating and cooling sources.

ACKNOWLEDGEMENT

This work was supported by the National Research Foundation of Korea (NRF) grant funded by the Korean government (No. 2012001927).

REFERENCES

Chung, T. W. 1994. Predictions of moisture removal efficiencies for packed-bed dehumidification systems. *Gas Separation and Purification* 8: 265–268.
Chung, T. W. & Ghosh, T. K. 1996. Comparison between Random and structured Packing for Dehumidification of Air by Lithium Chloride Solutions in a Packed Column and Their Heat and Mass Transfer Correlations. *Ind. Eng. Chem. Res* 35: 192–198.
Chung, T. W. & Luo, C. M. 1999. Vapor Pressures of the Aqueous Desiccant. *Journal of Chemical and Engineering Data* 44: 1024–1027.
Conde, M. R. 2004. Properties of aqueous solutions of lithium and calcium chlorides: formulations for use in air conditioning equipment design. *International Journal of Thermal Sciences* 43: 367–382.
Fumo, N. & Goswami, D. Y. 2002. Study of an Aqueous Lithium Chloride Desiccant System: Air Dehumidification and Desiccant Regeneration. *Solar Energy* 72: 351–361.
Jain, S. & Bansal, P. K. 2007. Performance analysis of liquid desiccant dehumidification system. *International Journal of Refrigeration* 30: 861–872.
Katejanekarn, T. et al. 2009. An experimental study of a solar-regenerated liquid desiccant ventilation pro-conditioning system. *Solar Energy* 83: 920–933.
Klein, S. A. et al. 2004. Engineering Equation Solver, EES Manual, Chapter 1: Getting started. *Solar Energy Laboratory, University of Wisconsin-Madison.*

Liu, X. H., Yi, X. Q. & Jiang, Y. 2011. Mass transfer performance comparison of two commonly used liquid desiccants: LiBr and LiCl aqueous solution. *Energy Conversion and management* 52: 180–190.

Longo, G. A. & Gasparella, A. 2005. Experimental and theoretical analysis of heat and mass transfer in a packed column dehumidifier / regenerator with liquid desiccant. *International Journal of Heat and Mass Transfer* 48: 5240–5254.

Martin, V. & Goswami, D. Y. 2000. Effectiveness of heat and mass transfer processes in packed bed liquid desiccant dehumidifier/regenerator. *HVAC&R Research* 6: 21–39.

Mechanical analysis on the tower-girder fixed region of Chongqing Jiayue bridge

Y. Pan
Shool of Civil Engineering and Architecture of Chongqing University of Science and Technology, Chongqing, P.R. China

Y. Deng
T.Y. Lin International Engineering Consulting (China) Co. Ltd, Chongqing, P.R. China

ABSTRACT: The Jiayue Bridge crosses the Jialing River at the core of the Chongqing Liangjiang New Area. It is a partially cable-supported girder bridge with a 250-meter main span. The girder section is a single-cell concrete box with wide wing slabs on both sides. Y-shaped tower and the girder are fixed together. Due to adopting double traffic system, the paths for pedestrians and bicyclists are placed at the crossbeam of the tower, which led to complex structure and mechanical behavior of the tower-girder fixed region. In order to analyze the stress of different working condition and verify the feasibility of the structure design, the three-dimensional finite element model of the tower-girder fixed region is established. The results of calculation have certain value for the similar bridge.

1 INTRODUCTION

1.1 General introduction

The partially cable-supported girder bridge is a new type of bridge structure. It has the form of normal cable-stayed bridge, but has lower towers. It very similar to an extradosed bridge, but has slightly taller towers. The advantage of a partially cable-supported girder bridge is that it utilizes the full capacity of both the girder and the cable-stayed system. It is very different from the normal cable-stayed bridge in cable arrangement, structure size and mechanical behavior. Tower-girder fixed system is usually used for partially cable-supported girder bridge with longer span and taller tower. The feature of the tower-girder fixed system is that tower bears unbalanced bending moment of the girder, the cable force is different remarkably on two sides of the tower and the structure has great integral rigidity.

In the tower-girder fixed region of the partially cable-supported girder bridge, the girder is tall, the distribution of steel bars and prestressed tendons is intensive. Therefore, this region is easy to go wrong. To insure the security and reliability of the bridge, this paper analyzes the local stress of the tower-girder fixed region in construction stage and service stage, discusses the three-dimensional mechanical behavior, verifies the feasibility of the structure design.

1.2 Introduction of the Jiayue bridge

The Jiayue Bridge is on the speedway loop of Chongqing and crosses the Jialing River at the town of Yuelai in the northern district of Chongqing. The Jialing River is a ship channel of national class III and goes through a deep valley at this location. The navigation requires 220 m horizontal clearance and 10 m vertical clearance. It fixes on the main span of the bridge to be 250 meters. The total length of the bridge is 756 m with the configurations of the 250 m main span and two 145 m side span, plus 66 m + 75 m + 75 m continuous approach, see Figure 1 and Figure 2.

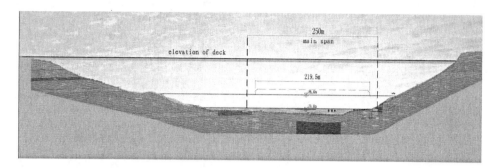

Figure 1. Bridge geological profile.

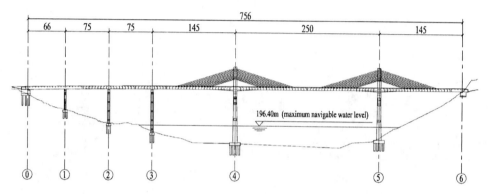

Figure 2. Bridge elevation.

The girder section is a single-cell concrete box with wide wing slabs on both sides. The cables are anchored at both sides of the girders and transfer cable forces to the box section through the diaphragms and top slab. Its section is 28 m width, consists of one 12 m width box and two 8.0 m width cantilever flanges. Two different thicknesses of diaphragms are set in the spacing of 5 m along the centerline of the bridge. Stronger one of 0.7 m thick is set at the cable stayed zone and 0.4 m thick diaphragms are set at the rest. The box girder is variable cross-section, from the pier segment to the segment of first stay cable, the depth is varied from 7 m to 5 m following a parabolic curve at the bottom of the box. It then becomes constant depth of 5 m.

The tower is designed as Y-shape, with 126.5 m total height and 32 m above the deck. The tower includes right and left legs. The two legs of the tower are connected by a crossbeam which is fixed with the girder. Vertical loads are transferred by the webs and diaphragms of the crossbeam. Due to adopting double traffic system, the paths for pedestrians and bicyclists are placed at the crossbeam of the tower, which led to complex structure and mechanical behavior of the tower-girder fixed region. In order to insure the reliability of the bridge, vertical, transverse and longitudinal prestressed tendons are arranged, see Figure 3.

2 ANALYSIS OF THE FINITE ELEMENT MODEL

2.1 Calculating idea

According to the Saint Venant's principle, the stress distribution of equivalent forces system don't agree with the actual situation near the equivalent area, while has high accuracy leave the equivalent area a bit far. Therefore, accurate three-dimensional unit simulation can just only used for the analysis of the tower-girder fixed region and the adjacent structure. Taking the displacement and

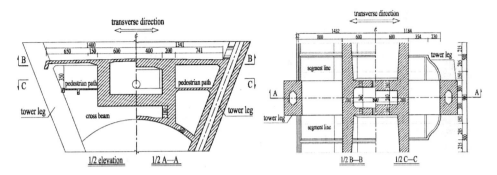

Figure 3. Bridge elevation structure of the tower-girder fixed region.

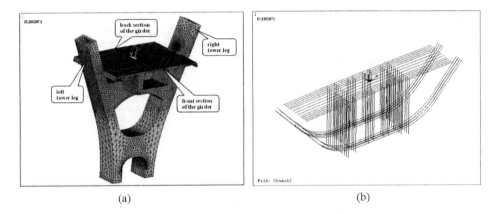

Figure 4. (a) 3D Finite element model, (b) Model of transverse and longitudinal prestressed tendons.

loads calculated by the integral truss model as boundary conditions and adding to the end of the three-dimensional model, the local stress of the tower-girder fixed region can be analyzed.

2.2 *Finite element model*

The analysis is focused on the tower-girder fixed region. The range of calculation model not only includes the tower-girder fixed region, but also extends at both ends appropriately. The tower above the deck 8 m, the tower below the deck 33 m and the girder around the tower center line on each side 9.5 m are taken into account. The concrete is simulated as tetrahedron element with 10 nodes, the prestressed tendons are simulated as 3D rod element by means of adding initial strain, see Figure 4.

The link between the prestressed tendons and the concrete is simulated by means of node coupling. Free meshes are used for dividing unit. At the haunches of the girder, as well as the juncture of the prestressed tendons and the concrete, the meshes are dense. There are 106845 elements and 27176 nodes in the finite element model. Origin of coordinates is set at the intersection point of the tower midline and the girder midline. The X axis is set as transverse, the Y axis is set as longitudinal and the Z axis is set as vertical.

Simulation of the boundary conditions:

The bottom of the tower is fixed.

Master nodes are established at the top tower and the centroid of the two girder section. The other force nodes work as slave nodes. Master nodes and corresponding slave nodes are connected

Table 1. The value of loads.

Loading cases	Location	Axial force (kN)	Bending moment (kN·m)		Shear force (kN)	Torque (kN)
			Mz	My		
Loading cases 1	Left side of the top tower	−70992	10323	30724	−1159	–
	Right side of the top tower	−70992	−10323	30724	−1159	–
	Front section of the girder	−445268	–	−53564	−5373	–
	Back section of the girder	441598	–	54834	−4916	–
Loading cases 2	Left side of the tower top	−74300	11755	38203	−2405	–
	Right side of the tower top	−71071	−10339	29748	−517	–
	Front section of the girder	−451214	–	2823	−8410	44281
	Back section of the girder	443549	–	25648	−7706	33719

by rigid joint. The forces calculated by two-dimensional truss are equivalently transformed into node loads adding to the centroid.

Two loading cases are taken into account. Loading cases 1 is the worst case in the construction. Loading cases 2 is the action of the eccentric loads plus dead loads, see Table 1.

3 ANALYSIS OF THE RESULTS

Basing on the three-dimensional finite element model, the stress distributions of the top slab, the webs, the diaphragms and the bottom slab are analyzed.

3.1 Stress distribution on top slab of the tower-girder fixed region

The upper surface and nether surface of the top slab are compressed under the action of cable forces and longitudinal prestress. The value of the normal pressure stress is from 0.4 MPa to 15 MPa. Because the cantilever is long, the stress distribution is affected by the transverse load distribution and decreases from the central box to the two sides. The maximum stress appears at the upper surface of the central position of the surface.

Under the action of vertical loads and cable forces transferred by the top tower, the tower-girder fixed region mainly bears bending moment and shear force at transverse direction. Because the paths for pedestrians are placed at the crossbeam of the tower, the transverse force of deck is similar to the continuous slab respectively supported by the top tower and the webs of girder. Through increasing the thickness and adding transverse prestress, the bearing capacity is improved. The results indicate that the whole top slab is compressed at transverse direction and the value of the normal pressure stress is from 0.2 MPa to 10 MPa. On both sides of the webs, the stress is the minimum. At the internal box and the end of the cantilever, the stress is the medium. At the junction of the top slab and the tower, the stress is the maximum, see Figure 5.

3.2 Stress distribution on webs of the tower-girder fixed region

Vertical loads of the girder are transferred by the webs to the crossbeam of the tower. Therefore, the webs are mainly subjected to the shear forces. The webs are compressed under the action of the external loads and vertical prestress. The tensile stress doesn't appear at the vertical direction and the value of pressure stress is less than10 MPa.There is no principal tensile stress on the webs and the value of principal pressure stress is from 0.6 MPa to 12 MPa. Local stress concentration appears at the junction of the webs and the tower legs.

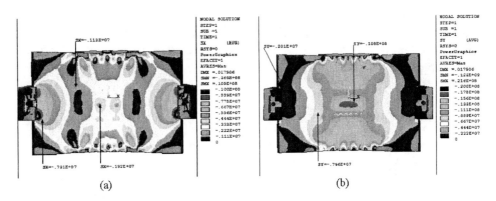

Figure 5. (a) X axial stress nephogram of the top slab, (b) Y axial stress nephogram of the top slab.

3.3 Stress distribution on diaphragms of the tower-girder fixed region

The front and back diaphragms at the tower-girder fixed region are key members for transferring vertical loads from the upper girder to the nether tower legs. Because the distance between the tower legs is large, in order to balance the bending moment of the crossbeam caused by the vertical loads, curve prestressed tendons are placed in the diaphragms. The calculation results indicate that under the action of vertical loads and prestress, the vertical tensile stress of the diaphragms is less than 1 MPa, the pressure stress is less than 10 MPa, the principal tensile is less than 1.5 MPa, the average stress is 0.6 MPa and the principal pressure stress is less than 12 MPa.

The calculation results of construction stage indicate, after the longitudinal, transverse and vertical prestressed tendons of the tower-girder fixed region stretched, there will be principal tensile stress about 1 MPa at the points along the transverse prestressed tendons, especially at the tower legs outside of the pavement. While at the service stage, principal tensile stress of the points with larger principal stress mentioned above will change to principal pressure stress.

3.4 Stress distribution on bottom slab of the tower-girder fixed region

Under the action of dead loads, X axial tensile stress of the bottom slab is less than 1.5 MPa, principal tensile stress is less than 1.6 MPa. Under the action of eccentric live loads, principal tensile stress at one side of the bottom slab is 2.5 MPa. Both the principal tensile stress doesn't exceed the bearing capacity of the concrete.

3.5 Summary of the stress analysis

The stress results of the tower-girder fixed region is seen Table 2.

4 CONCLUSIONS

The path of the loads transfer of the Jiayue bridge is clear. By strengthening the structural measures and designing reasonable prestressed system, the whole tower-girder fixed region is mainly subjected to pressure and all of the pressure stress doesn't exceed the demand of the codes. The design of the tower-girder fixed region is reasonable and feasible.

The cables are subjected to most of the vertical loads of the girder and transfer the loads by upper tower legs to the nether tower legs. Because the paths for pedestrians and bicyclists are placed at the crossbeam of the tower, the local stress is large at the junction of the top slab of the girder and the upper tower legs. By the structural measure of local chamfering, the local stress concentration is remitted.

Table 2. The stress results of the tower-girder fixed region.

Location and type of the stress		loading cases 1 (MPa)	loading cases 2 (MPa)
Transverse normal stress of top slab	$\sigma_{x\,max}$	−0.5	−0.2
	$\sigma_{x\,min}$	−10	−10.4
Longitudinal normal stress of top slab	$\sigma_{y\,max}$	−0.4	−0.3
	$\sigma_{y\,min}$	−15.2	−16.3
Transverse normal stress of bottom slab	$\sigma_{x\,max}$	1.1	1.4
	$\sigma_{x\,min}$	−13.2	−13.8
Longitudinal normal stress of bottom slab	$\sigma_{y\,max}$	−0.4	−0.1
	$\sigma_{y\,min}$	−15.4	−16.4
Vertical normal stress of diaphragm	$\sigma_{z\,max}$	0.5	0.7
	$\sigma_{z\,min}$	−4.5	−4.1
Maximum principal tensile stress of diaphragm	$\sigma_{1\,max}$	1.2	1.45
Maximum principal pressure stress of diaphragm	$\sigma_{3\,min}$	−9.3	−10.2
Vertical normal stress of web	$\sigma_{z\,max}$	−0.2	0.2
	$\sigma_{z\,min}$	−4.8	−3.5
Maximum principal tensile stress of web	$\sigma_{1\,max}$	1.3	1.6
Maximum principal pressure stress of web	$\sigma_{3\,min}$	−10.2	−11.2

Curve prestressed tendons placed in the diaphragm of the crossbeam effectively enhance the flexural capacity of the cross beam. In the construction process, the local principal tensile stress is large along the axes of the prestressed tendons, which can be improved by setting local anti-collapse steel bars.

REFERENCES

De-lan, Yin. & An-shuang, Liu. 2010. Jiayue Bridge, Its Design and Construction. *3rd$_{fib}$ International Congress*, Washington.
Jiang, Huang. & Cheng, Hu. 2012. Spatial Stress Analysis of Segment No. 0 of an Extradosed Cable-Stayed Bridge. *Journal of Hefei University of Technology* 35(8): 1097–1100.
Lu-song, Yu. & Dong-sheng, Zhu. 2008. Local Stress Analysis of Rigid Fixity Joint of Pylon, Girder and Pier of an Extradosed Bridge. *Bridge Construction* (1): 54–57.
Man-Chung, Tang. 2011. Poised for Growth. *Civil Engineering* (9): 65–72.
Yan, Wang. & Huai, Chen. 2008. Local Stress Analysis of Segment No. 0 of Long Span Continuous Beam Bridge Constructed With Cantilever Erection Method. *Journal of Rail Way Science and Engineering* 5(3): 23–27.
Yu, Deng & De-lan, Yin. 2012. Design and Construction of The Jiayue Bridge in Chongqing, China. *IABSE*, Seoul.

The development of intelligent aided system for safety assessment of buildings in post-earthquake based on web

Baitao Sun, Lei Zhang, Hongfu Chen & Xiangzhao Chen
Institute of Engineering Mechanics, China Earthquake Administration, Harbin, Hei Longjiang, China
Key Laboratory of Earthquake Engineering and Engineering Vibration of China Earthquake Administration, Harbin, Hei Longjiang, China

ABSTRACT: A computational aided system for safety assessment of buildings in post-earthquake based on Web is described in this article. Safety assessment of damaged buildings in post-earthquake is a significant and strong timeliness required job for earthquake emergency relief. The functional structure and functional logic of the system and workflow of safety assessment module are studied in depth. The input of the system is the subjective information of damaged buildings on site by inspectors which are possibly not experts of the safety assessment field. The system is based on Analytic Hierarchy Process (AHP) algorithm model and an effective tool for providing decisions about damage levels and safety of the post-earthquake buildings. Especially users can register and use the system once the computer connects with the Internet. Typical interfaces of the system are given and the system optimization is proposed.

1 INTRODUCTION

Safety assessment of buildings in post-earthquake indicates assessing the safety of damaged buildings in the expected earthquake, by means of examining the post-earthquake damage conditions and pre-earthquake seismic capability of the buildings during the post-earthquake emergency period. It requires strong timeliness, different from anti-seismic evaluation with seismic fortification intensity and dangerous building appraisal in pre-earthquake and post-earthquake. Assessing the safety of the damaged buildings on site fast, timely and efficiently can guarantee the safety of victims and properties in the disaster areas, thus reducing casualties and economic losses. It also has great significance to the maintenance of social stability, and provides the government with basis for the distribution of disaster relief supplies (Sun, B.T. et al. & Wang, D.M. 2003, Wang, X. 2007).

Research on safety assessment of buildings in post-earthquake in emergency work has already been launched abroad. In America, the Applied Technology Council (ATC) proposed *the Procedures for Post-Earthquake Safety Evaluation of Buildings* as a standard for the safety investigation of buildings. Once earthquake occurs, Federal Emergency Management Agency(FEMA) and local emergency departments will organize trained professionals to assess safety of post-earthquake buildings. The Japan Building Disaster Prevention Association published *Guideline for Post-earthquake Damage Evaluation and Rehabilitation* (1991, revised in 2001). And some Japanese experts proposed quantitative method to estimate the damage level of post-earthquake buildings (Nakano, Y. et al. 2004). In Spain and Colombia, Carreño M. L. and Cardona O. D. proposed assessment methods based on neural network and fuzzy mathematics, and developed computational tool for it. Some other countries (e.g., Mexcio; Italy) also carried out research in this field (Carreño, M. L. et al. 2010).

In China, in terms of the severe earthquake situation and urgent demands for safety assessment of post-earthquake buildings, the government and researchers have been thinking highly of safety assessment research and practice. The author has been committed to studies of safety assessment of post-earthquake damaged buildings in recent decades. In 2001, the author presided over the compilation of the national standard *Post-earthquake field works-Part 2: Safety assessment of buildings* (2001). Afterwards, quantitative method for safety assessment based on analytic hierarchy process (AHP) was proposed (Wang, D.M. 2003, Wang, X. 2007). The method was based on national standards, and combined with actual damages and expertise. Then a single-location version and a PDA version safety assessment system software had been developed successively (Chai, X.H. 2009). They had been used in seismic field of China, and achieved good effect of disaster reduction. However, some problems and deficiencies appeared in the practical application of those systems. For example, when users want to apply the single-location system, they must install the program first. Besides, managers can't update model parameters in time, and collection of basic data and sharing of assessment results were difficult.

Therefore, the author proposed the establishment of a Web-based safety assessment of buildings in post-earthquake intelligent aided system. The system makes it possible for the professionals and inspectors which are possibly not experts of the safety assessment field to complete safety assessment. And the experts' knowledge and experience can be shared by the general public in the field timely. This paper mainly describes the functional structure and functional logic of the system, and workflow design of safety assessment module, the global framework design, algorithm model and system development.

2 FUNCTION ANALYSIS OF THE SYSTEM AND WORKFLOW DESIGN OF SAFETY ASSESSMENT MODULE

2.1 Function analysis of the system

The core function of the system is that by means of data collection of earthquake, the building and its damages, the user assesses the safety of the building, and output the results and statistic analysis. The system is divided into four modules according to functions: Web module, assessment system module, forum module and background management module. The Web module publicizes the knowledge of earthquake and post-earthquake safety assessment. The assessment system module provides services of complete building safety assessment and statistical analysis of the results. The forum module is intended to provide a platform for experience exchange and resource sharing. And the background management module makes unified organization and management of the modules above (Zhang, X.H. 2010). The function structure is shown in Figure 1.

The assessment system module is core of the system. The function logic of this module is shown in Figure 2.

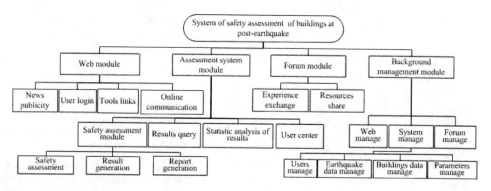

Figure 1. Function structure of the system.

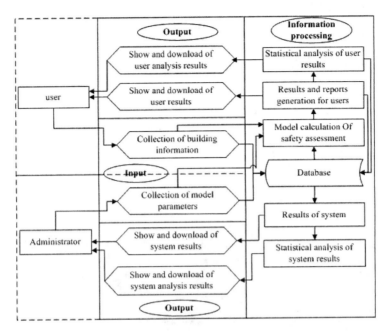

Figure 2. Function logic of the assessment system module.

2.2 *Workflow design of the safety assessment module*

The application flow of the safety assessment module can be devided into six steps sequentially as follows: (1) selecting earthquake; (2) inputing basic information of the building; (3) setting surrounding information; (4) setting expect earthquake effect; (5) inputing earthquake damage information; (6) generating assessment results.

The detailed workflow is shown in Figure 3.

3 FRAMEWORK DESIGN AND ALGORITHM MODEL OF THE SYSTEM

3.1 *Framework design of the system*

The system, adopting the design of the layered architecture, is divided into five layers (data layer, persistence layer, business layer, control layer and presentation layer from the top to down), as shown in Figure 4. Data layer is used for administration and maintenance of relational database. Persistence layer achieves mapping business between Java object and the forms in the database. Business layer incorporates all operations related to database, and encapsulate the operations involving database. Control layer completes transfer and control between the presentation layer and business layer. The presentation layer provides interface of the interaction between user and the system and offers display of data. By the guidance of the layered architecture development, the system realizes the high cohesion of inner modules and of inner layers, also implements low coupling among modules and among layers (Zhang, X.H. 2010).

3.2 *Algorithm model*

The algorithm model of the system mainly adopts the first model by Wang, X. (2007). It's based on analytic hierarchy process (AHP). Verified by earthquake damage examples, the model can

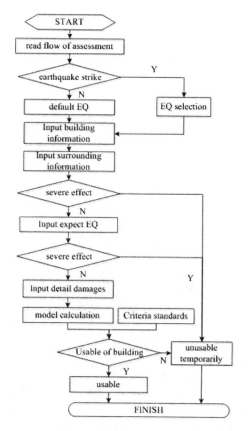

Figure 3. Workflow of safety assessment module.

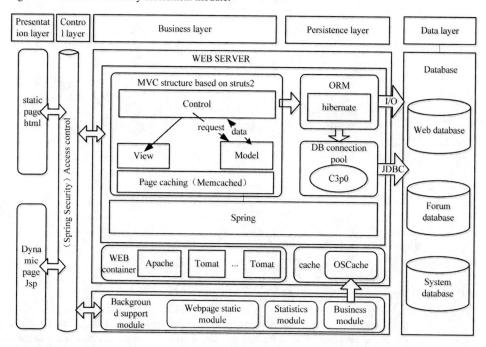

Figure 4. The layered architecture of safety assessment system.

perform the safety assessment of buildings on site accurately and reliably, and can replace the previous expert subjective evaluation methods well (Sun, B.T. et al. 2009).

4 THE DEVELOPMENT AND IMPLEMENTATION OF THE SYSTEM AND FUNCTIONAL TEST

4.1 *The development and implementation of the assessment system*

For the webpage development of the assessment system, technologies of HTML, JSP, CSS, Flash, JQuery and Ajax were adopted. The implementation of the logical part adopted J2EE technology including Struts2, Hibernate, Spring and Freemarker etc. The assess control of the system employed Spring Security technology. The output of the result report used the technology of iText and POI. And the usage of Google Map achieved obtaining of latitude and longitude, display of historical earthquakes and buildings distribution etc (Zhang, X.H. 2010).

The implementation of the webpage development adopted the layout of DIV&CSS. The system achieved the desired functions: data collection of expect earthquake, surrounding information, building information and detail damages, then safety assessment and results output etc. Convenient and rapid assessment is implemented by the development of the system, and it can be registered by the general public for free.

4.2 *Functional test*

For the functional test of the assessment system, the algorithm model involved in 2.2 was adopted, the model parameters and building examples were from paper by Chai, X.H. (2009), and the parameter α and β were specified 20 and 0.4 respectively.

According to the calculation of 53 multi-storey masonry structures, 26 multi-storey and high-rise reinforced concrete structures, 22 inside frame structures, and 12 single reinforced concrete column factories, and comparison with the actual investigation results, it is found that the accuracy of the system reached 90% or more. So it is demonstrated that the system has high accuracy and reliability. During the process of the test, the average time-consuming of safety assessment of each building is about two minutes, and the system is easy to operate. So the assessment system can offer reliable supports to safety assessment of buildings at post-earthquake in future.

5 CONCLUSIONS AND PROSPECT

The intelligent aided system is one of the major achievements that computer technology applies in earthquake field. Efficiency of safety assessment of buildings in post-earthquake will be greatly improved, and further development of earthquake field work will be propelled forward by the successful development of the system. By means of the system (1) the model parameters could be centrally managed and timely updated. And the safety assessment functions could be intellectualized and improved more; (2) The managers of the system could collect and share the buildings detail data more easily, while the data is important for the optimization of assessment models; (3) It is possible for the public to share the field achievements of safety assessment more conveniently, since the user could register and use the function through the Internet; (4) The experts could guide users online real-time through the online communication part. All those make contributions to the social public welfare service in earthquake field. The system adopts modular and hierarchical design principles in order to extend the system functions in the future, in addition to the system authority is developed.

In China, the built forms and seismic capacity of buildings from various regions are quite different because of vast territory and climate variation. Therefore the regional allocation of model parameters is a key research direction for model algorithm optimization. The system has some aspects which should be further studied as follows: (1) developing the GIS-Based information display and spatial analysis functions; (2) increasing the collection and automatic identification function of damage pictures and videos; (3) taking advantage of the newest computer technologies (e.g., cloud-computing); (4) developing the mobile intelligent terminal to serve the earthquake emergency work better.

ACKNOWLEDGEMENT

This paper was supported by International technical cooperation project of China-USA (Grant No. 2011DFA71100) and National natural science foundation (Grant No. 50938006).

REFERENCES

Carreño, M. L. & Cardona, O. D., Barbat, A. H. et al. 2010. Computational Tool for Post-Earthquake Evaluation of Damage in Buildings. *Earthquake Spectra*, 26(1): 63–86.
Chai, X.H. 2009. Safety Assessment of Buildings at Post-Earthquake Based on VB. Institute of Engineering Mechanics, China Earthquake Administration.
Nakano, Y., Maeda, M., Kuramoto, H. et al. 2004. Guideline for Post-earthquake Damage Evaluation and Rehabilitation of RC Building in Japan. 13th World Conference on Earthquake Engineering. Vancouver, Canada:124–139.
Sun, B.T. & Wang, D.M. 2003. Assessing safety of buildings on post-earthquake field quantitatively. *Earthquake Engineering and Engineering Vibration*, 23(5):209–213.
Sun, B.T. & Wang, X., Chai, X.H. et al. 2009. A quantitative method for safety assessment of buildings in post-earthquake field. *Journal of Harbin Institute of Technology*, 41(2):129–132.
Wang, D.M. 2003. Intelligence Aided System For Safety Assessment Of Buildings On Sit. Institute of Engineering Mechanics, China Earthquake Administration.
Wang, X. 2007. Study on the quantization-based method for safety assessment of buildings on seismic site and development of expert system. Institute of Engineering Mechanics, China Earthquake Administration.
Zhang, X.H. 2010. Research and Develop of J2EE based safety Assessment of Buildings at Post-Earthquake. Institute of Engineering Mechanics, China Earthquake Administration.

Impact of PV/T system on energy savings and CO_2 emissions reduction in a liquid desiccant-assisted evaporative cooling system

E.J. Lee & J.W. Jeong
Division of Architectural Engineering, College of Engineering, Hanyang University, Seoul, Republic of Korea

ABSTRACT: The main thrust of this research is to estimate the energy savings and CO_2 reduction potentials acquired by integrating photovoltaic/thermal (PV/T) system into the established liquid desiccant-assisted evaporative cooling system. The PV/T system provides the electric power and hot water simultaneously, which are required to operate the liquid desiccant unit in the established system. The required capacity of the PV/T system can be determined for the peak electric load or peak thermal load of the liquid desiccant-assisted evaporative cooling system. Consequently, energy simulations were performed for PV/T integrated system alternatives; the one is designed for the peak electric load, and the other is for the peak thermal load. The result shows that the PV/T system sized for the electric load may provide more than 10% of primary energy savings and CO_2 emissions reductions compared with the PV/T system sized for the peak thermal load.

1 INTRODUCTION

The conventional photovoltaic (PV) system applied into the building has generated electricity about 10%–15% of incident solar radiation. The rest of solar radiation is converted to the radiant heat, it causes the decrease of electric generation efficiency of the PV module by increasing the PV module surface temperature. The ratio of decreasing the generation efficiency of the PV module over the increase of PV module surface temperature is 0.5%/°C, the additional cooling system for PV module is required.

In order to overcome this problem, integration of PV system and solar thermal system (PV/T system), which generates both electricity and heat, is developed by Martin Wolf in 1976(**). This system could improve the use of solar radiation per unit area of a module, compared with a conventional PV system. However, the initial cost and system efficiency of the PV/T system are largely regarded depending on the design load, either electric load or thermal load of the building.

In this research, design process of PV/T system to obtain the optimum system efficiency is investigated. The energy saving potential is quantitatively estimated, when the PV/T system supplies the electric and thermal load of liquid desiccant and evaporative cooling-assisted-100% outdoor air system. And then, the CO_2 emissions reduction of the proposed system is also evaluated via conventional system.

2 LIQUID DESICCANT PILOT SYSTEM

From the previous research, one can be known that indirect and direct evaporative cooling-assisted-100% outdoor air system (IDECOAS) can save 40% of annual operating energy compared to the conventional VAV air-conditioning system. However, the IDECOAS may not achieve over 20% of operating energy savings, caused by decreasing the cooling performance of evaporative cooler under hot and humid climate. To solve this problem, integration of liquid desiccant (LD) system and

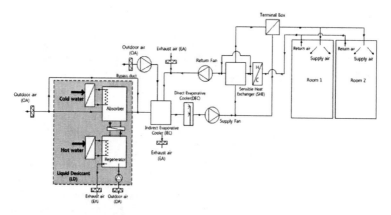

Figure 1. Schematic diagram of LD-IDECOAS.

the IDECOAS is suggested. This method could improve the cooling performance of evaporative cooling system by dehumidifying the hot and humid outdoor air.

As shown in Fig: 1, LD-IDECOAS is composed of the LD, IEC and DEC at upstream, and sensible heat exchanger (SHE) and heating coil (HC) is located in downstream. The LD is composed of absorber dehumidified by spraying desiccant solution and regenerator regenerated by heating desiccant solution. Heating coil is used to maintain 45~80°C in the regenerator and cooling coil helps continuous dehumidification to remove the heat at 15~30°C.

3 REQUIRED PV/T SYSTEM AREA PREDICTION PROCESS

3.1 Calculation of required load

PV/T system is one of combined heat and power (CHP) systems that generate electricity and heat simultaneously. The PV/T system is based on the required electric load and thermal load in calculation of the system capacity. From the previous research, one is reported that PV/T system save 25% of energy in following thermal load. On the other hand, CO_2 emissions reduce the amount compared with following electric load. That is, economic efficiency and eco friendliness of PV/T system depend on following operation strategy when designed.

Therefore, this section is calculation of electric and thermal load required LD-IDECOAS pilot system. Then energy saving rate and the CO_2 emissions reduction effect are quantitatively estimated when the PV/T system is applied instead of LD-IDECOAS pilot system.

(1) Calculation of required electric load

Load of the DC power based on maximum power capacity is 19.83 kW per day. The monthly-average daily electricity production amount is 178.50 kWh/day when inverter efficiency is assumed to be 90% and system operation time is 9 hours per day.

(2) Calculation of required thermal load

The temperature of heat source circulating the HC is 55°C~60°C at the regenerator It is essential for regeneration of desiccant solution (LiCl). The required thermal load is 43.47 kW per day using the mass flow, specific heat and temperature difference between inlet and outlet. The monthly-average daily thermal production amount considering operation time is 621.03 kWh/day.

3.2 Calculation of monthly-average daily total insolation on tilted surface

As shown in Eq. 1 for calculation of total insolation on tilted surface, Table 1 results the monthly-average daily total insolation on tilted surface from 3-year Seoul Weather Data. It uses following

Table 1. Seoul weather data (2010–2012).

Month	Monthly-average daily values			
	Ambient temperature (°C)	Clearness index	Insolation on horizontal surface (kW/m²day)	Total insolation on tilted surface (kWh/m²day)
Jun	23.17	0.48	0.836	8.350
Jul	25.27	0.32	1.062	4.708
Aug	26.47	0.34	0.938	4.644
Sep	21.53	0.45	0.676	5.333

Table 2. PV/T product characteristics.

Electrical characteristics	STC	Power (P_max)	250 Wp
		Voltage (V_mpp)	30 V
		Current (I_mpp)	8.34 A
		Module efficiency	15.4%
		Temperature coefficient	0.40%/°C
	NOCT	Temperature	45°C
Thermal characteristics		Operating temperature range	−40~85°C
		Heat removal fator × Transmittance-absorption ($F_R(\tau\alpha)_n$)	0.67
		Heat removal fator × Heat loss coefficient ($F_R U_L$)	4.56 W/m²°C
Size		Dimensions (H*W*D)	1637 × 992 × 50 mm
		Area (with frame)	1.62 m²

values; slope of 22e surface reflection rate of 0.3 and azimuth angle of 2u. The lowest total insolation on tilted surface is shown on August, because of short sunshine duration and rainy season.

$$\overline{H_T} = \overline{H_b} \overline{R_b} + \overline{H_d}\left(\frac{1+\cos\beta}{2}\right) + \overline{H}\rho_g\left(\frac{1-\cos\beta}{2}\right) \quad (1)$$

where, $\overline{H_T}$ = Monthly-average daily radiation on tilted surface [MJ/m²-day]; $\overline{R_b}$ = Geometric factor; β = Slope angle [°]; ρ_g = Ground reflectance.

3.3 Calculation of required PV/T area

In this research, PV/T system is configured by setting peak-load to be designed for greenhouse gas emission reduction as well as energy saving. In the result of weather data, it is maximum area that is designed on August. Therefore PV/T system is based on August-average daily values

As shown in Eq. 2, PV/T area is determined by total insolation on tilted surface and efficiency according to the temperature As result, PV/T area following electric load is 295.18 m² and following thermal load is 251.77 m². PV/T simulation model consisted of characteristics of a commercial PV/T product and Retscreen 4, as shown in Table 2.

$$A = \frac{P}{\eta \times \overline{H_T}} \quad (2)$$

where, A = Required PV/T area [m²]; P = Required load amount [kWh]; η = Efficiency.

Table 3. TPES and Emissions following peak electric load.

	Area (m^2)	Excess area (m^2)	TPES (TOE/m)	Emissions (tCO$_2$/m)
Reference	–	–	2.31	5.26
Jun	295.18	133.84	−2.13	−4.87
Jul	295.18	5.86	−0.33	−0.80
Aug	295.18	0.00	−0.26	−0.62
Sep	295.18	44.72	−0.34	−0.76

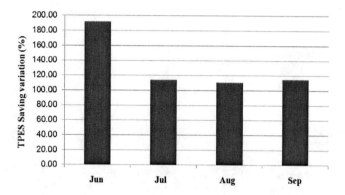

Figure 2. Energy saving rate compared with reference.

4 ENERGY SIMULATION

The conventional energy sources of electricity and LPG are convert primary energy (TOE). Assuming the operation period of 20day/month, the results are 0.82TOE for the electric load and 1.49TOE for the thermal load, for the total 2.31TOE/month. However, PV/T system installation show that energy saving rate is 94~112% compared with the conventional energy.

The PV/T system area following electric load requires more additional area of 43 m^2 but meets required load of LD-IDECOAS pilot system without additional energy sources. It is available heat source of 11~12% following electric load, while operation strategy following thermal load demands supplementary electricity of 6% in August.

As shown in Table 3, it is simulation result of total primary energy supply (TPES) CO$_2$ emission during summer-season operation period, designed for August-average daily values following peak electric load. The result in June shows that the system represented 45% of the excess area when compared to the conventional energy sources of electricity and LPG (References) The results in July and September show each 2% and 15% excess area. It is no such significant difference in part of the primary energy saving rate, as shown in Fig. 2.

5 CONCLUSIONS

This research calculates PV/T system area and analyzes saving rate of energy and greenhouse emissions to provide LD-IECOAS pilot system for required load. Area of PV/T system following electric load is required for additional area of 18% compared to area following thermal load. That is, rate of heat collection gain more energy than required thermal load. The excessive heat collection can be utilized for servicing hot water and other energy-saving means. Otherwise, in case of area following thermal load, lack of electricity is enforced to purchase commercial electricity.

Therefore, PV/T system following peak electric load save more than 110% of energy compared to the LD-IDECOAS pilot system.

Research results show that energy saving rate and carbon emission reduction rate save average 114% from July to September. The case in June results saving rate of 190%, but it only operates mere 10 days for LD-IDEOCAS pilot system. Consequently, PV/T system following electric load determines more effective way in terms of eco-friendliness and energy saving than thermal load.

Further research will obtain various weather data and actual PV/T system installation cases and review the suggested algorithm. Then it can be determined to predict the required area for the PV/T system installation at the design stage.

ACKNOWLEDGEMENT

This work was supported by the National Research Foundation of Korea (NRF) grant funded by the Korean government (No. 2012001927).

REFERENCES

Duffie Jone, A. & Beckman William, A. 1991. Solar Engineering of Thermal Process, *John Wiley & Sons Inc.*
Garg, H. P. & Agarwal, R. K. 1995. Some aspects of a PV/T collector/forced circulation flat plate solar water heater with solar cells. *Energy Conversion and Management* 36: 87–99.
Hawkes, A. D., & Leach, M. A. 2007. Cost-effective operating strategy for residential micro-combined heat and power. *Energy* 32: 711–723.
Hottel, H. C., & Whillier, A. 1955. Evaluation of flat-plate solar collector performance, *Trans. Conf. Use of Solar Energy* 3.
Jalalzadeh-Azar, A. A. 2004. A Comparison of Electrical- and Thermal-Load-Following CHP Systems. *ASHRAE Transactions* 110: 85–94.
Kalogirou, S. A., & Tripanagnostopoulos, Y. 2006. Hybrid PV/T solar systems for domestic hot water and electricity production. *Energy Conversion and Management* 47: 3368–3382.
Kim, M.H., Kim, J.H., Kwon, O.H., Choi, A.S. & Jeong, J.W. 2011. Energy Conservation Potential of an Indirect and Direct Evaporative Cooling Assisted 100% Outdoor Air System, *Building Services Engineering Research and Technology* 32: 345–360.
Kim, M. H., & Jeong, J. W. 2013. Cooling performance of a 100% outdoor air system integrated with indirect and direct evaporative coolers. *Energy* 59: 726–736.
Liu, B. Y. H., & R. C. Jordan 1960. The Interrelationship and Characteristic Distribution of Direct, Diffuse and Solar Radiation, *Solar Energy* 4: 1–19.
Mago, P. J., Chamra, L. M. & Ramsay, J. 2010. Micro-combined cooling, heating and power systems hybrid electric-thermal load following operation. *Applied Thermal Engineering* 30: 800–806.
Zondag, H. A., De Vries, D. W., Van Helden, W. G. J., Van Zolingen, R. J. C. & Van Steenhoven, A. A. 2003. The yield of different combined PV-thermal collector designs. *Solar energy* 74 : 253–269.

Study on multi-zone airflow model calibration process and validation

S.K. Han & J.W. Jeong
Division of Architectural Engineering, College of Engineering, Hanyang University, Seoul, Republic of Korea

ABSTRACT: The multi-zone network models simulating airflow and contaminants distribution inside a building are also useful tools to predict the air tightness of the building. However, the simulation outcomes of the multi-zone network models may have significant deviations from the actual building air tightness because the initial conditions entered for the simulation are commonly selected by subjective judgment of the engineer. Thus, the calibration of the multi-zone network models is required for getting more reliable and objective prediction of building air tightness. In this study, the model calibration process is proposed, and then verified for actual building cases. As a multi-zone network airflow modeling tool, CONTAMW3.1 is selected, and the simulation results acquired at each calibration step are compared to the actual test results. The result shows that the proposed model calibration process improves accuracy of the predicted air tightness of the building.

1 INTRODUCTION

The indoor air quality, ventilation system performance and energy-saving effect are evaluable by way of a simulation involving such environment in buildings such as airflow distribution and contaminants level and Stack Effect. With varied phenomenon are predictable over the design phase, multiple alternatives are available for comparison.

As it is analyzed to airflow distribution by CFD or multi-zone network models, the building performance is predicted by checking the simulation results for simplicity and time saving. Some input variables and assumptions are needed for configuring airflow simulation and have spatial limit for multi-zone network models. The simulation outcomes of the multi-zone network models, however, may have significant deviations from the actual building air tightness because the variables entered for the simulation are commonly selected by subjective judgment of the researcher. Also, the scope of available data is very limited. Proper calibration for input variables to better simulate what actually happens is thus necessary, which is why this study analyzes the effect of input variables on the results to evaluate validity of the calibration by using of experiment of design

2 METHODS

This study analyzed the effect of many variables to the predictable results and interaction among variables, using data and variables obtained from the existing literature. More influential variables were selected for measurement of the actual building tightness of the selected variables in design of experiment. The initial simulation model consists of input parameters obtained from the existing literature and performed the calibration process substituting the actual values for the specific valuables. As value of input parameters get calibrated, the air tightness performance and the simulated results were compared in checking validity of the calibration process.

2.1 Airflow model tuning method

To predict such phenomenon as ventilation performance in a building using a simulation demands various assumptions and preconditions, all input parameters established by the engineer's subjective judgment. According to Joseph Firrantello et al., the simulation results may substantially have a difference from what actually happens, where the input parameters should be chosen with special attention.

For more speed and simple simulation, this research incorporated 'Experiment of Design', one of the statistical approaches to simulate the phenomenon in an actual building involving least calibrations. Experiment of Design is of particular significance in that the probabilistic property of input parameters is involved in assessing the potential difference of the simulation in question, performing vary few calibrations for influential input variables to more accurately simulate what actually happens. From among these, this study incorporated '2-level factorial design', where the input parameters likely influencing the result are chosen for 2-level factorial designing. As for 'full factorial design' where the influence of every individual variable and interaction between variables can be understood, the reliability of the result is substantial and the significance of a variable may be assessed by analyzing the influence of individual variable or the interaction between variables.

3 2-LEVEL FACTORIAL DESIGN RESULT

Air tightness of randomly selected virtual buildings was predicted using CONTAM W3.1, in assessment of the influence [%] of variables to the air tightness. The input parameter and data, considering the building composition and leakage chariteristic by the size of building, both categorized into the maximum and minimum values, were obtained from the existing literature. A total of two virtual buildings were involved, with the two kinds of values, obtained from the existing study, applied to each simulation.

3.1 Building modeling

Modeling Building 1 and Modeling Building 2 are three-storied and as wide as 750 m^2, with WWR (Window to wall ratio) of 40% to the south; 20% to the east; 20% to the west and 20% to the north, to factor in the actual building structure.

3.2 Selecting input variables

Input parameters for 2-level Factorial Design were extracted, based upon existing studies in determining the influence to the air-tightness irrelevant to usage and dimension of the buildings.

According to Dickerhoff et al and Harrje and Born, the building mainly consists of the wall, windows, HVAC system, duct, fireplaces, etc. which the air leakage characteristic of each element is different. The wall accounts for 18~50% of the entire amount of air leakage, 35% on average. Ceiling accounts for from 18% to 30% of the entire amount. The door and window account for 15% of the total air leakage, which is set in multi-zone model by the size and shape. In this study, the input parameter have been considered through the existing studies including only exterior wall, windows, door, joints, and HVAC Ceiling penetration except for indoor air conditions.

3.3 Input parameters analysis for air-tightness simulation

This study is applied to 2-level factorial design, to select less influential parameters for least calibration processes. Refer to the following Formula 1 factoring in such individual variables as exterior wall, window, door, joints(wall-wall) and HVAC Ceiling penetration for assessment of the eventual dependent variable of 'air tightness of building'.

Building leakage rate = f(Exterior wall, Windows, Doors, Joints, HVAC penetration)   (1)

Table 1. Impact of Model Building-2 parameters.

Variable	Exterior wall	Window	Door	Joint	HVAC penetration
Contribution (%)	99.48	0.37	1.066E-005	0.14	6.874E-003
Variable	Exterior wall·Window	Exterior-Wall·Door	Exterior-wall·Joint	Exterior-wall·HVAC penetration	Window·Door
Contribution (%)	0.031	3.675E-006	1.719E-003	4.510E-004	3.384E-008
Variable	Window·Joint	Window·HVAC penetration	Door·Joint	Door·HVAC penetration	Joint·HVAC penetration
Contribution (%)	1.580E-005	4.151E-006	1.962E-009	5.157E-010	2.407E-007

Table 2. Results of the Object Leakage Test.

	Coefficient [C]	Exponent [n]	Flow@50Pa [m^3/h]	EqLA@10Pa [m^2]	Correlation [%]
Whole building leakage test	1469.42	0.70	23	0.6	92.99
Window leakage test-1	247.534	0.622	2820.78	0.0633	98.70
Exterior-wall leakage test-1	239	0.58	2282	0.265	99.97

Analysis of variance (ANOVA) test was performed for relevance (%) of and interaction among variables. Refer to the following Table 1 and check out Exterior wall, Windows and Joint, in order, influencing the most:

4 CALIBRATION PROCESS

Calibration Process was performed by substituting default value of parameters for value of actual measurement based upon the relevance of parameters. The initial model consist of the parameters that effect o based on the fundamental information from floor plan, and obtained parameters that may affect the results significantly. To verify the accuracy of model, the simulation results on each step were compared with actual results of experiment by calibrating the variables of model.

4.1 *Application of tuning process*

4.2 *Object building*

The subject building is a reinforced concrete structure and situates in a university in Seoul, 3-story and total floor area of 1474.515 m^2. The air tightness of the whole building was measured by blower door test based on ASTM E779-10 standard. As the measurement for the elements of building envelope, the window that comprised largest proportion of building envelope was measured first, and next the door was measured. The results of measurement is presented in Table 2.

4.3 *Modeling of actual building*

Variables for the initial simulation model, being air leakage through the external wall and window properties were obtained from existing literature, with the calibration variables assessed out of the results of '2 level factorial design'. The initial Mode 1 was set as Base Model, with the minimum

Table 3. Calibration Process for Parameters of Model.

Step	Model	Correlation	Correlation Parameter
Step.1	Model 1	X	–
Step.2	Model 2	O	Window-1
Step.3	Model 3	O	Exterior wall
Step.4	Model 4	O	Window-1, Exterior wall

Table 4. Result of Model and Deviation Rate using Minimum Value.

Model	Whole building leakage [m^3/h@50Pa]	C	n	Deviation rate [%]
Model 1	4564.55	357.05	0.65	78.52
Model 2	12361.45	1056.30	0.63	41.84
Model 3	15826.01	1609.73	0.58	25.54
Model 4	22112.73	2006.79	0.62	4.04

and maximum variables suggested by bibliographies considered together. Refer to the following Table 3 for breakdown of the calibration process. Model 1 generated by Step 1 process incorporates no other data than bibliographical sources, where no calibration processes were performed. As for Steps 2 and 3, one variable was calibrated for 'Window-1' and 'Exterior wall', respectively, with Step 4 performed calibration of two input parameters in generation of Model 4. The accuracy of results from each Model generated was confirmed by way of comparison with the result from the object building.

5 TUNING RESULT

The result from each model was represented in leakage, the flow coefficient 'C' and the flow exponent 'n' in recognition of air-tightness of building.

5.1 Predicted model and calibration results using minimum values

The calibration results of the model using the minimum value are presented in Table 4. The deviation rate for the results of model 1 which did not calibrated was 80%. It was verified that the results of model 4 was similar to the results of actual measurement (i.e. 21255 m^3/h) which undergone calibrated for two input parameters. Therefore, the validity of model was assessed by the foregoing flow coefficient and flow exponent in comparison between those of Models and those of the object building.

5.2 Predicted model and calibration results using maximum values

Table 5 shows the results of calibration performed by establishing the model of maximum default value. Like the previous calibration process, the initial Model 1 represented the deviation rate of 64%. As the input parameters are calibrated, it was found that the deviation rate of final model 4 decreased for 8.5%. The leakage characteristic of model 4 was gradually similar to that of the actual building.

Table 5. Result of Model and Error Rate using Minimum Value.

Model	Whole building leakage [m³/h@50Pa]	C	n	Deviation rate [%]
Model 1	34861.93	2741.63	0.65	64.01
Model 2	38396.78	3094.54	0.64	80.65
Model 3	17037.23	1684.38	0.59	19.84
Model 4	23060.13	2204.39	0.6	8.49

6 CONCLUSION

This research suggested a calibration process to improve accuracy of Model by way of air-tightness simulation of building. The variables of building elements influential to air tightness were suggested, modeling a randomly selected building by way of 2 level factorial design. The calibration process of model was performed by measuring the actual air tightness input to use for the existing Models. The result showed that, when simulation results were compared to the actual test results, a Model involving calibration processes for a couple of variable demonstrated gradual similarity to the leakage of the actual building, which evidences validity of calibration process of parameters. It is expected that other phenomenon within the building may also be predicted with the model getting more reliable. With the engineer's discretion factoring in the deteriorated state of building and environment as input parameters, the simulation models set out hereto may get much more accurate with deviation rate between actual building results and simulation results kept lower.

ACKNOWLEDGEMENT

This work was supported by the National Research Foundation of Korea (NRF) grant funded by the Korean government (No. 2012001927).

REFERENCES

ASHRAE Handbook (1997) American Society of Heating, Refrigerating and Air-conditioning Engineers, Inc., Chapter 25, Table 3.
Conference Thermal Performance of the Exterior Envelopes of Buildings VII, 829–837.
Dickerhoff, D. J., Grimsrud, D. T. and Lipschutz, R. D. (1982) Component leakage testing in residential buildings.
Emmerich, S.J. and Persily, A.K. (1996) Multizone Modeling of Three Residential Indoor Air Quality Control Options, NISTIR 5801, National Institute of Standards and Technology.
Firrantello, J., Bahnfleth, W. P., Musser, A., Freihaut, J. D. Jeong, J.-W. (2007) DA-07-069 Use of Factorial Sensitivity Analysis in Multizone Airflow Model Tuning. Transactions- American Society Of Heating Refrigerating And Air Conditioning Engineers. 113 p. 642–651.
Harrje, D. T., and Born, G. J. (1982) Cataloguing air leakage components in houses. In Proceedings of the ACEEE.
H. R. Trechsel and P. L. Lagus, Measured Air Leakage of Buildings, ASTM STP 904. American Society for Testing and Materials, p. 184–200.
Persily, A. K. and R. A. Grot. (1986) Pressurization Testing of Federal Buildings.
Persily, A.K. (1998) Airtightness of Commercial and Institutional Buildings: Blowing Holes in the Myth of Tight Buildings. DOE/ASHRAE/ORNL/BETEC/NRCC/CIBSE.
Persily, A.K., W.S. Dols, S.J. Nabinger and S. Kirchner. (1991) Preliminary Results of the Environmental Evaluation of the Federal Records Center. NISTIR 4643. National Institute of Standards and Technology.
Wang, L. L., Dols, W. S. and Chen, Q. (2010) Using CFD capabilities of CONTAM 3.0 for simulating airflow and contaminant transport in and around buildings. HVAC&R Research, 16(6), 749–763.

Impact of demand controlled ventilation on energy saving and indoor air quality enhancement in underground parking facilities

H.J. Cho & J.W. Jeong
Division of Architectural Engineering, College of Engineering, Hanyang University, Seoul, Republic of Korea

ABSTRACT: The main thrust of this paper is to verify possibility of indoor air quality enhancement and energy savings potentials in under-ground parking facilities by adopting the demand-controlled ventilation (DCV) strategy based on the real-time traffic load. For applying DCV approaches to under-ground parking facilities, the minimum ventilation rate per a single vehicle was determined. Two different DCV methods; DCV using the real-time vehicle detection system (i.e. VDS-DCV) and DCV based the variation of indoor CO concentration (i.e. CO-DCV) were proposed. By simulating transient ventilation flow rate and the indoor CO concentration variations for each DCV approach using a commercial equation solver program, applicability of the proposed DCV strategies was analyzed. And then a pilot ventilation system was installed in a real under-ground parking lot for verifying the simulation results. It shows that CO-DCV or VDS-DCV may provide significant reduction of fan energy consumption and good indoor air quality compared with the conventional parking facility ventilation approach.

1 INTRODUCTION

Demand control ventilation (DCV) method as a part of ventilation control system in building can be defined as an automatic control system for modulating the outdoor air flow rate. The supply air flow rate is varied based on the measuring real-time occupant number (Stipe 2003). DCV research has been developed over the past few years along with the advanced sensor technology. Its research is mostly concerned on the residential space whereas the research of non-residential spaces such as under-ground parking facilities has been rarely conducted. However, ITA Working Group is expected to be increase the construction of under-ground parking facilities (Godard 1995). Therefore, it is necessary for DCV researches to consider under-ground parking facilities as one of its subjects.

The existing ventilation control systems for under-ground parking facilities consists of constant air volume control (CAV) system and simply binary control system. In order to save the operating fan energy in the existing under-ground parking facilities using simple binary control system, relatively poor indoor air quality is inevitable. On the other hand, CAV system using a large single fan causes increasing energy consumption by over-ventilation rates.

Therefore, in order to improve fan energy savings potentials and indoor air quality, the DCV method is adopted to the under-ground parking facilities. The previous study was conducted first to DCV simulation (Cho 2013). And then its applicability to the proposed DCV strategies based on the real-time variation of the traffic load is verified by evaluating the real-time carbon monoxide concentration variation. As for the on-going study, the pilot ventilation system is located in a real under-ground parking lot for verifying the adaptability of the different control strategies.

2 PILOT VENTILATION SYSTEM FOR DCV STRATEGIES

Existing recommended ventilation standards is normalized for the unit floor area or unit volume of the underground parking lot. So applying current ventilation standard for under-ground parking lots to the DCV control may be difficult. Therefore, this research proposed ventilation rate required for a single vehicle which can be more easily integrated into the DCV control approach.

2.1 *VDS-DCV*

The vehicle detection system (VDS) DCV modulates the ventilation rate provided by monitoring data of the real-time traffic load. As for the applying the VDS-DCV, the required ventilation rate per a single vehicle was derived based on the total parking capacity and ventilation rate in the under-ground parking lot. In order to operate the VDS-DCV, the traffic loads has been monitored by the two vehicle detection sensors that installed at the entrance and the exit of the parking lot. And those sensors transmit the traffic loads data to the control server via wireless network system. Based on the real-time data, VDS-DCV can adjust automatically the required ventilation rate for 8 steps. The control server PC logs the real-time traffic loads and indoor CO level during the setup time.

2.2 *CO-DCV*

The CO-DCV uses CO detection system for estimating real-time variation of the traffic load by monitoring the indoor CO concentration variation, and modulating the ventilation rate. Even though CO level is the representative contaminant index in parking facilities, it is hard to emit artificially for the DCV tests. So in order to show the possibility of CO-DCV, the imitating test was conducted by injecting CO_2 tracer gas in CO-DCV case. The imitating test using CO_2 gas is also adjusted the ventilation rate for 8 steps based on the variation of CO_2 concentration.

2.3 *Pilot ventilation system*

As shown in Fig. 1, the schematic diagram of pilot ventilation system is consisted of the devices for operating control system in under-ground parking lot. For modulating ventilation rate, each

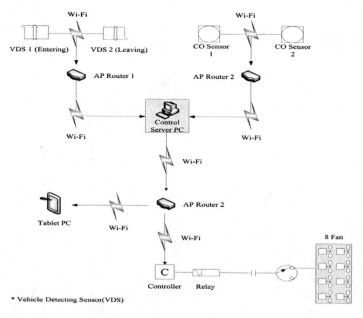

Figure 1. Schematic diagram of the pilot ventilation system.

supply and exhaust ventilation units have eight fans, and air flow rate for each fan is set to 6500 CMH, and total ventilation rate is 52000 CMH. So, it is possible to adjust variable air volume by controlling the number of fans from signal on the proposed DCV strategies. For minimization of impact from the outdoor conditions, the testing space was selected fourth floor in under-ground parking lot in Seoul, South Korea. The total floor area of the testing space is 3895 m^2 (13243 m^3) and the parking space is able to accommodate 125 vehicles.

3 EXPERIMENT RESULTS FOR DCV STRATEGIES

Fan energy consumption was measured by both DCV strategies and simply binary control The pilot ventilation system is set to activation from 9:00AM to 18:00PM. The static pressure for supply and exhaust fans is 200 Pa, and 550 W power is required per 6500-CMH air supply. The 10 minutes for minimum operating time is applied because variable frequency operation might cause unstable overload to the ventilation units.

3.1 VDS-DCV

The traffic load schedule is defined randomly as the testing condition for VDS-DCV. One important observation in Fig. 2 is that VDS-DCV can modulate proportionally the number of fan units by variation of traffic load data. It means that application of DCV based on real-time traffic load for under-ground parking lot is verified. For analyzing fan energy, the number of fan operation was estimated first to VDS-DCV test, and then the fan energy consumption was compared with CAV assumed in same condition. The result indicated that about 35% of fan energy can be reduced based on VDS-DCV compared with CAV control system.

3.2 CO-DCV & Simply binary control

For comparison, two different control system; imitated CO-DCV and simply binary control system were tested by using CO_2 tracer gas in similar condition. In case of CO-DCV, the number of vehicles in the space can be estimated indirectly from the transient variation of the contaminant concentration. Fig. 3 shows that the variation of ventilation air flow rate is significantly adjusted by the transient variation of CO_2 level. It was found that CO-DCV based on imitating test provides good energy saving performance compared with conventionally CAV system. Despite imitating test was performed alternatively by using CO_2, this result indicates that CO-DCV using CO index

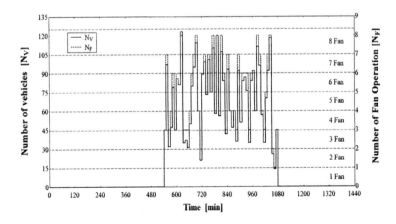

Figure 2. Fan operation profile for VDS-DCV.

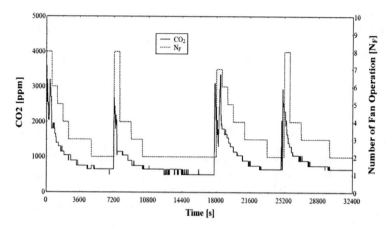

Figure 3. Fan operation profile for CO-DCV.

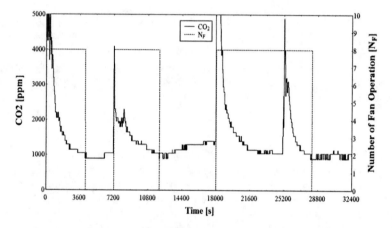

Figure 4. Fan operation profile for simply binary control.

can be implemented by sensing CO level. The measuring data for simply binary control has been estimated (Fig. 4). When the indoor CO concentration reaches at its upper limit level, constant volume ventilation units are activated and supply design flow until the CO concentration is back to normal level.

3.3 *Comparison of fan energy consumption*

Fig. 5 depicts the fan energy consumption for each ventilation strategy in under-ground parking lot. One may see that good energy saving potentials is provided by the proposed DCV strategies (VDS-DCV, CO-DCV) compared to the conventional CAV system. The test results indicated that a large amount of fan energy consumption can be reduced by imitated CO-DCV approach, in addition to the good indoor air quality. VDS-DCV approach would be good performance in terms of indoor air quality by supplying constant air volume continuously even though it is not available to compare with CO-DCV and simply binary control.

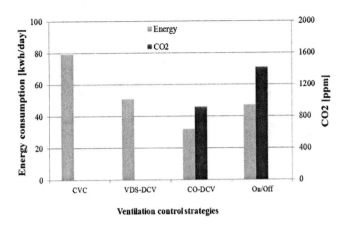

Figure 5. Impact of ventilation control strategies.

4 CONCLUSION

In order to improve fan energy savings potentials and indoor air quality of the conventional ventilation control system, the DCV strategies are applied to the under-ground parking facilities. The previous study was conducted first to DCV simulation. And then the pilot ventilation system has been installed for verifying applicability to the proposed DCV strategies in a real under-ground parking lot. The test results indicate that proposed DCV strategies can be used in real under-ground parking facilities by using CO level even though imitating test was conducted. The VDS-DCV approach based on real-time traffic load may also be a good choice to saving fan energy. Consequently, the continuous operation of ventilation units based on the CO-DCV or VDS-DCV may be a good choice for achieving energy savings and good indoor air quality simultaneously.

ACKNOWLEDGEMENT

This work was supported by the National Research Foundation of Korea (NRF) grant funded by the Korean government (No. 2012001927). This research was also supported by the MSIP (Ministry of Science, ICT & Future Planning), Korea, under the CITRC (Convergence Information Technology Research Center) support program (NIPA-2013-H0401-13-1003) supervised by the NIPA (National IT Industry Promotion Agency).

REFERENCES

ASHRAE Handbook HVAC Applications 2007. Chapter 13 Enclosed vehicular facilities. *American Society of Heating, Refrigerating, and Air-Conditioning Engineers.*
Cho, H. J. et al. 2013. Energy saving potentials of demand-controlled ventilation based on real-time traffic load in underground parking facilities. *Proceeding of AEI conference*: 524–533.
Fisk, W. J. et al. 1998. Sensor-based demand-controlled ventilation: a review. *Energy and Buildings* 29: 35–45.
Godard, J. P. 1995. Underground Car Parks: International case studies. *Tunneling and Underground Space Technology* 10, 3: 321–342.
Jeong, J. W. et al. 2010. Improvement in demand-controlled ventilation simulation on multi-purposed facilities under an occupant based ventilation standard. *Simulation Modelling Practice and Theory* 18, 1: 51–62.
Krarti, M. et al. 1999. Evaluation of fixed and variable rate ventilation system requirements for enclosed parking facilities. *Final report for ASHRAE Project 945-RP.*

Krarti, M. et al. 2001. Ventilation for Enclosed Parking Garages. *ASHRAE Journal* 43, 2: 52–57.

Lu, T. et al. 2011. A novel and dynamic demand-controlled ventilation strategy for CO_2 control and energy saving in buildings. *Energy and Buildings* 43: 2499–2508.

Martin, H. 2001. Demand-controlled ventilation in vehicle parks Technical Note-021-1. *Available via SenseAir AB*.

Mysen, M. et al. 2005. Occupancy density and benefits of demand-controlled ventilation in Norwegian primary schools. *Energy and Buildings* 37: 1234–1240.

Owen, Ng. M. et al. 2011. CO_2-based demand controlled ventilation under new ASHRAE Standard 62.1-2010: a case study for gymnasium of an elementary school at West Lafayette, Indiana. *Energy and Buildings* 43: 3216–3225.

Stipe, M. 2003. Demand-Controlled Ventilation: A Design Guide. *Oregon Office of Energy*.

Technique for detecting vertical prestress based on vibration principle

Xiaojuan Shu
School of Civil Engineering and Architecture, Central South University, Changsha, China

Mingyan Shen, Xingu Zhong, Tao Yang & Hongbing Chen
School of Civil Engineering, Hunan University of Science & Technology, Xiangtan, China

ABSTRACT: One of the important causes that result in cracking in webs of prestressed concrete bridge is that the prestress of many finely rolled thread steel bars do not meet the design requirements. Theoretical analysis and experimental tests of vibration behavior of the exposed segments of vertical prestressing bars showed that the values of effective tensions affect their vibration frequencies. The exposed segment was modeled as a cantilever beam with an elastically confined end, and the vibration frequency formula is deduced. Furthermore, influences of some factors such as diameter, linear density of the bar, mass of the sensor, confining rigidity of rotation are analyzed. The length of the exposed segment is modified to consider the influence of anchoring screws. Experimental results proved that there was a stable relationship between the confining rigidity of rotation and the effective tension force. Hence the effective prestress can be predicted from the measured vibration frequency.

1 INTRODUCTION

Prestressed concrete box girder is widely used in bridges such as continuous girder bridge, rigid frame bridge and cable stayed bridge. Most of large-span prestressed concrete box girder bridges adopt single-room section with wide wing and thin belly, and the web plate is designed with vertical prestress forced to improve the crack resistance of web plate. In the recent 30 years, most of domestic vertical prestressed steels adopt fining twisted steel vertical prestressing system, through which, however, the anticipated goals are not realized. Numerous engineers and researchers reach a consensus: over loss or even failure of vertical prestress is the important reason causing the box girder web crack [Zhong Xingu, 2002; Zhou Junsheng, 2000], this means lots of vertical prestressed steels are not stretched to design value, and the tension for prestressed steel is too random during construction. The prestress tension effect may be improved to a certain extent via methods like the second tension, lag tension and super tension [Shen Mingyan, 2007; Shen Mingyan, 2008], however, the ratio of forced steel bar under inspection is relatively small, unable to form effective monitoring function due to lack of relevant regulations for constraint as well as the limitation of detection means.

The current vertical prestressed tensioning bar of the construction side actually just simply pre-controls the controlling stress, and the general mode is to stretch the rebar to the controlled tonnage based on the oil pressure gauge and elongation, and twist the nut to the surface of the girder. The supervisor will also conduct monitoring mainly based on the above two indexes. It shows in the engineering practice that the construction side often won't attach great importance to the vertical prestressed bar just as they do to the longitudinal prestress tension, and there is great subjectivity during twisting nuts, which causes untight sticking or too large tension deflection angle between nut anchor plates, leading to the prestess loss of many rebars is significantly larger than the theoretical value [Zhou Junsheng, 2000; Shen Mingyan, 2007; Shen Mingyan, 2008], and the effective vertical prestressed bar is far different from the controlling stress. However, the second tension for all vertical prestressed bars will obviously enhance the construction strength, which will suffer contradiction from the construction side.

The most direct and most effective way to solve the failure of vertical prestressed bar tension reaching the designated position is to find an efficient, fast and economic nondestructive testing mode, which can detect the tensioning force of all prestressed bars after tension, and rebar without reaching the requirement can be timely added tension. There are many methods to measure the effective tensioning force of tension member; one way is to install pressure sensor [Chen Lu, 2006] for the anchored rebar, the number of vertical prestressed bar is too large, and it is unrealistic and uneconomic to install for a small proportion. As for the vibration frequency method [Chen Lu, 2006; Li Hui, 2006; Hao Cao, 2000; Fang Zhi, 1997] used to measure the tension of the guyed structure, it needs to place the sensor in the anchor point of the inhaul cable, which cannot be directly applied for vertical prestressed bar buried in the girder. The elastic magnetic flux method is still immature [Li Hui, 2006; Hao Cao, 2000]. Therefore, developing a detection method and formulating relevant detection standards can not only greatly constrain and regulate the prestressed construction behavior, make the tension quality with testability and controllability and thoroughly solve the over loss problem of vertical prestress, but also urge the construction unit to initiatively improve the construction process and the reliability of construction. This technique has a technical and economic significance in preventing the web plate crack of prestressed concrete box girder bridges and improving the durability of such bridge.

2 BASIC PRINCIPLES OF FREQUENCY METHOD AND NONDESTRUCTIVE TESTING FOR VERTICAL PRESTRESSED TENSION

The vibration frequency method to detect the tension of flexible tension members such as stay cable has been quite mature, and its principle is the inherent vibration frequency value of the stretched cable or truss is related to the tension [Fang Zhi, 1997; Wu Kangxiong, 2006]. Therefore, the tension can be speculated via measuring the lateral vibration frequency of the tension section of components. But due to the fact that the stretched area of the vertical prestressed bar is buried in the girder, it is impossible to install vibrating sensor, so other ways shall be considered.

2.1 Basic thinking of vibration frequency method for detection

Structure of fining twisted steel vertical prestressing at the tension end is shown in Figure 1. Its anchoring principle is to twist nuts to make a tight occlusion between the male screws of rebars with the female screws of nuts after tension of rebar with threads. Different tensioning force will cause different squeezing degrees of nuts for rebar threads. In other word, with different tensioning forces, the constraint effect of nuts for the exposed section of rebar will be different. So the fixed-end exposed rebar section formed by nuts should be regarded as elastic fixed-end cantilever girder. In case that the rotational restraint stiffness of elastic fixed end is different, the inherent vibration frequency of the exposed section is also different. It is discovered in the experiment that under different tension stress levels, the lateral vibration frequency of the exposed section will appear measurable floating. Based on this principle, if the relation of the tensioning force and fixed-end elastic rotational stiffness can be established, that means establishing the relation between the tensioning force and the natural vibration frequency of the exposed section, then the tensioning force can be directly gained through testing the vibration frequency of the exposed section theoretically. This testing thought is clear in mechanics principle, simple and fast in operation, easy for large-scale measurement and won't disturb the construction and produce any adverse influence on the structure.

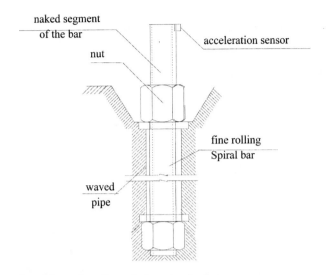

Figure 1. Constitution of the anchor of a vertical prestressing bar.

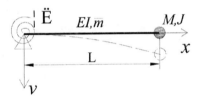

Figure 2. Model of a cantilever beam with an elastically confined end.

2.2 *Vibration model of exposed section of vertical prestressed bar*

Taking fining spiral steel with diameter of 32 mm- the most widely used vertical prestressed bar in engineering as an example, the exposed rebar section shown in Figure 1 is used as the reserved tension length, whose general length is about 8–20 cm, and its lateral vibration features meet the bending and vibration features of straight girder with equal section. As for the straight girder with equal section, the distribution mass is m, the constant of the section stiffness is EI, the length is L, and the differential equation of free bending vibration is [Ray W Clough, 1981]:

$$EI\frac{\partial^4 v}{\partial x^4} + \bar{m}\frac{\partial^2 v}{\partial t^2} = 0 \qquad (1)$$

Function for the general vibration form:

$$\varphi(x) = A_1 \sin \beta x + A_2 \cos \beta x + A_3 \sinh \beta x + A_4 \cosh \beta x \qquad (2)$$

A_1–A_4 is undetermined coefficient, determined by the boundary conditions. The calculation method for vibration frequency is:

$$\omega^2 = \frac{\beta^4 EI}{\bar{m}} \qquad (3)$$

Exposed prestresed bar section structure shown in Figure 1 can simplify the elastic fixed-end girder with elastic rotational restraint shown in Figure 2; stiffness of the elastic rotational restraint is λ, the terminal concentrated mass is M, and its boundary conditions are:

$$\begin{aligned}&\phi(0)=0\\&EI\phi''(0)-\lambda\phi'(0)=0\\&EI\phi''(L)-\omega^2\phi(L)J=0\\&EI\phi'''(L)+\omega^2\phi(L)M=0\end{aligned} \quad (4)$$

Linear system of equations (5) of A_1–A_4 will be gained via putting each derivative function of (2) gained into formula (4).

$$\begin{bmatrix} 0 & 1 & 0 & 1 \\ \lambda & 1 & \lambda & -1 \\ EI\beta & & EI\beta & \\ -(K_1+K_2)\sin\beta l & -(K_1+K_2)\cos\beta l & (K_1-K_2)\sin h\beta l & (K_1-K_2)\cosh\beta l + \\ -K_3\cos\beta l + & K_3\sin\beta l + & K_3\cosh\beta l + & K_3\sinh\beta l + \\ K_4\sin\beta l & K_4\cos\beta l & K_4\sinh\beta l & K_4\cosh\beta \end{bmatrix} \begin{Bmatrix} A_1 \\ A_2 \\ A_3 \\ A_4 \end{Bmatrix} = 0 \quad (5)$$

In the formula, $K_1 = EI\beta^2$, $K_2 = \omega^2 J$, $K_3 = EI\beta^3$, $K_4 = \omega^2 M$.

If the equation set is required with nontrivial solution, the determinant of coefficient shall be equal to 0, and the simplified frequency equation is shown in formula (6).

$$\begin{vmatrix} -\lambda\beta & -\lambda\beta & 2K_1 \\ (-K_1-K_2)G_1 & (K_1-K_2)G_3 & (K_1-K_2)G_4-(-K_1-K_2)G_2 \\ -K_3G_2+K_4G_1 & K_3G_4+K_4G_3 & (K_3G_3+K_4G_4)-(K_3G_1+K_4G_2) \end{vmatrix} = 0 \quad (6)$$

In formula (6), K_1–K_4 and G_1–G_4 are parameters related to ω, if the flexural rigidity of rebar EI, linear density m, concentrated mass M and J are known, the relation between the frequency ω and rotational restraint stiffness λ is in fact established in formula. That is to say, if the frequency ω is known, the corresponding λ can be calculated through formula (6), and the λ as mentioned before is related to the effective tensioning force. This is the basic principle of frequency-based nondestructive testing for effective vertical prestress discussed in this paper.

3 SENSIBILITY ANALYSIS FOR PARAMETERS INFLUENCING THE VIBRATION FREQUENCY OF EXPOSED SECTIONS

3.1 Machining error influence of rebar diameter and parameter calculation of rebar bending stiffness

Machining error of rebar diameter will influence the section stiffness and the rebar distribution mass, and the vibration frequency directly induced will be significantly changed. This item can be determined by trial calculation. Taking the fining twisted steel with $\phi 32$ as an example, Figure 3 shows the standard texture with 7800 kg/m³, and the frequency difference induced by different machining errors of rebar diameters. It indicates through calculation that machining errors will greatly affect the frequency of short rebar but slightly affect the absolute value of vibration frequency of long rebar. As for rebar with 8 cm, its frequency difference is close to 100 Hz, but when the length of rebar is larger than 16 cm, its influence tends to be stable. If the length of the exposed section is larger than 10 cm and the diameter difference is within ±0.5 mm, then the frequency difference caused by this error is less than 40 Hz.

Diameter specification of rebar produced in the same batch is basically constant, but there will be difference in diameter for rebar produced in different batches, so standardization for rebar diameter shall be accurately conducted in batches. Section shape of fining twisted steel is not standard circle,

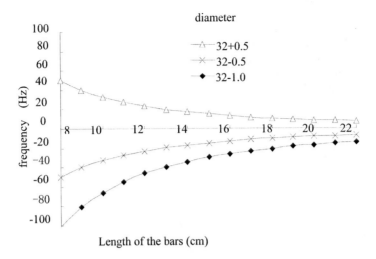

Figure 3. Frequency disparture caused by diameter errors of the bar.

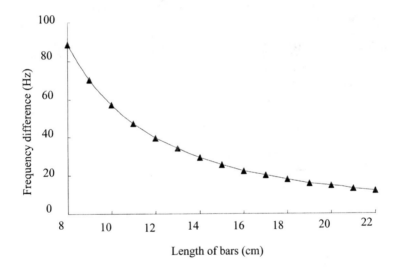

Figure 4. Difference of first vibration frequency for different linear density.

and the bending inertia moment can only be preliminarily estimated for the section directly as per the diameter. It can be determined via the testing method of static test.

3.2 *Influence of rebar linear density*

The standard linear density of fining twisted steel with $\varphi 32$ given in the regulation is generally 6.65 kg/m. Assume the diameter is unchanged, but the linear density changes, and the linear density error takes 0.5% (0.34 kg/m), the calculation result of frequency difference induced by this error is shown in Figure 4. Figure 4 shows decline of the rebar linear density causes the increase of vibration frequency of the structure; the linear density takes 0.5%, and the relative frequency increases about 2.6%. The absolute frequency difference shows an exponential function for progressive increase from 88 Hz of 8 cm to 12 Hz of 22 cm.

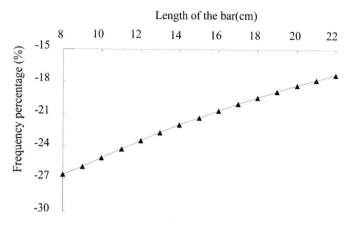

Figure 5. Error percent of relative frequency caused by sensor.

The material density of rebar produced in the same batch is basically constant, but there will be difference in material density for rebar produced in different batches, so standardization for rebar linear density shall be also accurately conducted in batches.

3.3 *Influence of vibration testing sensor*

It shows in Figure 5 that adding sensor at the end of cantilever structure will lead to the significant decline of vibration frequency of the structure. The relative decline percentage of frequency shows a line relationship to the rebar length. For rebar with length of 8 cm, the absolute value of frequency decline is 985.5 Hz, and compared with the original frequency, it declines 27%. Point mass formed by sensor needs to be marked accurately. During detection, the fixed sensor and layout method are suitably adopted and the centroid location of sensor shall be aligned to the end of rebar.

The sensor is a small cylindrical object, and can be regarded as homogeneous object, and the rotating inertia moment J can be calculated based on the geometric characteristics.

3.4 *Influence of rotational restraint stiffness*

The influence of rotational restraint stiffness on the vibration frequency is shown in Figure 6. It can be seen from Figure 6 that the most obvious area of rotational stiffness influencing the structure vibration frequency is 1×10^4 to 1×10^7 N.m. When the rotational restraint stiffness is larger than 1×10^7 N.m, there is no significant change for the vibration frequency of rebar, and its boundary conditions can be equivalent to complete solidification. For fining twisted steel, if the elastic rotational stiffness declines from 1×10^7 N.m to 1×10^4 N.m, the structure vibration frequency difference changes from about 300 Hz of rebar with 22 cm to about 2000 Hz of rebar with 8 cm.

It shows in the experiment that the effective tensioning force mainly influences the rotational restraint stiffness, but slightly affects the distribution of nut pressure. Therefore, if a stable regression relation is measured between the effective tensioning force and the rotational restraint stiffness, the tensioning force can be inversely calculated based on the rotational restraint stiffness. Via numerous experiments and tests by the research group of the author, they discover that there really exists a stable regression relation between the two. With the limitation of patent confidentiality, this paper won't give the specific regression function type or curve. Figure 6 shows the relationship between the tensioning force and vibration frequency of rebar with different lengths, indirectly reflecting the relationship between the tensioning force and the rotational restraint stiffness.

3.5 *Influence of nuts*

Literature [Wang Zhiqin, 2001] reported thatthe axial load distribution condition along the thread of nut nominal stress respectively with 70 MPa, 200 MPa and 300 MPa in load born by thread

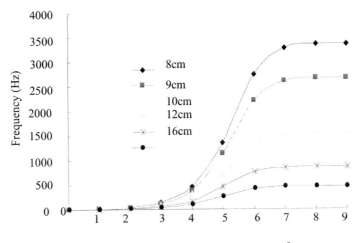

Figure 6. Relationship between frequencies and rotational rigidities of bars with different length.

teeth of 1–3 takes up 70–80% of the overall tension born by nuts. Therefore, length exposed of prestressed bar to nuts cannot be treated as the cantilever length, and there is certain length L_0 within the nut participating in the free vibration of cantilever. It also said that there is no obvious coupling relationship between the distribution rules of the rebar tension stress and nut stress, and the distribution of thread tooth is basically uninfluenced by the tensioning force. Based on this analysis, using correcting cantilever length to eliminate the influence of nuts is proposed. Figure 7 shows the relationship of tension tonnage and vibration frequency of rebar with different lengths, when the tension tonnage is 20T–50T, its vibration frequency shows a non-linear increase. When the prestress tension reaches 50T, the frequency slightly changes, indicating nuts tightly pressing the anchor plate, and its frequency is close to the vibration frequency of prestressed steel of fixed cantilever. On this basis, the vibration frequency measured with tensioning force of 50T is used to compare with the theoretic frequency without counting in the rebar length of nuts, thus inversely calculating the cantilever length of fixed cantilever girder, which is the correcting length. Via numerous actual measurements, the relationship of the correcting length and exposed length of rebar with different lengths is shown in Figure 8, and a linear regression equation (5) can be established between them.

$$L_m = \frac{11}{12}L + 4.8 \qquad (5)$$

4 EXAMPLE

Assumed question: there is a vertical prestressed steel on site, the standard diameter is 32 mm, measured length of the exposed section is 16 cm, and then a vibrating sensor with weight of 208 g is installed at its end. The first-order free vibration frequency is measured to be 450 Hz. Whether the tensioning tonnage has a design value of 50t?

Solution:

Step 1, the correcting length of the exposed section is calculated to be 19.4 cm with formula (3) applied.

Step 2, the measured value of the first-order free vibration frequency of the exposed section is measured via installing the sensor at the front of the free end.

Step 3, the fixed-end elastic rotational restraint stiffness value is inversely calculated via inputting the known parameters into the theoretical vibration model and applying formula (6).

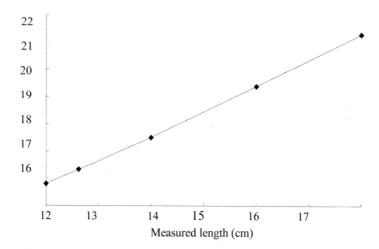

Figure 7. Relationship between tensile forces and frequencies.

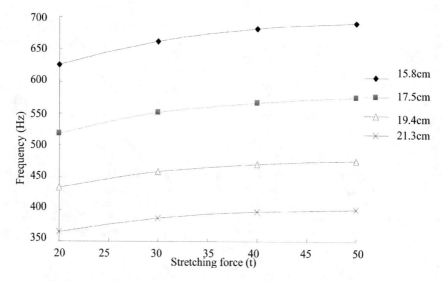

Figure 8. Relationship between modified length and measured real length.

Step 4, the actual effective tensioning force is evaluated based on the measured restraint stiffness and tensioning force.

The exposed length is 19.4 cm, diameter of the vertical fining twisted steel is 32 mm, the actual vibration frequency is 450 Hz, and the relevant effective tensioning force is 25T via referring to the curve in Figure 7, therefore, it fails to reach the design requirement, and added tension is required. The effective prestress can be calculated based on the effective prestress. Figure 9 shows the effective prestress detected in the test room.

5 CONSLUSIONS

(1) On the basis of analyzing the mechanical property of the anchored exposed section of vertical prestressed steel, the elastic fixed-end cantilever girder model of exposed section is proposed,

its elastic rotational restraint stiffness is related to the effective tensioning force, and the thought of detecting the effective tensioning force of the vertical prestressed steel is proposed based on this principle.
(2) The free end with concentrated mass and the natural vibration frequency equation of cantilever girder model of exposed rebar section for elastic fixed-end rotational restraint stiffness λ are inferred, thus establishing the relational expression of vibration frequency with λ.
(3) With combination of features of the vertical prestressed steel, the vibration frequency influence and sensitivity degree of parameters such as steel diameter, linear density, sensor and λ to the model is analyzed. Various calculation methods are discussed and analyzed. Based on the analysis of the distribution rules of nut load, correcting length is proposed to eliminate the influence of nut on the vibration frequency.
(4) A stable regression relation is determined between the tensioning force and λ, and the feasibility of frequency method detecting the effective tensioning force is established. The relation curve of the tensioning force and the vibration frequency is given for the convenience of calculation under the condition that the most common fining twisted steel with $\varphi 32$ is set with specific sensor.

REFERENCES

Chen Lu, Zhang Qilin, Wu Ming'er. 2006. Principle and method of measuring cable tension in a cable structures [J]. Industrial Construction 36 (sp):361–371.(in Chinese)
Fang Zhi, Zhang Zhiyong. 1997. Test of cable tension in cable-stayed bridges[J]. China Journal of Highway and Transport 10 (1): 51–58. (in Chinese)
Fang Zhi, Wang Jian. 2006. Vertical prestressing loss in the box girder of long-span PC continuous bridges [J]. China civil engineering journal 39(5):78–84.(in Chinese)
Shen Mingyan, Zhong Xingu, Shu Xiaojuan. 2008. Preliminary research on construction of vertical prestressed bars in webs of PC box beam [J]. Building Structure 38(12):73–76. (in Chinese)
Hao Cao, Pei Minshan, Qiang Shizhong. 2000. New method for testing cable tension in cable-stayed bridges-method of magnetic flux[J]. Highway 2000(11): 30–31. (in Chinese)
Li Hui, Liu Min, Ou Jinping, etal. 2006. Design and analysis of magnetorheological dampers with intelligent control systems for stay Cables [J]. China Journal of Highway and Transport 18(4):37–41. (in Chinese)
Ray W Clough, Joseph Penzien. 1981. Dynamics of structures [M]. Science Press, 204–208. (in Chinese)
Shen Mingyan, Zhong Xingu, Shu Xiaojuan. 2007. Research on tension technique and stretching time of vertical prestress in a bridge with box girder [J]. Construction Technology 36(10):38–40. (in Chinese)
Wang Zhiqin. 2001. Analysis of Stress-Strain and Fatigue Life of Bolts [J]. Aviation manufacturing technology 2011(2):44–46. (in Chinese)
Wu Kangxiong, Liu Keming, Yang Jinxi. 2006. Measuring system of cable tension based on frequency method [J]. China Journal of Highway and Transport 19(2): 62–66. (in Chinese)
Yu Xianlin, Ye Jianshu, Wu Wenqing. 2011. Cracking damage of long span prestressed concrete box girder bridges [J]. Journal of Harbin Institute of Technology 43(6):101–104. (in Chinese)
Zheng Hui, Fang Zhi, Cao Minhui. 2013. Experimental Study on the Shear Behavior of Reinforced Thin-wall Box Girders with Vertical Prestressed Tendon [J]. Journal of Hunan University (Natural Sciences) 40(1):1–8. (in Chinese)
Zhong Xingu. 2002. Analysis and Prevention of the cracks in continuous prestressed concrete box girders & Research on the concrete filled steel box composite beams [M]. Post-Doctoral Research Report, Hunan University, Changsha. (in Chinese)
Zhou Junsheng, Lou Zhuanghong. 2000. The status quo and developing trends of large span prestressed concrete bridges with continuous rigid frame structure [J]. China Journal of Highway and Transport, 13(1):31–37 (in Chinese)

Influence analysis of ground asymmetric load on spandrel with different bases of existing open trench tunnels

Shisheng Zeng
Faculty of Civil Engineering and Mechanics, Kunming University of Science and Technology, Kunming, Yunnan, China
Kunming Metro Construction Management Co. Ltd, Kunming, Yunnan, China

Jing Cao, Qiongxiang Pu & Haiming Liu
Faculty of Civil Engineering and Mechanics, Kunming University of Science and Technology, Kunming, Yunnan, China

ABSTRACT: Based on the theory of equivalent internal friction angle, the soil layers around the tunnel are equal to cohesionless homogeneous mass in this paper. On the basis of the Terzarghi unconsolidated mass theory, the local soil arching effect of the existing open trench tunnel which is located the side walls under the ground asymmetric load is discussed. Furthermore, the critical embedded depth formula of the backfill soil appearing the local soil arching effect is given. Finally, it is concluded that both the size and distribution of the ground asymmetric load have influence on the local soil arching effect of the existing open trench tunnel.

1 INTRODUCTION

Generally, due to the self-supporting capability, the rock-soil mass will produce the soil arching effect during the excavation process in the underground space (Yu 2008). The phenomenon of the soil arching effect, occurred in the underground engineering, has long been recognized by researchers. Initially Kovari proposed the arching effect during the process of tunnel excavation (Kovari 1994). He made an inference that the cohesionless material will appear arching effect, which is based on the research of the tunnel roof subsidence. Promojiyfakonov theory, which is established by a Russian scholar, made a further study on natural equilibrium arch theory in loose medium. Terzarghi confirmed the existence of the arching effect in the sand layer through tests, and the mechanical analysis is carried out (Terzarghi 1947). Moreover, through the movable door famous test (Jia 2003), he confirmed the existence of the soil arching effect in the field of the soil mechanics. At the same time, he obtained the existence condition of the soil arching effect, which is based on the description of the soil arching stress distribution. In recent years, considerable progresses have been made on the arching effect of tunnel. For example, Liang (Liang 2005) defined the pressure-arch, and determined the internal and external borders of the arch by the stress analysis. Then, he analyzed the relationship between the pressure-arch and the surrounding rock deformation (Liang 2008). It was discovered that as the inner boundary of pressure-arch close to the tunnel, the tunnel deformation become smaller and more stable. Whereas, the tunnel will appear large deformation. Jia (Jia 2003) explored the arch foot form in the "Several Soil Arching Effect Problems". Wang (Wang 2012) studied the soil arching effect of the deep circular tunnel, and drew a conclusion that different ground stress mode generates the different pressure-arch. So far, many researches have been done on soil arching effect. However, because the embedded depth of open trench tunnel is shallow relatively, the existence of the arching effect is almost neglected. Therefore, little information concerning the soil arching effect of the open trench tunnel has been published.

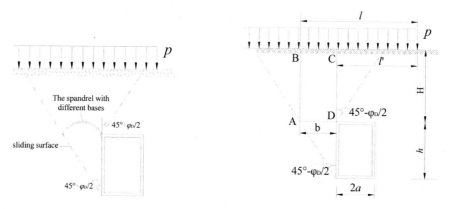

Figure 1. The spandrel with different bases.   Figure 2. The calculation chart ($l \geq b + H \tan(45° - \varphi_D/2)$).

2 THE BASIC ASSUMPTIONS AND SIMPLIFIED CONDITIONS

In order to facilitate the derivation, some simplifications and assumptions on the open trench tunnel are made as following:

1) It is based on an assumption that the surrounding rock pressure parallel to the cross section of the existing open trench tunnel. Then, the size and distribution of the surrounding rock pressure along the tunnel axis direction has no change. Therefore, this is a plane strain problem.
2) It is assumed that the two sliding surfaces start from the tunnel arch foot, and the two sliding surfaces form angles ($45° - \varphi_D/2$) with tunnel side walls respectively.
3) It is supposed that the sliding surface extends to the surface. Meanwhile, only the backfill soil, within the slip plane, is likely to decline. However, the backfill soil, outside the slip plane is stable, which is shown in Fig. 2 to Fig. 6.
4) The soil layers around the tunnel are regarded as the cohesionless homogeneous mass. The equivalent internal frictionangle φ_D is based on an equivalent substitution in which the shear strength is equal, and the formula as following:

$$\varphi_D = \arctan\left(\tan\varphi + \frac{c}{\gamma z}\right) \quad (1)$$

where z = the backfill soil depth (Zhu 2010); and c = the cohesion of backfill soil.
5) The ground asymmetric load diffuses outward along the angle between the ground asymmetric load and the horizontal plane which is ($45° + \varphi_D/2$).

3 DERIVATION OF THE LOCAL SOIL ARCHING EFFECT DEPTH ABOVE THE SIDE WALL

When the tunnel is shallow buried, the top of tunnel cannot generate the pressure arch. However, on account of the existence of ground asymmetric load induces the different force situation between the left side wall and right side wall of the existing open trench tunnel. It is necessary to determine whether it produce spandrel with different bases between the sliding surface and the ground loading side wall.

According to the ref. (Jia 2003), the backfill soil will form a spandrel with different bases which is supported by both the retaining structure and the stable soil on the sliding surfaces. In other words, one arch foot is located in the side wall of the tunnel, which is supported by the friction resistance between the soil and the side wall. Another arch foot is located in the interface between the active region and the stable soil, which is supported by the shear behavior of the stable soil.

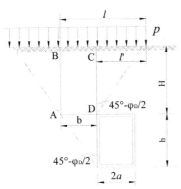

Figure 3. The calculation chart ($b \leq l < b + H\tan(45° - \phi_D/2)$).

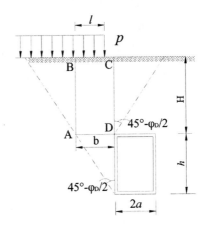

Figure 4. The calculation chart ($0 \leq l < b$).

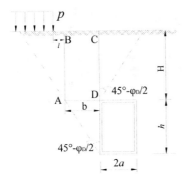

Figure 5. The calculation chart ($H\tan(45° - \varphi_D/2) \leq l < 0$)

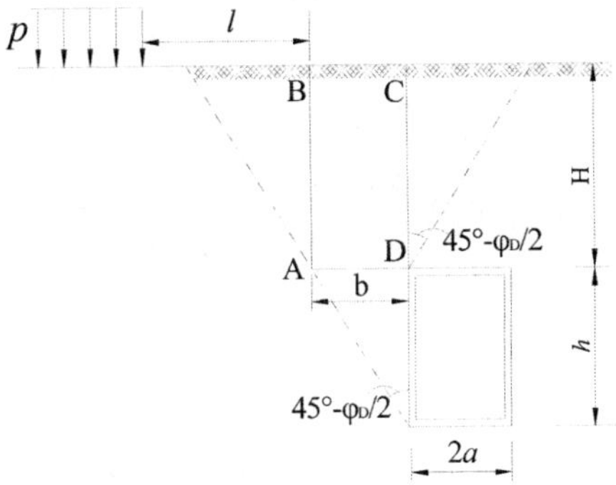

Figure 6. The calculation chart ($l < H\tan(45° - \varphi_D/2)$).

Therefore, it is assumed that a sliding surface starts from the arch foot of the tunnel, which forms an angle $(45° - \varphi_D/2)$ with tunnel side walls. Meanwhile, it is supposed that another sliding surface starts from the top of the side wall, which forms an angle $(45° - \varphi_D/2)$ with the vertical direction. What's more, it is considered that both the two sliding surfaces extend to the ground surface. Taking ABCD as the downward soil sliding, it is subjected to resistance along the AB and CD friction forces, where l is the distance from the rightmost of ground overload p to the AB side of sliding soil ABCD, and l' is the distance from the rightmost of ground overload p to the CD side of sliding soil ABCD.

When the position of the ground asymmetric load changed, it will produce different influence on the local soil arching effect of the existing open trench tunnel. If the point B is regarded as a reference, the process of analysis will be divided into the following categories

1) $l \geq b + H\tan(45° - \varphi_D/2)$

The total pressure Q acts on the underground structure, which can be expressed as the following equation:

$$Q = G + P - F_1 - F_2 \quad (2)$$

where $G =$ the gross weight of sliding soil ABCD; $P =$ the total ground overload acting on sliding soil ABCD; $F_1 =$ the friction force that G is subjected to AB side; and $F_2 =$ the friction force that G is subjected to CD side.

(1) The gross weight G of sliding soil $ABCD$ is given by:

$$G = \gamma Hh \tan(45° - \frac{\varphi_D}{2}) \tag{3}$$

(2) The total ground overload P acting on sliding soil $ABCD$ can be shown that:

$$P = p \cdot b = ph \tan(45° - \frac{\varphi_D}{2}) \tag{4}$$

(3) The friction force $F_1(F_2)$ that G is subjected to $AB(CD)$ side will be written as:

The total horizontal force $E_{AB}(E_{CD})$ is shown as follows:

$$E_{AB} = E_{CD} = (\frac{1}{2}\gamma H^2 + pH)\tan^2(45° - \frac{\varphi_D}{2}) \tag{5}$$

The friction resistance $F_1(F_2)$ is shown as follows:

$$F_1 = F_2 = E_{AB}\tan\varphi_D = (\frac{1}{2}\gamma H^2 + pH)\tan^2(45° - \frac{\varphi_D}{2})\tan\varphi_D \tag{6}$$

Through take Eq. 3, 4, 5, to the Eq. 2, the total vertical pressure values Q which acts on the top of the side wall will be obtained. The formula can be shown as follows:

$$Q = (\gamma H + p)h\tan(45° - \frac{\varphi_D}{2}) - (\gamma H^2 + 2pH)\tan^2(45° - \frac{\varphi_D}{2})\tan\varphi_D \tag{7}$$

The surrounding rock pressure collection degree q can be expressed as the following equation:

$$q = \frac{Q}{b} = (\gamma H + p) - \frac{(\gamma H^2 + 2pH)\tan^2(45° - \frac{\varphi_D}{2})\tan\varphi_D}{b} \tag{8}$$

where H = the tunnel depth; b = the span of the soil arching which is located in the top of the side wall; γ = the backfill soil gravity density, and φ_D = the equivalent internal frictionangle of backfill soil, as shown in Eq. 1.

Through the above analysis, it is implied that when the total vertical pressure that acts on the top of tunnel is equal to zero, the frictional resistance can overcome all the weight of sliding soil. Thus, the sliding soil $ABCD$ will not slip. So the tunnel under the $ABCD$ will not bear its weight. According to the Eq. 7, it is indicated that the total vertical pressure values Q that acts on the top of the side wall is changed with the variation of the tunnel buried depth H. It is assumed that Eq. 8 is equal to the zero. Then, the depth value H_p that the backfill soil can generate the spandrel with different bases at the top of the side wall will be attained. The formula has been established as follows:

$$H_p = \frac{b}{2\tan^2(45° - \frac{\varphi_D}{2})\tan\varphi_D} - \frac{p}{\gamma} + \sqrt{\frac{b^2}{4\tan^4(45° - \frac{\varphi_D}{2})\tan^2\varphi_D} + \frac{p^2}{\gamma^2}} \tag{9}$$

2) $b \leq l < b + H\tan(45° - \varphi_D/2)$

The calculation chart of case 2) is shown in Fig. 3. The derivation process is similar to case 1). But the friction resistance F_2 by CD surface is different. The formula will be written as:

$$F_2 = E_{CD}\tan\varphi_D$$

$$= [\frac{1}{2}\gamma H^2 + pl'\tan(45° + \frac{\varphi_D}{2})]\tan^2(45° - \frac{\varphi_D}{2})\tan\varphi_D \tag{10}$$

Similarly, in this case, the depth value H_p which the backfill soil can generate the spandrel with different bases at the top of the side wall will be attained. The formula has been established as follows:

$$H_p = \frac{b}{2\tan^2(45°-\frac{\varphi_D}{2})\tan\varphi_D} - \frac{p}{2\gamma} + \sqrt{\frac{b^2}{4\tan^4(45°-\frac{\varphi_D}{2})\tan^2\varphi_D} + \frac{p^2}{4\gamma^2} + \frac{bp}{2\gamma\tan^2(45°-\frac{\varphi_D}{2})\tan\varphi_D} - \frac{pl\tan(45°+\frac{\varphi_D}{2})}{\gamma}} \quad (11)$$

3) $0 \leq l < b$

The calculation chart of case 3) is shown in Fig. 4. The derivation process is similar to case 1). But the total ground overload P change into the following form:

$$P = pl \quad (12)$$

The friction resistance F_2 by CD surface is different and it is shown as follows:

$$F_2 = E_{CD}\tan\varphi_D = \frac{1}{2}\gamma H^2 \tan^2(45°-\frac{\varphi_D}{2})\tan\varphi_D \quad (13)$$

Likewise, in this case, the depth value H_p that backfill soil can generate the spandrel with different bases at the top of the side wall will be attained. The formula has been established as follows:

$$H_p = \frac{b}{2\tan^2(45°-\frac{\varphi_D}{2})\tan\varphi_D} - \frac{p}{2\gamma} + \sqrt{\frac{b^2}{4\tan^4(45°-\frac{\varphi_D}{2})\tan^2\varphi_D} + \frac{p^2}{4\gamma^2} + \frac{p(2l-b)}{2\gamma\tan^2(45°-\frac{\varphi_D}{2})\tan\varphi_D}} \quad (14)$$

4) $H\tan(45°-\varphi_D/2) \leq l_1 < 0$

The calculation chart of case 4) is shown in Fig. 5. The derivation process is similar to case 1). But the total ground overload P changed into zero. The friction resistance F_1 by AB surface is also changed into the following form:

$$F_1 = E_{AB}\tan\varphi_D$$
$$= \{\frac{1}{2}\gamma H^2 + p[H - l\tan(45°+\frac{\varphi_D}{2})]\}\tan^2(45°-\frac{\varphi_D}{2})\tan\varphi_D \quad (15)$$

The friction resistance F_2 by CD surface is different and it is shown as follows:

$$F_2 = E_{CD}\tan\varphi_D = \frac{1}{2}\gamma H^2 \tan^2(45°-\frac{\varphi_D}{2})\tan\varphi_D \quad (16)$$

In the same way, the depth value H_p that backfill soil can generate the spandrel with different bases at the top of the side wall will be attained. The formula has been established as follows:

$$H_p = \frac{b}{2\tan^2(45°-\frac{\varphi_D}{2})\tan\varphi_D} - \frac{p}{2\gamma} + \sqrt{\frac{b^2}{4\tan^4(45°-\frac{\varphi_D}{2})\tan^2\varphi_D} + \frac{p^2}{4\gamma^2} - \frac{bp}{2\gamma\tan^2(45°-\frac{\varphi_D}{2})\tan\varphi_D} + \frac{pl\tan(45°+\frac{\varphi_D}{2})}{\gamma}} \quad (17)$$

5) $l_1 < H\tan(45°-\varphi_D/2)$

The calculation graphic expression of case 5) is shown in Fig. 6. The derivation process is similar to case 1). But the friction resistance F_1 and F_2 is changed into the following forms:

$$F_1 = F_2 = \frac{1}{2}\gamma H^2 \tan^2(45°-\frac{\varphi_D}{2})\tan\varphi_D \quad (18)$$

In similar manner, the depth value H_p that backfill soil can generate the spandrel with different bases at the top of the side wall will be attained. The formula has been established as follows:

$$H_p = \frac{b}{\tan^2(45° - \frac{\varphi_D}{2})\tan\varphi_D} \tag{19}$$

In a word, a comparison among the Eq. 9, 11, 14, 17, and 19 is carried out. It indicates that as the ground asymmetric load distribution away from the sliding soil $ABCD$, the critical depth value H_p which is generate the spandrel with different bases at the top of the side wall will bigger and bigger.

4 CONCLUSION

This article is guided by the unconsolidated mass theory of *Terzarghi*. Typically, based on the theory of equivalent internal frictionangle, the soil layers around the tunnel are equal to cohesionless homogenate mass. Consequently, the failure model of existing open trench tunnels under ground asymmetric load is analyzed. Meanwhile, the soil arching effect of existing open trench tunnels under ground asymmetric load is discussed. In addition, the critical embedded depth formula of the local soil arching effect of the backfill soil is given.

(1) The existing open trench tunnels under ground asymmetric load will form two sliding surfaces respectively. A sliding surface which forms an angle $(45° - \varphi_D/2)$ with tunnel side walls starts from the arch foot of the tunnel. Meanwhile, another sliding surface which forms an angle $(45° - \varphi_D/2)$ with the vertical direction starts from the top of the side wall. In addition, it is considered that the two sliding surfaces extend to the surface.
(2) According to the failure model of existing open trench tunnels under ground asymmetric load, the soil arching effect of existing open trench tunnels under ground asymmetric load is discussed. At the same time, the critical embedded depth formula of the local soil arching effect of the backfill soil is obtained.

ACKNOWLEDGEMENTS

The authors would like to thank the Talents Training Project of Yunnan Province (No. 14118651) for providing the financial support to conduct this research.

REFERENCES

Jia, H.L., Wang, C.H. & Li, J.H. 2003. Several problems about the soil arching effect. Journal of Southwest Jiao Tong University 38(4): 398–402.

Kovari, K. 1994. Erroneous concepts behind the new austrian tunnelling method. Tunnels and Tunnelling 26(11): 38–42.

Liang, X.D., Liu, G. & Zhao, J. 2005. Definition and analysis of arching action in underground rock engineering. Journal of Hohai University 33(3): 314–317.

Liang, X.D., Song, H.W. & Zhao, J. 2008. Relationship between tunnel pressure-arch and surrounding rock deformation. Journal of Xi'an University of Science And Technology 28(4): 647–650.

Terzaghi, K. 1947. Theoretical soil mechanics. New York: John Wiley & Sons.

Wang, Y.C. & Yan, X.S. 2012. Study of deep circular tunnel pressure arch. Journal of underground space and engineering 8(5): 910–915.

Yu, B. & Wang, H.J. 2008. A research of pressure-arch theory and tunnel depth classification. Beijing: China railway publishing house.

Zhu, H.H. 2010. Underground structures. Beijing: China Building Industry Press.

Zhang, J.B. & Lian, W. 2009. Talk about the calculating formula of equivalent internal friction angle of rock mass. Geotechnical Engineering World 12(12): 16–17.

Sensitivity analysis of prestressing loss of long-curved tendons in tensioned phase

Jiangang Tian
School of Roads and Highway, Chang'an University, Xi'an, Shaanxi Province, China

Xiaoming Fang
Shaanxi Huashan Road and Bridge Engineering Company, Xi'an, Shaanxi Province, China

Long Qin, Wei Xie & Jiahuan Tong
School of Roads and Highway, Chang'an University, Xi'an, Shaanxi Province, China

ABSTRACT: In the modern field of bridge engineering, prestressing technology has been more and more widely applied with its unique advantages. Prestressing loss of long-curved tendons in the tensioned phase in post-tensioned concrete bridge accounts for the majority of total loss. The values of prestressing loss need to be carefully calculated, and it becomes one of the most important issues regarding the analysis of prestressed bridge. The sensitivity analysis of key parameters of the prestressing loss of long-curved tendons in tensioned phase was conducted in this paper. Based on the research of prestressing losses and elongation of test points at different locations of three prestressing tendons, the sensitivity of prestressing loss towards different parameters in the formula of codes such as μ, k, θ, L, and other factors are analyzed.

1 INTRODUCTION

Many factors can affect the tension of long-curved prestressing tendons. these include whether the applied initial stress values meet the design requirements or not, whether the embedded bellows is smooth and secure or not, if there is any leakage of slurry in concrete casting or not, if the bellow joints are smooth or not, if the bellows are suffered from compressional deformation or even being blocked during construction, and whether the elongation matches the stress value or not, etc. Theoretical research was conducted about the friction resistance of tension in the long-curved tendons and data were analyzed, as well as experienced research both home and abroad were referenced. Conclusion was summarized; a reasonable channel friction loss of prestressing research theory was generalized, and the effect of deviation coefficient k was evaluated. The finds provide the basis information to the stress and strain calculations in long-curved prestressing tendons, and the result plays an important role in the construction control on structures.

2 PROJECT PROFILE

Lishan Road overpass is located in the south of the road. As a result of the elevation limit, an upper cross is needed, thus forming a T intersection. The bridge starts with the stake number of K4 + 198.447 and ends up with K4 + 480.127. The total length of the bridge is 281.68 m. The total width of the cross section is 31 m. A single deck is used with the bi cross slope of 1.5%. The bridge is divided into three links with the main bridge at the second link. A prestressed concrete variable cross-section continuous box girder is used with 30 m + 45 m + 30 m. The first link is 4 × 25 m prestressed concrete variable cross-section continuous box girder and the third with a

Table 1. Sensitivity analysis of μ and L to the prestressing loss.

μ	0.14		0.16		0.18		0.20		0.22		0.25	
L (m)	Loss (MPa)	Scale (%)	Loss (MPa)	Scale (%)	Loss (MPa)	Scale (%)	Loss (MPa)	Scale (%)	Loss (MPa)	Scale (%)	Loss (MPa)	Scale (%)
25	179.3	12.9%	196.5	14.1%	213.5	15.3%	230.3	16.5%	246.8	17.7%	271.2	19.4%
35	197.4	14.1%	214.4	15.4%	231.1	16.6%	247.7	17.8%	263.9	18.9%	287.9	20.6%
45	215.2	15.4%	232.0	16.6%	248.5	17.8%	264.7	19.0%	280.8	20.1%	304.4	21.8%
55	232.8	16.7%	249.3	17.9%	265.5	19.0%	281.6	20.2%	297.4	21.3%	320.6	23.0%
65	250.1	17.9%	266.3	19.1%	282.3	20.2%	298.1	21.4%	313.7	22.5%	336.6	24.1%
75	267.1	19.1%	283.1	20.3%	298.9	21.4%	314.5	22.5%	329.8	23.6%	352.4	25.3%
85	283.9	20.4%	299.7	21.5%	315.2	22.6%	330.6	23.7%	345.7	24.8%	367.9	26.4%

3 × 25 m. The span is arranged as (4 × 25 m) + (30 m + 45 m + 30 m) + (3 × 25 m) of prestressed concrete variable cross-section continuous box girder. The first and third links use the prestressed concrete variable cross-section continuous box girder with single-box multi-room box section. The second link uses prestressed concrete variable cross-section continuous box girder with single-box multi-room box section. In this paper, the second link was analyzed.

3 SENSITIVITY ANALYSIS INFLUENCED BY TENSIONING

This project analyzes the prestressing loss in long-curved tendons under tensioned phase. Based on the research of prestressing losses and elongation of test points at different locations in the prestressing tendons N1, N2, and N3, different parameters such as μ, k, θ in the formula "L, $\sigma_{loss} = \sigma_{con}(1 - e^{-(\mu\theta + kx)})$" were evaluated for their influencing sensitivity towards the loss of prestress.

① Assuming k, θ as a constant, the value of μ and L can be changed. According to the range of μ in the codes, σ_{loss} is calculated with the μ taken as 0.14, 0.16, 0.18, 0.20, 0.22, and 0.25, and compared to the tensile stress control values, The percentage of their losses is obtained. Assuming L in the range from 25 m to 85 m, and the value changes every a 10 m, theoretical σ_{loss} is calculated and compared to the tensile stress control value, The percentage of their losses is obtained. Based on the above analysis of the calculated values, the sensitivity analysis of influence of μ and L on the prestressing loss are evaluated and shown in the table 1.

② Assuming μ, θ as constant, the value of k and L can be changed. According to the range of k in the codes, σ_{loss} is calculated with K taken as 0.0015, 0.0020, 0.0025, 0.0030, 0.0035 and 0.0040, and compared to the tensile stress control values, The percentage of their losses is obtained. Assuming L be in the range from 25 m to 85 m, and the value changes every a 10 m, theoretical σ_{loss} is calculated and compared to the tensile stress control value. The percentage of their losses is obtained. Based on the *above* analysis of the calculated values, the sensitivity analysis of influence of k and L on the prestressing loss are evaluated and shown in the table 2.

③ Assuming $\mu = 0.20$, $k = 0.0015$, the value of θ and L the value can be changed. Make θ in the range from 0.5 to 1.5rad, the value changes every a 0.2 rad, theoretical σ_{loss} is calculated and compared to the tensile stress control value. The percentage of their losses is obtained. Based on the above analysis of the calculated values, the sensitivity analysis of influence of θ and *L* on the prestressing loss are evaluated and shown in the table 3.

④ Assuming $\mu = 0.25$, $k = 0.0040$, the value of θ and L the value can be changed. Make θ in the range from 0.5 to 1.5 rad, the value changes every a 0.2 rad, theoretical σ_{loss} is calculated and compared to the tensile stress control value. The percentage of their losses is obtained. Based

Table 2. Sensitivity analysis of k and L to the prestressing loss.

k	0.0015		0.0020		0.0025		0.0030		0.0035		0.0040	
L (m)	Loss (MPa)	Scale (%)	Loss (MPa)	Scale (%)	Loss (MPa)	Scale (%)	Loss (MPa)	Scale (%)	Loss (MPa)	Scale (%)	Loss (MPa)	Scale (%)
25	271.2	19.4%	285.2	20.4%	298.9	21.4%	312.6	22.4%	326.0	23.4%	339.3	24.3%
35	287.9	20.6%	307.1	22.0%	326.0	23.4%	344.5	24.7%	362.8	26.0%	380.7	27.3%
45	304.4	21.8%	328.7	23.6%	352.4	25.3%	375.6	26.9%	398.3	28.6%	420.5	30.1%
55	320.6	23.0%	349.8	25.1%	378.1	27.1%	405.7	29.1%	432.6	31.0%	458.7	32.9%
65	336.6	24.1%	370.5	26.6%	403.2	28.9%	435.0	31.2%	465.7	33.4%	495.4	35.5%
75	352.4	25.3%	390.8	28.0%	427.7	30.7%	463.3	33.2%	497.6	35.7%	530.7	38.0%
85	367.9	26.4%	410.7	29.4%	451.6	32.4%	490.9	35.2%	528.5	37.9%	564.5	40.5%

Table 3. Sensitivity analysis of θ and L to the prestressing loss ($\mu=0.2, k=0.0015$).

θ (rad)	0.50		0.70		0.90		1.10		1.30		1.50	
L (m)	Loss (MPa)	Scale (%)	Loss (MPa)	Scale (%)	Loss (MPa)	Scale (%)	Loss (MPa)	Scale (%)	Loss (MPa)	Scale (%)	Loss (MPa)	Scale (%)
25	179.2	12.8%	226.9	16.3%	272.7	19.5%	316.7	22.7%	359.0	25.7%	399.6	28.6%
35	197.3	14.1%	244.3	17.5%	289.4	20.7%	332.7	23.9%	374.4	26.8%	414.4	29.7%
45	215.1	15.4%	261.4	18.7%	305.9	21.9%	348.6	25.0%	389.6	27.9%	429.0	30.8%
55	232.7	16.7%	278.3	19.9%	322.1	23.1%	364.1	26.1%	404.6	29.0%	443.4	31.8%
65	250.0	17.9%	294.9	21.1%	338.0	24.2%	379.5	27.2%	419.3	30.1%	457.6	32.8%
75	267.1	19.1%	311.3	22.3%	353.8	25.4%	394.6	28.3%	433.8	31.1%	471.5	33.8%
85	283.9	20.3%	327.4	23.5%	369.3	26.5%	409.5	29.4%	448.1	32.1%	485.3	34.8%

Table 4. Sensitivity analysis of θ and L to the prestressing loss ($\mu=0.25, k=0.004$).

θ (rad)	0.50		0.70		0.90		1.10		1.30		1.50	
L (m)	Loss (MPa)	Scale (%)	Loss (MPa)	Scale (%)	Loss (MPa)	Scale (%)	Loss (MPa)	Scale (%)	Loss (MPa)	Scale (%)	Loss (MPa)	Scale (%)
25	281.1	20.1%	335.4	24.0%	387.1	27.7%	436.2	31.3%	483.0	34.6%	527.5	37.8%
35	324.7	23.3%	376.9	27.0%	426.6	30.6%	473.8	34.0%	518.8	37.2%	561.5	40.2%
45	366.7	26.3%	416.9	29.9%	464.6	33.3%	509.9	36.6%	553.1	39.6%	594.2	42.6%
55	407.0	29.2%	455.2	32.6%	501.1	35.9%	544.6	39.0%	586.1	42.0%	625.6	44.8%
65	445.8	32.0%	492.1	35.3%	536.1	38.4%	578.0	41.4%	617.8	44.3%	655.7	47.0%
75	483.0	34.6%	527.5	37.8%	569.8	40.8%	610.0	43.7%	648.3	46.5%	684.7	49.1%
85	518.8	37.2%	561.5	40.2%	602.1	43.2%	640.8	45.9%	677.6	48.6%	712.6	51.1%

on the above analysis of the calculated values, the sensitivity analysis of influence of θ and L on the prestressing loss are evaluated and shown in the table 4.

From the above table, it can be seen that μ, k, θ, and L all have the impact on the loss of prestressing during tensioning of tendons based on the formula "$\sigma_{loss}=\sigma_{con}(1-e^{-(\mu\theta+kx)})$". But they have influenced at different extent. In the range values of μ, k, θ, and L, six values were taken equally, and the percentage values were calculated in order to analyze the sensitivity of various factors. According to the figure 1 to 3, L were taken 25 m, 55 m, and 85 m, the influencing factors increase its growth rate with value changing. (Series 1 changes due to the changing value of μ; Series 2 changes due to the changing value of k; Series 3, 4 are based on that $\mu=0.20$, $k=0.0015$ and $\mu=0.25$, $k=0.0040$, which were due to the changing value of θ respectively.)

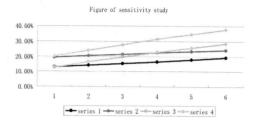

Figure 1. Influencing rate of μ, k and θ regarding the σ_{loss} for span of 25 m.

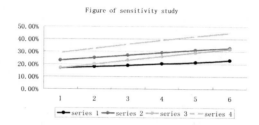

Figure 2. Influencing rate of μ, k and θ regarding the σ_{loss} for span of 55 m.

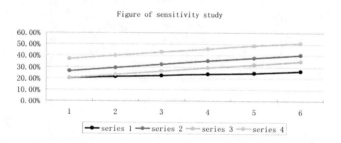

Figure 3. Influencing rate of μ, k and θ regarding the σ_{loss} for span of 85 m.

4 TEST RESULT

According to table 1 to table 4, and figure 1 to figure 3, it can be seen that μ, k, θ, and L all have the impact on the loss of prestressing during tensioning of tendons based on the formula "$\sigma_{loss} = \sigma_{con}(1 - e^{-(\mu\theta + kx)})$". But they influence at different extent. With the increase in span, the increasing of value μ have less impact on σ_{loss}; the increasing of value k have more impact on σ_{loss}; and the increasing of value θ have less impact on σ_{loss}. With the increase in span, the change of value k has more impact on the growth rate of σ_{loss} than μ, which indicates the value of k has greater impact on the sensitivity. With the constant value of μ and k, and with the increase in span, the change of value θ has less sensitivity influence. In a small span condition, θ has greater impact on the sensitivity. With the increase in span, k has greater impact on the sensitivity.

5 CONCLUSION

The paper studies the loss of prestressing in long-curved tendons during tensioning phase. By analyzing the equation of $\sigma_{loss} = \sigma_{con}(1 - e^{-(\mu\theta + kx)})$, the sensitivity of μ, k, θ, and L affecting the loss of prestressing were calculated. All of the factors have impacts on the prestressing loss, but they function in different degree. However, only a part of the conclusion can be drawn from this study.

1) With the increase in span, the increasing of value μ and, θ has less impact on σ_{loss}, and the increasing of value k has more impact on σ_{loss}.
2) With the increase in span, the change of value k has more impact on the growth rate of σ_{loss} than μ, which indicates the value of k has greater impact on the sensitivity.
3) With the constant value of μ and, k and with the increase in span, the change of value θ has less sensitivity influence.
4) In a small span condition, θ has greater impact on the sensitivity. With the increase in span, k has greater impact on the sensitivity.

More experimental research on the principle of prestressing loss is needed in order to build more accurate verification. For that reason, assessing methods of tensioned prestressing loss can be enhanced and further used in the prestressed construction phase.

REFERENCES

Cheng, C. 2011. Theoretical analysis and experimental study of effective prestress at curved channel in large span PC Bridge, [D]. Wuhan University of Technology

Feng, H. 2007. *Research about the stress performance in post-tensioned concrete beam specified by the HRB500 reinforced bars, [D]*. Zhengzhou University

Lv, J. 2007. *Research about stress performance and prestress loss in prestressed concrete curved beam bridge, [D]*. Dalian University of Technology

Su, H. 2007. *Research about prestressing loss in webs of prestressed concrete box girder, [D]*. Wuhan University of Technology

Ye, J. Z. 2008. *Structural design principles, [M]*. Beijing: People's Communications Press.

Xu, Y. Z. 2003. *Research about prestress loss in large span continuous prestressed concrete beam bridge, [D]*. Wuhan University of Technology

Numerical analysis of the wind field on a low-rise building group

C. Zhang & W.B. Yuan
College of Architecture and Civil Engineering, Zhejiang University of Technology, Hangzhou, China

ABSTRACT: From the perspective of complex wind effects, the wind field of unequal spacing low-rise building group was computed numerically. Then the interaction between building are studied in detail. The results showed that the eddy current has a local maximum in the first lee area when the spacing is small, which is also the most dangerous place. There is a great impact on the wind field under different distance between buildings, When spacing is proper, we can guarantee a minimum pressure on the rear housing. These results can help us to provide a reasonable basis for the layout of the building complex.

1 INTRODUCTION

The lower gable roof houses is the most large wide range of housing types in China's southeast coastal areas, this type of house often cause damage or even collapse in the coastal frequent activities of the typhoon. Accurately predicting the distribution of wind load in order to propose effective mitigation measures are necessary.

Wind tunnel model test and full-size field test is the primary method of forecasting surrounding wind field and surface pressure of buildings (Surry, 1999). Research on the particular situation relies mainly on wind tunnel model test, and this method has many disadvantages such as long period, high cost and susceptible precision. Since the 1980s, numerical simulation has become a new and effective method of forecasting wind load of building surface. It makes the study of wind field of building complex possible, this proved to be a simple and reliable way (Murakami, 1997). In the late 1800s foreign scholars carried out numerical simulation to wind farms and wind environmental of building groups. For example, Stathopoulos & Baskaran (Stathopoulos & Baskaran, 1996/1989) used the orthogonal discrete grid and standard k-ε model to simulate a buildings wind field of blocks of square, they focused on the simulation of wind flow between the building and the change of wind speed, does not involve building more complex figure. China's Wang Hui and Chen Shui fu (Wang & Chen. & Tang, 2003) used the same six figure lower house to simulate wind environment, but they did not consider changes in the spacing of buildings.

This article will use the widespread adoption of low-rise housing groups of gable roof with a cornice. Using tetrahedral discrete grids with good adaptive meet the requirements of the sloping roof of the body. Numerical simulation will be focused on the changes of wind farms and the distribution of surface pressure in the group effect.

2 THE BASIC EQUATIONS AND TURBULENCE MODEL

Near-ground wind is a nearly incompressible turbulent flow. Under the steady-state effect, the differential equations based on Reynolds average equation and standard k-ε model see equation 1–4 (Zhou & Stathopulos, 1997).

$$\frac{\partial U_j}{\partial x_j} = 0 \qquad (1)$$

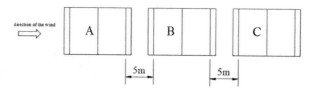

Figure 1. Buildings arrangement.

$$U_j \frac{\partial U_i}{\partial x_j} = -\frac{\partial p}{\partial x_i} + \frac{\partial}{\partial x_j}\left[(v+v_t)\frac{\partial U_i}{\partial x_j}\right] + \frac{\partial}{\partial x_j}\left[v_t \frac{\partial U_j}{\partial x_i}\right] \quad (2)$$

$$U_j \frac{\partial k}{\partial x_j} = \frac{\partial}{\partial x_j}\left[\left(v+\frac{v_t}{\sigma_k}\right)\frac{\partial k}{\partial x_j}\right] + v_t\left[\frac{\partial U_i}{\partial x_j}+\frac{\partial U_j}{\partial x_i}\right]\frac{\partial U_i}{\partial x_j} - \varepsilon \quad (3)$$

$$U_j \frac{\partial \varepsilon}{\partial x_j} = \frac{\partial}{\partial x_j}\left[\left(v+\frac{v_t}{\sigma_\varepsilon}\right)\frac{\partial \varepsilon}{\partial x_j}\right] + C_1 \frac{\varepsilon}{k}v_t\left[\frac{\partial U_i}{\partial x_j}+\frac{\partial U_j}{\partial x_i}\right]\frac{\partial U_i}{\partial x_j} - C_2 \frac{\varepsilon^2}{k} \quad (4)$$

where <Ui> (i = 1, 2, 3) representing average velocity component of orientation X, Y, Z. <P> is the average pressure. <k> is turbulent kinetic energy. <ε> is turbulent kinetic energy dissipation rate. <ν> is the air dynamic viscosity coefficient. $v_t = C_\mu k^2/\varepsilon$ is turbulent eddy viscosity coefficient. $C_\mu = 0.09$, $C_1 = 1.44$, $C_2 = 1.92$, $\sigma_k = 1.0$, $\sigma_z = 1.3$.

After the dimensionless equation 1–4 can be written as a unified expression:

$$\frac{\partial}{\partial x_j}\left[U_j \Phi - \Gamma \frac{\partial \Phi}{\partial x_j}\right] = S \quad (5)$$

where <Φ> representing the variable <Ui> (i = 1, 2, 3), k , ε. <Γ> is the effective diffusion coefficient. <S> is the source term.

3 RESEARCH CONTENT

Using three pairs of sloping roof building groups to research calculation, the arrangement form is shown in Figure 1. Structure used in this paper is the roof slope of 25° of ordinary houses, flat size 8 m × 10 m, eaves height of 9.6 m, cornice length 1.0 m. Three situations were simulated respectively, that house pitch: 5m, 10 m, 20 m. Computational domain of three cases are three-dimensional rectangular area of 280 m × 120 m × 80 m, 300 m × 120 m × 80 m, 320 m × 120 m × 80 m. This article uses the automatic generation of tetrahedral mesh CFD, housing surface mesh is shown in Figure 2. Finally the calculation area of 5 m spacing formed 616,111 units and 104,327 tetrahedral nodes, the calculation area of 10 m spacing formed 663,730 units and 112,324 tetrahedral nodes, the calculation area of 20 m spacing formed 696,061 units and 117,811 tetrahedral nodes.

In this paper, the gas flow pressure values are calculated at the height of the building or at the highest point above the undisturbed pressure as a pressure reference value and by the dimensionless are given in the form of pressure coefficients. Taking into account the low-rise building most at the height of about 15 to 20 meters, the pressure on the surface is less affected by the wind speed changes of atmospheric boundary layer, so setting the wind flow is uniform flow (Huang, 2008). In this paper wind-level is eleven, that is V = 29.0 m/s.

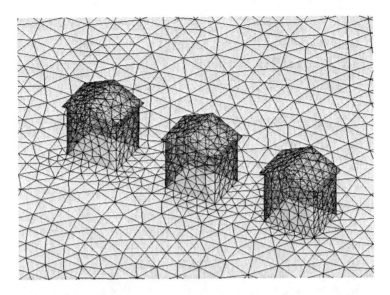

Figure 2. Housing surface mesh.

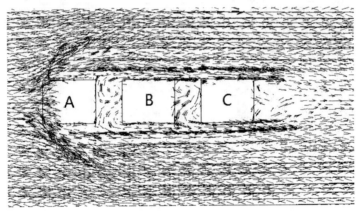

Figure 3. The wind vector of 4.8 m height horizontal section by 5 m spacing.

4 RESULTS

4.1 *Wind vector simulation results*

Figures 3–5 is the 4.8 m height horizontal section of the local velocity vector of conditions: 5 m, 10 m, 20 m. Each vector segment of length indicates the relative size of the average velocity, the direction of arrows indicates the direction of speed. The figure shows that leeward areas where flow is impeded results in varying degrees of vortex as wind flow along. The vortex behind the house is the strongest when the spacing is 5 m, and with the housing spacing increases the intensity of the leeward vortex decreases, moreover vortex intensity of the first leeward areas is generally much stronger than the second leeward areas.

4.2 *Surface pressure simulation results*

Figures 6–8 is the 4.8 m height horizontal section of surface pressure coefficients of three conditions. Based on analysis and computation, when the spacing is 5 m, windward of the house suffers the

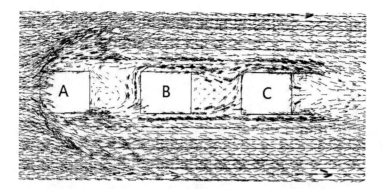

Figure 4. The wind vector of 4.8 m height horizontal section by 10 m spacing.

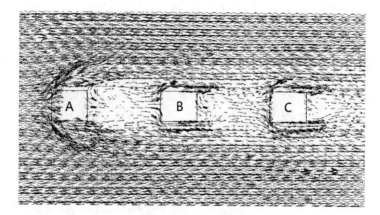

Figure 5. The wind vector of 4.8 m height horizontal section by 20 m spacing.

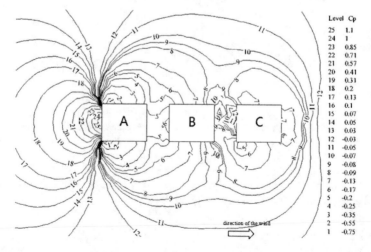

Figure 6. The surface pressure coefficients of 4.8 m height horizontal section by 5 m spacing.

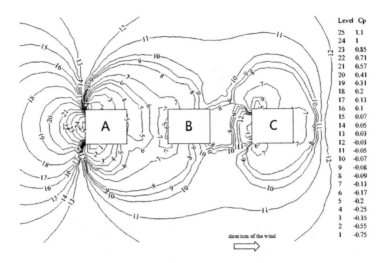

Figure 7. The surface pressure coefficients of 4.8 m height horizontal section by 10 m spacing.

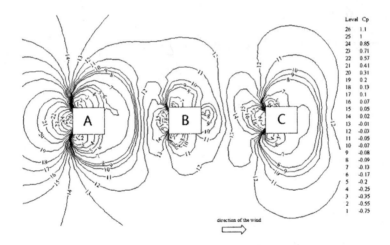

Figure 8. The surface pressure coefficients of 4.8 m height horizontal section by 20 m spacing.

largest positive pressure in the direction of the wind and its value is 1.1 (The following pressure values refer to the pressure coefficient, comparison of absolute pressure), but windward of the rear housing B and C are subjected to negative pressure, and the negative pressure decreases from B to C and respectively they are -0.17 and -0.03. In the leeward direction, the pressure are all negative, and among minimum value is -0.08 in the leeward side of B, by the way the negative pressure of A is about twice than that of B. All sides of the house show negative pressure, and the pressure of the windward front corner area of A is the largest, then has a tendency to decrease from A to C, but the pressure of C at the front corner area is little higher than that of B at the posterior horn, and it is about one fifth of values of A front corner, and minimum value is -0.03 at the posterior horn of C; When the spacing is 10 m, in the direction of the wind windward of B still suffers negative pressure, but the values has been less than -0.1, On the contrary the pressure of windward side of C has a tendency from negative to positive. In the leeward direction, the pressure of A and C are all negative, and the value of A is still twice than that of B, at the same time the pressure of B leeward area is significantly weakened. The side pressure value from A to B decreases from -0.75

to −0.08, but the pressure of the C front corner area sharply increases to −0.2 about a quarter of A, then slowly decreases to about −0.03; When the spacing is 20 m, A suffer positive pressure unchanged in the direction of the wind, while there is a large change on the pressure of B and C, and they are all positive, incidentlly the windward pressure of B is only 0.1 that it is half of C. In the leeward direction, the pressure of A and C are substantially all −0.13, and the leeward zone pressure value of house B is about half of the front row house. The sides of three houses are all negative pressure and the pressure changes shows significant regularity, generally the pressure values of front corner area of rear house are higher than that of rear corner area of front house, which the front corner area of B is the minimum about −0.17, and A of C is the third at the front corner, meanwhile pressure value of the rear corner of three houses are reduced to −0.03.

5 CONCLUSION

(1) When the building space is smaller the windward side of rear housing will have a greater negative, and the pressure value decreases along the wind. However as the spacing increases it helps the pressure of windward side of the rear housing from negative to positive. As can be seen from this article where appropriate spacing we can ensure smaller pressure on rear housing.
(2) The numerical simulations shows that the spacing between two buildings has little effect on the leeward side of the house.
(3) Under groups effect the influence on the front housing is smaller than the rear housing.
(4) In this paper, the numerical simulation is carried out only for 3 simple housing of different spaces, it requires further research to simulate more groups at different spacing.

REFERENCES

BC, Huang. 2008. *Principle and application of structural wind resistance analysis.* ShangHai: Tongji University Press.
Baskaran, A. & Stathopoulos, T. 1989. Computational evaluation of wind effects on buildings, *Bldg Environ* 24(4): 325–333.
Hui, Wang. & S.F., Chen. & J.C., Tang. 2003. Numerical simulation of wind pressures on a low-rise building complex with gable roofs, *Engineering Mechanics* 20(6): 135–140.
Murakami, S. 1997. Current Status and Future Trends in Computational Wind Engineering, *J. Wind Eng* (67–68): 3–34.
Stathopoulos, T. & Baskaran, A. 1996. Computer simulation of wind environmental conditions around buildings, *Engineering Structures* 18(11): 876–885.
Surry, D. 1999. Wind loads on low-rise buildings: past, present and future, *Wind Engineering into the 21st Century—Proceedings of the 10th International Conference on Wind Engineering.* Copenhagen, Denmark: 105–114.
Zhou, Y. & Stathopulos, T. 1997. A new technique for the numerical simulation of wind flow around buildings, *J Wind Eng Indus Aerodyn* 72: 137–147.

Modeling of a tunnel form building using Ruaumoko 2D program

N.H. Hamid & S.A. Anuar
Faculty of Civil Engineering, Universiti Teknologi Mara, Shah Alam, Selangor, Malaysia

ABSTRACT: Modeling of hysteresis loops for a three-storey precast tunnel form building using Ruaumoko 2D program under in-plane lateral cyclic loading is performed. A one-third scale of three-storey RC building is designed, constructed and tested under quasi-static lateral cyclic loading. The hysteresis loops are obtained from experimental work by plotting the graph load versus displacement using four linear potentiometers. Comparisons of hysteresis loops between experimental and modelling were performed and a good agreement them was obtained. Based on these models and experiment results, the percentage differences between them are calculated for lateral strength, stiffness and ductility. As a conclusion, this type of building can carry gravity loading together with lateral loading and it is safe under seismic loading because the ductility factor lies within the acceptable standard seismic codes of practice.

1 INTRODUCTION

Seismic performances of tunnel form buildings have been observed during the 1999 Izmit Earthquake and the 2003 Bingol Earthquake in Turkey. These buildings experienced only minor nonstructural damage in the form of separation of precast panels from floors (Yakut and Gulkan, 2003). The occurrences of Kocaeli ($M_w = 7.4$) and Duzce ($M_w = 7.1$) earthquakes in Turkey 1999 once again demonstrated that there were nonstructural damage and excellent performance conditions of RC shear walls which built using tunnel form technique (Kalkan and Balkaya, 2003). Multi-storey reinforced concrete tunnel buildings are the common structural types which normally constructed from low to high seismic regions. Despite their good performance during the 1999 Izmit Earthquake, abundance of these structures are scattered around the world including Malaysia, Indonesia, Italy, Spain, Japan and others countries. The current seismic codes of practice and design provisions provide insufficient guidelines for seismic design of tunnel form buildings (Kalkan and Balkaya, 2004). A similar good seismic performance of these types of buildings was also experienced in Romania during the 1977, 1986 and 1990 earthquakes (Moroni, et al, 2004).

There are very limited research had been conducted regarding the seismic behaviour and modeling of reinforced tunnel form building under earthquake excitations. One of the experimental investigations on the inelastic lateral behavior of four-storey tunnel form buildings under quasi-static in-plane lateral cyclic loading were conducted by Bahadir and Kalkan (2007). Results of their study will augment the literature with the tests involving 3D wall configurations where the response is dominated by flexure triggered brittle mechanism (Yuksel, 2003). This failure mechanism is similar to the visual observation in an eight-story heavily damaged shear-wall dominant building (EL-Faro building) during the 1985 Chile Earthquake (Wood, et al, 1991). The load-carrying lateral capacity, stiffness, ductility and failure mechanism of tunnel form wall system depend on the amount of vertical reinforcement ratio, wall's thickness and boundary of reinforcement in the wall (Kalkan and Yuksel, 2009).

Even though, there are very little research had been conducted on the seismic performance of tunnel form buildings under long-distant earthquake excitation and near-field earthquakes but it is important to predict the safety of these buildings in Malaysia. For that reason, a comprehensive research needs to be carried out in order to determine the seismic behavior of tunnel form building under various level of earthquake attack. Hence, a one-third scale of a three-storey building was designed, constructed in Heavy Structure Laboratory, Universiti Teknologi MARA and tested under quasi-static out-of-plane lateral cyclic loading. Hysteresis loops are measured at different locations along one side of the building using linear potentiometers. Ruaumoko 2D program is used to model this hysteresis loops which obtained from experimental results. Then, the experimental results of hysteresis loops are compared with Ruamoko 2D modeling and some similarities between them were obtained. The percentage differences of lateral strength, stiffness and ductility were compared between them.

2 CONSTRUCTION OF PROTOTYPE BUILDING

A prototype of one-third scale three-storey precast tunnel form building which consists of shear wall panels, floor slabs and foundation beam were constructed in Heavy Structural Laboratory, Universiti Teknologi Mara, Malaysia (Samsudin, 2010). The building was constructed with different height of shear wall based on actual project of Kolej Sains Kesihatan Bersekutu, Ulu Kinta Perak. In this project, the tunnel form building was designed according to BS 8110 and constructed for hostel and residential block. The floor height for the ground floor is 450 mm, the intermediate floor is 750 mm and the top floor is 1000 mm. The wall thickness for each floor is limited to 75 mm thickness with the width of 1000 mm. Fabric wire mesh of a single layer (BRC-A7) was used to replace the longitudinal and transverse reinforcement bars for wall panel and floor slab. Slab dimension for each floor is similar to the wall panel with 640 mm width and 75 mm thick. Figure 1 shows the isometric view together with dimensions of three-storey tunnel form building which had been constructed in the laboratory.

Figure 2 shows the systematic arrangement of the nine (9) linear potentiometers located along left-hand side and load actuator was at right-hand side of the tunnel form building. These potentiometers were used to measure the deformation of tunnel form reinforced concrete building when subjected to out-of-plane lateral cyclic loading.

Figure 3 shows the real location of linear potentiometers along one side of tunnel-form wall building. These linear potentiometers were employed to measure the lateral displacement of the building when out-of-plane lateral cyclic loading was applied at top of the building. Load actuator was applied to the wall panel using two welded steel which bolted to each other using 30 mm diameter bolt. Tunnel-form building was applied at ±0.1%, ±0.2%, ±0.3%, ±0.4%, 0.5%, ±0.6%, ±0.7%, ±0.8% and ±0.9% drift. For each drift, two cycles of similar drift was applied to the

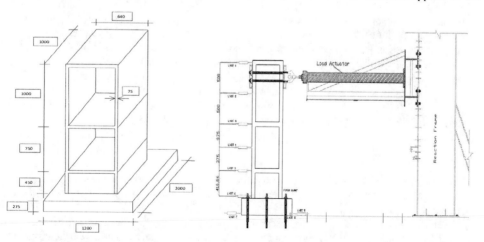

Figure 1. Isometric view of one-third scale of 3-storey tunnel form building.

Figure 2. Systematic arrangement of linear potentiometers and load cell.

Figure 3. Location of linear potentiometers along one side of tunnel-form wall in the laboratory.

Figure 4. Three-storey building tested under quasi-static lateral cyclic loading.

3-storey building. Figure 4 shows the 3-storey precast building was imposed by lateral cyclic loading using double actuator with load cell. The structure experienced the first mode shape under dynamic loading with control displacement method. From visual observation, the prototype building performed well under out-of-plane lateral cyclic loading with minor damage at top of the wall. No cracks were observed at connection of wall-foundation interface.

3 COMPARISON BETWEEN EXPERIMENTAL RESULTS AND MODELING

Modeling of the hysteresis loops for 3-storey tunnel-form building is using HYSTERES PROGRAM in Ruaumoko 2D Programming. Ruaumoko 2D Programming is the inelastic dynamic analysis program which can be run based on the database from past earthquake records. The programmer of this program is Prof Athol Carr from University of Canterbury and it was written in Fortran Language. The steps and procedures of getting Modified Takeda Hysteresis model are written and well-explained in the manual (Carr, 2004). Modified Takeda Hysteresis model was chosen from the manual and this hysteresis model is similar to the experimental hysteresis loops. Modified Takeda Hysteresis model from Ruaumoko 2D was selected based on the similar shape that produced by the experimental results obtained from experimental work. The comparison of hysteresis loops (load versus displacement) obtained from experimental work and modeling are made based on the measure parameter of displacement and loading.

Figure 5 shows the comparison of hysteresis loops between experiment and modeling result for LVDT 1. The graph of modeling result is outlined as red color and experimental result is marked as blue color. There are some similarities in term of the shape between the experiment and model. The maximum displacement of LVDT 1 is 52.52 mm and maximum lateral load is 15.52 kN, while the maximum displacement of LVDT 1 from Modified Takeda model is 58.0 mm at 16.0 kN. Figure 6 shows the comparison of hysteresis loops between experimental result and modeling result which are measured by LVDT 3. A similar color of notation used for hysteresis loops for experimental results and modeling results. The maximum displacement of LVDT 3 for experimental result is 23.20 mm at 15.52 kN, while the maximum displacement in LVDT 3 from Modified Takeda model is 26.0 mm at 16.0 kN. Therefore, these values shows good agreement between them and it is acceptable to use this model for different numbers of storey and height of the buildings.

Figure 7 shows the comparison of hysteresis loops between experimental result and modeling for LVDT 5. The maximum displacement of LVDT 5 for experimental result is 20.50 mm at 15.52 kN

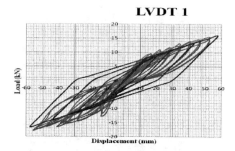

Figure 5. Comparison of hysteresis loops between experimental result and modeling result for LVDT 1.

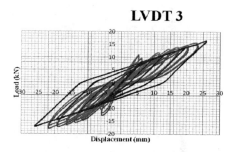

Figure 6. Comparison of hysteresis loops between experimental result and modeling result for LVDT 3.

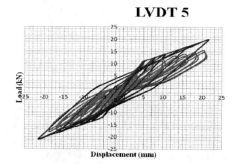

Figure 7. Comparison of hysteresis loops between experimental result and modeling result for LVDT 5.

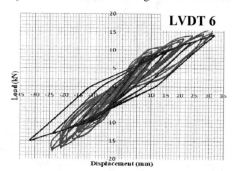

Figure 8. Comparison hysteresis loops between experimental result and modeling result for LVDT 6.

Table 1. Comparison between experimental and modeling results for lateral strength, stiffness and ductility for LVDT 1.

LATERAL STRENGTH			STIFFNESS			DUCTILITY	
Experiment Results	Modeling Results	Percentage Difference	Experiment Results	Modeling Results	Percentage Difference	Experiment Results	Modeling Results
$\Delta_y = 16.53$	$\Delta_y = 13.0$	27%	$k = 0.59$	$K = 0.62$	5%	3.18	4.46
$F_y = 9.70$	$F_y = 8.0$	21%	$k_{secant} = 0.16$	$k_{secant} = 0.18$	11%	–	–
$\Delta_{max} = 52.52$	$\Delta_{max} = 58.0$	9%	–	–	–	–	–
$F_{max} = 15.52$	$F_{max} = 16.0$	3%	–	–	–	–	–

and the maximum lateral displacement for LVDT 5 based on Modified Takeda Hysteresis Model is 21.6 mm at 20.02 kN. Figure 8 shows the comparison of hysteresis loops between experimental and modeling result for LVDT 6. The maximum lateral displacement for LVDT 6 for experimental result is 31.0 mm and maximum lateral load is 15.52 kN.

4 COMPARISON OF LATERAL STRENGTH, STIFFNESS AND DUCTILITY

From experimental results of hysteresis loops, the lateral strength can be obtained from four LVDTs which located at different levels of tunnel form building. However, the highest deformation shape located at the top of the building which is measured by LVDT 1. Table 1 tabulates the comparison between experimental result and modeling result of the lateral strength, stiffness and ductility of the hysteresis loops for LVDT 1. The variations of differences of percentage lie between 3% and 27%.

5 CONCLUSION

Based on the experimental results and modelling of hysteresis loops using Ruaumoko 2D programming, the following conclusion can be drawn as follow:

1) The modeling of hysteresis loops using HYSTERES program has similar shape and pattern as hysteresis loops obtained from experimental work.
2) There are some differences in term of percentages with range of 3% to 27% for yield displacement, ultimate displacement, yield strength and ultimate strength.
3) The difference between modelling and experimental result for elastic stiffness is 5% and secant stiffness is 11%.
4) The ductility calculated using experimental work is 3.18 and ductility using modelling is 4.46. This due to the modelling has higher value for maximum displacement and smaller value for yield displacement.
5) This type of building will be survived under low seismic loading and long-distant earthquake excitations.

ACKNOWLEDGEMENT

Special thanks goes to FRGS (Fundamental Research Grant Scheme), MOHE (Ministry Of Higher Education), Putrajaya, Malaysia and RMI (Research Management Institute) of Universiti Teknologi Mara, Shah Alam, Malaysia for providing fund for this research. Gratitude and appreciation to the Faculty of Civil Engineering, UiTM Malaysia in providing the experimental facilities and technicians/laboratory staff members who provided invaluable assistance during the course of this experimental work.

REFERENCES

Balkaya, C. and Kalkan, E., (2004), Seismic Vulnerability, behavior and Design of Tunnel Form Builiding Structures, Journal of Engineering Structures, Elsevier, 26, 2081–2099.
Carr, A.J. (2004). Ruaumoko: Inelastic Dynamic Analysis Program. Department of Civil Engineering, University of Canterbury, New Zealand.
Kalkan, E. and Balkaya, C. (2003), Nonlinear Seismic Response Evaluation of Tunnel Form Building Structures, Computers and Structures Journal, Pergamon, Science Direct, Elsvier, 81(2003), 153–165.
Kalkan, E. and Balkaya, C. (2004), Seismic Vulnerability Behaviour and Design of Tunnel Form Building Structures, Engineering Structures Journal, Elsevier, 26(2004), 2081–2099.
Kalkan, E. and Yuksel, S.B. (2009), Pros and Cons of Multi-Storey RC Tunnel-form (Box-Type) Buildings, The Structural Design of Tall and Special Building Journal, 8(2007), 109–118.
Moroni, M.O., Astroza, M., and Acevedo, C. (2004), Performance and Seismic Vulnerability of Tunnel Form Types Used in Chile, Journal of Performance of Constructed Facilities, ASCE, Vol. 18, No. 3, August 1, 2004, ISSN 0887-38-28, pg 173–179.
Samsudin, A.H. (2010), Seismic Performance of 3-storey Precast Tunnel Form Buildings Using IBS Under Quasi-static Lateral Cyclic Loading, Master Dissertation, Faculty of Civil Engineering, Universiti Teknologi MARA, 40450 Shah Alam, Selangor, Malaysia.
Yakut, A. and Gulkan, P. (2003), Housing Report on Tunnel Form Building, World Housing Encylopedia, EERI Report, 101, pg 14.
Yuksel, S.B. and Kalkan, E. (2007), Behaviour of Tunnel Form Buildings Under Quasi-static Cyclic Lateral Loading, Structural Engineering and Mechanics, Vol. 27, No.1, (2007), 9, 178–182.
Yuksel, S.B. (2003), Experimental Investigation of The Seismic Behaviour of Panel Buildings, PhD Thesis, Middle east technical University, Ankara.
Wood, S.L., Stark, R., and Greer, S.A. (1991), Collapse of Eight-storey RC Building During 1985 Chile Earthquake, Journal of Structural Engineering, ASCE 117(2): 600–619.

A simplified design method for CFST frame-core wall structure with viscous dampers

Fengming Ren & Congxiao Wu
School of Civil Engineering, Guangzhou University, Guangzhou, Guangdong, China

Jianwei Liang
School of Civil and Transportation Engineering, Guangdong University of Technology, Guangzhou, Guangdong, China

ABSTRACT: Based on the limited demands for the inter-story drift of the concrete filled steel tubular (CFST) frame-core wall structure in the *technical specification for concrete structures of tall building (JGJ3-2010)* and the *code for seismic design of buildings (GB 50011-2010)*, and combined with the response spectrum of displacement and the formula of equivalent viscous damping ratio, the calculation formula of reasonable damping coefficient and simplified design method for viscous dampers in the CFST frame-core wall structure were derived. Finally, an example structure was studied by utilizing this method. The results show that the proposed formulas and method are valid and can be a reference for preliminary design of CFST frame-core wall structure with viscous dampers.

1 INTRODUCTION

All the top ten high-rise buildings to be built or under construction in the next ten years are located in Asia, which five of them are located in China, and 70% of these high-rise buildings are hybrid structure (CTBUH 2012). As an important part of the hybrid structures, CFST frame-core wall structures have been widely used, and many researches have been carried out by scholars around the world. Han Lin-hai et al. (Han et al. 2009) performed shaking table test and finite element analysis on a CFSST frame-core wall structure and a CFCST frame-core wall structure. Huang Zhong-hai (Huang et al. 2009) performed elasto-plastic time-history analysis on a super high-rise CFST frame-core wall structures with strengthened stories under rare earthquake action. The choice of structural system, the whole stability analysis and dynamic elastic-plastic analysis according to a CFST frame-core wall structure were presented by Yao Guo-huang (Yao et al. 2011). The mechanical characteristics of the CFST frame-core wall structure system are extremely complicated, so Zhong Shan-tong (Zhong 2006) pointed out that the systematic researches on the seismic performance of CFST hybrid structures are necessary urgent.

The technology of energy dissipation has been widely used in structural engineering in recent years, Soong T T (Soong et al. 1997) and Dargush GF (Dargush et al. 2002) had studied on the design method of the structure with energy dissipation equipments in earlier years. C.P. Providakis (Providakis 2008) performed static elastic-plastic analysis on a base-isolated steel-concrete composite structure under strong earthquake action. The shaking table test and elastic-plastic time-history analysis of a CFST frame with viscous dampers were performed by Lu Xi-lin (Lu et al. 2006). Ren Feng-ming and Zhou Yun (Zhou et al. 2011) adopted the energy dissipation technology on a super high-rise CFST frame-core wall structure, and the shaking table test of this hybrid structure with dampers was performed and analyzed.

As a new structural system, the specifications of China have not involved the design methods of the CFST frame-core wall structures with dampers. The development of the hybrid structure with dampers was limited because of the lack of design reference. In this paper, the value of the reasonable

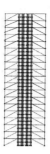

Figure 1. Simplified analysis model.

damping coefficient of the viscous dampers set in the CFST frame-core wall structures and the preliminary design method of the CFST frame-core wall structures are studied, which combined with the limited demands for the inter-story drift of the CFST frame-core wall structures.

2 DERIVATION OF THE DAMPING COEFFICIENT FOR VISCOUS DAMPERS

2.1 Structure model

In this paper, the analysis model was a CFST frame-core wall structure, which the inter-story drift angle of the structure exceeded the limit of the specification, so the viscous dampers were set in the structure to improve the structural performance. The simplified analysis model is shown in Fig. 1.

2.2 Top displacement of the structure

Motion equations of the multi-particle structure with viscous dampers are:

$$M\ddot{u}(t)+(C_s+C_a)\dot{u}(t)+Ku(t)=-M\ddot{u}_g(t) \qquad (1)$$

$$\ddot{u}(t)+2(\zeta_s+\zeta_a)\omega_n\dot{u}(t)+\omega_n^2 u(t)=-\ddot{u}_g(t) \qquad (2)$$

where $C_a =$ the linear viscous coefficient matrix or equivalent viscous coefficient matrix of the viscous damper, $C_s =$ the damping coefficient matrix of the structure without dampers, $\zeta_a =$ the additional damping ratio of the viscous damper, $\zeta_s =$ the damping ratio of the structure without dampers, $M =$ the mass matrix of the structure, $K =$ the stiffness matrix of the structure and $\ddot{u}(t), \dot{u}(t), u(t) =$ the acceleration vector, velocity vector and displacement vector of the structure respectively.

In order to make the equation (2) calculated by mode-superposition method, the non-proportional damping matrix of the structure should be diagonalizable. The assumption is shown in the equation (3):

$$\overline{C}_e = diag(2\zeta_{a1}\omega_1 M_1, 2\zeta_{a2}\omega_2 M_2, ... 2\zeta_{an}\omega_n M_n) \qquad (3)$$

where $M_1 \ldots M_n =$ the generalized mass of each vibration mode of the structure.

The top displacement of the simplified CFST frame-core wall structure with viscous dampers can be obtained from the motion equations and the response spectrum method, as shown in equation (4).

$$\left. \begin{array}{l} y'_{j1} = S_{jd}\gamma_{j1}X_{j1} \\ y_1 = \sqrt{\sum y_{j1}^2} \end{array} \right\} \qquad (4)$$

where $y'_{j1} =$ the top displacement of jth vibration mode, $S_{jd} =$ the spectral displacement of the structure, $\gamma_{j1} =$ the mode participation coefficient, $X_{j1} =$ the top horizontally relative displacement of jth vibration mode and $y_1 =$ the total top displacement of the structure with dampers.

In order to calculate the top displacement of CFST frame-core wall structure with dampers, the displacement response spectrum of structure should be derived. The displacement response spectrum was obtained by transforming the acceleration response spectrum of the *code for seismic design of buildings (GB 50011-2010)*, as shown in equation (5).

$$S_d = \frac{1}{\omega^2(1-\zeta^2)}S_a \Rightarrow S_d = \frac{T^2}{2\pi^2(1-\zeta^2)}S_a = \frac{T^2}{2\pi^2(1-\zeta^2)}\alpha g \quad (5)$$

where S_d = the displacement response spectrum, S_a = the acceleration response spectrum, ω = the natural frequency of the structure, ζ = the damping ratio of the structure, α = the earthquake affecting coefficient.

Generally the CFST frame-core wall structures are high-rise buildings, and the natural period of the structure is in the range of T_g to 6s. The equations of the displacement response spectrum curve are obtained by transforming the acceleration response spectrum in the *code for seismic design of buildings (GB 50011-2010)*, as shown in equation (6).

$$\left. \begin{array}{l} T_g < T \leq 5T_g \\ S_d = \dfrac{T^2}{(2\pi)^2\sqrt{(1-\zeta^2)}}(\dfrac{T_g}{T})^{(0.9+\frac{0.05-\zeta}{0.3+6\zeta})}(1+\dfrac{0.05-\zeta}{0.08+1.6\zeta})\alpha_{max}g \\ 5T_g < T \leq 6s \\ S_d = \dfrac{T^2}{(2\pi)^2\sqrt{(1-\zeta^2)}}\left[(1+\dfrac{0.05-\zeta}{0.08+1.6\zeta})0.2^{(0.9+\frac{0.05-\zeta}{0.3+6\zeta})} - \left(0.02+\dfrac{0.05-\zeta}{4+32\zeta}\right)(T-5T_g)\right]\alpha_{max}g \end{array} \right\} \quad (6)$$

2.3 Calculation of viscous damping coefficient

The additional damping ratio of the viscous dampers could be derived by equation (1) to equation (6). The inter-story drift and story shear force of the structure are obtained by performing response spectrum analysis on the CFST frame-core wall structure whose damping ratio had been adjusted, as shown in equation (7). Because of the effect of the gap during the process of installing the damper, the value of displacement should be reduced when the displacement between the two ends of the dampers was taken as the inter-story displacement.

$$\left. \begin{array}{l} W_c = (2\pi^2/T_1)\sum C_e \cos^2\theta_j \Delta_j^2 \\ W_s = \dfrac{1}{2}\sum F_i u_i \\ \zeta_a = \dfrac{W_c}{4\pi W_s} \end{array} \right\} \quad (7)$$

where W_c = the total energy dissipated by all of the dampers during a loading circle, W_s = the total strain energy of the structure, Δ_j = the displacement between the two ends of the *j*th damper, θ_j = the angle between the axis direction of the *j*th damper and horizontal direction, T_1 = the fundamental natural period of the structure.

The equivalent linear viscous damping coefficient C_e of the viscous dampers in the structure could be obtained from the equations above.

3 PROCEDURE OF THE SIMPLIFIED DESIGN METHOD

According to the analysis above, the procedure of the simplified design method of the CFST frame-core wall structure with viscous dampers can be listed below:

1) If the performance indexes exceed the limits of the specification, the CFST frame-core wall structure with viscous dampers was adopted;

2) The stories (the inter-story drift exceeds the limits of the specification) were found out. Combined with the result of step 1), the top displacement of the structure with dampers was calculated from the equation (4) according to the regulation and stipulation of the code.
3) The damping ratio of the structure with dampers and the additional damping ratio from the viscous dampers were calculated respectively from equation (6);
4) Substituting the additional damping ratio calculated in the step 3), then the equivalent linear viscous damping coefficient C_e of the viscous dampers were obtained from equation (7);
5) Determining the number and the specific parameters of the viscous dampers set in the CFST frame-core wall structure according to the value of C_e.

4 EXAMPLES

4.1 Engineering situation

The plane layout of a ten stories CFST frame-core wall structure is shown in Fig. 2. The story height of every floor is 4 m. The sections of the components are shown in Table 1. The live load of the floor is $2.0\,kN/m^2$. The structure was designed in the 1st classification of design earthquake, the site classification of II and the 8 fortification intensity.

4.2 Analysis result

Combining with the each inter-story drift of the original structure obtained from the analysis of response spectrum, the top displacement satisfied the demand of the specification could be calculated, and then the damping ratio of the CFST frame-core wall structure with dampers and the additional damping ratio of viscous dampers could be obtained from the equation (5) above. As a

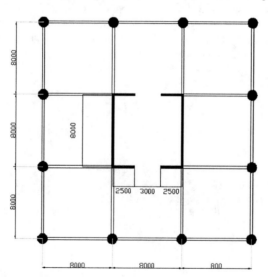

Figure 2. Layout of the model.

Table 1. Structural parameters.

	Beam	Column	Core wall	Slab
Section (mm)	400 × 200 × 18 × 10	450 × 5	200	100
Concrete	–	C30	C30	C30
Steel	Q345	Q345	–	–

result, each story was set two dampers, and the damping coefficient of the dampers is $16\ kN \cdot s/mm$ according to the calculated result. The arrangement of the dampers is shown in Fig 3, and the monoclinic arrangement of dampers is chosen.

The original structure and the structure with viscous damper were analyzed by the software of ETABS. Because the viscous dampers could not provide the additional stiffness for the structure, the natural periods of the two structures were the same. The inter-story drift angle and the story shear force obtained from the analysis of response spectrum analysis and time-history analysis are shown in Fig. 4 and Fig. 5 respectively.

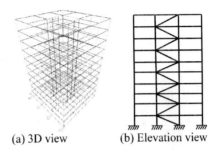

(a) 3D view (b) Elevation view

Figure 3. Arrangement of the dampers.

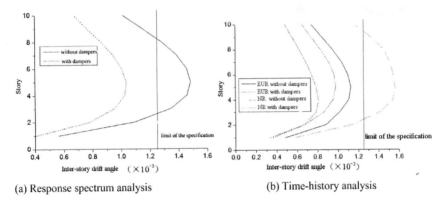

(a) Response spectrum analysis (b) Time-history analysis

Figure 4. Inter-story drift angle of structure.

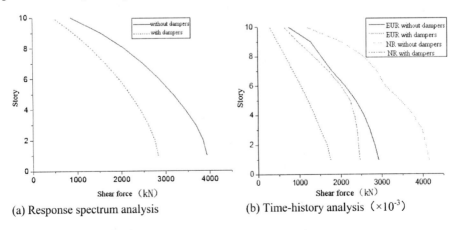

(a) Response spectrum analysis (b) Time-history analysis $(\times 10^{-3})$

Figure 5. Story shear force of structure.

From the Fig. 4 and Fig. 5, it can be found that the story shear force of the CFST frame-core wall structure with viscous dampers is less than that of the CFST frame-core wall structure significantly, and the maximum of the shear force is decreased by 45% for response spectrum analysis and 50% for time-history analysis respectively. The inter-story drift angle of structure with dampers is also less than that of the structure without dampers significantly and satisfied the demand of the specification. The seismic performance of the CFST frame-core wall structure is improved because of the effect of the dampers.

5 CONCLUSIONS

According to the study on the reasonable damping coefficient of the CFST frame-core wall structure with viscous dampers, it can be concluded that the CFST frame-core wall structure with viscous dampers could be satisfied the demand of the specification after it was designed by the method proposed in this paper. It indicates that this simplified design method is feasible for the CFST frame-core wall structure with viscous dampers, and the method is simple and practical, so it can guide the preliminary design of the CFST frame-core wall structure with viscous dampers.

ACKNOWLEDGEMENTS

The research reported in the paper is part of the Project 51108095 and the Project 51278130 sponsored by the National Natural Science Foundation of China and the Project 2012KB11 sponsored by the State Key Laboratory of Subtropical Architecture Science. Their financial support is highly appreciated.

REFERENCES

Constantinou M C, Dargush G F, Lee G C, et al. 2002. Analysis and Design of Buildings with Added Energy Dissipation Systems. MCEER.
Han Lin-hai, Li Wei & Yang You-fu. 2009. Seismic behaviour of concrete-filled steel tubular frame to RC shear wall high-rise mixed structures. *Journal of Constructional Steel Research* 65:1249–1260.
http://buildingdb.ctbuh.org/index.php
HUANG Zhonghai, LIAO Yun, LI Zhishan & WANG Shaohe. 2009. Elasto-Plastic Time-History Analysis for a Frame-Tube Tall Building in Zhu Jiang Xin Cheng J2 - 5 Plot under Expected Rare Earthquakes. *Structural Engineers* 25(2): 91–97.
Lü Xilin, Meng Chunguang & Tian Ye. 2006. Shaking table test and elasto-plastic time history analysis of a high-rise CFRT frame structure with dampers. *JOURNAL OF EARTHQUAKE ENGINEERING AND ENGINEERING VIBRATION* 26(4):231–238.
Providakis, C.P. 2008. Pushover analysis of base-isolated steel-concrete composite structures under near-fault excitations. *Soil Dynamics and Earthquake Engineering* 28(4):293–304.
Soong T T & Dargush G F (eds). 1997. *Passive Energy Dissipation Systems in Structural Engineering.* USA: State University of New York at Buffalo.
Zhong Shantong (ed). 2006 *CFST unified theory.* Beijing: Tsinghua University Press.
Yao Guo-huang, Chen Yi-yan, Guo Ming & Pan Dong-hui. 2011. Summarization of Structural Design for a High-rise CFST Frame-tube Structure. *Earthquake Resistant Engineering and Retrofitting* 33(4):66–72.
Zhou Yun, Ren Fengming, Fan Huabing & Zhang Jichao. 2011. Shaking table experimental study on a CFST frame-RC core structure with energy dissipating devices. *JOURNAL OF EARTHQUAKE ENGINEERING AND ENGINEERING VIBRATION* 31(2):130–137.

Advanced Engineering and Technology – Xie & Huang (Eds)
© 2014 Taylor & Francis Group, London, ISBN 978-1-138-02636-0

Experiment research on the bending behavior of stainless steel reinforced concrete slabs

Ying Zhang
Guangdong University of Technology, Guangzhou, Guangdong, P.R. China

Guoxue Zhang & Xiwu Zhou
Foshan University, Foshan, Guangdong, P.R. China

Ziqing Chen
Guangdong University of Technology, Guangzhou, Guangdong, P.R. China

ABSTRACT: To study the design methods and application of stainless steel concrete slab, bending tests of Stainless Steel Reinforced Concrete Slabs (SSRCS) and Ordinary Steel Reinforced Concrete Slab (OSRCS) were carried out. Results show that, compared to OSCRS, ultimate bearing capacity of SSRCS has been improved. The stiffness degradation of SSRCS is slower than OSRCS, the crack patterns, deflections and strains of reinforcement in SSRCS are significantly different from OSRCS. It shows that the crack width of SSRCS can meet the requirement of the maximum crack width in Code for Design of Concrete Structures (GB50010-2010). Based on the results found, it is suggested that the ultimate bearing capacity and mid-span deflection need to be corrected for reasonable estimation.

1 INTRODUCTION

The reinforced concrete bridges used all over the world shows that corrosion is the main factor of durability degradation for structure, and using stainless steel rebar can solve the problem effectively, so as to prolong the service life of concrete structure. Stainless steel rebar has been widely applied in the bridge engineering and port engineering in recent years. The Hong Kong-Zhuhai-Macao Bridge uses 3462 tons of duplex stainless steel rebar in the pile caps, tower, pier shafts and other elements. Researchers have studied stainless steel reinforced concrete (García-Alonso et al. 2007, Zhang et al. 2008, 2010, 2011, Zhou et al. 2008), but the research of concrete bridge decks reinforced with stainless steel rebar flexural performance have few report. This paper experimentally compares the bending capacity of stainless steel reinforced concrete slabs and ordinary steel reinforced concrete slab.

2 EXPERIMENT SCHEME AND DESIGN

Test simulates simply-supported single span slab. 3 specimens were made in this study. The diagram of loading is shown in Figure 1. Design parameters of specimens as shown in Table 1. The thickness of longitudinal reinforcement protective layer is 20 mm.

The 1.4362 duplex stainless steel rebar was provided by UGITECH. Ordinary reinforced was hot rolled steel. The Mechanical properties of carbon steel and stainless steel rebar are shown as Table 2. And the material performance metrics of concrete is shown in Table 3.

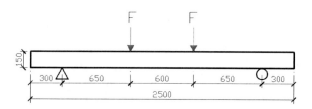

Figure 1. Diagram of loading.

Table 1. Design parameters of specimens.

Specimen number	Section size (mm × mm × mm)	Longitudinal rebar	Distributing rebar	Reinforcement ratio
B-1-0	150 × 500 × 2500	4 S10*	Φ8@200	0.5%
B-2-0	150 × 500 × 2500	4 S14*	Φ8@200	0.98%
B-3-0	150 × 500 × 2500	4 Φ10	Φ8@200	0.5%

*1.4362 duplex stainless steel rebar.

Table 2. Mechanical properties of carbon steel and stainless steel rebar.

Steel types	Yield strength (MPa)	Tensile strength (MPa)	Elasticity modulus (MPa)
HRB335	335	455	2.0×10^5
1.4362	750	880	1.93×10^5

Table 3. Material performance metrics of concrete.

Concrete grade	Slump (mm)	Compressive strength (MPa) 7d	Compressive strength (MPa) 28d
C30	125	22.4	37.9

3 EXPERIMENT PROCEDURE AND RESULTS

3.1 Deflection

The load-deflection curve of specimens is shown in Figure 2. In the initial stages of small load, three pieces of slab show similar mechanical characteristics. During this phase, the cracks did not appear in specimens. The curve of load-deflection is approximately linear. There are no obvious inflection point and have greater slope. It reveals that the slabs are in elastic stage. The tensile stress of reinforcement is small, and the deformation of specimens is not obvious. The first vertical crack is observed when the load is added to cracking load. This phase is classified as elastic-plastic stage. At this moment, the cracking load of stainless steel reinforced concrete slab is greater than ordinary reinforced concrete slab and the deflection of specimens grows slowly. Before the steel bar reaches the yield point, the deflection of ordinary reinforced concrete slab is greater than the stainless steel reinforced concrete slab in the case of the same reinforcement ratio. With the increase of reinforcement ratio, the deflection also increases. Load increases slightly but the mid-span deflection increases sharply. The deflection of stainless steel reinforced concrete slab has a significantly greater growth than ordinary reinforced concrete slab when specimen goes into the

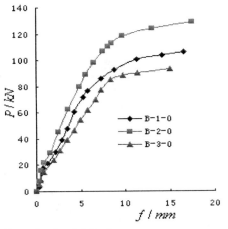 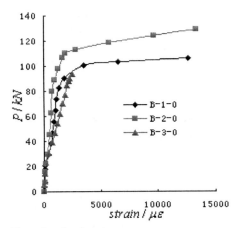

Figure 2. Load-deflection curve.   Figure 3. Load-strain curve.

failure stage. When failure load is reached, concrete suffer compression failure at the supports for both stainless steel reinforced concrete slab and ordinary reinforced concrete slab.

3.2 Strain of reinforcement

The load-strain curve of stainless steel reinforced concrete slabs and ordinary reinforced concrete slab are shown in Figure 3. The tensile stress of slabs is mainly borne by concrete and strain is very small for both two kinds of reinforcement before the normal section cracks. When the load is increased to the occurrence of crack, the tension zone of the fracture cross-section of concrete is majority out of work. At this time, as the bond between steel bar and concrete produces local bond stress, the tensile stress in the tension zone was transferred to the steel bar, so the stress and strain of the reinforcement have a sharp increase before crack occurs, and the load-strain curve shows the first turning point. Before the steel bar reaches yield point, the strain of stainless steel is close to ordinary steel. When specimen goes into the failure stage, although the load will slightly increase, the specimen was actually damaged after the strain of ordinary steel reinforcement suffers a minor growth. The strain of stainless steel reinforced rapidly increase after the steel yields. Test results show that when two kinds of slab with the same reinforcement ratio become damaged, the strain of stainless steel bar and ordinary steel bar is 13329 and 2508, respectively. The strain of stainless steel bar is about 5.3 times of ordinary steel bar.

3.3 Result analysis

Before the specimen reach cracking load, stiffness is governed by concrete, so the mechanical properties of two kinds of slab is similar. When the first crack appears, the load of stainless steel reinforced concrete slab B-1-0 is 17.95 kN, B-2-0 is 22.34 kN, and the ordinary reinforced concrete slab B-3-0 is 14.75 kN. Results indicate that the bonding between stainless steel and concrete is higher than ordinary and concrete in the elastic stage. It is beneficial to delay the early cracking time of specimens, which increase the crack resistant capacity.

The load increases after the cracking of the specimen, with the increase of load, the original cracks propagate and the new cracks occur, and the stiffness of slab begins to degrade. The load-deflection curve of stainless steel reinforced slab is bent and there is a turning point when $P = P_{cr}$. It shows that the stiffness degradation process of stainless steel bar is a gradient process. The ordinary reinforced concrete slab load-deflection curves are approximately linear. There are two inflection points on the curve at $P = P_{cr}$ and $P = P_u$. The rate of ordinary reinforced concrete slab stiffness degradation is higher than stainless steel reinforced concrete slab after cracking, which leads to that the stiffness of the stainless steel reinforced concrete is greater than the ordinary reinforced concrete slab. The mid-span deflection, crack width and rate of steel strain growth of stainless steel reinforced concrete are smaller than the ordinary reinforced concrete slab in this period.

After the tensile reinforcement yields, strain of reinforcement and deflection grows rapidly but load grows slowly. The Figure 2 shows that the stainless steel reinforced concrete load-deflection curve has no obvious turning point after the tensile reinforcement yield and the tangent modulus of stainless steel reinforced is gradually changing. However, the tangent modulus changes obviously after ordinary reinforced yields, which results in the load-deflection curve of ordinary reinforced concrete slab has an obvious turning point.

3.4 Result comparison

According to the formula calculation of Code for Design of Concrete Structures (GB50010-2010), the bearing capacity of normal section (P), maximum crack width (w) and mid-span deflection (f) of specimen and experimental value are determined as shown in Table 4, Table 5 and Figure 4.

As can be indicated from Table 4, the experiment result of the bearing capacity of normal section of reinforced concrete slabs with stainless steel rebar and carbon steel rebar is greater than the value calculated by the Code (GB50010-2010). Table 5 suggests that the results which use ordinary

Table 4. Bearing capacity of normal section.

Specimen number	P_E/kN	P_C/kN	P_E/P_C
B-1-0	106.49	85.92	1.239
B-2-0	128.93	104.65	1.232
B-3-0	93.58	79.84	1.172

*P_E/experimental value, P_C/calculation value.

Table 5. Maximum crack width.

Specimen number	W_E/mm	W_C/mm	W_E/W_C
B-1-0	0.50	0.52	0.967
B-2-0	0.46	0.45	1.023
B-3-0	0.61	0.64	0.958

*W_E/experimental value, W_C/calculation.

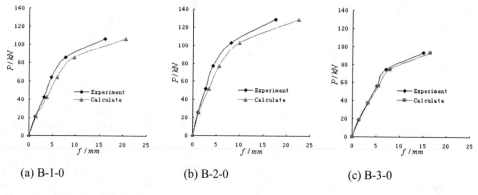

(a) B-1-0   (b) B-2-0   (c) B-3-0

Figure 4. Mid-span deflection.

reinforced concrete flexural member of crack calculation formula (GB50010—2010) are close to the experimental results. And the crack calculation formula of ordinary reinforced concrete can be used to calculate the maximum crack width of stainless steel reinforced concrete slab. As shown in Figure 4, the stainless steel reinforced concrete slab has great gap between calculated value and experimental value of maximum mid-span deflection. Therefore, the calculation formula of ultimate bearing capacity and maximum mid-span deflection need to be corrected for the stainless steel reinforced concrete slab.

4 CONCLUSION

Stainless steel bar can delay the occurrence of cracking time of specimens. The stiffness of stainless steel reinforcement concrete degrades later than the ordinary reinforced concrete. The results obtained using ordinary reinforced concrete flexural member of crack calculation formula are close to experiment results, while the ultimate bearing capacity and mid-span deflection of the stainless steel reinforced concrete slab remains to be further corrected.

ACKNOWLEDGEMENT

The authors wish to acknowledgement the financial support of Guangdong Natural Science Foundation (S2011010001225), and Center of Guangdong Universities Civil Engineering and Technology Development.

REFERENCES

García-Alonso, M.C., González, J.A., Miranda, J., Escudero, M.L., Correia, M.J., Salta, M. & Bennani, A. 2007. Corrosion behavior of innovative stainless steels in mortar. *Cement and Concrete Research* 37(11):1562–1569.
GB50010-2010, Code for Design of Concrete Structures. (in chinese)
Zhang Guoxue & Wu Miaomiao 2008. Experimental research on the bond-anchorage properties of stainless steel reinforced concrete. *Advances in Concrete structural Durability: Proceedings of the International Conference on Durability of Concrete Structures, ICDCS 2008*: 681–686.
Zhang Guoxue, Zhang Zhihao & Wang Changwei 2011. Seismic performance of stainless steel reinforced concrete columns, *Advanced Science Letters* 4(8/10):3119–3123.
Zhang Guoxue, Zhao Feng, Zhang Zhihao, Zhang Huaying, Zhou Xiwu & Rao Dejun 2010. Experimental research on seismic performance of stainless steel reinforced concrete beams, *China Railway Science* 31(5):35–40. (in chinese)
Zhou Yihui, Ou Yu-Chen, George C. Lee & Jerome S. O'Connor 2008. A pilot experimental study on the low cycle fatigue behavior of stainless rebar for earthquake engineering applications. University at Buffalo.

Effect of geogrid-confinement on swelling deformation of expansive soil

Jinhua Ding
Zhejiang University, Hangzhou, Zhejiang, China;
Changjiang River Scientific Research Institute, Wuhan, Hubei, China

Renpeng Chen
Zhejiang University, Hangzhou, Zhejiang, China

Jun Liu
Changjiang River Scientific Research Institute, Wuhan, Hubei, China

ABSTRACT: Large-dimension expansion model tests were conducted to study the swelling deformation behavior of expansive soil and the effect of geogrid confinement. The results indicate that under the water-infiltration conditions, geogrid confinement substantially decreases the lateral swelling deformation of expansive soil, but it has little influence on vertical displacement. The confinement degree on lateral swelling deformation is related to geogrid type and strength. In comparison with expansive soil without geogrid, the uniaxial HDPE50 geogrid, the HDPE80 geogrid, and the biaxial SS30 geogrid reduces lateral expansion by approximately 37% to 70%, 65% to 83%, and 46% to 80%, respectively.

1 INTRODUCTION

Expansive soil has several special properties, such as swell-shrinkage behavior, fissure presence, and overconsolidation effect (Liu 1997; Zhan 2003); these properties trigger a great deal of engineering damages to geotechnical structures. China has a wide distribution of expansive soil, and expansive soil can be found in more than 380 km of channels involved in the Middle Route Project of South-to-North Water Transfer Project (SNWTP). Thus, such problems in engineering can be effectively resolved by understanding the mechanical behavior of expansive soil.

A great number of recent studies and engineering cases have established that geogrid reinforcement improves soil strength and control deformation; however, its effectiveness for solving engineering problems relating to expansive soil remains uncertain. Moreover, limited study has been conducted on geogrid-reinforced expansive soil. Sharma et al. (2005) conducted laboratory model testing on expansive clay beds reinforced by geopiles, which are formed by filling geogrid cylinders with different geomaterial; they found that heaves could be controlled because of the friction mobilized at the interface among the filling material, the geogrid, and the expansive soil. Gupte et al. (2006) introduced a design and construction methodology of a demonstration project that involved the use of polypropylene woven geotextile as separator cum reinforcement of expansive soil subgrade in India; the monitoring data showed that no signs of visible distress have occurred after 18 months. Kameshwar et al. (2008) studied an integrated soil subgrade system by using geogrids as horizontal reinforcing elements; the laboratory experiment results demonstrated that geogrid-reinforced soft black cotton soil subgrade can safely withstand and sustain traffic loads and minimize the rut depths for the designed traffic loading.

To a certain extent, these previous findings illustrate that geogrids can be used to control the swelling deformation of expansive soil. However, limited information is available with regard to

Table 1. Index Properties of Expansive Soil.

Property	Value	Property	Value
Natural field conditions		*Percentage of total non-clay minerals (%)*	
Specific gravity, Gs	2.72	Quartz	19
Water content, ω (%)	14.7	Alkali	2
Dry unit weight, γd (kN/m^3)	16.7	Anorthose	3
Void ratio, e	0.60	Calcite	51
Consistency limits		*Percentage of total clay minerals (%)*	
Liquid limit (%)	46.1	Montmorillonite	12
Plastic limit (%)	19.1	Kaolinite	5
Plasticity index (%)	27.0	Illite	8
Grain size		*Expansion*	
Silt (>0.005 mm) (%)	47.5	Free swell ratio (%)	57
Fine silt (<0.005 mm) (%)	52.5	Degree of expansion	Weak
Clay (<0.002 mm) (%)	29.2	*Saturated consolidation shear strength*	
Compaction (Light)		Cohesion (kPa)	55.7
Maximum dry unit weight, $\gamma dmax$	17.5	Friction angle (°)	33.9
Optimum water content, ωop (%)	18.7	Soil Classification	CH

the swelling expansion mechanism of expansive soil under geogrid-confined conditions. This paper presents the results of laboratory expansion tests of geogrid-confined expansive soil under water infiltration conditions. A series of model tests were performed to study the effect of geogrids confinement on controlling swelling deformation, particularly the lateral swelling deformation of expansive soil.

2 PROPERTIES OF COMPACTED EXPANSIVE SOIL AND GEOGRIDS

2.1 Expansive soil

The expansive soil sample used in laboratory model tests was obtained from the channel of SNWTP in Xinxiang, Henan Province, China. The physical and mechanical properties of the expansive soil sample are presented in Table 1. Its free swell ratio is approximately 57%, as obtained by free swelling test. According to the Chinese national standard "Specification of the construction technology in expansive soil area" (GBJ112-87), the sample can be classified under "weak expansive soil" because its free swell ratio is smaller than 60%.

2.2 Geogrids

Three types of geogrids were chosen as confinement in laboratory model tests of expansive soil: two high-density polyethylene (HDPE) uniaxial geogrids (HDPE50 and HDPE80) and one polypropylene (PP) biaxial geogrid (SS30). The basic tensile properties of these geogrids are presented in Table 2.

3 OVERVIEW OF EXPANSION MODEL TESTS

3.1 Device of model tests and monitoring arrangement

The specimen dimension of expansive soil in the conventional expansion tests using a consolidometer is only $\varphi 61.8$ mm and a height of 20 mm, and the size is too small for placing the geogrid into the cutting ring. Thus, in consideration of the deformation of expansive soil with water infiltration and

Table 2. Mechanical Properties of Geogrids.

Geogrid Type		Ultimate Elongation (%)	Tensile Strength (kN/m)		
			Ultimate	Corresponding to Elongation 2%	Corresponding to Elongation 5%
HDPE50		11.98	52.32	15.81	29.83
HDPE80		10.55	85.10	30.37	55.23
SS30	Longitudinal	15.87	31.42	9.08	19.95
	Transverse	10.29	35.00	14.03	26.16

(a) expansive soil   (b) put the geogrid   (c) geogrid-confined model soil sample

Figure 1. Photo of the model tank.

the influence of different geogrids on swelling deformation, a model test container with dimensions of 630 mm × 400 mm × 400 mm (length × width × height) is constructed by using steel plates on three sides and a special glass on one side. To observe and measure lateral deformation, the dimensions of compacted soil sample are limited to a length of 462 mm (by using a lateral rigid baffle), a width of 400 mm, and a height of 360 mm. First, the soil sample was prepared according to the specified water content and then compacted to a specified dry density by means of layered roller compaction method. Then, the sample was covered with plastic film to keep the water content constant for 24 h. Finally, the lateral rigid baffle was dismantled, and a double-ring infiltrometer was set. The water level in both rings was kept constant during the entire test process. For geogrid-confined model tests, the geogrid was laid down at the specific depth first, and then folded up and packed with expansive soil to form the model soil sample (Shown in Fig. 1).

The model tests focus on the deformation mechanism and water content distribution of expansive soil after water infiltration; thus, the selected monitoring variables were water content and displacement (internal and external). A typical monitoring arrangement is shown in Fig. 2.

Two ML2x-KIT ThetaKit transducers were used to automatically measure the water content of soil; these transducers were individually embedded at two designated locations (points W1 and W2 in Fig. 2) at the surface and at a depth of 12 cm.

The internal vertical displacements at different depths of 12 and 24 cm were monitored by two self-regulating settlement poles with dial indicators (points S1 and S2). Three dial indicators were set at points B1, B2, and B3 to measure the surface settlement.

To measure the lateral displacements at different depths of 6, 18, and 30 cm at the lateral free face of the soil sample (measurement points from L1-1 to L3-2), eight dial indicators were fixed in a customized iron shelf. This shelf was placed after the lateral rigid baffle was dismantled. The probe bottoms of the eight dial indicators must be kept close to the lateral free face of the soil sample.

All the measuring instruments were calibrated prior to testing. After supplying water, the instrument readings were recorded every 30 minutes. When the monitoring displacement rate is less than 0.005 mm/h, the soil deformation is assumed to be stable; thus, the test can be terminated.

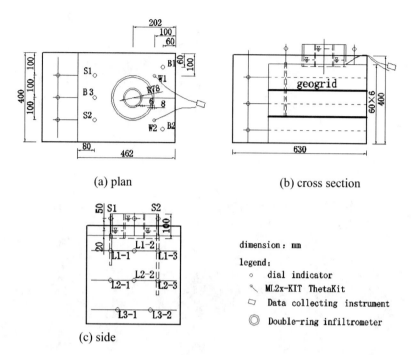

(a) plan

(b) cross section

(c) side

dimension: mm

legend:
∘ dial indicator
✎ ML2x-KIT ThetaKit
▫ Data collecting instrument
◎ Double-ring infiltrometer

Figure 2. Schematic diagram of model test.

Table 3. Model test scheme.

Test Number	Expansive Soil		Geogrid	
	Water Content (%)	Dry Density (kN/m^3)	Type	Number of Layer*
No1-1	14%	17.5	–	–
No1-2	18%	17.5		
No1-3	18%	16.5		
No2	14%	17.5	HDPE50	Double
No3-1			HDPE80	Single
No3-2			HDPE80	Double
No4			SS30	Double

*Single layer: one layer of geogrid embedded only at depth of 24 cm of model soil sample
Double layer: two layers of geogrid embedded at depth of 12 cm and 24 cm of the model soil sample

3.2 Model test scheme

To study the swelling deformation of expansive soil and the effect of geogrid confinement, a test scheme was established in view of several influencing factors, including water content, dry density of expansive soil, and the strength, type, and number of geogrid layers (Table 3).

4 TEST RESULTS AND ANALYSIS

4.1 Influence of water content and dry density of expansive soil

Fig. 3 shows the lateral displacement-time curves of expansive soil with water infiltration under different conditions of initial water content and dry density. These results demonstrate that swelling deformation decreases with increasing initial water content.

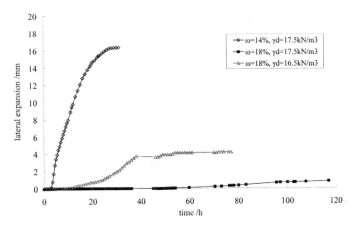

Figure 3. Influence on lateral deformation of initial water content and dry density (at a depth of 6 cm).

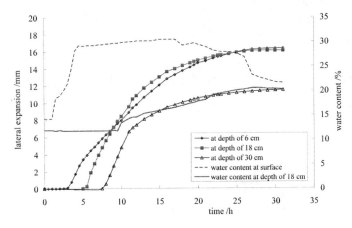

Figure 4. Lateral deformation vs. time at different depths of expansive soil (Test NO1-1).

The test NO1-1 was considered as an example to illustrate the expansion process of expansive soil with water infiltration. At a water content of 14% and a dry density of 17.5 kN/m³, the vertical expansion displacement at the surface reaches up to 10 mm. The lateral swelling deformation at the upper part of the soil is greater than that in the deep portion (Fig. 4), and the greatest lateral displacement is 16.4 mm, which is measured at a depth of 6 cm. The water content at different depths measured by ML2x-KIT ThetaKit transducers are described too in Fig. 4.

4.2 Influence on swelling deformation of geogrid confinement

4.2.1 Pattern of swelling deformation of geogrid-confined expansive soil

The influence on expansion of different geogrids is shown in Fig. 5, which shows that the lateral swelling deformation is noticeably decreased, but vertical displacement is almost unaffected.

Moreover, the expansion pattern of geogrid-reinforced expansive soil was also changed. Without geogrid, the lateral swelling deformation at the upper part of expansive soil sample was greater than that at the bottom because several transverse perfoliate cracks occurred near the volley face. With geogrid-confinement, the lateral swelling deformation was greater as the depth of the soil sample deepened. In other words, geogrid-confinement significantly influences the lateral deformation at the upper part of the soil. For example, at a shallow depth of 6 cm, the lateral displacement of expansive soil with HDPE50 geogrid-confinement is 4.994 mm, which is a decrease by approximately

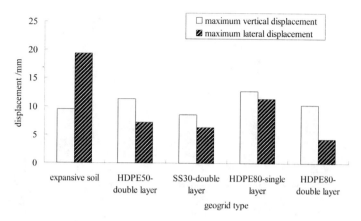

Figure 5. Influence of different geogrids on swelling deformation of expansive soil.

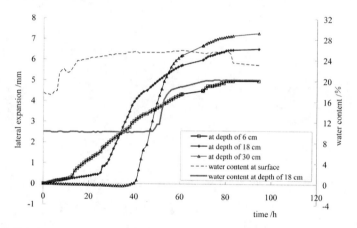

Figure 6. Lateral deformation and water content vs. time at different depths (HDPE50 geogrid, double layers).

70% from the lateral displacement of expansive soil without geogrid (16.361 mm). However, at a deep depth of 30 cm, the corresponding lateral deformation decreases only from 11.512 mm to 7.302 mm. Fig. 6 shows the entire lateral deformation process and the water content at different depths of the test NO$_2$ with HDPE50 geogrid-confinement.

4.2.2 *Type of geogrid*

Considering the test results of expansive soil without geogrid as the basis for comparison, Fig. 7 contrasts the lateral deformation at a depth of 6 cm with different types of geogrid-confinement. Despite its low strength, the biaxial SS30 geogrid effectively decreased the lateral expansion at the upper part of the soil during the early stage of testing. The uniaxial HDPE80 geogrid reduced the lateral deformation more effectively than the HDPE50 geogrid. Compared with the measured deformation at different depths in the test without geogrid, the HDPE50, HDPE80, and SS30 geogrids reduced lateral expansion by approximately 37% to 70%, 65% to 83%, and 46% to 80%.

4.2.3 *Layer numbers of geogrid*

In this paper, the lateral expansion at a depth of 6 cm of single- and double-layer HDPE80 geogrid-confined expansive soil model tests was compared (Fig. 8). When the geogrid was embedded at a depth of 24 cm, the lateral expansion at a depth of 6 cm was reduced to 9.489 mm from 16.361 mm

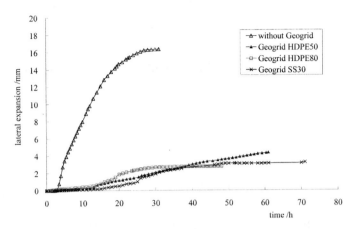

Figure 7. Lateral deformation vs. time curves of different types of geogrid (double layer, at depth of 6 cm).

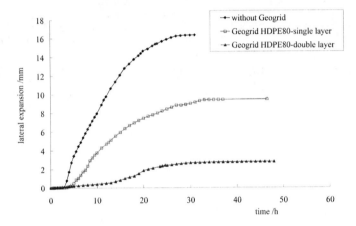

Figure 8. Lateral deformation vs. time curves of different layers of geogrid (at depth of 6 cm).

(without geogrid). These findings indicate that geogrids can change the lateral expansion in the range of 18 cm at least.

5 SUMMARY AND CONCLUSIONS

This paper investigates the influence of geogrid confinement on swelling deformation of expansive soil induced by water infiltration. The following conclusions were drawn from the results of model tests.

(1) Swelling deformation of expansive soil is related to initial water content and dry density. Greater water content corresponds to less expansion of expansive soil.
(2) The factors influencing the expansion of expansive soil with geogrid confinement include geogrid type, strength, and number of geogrid layers. A biaxial geogrid with low strength can achieve almost the same restraint effect that is achieved by high-strength uniaxial geogrids.
(3) The lateral swelling deformation of expansive soil with geogrid confinement is substantially reduced; however, the restriction effect on vertical displacement remains weak.
(4) The swelling deformation pattern of geogrid-confined expansive soil varies. For expansive soil without geogrid confinement, the lateral deformation at the upper part is large. However, with

geogrid-confinement, lateral expansion at the upper part is significantly reduced. In addition, the expansion at the deep part of soil is greater than that in the upper part.
(5) The results of the model tests reveal that the HDPE50, HDPE80, and SS30 geogrids reduced lateral expansion by approximately 37% to 70%, 65% to 83%, and 46% to 80%.

ACKNOWLEDGEMENTS

This research is financially supported by the Chinese national natural science fund (Project 51225804 and Project U1234204).

REFERENCES

Cheng Z. L, Li Q.Y, Guo X.L, Gong B.W. 2011. Study on the stability of expansive soil slope. Journal of Yangtze River Scientific Research Institute, Vol. 28(10): 03–11.
Ding J.H, Bao C.G, Ding H.S. 2008. Pullout test study on the interaction between geogrid and swelling rock. South-to-North Water Transfers and Water Science & Technology, Vol. 6(1): 52–56.
Gupte A, Satkalmi V, Bhonsle S, Viswanadham B.V.S. 2006. Performance of road reinforced with polypropylene geotextile-a case study. Proceedings of the 8th International Conference on Geosynthetic, Japan, 821–824.
Kameshwar Rao Tallapragada, Anuj Kumar Sharma. 2008. Experimental study of use of geogrids as horizontal reinforcing element in swelling subgrades. Proceedings of the 4th European Geosynthetics Conference, Edinburgh, Scotland.
Liu T.H. 1997. Expansive soil problems in engineering construction. China Architecture and Building Press, Beijing.
Sharma RS, Phanikumar BR. 2005. Laboratory study of heave behavior of expansive clay reinforced with geopiles. Journal of geotechnical and geoenvironmental engineering, Vol. 131(4): 512–520.
Zhan L.T, NG.W.W Charles, Bao C.G, Gong B.W. 2003. Artificial rainfall infiltration tests on a well Instrumented unsaturated expansive soil slope. Rock and Soil Mechanics, Vol. 24(4): 152–158.

A micro-mechanics study of saturated asphalt concrete

Y. Yuan
Institute of Road and Bridge Engineering, Dalian Maritime University, Dalian, Liaoning, China
Department of Transportation Highway Administration of Liaoning Province, Shenyang, Liaoning, China

Y.H. Zhao
Institute of Road and Bridge Engineering, Dalian Maritime University, Dalian, Liaoning, China

ABSTRACT: Water existing in the asphalt concrete has become the main factor which significantly influences the mechanical properties and service life of the asphalt pavement. A multiphase composite material model is developed for the viscoelastic saturated asphalt concrete, and elastic modulus, secant modulus and effective complex modulus are derived applying viscoelastic theory and micromechanics methods under the Burgers model of the asphalt concrete. The effect of pore-water pressure on the porosity and the effective modulus are studied under Sine function and pulse wave traffic loads. The excess pore-water pressure of asphalt pavement is suggested as a major reason of water damage in pavement.

1 INTRODUCTION

The rapid growing traffic in China causes the increasingly development of serious pavement diseases such as frost boiling, cracking, upheaval etc. It is found that many diseases are related to the effect of pore-water. Therefore, researches on water damage of asphalt concrete have become imperative.

Most of the studies of water damage of asphalt pavements are related to experimental methods. Another major part of researches are of finite element methods with 2-D numerical analysis, and a few focus on theoretical principles by using Biot consolidation theory. There is little result about constitutive properties. In this paper a micromechanics method is applied to predict the mechanical behavior of saturated asphalt concrete based on viscoelasticity theory.

Micro-mechanics theory has been widely used in the research of composite materials. As the important references of this research several literatures are introduced here. Weng & Zhao (1990) derived the effective elastic modulus of porous materials based on the Eshelby's equivalent inclusion theory and Mori-Tanaka method. Li & Weng (1995) predicted the creep behavior and complex modulus of fiber-reinforced polymer matrix composites by using Mori-Tanaka method in the Laplace Transform Domain. M-T method had gradually applied in the study of creep and relaxation modulus of asphalt concrete (Huang et al. 2009).

On the foundation of the above works, in this study the effective secant modulus of viscoelastic saturated asphalt concrete are established first and so the time dependent stress-strain relations are derived. Then the bulk complex modulus is developed for the analysis of high frequency harmonic load. Traffic loads are considered in the analysis of the pore-water pressure effect on porosity change as well as bulk modulus of the damaged materials. Finally an excess pore water pressure is investigated for valuing its influence on the material properties.

2 MICRO-MECHANICAL PROPERTIES OF SATURATED ASPHALT CONCRETES

The saturated asphalt concrete is assumed to be a two-phase composite (Fig. 1) with spherical saturated pores as phase 1 and asphalt concrete phase 0, which is undrained under homogeneous

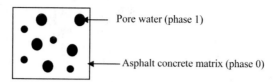

Figure 1. Schematic of saturated asphalt concrete system.

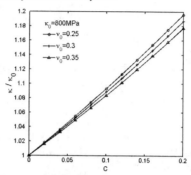

Figure 2. Effective bulk modulus of saturated asphalt concrete.

boundary and load condition. The viscoelastic properties of the system are analyzed based on micro mechanics theory for composite materials.

2.1 The effective bulk modulus of a 2-phase composite

For a 2-phase isotropic composite with elastic inclusion and matrix the effective bulk modulus of are provided explicitly by Weng (1984) as Equation 1, where the bulk modulus of rth phase are denoted by κ_r.

$$\frac{\kappa}{\kappa_0} = 1 + \frac{c(\kappa_1 - \kappa_0)}{(1-c)(\kappa_1 - \kappa_0)\alpha + \kappa_0} \quad (1)$$

where c is volume fraction of the inclusions, $\alpha = (1 + \nu_0)/[3(1 - \nu_0)]$, $\nu_0 =$ Poisson's ratio of matrix.

In the case of saturated asphalt concrete the pore-water sustains only compression. Under 20°C and constant pressure, the compressive bulk modulus of the pore-water given by Li. (2006) is 2190 MPa. The bulk modulus and Possion's ratio of asphalt concrete matrix are enactment by practical engineering. The variations of the effective modulus as a function of c under different conditions are shown in Fig. 2.

It is suggested in the Fig. 2 that the effective bulk modulus increases with the rise of saturated porosity when all the porous are non-penetrate and undrained.

2.2 The bulk secant modulus of saturated asphalt concrete

Asphalt concrete is a typical viscoelastic material in which the constitutive property depends on the temperature and the loading time. The ratio of stress to strain is denoted as the bulk secant modulus. The bulk stress-strain relation of the matrix in the Laplace Transform domain can be recast into (Li & Weng 1995)

$$\hat{\sigma}_{kk}^{(0)}(s) = 3\kappa_0^{TD} \hat{\varepsilon}_{kk}^{(0)}(s) \quad (2)$$

in terms of its bulk stress $\hat{\sigma}_{kk}^{(0)}(s)$ and strain $\hat{\varepsilon}_{kk}^{(0)}(s)$ of matrix, where κ_0^{TD} is bulk secant modulus in the Laplace transform domain. Under a constant strain rate, take inversion of Equation 2.

$$\sigma_{kk}^{(0)}(t) = \kappa_0(t) \times \dot{\varepsilon}_{kk}^{(0)} t = d[\frac{\eta_M^2 (E_M + E_K - E_M e^{E_K/\eta_K})}{Me^{E_M E_K t/M}(6\nu_0 - 3)} - \frac{\eta_M}{6\nu_0 - 3}]/dt \times \varepsilon_{kk}^{(0)} \quad (3)$$

Table 1. Viscoelastic coefficients of five kinds of asphalt concrete at 20°C.

Grading types	E_M	η_M	E_K	η_K
SAC-16	190.87	115.56	255.27	125.52
AC-16	148.69	96.88	200.80	79.01
AC-13	131.69	91.80	192.34	77.26

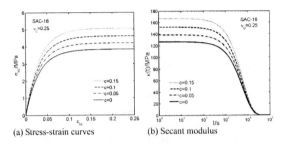

(a) Stress-strain curves (b) Secant modulus

Figure 3. Stress-strain curves and bulk secant modulus with different porosity.

where $\tau = \eta_K/E_K$ = relaxation time. η_M = coefficient of viscosity, η_K = coefficient of elastic delay viscosity, E_M = Young's modulus of immediate elasticity, E_K = modulus of delayed elasticity.

The secant modulus of matrix can be derived from Equation 4 that

$$\kappa_0(t) = \frac{E_M E_K [\eta_M^2 (E_M + E_K - E_M e^{E_K/\eta_K})]}{M^2 e^{E_M E_K t/M} (6v_0 - 3)} \quad (4)$$

According to the elastic-viscoealstic correspondence principle, the effective modulus in transform domain can be written from their elastic counterparts as

$$\frac{\kappa^{TD}}{\kappa_0^{TD}} = 1 + \frac{c(\kappa_1^{TD} - \kappa_0^{TD})}{(1-c)(\kappa_1^{TD} - \kappa_0^{TD})\alpha + \kappa_0^{TD}} \quad (5)$$

where $\alpha = (1 + v_0^{TD})/[3(1 - v_0^{TD})]$. The inversion under $v_0^{TD} = v_0$ leads to the bulk secant modulus of saturated asphalt concrete. The paramount of AC-13, AC-16 and SAC-16 asphalt concrete at 20°C are provided in Tab. 1 (Wang. 2011).

The following figure is the bulk stress-strain relationship curves of the SAC-16 saturated asphalt concrete at the strain rate 5×10^{-6} s^{-1}.

It is displayed in Fig. 3a, the value of the stress will reach constant related to η_K, c and strain rate. And bulk modulus of SAC-16 shown in Fig. 3b display a downward trend, and finally become 0. This is coincidence with the performance of Burgers model. When the strain rate is very slow, the theoretic result of the stress-strain relations are coincided with the laboratory tests (Ye. 2009).

2.3 The complex modulus of asphalt concrete

The linear viscoelastic property of saturated asphalt concrete can be provided by quasi-static tests. In practice engineering, asphalt pavement is subject to harmonic loads. Under this kind of loads, mechanical properties of saturated asphalt concrete can be expressed in effective complex modulus.

The complex modulus of matrix denoted as κ_0^* derived from creep compliance or relaxation modulus of the asphalt concrete can be decomposed as

$$\kappa_0^* = \kappa_0^R - i\kappa_0^I \quad (6)$$

where κ_0^R = real part of the complex bulk modulus, κ_0^I = imaginary part of the complex bulk modulus. Replace TD by $*$ in Equation 6.

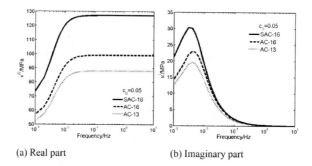

(a) Real part (b) Imaginary part

Figure 4. The real and imaginary parts of the effective complex modulus.

Consider Burgers model as viscoelastic model of asphalt concrete matrix. The complex modulus curves can be given as follow under the assumption of $v_0^* = v_0$.

It is evident from Fig. 4 that, as frequency increase, the real part of effective complex modulus will approach to the transient elastic modulus. The imaginary part will increase from 0 to maximum, then decrease to zero as frequency increase. When the frequency is above 1 Hz, the complex modulus of the asphalt concrete will exhibit transient elastic property.

3 THE EFFECT OF PORE-WATER PRESSURE ON EFFECTIVE MODULUS

3.1 *The effect of pore-water pressure on porosity and bulk modulus*

As a hydrodynamic media, water existing in the porous will generate pore-water pressure. While under the external loading, the variation of pore-water pressure causes the change of porosity. So the investigation of the effect of water pressure on the modulus of saturated asphalt concrete is the efficient route to study the mechanism of water damage. Li et al. (2003) established the relationship between porosity c and pore-water pressure ΔP.

$$c = \frac{c_0 + \varepsilon_v + (1-c_0)\Delta P / \kappa_0}{1+\varepsilon_v} \tag{7}$$

where c_0 = initial porosity, ε_v = total strain of saturated asphalt concrete. With the saturated porosity very small, we have $\varepsilon_v = \varepsilon_0 + \varepsilon_1$. When the bulk modulus of matrix changes with time It then follows from Equation 7 that

$$c(\Delta p, \kappa_0(t)) = \left\{ c_0[1 - \frac{\Delta P}{\kappa_0(t)}] + \frac{\Delta P}{\kappa_1} \right\} / \left[1 - \frac{\Delta P}{\kappa_0(t)} + \frac{\Delta P}{\kappa_1} \right] \tag{8}$$

As a result of changed porosity, the effective modulus then becomes

$$\kappa(t) = \kappa_0(t)[1 + \frac{c(\Delta P, \kappa_0(t))[\kappa_1 - \kappa_0(t)]}{[1 - c(\Delta P, \kappa_0(t))][\kappa_1 - \kappa_0(t)]\alpha + \kappa_0(t)}] \tag{9}$$

3.2 *Sine function pore-water pressure*

The pore-water pressure can be predicted by a sine function as:

$$P = \sigma_0 \sin(2\pi f t) \tag{10}$$

where σ_0 is the maximum of pore-water pressure, in another word, the amplitude of the sine function, f is the frequency. The effective modulus and the pore-water pressure of AC-13 asphalt concrete under different porosity and frequency are displayed in Fig. 5.

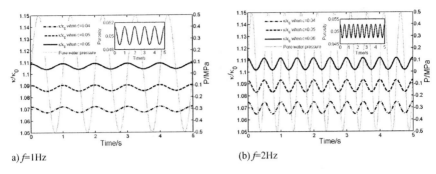

a) $f=1$Hz (b) $f=2$Hz

Figure 5. Effective bulk modulus at $\sigma_0 = 0.5$ MPa.

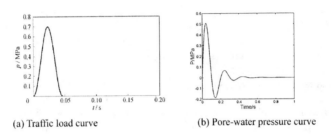

(a) Traffic load curve   (b) Pore-water pressure curve

Figure 6. Traffic load and pore-water pressure curves.

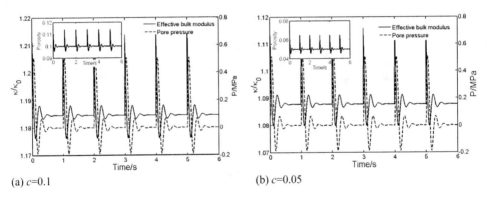

(a) $c=0.1$ (b) $c=0.05$

Figure 7. Effective bulk modulus curves under excess pore-water pressure.

The fluctuation curves of effective bulk modulus and porosity differs by one-fourth phase from pore-water pressure. And the amplitude of porosity and effective modulus increase with the growth of the frequency.

3.3 Excess pore-water pressure

The pore-water in the asphalt concrete will cause excess pore-water pressure under the high speed and repeated travel loads. Dong et al. (2007) have obtained the excess pore-water pressure curves with the penetration rate and vehicle speed under a pulse wave traffic load (Fig. 6).

Taking the initial porosity of the AC-13 asphalt concrete 0.05 and 0.1 respectively, the changes of excess pore-water pressure, porosity and effective bulk modulus under the traffic load are depicted in Fig. 7.

Compare Fig. 7a with Fig. 7b, the variation range of porosity and effective modulus with the different initial porosity is basically undiversified. The effective modulus relative to the changes of pore-water pressure reflects the retardation phenomenon. It is shown under the action of traffic loads that, the effect of the excess pore-water pressure on porosity and effective modulus of asphalt road can not be ignored, which not only cause internal stress concentrations in the weaknesses of pavement, but also resulting in accelerated destruction of pavement.

It is known that the driving speed and traffic volume determine the load frequency, and the axle-loads decide the maximum of traffic loads. Therefore, it is significant to investigate the three factors which affect the seriousness of water damage in asphalt pavement.

4 CONCLUDING REMARKS

(1) The micro-mechanical models of the saturated asphalt concrete are proposed by assuming asphalt concrete as elastic or viscoelastic medium and saturated pores elastic but sustain only compression, and the effective elastic modulus, secant modulus are predicted by using viscoelastic theory and micromechanics method.
(2) Base upon the elastic-viscoealstic correspondence principle, the explicit solutions of secant modulus and constitutive equations of the matrix is deduced under constant strain rate. In addition, the effective complex modulus is derived. The complex modulus is used to exhibit transient elastic property under high frequency harmonic load.
(3) The variation of pore-water pressure causes the changes of porosity and effective bulk modulus. Under the quick and cycling traffic load, the excess pore-water pressure of asphalt pavement is squeezed, and effective bulk modulus as well as porosity instantaneously increase or decrease.

REFERENCES

Eshelby, J.D. 1957. "The determination of the elastic field of an ellipsoidal inclusion and related problems". *Proc. R. Soc. London,* Vol. A, 241: 376–396.

Mori, T & Tanaka, K. 1973. Average stress in matrix and average elastic energy of materials with misfitting inclusions. *Acta Metall,* 21: 571–574.

Zhao, Y. H. & Weng, G. J. 1990. "Effective elastic modulus of ribbon-reinforced composites". *Transaction of the ASME,* Vol. 57, 158–167.

J. Li. & G. J. Weng. 1994, Strain-Rate Sensitivety, Relaxtion Behavior, and Complex Modulus of a Class of Isotropic Viscoelastic Composites. *Journal of Engineering Materials and Technology,* Vol. 116:495–504.

Huang, X. M. et al. 2009. "Visco-elastic Micromechanics analysis of asphalt mixture considered coarse aggregate and porosity". *Journal of South China University of Technology,* Vol. 37(7): 31–36.

Weng, G. J. 1984. "Some elastic properties of reinforced solids, with special reference to isotropic ones containing spherical inclusions". *Int. J. Eng. Sci.,* 22: 845–856.

Li, F. et al. 2006. "Research on Numerical of Micro-mechanics Damage numerical simulation of asphalt concrete". *Transportation Science & Technology,* Vol. 214: 56–58.

Huang, Z. Q. et al. 2011. "Simulation of effective elastic properties of solids with fluid-filled pores using boundary element method". *China Academic Journal Electronic Publishing House,* Vol. 51 (4): 471–477.

Kachanov, M 1993. "Elastic solids with many cracks and related problems". *Journal of Advances in Applied Mechanics,* Vol. 30: 259–445.

Wang, Z. C. 2011. *Study on Viscoelastic Properties of Asphalt Mixtures Based on Micromechanics.* Dalian: Dalian Maritime University.

Ye, Y. 2009, *Experimental researches on viso-elastoplastic constitutive model of asphalt mixture.* Wuhan: Huazhong University of Science and Technology.

Li, P. C. el al. 2003. "Mathematical modeling of flow in saturated porous media on account of fluid-structure coupling effect". *Journal of Hydrodynamics.* Ser. A, Vol. 18, No. 4: 439–426.

Dong, Z. J. et al. 2007. "Research on pore pressure within asphalt pavement under the coupled moisture-load action". *Journal of Harbin Institute of Technology.* Vol. 39, No. 10: 1614–1617.

Numerical analysis of SRC columns with different cross-section of encased steel

H.J. Jiang & Y.H. Li
State Key Laboratory of Disaster Reduction in Civil Engineering, Tongji University, Shanghai, P.R. China

Z.Q. Yang
The Architectural Design & Research Institute of Tongji University, Shanghai, P.R. China

ABSTRACT: The finite element analysis models of Steel Reinforced Concrete (SRC) columns with different cross-section of encased steel were constructed with the aid of ABAQUS program. The load-displacement curves of the columns under monotonic loads were obtained. In addition, the effects of axial compression ratio, the steel ratio and the arrangements of encased steels on the seismic behavior of the SRC columns were investigated. The results indicate that both the axial compression ratio and the steel ratio have significant effects on the seismic behavior of SRC columns while the effect of cross-section type of encased steel is not significant. With the increase of the steel ratio, the ultimate strength of SRC columns rises up roughly linearly. The effect of the steel ratio on the ultimate strength is more apparent when the axial compression ratio remains a low level, while the influence is inconspicuous when the axial compression ratio is relative high. With the increase of the axial compression ratio, the ductility of SRC columns rises up, and the ultimate strength increases at first and then decreases.

1 INTRODUCTION

Steel reinforced concrete (SRC) columns which have been widely used in high-rise buildings have the advantages of steel columns and reinforced concrete columns, such as high strength, stiffness and ductility. In addition, the performance of fire resistance and durability of SRC columns is also excellent. SRC columns can reduce the space of columns in buildings effectively. A lot of researches have been carried out on the structural behavior of SRC columns. However, most of them focused on SRC columns with simply shaped steel, such as cross-shaped (Chen et al. 2009) and I-shaped (Liu et al. 2011), and normal steel ratio. In recent years, the super SRC columns with huge cross sections and higher steel ratios have been used in super tall buildings. For super columns, due to the huge cross sections, there can be multiple choices to arrange the encased steel. Meanwhile, the steel ratio can have significant influence on the seismic behavior of SRC columns. Up to now, few researchers have studied the effects of the arrangement and the steel ratio of the encased steel on the seismic behavior of SRC columns. As the first part of the study on super SRC columns, five scaled specimens with different cross-section of encased steel and different steel ratio were designed based on the super columns from the actual engineering, which will be tested under cyclic loading. The seismic performance of the specimens were analyzed with the aid of ABAQUS program.

2 DESIGN OF SPECIMENS

Five 1/8-scaled specimens with three types of cross-section of encased steel and three steel ratios were designed. The cross-sections of the encased steel are show in Figure 1. The dimensions of

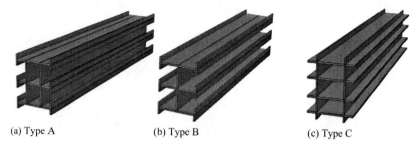

(a) Type A (b) Type B (c) Type C

Figure 1. Three types of cross-sections of encased steel in SRC columns.

Table 1. Main parameters of specimens.

Specimen number	ρ_{ss} (%)	ρ_s (%)	ρ_v (%)	n	n_f
A4	4	1.35	0.83	4.54	0.51
A8	8	1.35	0.83	4.54	0.51
A12	12	1.35	0.83	4.54	0.51
B8	8	1.38	0.76	4.54	0.51
C8	8	1.40	0.88	4.54	0.51

Notes: The specimen number A4 represents that the cross-section of encased steel is Type A and its steel ratio is 4%; ρ_{ss} represents the steel ratio; ρ_s represents the longitudinal steel bar ratio; ρ_v represents the transverse reinforcement ratio; n represents the shear span ratio and n_f represents the axial compression ratio.

specimens are identical. The cross section is the rectangle with the height of 535 mm and the width of 660 mm, and the length is 3 m. The concrete used in the specimens was specified with cubic compressive strength of 40 MPa and modulus of elasticity of 3.0×10^4 MPa. The encased steel was specified with yield stress of 248 MPa and modulus of elasticity of 2.1×10^5 MPa. The reinforcement bars were specified with yield stress of 353 MPa and modulus of elasticity of 2.1×10^5 MPa. The prototype of these SRC columns were designed according to Chinese code. The main parameters of each specimen are shown in Table 1.

3 FINITE ELEMENT ANALYSIS AND VERIFICATION

The five specimens designed above were analyzed with the aid of ABAQUS program to study the seismic behavior of SRC columns. In order to verify the numerical model used in this study, several comparisons were implemented between the simulation and the experiment.

3.1 Finite element model

Bilinear constitutive relationships are adopted for the steel and reinforcement rebar. The post-yielding stiffness ratio is taken as 0.01. The reinforcement bars are modelled by truss elements. They are coupled to the concrete elements using the embedded modelling technique. The steel is modelled by shell element or solid element.

The concrete damage plasticity model is used to simulate the nonlinear behavior of concrete. The damage factors proposed and verified by Birtel & Mark (2006) are applied here for the damage plasticity model. Two constitutive relationships of concrete are applied according to the constraint condition of the concrete. The constitutive relationships proposed by Han (2007) are applied for concrete in shadow area with high constrain (as shown in Fig. 2) and the constitutive relationships

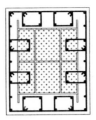

Figure 2. Division of cross section.

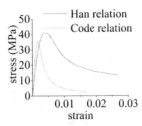

Figure 3. Comparison of two constitutive relationships.

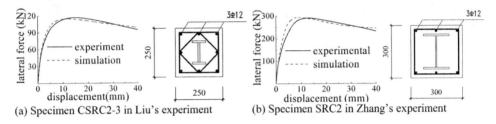

(a) Specimen CSRC2-3 in Liu's experiment   (b) Specimen SRC2 in Zhang's experiment

Figure 4. Comparison of simulation results and experimental results.

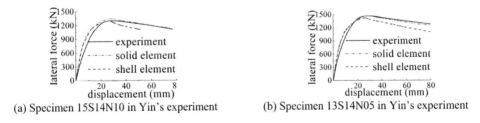

(a) Specimen 15S14N10 in Yin's experiment   (b) Specimen 13S14N05 in Yin's experiment

Figure 5. Comparison of simulation results and experimental results with different element.

proposed in Chinese code are used for the other areas. The comparison of two constitutive relationships is shown in Figure 3.

3.2 *Verification of finite element model*

Four SRC columns in Yin's (2012) experiment, three in Liu's (2011) experiment, two in Zhang's (2007) experiment and one in Chen's (2009) experiment were applied to verify the finite element model in this study. The comparison of load-displacement curves between the test results and simulation results for some specimens is shown in Figure 4. They agree well.

For the encased steel, there are two types of element can be adopted as mentioned above. To make sure which one is better, two SRC columns in Yin's experiment named 15S14N10 and 13S14N05 were simulated with the two types of element. The results as shown in Figure 5 indicate that the

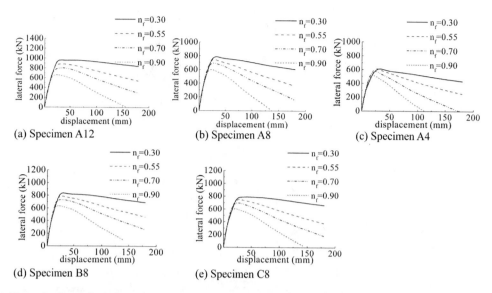

Figure 6. Load-displacement curves of specimens with different axial compression ratio.

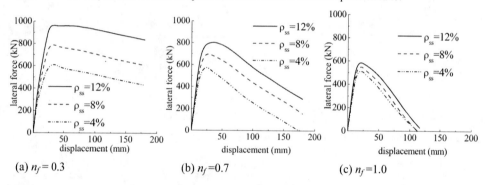

Figure 7. Comparison of load-displacement curves between SRC columns with different steel ratio.

solid element model can match the results better than the shell element model. In the following study, the solid element is used for the encased steel.

4 ANALYSIS OF SEISMIC BEHAVIOR OF SRC COLUMNS

4.1 Load-displacement curves

The numerical models of the five specimens designed above were constructed. The load-displacement curves of the specimens under the constant vertical load and monotonically increased lateral load at the top were obtained as shown in Figure 6.

From Figure 6 it can be found that the slop of descending branch is small so that the ductility is good when the axial compression ratio is small. With the increase of the axial compression ratio, the peak load and corresponding displacement decrease, and the descending branch becomes steeper so that the ductility decreases.

4.2 Parameter study

The load-displacement curves for the three SRC columns with the steel cross section type of A and different steel ratio are shown in Figure 7. With the increase of steel ratio, the initial stiffness,

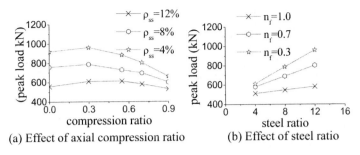

Figure 8. Effects of axial compression ratio and steel ratio on the ultimate strength.

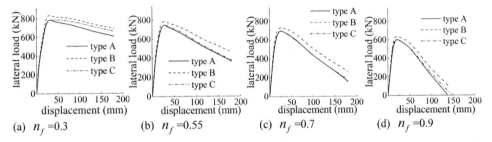

Figure 9. Comparison of SRC columns with different cross-section of encased steel.

the peak load and the corresponding displacement increase. The effect of steel ratio becomes insignificant when the axial compression ratio is large.

The effects of axial compression ratio and steel ratio on the ultimate strength are shown in Figure 8. The ultimate strength rises up slowly at first and then drops down with the axial compression ratio increasing. For the columns with high steel ratio, the rate of descending part is higher than the columns with low steel ratio. With the increase of steel ratio, the ultimate strength rises up roughly linearly. In addition, the effect of the steel ratio on the ultimate strength is more apparent when the axial compression ratio is small.

The load-displacement curves of SRC columns with different cross-section of encased steel are shown in Figure 9. For these columns with the same steel ratio, the difference is not significant. Generally, the columns with the steel cross-section type of B have the largest ultimate strength and ductility while the columns with the steel cross-section type of A have the smallest ultimate strength and ductility.

5 CONCLUSIONS

The seismic behavior of SRC columns with different cross-section of encased steel is evaluated by finite element analysis with the aid of ABAQUS program. The steel ratio and axial compression ratio influence the seismic behavior of SRC columns significantly. With the increase of steel ratio, the ultimate strength of SRC columns rises up effectively. The influence of steel ratio is more significant when the axial compression ratio is low. With the increase of the axial compression ratio, the ductility rises up, and the ultimate strength increases at first and then decreases. The effect of the cross-section type of encased steel is not significant.

ACKNOWLEDGEMENTS

The financial supports from the Ministry of Science and Technology of China through Grant No. 2012BAJ13B02 and National Natural Science Foundation of China through Grant No. 91315301-4 are gratefully appreciated.

REFERENCES

Birtel, V. & Mark, P. 2006. Parameterized finite element modeling of RC beam shear failure, *Proceedings of the 19th Annual International ABAQUS Users' Conference*, Boston, 23–25 May, 2006: 95–108.
Chen, X.G., et al. 2009. Experimental study on the seismic behavior of steel reinforced concrete columns. *Journal of University of Science and Technology*, 31(12): 1516–1524.
Han, L.H. 2007. *Concrete filled steel tublar structures: theorem and application, second ed.*, Beijing: Science Press.
Liu, Y., et al. 2011. Experimental investigation and numerical simulation of seismic behavior of CSRC columns. *Journal of Huaqiao University (Nature Science Edition)*, 32(1): 72–76.
Yin, X.W. 2012. *Seismic Performance of SRC Columns with High Ratio of Encased Steel*. Shanghai: Tongji University.
Zhang Z.W. 2007. *Experiment study on deformation performance and hysteretic characteristic of SRC columns*. Xiamen: Huaqiao University.

Modeling hysteresis loops of corner beam-column joint using Ruaumoko program

A.G. Kay Dora & N.H. Hamid
Faculty of Civil Engineering, Universiti Teknologi MARA, Selangor, Malaysia

ABSTRACT: This paper presents the modeling of precast RC beam-column corner joint with corbels using Ruaumoko Programming. Hysteresis loops (load versus displacement) curve is compared between experimental and modeling results. The experimental work is carried out on a full-scale specimen which representing the precast RC beam-column corner joint with corbels. This specimen is tested under reversible lateral cyclic loading up to ±1.35% inter-story drift. Modeling work is carried out using Hysteres Program in Ruaumoko 2D. A curve of hysteresis loop for beam-column corner joint is generated using Pampanin Reinforced Concrete Beam-Column Joint Hysteresis Model with reloading slip factor. The modeling result is compared with experimental work. It was found that the hysteresis loops exhibited good agreement between experimental and modeling results. The effective stiffness, displacement ductility and equivalent viscous damping for the specimen are also discussed and compared in this paper.

1 INTRODUCTION

Malaysian earthquake event is not anxiously seen in history and seismic assessment for buildings in Malaysia has never been done previously. As recorded, the nearest distance of earthquake epicenter from Malaysia is approximately 350 kilometers. The tremors due to Sumatra earthquakes had been reported several times in Peninsular Malaysia although it is located in the stable Sunda Shelf with low to medium seismic activity level. For instance, there were two large earthquakes near Sumatra which occurred at the end of 2002 ($M_w = 7.4$) and early 2003 ($M_w = 5.8$) (Adnan, et al., 2005). Major earthquake with M8.6 that struck East Indian Ocean on April 11, 2012, after deadliest M9.2 Sumatra Earthquake in 2004, have altered global seismicity patterns. The effects of those earthquakes to building depend on the natural frequency of the buildings. According to the data analysis, the maximum effect of the motion will occur on 1 to 10-storey buildings in Penang and Kuala Lumpur. Based on the facts, seismic hazard assessment for Malaysia is essential in order to mitigate the effects of potential large earthquake that may occur in the future.

Ruaumoko program is a finite element application which designed to provide a step-wise time history nonlinear analysis for 2D and 3D structures to ground accelerations and to time varying force excitations. Ruaumoko Programming has been used throughout the world by academic researchers and professionals for modeling the concrete frame structures under earthquake loading (Balendra, et al., 2002; Elmas and Guler, 2010; and Spieth, et al., 2004). The dynamic performance of concrete, steel and timber structures and bridges can be determined using Ruaumoko program as well as pushover analyses for monotonic and cyclic loading of structures. The results obtained from Ruaumoko program can be used to determine the mode shape, hysteresis loops, moment-rotation, damage indices and energy absorption. The key advantage of Ruaumoko is it allows a wide range of modeling variables to be utilized includes of hysteresis rules, mass of the structure, damping and stiffness factors Carr (2007). In this study, Ruaumoko program will be used to model the seismic behavior of precast reinforced concrete of two-story school building which had been constructed in Malaysia. The beam-column corner joint is designed by using BS8110 without considering

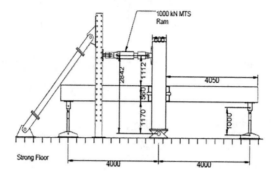

Figure 1. Test setup of subassembly test.

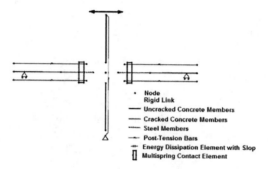

Figure 2. Ruaumoko subassembly model of beam-column joint.

earthquake loading. Hysteresis Program in Ruaumoko Program is used as the preliminary program to model the inelastic behavior of beam-column joint, before modeling the whole building. In this study, Pampanin Reinforced Concrete Hinge Hysteresis rule with reloading slip factor was chosen to model the inelastic member behavior of the sub-assemblage corner beam-column joint.

2 PRINCIPLE APPROACH

The principle approach of the new developed multispring contact element is presented using the model of a sub-assembly test which was performed at the University of Canterbury (Mander, 2008). The test represents an inner node of a precast post-tensioned concrete frame. Each beam was joined by post-tensioned high strength steel bars with the column element. This leads to two rocking connections at the beam-column interfaces. The experiment setup is shown in Figure 1.

A principle drawing of the Ruaumoko model is shown in Figure 2. The beams and the column were modeled with linear elastic Giberson frame members. The model was tested with displacement control at the top node.

3 HYSTERESIS PROGRAM

The HYSTERESIS program consists in Ruaumoko 2D program. This program can be used to model the beam-column joint hysteresis loops by putting some data which obtained from analysis. The parameters for the hysteresis loops are calculated based on the experimental data. Figure 3 shows the Pampanin Hysteresis loops with reloading slip factor for modeling purpose. The data of elastic initial stiffness, K_0, is calculated using Equation q and bi-linear factor, r, is calculated using

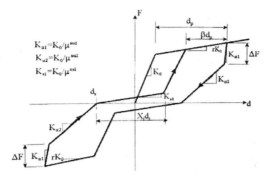

Figure 3. Pampanin Hysteresis with reloading slip factor (Carr, 2008).

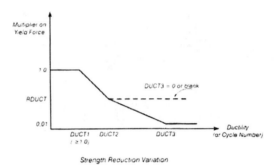

Figure 4. Degrading Strength Rule (ILOS).

Equation 2.

$$K_0 = \frac{(Y_2 - Y_1)}{(X_2 - X_1)} \qquad (1)$$

$$r = \frac{(Y_2 - Y_1)}{(X_2 - X_1)} \times \frac{1}{K_0} \qquad (2)$$

where, X_1 is the coordinate at Node 1 in x-x axis; X_2 is the coordinate at Node 2 in x-x axis; Y_1 is the coordinate at Node 1 in y-y axis; and Y_2 is the coordinate at Node 2 in y-y axis.

As specified by Carr (2007), to allow for strength degradation, the yield levels in the interaction diagrams may be reduced as either a function of the ductility or the number of load reversals from the backbone of the hysteresis rule as shown in Figure 4. DUCT1 is the ductility at which degradation begins; DUCT2 is the ductility at which degradation stops; DUCT3 is the ductility at 0.01 initial strength and RDUCT is the residual strength as a fraction of the initial yield strength.

The stiffness degradation parameters which are known as AlfaS1(αs1); AlfaU1(αu1); AlfaU2(αu2); DeltaF and Beta were calculated using Eq. 3, Eq. 4, Eq. 5, based on the experimental hysteresis loops before input them into Pampanin Joint Hysteresis Rule.

$$K_{\alpha 1} = K_0 / K^{au1} \qquad (3)$$

$$K_{\alpha 2} = K_0 / K^{au2} \qquad (4)$$

$$K_{s1} = K_0 / K^{as1} \qquad (5)$$

where, $K_{\alpha 1}$ is the initial unloading stiffness, $K_{\alpha 2}$ is the final unloading stiffness, K_{s1} is the slip stiffness, μ_{au1} is the ductility with initial unloading power factor, μ_{au2} is the ductility with final unloading power factor, and μ_{as1} is the ductility with reloading power factor.

4 COMPARISON BETWEEN MODELING AND EXPERIMENTAL RESULTS

The experimental results were recorded using data logger and the graph were plotted on the computer screen to monitor the seismic performance of the precast beam-column corner joint with corbels as described in (Kay Dora and Abdul Hamid, 2012). The comparison between experimental and modeling results was shown in Figure 8. The hysteresis loops from experimental results represents by dotted lines and the black dark curve represents modeling results obtained from Hysteresis Program. Linear variable displacement transducers were used to measure the lateral displacement of the sub-assemblage in millimeter. For experimental work, ten sets of history drifts were applied at top of column (±0.01%, ±0.05%, ±0.1%, ±0.2%, ±0.5%, ±0.75%, ±1.0%, ±1.25%, and ±1.35% drift). Two numbers of cycles were applied for each drift level. However, Hysteresis Program plotted the inelastic behavior of the beam-column joint for one cycle only at ±1.0%, ±1.25% and ±1.35% drift. The modeling results were exhibits similar pattern with the experimental results as shown in Figure 5.

From the results, comparison of strength, effective stiffness, displacement ductility and equivalent viscous damping between the modeling and experimental were made. The comparison was considered at ±1.0%, ±1.25% and ±1.35% drift at both pushing (positive) and pulling (negative) direction. Table 1 shows the comparison of strength values between experimental and modeling results. The measurements of strength were made based on maximum force value at each of interstory drift. Table 2 shows the comparison for effective stiffness between experimental and modeling results. Most of the values show that the experimental results exhibit slightly higher values as compared to the modeling results. On the other hand, Table 3 shows the comparison of displacement

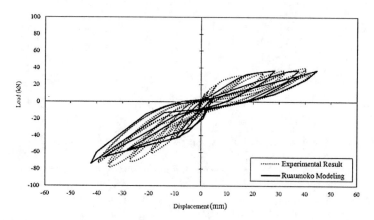

Figure 5. Hysteresis loops comparison between experimental and modeling results.

Table 1. Comparison of lateral strength between experimental and modeling results.

Target drift (%)	Positive Direction			Negative Direction		
	Model (kN)	Exp (kN)	% Diff	Model (kN)	Exp (kN)	% Diff
1.00	37.09	36.95	0.38	−57.07	−71.36	20.03
1.25	36.93	40.32	8.41	−65.91	−77.93	15.42
1.35	36.97	36.10	2.35	−73.50	−72.93	0.78

Table 2. Effective stiffness comparison between experimental and modeling results.

Target drift (%)	Positive Direction			Negative Direction		
	Model (kN)	Exp (kN)	% Diff	Model (kN)	Exp (kN)	% Diff
1.00	1.32	1.16	12.07	2.03	2.64	23.12
1.25	0.98	1.00	1.72	1.76	2.21	20.47
1.35	0.83	0.82	1.23	1.74	1.85	5.83

Table 3. Displacement ductility comparison between experimental and modeling results.

Target drift (%)	Positive Direction			Negative Direction		
	Model (kN)	Exp (kN)	% Diff	Model (kN)	Exp (kN)	% Diff
1.00	1.47	1.67	11.74	1.21	1.67	27.54
1.25	1.96	2.11	6.81	1.61	2.10	23.47
1.35	2.33	2.28	2.40	1.81	2.19	17.14

Table 4. Equivalent viscous damping comparison between experimental and modeling results.

Target drift (%)	Modeling (kN)	Experimental (kN)	% Different
1.00	1.47	1.67	11.74
1.25	1.96	2.11	6.81
1.35	2.33	2.28	2.40

ductility between the experimental and modeling results. Table 4 shows the comparison of equivalent viscous damping between the experimental and modeling results. These values also exhibit good deals amongst each other. Overall values show that modeling results have a good agreement with the experimental results.

5 CONCLUSION

In this paper, the comparison between experimental and modeling result using Ruaumoko Hysteresis Program was successfully presented. The load versus displacement curve or hysteresis loops of the specimen were modeled and exhibits good agreement with the hysteresis loops obtained from experimental results. The different percentage of maximum strength, effective stiffness, displacement ductility and equivalent viscous damping between experimental and modeling results estimated between 1.20 to 6.63%, demonstrated the use of Ruaumoko Hysteresis Program in verifying the experimental results. Further task is to implement other associated program in Ruaumoko such as RUAUMOKO2D and DYNAPLOT to give more analysis on inelastic behavior of precast reinforced structures under earthquake excitation.

ACKNOWLEDGEMENT

The authors would like to thank the Research Management Institute, University Teknologi MARA, Malaysia and FRGS (Fundamental Research Grant Scheme), MOHE (Ministry Of Higher Education), Malaysia for the funding this research work. Nevertheless, the authors also want to express their gratitude to the technicians of Heavy Structures Laboratory and also postgraduate students of Faculty of Civil Engineering, UiTM, for their great commitment toward this research work.

REFERENCES

Adnan, A., Hediyawan, Mato, A., and Irsyam, M. (2005). Seismic Hazard Assessment for Peninsular Malaysia using Gumbel Distribution Method. Journal Teknologi, 42(B): 57–73, Universiti Teknologi Malaysia.

Balendra, T., Lam, N.T.K., Wilson, J.L., and Kong, K.H. (2002). Analysis of Long-Distance Earthquake Tremors and Base Shear Demand for Buildings in Singapore. Engineering Structures, Vol. 24, No.1, pp 99–108.

Carr, A.J. (2007). RUAUMOKO Manual (Volume 1: Theory). University of Canterbury, Christchurch, New Zealand.

Carr, A.J. (2008). RUAUMOKO Manual (Volume 5: Appendices). University of Canterbury, Christchurch, New Zealand.

Elmas, A.Z. and Guler, K. (2010). Seismic Performance Evaluation of An Existing RC School Building. Paper presented at the 14th European Conference on Earthquake Engineering, August 30–September 3, Ohnid, Republic of Macedonia.

Kay Dora A.G. and Abdul Hamid, N.H. (2012). Comparing The Seismic Performance Of Beam-Column Joints With And Without SFRC When Subjected To Cyclic Loading, Advanced Materials Research, 626, 85.

Mander, J.B. (2004). Beyond ductility: The Quest Goes On. Bulletin of the New Zealand Society for Earthquake Engineering, 37(1), pp 35–44.

Spieth, H.A., Arnold, D., Davies, M., Mander, J.B., and Carr, A.J. (2004). Seismic Performance of Post-Tensioned Precast Concrete Beam to Column Connections with Supplementary Energy Dissipation, Proceedings of NZSEE Conference, Rotorua, New Zealand, March 19–21.

Advanced Engineering and Technology – Xie & Huang (Eds)
© 2014 Taylor & Francis Group, London, ISBN 978-1-138-02636-0

Prediction of ship resistance in muddy navigation area using RANSE

Zhiliang Gao, Hua Yang & Qixiu Pang
Tianjin Research Institute for Water Transport Engineering, M.O.T., Tianjin, China

ABSTRACT: The computational fluid dynamics method, based on solving the Reynolds averaged Navier-Stokes equations, was used to simulate the flow field around a ship navigating in the shallow water area with fluid mud layer. The rheological behaviour of fluid mud was characterized by the Bingham plastic model. The water-mud interface was captured using the volume of fluid method. Benchmarking test of the KVLCC2 tanker advancing in the deep and shallow water was conducted. The computed resistance was validated against experimental data. To investigate the effect of fluid mud on the ship resistance, the tanker advancing with different under keel clearance in the muddy area were simulated. The ship resistance and undulation of water-mud interface in different navigation conditions were compared. The resistance characteristic is briefly discussed.

1 INTRODUCTION

For muddy harbour, regular dredging is necessary to guarantee the navigability of approach channel. Due to the high cost of maintenance, it is also desirable to avoid unnecessary dredging. The question then arises: how much of the mud needs to be dredged. The answer could be found following the nautical bottom concept introduced by PIANC (1997). The nautical bottom is the level where physical characteristics of the mud reach a critical limit. If the ship draft exceeds the nautical bottom, the ship might suffer either damage or unacceptable effects on controllability and manoeuvrability. Accordingly, it is possible to reduce the maintenance cost by utilizing the fluid mud layer as a part of navigable depth of channel. This utilization requires a clear understanding of the hydrodynamic performance of a ship advancing in fluid mud area.

The resistance is one of the most important criterions for the ship hydrodynamic performance. Its characteristic is crucial for assessing the navigability of fluid mud layer. The method of model test is traditionally employed for ship resistance prediction. But this method has the disadvantages of high expense, low efficiency and system errors. Those disadvantages become more notable for the test conducted in specific muddy conditions (Vantorre, 1994; Delefortrie, 2007). Numerical simulation of the flow field around a ship based on the computational fluid dynamics (CFD) method is an alternative approach to study the ship resistance. Among a variety of CFD methods, the method of solving the Reynolds averaged Navier-Stokes equations (RANSE) with the free surface capturing technique is most widely used for such simulation (Larsson et al., 2003; Gu and Wu, 2005; Carrica et al., 2007; Ahmed, 2011). In the area of resistance prediction, RANSE simulation has become a supplement to traditional model test and even replaced part of the experimental work. Therefore, its application to prediction of ship resistance in muddy navigation area is viable and promising.

In this study, a commercial multiphase RANSE solver is adopted to simulate flow field around a ship advancing in shallow water area with fluid mud layer. Preliminary benchmarking test is conducted to validate the present method. The effect of fluid mud on ship resistance is discussed based on the numerical results.

2 METHODOLOGY

2.1 *Mathematical model*

The present method considers incompressible multiphase flow. The fluid motion is governed by the continuity and RANS equations. The renormalization group (RNG) k-ε model proposed by Yakhot and Orszag (1986) is utilized for turbulence modeling. The interface between different fluids is captured using the volume of fluid (VOF) method (Hirt and Nichols, 1981). The governing equations described in the Cartesian coordinate system are as follows:

$$\frac{\partial u_i}{\partial x_i} = 0 \tag{1}$$

$$\frac{\partial}{\partial t}(\rho u_i) + \frac{\partial}{\partial x_j}(\rho u_i u_j) = \frac{\partial}{\partial x_j}\left[\mu_{\text{eff}}\left(\frac{\partial u_i}{\partial x_j} + \frac{\partial u_j}{\partial x_i}\right)\right] - \frac{\partial P}{\partial x_i} + \rho g_i \tag{2}$$

$$\frac{\partial}{\partial t}(\rho k) + \frac{\partial}{\partial x_j}(\rho u_j k) = \frac{\partial}{\partial x_j}\left(\alpha_k \mu_{\text{eff}} \frac{\partial k}{\partial x_j}\right) + G_k - \rho\varepsilon \tag{3}$$

$$\frac{\partial}{\partial t}(\rho\varepsilon) + \frac{\partial}{\partial x_j}(\rho u_j \varepsilon) = \frac{\partial}{\partial x_j}\left(\alpha_\varepsilon \mu_{\text{eff}} \frac{\partial \varepsilon}{\partial x_j}\right) + C_{1\varepsilon} G_k \frac{\varepsilon}{k} - C_{2\varepsilon}^* \rho \frac{\varepsilon^2}{k} \tag{4}$$

$$\frac{\partial \alpha_q}{\partial t} + \frac{\partial}{\partial x_j}(\alpha_q u_j) = 0 \tag{5}$$

with

$$\mu_{\text{eff}} = \mu + \mu_t \tag{6}$$

$$G_k = \mu_t \left(\frac{\partial u_i}{\partial x_j} + \frac{\partial u_j}{\partial x_i}\right) \frac{\partial u_i}{\partial x_j} \tag{7}$$

$$C_{2\varepsilon}^* = C_{2\varepsilon} + \frac{C_\mu \eta^3 (1 - \eta/\eta_0)}{1 + \beta \eta^3} \tag{8}$$

$$\eta = \left[\frac{1}{2}\left(\frac{\partial u_i}{\partial x_j} + \frac{\partial u_j}{\partial x_i}\right)\left(\frac{\partial u_i}{\partial x_j} + \frac{\partial u_j}{\partial x_i}\right)\right]^{1/2} \frac{k}{\varepsilon} \tag{9}$$

where t is the time; u_i ($i = 1, 2, 3$) is the mean velocity component in the x_i-direction; $\rho = \Sigma \alpha_q \rho_q$ is the mixture density; ρ_q is the density of the q-phase fluid; α_q is the volume fraction of the q-phase fluid and satisfies $\Sigma \alpha_q = 1$; P is the pressure; g_i is the component of gravitational acceleration in the x_i-direction; μ_{eff} is the effective viscosity; $\mu = \Sigma \alpha_q \mu_q$ is the mixture viscosity; μ_q is the apparent viscosity of the q-phase fluid; $\mu_t = \rho C_\mu k^2 / \varepsilon$ is the turbulent viscosity; k is the turbulence kinetic energy; ε is the turbulence dissipation rate; G_k is the turbulence production; the constants are given as: $C_\mu = 0.085$, $C_{1\varepsilon} = 1.42$, $C_{2\varepsilon} = 1.68$, $\alpha_k = 1.393$, $\alpha_\varepsilon = 1.393$, $\eta_0 = 4.38$, $\beta = 0.012$.

The apparent viscosity is a constant for Newtonian fluid such as water and is a function of the shear rate for non-Newtonian fluid such as fluid mud. In this study, the Bingham plastic model is used to determine the rheology of fluid mud. For low shear stress, the fluid mud behaves as a rigid body. Once the shear stress exceeds the yield stress, the fluid mud flows as a viscous fluid and its apparent viscosity reads:

$$\mu = \left(\frac{\tau_c}{\dot{\gamma}} + \mu_d\right) \tag{10}$$

where $\dot{\gamma}$ is the shear rate; τ_c is the yield stress; μ_d is the dynamic viscosity.

Table 1. Main particulars of KVLCC2.

Size	Full scale	Model scale
Scale	1	45.714
Length between perpendiculars (m)	320.0	7.0
Breadth of waterline (m)	58.0	1.269
Draft (m)	20.8	0.455
Depth (m)	30.0	0.6563
Displacement (m^3)	312622	3.272
Block coefficient	0.81	0.81

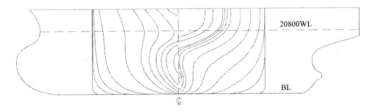

Figure 1. Body plan of KVLCC2.

In order to completely specify the mathematical model, it is necessary to define the boundary condition of the flow domain. The definition depends on the specific simulation cases and will be given in Section 3.

2.2 *Numerical method*

The commercial CFD software, FLUENT 12.0 (Fluent User's Guide, 2008), is employed to solve the governing equations (1) to (5). These equations are discretised by the finite volume method on a collocated grid. The Euler implicit scheme is used for the temporal discretisation. The convective and diffusive terms in Eqs. (2) to (4) are discretised by the second order upwind and central differencing schemes, respectively. The convective term in Eq. (5) is discretised by the modified high resolution interface capturing (HRIC) scheme. The enhanced wall treatment is used for modeling the fluid motion in the near-wall region. The Y plus that measures the distance from the wall to the centroid of the first cell is around 30. The SIMPLE algorithm is employed for pressure-velocity coupling. The Bingham plastic model is implemented in FLUENT utilizing the Herschel-Bulkley model provided by the software.

3 NUMERICAL SIMULATION

3.1 *Test ship*

The KRISO Very Large Crude Carrier 2, which is called KVLCC2, is adopted as the test ship in this study. The main particulars of the ship are summarized in Table 1 and its body plan is illustrated in Figure 1. All the numerical simulations in this study are performed in model scale.

3.2 *Benchmarking test*

To test its performance for ship resistance prediction, the present method is used to simulate flow field around a ship advancing at low speed in the deep and shallow water. The water depths in the simulation are selected as 1.2 and 10 times of the ship draft. The advance speed of the ship is 0.533 m/s, corresponding to the Froude number of 0.064. Since the ship speed is low, the undulation of free surface is very small and is ignored in the simulation. Additionally, the flow is symmetrical

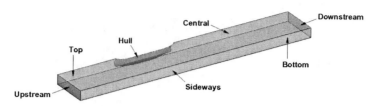

Figure 2. Computational domain for the shallow water case.

Table 2. Comparison of ship resistance obtained with experiment and computation.

Water depth	Experiment (N)	Computation (N)	Error (%)
Shallow	14.44	13.26	8.17
Deep	7.82	7.96	1.79

about the ship's longitudinal section in center plane, so only half of the flow field needs to be considered.

The computational domain for the shallow water case is shown in Figure 2. It is extended one ship length in front of bow, two ship length behind stern and half ship length away from ship side. The velocity inlet and pressure outlet conditions are imposed on the upstream and downstream boundaries, respectively. The symmetry condition is specified at the top and central boundaries. The condition of no-slip rigid wall is specified at the hull surface. The condition of no-slip moving wall, whose moving speed is equal to the flow speed specified at the upstream condition, is imposed on the bottom and sideways boundaries. The computational domain and boundary conditions for the deep water case are similar to those for the shallow water case. The total numbers of cell element for the shallow and deep water cases are approximately 0.49 million and 0.67 million, respectively.

Table 2 shows the comparison of ship resistance obtained with experimental measurement (SIMMAN, 2008) and numerical simulation. The numerical results agree with the experimental data. Their difference is less than 10%. For the case of shallow water, the difference between computed and experimental results is clear and mainly attributed to the neglect of free surface and ship's sinkage and trim in the computation. The defect of experimental equipment may also cause such difference, as reported in the literature (Simonsen et al., 2006).

3.3 Case study

To study the effect of fluid mud on the ship resistance, the present method is used to simulate the case of KVLCC2 tanker advancing in the shallow water area with muddy bottom, as shown in Figure 3. The water density (ρ_{water}) is 998.2 kg/m³. The fluid mud used in this study is dredged from an approach channel. The rheological test shows that the critical value of mud density (ρ_{mud}) for definition of nautical bottom is 1220 kg/m³. This value of density is adopted in the simulation. The ship speed (V_{ship}) in model scale is selected as 0.533 m/s, which corresponds to 7 knot in full scale and is the typical speed for ship entering or departing from the harbour. The sea bottom level (H_{bed}) is 1.2 times of the ship draft (D). The mud depth (H_{mud}) ranges from 0 to 0.4D. The ship's under keel clearance (H_{ukc}) is between −0.2D and 0.2D. Table 3 summarizes the navigation conditions for the numerical simulation. The undulation of air-water interface is not taken into account in the simulation. The computational domain, boundary condition and mesh arrangement for the current studied cases are identical to those for the shallow water case in the previous benchmarking test.

Figure 4 shows the ship resistance predicted with present numerical method. For the case of $H_{ukc} > -0.1D$, the increase of resistance as a function of decrease of H_{ukc} is almost linear. For the case of $H_{ukc} < -0.1D$, the resistance increases abruptly with further decrease of H_{ukc}. The computed resistance of the ship advancing at its service speed (1.177 m/s) in deep water is 34.73 N and is regarded as a reference. The resistances for cases 01, 02 and 03 are lower than the reference value,

Figure 3. Illustration of ship advancing in muddy area.

Table 3. Navigation conditions for ship advancing in muddy area.

Case	V_{ship} (m/s)	ρ_{water} (kg/m^3)	ρ_{mud} (kg/m^3)	H_{bed}	H_{mud}	H_{ukc}
01	0.533	998.2	1220	1.2D	0	0.2D
02	0.533	998.2	1220	1.2D	0.1D	0.1D
03	0.533	998.2	1220	1.2D	0.2D	0
04	0.533	998.2	1220	1.2D	0.3D	−0.1D
05	0.533	998.2	1220	1.2D	0.4D	−0.2D

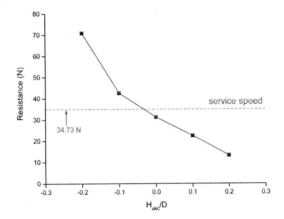

Figure 4. Comparison of ship resistance in different cases.

Figure 5. Comparison of water-mud interface in different cases (from left to right: case 02 to case 05).

whereas those for cases 04 and 05 are nearly 23% and 100% higher than the reference value. From the resistance point of view, the ship can navigate with negative under keel clearance provided that the penetration of mud layer is less than 10% of the ship draft. Figure 5 shows the undulation of water-mud interface obtained in the numerical simulation. It is clear that the undulation increases as the mud depth increases. In case 05, the maximum wave trough and crest are located at the bow area. Their heights are approximately 1% of the ship length. The undulation decays towards the stern. In cases 03 and 04, the characteristics of undulation are similar to those in case 05 except for the undulation amplitude. In case 02, the undulation is unobservable. Since the undulation of water-mud interface is very small in these cases, its contribution to the increase of ship resistance is small. Therefore, the increase of resistance for a ship advancing at low speed in muddy area is mainly attributed to the increase of viscous component. Before the ship contacts with the mud layer, the large viscosity of fluid mud reduces the effective depth of water and consequently enhances the blockage effect below the ship bottom. This leads to the increase of viscous pressure resistance. After the ship penetrates the mud layer, the viscosity of fluid that contacts the hull surface increases. This leads to the increase of frictional resistance and further increase of total resistance.

4 CONCLUSIONS

A commercial RANSE solver with the VOF method is used to predict the ship resistance in the shallow water area with fluid mud layer, whose rheological behaviour is determined by the Bingham plastic model. The present method is first validated by the test of the KVLCC2 tanker advancing in deep and shallow water. The benchmarking results confirm that the adopted mathematical model and numerical method are valid. The method is then used to simulate the tanker advancing at low speed with different under keel clearance in muddy area. It is found that the ship resistance and undulation of water-mud interface increase as the under keel clearance decreases. The increase of total resistance mainly results from the increase of viscous resistance and is in the normal range if the ship's penetration into the mud layer is limited, such as less than 10% of the ship draft found in this study. Future work will be carried out to investigate the squat of a ship advancing in muddy area.

ACKNOWLEDGEMENT

This work is supported by the Fundamental Research Funds for the Central Public Research Institutes of China (Grant No: TKS130201) and the Technology Foundation for Selected Overseas Chinese Scholar (granted by Ministry of Human Resources and Social Security of China).

REFERENCES

Ahmed, Y.M. 2011. Numerical simulation for the free surface flow around a complex ship hull form at different Froude numbers. *Alexandria Engineering Journal* 50 (3): 229–235.
Carrica, P.M., Wilson, R.V. & Stern, F. 2007. An unsteady single-phase level set method for viscous free surface flows. *International Journal for Numerical Methods in Fluids* 53 (2): 229–256.
Delefortrie, G. 2007. Manoeuvring behaviour of container vessels in muddy navigation areas. *Ph.D. Thesis, Ghent University.*
Fluent User's Guide, 2008. ANSYS-Fluent Inc, Lebanon, New Hampshire.
Gu, M. & Wu, C. 2005. CFD calculation for resistance of a ship moving near the critical speed in shallow water. *Journal of Ship Mechanics* 9 (6): 40–47.
Hirt, C.W. & Nichols, B.D. 1981. Volume of fluid method for the dynamics of free boundaries. *Journal of Computational Physics* 39 (1): 201–225.
Larsson, L. Stern, F. & Bertram, V. 2003. Benchmarking of computational fluid dynamics for ship flows: the Gothenburg 2000 workshop. *Journal of Ship Research* 47 (1): 63–81.
PIANC, 1997. Approach channels – A guide for design. Final report of the joint working group PIANC and IAPH, in cooperation with IMPA and IALA. Supplement to PIANC Bulletin, No. 95.
SIMMAN, 2008. *In Proceedings of the Workshop on Verification and Validation of Ship Maneuvering Simulation Method, Copenhagen, 14–16 April 2008.*
Simonsen, C., Stern, F. & Agdrup, K. 2006. CFD with PMM test validation for maneuvering VLCC2 tanker in deep and shallow water. *In Proceedings of the International Conference on Marine Simulation and Ship Maneuverability, Terschelling, 25–30 June 2006.*
Vantorre, M. 1994. Ship behaviour and control in muddy areas: state of the art. *In Proceedings of the 3rd Conference on Manoeuvring and Control of Marine Craft, Southampton, 7–9 September 1994.*
Yakhot, V. & Orszag, S.A. 1986. Renormalization group analysis of turbulence. I. Basic theory. *Journal of Scientific Computing* 1 (1): 3–51.

Numerical analysis of depth effects on open flow over a horizontal cylinder

Yehui Zhu, Liquan Xie & Xin Liang
College of Civil Engineering, Tongji University, Shanghai, China

ABSTRACT: Numerical models were built to investigate the effects of flow depth on open flow over a horizontal cylinder. Results show the abrupt change of flow surface at the cylinder, backup at the upstream side and drawdown at the downstream side. The center of drawdown curve is somewhere downstream of the cylinder center, which coincides with previous experimental results. With the decrement of flow depth, the drawdown effect becomes more and more obvious. Compared with the inlet velocity, flow beneath and downstream of the cylinder accelerates. The maximum bed shear stress shows up under the cylinder. The bed shear stress goes up with the flow depth, but the climbing rate slows down. The pressure on the bed increases with the flow depth. The pressure difference upstream and downstream the cylinder decreases with the decrement of flow depth.

1 INTRODUCTION

The study of unidirectional flow submerging a horizontal cylinder has attracted much attention. An important application of these studies is to analyze the response of marine pipelines and the permeable bed underneath subjected to high flows. Many previous studies have investigated the flow field around a cylinder based on experiments. Bearman & Zdravkovich (1978) studied the influence of gap size on pressure distribution around the cylinder and the flow structures. Mao (1988) found a large jump of pressure upstream and downstream the cylinder in unidirectional currents and also found the three-vortex structure near the horizontal cylinder (see Fig. 1). Jensen et al. (1990) carried out a series of experiments in a flume to study the flow field characteristics over a cylinder and transient bed profiles in the period of local scouring around pipeline. Wu & Chiew (2013) analyzed the three-dimensional flow structure near scour hole in unidirectional current, and found a strong and concentrated flow at the span shoulder due to the pipeline and the onset of scour, which plays an important role in the development of scour hole.

Due to the high expenses associating with the physical model tests in lab, numerical simulation gains popularity as a perfect alternative way. Smith & Foster (2005) simulated the flow field above the frozen scour bed under a pipeline with k-ε model and Large Eddy Simulation (LES) model, and verified the numerical results with experimental results by Jensen et al. (1990). Liang & Cheng (2005) and Liang et al. (2005) conducted numerical studies on the flow structures around a cylinder in unidirectional current. Lu et al. (1997) examined the oscillating flow over a circular cylinder

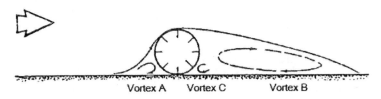

Figure 1. The three-vortex system (Mao, 1988).

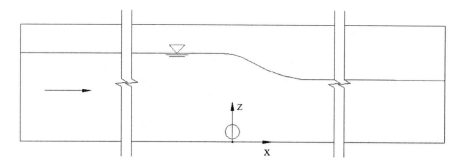

Figure 2. Layout of the numerical flume.

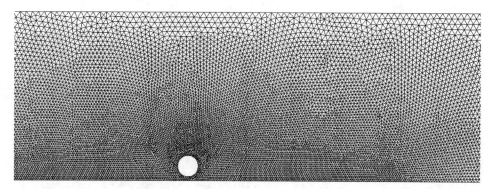

Figure 3. Triangular grid meshes adjacent to the circular cylinder.

with LES model. Zhang and Dalton (1999) studied the oscillating flow over a circular cylinder with finite difference method. They found the critical transition from two-dimensional flow to three-dimensional flow around the cylinder when $KC < 4$.

The present study aims to investigate the influence of flow depth on water surface profile, velocity distribution, the shear stress and pressure on the bed in open channel flow over a horizontal cylinder.

2 MODEL SETUP

A two-dimensional numerical flume with a length of 5.0 m and a height of 0.4 m was built shown in Figure 2. The diameter of the cylinder was 0.05 m and the distance between the cylinder and the inlet boundary was 2.7 m. The flow depths at the inlet were set as $z_0 = 0.08$ m, 0.12 m, 0.16 m and 0.32 m respectively in four cases. The depth averaged velocity at the inlet boundary was specified as 1.0 m/s, and kept constant for every case of flow depth. The gap between the bottom of the cylinder and the bed of the flume is set 0.01 m for all cases examined.

The boundary conditions were set as follows: the inlet of fluid was defined as velocity inlet, and the velocity was 1.0 m/s; the bed of the flume and the cylinder were defined as wall; the outlet of fluid was set to be pressure outlet; the free surfaces were set to be pressure inlet. In addition, the operation pressure was defined as standard atmospheric pressure, i.e. 101,325 Pa. The reference pressure location was at the top right corner of the flume, and the operation density at this point was 1.225 kg/m^3, the density of air in standard conditions. The turbulent kinetic energy and turbulent dissipation rate at each boundary were calculated by empirical formula. Triangular elements were chosen to mesh the model (Fig. 3) and the total domain was meshed into 39,000 cells. The software Fluent was used in the simulations. Unsteady simulations were conducted until converged. The time step was 0.01 s, and the total simulation time was 300 s.

The governing equation of incompressible flow is the unsteady Reynolds-averaged Navier-Stokes equations, expressed in a Cartesian tensor form as

$$\frac{\partial(\rho U_j)}{\partial X_j} = 0 \qquad (1)$$

$$\frac{\partial(\rho U_i)}{\partial t} + \frac{\partial(\rho U_i U_j)}{\partial X_j} = -\frac{\partial P}{\partial X_i} + \frac{\partial}{\partial X_j}\left(\mu\frac{\partial U_i}{\partial X_j} - \rho\overline{u_i u_j}\right) \qquad (2)$$

in which U = mean velocity; u = fluctuating velocity; X_i = coordinates, for 2D problems, i = 1, 2; $-\rho\overline{u_i u_j}$ = Reynolds stress; ρ = fluid density; $\mu = C_\mu \rho k^2/\varepsilon$ = coefficient of kinematic viscosity, k = turbulent kinetic energy, and ε = turbulent dissipation rate. In the present study, the standard k-ε model (Launder & Spalding, 1974) was used. k-ε model has been tested and verified in many simulations, and the results were satisfying. The transportation equations of k and ε are

$$\frac{\partial(\rho k)}{\partial t} + \frac{\partial(\rho U_j k)}{\partial X_j} = \frac{\partial}{\partial X_j}\left[\left(\mu + \frac{\mu_t}{\sigma_k}\right)\frac{\partial k}{\partial X_j}\right] + G - \rho\varepsilon \qquad (3)$$

and

$$\frac{\partial(\rho\varepsilon)}{\partial t} + \frac{\partial(\rho U_j \varepsilon)}{\partial X_j} = \frac{\partial}{\partial X_j}\left[\left(\mu + \frac{\mu_t}{\sigma_\varepsilon}\right)\frac{\partial \varepsilon}{\partial X_j}\right] + C_{\varepsilon 1}\varepsilon G/k - C_{\varepsilon 2}\rho\varepsilon^2/k \qquad (4)$$

The coefficients were set as default, i.e. $C_\mu = 0.09$, $\sigma_k = 1.0$, $\sigma_\varepsilon = 1.3$, $C_{\varepsilon 1} = 1.44$, $C_{\varepsilon 2} = 1.92$.

A finite volume method (FVM) was introduced to solve the governing equations in this mathematic model. The Volume of Fluid (VOF) model was used to trace the interface of water and air. The pressure-based segregated algorithm was used to solve the governing equations. A first-order upwind interpolation was used for the discretization of momentum, turbulent kinetic energy and turbulent dissipation rate, and the PRESTO! scheme for pressure.

3 RESULTS AND ANALYSIS

A 2D Cartesian coordinate system in the simulation was defined as x = streamwise, z = bed-normal (see Fig. 2). The z axis goes through the center of the cylinder and the origin of z axis is on the bed of flume. Due to the installation of a cylinder in the bottom flow, some changes of the flow could be represented such as water surface transition, vortexes around the cylinder, shear stress and hydrodynamic pressure changes on the bed.

3.1 *The interface of water and air*

Figure 4 illustrates the interfaces of water and air with respect to the same gap size and different flow depths. Due to the choking of flow and weir flow at the cylinder, the water level rose on the upstream side of the cylinder and fell on the downstream side. The profiles in Figure 4 also show that the drawdown curve is asymmetry about the line of $x = 0$. The center of drawdown curve is at somewhere downstream of the $x = 0$ line, which agrees with the experimental result of Chiew (1991). Figure 5 shows the same results by introducing z/z_0 as a non-dimensional parameter. Figure 4 and 5 indicates that the decreased flow depth leads to a steeper water surface profile, which means that the effect of flow depth becomes greater in shallower flows.

3.2 *Velocity field*

Figure 6 shows the contour plot of velocity magnitude when $z_0 = 0.32$ m. The contour plots of other flow depth are quite similar. As the coming flow approaches the cylinder, it separate into two

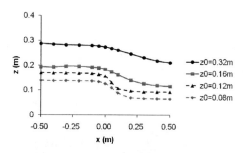
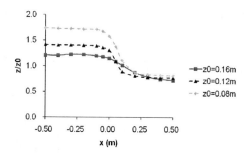

Figure 4. Water surface profile with respect to different flow depths.

Figure 5. Water surface profile at the cylinder with z/z_0.

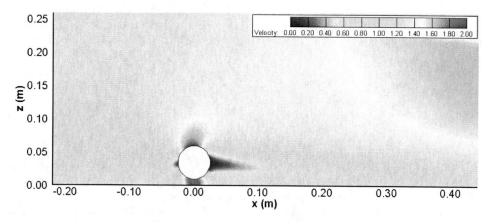

Figure 6. Velocity magnitude contour when $z_0 = 0.32$ m (m/s).

parts. One goes upward, climbing over the crest of the cylinder, while the other goes downward, running through the gap between the cylinder and the bed. Both pieces of flow accelerate, and have similar speed. Figure 6 also indicates that the flow velocity downstream of the cylinder climbs up. This phenomenon could be attributed to the flow depth decrease due to the drawdown phenomenon which is mentioned in the previous section.

The high flow beneath the marine pipeline is an important hydrodynamic force for the expansion of local scour hole under the pipelines. The results show the gap flow velocity beneath the cylinder changing with open flow depth. The maximum gap flow velocity is 2.17 m/s, 1.72 m/s, 1.58 m/s and 1.45 m/s, respectively corresponding to the flow depth of 0.32 m, 0.16 m, 0.12 m and 0.08 m. It can be concluded that the gap flow velocity increases with the increase of flow depth under shallow water conditions.

3.3 *Shear stress*

It is of great importance to quantify the bed shear stress around a horizontal cylinder since it is directly associated with the local scour of marine pipelines. Local scour underneath the pipeline may lead to the vibration due to hydrodynamic forces, which may further lead to the failure of the pipeline. Curves in Figure 7 show the bed shear stress with respect to different flow depths and identify an obvious tendency of an increase of shear stress with the increase of flow depth around the cylinder. It is easy to observe a sharp increase of shear stress when the upstream flow approaches the cylinder and the shear stress rose more than three times within distance of several centimeters on the upstream side of the cylinder. This sharp increase may be attributed to the

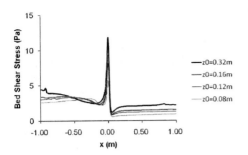

Figure 7. Bed shear stress distribution with z_0 as a third parameter.

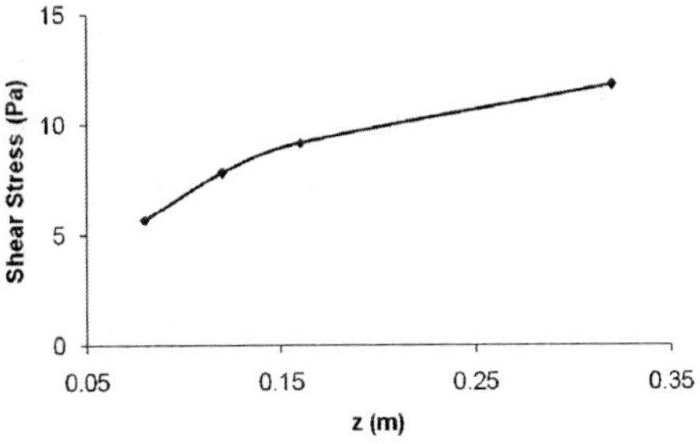

Figure 8. Peak shear stress of different flow depths.

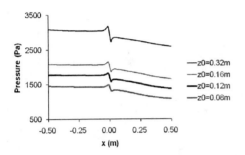
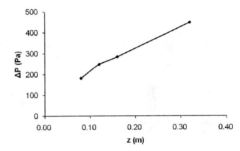

Figure 9. Bed pressure distribution with z_0 as a third parameter.

Figure 10. Pressure difference on two sides of the cylinder corresponding to different flow depths.

increase of velocity. High shear stress values may lead to the movement of sediments, and thus cause local scour of the pipelines.

A closer look into the peak shear stress of four different flow depths is shown in Figure 8. The result curve indicates that the peak value gradually increases with the flow depth increase, but the increasing rate becomes lower. There is a reasonable explanation for this phenomenon: there may exist a critical flow depth when the gap size is constant, that is, the peak shear stress value keeps constant with the flow depth increase when the flow depth is larger than the critical value.

3.4 Pressure on the bed

Figure 9 shows the pressure distribution on the bed of flume with respect to different flow depths. The curves indicate that the pressure on the flume bed decreases with the decrement of flow depth. All four curves show a gentle downward slope. This may be attributed to two facts. On one hand, the drawdown of water surface mentioned above will certainly cause a loss on the static water pressure; on the other hand, the wall boundaries are hydraulically rough, which also lead to loss in dynamic pressure.

From Figure 9 it is quite apparent that bed pressure fluctuates at the location of the cylinder, i.e. a high pressure upstream of the cylinder and a low pressure downstream. The difference of pressure on the two sides of the cylinder may lead to piping effect in the bed soil. In fact, the high pressure gradient when the gap size is zero is an important fact in the onset of local scour beneath the marine pipelines. Chiew (1990) linked the onset of the pipeline local scour straightly to the seepage in the sea bed. The simulation results in Figure 9 shows that the pressure difference still exists when gap size is not zero. That is to say, the piping effect may also happen after the onset of scour. Figure 10 shows the pressure difference on two sides of the cylinder. It is clear that the pressure difference drops down with the decrease of flow depth.

4 CONCLUSIONS

The following conclusions can be drawn based on the numerical results above:

(1) The water surface curves indicate the abrupt change of flow surface around the cylinder, backup at the upstream side and drawdown at the downstream side, due to the choking of flow and weir flow at the cylinder. The center of drawdown curve is somewhere downstream of the center of cylinder, which agrees with the experimental results of a previous study. With the decrement of flow depth, the drawdown effect becomes more and more obvious, and the drawdown curve becomes much steeper.

(2) Flow velocity climbs up downstream of the cylinder, due to the backup and drawdown effect. The velocity both beneath and over the cylinder rises up magnificently compared with the inlet velocity.

(3) Bed shear stress underneath the cylinder rockets up compared with undisturbed areas. When the flow depth goes up, the maximum shear stress generally goes up. However, the increment rate gradually drops down, showing the existence of a critical value for peak bed shear stress. When the flow depth is over the critical value, the peak shear stress no longer climbs up.

(4) The pressure on the bed climbs up with the flow depth. The pressure shows a high peak and a low peak on two sides of the cylinder. The pressure difference gradually decreases with the decrement of flow depth.

ACKNOWLEDGMENTS

This work was financially supported by the National Natural Science Foundation (No. 11172213) and the Fundamental Research Funds for the Central Universities (No. 20123106).

REFERENCES

Bearman, P. & Zdravkovich, M. 1978. Flow around a circular cylinder near a plane boundary. *Journal of Fluid Mechanics*, 89, 33–47.

Chiew, Y. 1990. Mechanics of Local Scour around Submarine Pipelines. *Journal of Hydraulic Engineering*, 116, 515–529.

Chiew, Y. 1991. Flow around Horizontal Circular Cylinder in Shallow Flows. *Journal of Waterway, Port, Coastal, and Ocean Engineering*, 117, 120–135.

Jensen, B. L., Sumer, B. M., Jensen, H. R. & Fredsøe, J. 1990. Flow Around and Forces on a Pipeline Near a Scoured Bed in Steady Current. *Journal of Offshore Mechanics and Arctic Engineering*, 112, 206–213.

Launder, B. E. & Spalding, D. 1974. The numerical computation of turbulent flows. *Computer methods in applied mechanics and engineering*, 3, 269–289.

Liang, D. & Cheng, L. 2005. Numerical modeling of flow and scour below a pipeline in currents: Part I. Flow simulation. *Coastal Engineering*, 52, 25–42.

Liang, D., Cheng, L. & Li, F. 2005. Numerical modeling of flow and scour below a pipeline in currents: Part II. Scour simulation. *Coastal engineering*, 52, 43–62.

Lu, X., Dalton, C. & Zhang, J. 1997. Application of large eddy simulation to an oscillating flow past a circular cylinder. *Journal of fluids engineering*, 119, 519–525.

Mao, Y. Seabed scour under pipelines. Proceedings of 7th International Symposium on Offshore Mechanics and Arctic Engineering (OMAE), Houston, 1988. 33–38.

Smith, H. & Foster, D. 2005. Modeling of Flow around a Cylinder over a Scoured Bed. *Journal of Waterway, Port, Coastal, and Ocean Engineering*, 131, 14–24.

Wu, Y. & Chiew, Y. 2013. Mechanics of Three-Dimensional Pipeline Scour in Unidirectional Steady Current. *Journal of Pipeline Systems Engineering and Practice*, 4, 3–10.

Zhang, J. & Dalton, C. 1999. The onset of three-dimensionality in an oscillating flow past a fixed circular cylinder. *International journal for numerical methods in fluids*, 30, 19–42.

Choice of azimuth axis and stability analysis of surrounding rock of Lianghekou underground powerhouse

Z.Z. Wang, J.H. Zhang & H.Y. Zeng
State Key Laboratory of Hydraulics and Mountain River Development and Protection, College of Water Resources & Hydropower, Sichuan University, Chengdu, Sichuan, China

C.G. Liao & D.Q. Hou
Chengdu Investigation and Design Institute, China Hydropower Engineering Consultation Group Company, Chengdu, Sichuan, China

ABSTRACT: Lianghekou hydropower station locate in deep valley with slopes up to 500 m~1000 m. The geo-stress measurement in range of PD12 foot-rill in the underground powerhouse show that the maximum principal stress value is about 21.57~30.44 MPa, obvious peeling phenomenon is found on the cave wall of $T_3lh^{1(5)}$ sandstone. The azimuth axis of the power house cavern must be chosen in consideration of both hydraulic characteristics of passageway and stability of surrounding rock. Three-dimensional nonlinear finite element analysis are carried out for both axis N3°E and N7°W respectively, then the comparison of the stress-strain, the stability of surrounding rock and supporting parameters in the two different axes are made. As the analysis shows, compared with N7°W axis scheme, the N3°E axis plan avoids stress release better, which significantly limits the deformation and the development of plastic zone. So the N3°E solution is recommended.

1 INTRODUCTION

The Lianghekou hydropower station locates in Yalong River in Sichuan province. The horizontal bury depth of the underground powerhouse is about 368 m, while vertical bury depth about 439 m. And the elevation of the underground powerhouse arch is 2635 m, the floor is at El.2587 m. Span of the powerhouse is 29.2 m; height of the machine hall is 63.8 m. According to the data of PD12 adit and above PD12, the lithological formation of the powerhouse is meta-sandstone, meta-morphic siltstone and silty slate, belong to $T_3lh^{1(3)}$—$T_3lh^{1(5)}$ and $T_3lh^{2(1)}$—$T_3lh^{2(3)}$. Rock mass of the powerhouse is slightly weathered, and hard rocks, saturated uni-axial compressive strength is 60~100 MPa, rock acoustic wave velocity is generally greater than 5000 m/s. Obvious peeling phenomenon is found on the cave wall of $T_3lh^{1(5)}$ sandstone and blocky structure can be seen in σ_{PD12up} stress release hole during the construction of adit above PD12. Seven groups of geo-stress measurement are made within the scope of PD12 adit, which show maximum principal stress ranges from 21.57 to 30.44 MPa, belonging to high stress area. Difference of direction ranges from N34.7~57.7°E. Maximum σ_1 is 30.44 MPa, azimuth angle is 35.5°, dip angle is −10.6°. The average value of rock's saturated uni-axial compressive strength is 80MPa, integrity coefficient of the rock mass is 0.65, strength-stress ratio of surrounding rock is 2.08.

As PD12 adit reveals, the powerhouse does not have regional faults, and its main structural planes are small faults and fissures. Small faults mainly developed in $T_3lh^{1(3)}$ and $T_3lh^{1(4)}$, basically identical with the attitude of stratum, width of fracture zone is small, range from a few centimeters to 50 centimeters; most of fillings in fault is schistose, while carbonization and mylonitization is severe. Statistics of fissures show that there are four dominant groups: ①N60~70°W/SW ∠60°~70°; ⑥-1N0~30°E/SE ∠10~30°; ②-1N0~30° E/SE∠40~60°; ④N0~30°E/SE (NW) ∠70~90°. Both of

① and ⑥-1 fissures are developed relatively. Surface of fissure is fresh, rough and closed with no filling.

Groundwater of the powerhouse is not active, water content of rock mass is not abundant, and water permeability is weak. Wall of the PD12 adit is more wet, part of seepage flows along the fissures. Part of bedrock fissure water is confined aquifer.

Because of the influence of terrain and geological conditions, the azimuth axis of the powerhouse cavern must be chosen in consideration of both hydraulic characteristics of passageway and stability of surrounding rock. Three-dimensional nonlinear finite element is carried out for axis N3°E and N7°W respectively, then the comparison of the stress-strain, the stability of surrounding rock and supporting parameters in the two different axes programs are made.

2 THREE-DIMENSIONAL FINITE ELEMENT ANALYSIS

3D nonlinear finite element analysis program NASGEWIN was used in this paper. Zhang jianhai who is professor of geotechnical engineering institute in Sichuan University has perfected this program by many years. Serial registration number of the computer software copyright is 2009SR027603. The program has convenience user interface and visual interface in pretreatment and post-processing. NASGEWIN has successfully applied in the national research project and more than 20 important research projects such as Ertan, Jinping-1, Shapai, Zipingu, Guandi, Xiluodu, Xiaowan, Baise, Pubugou, Huangjinping, and Houziyan. It has obtained good application effect, and continuously be improved in the engineering practice. Now, it has powerful function for engineering problems such as deformation and stability analysis of underground power house.

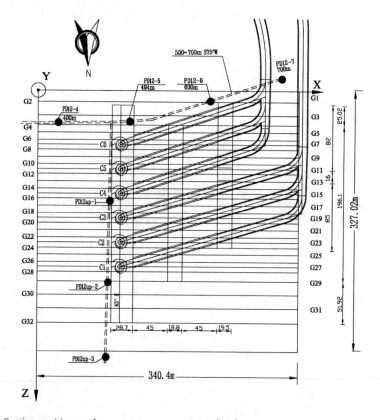

Figure 1. Section positions and geo-stress measurement points layout.

2.1 Calculation range and finite element model

As shown in Fig. 1, the X direction of computational model points from main powerhouse to tailrace surge chamber, with total length being 340.4 m; the Y direction is vertical from bottom elevation 2500.0 m to mountain surface. The Z direction is from unit 6# to 1#, with a total length of 327.02 m. Fig. 1 also shows geo-stress measurement points. The geological characteristics that rock interface, terrain, fault and excavation of the 3 chambers were simulated accurately in 3D modeling. The finite element model has 199631 nodes and 167442 elements in total. The model is consists of 134167 nodes and 134383 element for solid element, and consists of 65464 nodes and 33059 element for anchor rod and anchor rope. Fig. 2 shows the 3D grid of the research area and excavation zone, Fig. 3 shows the arrangement of 3D anchor rod and anchor rope.

2.2 Physical and mechanical parameters

Table 1 shows the physical and mechanical parameters of surrounding rock in 3D nonlinear finite element analysis.

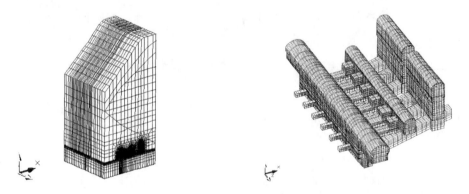

Figure 2. 3D grid of Lianghekou underground powerhouse.

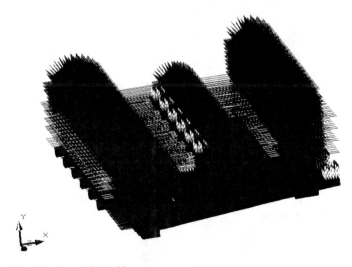

Figure 3. 3D anchor rod and anchor cable arrangement.

Table 1. Rock mass physical mechanics parameter.

Code	Modulus of deformation E(GPa)	Poisson's ratio μ	Unit weight (t/m^3)	Shearing strength	
				Internal friction coefficient f'	Cohesion C' (MPa)
III$_1$	10	0.25	2.72	1.1	1.25
III$_2$	6.5	0.28	2.70	0.9	0.85
IV	3.5	0.31	2.675	0.70	0.50
V	1.75	0.38	2.625	0.50	0.12
f	1.50	0.35	1.80	0.35	0.07
g	3.5	0.31	2.50	0.70	0.50

2.3 Research scheme

(1) Based on the measured in-situ stress date, 3D ground stress in the powerhouse was calculated through multiple argument regression algorithm.
(2) For the two axes under study, the entire chamber group consists of 11-stage excavation, and reinforcement effect is considered in analytical calculation.

2.4 Boundary constraint

In the research scope of the foundation, the method of normal bar chain constraint is used around the boundary surface; and the slope surface is left as free boundary.

3 CALCULATION RESULTS AND ANALYSIS

3.1 Results and discussion of geo-stress regression

Using multiple argument regression algorithms, regressed geo-stress state of the two axes (measuring point arrangement is shown in Fig. 1) are shown in Table 2 and Table 3.
By the results, the geo-stress has the following characteristics:

(1) Geo-stress regression correlation coefficient of the N3°E scheme is 0.9454 and the N7°W scheme is 0.9466. It shows that the regressed geo-stress fields are quite acceptable.
(2) The maximum value of first principal stress in each measuring point is fluctuating between 21.57 MPa to 30.44 MPa, generally increase with burial depth increasing. The fluctuation of first principal stress value in main powerhouse is small, from 23.61 MPa of σPD12-5 to 24.1 MPa of σPD12up-2. Components stress are compared between N3°E scheme and N7°W scheme, it shows horizontal component stress σx of N3°E scheme is less than N7°W scheme. Obviously, although N7°W is only 10 degrees away from N3°E, the release of geo-stress is increased by nearly 1/4, stress increment is from 1.8 MPa to 3.7 MPa. In N3°E scheme, the release stress σx is greater than axial stress σz only in point σPD12-7; but in N7°W scheme, there are four points. It shows that compared with N3°E scheme, N7°W scheme has not avoided geo-stress release very well.
(3) Compared with the normal stress, components of shear stress of measured geo-stress is smaller.
(4) Regularity of transverse and longitudinal measured values of stress in point σPD12-4 is different significantly with others. Errors of fitting in point σPD12-7 and σPD12up-3 are greater. The fitting geo-stress field of other points in two schemes retained the basic characteristics of measured geo-stress field.

Table 2. Measured value and fitted value of geo-stress (N3°E scheme, stress is positive to pressure).

Point		Principal (MPa)			Component of stress (MPa)					
		σ_1	σ_2	σ_3	σ_x	σ_y	σ_z	τ_{yz}	τ_{zx}	τ_{xy}
σ_{PD12-4}	measured	21.57	10.16	5.67	8.02	9.10	20.28	0.14	−4.15	1.96
	fitting	21.14	10.02	5.69	8.50	8.90	19.45	−0.14	−4.53	2.1
	error	0.43	0.14	−0.02	−0.48	0.20	0.82	0.28	0.37	−0.13
σ_{PD12-5}	measured	23.61	13.4	7.5	16.82	10.45	17.25	−2.61	−4.2	5.24
	fitting	23.87	12.93	7.45	16.73	10.33	17.19	−2.63	−4.64	5.16
	error	−0.26	0.47	0.05	0.08	0.11	0.06	0.01	0.43	0.07
σ_{PD12-6}	measured	24.56	12.39	7.57	15.62	10.14	18.77	−0.79	−6.56	3.71
	fitting	25.13	12.27	7.62	16.01	10.18	18.82	−0.9	−6.92	3.76
	error	−0.57	0.12	−0.05	−0.39	−0.04	−0.05	0.1	0.35	−0.05
σ_{PD12-7}	measured	23.9	12.03	8.78	18.19	11.9	14.62	0.73	−6.31	−3.59
	fitting	25.15	12.7	10.31	18.05	12.36	17.75	−0.39	−7.24	−0.79
	error	−1.25	−0.67	−1.53	0.14	−0.46	−3.13	1.13	0.92	−2.8
$\sigma_{PD12up-1}$	measured	25.17	10.3	5.61	13.35	11.65	16.09	−3.7	−8.93	2.39
	fitting	25.35	10.13	5.41	13.47	11.49	15.92	−3.68	−9.15	2.5
	error	−0.18	0.17	0.2	−0.12	0.15	0.16	−0.01	0.22	−0.11
$\sigma_{PD12up-2}$	measured	24.1	16.08	5.93	10.77	15.57	19.78	−0.57	−6.64	4.25
	fitting	24.33	15.27	5.98	11.13	14.94	19.52	−0.79	−7.01	4.08
	error	−0.23	0.81	−0.05	−0.35	0.63	0.26	0.22	0.37	0.16
$\sigma_{PD12up-3}$	measured	30.44	12.07	5.06	14.68	9.95	22.94	−5.00	−10.13	−0.83
	fitting	28.49	10.8	6.20	13.98	10.11	21.4	−4.10	−9.41	0.2
	error	−1.95	1.27	−1.14	0.70	−0.15	1.54	−0.89	−0.71	−1.04

Table 3. Measured value and fitted value of geo-stress (N7°W scheme, stress is positive to pressure).

Point		Principal (MPa)			Component of stress (MPa)					
		σ_1	σ_2	σ_3	σ_x	σ_y	σ_z	τ_{yz}	τ_{zx}	τ_{xy}
σ_{PD12-4}	measured	21.56	10.16	5.67	9.81	9.1	18.49	0.48	−5.99	1.91
	fitting	21.2	9.99	5.68	10.34	8.98	17.63	0.21	−6.15	2.07
	error	0.36	0.17	−0.01	−0.52	0.2	0.85	0.26	0.15	−0.16
σ_{PD12-5}	measured	23.62	13.41	7.5	18.27	10.45	15.8	−1.67	−4.02	5.61
	fitting	23.83	12.95	7.47	18.22	10.33	15.7	−1.67	−4.49	5.49
	error	−0.21	0.46	0.03	0.04	0.11	0.1	0	0.46	0.12
σ_{PD12-6}	measured	24.55	12.4	7.58	17.96	10.14	16.43	−0.14	−6.7	3.79
	fitting	25.11	12.26	7.62	18.34	10.18	16.47	−0.23	−7.01	3.82
	error	−0.56	0.14	−0.04	−0.37	−0.04	−0.04	0.09	0.3	−0.02
σ_{PD12-7}	measured	23.9	12.03	8.77	20.24	11.9	12.57	0.1	−5.32	−3.67
	fitting	25.1	12.68	10.3	20.36	12.34	15.37	−0.52	−6.78	−0.7
	error	−1.2	−0.65	−1.53	−0.12	−0.44	−2.8	0.62	1.45	−2.96
$\sigma_{PD12up-1}$	measured	25.16	10.3	5.62	16.49	11.65	12.95	−3.22	−8.85	3
	fitting	25.34	10.13	5.44	16.58	11.47	12.86	−3.17	−9.03	3.08
	error	−0.18	0.17	0.18	−0.08	0.17	0.08	−0.04	0.17	−0.08
$\sigma_{PD12up-2}$	measured	24.1	16.08	5.94	13.31	15.57	17.24	0.17	−7.78	4.29
	fitting	24.36	15.25	5.99	13.73	14.91	16.95	−0.09	−8.04	4.13
	error	−0.26	0.83	−0.05	−0.42	0.65	0.28	0.26	0.25	0.15
$\sigma_{PD12up-3}$	measured	30.44	12.07	5.06	18.40	9.95	19.23	−5.07	−10.93	0.04
	fitting	28.47	10.82	6.24	17.34	10.17	18.03	−3.97	−10.1	0.9
	error	1.97	1.25	−1.18	1.06	−0.21	1.19	−1.09	−0.82	−0.86

Table 4. Calculation results of deformation in 4 # section.

Position		Scheme	
		N3°E	N7°W
Main powerhouse	roof (mm)	10.50	9.38
	upstream arch foot (mm)	43.56	53.66
	upstream crane beam (mm)	101.45	115.19
	middle of upstream sidewall (mm)	106.22	123.30
	downstream arch foot (mm)	20.66	30.93
	downstream crane beam (mm)	64.47	84.00
	middle of downstream sidewall (mm)	83.47	116.79
Main transformer chamber	roof (mm)	16.14	18.06
	middle of upstream sidewall (mm)	14.27	11.76
	middle of downstream sidewall (mm)	5.39	5.56
Tailrace surge chamber	roof (mm)	15.85	16.85
	middle of upstream sidewall (mm)	67.39	76.84
	middle of downstream sidewall (mm)	71.78	83.07

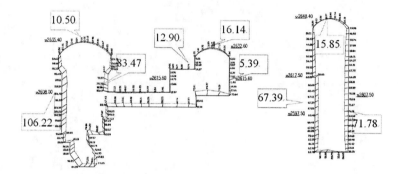

Figure 4. Deformation around the tunnel of N3°E scheme.

3.2 Results and discussion of excavation and support

On the basis of geo-stress regression, excavation process on N3°E and N7°W axes scheme are simulated step by step. Deformation, stress and plastic zone of surrounding rock are calculated and analyzed in reinforced condition.

For simplicity, the 4# unit was selected as representative section for comparison of calculation results. Table 4 to Table 7 list deformations around the tunnel, anchor rod and anchor cable stress, depth of plastic zone in 4# unit.

(1) According to Table 4, after reinforcement, surrounding rock deformation of N3°E is smaller than N7°W. For example, the deformation of upstream sidewall in N3°E scheme is 106.22 mm, which is smaller than 123.30 mm of N7°W scheme, with a 17.08 mm decrease; the deformation value of downstream sidewall in N3°E scheme is 83.47 mm, smaller than 116.79 mm of N7°W scheme, with a 33.32 mm decrease, as shown in Fig. 4 and Fig. 5.

(2) According to Table 5, in general, the tensile and compressive stress of anchor rod on host in N3°E scheme is smaller or close to N7°W scheme, extreme value appeared on the middle of sidewall both upstream and downstream, the values reached 357.0 MPa and 388.74 MPa in N3°E scheme, and in N7°W scheme are 335.54 MPa and 388.91 MPa.

(3) According to Table 6, in a general way, internal force of anchor rope in N3°E scheme is smaller than N7°W scheme. For example, the force of downstream sidewall on host in N3°E scheme is

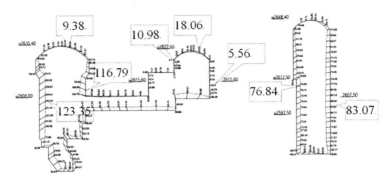

Figure 5. Deformation around the tunnel of N7°W scheme.

Table 5. Calculation results of anchor rod stress in 4 # section.

Position		Scheme	
		N3°E	N7°W
main powerhouse	roof (MPa)	−2.28	11.07
	upstream arch foot (MPa)	−2.43	11.38
	upstream crane beam (MPa)	237.02	212.08
	middle of upstream sidewall (MPa)	357.0	335.54
	downstream arch foot (MPa)	113.32	134.45
	downstream crane beam (MPa)	329.15	365.20
	middle of downstream sidewall (MPa)	388.74	388.91
main transformer chamber	roof (MPa)	60.10	102.75
	middle of upstream sidewall (MPa)	143.13	125.35
	middle of downstream sidewall (MPa)	161.59	155.42
tailrace surge chamber	roof (MPa)	29.46	−4.64
	middle of upstream sidewall (MPa)	176.94	218.11
	middle of downstream sidewall (MPa)	234.06	271.30

Table 6. Calculation results of anchor rope stress in 4 # section.

Position		Scheme		Remark
		N3°E	N7°W	
main powerhouse	upstream sidewall (kN)	2370.10	2345.86	design force:2000t
	downstream sidewall (kN)	2793.82	3019.50	design force:2000t
main transformer chamber	upstream sidewall (kN)	2302.36	2370.39	design force:2000t
	downstream sidewall (kN)	2285.53	2341.45	design force:2000t
tailrace surge chamber	upstream sidewall (kN)	1794.98	1818.05	design force:1750t
	downstream sidewall (kN)	1868.69	1869.24	design force:2000t

2793.82 kN, and N7°W scheme is 3019.50 kN. By contrast, N3°E scheme could reduce anchor rope internal force by 8.07%.

(4) According to Table 7, depth of plastic zone in N7°W scheme is larger than N3°E scheme, because its stress release and deformation of surrounding rock are larger. The depth of plastic zone in the middle of upstream sidewall on host in N3°E scheme is 18.12 m, which is smaller than 18.37 m in N7°W scheme, with a 0.25 m decrease. The depth of plastic zone in the middle of downstream sidewall on transformer chamber in N3°E scheme is 7.12 m, which is smaller

Table 7. Calculation results of the plastic zone.

Position			Scheme N3°E	Scheme N7°W	Remark
depth of plastic zone	main powerhouse	roof (m)	9.29	10.61	throughout bus bar hole
		upstream arch foot (m)	2.46	2.46	
		upstream crane beam (m)	15.06	15.51	
		middle of upstream sidewall (m)	18.12	18.37	
		downstream arch foot (m)	2.30	2.30	
		downstream crane beam (m)	33.45	33.45	
		middle of downstream sidewall (m)	13.93	14.05	
	main transformer chamber	roof (m)	6.73	8.53	
		middle of upstream sidewall (m)	12.19	12.20	
		middle of downstream sidewall (m)	7.12	7.81	
	tailrace surge chamber	roof (m)	8.67	8.78	
		middle of upstream sidewall (m)	9.46	10.17	
		middle of downstream sidewall (m)	9.88	10.48	
plastic volume	overall underground caverns	$(1.0 \times 10^4 \text{ m}^3)$	123.3	216.1	the three chambers

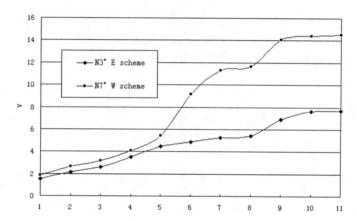

Figure 6. Plastic volume of per unit width with 11 excavation stage.

than 7.81 m in N7°W scheme, with a 0.69 m decrease. The depth of plastic zone in the middle of upstream on tailrace surge chamber in N3°E scheme is 9.46 m, which is smaller than 10.17 m in N7°W scheme, with a 0.71 m decrease. Plastic volume of N3°E scheme is 123.3 × 104 m³, which is smaller than 216.1 × 104 m³ of N7°W scheme, with a 75.2% decrease.

As Fig. 6 shows, in all different excavation stage, the plastic volume of per unit width in N7°W scheme greater than N3°E scheme, and with proceeding of excavation, the differences of plastic volume trend to increasing gradually. The volume of first excavation on 4# unit in N7°W scheme greater 26.41% than N3°E scheme, to the end of excavation, it more than 82.94%. The differences of plastic zone are obvious in machine fossa area, draft tube and the middle of downstream sidewall on host.

4 CONCLUSIONS AND SUGGESTIONS

Because of Lianghekou underground station locates in high geo-stress areas, additionally, for the specific geological conditions of surrounding rock and the limitation of water facilities, two axes schemes (N3°E and N7°W) are compared and analyzed. The results show:

(1) Geo-stress regression analysis shows that release stress of sidewall on host in N3°E scheme is decreased by 8.6%~25.34% than N7°W scheme, or stress decrease by 1.8 MPa~3.7 MPa.
(2) Maximum deformation of upstream sidewall on host in N3°E scheme is decreased by about 1.0 cm than N7°W scheme, whereas the displacement of downstream sidewall is decrease by about 4.0 cm, internal force of anchor rope is decreased by 8.0%.
(3) Plastic volume of per unit width in N3°E scheme is smaller than N7°W scheme for all units. In earlier stage of excavation, differences of plastic volume in two schemes are small. In later stage, plastic volumes develop quickly on machine fossa area and draft tube in N7°W scheme, and the differences of two schemes reach maximum after the excavation finished. It shows that the N3°E scheme observably limit the deformation and development of plastic zone, especially in machine fossa area and around the draft tube.
(4) According to the above statement, the N3°E solution is recommended.
(5) Because the limits of tail-water facilities, axis of underground powerhouse is difficult to achieve optimal direction to avoid release of high geo-stress. It brings about some problems such as depth of damage zone on high sidewall is larger. And there is partially penetration destruction of rock pillar between main powerhouse and main transformer chamber. And serious damage to the lower part of tailrace surge chamber is also found. So it is necessary to adopt pre-stressed anchorage rope and other measures in design for reinforcement between main powerhouse and main transformer chamber. Penetration plastic destruction appears in rock plates between bus bar hole and tailrace tunnel in many zones, so additional support is propose for strengthening.

REFERENCES

Chen, xiutong & Li, lu. 2008. Stability analysis of surrounding rock of large underground powerhouse cavern group. Chinese Journal of Rock Mechanics and Engineering. 27(supp. 1): 2866–2872.

Gao, wei & Zheng, yinren. 2000. Development of research on uncertain back analysis in geotechnical engineering. Underground Space. 20 (2): 81–85.

Han, rongrong & Zhang, jianhai & Zhang, xiao & etal. 2010. Feedback analysis of excavation monitoring to 7th excavation at right bank of Xiluodu underground powerhouse. Chinese Journal of Underground Space and Engineering. 6(3): 552–558.

Han, rongrong & Zhang, Jianhai & Zhang, Xiao & et al. 2008. An improved algorithm of inversion regression analysis for groundstress field. Sichuan Water Resource. (4): 72–75.

Wang, huijuan & Zhang, jianhai & Zhao, xiaofeng & et al. 2011. Study on deformation monitoring feedback and lining deformation stability of the outgoing well. Chinese Journal of Underground Space and Engineering. 7(6):1227–1232.

Zhang, jianhai. & et.al. 2001. Static and dynamic stability assessment of rock slopes and dam foundations using rigid body-spring element method. Int.J.Rock Mechs& Mining Scis. 38(8):1081–1090.

Zhang, jianhai & Hu, zhuxiu & Yang, yongtao & et al. 2011. Acoustic velocity fitting and monitoring feedback analysis of surrounding rock loosing zone in underground powerhouse. Chinese Journal of Rock Mechanics and Engineering. 30(6): 1191–1197.

Zhang, jianhai & He, jiangda. 2000. Dynamic stability analysis of Xiaowan arch dam shoulders YUNNAN WATER POWER. 16 (1):58–61.

Zhang, zhiguo & Xiao, ming. 2009. Inversion of monitored displacement field and evaluation of surrounding rock stability of underground caverns. Journal of Rock Mechanics and Engineering. 28(4): 813–818.

Experimental study on the chloride-permeation resistance of marine concrete mixed with slag and fly ash

Subin Cheng
Collage of Civil Engineering, Tongji University, Shanghai, China

Su Cheng
Wuxi Municipal Development Group, Wuxi, China

Chengfei Hu
Collage of Civil Engineering, Tongji University, Shanghai, China

ABSTRACT: Abstract: In the marine environment, destruction early by reinforcement corrosion is the most outstanding disaster of the marine structure, and durability of the concrete structure impairment mainly is the carbonization and the reinforcement corrosion problem of the concrete. This paper explores lots of test to study the influences of combinations of slag and fly ash at different rations on the mechanical properties and durability of high performance marine concrete for ocean structure. The results show that the marine concrete, with a proper mix ratio with admixture as slag and fly ash, can improve its resistance of chloride permeability performance.

1 INTRODUCTION

For marine environment even worse than on land, concrete construction suffer with seawater salts, especially chloride erosion, resulting in corroded, and leaving the premature corrosion of concrete structures damaged or lost durability (Anna, 2009). It is widely accepted that the most outstanding disaster of the marine structure is reinforcement corrosion, and the reinforcement corrosion problem induced by chloride-permeation is critical to the life duration of the concrete engineering (Song, 2008) Thus a serious threat to normal life of marine concrete, therefore, to take effective measures to ensure that marine concrete resistance to chloride ion penetration is extremely important. China also has a lot of marine structure appearing steel corrosion damaged or destroyed.

Therefore, it is of great practical significance in analyzing how to improve the resistance to chlorideion permeability and the durability of marine concrete using the slag and fly ash.

2 THE MECHANISM OF CHLORIDE-PERMEABILITY

The steel corrosion induced by the chloride ion is a complex electrochemical process in marine structure, and it shows that the destructive erosion may take place when the weight of chlorine salt represents up to 0.1%~0.2% of the weight of concrete, which will impact the durability and security of reinforced concrete structure seriously. For chlorine ion being absorbed in the defective place of marine structure while chloride ion concentrating in concrete pore fluid around the reinforced structure reaches a certain degree, it makes the indissoluble hydrogen ferric oxide into a soluble $FeCl_2$ and destroys passivation membrane on the surface of reinforcing bar partly (Liu, 2011).

The main response equations in the process of corrosion are shown as below.

$$Fe \rightarrow Fe^{2+} + 2eFe^{2+} + 2Cl^- + 4H_2O \rightarrow FeCl_2 \cdot 4H_2O \quad (1)$$

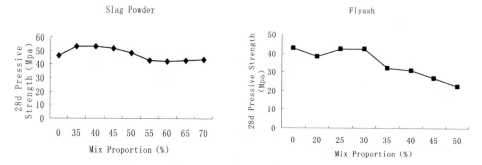

Figure 1. Compressive strength of Marine concrete in the 28 days.

$$FeCl_2 \cdot 4H_2O \rightarrow Fe(OH)_2 \downarrow +2Cl^- + 2H^+ + 2H_2O \quad (2)$$

$$4Fe(OH)_2 \downarrow +O_2 + 2H_2O \rightarrow 4Fe(OH)_3 \downarrow \quad (3)$$

Theoretical the volume of rust material can come to 2~7 times, at least 1.5~3 times, of that of primary steels in oxygen-free environments. The expansion of cracks and the cracking and stripping of concrete cover destroy the structural integrity of marine concrete, that also weaken the reinforced performance which will lead to a very short service life of structure.

A ton of research shows that, mixing a certain amount of admixture such as silane and coal ash will improve the resistance to chloride ion permeability by preventing the penetration path of Cl^- and reduce the penetration speed of Cl^-. But the great proportion only mixes the pulverized coal ash and only mixes the gangue to be possible through the volcanic ash effect production hydration calcium silicate gelatin deposition in the contact surface crevice, enhanced the concrete caking intensity and the impermeability. This paper discusses experimental analysis on the chloride-permeation resistance of marine concrete mixing content as slag and fly ash.

3 CHLORIDE-PERMEATION RESISTANCE TEST OF MARINE CONCRETE MIXED WITH SLAG AND FLY ASH

3.1 Design of experiment

Based on the study above, the experiment plans to test the influence of different admixture as slag and fly ash in the improvement of anti-chloride-permeability performances of marine concrete. Main material contain: the Conch Portland cement of 42.5 level, 5~30 mm continuous gradation coarse aggregate, fine aggregate, S95 Baotian ground fine mineral powder and GGBF, II level Baotian coal cash, FDN naphthalene series high range water reducing agents.

Nine groups test of slag admixture in different proportion and eight groups of concrete with fly ash admixture in different proportion, which are cultivated according to the procedures of marine concrete. After 28 days cultivation, strength test is made for samples. Then the concretes samples are divided into two groups in terms of different admixture of fly ash and slag in different proportion. In every group half of the test cubes are immersed totally and the rest immersed partly. Then, the depth of chloride-permeation in concrete cube is measured by silver nitrate color-developing method.

3.2 Experimental result of the improvement of anti-chloride-permeability performances of concrete caused by admixtures (slag/fly ash)

3.2.1 The analysis of experiment strength

The 28 days strength of concrete test cube with mineral powder and fly ash in different proportion is shown in the figure 1.

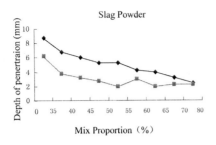 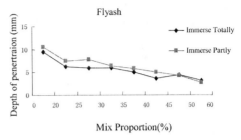

Figure 2. Depth of chloride-permeation in only mixed Slag Powder or fly ash concrete.

According to the compressive strength of concrete 28D., mixed by only slag powder, does not sharply descend. Further, it is increased by the slag powder when blend ratio is between 35% and 50%. In addition, even though the compressive strength is the weakest, with the blend ratio being 60%, it exceeds C40, which is capable for the design requirement of most projects.

The compressive strength of concrete, with fly ash as admixture, drop gently. As the blend ratio is from 25% to 30%, the decrease is not significantly. However, when the blend ratio increases up to 40%, the strength fall slightly, which is under C30. Therefore, in terms of the fly ash as the admixture, it should be cautious to take fly ash with high quality as well as small amount, therefore meeting the strength requirement.

3.2.2 *Experimental result of the improvement of anti-chloride-permeability performances of concrete caused by single/various admixture*
(a) Single admixture (slag/fly ash)

The improvement of anti-chloride-permeability performances of marine concrete is a vital indicator to assess the durability of marine concrete. Different amounts of Slag and fly ash have different impacts on it. In experiment, the concrete cubes are immersed by *Nail* solution with 0.4% concentration and the depth permeation of chloride ion in cube is measured as figure 2.

According to the depth of chloride-permeation, the result of samples immersing partly is deeper than that in the condition of admixture immersing totally no matter what the admixture is slag or fly ash. When the admixture is only slag, the difference is 150%. However, when the admixture is fly ash, the difference is 29%. Due to the influence of wave action, it is imperative to take actions to protect concrete containing corroding reinforcing steel bar at tidal and splash zone.

The marine construction mixed by slag or fly ash would decrease the depth of chloride-permeation. Further, as the proportion of slag and fly ash rise, the depth of chloride-permeation in concrete has downward trend no matter what the condition is immersing partly or totally. As to slag, when the proportion is 70%, the depth of chloride-permeation of concrete would decrease to 25%~30% of normality in the condition of immersing partly. When the proportion is 50% and 60%, the depth of chloride-permeation of concrete would decrease to 30%~35% of normality in the condition of immersing totally. When it comes to fly ash, when the proportion is 50%, the depth of chloride-permeation of concrete would decrease to 20%~35% of normality in the condition of immersing partly. When the proportion is 50%, the depth of chloride-permeation of concrete would decrease to 30%~35% of normality in the condition of immersing totally.

Taking strength and the depth of chloride-permeation into consideration, submarine engineering works had better to choose the concrete with 50%~70% slag and the concrete with 30% fly ash while engineering works in which water level changes are supposed to choose the concrete with 50%~60% slag and with 30% fly ash.

(b) Mixed admixture
The chloride diffusion effect with different admixture including fly ash and slag is shown in table 1.

Table 1. The chloride diffusion effect with different admixture including fly ash and slag.

Group number	Water-binder ratio	Cement (kg)	Slag Powder (kg)	Fly ash (kg)	Admixture (kg)	Chloride Diffusion Coefficient ($10^{-12} cm^2/s$)
1	0.31	310	90	70	5.64	0.62
2	0.34	225	90	130	4.45	1.08
3	0.35	220	126	74	4.2	2.17
4	0.35	170	140	110	4.2	1.41
5	0.35	420	0	0	4.32	3.24

In group 1, 2 and 3, the coefficient of chloride diffusion in concrete increases with the ratio of water to binder rising up. The reason is that after the cement hydrate, it causes tons of capillary holes in the interior of concrete. As the water to binder ratio grows up, the amount of the capillary hold would increase. The chloride ion could go across the hole to penetrate the interior of concrete, which leads to the anti-chloride-permeability performances of concrete decrease.

From the outcome of group 3, 4 and 5, it is clear that group 3 and 4 use plenty of slag and fly ash instead of cement mix into concrete, which could enhance the anti-chloride-permeability performances of concrete. As the amounts of mineral admixture rise, the anti-chloride-permeability performances of concrete increase, diffusion coefficient of chloride ion would drop. When the admixture is 33% slag and 26% fly ash, diffusion coefficient of chloride ion decease to 56.5%. Therefore, in the construction of ocean engineering and coastal engineering, using the marine concrete mixed by both slag and fly ash could increase the anti-chloride-permeability performances of concrete significantly.

4 CONCLUSION

Slag and fly ash as composite incorporation can refine overall surface particles of cement paste and increase the specific surface area, so the physical adsorption capacity has been significantly enhanced. Thus, the marine concrete mixed with slag and fly ash, whether single or complex admixture, can improve anti-erosion ability to a certain extent. When concrete is mixed by both slag and fly ash, the anti-chloride-permeability performances of concrete would be enhanced markedly. As a result, mixing both some slag and fly ash in the marine concrete could effectively improve the anti-chloride-permeability performances of concrete, therefore lengthening the life duration of the project. The chloride-permeability resistance of concrete works better in the condition of concrete immersing partly than that of concrete immersing totally. Thus, marine concrete with slag and fly ash would be widely applied to improve the anti-chloride-permeability performances for certain submerged marine structure, such as jetty. Taking the strength of marine concrete into consideration, mixing slag would be more realistic and profitable than mixing fly ash. Furthermore, insofar as the permission of projects, the more the amounts of slag is used, the better the effects take.

REFERENCES

K.Y. Anna, J.H. Ahnb, J.S. Ryouc. 2009. The importance of chloride content at the concrete surface in assessing the time to corrosion of steel in concrete structures. *Construction and Building Materials* 23(1): 239–245

Ha-Won Song, Chang-Hong Lee, Ki Yong Ann. 2008. Factors influencing chloride transport in concrete structures exposed to marine environments. *Cement and Concrete Composites* 30(2): 113–121

Liu Bingyun, Zhang Sheng, Li Kai. 2011. The corrosion mechanism and anticorrosion method of marine concrete construction. *Construction & Design for Project*, 1: 88–92

Experiments on coupling effect of upward seepage on slope soil detachment

Xin Liang, Liquan Xie & Yehui Zhu
Department of Hydraulic Engineering, Tongji University, Shanghai, China

ABSTRACT: With respect to general consensus of significant effects of seepage flow on slope erosion and few published experimental results or mature theory to explain the erosion mechanism quantitatively, quantitative experimental studies were initiated by new proposed apparatus in this paper. The model tests based on new experimental techniques of proposed apparatus, were conducted to represent the scouring process of slope soil detachment by shallow flow, revealing the coupling mechanism of seepage on slope erosion. According to the experimental analysis by different combinations of surface flow and seepage flow, performance of the proposed apparatus was confirmed, and the seepage coupling effects on slope erosion were quantified in detail. The results can be introduced into the theories of slope erosion to consider the coupling effect of upward seepage on slope soil detachment, and be applied in the design of erosion protection.

1 INTRODUCTION

Soil particles are susceptible to being detached and carried away with concentrated runoff streams, forming rills and gullies that grow with time. The slope integrity is diminished, eroding the slope face and becoming a larger environmental and safety issue. With respect to the erosion mechanism, water erosion for slope soil usually results from the surface flow on the soil or the seepage in the soil, or both of them. Erosion by surface flow primarily features physical disturbances, such as transportation of slope soil. Erosion by seepage primarily features dispersing and suspension of slope soil.

There has been a lot of studies on both kinds of erosive force, i.e. the surface flow on the soil and the seepage in the soil. To study the former kind of slope erosion, the research interests include statistical analysis on the primary factors causing slope soil's movement, physical experiments and numerical simulations (Abrahams A D, & Parsons G L, 1996, Chaplot V A, M, & Bissonnais Y L, 2003, Flanagan D C, & Nearing M A, 2000, Foster G R, & Huggins L F, & Meyer L D, 1984, Nearing M, & Bradford M, & Parker C, 1991) based on the combination of surface flow erosion and splash erosion from rainfall, theoretical analysis considering the effect of slope length (Kinnell P I A, 2000), vegetation (Braud I, & VichA I J, & Zuluaga J, & Fornero L, & Pedrani A, 2001) and the angle of force on the soil particles. Reports from Dunne (Dunne, T, 1990), Howard and McLane's (Howard, A. D., & McLane, C. F. 1988) indicated that the slope soil was moved away from ground at the slope toe by the upward seepage force, causing slope collapse when the support of slope toe disappeared, or even causing serious retreat of slope.

A new, special equipment was applied to study the coupling effect of upward seepage on slope soil detachment quantitatively in this paper.

2 EXPERIMENTS SETUP

The equipment contains four parts: a soil-bin, a set of water supplying devices, a set of seepage supplying devices and outlet water receiver (see Fig. 1). The soil-bin and the seepage supplying devices are shown in Fig. 2. The seepage supplying devices mainly contains a seepage head

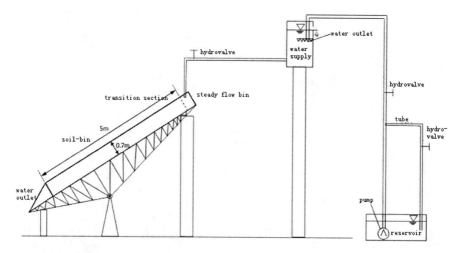

Figure 1. Experimental set-up.

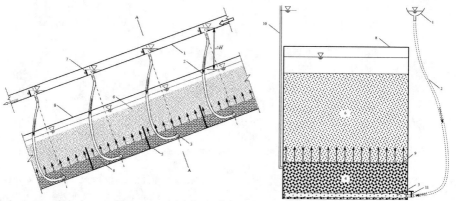

1.the seepage head controlling groove 2.joint-to-joint tubes 3.permeable tubes 4.sand gravel 5.seepage cabin walls 6.sand 7.seepage walls 8.the soil-bin 9.stainless-steel net

Figure 2. Sketch of the uniform seepage supplying devices.

controlling groove, joint-to-joint tubes, permeable tubes, sand gravel and stainless-steel nets. A stainless-steel net is installed between sand gravel and slope soil to prevent the upper soil from dropping into gravel. Coarse sand is used to smooth the surface of sand gravel so that it would be easier to install the stainless-steel net. The soil-bin is a double-cell flume (see Fig. 3) which is divided symmetrically by an organic glass wall. Experiments can be done in the two cells at the same time to contrast the difference of slope erosion directly. The slope soil may be taken away while surface flow running over. The slope at the upper part of 0.5 meter is covered by plastic cloth during the model tests in order to avoid soil subsidence due to the boundary effect between the current stabilization groove and the soil-bin. And it can also make the current more uniform and stabilized. The soil-bin has a height of 30 cm, a width of 30 cm, a length of 180 cm and an inclination angle of 14.7°. It is divided into five parts by seepage cabin walls. Seepage cabin walls have a height of 14 cm. 8 cm of sand gravel is put at the bottom of the soil-bin and 12 cm of soil is put upward. The experiments in this paper would all contrast the effect of seepage on slope erosion directly. Thus, only one side of the soil-bin has seepage. The posticum is on the downside of the soil-bin with the outlet water receiver.

Figure 3. Structure of the soil-bin.

Table 1. Test condition: washing time (s).

	Flow		
Seepage intensity	Group 1: 0.438 L/s	Group 2: 0.370 L/s	Group 3: 0.205 L/s
0.36 with seepage without seepage	64	38	55

3 DATA COLLECTION AND DISCUSSION

Three contrastive groups of experiments were conducted by changing surface flow rate with the same seepage intensity (see Table 1). In the tests, the right cell of the soil-bin was supplied with seepage condition, and the video of scouring process illustrated the obvious difference of soil erosion initiation and development in both of the cells. The erosion initiation was earlier and the erosion intensity was higher for the condition of seepage coupling in the right cell. Meanwhile, small scale of landslide in some area also occurred due to the seepage. Water flow samples were collected at the bottom end of the soil-bin for the calculation of the sediment concentration. Seepage intensity equals the vertical height difference between the water level of the seepage head controlling groove and the level of the surface flow ΔH divided by the soil depth. $J = \Delta H/H$.

Figure 4 contrastively shows terrain photos of the slope after erosion in both of the cells, from which we could see the significant effect of upward seepage on slope soil detachment. Measuring the landform could be used to analyze the coupling of surface flow and seepage effect quantitatively. And the measured results of slope erosion are shown in Figure 5. The figure shows that the slope erosion would be worse and the scour pits would be deeper if seepage was coupled in the water erosion, and erosion intensity differs from 15.1% to 40.8%. The upper part of slope was eroded

Figure 4. Contrast of slope erosion (the left without seepage, the right with seepage: Group 1.

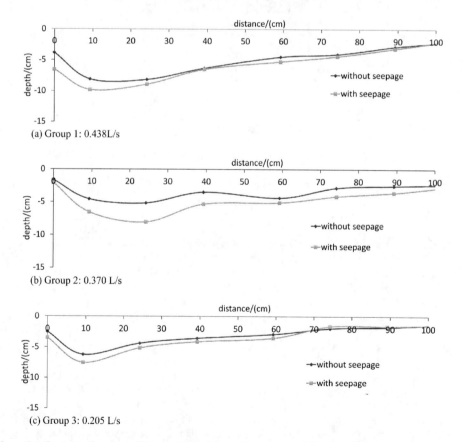

(a) Group 1: 0.438L/s

(b) Group 2: 0.370 L/s

(c) Group 3: 0.205 L/s

Figure 5. Contrast of seepage effect on slope erosion with respect to different surface flow.

more seriously than the lower part, the reason of which is that the flow is clear and has a much high sediment carrying capacity when water enters the soil-bin from the inlet. When water flows down, plenty of sediments were involved in the flow and the sediment carrying capacity of water becomes lower.

4 CONCLUSION

As far as we know, most of researches have been conducted on erosion by either surface flow or seepage. The coupling effect of seepage and surface flow on slope erosion has been confirmed qualitatively, but the quantitative study is so difficult to carry out because of lack of corresponding experimental setup that few quantitative results have been published up to now. Based on new experimental techniques of proposed apparatus, the model tests were carried out to study the scouring process of slope soil detachment by shallow flow, revealing the coupling mechanism of seepage on slope erosion. The results of this study highlight important similarities and differences of slope soil erosion between erosion by surface flows over mobile bed forms and erosion by coupling of surface flows and seepage flows in the slope soil.

The following conclusions can be drawn based on the results above:

(1) The performance of the proposed apparatus and experimental methods was effective to consider the coupling behavior of seepage and surface flow.
(2) The results indicated that the coupling effect will increase the rate of slope erosion (from 15.1% to 40.8%) and has a serious influence on the stability of slope.
(3) The erosion initiation is much earlier for the condition of seepage coupling.

ACKNOWLEDGMENTS

This work was financially supported by the National Natural Science Foundation (No. 11172213) and supported by the Fundamental Research Funds for the Central Universities (No. 20123106).

REFERENCES

Abrahams A D, & Parsons G L. 1996. Rill hydraulics on a semiarid hill slope, southern Arizona [J]. *Earth surface Processes and Landforms*, 21: 35–47.
Braud I, & Vich A I J, & Zuluaga J, & Fornero L, & Pedrani A. 2001. Vegetation influence on runoff and sediment yield in the Andes region: observation and modeling [J]. *Journal of Hydrology*, 254(1–4): 124–144.
Chaplot V A M, & Bissonnais Y L. 2003. Runoff features for inter rill erosion at different rainfall intensities, slope lengths, and gradients in an agricultural loessialhill slope [J]. *Soil Science Society of America Journal*, 67(3): 844–851.
Dunne, T. 1990. Hydrology, mechanics and geomorphic implications or erosion by subsurface flow. p. 11–28. In C.G. Higgins and D.R. Coates (ed.) *Groundwater geomorphology: The role of subsurface water in earth-surface processes and landforms*. Geol. Soc. Am., Boulder, CO.
Flanagan D C, & Nearing M A. 2000. Sediment particle sorting on hill slope profiles in the WEPP model [J]. *Transactions of the ASAE*, 43(3): 573–583.
Foster G R, & Huggins L F, & Meyer L D. 1984. A laboratory study of rill hydraulics I: Velocity relationships [J]. Transactions of ASAE, 27(3): 790–796.
Howard, A. D., & McLane, C. F. 1988. Erosion of cohesionless sediment by groundwater seepage [J]. *Water Resour. Res.*, 24(10), 1659–1674.
Kinnell P I A. 2000. The effect of slope length on sediment concentrations associated with side-slope erosion [J]. *Soil Science society of America Journal*, 64(3): 1004–1008.
Nearing M, & Bradford M, & Parker C. 1991. Soil detachment by shallow flow at low slopes [J]. *Soil Science society of America Journal*, 55(2): 339–344.

Seismic safety evaluation of gravity dam in aftershocks

Shuli Fan & Jianyun Chen
State Key Laboratory of Coastal and Offshore Engineering Dalian University of Technology, Dalian, Liaoning, China

Shaolan Lv
Zhejiang Zheneng Yanshan Construction & Development Co. Ltd., Shanghai, China

ABSTRACT: Taking a RCC gravity dam for example, nonlinear frictional model developed in ABAQUS software was used to simulate the penetrated crack surface under different probability levels of aftershocks. The dynamic response of gravity dam with a penetrated crack at the slope change on the downstream face under aftershock was studied. The residual displacement control algorithm on the dam crest was taken as a criterion to decide whether the dam was stable, and the largest aftershock that the dam could withstand was obtained from the results of numerical analyses. Results also show reservoir water level has great effect on the safety of dam in aftershock. It is necessary to drop water level below the crack location after main shock.

1 INTRODUCTION

For the huge damage to industrial and agricultural production as well as lose of life and property, earthquake-resistant design of dams is particularly important. Although concrete gravity dams are, in general, designed to perform satisfactorily under all kinds of loading, but in lifecycle, hydraulic structure may be attacked by large earthquakes exceeding design standard for earthquake's complexity and indeterminacy. Such as the Koyna dam had suffered serious earthquakes which were stronger than their design earthquakes, and was damaged seriously. These seismic events demonstrate that concrete gravity dams may be damaged seriously in earthquakes and experience a crack at the upper part which passed through the entire thickness at the slope change on the downstream face of gravity dam. At same time, many aftershocks follows a strong main shock, especially near the epicenter, Especially for a homogeneous material and uniform external stress, aftershocks typically begin immediately after main shock (Alliard and Leger, 2008). Sometime aftershock may be more severe than the main shock in one direction such as the Valparaiso earthquake occurred in 1730. These aftershocks cause serious secondary damage to the structures damaged in the main shock and increase the risk of structural collapse. So in this condition, damage is not repaired immediately, the cracked gravity dam may be collapsed or become completely unusable under strong aftershocks.

Numerous investigations have been conducted to study the seismic response of damaged concrete dam. Almost all the relevant studies have focused on the damage or the propagation of cracks in the dams under main shock. Several scholars have carried out researches on the dynamic response of the gravity dam with an existing crack in aftershock. Sherong Zhang (Sherong et al., 2013) evaluated the nonlinear dynamic response of Koyna dam under main shock-aftershock seismic sequences using a concrete damaged plasticity model. O.A. Pekau (O. A. and Zhu, 2006) implemented IDCE model to deal with dynamic contact surface in the seismic behaviour analysis of cracked concrete gravity dams. But the investigation assumed the dam to be founded on rigid foundation. Alliard (Alliard and Leger, 2008) studied the effect of the drainage system dimensions and model parameters on aftershock response of gravity dam.

In this paper, a mainshock-damaged gravity dam with a crack at the level of downstream slope change is selected as an example to study its dynamic response and the safety assessment in aftershock condition. Nonlinear frictional model proposed by Clough and Duncan is developed in ABAQUS software and used to simulate the penetrated crack surface under different probability levels of aftershocks. The validity of the joint element and numerical algorithm has been checked using the available numerical results. The applicability of the proposed model in simulating the non-linear behavior of typical cracked gravity dam has been shown to research the dynamic response characteristics and seismic safety of cracked gravity dam in aftershock.

2 MODEL APPLIED IN ANALYSES

It is an important subject to study the mechanical properties of contact surface and its numerical realization. Through a long term research, many interface models have been proposed such as ideal model, nonlinear elastic model, and damage model (Zhang and Zhang, 2005). The Mohr-Coulomb friction law used to govern the contact conditions of sliding or not is a ideal elastic-plastic constitutive model. This model underestimates the deformability of interface. The results of displacement and safety factor are relatively small. The nonlinear elastic model developed by Clough and Duncan (Clough G W and Duncan J M, 1973) has been used widely in the analyses of soil-structure interface behavior. The hyperbolic relationship of shear stress and relative displacement satisfies the following relation according to reference (Clough G W and Duncan J M, 1973):

$$\Delta \tau = k_s \Delta \gamma \tag{1}$$

The coefficient of tangential stiffness k_s can be expressed as:

$$k_s = \left(1 - R_f \frac{\tau}{\sigma_n \tan \delta}\right)^2 K \gamma_w \left(\frac{\sigma_n}{P_a}\right)^n \tag{1}$$

where γ_w is unit weight of water; σ_n is the normal stress; τ is the shear stress; P_a is atmospheric pressure; δ is the friction angel; K is modulus number, this dimensionless parameter represents Young's modulus; n is modulus exponent; R_f is the failure ratio.

3 SEISMIC INVESTIGATION OF CRACKED CONCRETE GRAVITY DAM

3.1 *Model of cracked dam, material parameters and loadings*

The gravity dam considered in this analysis is 114 m high and 12 m wide at the dam top, 92 m wide at the base. The designed normal water level is 109 m at the up-steam face of the dam. Figure 1 shows the shape of the dam partially and the finite element model which includes 1294 unit elements and 1400 nodes. A crack penetrated the monolith is assumed to be horizontal and emanating from the point of slope change on the downstream face as shown in Figure 1(a).

The material parameters of dam concrete are as follows: the elasticity modulus $E = 2.4 \times 10^4$ MPa, Poisson's ratio $v = 0.17$, mass density $\rho = 2400$ kg/m³. The energy dissipation of the monolith is considered by the Rayleigh damping method with 5% damping ratio. The reservoir water level is 109 m. In addition to self-weight of the dam, hydrostatic, and earthquake forces, Westergraard added masses (HM, 1933) are employed to represent the hydrodynamic effect. The static solutions of the dam due to its gravity loads and hydrostatic loads are taken as initial conditions in the dynamic analyses of the system.

The time history of aftershock is shown in Figure 2. The record has been modified to match a given target design spectrum. The dam was excited in a single axis corresponding to a horizontal motion along the upstream-downstream axis. In overloading analyses, nineteen different levels of

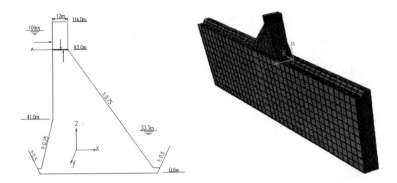

Figure 1. Non-overflow section of Longkaikou gravity dam: (a) geometric model of the model with a crack; (b) finite element mesh of the gravity dam.

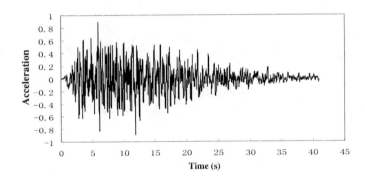

Figure 2. The seismic record for upstream/downstream motion.

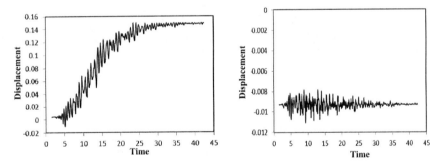

Figure 3. Displacement history of dam crest under PGA 1.45 g: (a) horizontal direction; (b) vertical direction.

PGA are considered for the input motions varying linearly from 0.154 g to 1.54 g. A 5% damping ratio has been assumed in all dynamic analyses.

3.2 *Dynamic response analyses of cracked gravity dam*

The time histories of the dam crest displacement in horizontal and vertical directions, caused by aftershocks with a PGA of 0.15 g, are shown in Figure 3. The variation of displacements along

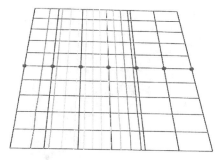
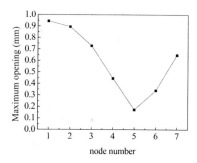

Figure 4. The Maximum joint openings in aftershock: (a) node number and location; (b) maximum opening of each node.

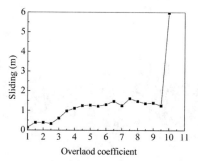

Figure 5. Residual displacements of dam crest under different aftershock levels.

two directions suggests that the upper block lost contact and rotated in aftershocks in the x-o-z plane. The maximum horizontal displacement (0.14 m) on the dam crest sliding to downstream was obtained after aftershock. The Maximum joint openings in aftershock are illustrated in Figure 4. The opening is quite limited in aftershocks, just 10^{-4} m, when the PGA is 0.154 g. So for the gravity dam with horizon broken area in upper region, the failure mode is sliding to downstream and the trend of falling into the river, not overturning down to reservoir area.

3.3 *Overloading analysis of gravity dam subjected to aftershocks*

For the parameter uncertainties of earthquake, in the second part of this study, the slip deformation and the joint opening were calculated and discussed in different levels of PGA changing from 0.154 g to 1.54 g to determine the security of cracked gravity dam in aftershocks. Using the seismic record shown in Figure 2, the PGA in the upstream-downstream direction of the dam increases until failure occurs.

The residual displacements of gravity dam crest under different aftershock levels are shown in Figure 5. The results indicate that the residual displacement presents increasing trend with the creasing of aftershock overload coefficient. The overall sliding can be neglected for the case of crack at the height when the peak acceleration of earthquake is scaled to 0.4 g. Nevertheless, when the peak acceleration increases to 0.8 g, the sliding at the height is 35.7 mm.

The residual sliding displacement is 1.28 m under aftershock load with a PGA of 1.45 g. The top block can be observed to be progressively sliding down the crack surface in Figure 6. When the overload coefficient reaches 10, the PGA is 1.54 g, the residual sliding displacement of dam crest varies tremendously, more than 6 m, and exceeds the half width of contact surface. So the overload

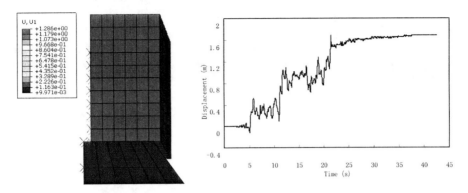

Figure 6. Sliding displacement of contact surface under PGA of 1.45 g: (a) sliding nephogram; (b) time history curve.

coefficient of the crack gravity dam is about 9.5, which the peak of horizontal acceleration is 1.45 g. The concrete gravity dam has adequate sliding and overturning stability in aftershock condition.

3.4 The influence of water level to the dynamic response of gravity dam in aftershock

In earthquake, hydrostatic and hydrodynamic pressure pushes groin head sliding to downstream. In this section, the influence of water level on the residual displacement of gravity dam in aftershock is discussed. From above analyses, we know, the maximum displacement to down stream exceeds the centroid of dam upper part and the top of the dam toppled from the dam during aftershock with a PGA of 1.54 g, at this time the water level is 109 m. If the water level is under the crack face formed in main shock, the residual displacement decreases to 1.6 m when the dam suffers the same aftershock, as the water depth is 92 m, and upper part do not fall down. This illustrates the great influence of water level on the residual displacement due to aftershock. Especially, water level has no effect on the residual displacement when water level is under the crack surface. So we can draw off the reservoir to make water level being under crack surface after main shock to prevent upper part falling down in strong aftershock.

4 SUMMARY AND CONCLUSION

Nonlinear frictional model is used to simulate the dynamic response of the penetrated crack under different probability levels of aftershocks. Using the nonlinear frictional model, several conclusions were obtained from overloading analyses:

1. The time displacement curve of dam crest shows that for cracked dam, in the strong aftershock groin head slips back and forth along stream direction and stops when earthquake is end, keeping the residual displacement.
2. The residual sliding displacement increases with the overload coefficient of earthquake. When increasing of overload coefficient attains or surpasses certain extent, the upper part will disengage and fall down from dam body.
3. The curves of dam crest residual displacement show that the gravity dam can bear strong earthquakes and has big safety margin after the dam appear penetrated cracks on groin head.
4. Reservoir water level has great influence on the residual displacement. If the reservoir water level is drop to slide plane after main earthquake, the dam stability will be increased during aftershock.

ACKNOWLEDGMENTS

Project supported by the Fundamental Research Funds for the Central Universities (No. DUT13LK11) and the State Key Development Program for Basic Research of China (Grant No. 2013CB035905), is acknowledged.

REFERENCES

Alliard, P. M. & Leger, P. (2008) Earthquake safety evaluation of gravity dams considering aftershocks and reduced drainage efficiency. *JOURNAL OF ENGINEERING MECHANICS-ASCE*, 134, 12–22.
Clough, G. W. & Duncan, J. M. (1973) Finite element analyses of retaining wall behavior. *Journal of the Soil Mechanics and Foundations Division, ASCE*, 97, 1657–1673.
HM, W. (1933) Water pressures on dams during earthquakes. *Transactions of the American Society of Civil Engineers*, 98, 418–433.
O. A., P. & Zhu, X. (2006) Seismic behaviour of cracked concrete gravity dams. *Earthquake Engineering & Structural Dynamics*, 35, 477–495.
Zhang, G. & Zhang, J. (2005) Application of elasto-plasticity damage interface model in stress-strain analysis of CFRD. *Journal of Hydroelectric engineering*, 25, 5–10 (In Chinese).

Research on the features of the wide range continuous rainstorms in Guangxi in the conditions of three types of trough

Weiliang Liang, Mingce Huang & Meifang Qu
Guangxi Meteorological Observatory, Nanning, Guangxi, China

Xianghong Li
Guilin Meteorological Observatory, Guilin, Guangxi, China

ABSTRACT: By using the daily precipitation data of Guangxi and ERA-interim reanalysis data from 1981 to 2010 and the classification and statistical methods, the wide range continuous rainstorm cases in the conditions of plateau trough (PT), south branch trough (SBT) and north China trough (NCT) were analyzed separately. The circulations and physical quantities in the same type of trough were composed. The results show that the wide range continuous rainstorms could occur in Guangxi between January and September, mostly between May and July. The features of circulation and physical quantities in different condition were particular and different. The conceptual models of the wide range continuous rainstorms were found.

1 INTRODUCTION

Guangxi takes place in the southeast side of plateau. Indian monsoon, South China Sea monsoon, front and troughs from mid-latitude interact regularly in Guangxi. Rainstorms usually occur after the summer monsoon set up, lead to flood and geological disasters.

Lots of researches have been done for the features, causes, and forecast methods, such as the short wave trough which moved along the blocking high was the main cause in Songhuajiang river and Nenjiang river basin, in the year 1998 (Zhang et al. 2001). As same as the stable south Asia high, the stable upper level jet and low level jet were important to rainfall in south China (Ding et al. 2008). The stable shear line and vortex could provide moisture condition and uplift condition for continuous rainfall (Huang et al. 2008). Some researchers studied on continuous rainstorms in single station and got common results, such as a stable circulation was benefit to the generation, retention, and regeneration of rainstorm systems (Hu et al. 2007 Zhang et al. 2004 Jiang et al. 2006).

Gao et al. (1999) and Liu et al. (2013) separately announced definitions of Guangxi continuous rainstorms for single station, according to the experience of daily forecasting. But the wide range continuous rainstorms in Guangxi were seldom researched.

This article focus on the features and the anomalies of the circulations, use statistical and composite methods, find the particular features in the condition of PT, SBT and NCT, provide basis for daily forecasting.

2 DATA AND METHODS

The data included such as daily precipitation data of Guangxi, daily ground based and sounding observations in Guangxi, ERA-interim reanalysis data from EC. All the data above were from 1981 to 2010, 30a totally. ERA-interim reanalysis data was 4 times everyday. The resolution was $0.75° * 0.75°$ horizontally and 37 levels vertically. It was the highest resolution for reanalysis data all around the world.

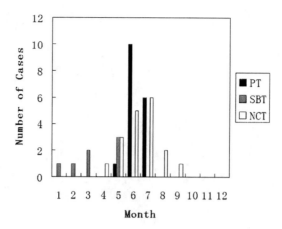

Figure 1. Monthly distribution of wide range continuous rainstorms in Guangxi.

In order to select cases as more as possible, and the rainstorms should cover a wide range of Guangxi, we announced that a wide range continuous rainstorm case was that daily precipitation, $R \geq 50$ mm in 20 observing stations or more in one day, and 10 or more in the day before or after (There are 89 observing stations in Guangxi totally).

Research method was that selecting cases base on 30a precipitation data, analyzing the weather systems in every case using observations and ERA-interim data, classifying the cases according to the systems (PT, SBT or NCT), compositing and analyzing the circulation and physical quantities of each classification.

3 STATISTIC OF THE WIDE RANGE CONTINUOUS RAINSTROMS

Except tropical storms (typhoons), there were 42 cases in the years from 1981 to 2010, including 17 PT cases, 7 SBT cases and 18 NCT cases.

PT cases occurred in the period between May and July, mostly in June. It was the most centralized in three classifications. SBT cases occurred from January to May. It was the only one that occurred in winter. It indicated that SBT should be paid close attention to in winter. NCT cases occurred from April to September, spread all over the flood season in Guangxi. It centred in June and July, single peak distributed (Fig. 1). Generally, the wide range continuous rainstorms occurred from January to September. 80% cases occurred from May to July.

4 COMPOSITE ANALYSIS OF CIRCULATION AND PHYSICAL QUANTITIES

Select the day as a benchmark, in which the number of rainstorm ($R \geq 50$ mm) stations was the most. By composing ERA-interim reanalysis data, we found the features and anomalies of each classification.

4.1 *Plateau trough*

The circulation of 200 hPa was stable before rainstorms. The south Asia high was at the south of Tibet Plateau. The ridge line took place between 23°N and 25°N. It was a bypass region in Guangxi. In the north China, about 35°N, the upper level westerly jet made the divergence stronger in Guangxi. South Asia high and jet caused the divergence and uplift motion in Guangxi. It was benefit to the occurrence and strengthens of rainstorms.

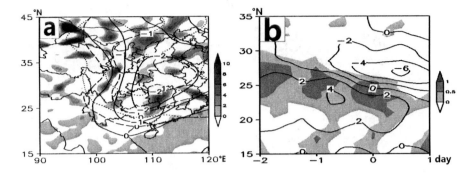

Figure 2. Composite Circulation and Anomaly of PT Cases a. the variation of geopotential height anomaly (dotted: −2 day, dashed: −1 day, solid: 0 day, unit: dagpm) and the vorticity advection (shaded, unit: $10^{-5}s^{-2}$) of 500 hPa. b. the time section of v-wind anomaly (contour, unit: $m \cdot s^{-1}$) and specific humidity anomaly (shaded, unit: $g \cdot kg^{-1}$) along 108°E.

On 500 hPa, there was a trough on the northeast side of plateau. It moved southeastward and became stronger in 2 days before rainstorms in north Sichuan province. The position of trough line was about (35°N, 115°E) to (24°N, 105°E) at 12:00 UTC, 1 day before rainstorms. Vorticity and vorticity advection were both positive in north Guangxi (Fig. 2a). Positive vorticity advection was an effective dynamic starting factor for rainstorms. It appeared before the though and corresponded to the rainstorms (Xiao et al. 2013 Zhang et al. 2009). It meant that when PT strengthened and moved southeastward, the positive vorticity and vorticity advection were the important introductions for Guangxi (especially for north Guangxi) rainstorms. The variation of geopotential height anomaly (Fig. 2a) showed that the negative signal spread southeastward and the geopotential height was lower than usual in Guangxi before rainstorms.

Shear line (or vortex) of 850 hPa existed in 82% PT cases. The shear line moved southward slowly before rainstorms. At 00:00 UTC in the rainstorm day, the shear line located at the north boundary of Guangxi. Strong southwesterly dominated all of Guangxi. In 59% cases the southwesterly increased to be low level jet ($\geq 12\, m \cdot s^{-1}$). In the rest cases the velocity was close to $12\, m \cdot s^{-1}$. Low level jet was also a dynamic starting and increasing factor for rainstorms. The low level jet coupled with the upper level jet, strengthened the uplift motion and precipitation by positive feedback effect.

Water vapor converged along shear line. Moisture flux divergence center moved eastward along shear line and increased above $-5*10^{-6}\, g \cdot cm^{-2}\, hPa^{-1}\, s^{-1}$. The convergence of water vapor was intense. The wind anomaly showed an anomalous vortex in north Guangxi. Moisture flux convergence anomaly located at the southeast of the anomalous vortex. The time section of v-wind component anomaly and specific humidity anomaly along 108°E reflected that water vapor convergence region companioned the shear line (fig. 2b). It was moister at the south of shear line in rainstorm cases than usual. More quality convergence and water vapor convergence was a characteristic of rainstorms.

The circulation which is benefit to wide range continuous rainstorms in PT condition is: south Asia high locates at south Tibet plateau, there is upper level jet in north China, plateau trough strengthens and moves southeastward, southwesterly is strong (close to be jet at least), the shear line moves slowly in Guangxi, quality and water vapor converge along the shear line.

4.2 South branch trough

The cases of SBT occur in winter and spring. In these seasons, the south Asia high was still nearby Philippine islands. The upper level west-southwesterly jet was in east China, containing a core which exceeded $60\, m \cdot s^{-1}$. Guangxi was at the right side of the entrance of upper level jet. Upper level circulation was advantageous for rainstorms.

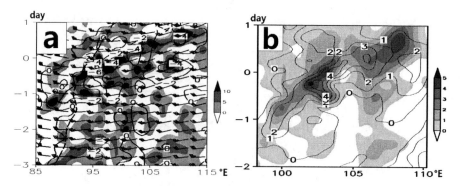

Figure 3. Compose in the SBT condition. a. the time section of wind (barb), vorticity advection (shaded, unit: $10^{-5}\mathrm{s}^{-2}$), and vertical velocity (contour, unit: $10^{-3}\,\mathrm{hPa\cdot s^{-1}}$) of 500 hPa along 23°N. b. the time section of vorticity (contour: 700 hPa, shaded: 850 hPa, unit: $10^{-5}\mathrm{s}^{-2}$).

The ridge line of west Pacific subtropical high of 500 hPa stagnated at 15°N. Empirically, when the west Pacific subtropical high weakened and shift southward to south of 18°N, the front could cause rainstorms, even wide range rainstorms in high probability while the shear line (or vortex) was interacting.

The SBT increased and moved across 95°E at 12:00 UTC the day before rainstorms, from west of 90°E. At this time, the SBT had large amplitude that covering the region from 5°N to 25°N. Thus it was capable to transport and provide water vapor and heat to maintain the rainstorms in Guangxi. In front of SBT, it was the straight southwesterly and positive vorticity advection. Figure 3a showed the process that positive vorticity advection transferred eastward, following the SBT. The trough line moved from west of 90°E to east of 97°E progressively in 3 days. The positive vorticity advection arrived at Guangxi (104–112°E) in the day before rainstorms and lasted for a long time. It led to continuous uplift motion and continuous precipitation.

Different from the other two classifications, it was warm shear line at the north boundary of Guangxi in SBT condition. There was no low level jet, but the warm and moist southwesterly from Indochina peninsula converged with the southeasterly at the edge of west Pacific subtropical high. That also caused large quantity of water vapor convergence and provided abundant water vapor for rainstorms in Guangxi. A time section along the line between (95°E, 15°N) and (110°E, 25°N) indicated that an anomalous vortex, which came from Bay of Bengal 2 days before, kept in Guangxi throughout the rainstorms (Fig. 3b). The anomalous convergence and vortex were obvious on both 700 hPa and 850 hPa. They caused the anomalous uplift. The rainstorms increased abnormally.

The circulation that is benefit to wide range continuous rainstorms in SBT condition is: the south Asia high is on Philippine islands, strong upper level jet locates at north China, the amplitude of SBT becomes larger and the trough line moves eastward across 95°E, the west Pacific subtropical high shifts southeastward to 15°N, the 850 hPa warm shear line maintain in north Guangxi, the southwesterly and southeasterly converge in Guangxi.

4.3 North China trough

The south Asia high located at Indochina peninsula. Ridge line was along 25°N. The bypass region provided divergence on upper level. The circulation of 200 hPa was stable.

The subtropical high was at the south of 15°N. No troughs but the zonal westerly occupied the south China. 1day before rainstorms, a trough, which took place from (115°E, 45°N) to (105°E, 30°N) in north China, moved eastward at a high speed.

The NCT did not affect directly, but it could not be ignored in Guangxi rainstorms because of the forcing effect. The temperature behind the NCT dropped rapidly. The negative temperature

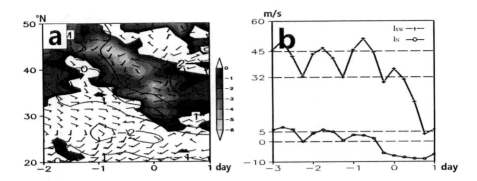

Figure 4. Composite of NCT cases. a. time section of temperature advection (shaded, unit: 10^{-5} K·s^{-1}) of 500 hPa, geopotential height anomaly (contour, unit: dagpm) and wind (barb) of 850 hPa along 108°E. b. south wind index (I_{SW}) and north wind index (I_N).

advection was larger than $-4*10^{-5}$ K·s^{-1}. It led to a large anomalous geopotential height on low levels and anomalous high pressure on surface, which forced the shear line and front to shift southward rapidly. Fig. 4a showed the synchronous process clearly. The force from 500 hPa aroused geopotential height anomaly of 850 hPa. The anomalous high was at the north of shear line, and the anomalous low was at the south. It meant more convergence and higher speed of shear line. Actually, it moved about 10 latitudes per day. The shear line moved into Guangxi at 00:00 UTC in the rainstorm day. Wind convergence and water vapor convergence provided uplift and vapor for rainstorms.

South and North Wind Indices were useful empirical forecasting methods in Guangxi. It judged the position and intensity of shear line by calculating the v-wind component (Chen et al. 2012). Added v-wind at the 9 grids in region 20°–25°N, 105°–110°E (grid interval 2.5°), obtained the South Wind Index:

$$I_{SW} = \sum_{i=1}^{m}\sum_{j=1}^{m} v_{850}(i,j), m=3$$

Added v-wind at the 3 grids in region 27.5°N, 105°–110°E (grid interval 2.5°), obtained the North Wind Index:

$$I_N = \sum_{i=1}^{n} v_{850}(i), n=3$$

$I_{SW} \geq 32$ m·s^{-1}, $I_N \leq 5$ m·s^{-1} could be used as indices for regional rainstorms in Guangxi (10 stations or more), and $I_{SW} \geq 45$ m·s^{-1}, $I_N \leq 0$ m·s^{-1} for wide range rainstorms (30 stations or more) (Chen et al. 2013).

As showed in Fig. 4b, I_{SW} was greater than 32 m·s^{-1} previously, indicating that southerly was stable and strong. I_N reduced to less than 5 m·s^{-1}. The northerly was replacing southerly in Guizhou and Hunan province, the shear line shift close to Guangxi progressively. When the rainstorms were coming, I_N reduced to less than 0 m·s^{-1}. It was northerly in Guizhou and Hunan province, while the southerly in Guangxi was still strong. The convergence along shear line was violent enough to cause wide range rainstorms. After shear line shift southward, I_{SW} reduced rapidly, indicating that northerly was expanding southward rapidly.

The circulation that is benefit to wide range continuous rainstorms in NCT condition is: south Asia high is on the north of Indochina peninsula, it is zonal westerly at low latitudes on 500 hPa, the temperature advection is strong behind the NCT, the shear line on 850 hPa moves southward into Guangxi.

5 SUMMARY

The wide range continuous rainstorms in Guangxi distribute from January to September. 80% of them distribute from May to July.

When the PT increases and moves southeastward, the position of south Asia high and upper level jet is appropriate, the shear line and low level jet can provide continuous uplift and vapor, there will be wide range continuous rainstorms probably.

The SBT is the only system which leads to wide range continuous rainstorms in winter. When the SBT increases and moves eastward, the south Asia high and upper level jet is benefit, the water vapor is abundant, the wide range continuous rainstorms could occur.

When the divergence on high level and water vapor convergence on low level are benefit, the NCT can force shear line and front to shift southward rapidly and arouse incense uplift. It could also cause wide range continuous rainstorms in Guangxi, although there is no trough in south China.

ACKNOWLEDGEMENT

Project funding: Southern China Regional Meteorological Center Project (CRMC2012M07), 2013 Guangxi Natural Science Fund Project (2013GXNSFAA019288), Guangxi Scientific Research and Technology Development Project (10123009-8).

REFERENCES

Zhang Qingyun, Tao Shiyan & Zhang Shunli. 2001. A Study of Excessively Heave Rainfall in the Songhuajiang-Nenjiang River Valley in 1998. China Journal of Atmosphere Sciences 25(4): 567–575.

Ding Zhiying, Chang Yue & Zhu Li. et al. 2008. Research on the Reason of the Double Rain-Bands' Forming in A Sustaining Storm Rainfall of South China. Journal of Tropical Meteorology 24(2): 117–126.

Huang Zhong, Wu Naigeng & Feng Yerong. et al. 2008. Causality Analysis of the Continuous Heavy Rain in Eastern Guangdong in June 2007. Meteorological Monthly 34(4): 53–60.

Hu Liang, He Jinhai & Gao Shouting. 2007. An Analysis of Large-Scale Condition for Persistent Heavy Rain in South China. Journal of Nanjing Institute of Meteorology 30(3): 345–351.

Zhang Xiaoling, Tao Shiyan & Zhang Shunli. et al. 2004. A Case Study of Persistent Heavy Rainfall over Hunan Province in July 1996. Journal of Applied Meteorological Science 15(1): 21–30.

Jiang Lijuan, Li Xianghong & Xue Rongkang. 2006. The Monsoon Perturbation and Dynamic Condition Analysis for Guilin Continuous Rainstorm in June 2005. Journal of Guangxi Meteorology 27(2): 1–4.

Gao Anning, Liang Zhihe & Wu Rencai. 1999. Research on the Long-Term Forecasting of Wide Range Continuous Rainstorms in Flood Season in Guangxi. Journal of Guangxi Meteorology 20(1): 2–7.

Liu Guozhong, Huang Kaigang & Luo Jianying et al. 2013. Research on the Short-Term Forecasting Technique of Persistent Rainstorm with Conceptual Model and Ingredients-Based Method. Meteorological Monthly 39(1): 20–27.

Xiao Dixiang, Xiao Dan & Zhou Changchun et al. 2013. Analysis on Dynamic Effects of Low-Level Southerly Airflows on One Rainstorm Process and the Numerical Simulation. Meteorological Monthly 39(3): 281–290.

Zhang Xiaohui, Xu Aihua & Liu Xiaohui. 2009. A Diagnostic Case Study on the Interaction between Mid-Latitude Trough and Tropical Depression "NEOGURI". Journal of Tropical Meteorology 25(B12): 29–38.

Writing Group. 2012. Weather Forecasting Technique and Method of Guangxi. Beijing: China Meteorological Press: 91.

Chen Jian, Liu Chouhua & Gao Anning et al. 2012. Application of Low Latitude South and North Wind Indices to Prediction of Rainstorm in Guangxi. Meteorological Monthly 38(11): 1348–1354.

Chen Jian, Gao Anning & Tang Wen. 2013. Forecast Method Research and Occurrence Characteristics of Large Range of Frontal Rainstorm in Guangxi. Journal of Meteorological Research and Application 34(1): 7–12.

Precipitation analysis in Wuhan, China over the past 62 years

S. Deng
College of Hydrology and Water Resources, Hohai University, Nanjing, China

B.H. Lu
College of Hydrology and Water Resources, Hohai University, Nanjing, China
State Key Laboratory of Hydrology Water Resources and Hydraulic Engineering, Hohai University, Nanjing, China

H.W. Zhang & E. Akwei
College of Hydrology and Water Resources, Hohai University, Nanjing, China

ABSTRACT: This paper studied the variability of the statistical structures of precipitation in Wuhan, China, by analyzing time series of annual and monthly precipitation of Wuhan station from the period 1951 to 2012. Mann–Kendall test and Precipitation Concentration Degree (PCD) were used to detect precipitation concentrations and the characteristics of variation. The results show that no significant trend of annual precipitation was found, but the number of annual precipitation days was significantly reduced at the 0.05 significance level. Annual PCD of Wuhan shows an increasing trend but not significant, which was mainly attributed to the increasing proportion of the flood season precipitation in the annual precipitation, especially in July. While the proportion of precipitation in the non-food season decreases, especially in March. This research seeks to contribute to the study of the hydrology of a catchment in order to ensure sustainable water resource management.

1 INTRODUCTION

Global warming and its effects have received increasing attention from human society (Zhang et al., 2008). Precipitation intensity, amounts and patterns are expected to change. Precipitation extremes accounts for high percentages of the annual total, are directly responsible for flood occurrences and will bring more frequent disasters for human society. A significant decreasing number of rainy days and significant increasing in precipitation intensity values have been identified in many places in the world, such as America (Karl et al., 1996) and China (Zhai et al., 2005). Since the 1990s, there has been high frequency of flood occurrence in the seven big river valleys in China, and both flood and geological disasters have increased due to the increase of intense precipitation events and the consequent increase in their degree of concentration. The changes in spatial and temporal concentration degree and variation of heavy precipitation in China were the major possible reasons for the frequent flood disasters (Jishun et al., 2000).

Higher precipitation concentration, represented not only by higher percentages of the annual total precipitation in a few rainy days, but also by the time and degree of concentration of the annual total precipitation within a year, has the potential to cause droughts and floods, which are expected to put considerable pressure on water resources (Zhang and Qian, 2003). In this paper, the statistical characteristics of annual and monthly precipitation were analyzed with Mann–Kendall test, precipitation concentration degree (PCD), and other methods.

2 STUDY REGION AND DATA

Wuhan, the largest city in central China, lies at the river mouth where the Yangtze River is joined by its main tributary, the Hanjiang River. As the capital of Hubei Province, Wuhan is the largest mega-city in central China and in the middle reaches of Yangtze River. It has a population of eight million and an area of 8467.11 km^2. With abundant water resources, Wuhan depends heavily on utilizing surface water but little care has been taken to protect them.

In this study, monthly precipitation data and precipitation days from 1951 to 2012 of Wuhan station were analyzed. The rain-gauge data were collected and quality-controlled by the National Meteorological Information Center (NMIC) of China Meteorological Administration (CMA).

3 METHODOLOGY

3.1 Mann–Kendall test

The Mann–Kendall test is a robust, nonparametric procedure for randomness against trend. It works without requiring normality or linearity, and is therefore widely used for trend detection (Zhang et al., 2010). The basic requirement of MK test is that the number of observed measurements is no less than ten (Su et al., 2011). The procedure is described as follows:

Let $x_1, x_2, \ldots x_n$ represents n data points, where x_j is the data point at time j. Then the MK statistic (S_k) is expressed by Equations (1).

$$S_k = \sum_{k=1}^{n-1} \sum_{j=k+1}^{n} \text{sgn}(x_j - x_k), \quad k = 2, 3, 4 \ldots n \tag{1}$$

$$\text{where } \text{sgn}(x_j - x_k) = \begin{cases} +1, & x_j - x_k > 0 \\ 0, & x_j - x_k = 0 \\ -1, & x_j - x_k < 0 \end{cases} \tag{2}$$

Compute statistic UF_k using following equation:

$$UF_k = \frac{[S_k - E(S_k)]}{[Var(S_k)]^{1/2}}, \quad (k = 1, 2, 3 \ldots n) \tag{3}$$

where $UF_1 = 0$, $E(S_k)$ and $Var(S_k)$ are the mean and variance of S_k. Compute the probability and decide on a probability level of significance. The trend is assumed as downward if UF_k is negative and the computed probability is greater than the level of significance. The trend is regarded as upward if UF_k is positive and the computed probability is greater than the level of significance. If the computed probability is less than the level of significance, there is no trend.

When MK test is used for mutation analysis, we can apply the above procedure to the reverse sequence of x_n and calculate statistic UB_k, where $UB_k = -UF_k$, $k = n, n-1, \ldots, 1$; $UB_1 = 0$. Then the cross-point between curve UF_k and curve UB_k is assumed as catastrophe point.

3.2 Precipitation concentration degree (PCD)

The basic principle for calculating the PCD is based on the vector of monthly total precipitation. Based on the assumptions that monthly total precipitation is a vector quantity with both magnitude and that the direction for a year can be seen as a circle (360°), the annual PCP and PCD for a location can be defined as follows:

$$R_x = \sum_{i=1}^{12} R(i)\cos\theta_i, \quad R_y = \sum_{i=1}^{12} R(i)\sin\theta_i \tag{4}$$

$$PCD_i = \sqrt{R_x^2 + R_y^2} \Big/ \sum_{i=1}^{12} R(i) \tag{5}$$

where i represents the month ($i = 1, 2, \ldots, 12$). R_i stands for monthly total precipitation in the i th month, and θ_i is the azimuth of the i th month. PCD_i represents the degree that the total precipitation of the ith month concentrates in 12 months.

Based on Equations (5), annual PCD can reflect the degree that total precipitation is distributed in 12 months. The annual PCD ranges from 0 to 1. If annual total precipitation all concentrates on one specific month, the maximum 1 can be obtained. If total precipitation of each month within a year is evenly distributed, annual PCD can reach 0 (Zhang and Qian, 2003).

4 RESULTS

4.1 Inter-annual variations of precipitation

By calculating and analyzing precipitation data from 1951 to 2012 in Wuhan, the average annual precipitation is 1257 mm and maximum annual rainfall occurring in 1954 is 2057 mm, while minimum precipitation taking place in 1966 is 727 mm. Extreme ratio is 2.8. The inter-annual variability is large. The rate of precipitation increases slowly—approximately 6.1 mm/10 a.

Curve of cumulative departures of precipitation in Wuhan is shown in Fig. 1. The value of cumulative departures is wavelike and rises first, then become sharply decrease at 1964, reaching the lowest point at 1979; then rises sharply, reaching peak at 2004, and then come to decreases.

MK test was applied for visualizing the trends and catastrophe point of annual precipitation and precipitation days. The UF value of annual precipitation was 0.513, indicates that Wuhan characterized by increase trends but not significant at the 0.1 significance level; Figure 2(a) exhibited 1980 was identified as catastrophe point, denoting that the annual precipitation increased after

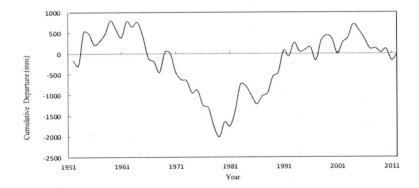

Figure 1. Cumulative departure curve of annual precipitation.

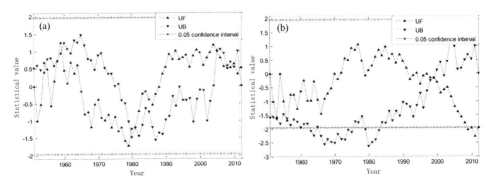

Figure 2. (a) Results of MK test for annual precipitation during 1951–2012. (b) Results of MK test for precipitation days during 1951–2012.

1980. However the UF value of annual precipitation days was −1.97, showing downward trend at the 0.05 significance level. As exhibited in Figure 2(b), 1989 was identified as catastrophe point, denoting that annual precipitation days were reduced after 1989.

These results signify that during 1951–2012, the number of the rainy days decreased notably while the annual total precipitation did not have a notable extreme changing tendency, and would result in precipitation intensity increasing.

4.2 *Analysis of precipitation during the year*

The distribution of precipitation in Wuhan was uneven. In non-flood season (October to April of the following year), average precipitation was 491 mm, taking about 39% of annual precipitation; average rainfall reached up to 766 mm in flood season (May–September), accounting for 61%. As displayed in Figure 3(a), the rainfall distribution had a single peak in June, and average peak value was 218 mm. Precipitation decreased to 41 mm in January and 30 mm in December.

4.2.1 *Analysis of precipitation concentration degree (PCD)*

In order to analyze the concentration of precipitation during 1951–2012 in Wuhan, annual PCDs of rainfall data from 1951 to 2012 were calculated, the average of PCDs was 0.2, annual PCDs were displayed in Figure 3(b), showing that annual PCDs were of volatility and were less than 0.45.

For further analysis on the trend of PCDs, MK test of 1951–2012 PCD series was taken, results showed that the UF value of PCDs from 1951 to 2012 was 0.5, indicating an increasing trend, but it was not significant at the 0.1 significance level.

In order to analyze the relationship between PCD and summer precipitation, MK trend test was used. The UF value of proportion of flood season (May-September) precipitation from 1951 to

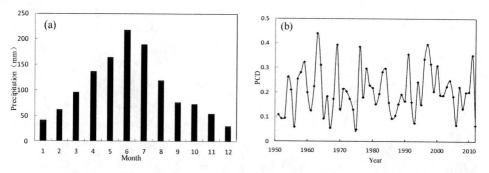

Figure 3. (a) Distributions of monthly precipitation in Wuhan. (b) PCD in Wuhan during 1951–2012.

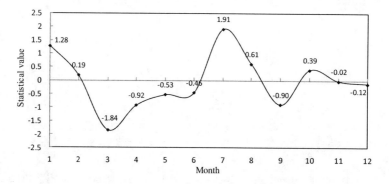

Figure 4. Results of MK trend test for monthly precipitation ratio during 1951–2012.

2012 is 0.96. It showed an increasing trend, but the trend is not significant. The result was basically the same with PCD analysis.

4.2.2 Analysis of monthly precipitation

For further study of monthly precipitation, MK trend test was used to analyze the trend of proportion of average monthly precipitation from 1951 to 2012. Results are shown in Figure 4, indicating that, during the flood season (May–September), the UF value of July was 1.91, showed upward trend at the 0.1 significance level, while other months did not show a significance trend. During non-flood season (October to April of the following year), the UF value of March was −1.84, showing downward trend at the 0.1 significance level, while other months did not show a significance trend.

5 CONCLUSIONS

Wuhan is the largest mega-city in middle reaches of Yangtze River. The statistical analysis of precipitation databases has scientific and practical value not only in climate change assessment but also in water resource management, agricultural planning and irrigation in Wuhan. PCD and MK test were used in this study to investigate the statistical structure of precipitation rates based on monthly precipitation datasets. Main findings are summarized as follows.

(1) The inter-annual precipitation in Wuhan varies, extreme ratio is 2.8, and precipitation increasing rate is 6.1 mm/10 a. No significant trend was found of annual precipitation, but the number of annual precipitation days has significantly reduced at the 0.05 significance level. 1980 was the catastrophe point of annual precipitation, as it increased after 1980. 1989 was the catastrophe point of precipitation days, while it came to decrease after 1989.

(2) Annual PCD of Wuhan shows a not significant increase trend, which was mainly attributed by the increasing proportion of flood season precipitation, especially in July (at the 0.1 significance level); while the proportion of precipitation in the non-flood season exhibited decreasing trend, especially in March (at the 0.1 significance level).

ACKNOWLEDGEMENTS

The study is sponsored in part by the Ministry of Water Resources, Public Welfare Fund 201201026.

REFERENCES

Cun, C. & Vilagines, R. 1997. Time series analysis on chlorides, nitrates, ammonium and dissolved oxygen concentrations in the Seine river near Paris. *Science of the total environment*, 208, 59–69.

Jishun, L., Angsheng, W. & Jiatian, C. 2000. Spatial-temporal variation of precipitation leads to the serious flood and drought in Yangtze River Valley. *Disaster Reduction in China*, 9, 27–40.

Karl, T. R., Knight, R. W., Easterling, D. R. & Quayle, R. G. 1996. Indices of climate change for the United States. *Bulletin of the American Meteorological Society*, 77, 279–292.

Su, S., Li, D., Zhang, Q., Xiao, R., Huang, F. & Wu, J. 2011. Temporal trend and source apportionment of water pollution in different functional zones of Qiantang River, China. *Water research*, 45, 1781–1795.

Zhai, P., Zhang, X., Wan, H. & Pan, X. 2005. Trends in total precipitation and frequency of daily precipitation extremes over China. *Journal of Climate*, 18, 1096–1108.

Zhang, L. & Qian, Y. 2003. Annual distribution features of precipitation in China and their interannual variations. *Acta Meteorologica Sinica*, 17, 146–163.

Zhang, Q., Xu, C.-Y., Tao, H., Jiang, T. & Chen, Y. D. 2010. Climate changes and their impacts on water resources in the arid regions: a case study of the Tarim River basin, China. *Stochastic Environmental Research and Risk Assessment*, 24, 349–358.

Zhang, Q., Xu, C.-Y., Zhang, Z., Ren, G. & Chen, Y. 2008. Climate change or variability? The case of Yellow river as indicated by extreme maximum and minimum air temperature during 1960–2004. *Theoretical and Applied Climatology*, 93, 35–43.

Climatological anomalies of 500 hPa height fields for heavy rainfalls over the Huaihe River Basin

J. He, G.H. Lu, Z.Y. Wu & H. He
College of Hydrology and Water Resources, Hohai University, Nanjing, Jiangsu, China

Y. Yang
Bureau of Hydrology, Ministry of Water Resources, Beijing, China

J.Y. Huang
School of Physics, Peking University, Beijing, China

ABSTRACT: The signal field method was adopted in an effort to effectively extract climatological anomalies from the 500 hPa height fields for the heavy rainfalls over the Huaihe River Basin (HRB) in 2003, 2005 and 2007. The results show that the heavy rainfalls were all associated with a similar anomaly pattern of the 500 hPa height signal fields: both the west of the Lake Baikal and the south of the HRB were dominated by significant positive signals; the vicinity of the HRB was covered by negative signals; the positive-negative height signals established a tripole of "+ − +" from mid-high to low latitudes. In addition, the forming and collapsing of this tripole to some extent corresponded to the heavy rainfall processes of the HRB.

1 INTRODUCTION

The Huaihe River Basin (HRB) is one of the most flood-prone basins in China for its transitional climate from humid zone to semi-humid zone and from subtropical zone to warm temperate zone. Heavy rainfalls contribute a large portion of total flood season (June–September) precipitation to the HRB and therefore play an important role in causing floods. Historically, the HRB severely suffered from floods triggered by heavy rainfalls, such as the catastrophic floods in the summers of 1954, 1975 and 1991. Just in the first 10 years of the 21st century, regional or basin-wide floods induced by heavy rainfalls continually inundated the HRB in 2003, 2005 and 2007. Heavy rainfalls and ensuing floods over the HRB have been critical troubles impacting on local economy and human, and attracted great research interest of lots of Chinese meteorologists and hydrologists.

Many of previous investigations on the heavy rainfall over the HRB were case studies, which adopted methods such as synoptic analysis, dynamic diagnosis and numerical simulation. Cheng et al. (2004) noted that the abrupt turning of the Ural Mountain blocking high from strengthening period to weakening period and the developing and westward landing of the western Pacific subtropical high (WPSH) were two major preludes to the heavy rainfall over the HRB in summer 2003. Song et al. (2006) simulated the heavy rainfall of 4–6 July 2003 with the mesoscale model Weather Research and Forecasting (WRF), which revealed the forming, developing, intensifying and declining of the related mesoscale synoptic systems. Zhao et al. (2007) pointed out that the heavy rainfall producing the severe flood over the HRB during June and July 2007 resulted from the interaction of blocking highs at high latitudes, persisting of the WPSH, deepening of troughs in the westerlies, coupling of upper level with low level jet (LLJ), and forming and developing of mesoscale vortexes. Most of summer heavy rainfalls over the HRB, including those that occurred in 2003, 2005 and 2007, were closely associated with the Mei-Yu front located in East China. Tao et al. (2008) found that with the development of cyclogenesis and frontogenesis induced by deepening upper troughs, rapidly upward motions along the Mei-Yu front resulted in the heavy rainfall over the HRB in middle of July 2007.

Since the heavy rainfall is a type of disruptive event of small probability, the related meteorological fields should exhibit significant anomalies, or departures from climatology, which might be used as clues for identifying heavy rainfall events. However, the conventional methods mentioned above are insufficient to evaluate how much a specific meteorological field departs from climatology objectively, for the existence of the regional and temporal variability with that field. In this study, the signal field method was introduced to help determine whether anomalies relative to climatology for a specific meteorological field were statistically significant or within normal fluctuations of climatology. Then an analysis was conducted, using the signal field method, on the 500 hPa height fields for the heavy rainfalls over the HRB in the flood seasons of 2003, 2005 and 2007. The goal is to demonstrate the potential utility of the signal field method in assessing climatological anomalies of meteorological fields, and to look for 500 hPa height signals favoring heavy rainfalls over the HRB.

2 METHODOLOGY

The Hydrological Information Annual Reports for 2003, 2005 and 2007 compiled by Ministry of Water Resources (MWR) Bureau of Hydrology were collected to determine periods of heavy rainfalls over the HRB in the three years. Three typical heavy rainfall events of 28 June–17 July 2003, 4–11 July 2005 and 29 June–9 July 2007, all of which occurred in the Mei-Yu seasons, were fixed to be analyzed due to their long duration and high severity. The heavy rainfalls directly induced by tropical cyclones, which sometimes occurred in August and September, were beyond the scope of this study. Additionally, areal precipitation data for the HRB mentioned in this paper was also from MWR Bureau of Hydrology.

The NCEP/NCAR reanalysis dataset as described in Kalnay et al. (1996) was used for this study. This global dataset has a resolution of 2.5° × 2.5° at up to 17 pressure levels, extending from 1948 to present, and is updated daily. For a fixed day of the three events, the 500 hPa height field was plotted and examined over a range of 30°E–180°E and 20°N–80°N. In addition, a 60-yr (1948–2007) time-series of 500 hPa height fields was treated as the local climatology. The mean and standard deviation of the time-series, which characterized the average and average variation of the local climatology, were then calculated. Thus, the 500 hPa height signal field was obtained in a method similar to Huang et al. (2002) and Yang et al. (2005) by:

$$s = \frac{h - \mu_h}{\sigma_h} \tag{1}$$

where h is the 500 hPa height field for that day, μ_h is the mean of the local climatology, and ∂_h is the standard deviation from that mean. Based on the hypothesis that the distribution of the local climatology is normal, the process described by (1), in fact, converts a general normal distribution toward a standard normal distribution. That is, the 500 hPa height signal field is composed of signals which are virtually standardized anomalies from the local climatology. According to the hypothesis testing, signals with absolute values no less than 1.96 are statistically significant under the significance level of 5%. In an absolute sense, the term significant throughout this paper refers to signals equal to or greater than 1.5 for convenience.

3 RESULTS

3.1 *Investigation of the 500 hPa height signal fields for the 2005 event*

In this section, similarities and differences between the signal field method and the conventional methods in analyzing heavy rainfalls are illuminated with the heavy rainfall from 4 to 11 July 2005 taken as an example.

Large-scale conditions associated with this event have been examined by Chen et al. (2007) and Zhang et al. (2008) in detail. The composite 500 hPa height field for the heavy rainfall, as shown in

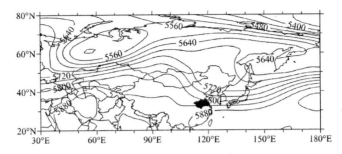

Figure 1. Composite 500 hPa height field (10 gpm) for 4–11 July 2005. The filled polygon is the HRB.

Figure 1, indicated that the circulation pattern over mid-high latitudes of Eurasia was "trough-ridge-trough". Two deep troughs both extending from the northeast to the southwest appeared around the Ural Mountain (60°E, 60°N) and the Sakhalin (140°E, 50°N) respectively, with the latter reaching southward to the vicinity of the HRB. The broad region between the two troughs was dominated by a large ridge. Cold air from high latitudes intruded into the HRB by two paths: one was behind the Sakhalin trough, cold air spreading southward, along Northeast and North China; the other was the northwest path, cold air that split away from the Ural Mountain trough pressing southeastward, through the Balkhash Lake (75°E, 45°N) and the Gansu Corridor (100°E, 40°N) west of the Yellow River in sequence. On the other hand, the enhanced WPSH was situated south and southeast to the HRB, with its ridge lying horizontally near 24.5°N and the 5880 gpm contour (characteristic height of the WPSH) stretching westward to 110°E. A long-term confrontation over the HRB between warm moist southwesterlies at the northwest edge of the WPSH and cold air from the north resulted in this heavy rainfall.

Troughs and ridges in the westerlies and the WPSH at low latitudes are quite common circulation systems. How to associate these seemingly ordinary circulation systems with extraordinary rainfall events such as that in 2005? Since the 500 hPa height field was unlikely to provide more information in answering this question, the 500 hPa height signal field for the period 4–11 July 2005 were then calculated, plotted and added to this investigation.

Noting the region of significant positive signals to the west of the Lake Baikal (108°E, 54°N) in the 500 hPa height signal field for 6 July (Fig. 2c), it remained almost in the same location and grew steadily in intensity in the next two days, of which the maximum value arrived at around 2.8 (96°E, 51°N) on 8 July (Fig. 2e). Although slightly weakened as it shifted eastward, it was still one of the most noteworthy features in the signal fields for 9–10 July (Fig. 2f–g). As viewed from 500 hPa height fields, these positive signals, which represented significant height anomalies above the 60-yr climatological heights, could be linked to the anomalously developing ridge over the central and western Mongolia.

Another noticeable feature was present over the western North Pacific between 20°N and 30°N. Significant signals greater than 1.5 broke out over the southeast of the Taiwan Strait (120°E, 25°N) first on 7 July (Fig. 2d), further spread to the southeast coast of China mainland and peaked at about 2.6 around (130°E, 20°N) on 9 July (Fig. 2f). However, the center of the positive signals was less moving during its intensifying and expanding. These suggested that the heavy rainfall was associated with the relatively stable and anomalously strong WPSH at proper location allowing for more warm moisture to be pulled northeastward into the HRB.

As for negative signals, those distributed from 110°E to 140°E and from 30°N to 60°N were particularly worth noting, even though they were not significant during much of the event. These negative signals extended southwestward across the HRB from 7 to 9 July (Fig. 2d–f), which revealed the direct impacts of the Sakhalin trough on the heavy rainfall.

Thus, the positive-negative signals over East Asia formed a tripole of "+ − +" from mid-high to low latitudes, as denoted by the crosses in Figure 2f. The resulting gradient between the negative pole and the positive pole to its northwest was likely to be associated with southward intrusion of northerly cold air, and that between the negative pole and the positive pole to its southwest with the

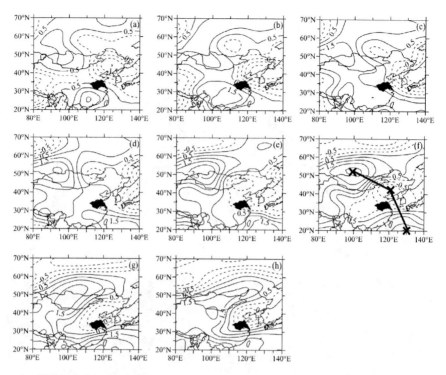

Figure 2. 500 hPa height signal fields from (a) 4 July to (h) 11 July 2005. The filled polygon is the HRB. The crosses represent the tripole.

Mei-Yu front in the vicinity of the HRB. During the period of 6–10 July, when the tripole established and then stabilized, the persistent heavy rainfall raged across the HRB, with the accumulated areal precipitation over 126 mm.

As shown in Figure 2a–b, the signal fields for 4–5 July had a tripole pattern similar to that on 9 July (Fig. 2f). However, there were subtle differences between them. The heights at mid-high latitudes were within the normal range of climatology due to the corresponding positive signals less than 1.5. And the significant signals centered in Southwest China were much further to the west and north than those to the southeast of the HRB during 6–10 July. Under these conditions, the HRB received relatively small areal precipitation of 8.7 mm and 5.2 mm on the two days respectively. On 11 July, the marked shrinking of the significant positive signals at mid-high latitudes implied that the associated anomalous ridge weakened abruptly, and the westward moving of the significant positive signals at low latitudes suggested the continued advance of the WPSH to the west (Fig. 2h). Meanwhile, the heavy rainfall was suspended.

With the signal method applied to the 500 hPa height fields for the 2005 event, the anomalous activities of the associated circulation systems were well visualized, easily captured and objectively evaluated by means of the derived 500 hPa height signal fields. Compared with conventional anomaly fields that are composed of departures from climatology, signal fields focus on diagnosing the statistical significance of the departures, as well as their patterns.

3.2 *Regularity of the 500 hPa height signal fields for the 2003, 2005, 2007 events*

To understand whether it was coincidence or general rule that associated with the anomaly characteristics of the 500 hPa height signal fields for the 2005 event, this paper further examined the 500 hPa height signal fields for the 2003 and 2007 events.

As shown in Figure 3, the anomaly pattern found in the 2005 event also appeared in the 500 hPa height signal fields for 7 July 2003 and 8 July 2007. The positive signals largely distributed in

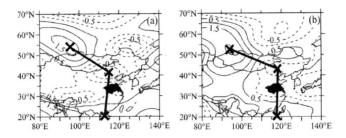

Figure 3. 500 hPa height signal fields for (a) 7 July 2003 and (b) 8 July 2007. The filled polygon is the HRB. The crosses represent the tripole.

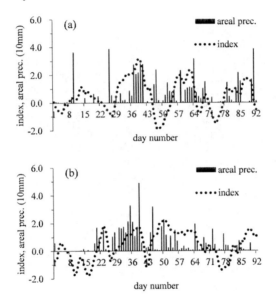

Figure 4. Index and areal precipitation (10 mm) from June to August in (a) 2005 and (b) 2007.

the west of the Lake Baikal with the significant center located around (95°E, 54°N). The negative signals originated from high latitudes, and extended through the HRB to the south and west. Together with the positive signals at low latitudes, the 500 hPa height signal fields for the 2003 and 2007 events exhibited a triple similar to that in the 2005 event.

Based on 500 hPa height signal fields, a simple index was derived as below to quantify the tripole aforementioned above:

$$index = m_a + m_c - m_b \qquad (2)$$

where m_a, m_b and m_c represent spatial averages of 500 hPa height signals over areas of (80°E–100°E, 40°N–60°N), (110°E–130°E, 30°N–50°N) and (110°E–130°E, 20°N–30°N) respectively. In general, the higher the index achieves, the better the tripole is.

According to Equation (2), the daily index was calculated from June to August in every of the three years, and shown accordingly with daily areal precipitation during the same period in Figure 4. It was found that the index increased from negative value to maximum value before and during the heavy rainfalls, and then decreased rapidly to minimum value at the end of the heavy rainfalls. Take the 2005 event as an example (Fig. 4a), the index increased from −1.2 on the 27th day (27 June), kept increasing to the maximum value 3.2 on the 39th day (9 July), maintained in high level during the following two days, and declined to the minimum value −2.2 on the 48th day (18 July) within a short period. Accordingly, the heavy rainfall occurred when the index was high from

4 to 11 July. A similar index fluctuation was observed from 26 June to 17 July in 2007. The index started to increase from -0.8 on the 26th day (26 June), after which the heavy rainfall started on the third day. The peak of the index was 1.9 on the 29th day (9 July), which was the last day of that heavy rainfall. The index declined to the minimum value -1.0 on the 44th day (12 July) shortly after the heavy rainfall. Moreover, the variation of the index during the 2003 event is similar yet not as obvious as that in 2005 or 2007.

4 CONCLUSIONS

This study analyzed the climatological anomalies of the 500 hPa height fields for the heavy rainfalls over the HRB in 2003, 2005 and 2007, which led to the following conclusions:

- Compared with conventional anomaly fields, the climatological anomalies are well illuminated by signal fields, which is a powerful tool to analyze small probability events with high mutability in climatic background, such as heavy rainfalls.
- Certain similarities have been observed in the distribution of the climatic anomalies in meteorological fields over the HRB for the heavy rainfalls in 2003, 2005 and 2007. Specially, the positive signals appeared to the west of the Lake Baikal and the southern of the HRB, while negative signals dominated the HRB and adjacent area. The positive-negative signals formed into a tripole of "+ − +" from mid-high to low latitudes.
- There are corresponding relations between heavy rainfall and the tripole of "+ − +" of the 500 hPa height fields in 2003, 2005 and 2007 over the HRB.

ACKNOWLEGEMENTS

This study was supported by the National Basic Research Program of China ("973" Program, Grant No. 2010CB428405), the Special Public Sector Research Program of Ministry of Water Resources (Grants No. 201301040), the Natural Science Foundation of Jiangsu Province of China (Grant No. BK20131368), the Foundation for the Author of National Excellent Doctoral Dissertation of P.R. China (Grant No. 201161), and the Qing Lan Project and Program for New Century Excellent Talents in University (Grant No. NCET-12-0842).

REFERENCES

Chen, X.H., Yu, J.L., Qiu, X.X & Zhang, J. 2007. Vapor investigation of a heavy rainfall event in the Huaihe River Basin in 2005. *Meteorological Monthly*. 33(4): 47–52.

Cheng, H.Q. & Chen, J.Y. 2004. Large scale circulation background of heavy rainstorm processes in the Huai River Valley in summer of 2003 and its precursor signal. *Progress in Geophysics*. 19(2): 465–473.

Huang, J.Y., Yang, Y. & Zhou, G.L. 2002. A study of the 500 hPa signal field about heavy rainfall in China. *Chinese Journal of Atmospheric Sciences*. 26(2): 221–229.

Kalnay, E. et al. 1996. The NCEP/NCAR 40-year reanalysis project. *Bulletin of the American Meteorological Society*. 77(3): 437–471.

Song, Q.Y., Wei, F.Y. & Xu, C.H. 2006. Numerical simulation and diagnostic analysis of a heavy rainfall process over the Huaihe River Valley. *Journal of Nanjing Institute of Meteorology*. 29(3): 342–347.

Tao, S.Y., Wei, J. & Zhang, X.L. 2008. Large-scale features of the Mei-Yu front associated with heavy rainfall in 2007. *Meteorological Monthly*. 34(4): 3–15.

Yang, Y., Zhou, G.L, Qi, J.G. & Huang, J.Y. 2005. Background of climatic anomalies about heavy rain in middle reaches of the Yangtze River in China. *Advances in Water Science*. 16(4): 546–551.

Zhang, J., Wang, D.Y., Tian, H., Zhu, H.F & Chen, X.H. 2008. Comparative analysis of atmospheric circulation characteristic over Huaihe River during the severe precipitation between 2003 and 2005. *Scientia Meteorologica Sinica*. 28(4): 402–408.

Zhao, S.X., Zhang, L.S. & Sun, J.H. 2007. Study of heavy rainfall and related mesoscale systems causing severe flood in Huaihe River Basin during the summer of 2007. *Climatic and Environmental Research*. 12(6): 713–727.

A method on long-term forecasting of meteorological droughts in the southwest of China

G.H. Lu, L. Dong & Z.Y. Wu
College of Hydrology and Water Resources, Hohai University, Nanjing, Jiangsu, China

ABSTRACT: Based on the Standardized Precipitation Index (SPI) and circulation parameters, in consideration of the nonlinear relationship between the regional drought and drought-formation factors, circulation parameters were selected as predictors to meet different linetypes by the correlation analysis. Then, four autumn droughts forecasting models for the southwest of China have been established, using multiple linear regression method. The results show that, in the calibration period from 1961 to 2004, all models pass the F test and the simulation effect of the exponential type regression model is the best. Nonlinear models hindcast the autumn droughts in the southwest of China in 2005, 2007, 2009 successfully and perform better than the linear model for hindcasting droughts in the validation period from 2005 to 2012, with a forecast period of 6 months. The nonlinear method for predictors is suitable for drought forecasting in the southwest of China, which has potential practical application in prospect.

1 INTRODUCTION

The southwest of China locates on the east of Tibetan Plateau, affected by the landform and the Southwest Monsoon. Thus, temporal and spatial distribution of precipitation is uneven, with great variations from year to year, which induces frequently occurred meteorological droughts. In recent ten years, drought-stricken area has been spreading (Dong et al., 2012), the Sichuan-Chongqing region suffered severe drought in 2006, and the whole southwest of China was hit by an extreme drought from autumn 2009 to spring 2010, which seriously influenced the industrial and agricultural production and people's living. Therefore, distinguishing the precursor signals of regional droughts and forecasting the tendency of future's droughts or floods have fundamental significance in providing scientific basis for droughts or floods prevention and disaster relief.

Statistical method is mainly used in the climate forecasting research in China, of which scholars studied statistical relation between factors and predictand, and have made some achievement in precipitation (Chen, 2013, Fan et al., 2008) and runoff (Ge et al., 2006, Liu et al., 2003) forecasting. By establishing mathematic models based on observational data and forecasting drought tendency in future, scholars have made some attempts to forecast droughts in different areas (Cheng et al., 2010, He et al., 2008, Li et al., 2007, Wei, 2003). As for the southwest of China, there are large amounts of achievement in precipitation affecting mechanization and drought formation (Hua, 2003, Ju et al., 2011, Wang et al., 2012, Yang et al., 2012, Yang et al., 2011). However, the hydrological and meteorological conditions in temporal and spatial are complicated, which causes great difficulties in drought forecasting, thus there is seldom research on drought forecasting in the southwest of China.

Therefore, it is vital to seek the key precursor signals of regional droughts by meteorological statistical analysis in the southwest of China. According to its distinctive location and complicated synoptic-climatological cause, the drought predictand and previous circulation parameters are nonlinearly processed in this paper to better characterize the relationship of different linetypes between predictors and predictand. Then, the autumn drought forecasting models in the southwest of China are established based on the multiple regression method, which has a great value in long-term forecasting of regional droughts.

2 MATERIALS AND METHODS

2.1 Regional drought index

Using the data set of "Chinese surface climate monthly precipitation in grid 0.5° × 0.5°" from China Meteorological Data Sharing Service System, taking November as a precursor signal for the period of winter and autumn drought in the southwest of China, regional autumn drought index in the southwest of China (RSPI) is computed as predictand. The specific methods are as follows: using Standardized Precipitation Index (SPI) of 576 grids in the southwest of China in November from 1961 to 2012, RSPI is computed by area-weighted average method, which is as follows:

$$RSPI = \frac{\sum SPI_i \times A_i}{A} \quad (1)$$

where A_i is the area of each grid, A is the area of the southwest of China, SPI_i is a Standardized Precipitation Index in each grid. The SPI3 is used in this study, which represents 3-month accumulative precipitation in advance (from September to November). The time scale reflects the influence of accumulative precipitation on the whole autumn droughts. The RSPI can appropriately describe the typical autumn droughts in the southwest of China in 1974, 1992, 2003, 2009 etc.

2.2 Predictors

Through drought mechanism analysis in the southwest of China, initial circulation characteristic indexes are selected, using the 74 circulation parameters extracted from the monthly data of 1960~2012 provided by the Climate Diagnostics and Prediction Division of the National Climate Center. Thirteen parameters in six zones are selected from the subtropical anticyclone category in eleven zones; six parameters in three zones are picked from the polar vortex category in five zones. Atlantic and Europe pattern C, Meridional index over Eurasian continent (IM, 0-150E) and Meridional index over Asia (IM, 60E-150E) are selected from the circulation category; the Tibetan Plateau index (25N-35N, 80E-100E) and Index of the strength of the India-Burma trough (15N-20N, 80E-100E) from Slot category and other categories indicators are also selected.

2.3 Processing methods

Firstly, in consideration of the nonlinear relationship between the predictand and predictors, function transformation is conducted to make nonlinear relationship to satisfy the linear modeling requirements. Supposing three nonlinear relationships (Huang, 2007):

- Logarithmic function type: $y = a + b \lg x$. Set $y' = y, x' = \lg x$;
- Power function type: $y = dx^b$. Set $y' = \lg y, x' = \lg x$;
- Exponential function type: $y = de^{bx}$. Set $y' = \ln y, x' = x$.

where y, x are predictand RSPI and candidate circulation factors respectively; y', x' are predictand and circulation factors supposed to match linear relationship better after the functional transformed; a, b and d are parameters.

Then, the linear correlation analysis can be made after the nonlinear relationship between predictand and indicators transformed to be linear through the functions. The circulation parameters of thirteen months from March of the current year to March of the year before are treated as preceding parameters and the correlation coefficients with RSPI are calculated individually, then the t-test (Huang, 2007) is carried out.

We need to conduct the collinearity diagnostics among the picked candidate parameters to ensure that predictors in modeling have major contributions to regression equation (Fan, 1999). The correlation coefficient is computed between the selected parameters, then those who cannot pass significant test of 0.05 do not share a collinearity.

After the three steps, the final selected preceding circulation parameters can share different linear relationship with predictand. So the selected parameters can be used to build multiple linear regression model.

2.4 Model evaluation methods

The model simulation results in verification period and calibration period are evaluated by the RMS error, the multiple correlation coefficient, the F test and the, respectively. The successful rate of forecasting method is illustrated as follows:

$$Q = N'/N \qquad (2)$$

In the formula, N is the number of years in forecasting model, N' is the number of years which has the same flood or drought trend reflected by simulation and actual values.

3 DROUGHT FORECASTING MODEL

The model calibration and verification period are 1961~2004 and 2005~2012, respectively. Twenty-five circulation indexes are selected as candidate predictors. We assume there exists linearity, logarithmic, power and exponential relations between predictand and each predictor, respectively. Afterwards, through factors screening in section 2.3 we pick out forecasting factors with satisfied linear or nonlinear relationship, and establish four multiple regression forecasting models. Among them, the predictands forecasted by power and exponential models need to be conducted with exponential transformation with 10 and e as the base, respectively, then predictand RSPI can be obtained in the southwest of China.

3.1 Forecasting model

Multiple regression forecasting models and predictors are listed in the Table 1.

3.2 Model evaluation

According to the above simulated results by four types of regression models, combined with exponential transformation of fitted values obtained from power and exponential models, the estimated predictand RSPI corresponding to four models can be obtained. The estimated and the actual values are compared through multiple methods and then model evaluation results are obtained.

In calibration period from 1961 to 2004, model fitting effects are estimated by the RMS error, the multiple correlation coefficient and the F test. The results are shown in the Table 2.

Under the condition of significance level $\alpha = 0.01$, numerator freedoms equal 8, 7 and 6, denominator freedoms are 35, 36 and 37, corresponding to the critical values of F, $F\alpha$ are 3.069, 3.183 and 3.334, respectively. Thus, in calibration period, it shows that F test value of each regression equation is far greater than the critical value, so that the fitting effect is remarkable. From the perspective of RMS error, model simulations of historical droughts are within the allowed error range and thus

Table 1. Multiple regression forecasting models.

Linetype	Forecasting models
1 Linear	$y = 0.454 + 0.014x_1 - 0.014x_2 + 0.005x_3 - 0.018x_4 - 0.041x_5 - 0.029x_6 - 0.018x_7 - 0.023x_8$
2 Logarithm	$y = 48.407 + 2.407 \lg x_1 - 1.627 \lg x_2 - 1.998 \lg x_3 + 3.338 \lg x_4 - 1.665 \lg x_5 - 18.659 \lg x_6 - 1.023 \lg x_7$
3 Power	$\lg y = 14.104 + 0.525 \lg x_1 - 0.378 \lg x_2 + 0.579 \lg x_3 - 0.583 \lg x_4 + 0.546 \lg x_5 + 0.244 \lg x_6 - 6.01 \lg x_7$
4 Exponential	$\ln y = 8.259 - 0.022x_1 - 0.012x_2 - 0.009x_3 + 0.048x_4 - 0.027x_5 + 0.007x_6$

Table 2. Significance test of the models.

Model	RMS error	Multiple correlation coefficient	F	Fα
Linear	0.267	0.796	7.557	3.069
Logarithm	0.275	0.782	8.093	3.183
Power function	0.290	0.805	9.469	3.183
Exponential	0.291	0.828	13.446	3.334

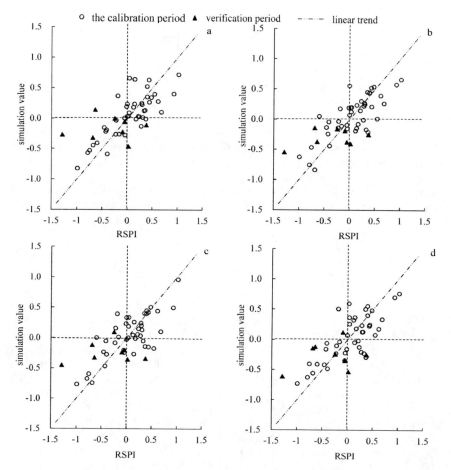

Figure 1. Simulation and forecasting of the models. a represents the linear model, b represents the logarithm model, c represents the power function model and d represents the exponential model.

the effect is satisfactory. The multiple correlation coefficient of exponential regression equation is the highest and it fits the best simulation results.

Using four models, autumn droughts are hindcasted for during 2005~2012 in the southwest of China. The results are shown in the Figure 1.

The closer the scatters in the Figure to the linear trend, the more accurately the models fits. It can be seen from the four figures that, most of the scatters are in the first and the third quadrants, which indicates that the simulation value and predictands have the same sign, suggesting that they are accurately simulated. However, the value of RSPI of the other scatters are between −0.5 and

Table 3. The comparison between the values of RSPI and forecast of three typical droughts.

Year	RSPI	Linear	Logarithm	Power function	Exponential
2005	−0.691	−0.317	−0.140	−0.114	−0.142
2007	−0.644	0.143	−0.370	−0.325	−0.120
2009	−1.291	−0.267	−0.547	−0.451	−0.613

+0.5, which indicates that the errors of simulated trend in normal state years of atypical droughts are in the reasonable ranges. The scatters in verification period intensively distribute closer to the trend line, with better simulations.

Among the four models, the success rate of logarithm model is up to 75%, while the others are 62.5%. The RMS error of logarithm model is 0.45, less than exponential function model (0.483) and power function model (0.494). It can also be found that the deviation between simulation values and actual ones of linear model, whose RMS error is 0.541, is larger than that of nonlinear model.

The values of RSPI and forecast of three typical autumn droughts in the southwest of China (the year of 2005, 2007 and 2009) are shown in the Table 3.

The three nonlinear models successfully hindcast these droughts, while the linear model fails to recognize the drought occurred in autumn 2007.

3.3 Predictors analysis

Among the 11 zones of subtropical high, the zones possessing significant influence on autumn droughts in the southwest of China are located in the Western Pacific (110E-180), the North American-Atlantic (110W-20W), the North African, Atlantic and North American (110W-60E), the Indian Ocean (65E-95E), the North African (20W-60E) and the Atlantic (55W-25W). And there is a significant nonlinear relationship between the North Africa or the Atlantic Ocean subtropical high and the autumn droughts in the southwest of China, while others represent the great linear relationship. Among the 5 polar vortex zone, the Atlantic European zone (the fourth, 30W-60E) and the Northern Hemisphere polar vortex (the fifth, 0-360) make the greatest contribution to regression models. The Meridional index over Eurasian continent (IM, 0-150E) and over Asia (IM, 60E-150E) appear abnormal strong signals ahead of around 8 months and one and a half years, respectively, affecting autumn droughts in the southwest of China. Also, among the circulating tanks indexes, the Tibetan Plateau (25N-35N, 80E-100E) 500 hPa geopotential height contributes greatly to the non-linear models, and there is a strong predictive signal in summer of the previous year.

4 CONCLUSION

Considering the predictand and circulation parameters should meet different linear correlations, the predictors that meet the linear, logarithmic type, the power function type and exponential function type are selected by using correlational analysis method. After function transformation, four types of autumn drought forecasting models in the southwest of China have been established, based on multiple linear regression methods. Conclusion are drawn as follows:

- In the calibration period from 1961 to 2004, all models pass though the 0.01 significance level of F test and the fitting effect is significant. Exponential regression equation is the best for simulating the historical droughts.
- According to the hindcast results in the validation period from 2005 to 2012, the forecast accuracy of three nonlinear models is better than linear model. And the logarithmic model has the highest success rate and the best trend simulation. All models are successful in forecasting three typical autumn droughts except the linear model, which fails in forecasting autumn drought in 2007.

– Overall, the nonlinear models perform better than the linear model, which proves that drought factors and autumn drought is not in a simple linear relationship. This paper establishes regression models by dealing with factors with nonlinear functions. This approach of forecasting widespread regional droughts with a forecasting period of more than half year has significant guiding meaning and practical application value.

ACKNOWLEDGEMENTS

This work is supported by the Special Public Sector Research Program of Ministry of Water Resources (Grant No. 201301040), the Natural Science Foundation of Jiangsu Province of China (Grant No. BK20131368), the National Natural Science Foundation of China (Grant No. 41001012), the Foundation for the Author of National Excellent Doctoral Dissertation of PR China (Grant No. 201161), and the Qing Lan Project and Program for New Century Excellent Talents in University (Grant No. NCET-12-0842).

REFERENCES

Chen, H. 2013. Statistical prediction model for summer extreme precipitation events over Huaihe River valley (in Chinese). Climatic and Environmental Research, 18, 221–231.
Cheng, Y.Q., Zhang, S.W. & Xu, Y.Q. 2010. Study on the Forecasting Method of Drought Intensity over Chifeng, Inner Mongolia in Summer (in Chinese). Meteorological Monthly, 1, 49–53.
Dong, L., Lu, G.H. & Wu, Z.Y. 2012. Analysis on Evolution Characteristics of Meteorological Drought in Southwest China during Recent 50 Years (in Chinese). New Development of Hydrological Science and Technology in China – 2012 China Hydrological Symposium Nanjing, Jiangsu Province, China.: Hohai University Press.
Fan, K., Lin, M.J. & Gao, Y.Z. 2008. Forecasting north China's summer rainfall (in Chinese). Sci China, 38, 1452–1459.
Fan, Z.X. 1999. The Mid-long Term Hydrological Forecasting (in Chinese), Nanjing, Jiangsu Province, China., Hohai University Press.
Ge, Z.X., Gu, Y.H., Cao, L.Q. & Qiang, X.M. 2006. Influences of sea surface temperature and meteorological factors on monthly runoff of Hongjiadu Hydropower Station (in Chinese). Journal of Hohai University (Natural Sciences), 34, 606–609.
He, X.X., Wu, H.B. & Chen, X.L. 2008. Ensemble Canonical Correlation Forecasting of Summer Drought Index over Southeastern China (in Chinese). Journal of Nanjing Institute of Meteorology, 31, 10–17.
Hua, M. 2003. Analysis and Simulation Study on the Influence of Heat Condition over Qinghai-Xizang Plateau on Climate over South-West China (in Chinese). Plateau Meteorology, 22, 152–156.
Huang, J.Y. 2007. Meteorological Statistical Analysis and Forecasting Method (the Third Edition), Beijing, China, China Meteorological Press.
Ju, J.H., Lv, J.M., Xie, G.Q. & Huang, Z.Y. 2011. Studies on the Influences of Persistent Anomalies of MJO and AO on Drought Appeared in Yunnan (in Chinese). Journal of Arid Meteorology, 29, 401–406.
Li, X.J., Zeng, Q., Liang, J., Ji, Z.P. & Xie, D.S. 2007. Climatic Forecasting of Droughts in South China (in Chinese). Meteorological Science and Technology, 35, 26–30.
Liu, L. P., Chen, C. M., Chen, J., Zhang, M. L. & Wang, Y. S. 2003. Principal components and prediction of annual maximum discharge in Beijiang River in Guangdong province. Advances in Water Science, 14, 385–388.
Wang, X.M., Zhou, S.W. & Zhou, B. 2012. Causative Analysis of Continuous Drought in South-west China from Autumn 2009 to Spring 2010 (in Chinese). Meteorological Monthly, 38, 1399–1407.
Wei, F.Y. 2003. A Predictingdrought Model with an Integration of Multi-scale in North China (in Chinese). Journal of Applied Meteorological Science, 14, 583–592.
Yang, H., Song, J., Yan, H.M. & Li, C.Y. 2012. Cause of the Severe Drought in Yunnan Province during Winter of 2009 to 2010 (in Chinese). Climatic and environmental research, 17, 315–326.
Yang, S.Y., Zhang, X.N., Qi, M.H. & Niu, F.B. 2011. Significant climate characteristic analysis on extremely less rain over Yunnan in autumn, 2009 (in Chinese). Journal of Yunnan University (Natural Sciences), 33, 317–324.

Comparison on river flow regime between pre- and post-dam

X.S. Ai & J.J. Tu
State Key Laboratory of Water Resource and Hydropower Engineering Science, Wuhan University, Hubei, China

Z.Y. Gao
Hubei Urban Construction Vocational and Technological College, Wuhan, Hubei, China

S. Sandoval-Solis
Department of Land Air and Water Resource, University of California, Davis, USA

Q.J. Dong
State Key Laboratory of Water Resource and Hydropower Engineering Science, Wuhan University, Hubei, China

ABSTRACT: The nature flows are the suitable processes for the river form and ecosystem, whereas the big dam built on the river great changed the nature flows regimes. This paper analyzes the characteristics of daily runoff series from 1952 to 2010 of Dongjiang section on Leishui River in china. Based on the analyzing results of IHA (Indicators of Hydrologic Alteration), the paper researches the characters of the flows before the Dongjiang dam build, and outflows after the dam operation. The paper uses two main factors, basic monthly flow and special daily flow. Basic monthly flow adopts the median flow of each month; the special daily flow considers four benchmarks and five especially flows. There are extreme flood events, small floods, high flows, low flows and extreme low flows. For large flood and small flood events, peak flows, frequency, return periods and the shape of flood are concerned with. For high flows, the magnitude, timing and frequencies are obtained. The case study shows that the median flows are reversed after the dam operation, and the large flood and small flood are not occurred post-dam. The results of comparing pre- and post-dam shows that the Dongjiang dam building has greatly changed the rivers nature flow regimes, we must pay more attention to the dam's reoperation in order to protect the river ecological system.

1 INTRODUCTION

A river is not only channel flow but also a set of structures, processes and interactions that can provide services to social and economic activities (Postel and Ritcher, 2003). From a hydrologic perspective, rivers play a central role in the global cycling of water between the sea, air, and land. Adequate stream flow regimes protect water quality, filter and decompose pollutants, and help maintain soil fertility (Thompson et al., 2012). The variations in river discharge determine flow velocity and bed morphology, as well as timing and duration of inundation of levees and floodplains. These conditions trigger ecosystems responses, such as migration or spawning (Poff et al., 2010). However, human actions alter rivers in numerous ways, one threat to river health looms over the others, a force of ecosystem decline that has quite literally reached geologic proportions, the alteration of natural river flows by dams, diversions, levees, and other infrastructure. The regulation of reservoirs, which modulates the natural water flow downstream of these dams, has often been a major cause of ecological impacts (Bunn and Arthington, 2002).

Dongjiang reservoir is located in Zixing City, Hunan Province, China. It is a big reservoir with the tasks of hydropower electricity, shipping, water available, etc. There is not considered the environment flow as an integral part of the water management in its operation figure. The main objective of this study is characterizing and comparing extreme flood events, small floods, high flows, low flows and extreme low flows in the Dongjiang reach for two periods, prior the dams alteration (pre-1992) and after (post-1993).

2 COMPARISON OF THE RELEASE BETWEEN PRE- AND POST-DAM OF DONGJIANG

2.1 Thresholds of the daily flows

The dam of Dongjiang was used from 1993, so before this year, under stream flows are in nature state of the runoff, and after that time, the runoffs are controlled by the dam and the director's method. So the two hydrologic periods have been analyzed: 1) pre-1992, a 41 year period of analysis (Mar/1952 to Feb/1993); and 2) post-1993, which data for 18 years and Mar/1993 to Feb/2011. For the post-1993 the last eighteen years period has been chosen with the aim of analyze the hydrological characteristics of the immediate hydrology.

In order to characterize different conditions, the mean daily flows at Dongjiang have been analyzed. Median monthly flow is the median value of the mean daily flows for each month. The daily flows have been categorized large floods, small floods, high flows, low flows and extreme low flows according to the following *benchmarks*.

- Extreme low flows with a peak under 10% (29.6 m^3/s).
- Low flows are flows with a peak between 10% (29.6 m^3/s) and 75% (143 m^3/s).
- High flows are flows with a peak between the 75% (143 m^3/s) and 2 year return interval event (1320 m^3/s) of the mean daily flows for the pre-1992 period.
- Small floods are flows with a peak between the 2 year return interval event (1320 m^3/s) and 10 year return interval event (2192 m^3/s) of the mean daily flows for the pre-1992 period.
- Large floods are flows with a peak above 10 year return interval event (2192 m^3/s). These large floods will definitely threaten the safety of the riversides.

The thresholds for the benchmarks have been used in the pre-1992 data in order to: 1) determine the characteristics (frequency, peak flows and duration) of the benchmarks that occurred prior the alteration of the channel, and 2) compare the frequency of these benchmarks prior alteration (pre-1992) with the post alteration period (post-1993). The median values have been used to determine characteristics hydrographs of high flows and small floods for each month and period of analysis (pre-1992 and post-1993).

The Indicators of Hydraulic Alteration (IHA) software developed by The Nature Conservancy (The Nature Conservancy, 2009) has been used to label the different flows along the periods of analysis. High flows and small floods have been characterized by determining their hydrograph for each period of analysis. Return periods for high flows and small floods have been calculated using the frequency of these benchmarks in the respective period of analysis.

A post-processing analysis was required to determine the typical hydrograph of pulses, small floods and large floods. For pulses, each pulse event was tagged and grouped for each month. Within each month pulses were arranged and centered around their peak; the typical pulse hydrograph for each month is composed of the median values for each day. The pulse duration (in days) was determined as the period with flows greater than the *normal flow* for that particular month. The same procedure was applied for small floods and large floods and for pre- and post-alteration periods.

2.2 Flow characters of pre-1992

2.2.1 Medium flows

Prior to 1992, medium flows of every month varied from 38 m^3/s in December to 209 m^3/s in May. The highest medium month flow is in May, and then June and April, it is decreased separately to December and January.

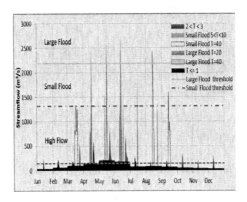

Figure 1. Pre-1992 annual hydro-characteristics.

2.2.2 *High flows*

From the pre-1992 hydrographs a total of 16 high flow pulses ($T = 1$ and $2 < T < 3$) were identified. On average pulses with $T = 1$ occurred once per month between February and September and twice in April. Pulses with a return period of 2–3 years ($2 < T < 3$) occurred mainly in the months January, March, May, June, October, November and December. The high flows occurred in May, June and April are larger than others.

2.2.3 *Small flood*

Prior 1992, small floods typically occur in March and May through October. There are fix small floods occurred, shown in figure 1. The four large floods occurred from April to September have a time period of 5–10 years, the others occurred in March and October have a time period of 40 years. durations range from 2 to 6 days and peak daily flows from 1,342 to 1,930 m^3/s; the total flow in any small flood event is from 13.0×10^6 m^3 to 45.5×10^6 m^3.

Small floods provide migration and spawning cues for fish, trigger new phase in life cycle (i.e insects), enable fish to spawn in floodplain, provide nursery area for juvenile fish, provide new feeding opportunities for fish, waterfowl, recharge floodplain water table, maintain diversity in floodplain forest types through prolonged inundation (i.e. different plant species have different tolerances), control distribution and abundance of plants on floodplain, deposit nutrients on floodplain etc.

2.2.4 *Large flood*

Large floods of pre-1992 have been analyzed at the Dongjiang gauge station. These events typically occur in April through June and August. There are four large floods occurred, shown in figure 1. The two large floods occurred in April and June have a time period of 20 years, two others occurred in May and August have a time period of 40 years. durations range from 2 to 6 days and peak daily flows from 2,440 to 2,740 m^3/s; the total flow in any large flood event is from 7.3×10^6 m^3 to 39.6×10^6 m^3.

These occurrences, along with frequent smaller floods moved sediment and prevented any substantial accumulation in the canal, maintaining a wide, sandy multi-threaded river with regular resetting events. Large floods apply to small and large floods, Maintain balance of species in aquatic and riparian communities, Create sites for recruitment of colonizing plants, Shape physical habitats of floodplain, deposit gravel and cobbles in spawning areas, flush organic materials (food) and woody debris (habitat structures) into channel, purge invasive, introduced species from aquatic and riparian communities, disburse seeds and fruits of riparian plants, drive lateral movement of river channel, forming new habitats (secondary channels, oxbow lakes), Provide plant seedlings with prolonged access to soil moisture etc.

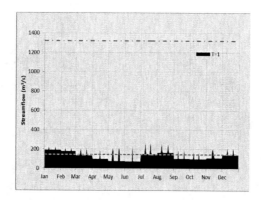

Figure 2. Post-1993 annual hydro-characteristics.

2.3 Flow characters of post-1993

Post-1993, median monthly values varied from 75 m^3/s in June to 191 m^3/s in January. The highest month is January, and the lowest month is June, from January to June the medium flow of every month is decreased step by step, and In July and August the medium flows is higher than 150 m^3/s, in December it is 141 m^3/s. A median of 21 high flows occurred each month per year; and twice in every month except June, August and January.

There are no small floods or large floods occurred at this period. Figure 2 shows a proposed annual hydrograph for the post-1993 conditions; it includes the benchmarks for 1 year return period. Values for the median monthly flow and the peak flows of the benchmarks are not occurred in the figure.

2.4 Compare the flows characters of pre- and post-dam

Comparing both hydrographs it can be concluded that part of what happened every year in the pre-1992 period is not happening in the post-1993 period. In the post-1993 hydrograph there is a proliferation of high flows but not of small floods. After analyzing different annual hydrographs for different return periods, these are the results:

- Small floods in the pre-1992 conditions were frequent (medium peak flow = 1612 m^3/s), once every two years; they are happened in April, June, July, August, September or October. There is no small flood in the post-1993 conditions.
- Large floods that occurred every seven years (T = 7) in the pre-1992 conditions happen every 7 years; they are happened in May, June, July and October. There is no large flood in the post-1993 conditions.
- High flows occurred every month between February and September each year in the pre-1992 period and happen almost two times every month each year, there are more frequently than before.

Medium flows of each month have changed a lot, especially the hydrological regime, the flow processes have inversed the sequence of the nature states totally.

3 CONCLUSIONS

The characterization of flood events, and the comparison of their results has shown how significantly different is the hydrology prior and after alteration. The change in the hydrology is the result of the infrastructure and human water management that has completely modified the hydrologic characteristics of the river. The quantity, frequency, and timing of the high flows have been severely altered, there are no flood occurred after the dam operation. As a result, the environment has been deteriorated.

Although hydrologists and ecologists have done some researches on environmental restoration flows and on how to operate the system in order to meet these flows, but there still has not an effective and acceptable method to control the release of the dam. This research intends to be the foundation of discussions and constructive opinions from academic, scientists and the environmental community. We need more and more people enter into this field and consider how to protect our river even there are some dams on the river.

ACKNOWLEDGMENTS

This research was funded by the National Natural Science Foundations of China (NSFC) (No. 51179130 and 51190094).

PREFERENCES

Bunn, S.E., Arthington, A.H., 2002. Basic principles and ecological consequences of altered flow regimes for aquatic biodiversity. Environmental Management 30, 492–507.

Poff, N.L., Richter, B.D., Arthington, A.H., Bunn, S.E., Naiman, R.J., Kendy, E., Acreman, M., Apse, C., Bledsoe, B.P, Freeman, M.C., Henriksen, J., Jacobson, R.B., Kennen, J.G., Merritt, D.M., O'Keeffe, J.H., Olden, J.D., Rogers, K., Tharme, R.E., Warner, A. 2010 The ecological limits of hydrologic alteration (ELOHA): a new framework for developing regional environmental flow standards. Freshw Biol 55: 147–170.

Postel, S., Richter, B. 2003. Rivers for Life: Managing Water for People and Nature, Island Press, Washington, D.C.

The Nature Conservancy. 2009. Indicators of hydrologic alteration software.

Thompson, L., Escobar, M., Mosser, C.M., Purkey, D.R., Yates, D., Moyle, P.B. 2012. Water management adaptations to prevent loss of spring-run Chinook salmon in California under climate change. Water Resources Planning and Management 138(5):465-478. ISSN:0733-9496.

Spatial-temporal variation and trend of meteorological variables in Jiangsu, China

Z.Y. Wu, Y. Mao, G.H. Lu & Y. Yang
College of Hydrology and Water Resources, Hohai University, Nanjing, China

J.H. Zhang
Department of Water Resources of Jiangsu Province, Nanjing, China

ABSTRACT: The Spatial-temporal variation and trend of meteorological variables were analyzed using the Mann-Kendall and Morlet wavelet methods. The daily data from 32 meteorological stations and 403 rainfall stations in Jiangsu Province from 1956 to 2011 were used for analysis. The results showed that the spatial and temporal distributions of meteorological variables were uneven in Jiangsu Province. In space, precipitation gradually increased from northwest to southeast, showing an increasing trend in southern but an opposite trend in northern. Temperature showed a rising trend in all regions of the province. Pan evaporation decreased from south and north to the central region with a decreasing trend in the whole province. In time, the inner-annual distribution of precipitation was uneven and concentrated in summer. There was a notable increasing trend in temperature from 1956 to 2011 and decreasing trend in pan evaporation from 1980 to 2011.

1 INTRODUCTION

Jiangsu is an eastern coastal province of China, locating in the transitional zones of a humid subtropical climate and a humid continental climate. Under the background of global warming in recent years, the climatic characteristics have changed gradually, leading to a high frequency of meteorological disasters such as extreme rainstorms and droughts, impacting on the sustainable development of social economy severely. As the main meteorological variables, precipitation, temperature and evaporation reflect the regional climate characteristics directly. Thus, analyzing the spatial-temporal variation and the trends is significant to use the regional water resources reasonably and promote the economic and social development.

In recent years, the studies in variation characteristics of meteorological variables in Jiangsu Province focused on temperature and precipitation, whereas there are only a few studies in evaporation. The studies of temperature focuses on the spatial-temporal variation of temperature abnormality (Sun et al., 2007, Zhu et al., 2013, Zhou et al., 2006, Sun et al., 2009, Wu et al., 2003) and that of precipitation concentrates upon the spatial-temporal variation, the anomaly characteristics and the trend of precipitation (Chen et al., 2003, Deng et al., 2004, Qin, 2012, Qiu et al., 2008, Wang, 1992). However, as for that of evaporation, only pan evaporation capacity is mentioned (Min et al., 2010, Wan and Ma, 2005, Zeng et al., 2007). A lot of results has been obtained in these studies, which builds a solid foundation of intensive study on hydrological and meteorological variables in Jiangsu Province. However, these studies were only based on the monthly data in about 60 meteorological stations covering a total area of 100,000 square kilometer, which cannot reflects the spatial non-uniformity of meteorological variables, especially for precipitation, due to the few and scattered stations. In this study, we combined 32 meteorological stations and 403 rain gauge stations to generate the new daily temperature, precipitation, and pan evaporation data in Jiangsu Province. Based on the new data, the spatial-temporal variation are analyzed in 1956–2011, and some new variation characteristics of meteorological variables in Jiangsu Province of recent years are found.

Figure 1. The distribution of meteorological stations and rain gauge stations in Jiangsu Province.

2 MATERIALS AND METHODS

2.1 *Materials*

This study used the daily temperature and precipitation data from 32 meteorological stations during 1956–2011, the daily precipitation data from 403 rain gauge stations during 1956–2011 and daily pan evaporation data gauged by E601 from 21 stations during 1980–2011. The detailed information for the evaporation data can be found in Wan and Ma (2005). The distribution of stations is shown in the Figure 1. The data was interpolated into 728 grids of 0.125° × 0.125° in Jiangsu Province with inverse distance method, then the regional average temperature, precipitation and pan evaporation were computed based on grid data.

2.2 *Methods*

2.2.1 *Mann-Kendall (MK) non-parametric test*

MK trend test is a nonparametric rank correlation statistical testing method, it is one of the most efficient tools to extract the sequence trends. MK trend test has been widely used in the analysis of climate parameters and significant trends in hydro-meteorological time series. Following equations can be found in Qiu et al. (2008).

We can judge the value of MK (U_{MK}). While n > 10, U_{MK} converges to normal distribution. When $|U_{MK}| < U_{\alpha/2}$, previous assumption accepted which means the variance trend is not significant. If $|U_{MK}| > U_{\alpha/2}$, then we deny the given assumption, which means the variance trend is significant (Zhou, 2005).

2.2.2 *Morlet wavelet analysis*

Given many deficiencies in traditional methods to multiscale analysis for hydrological series, Morlet, inheriting and developing the Fourier transform, proposed Wavelet analysis which includes wavelet function and wavelet transform (Wang et al., 2002). Formulas and Principles can be found in Hu et al. (2006).

3 RESULTS AND ANALYSIS

3.1 *Distribution characteristics of precipitation, air temperature and pan evaporation*

Figure 2 is the distribution of annual mean precipitation, air temperature and pan evaporation for Jiangsu province. It can be found that temperature showed increasing trend from south to north, while precipitation was increasing gradually from northwest to southeast and more pan evaporation capacity was showed in Xuzhou, Taizhou, Zhenjiang, Yangzhou etc.

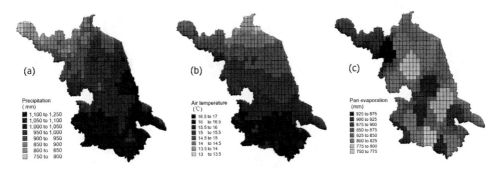

Figure 2. The distributions of annual mean precipitation, temperature and pan evaporation in Jiangsu province.

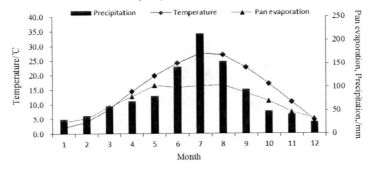

Figure 3. Monthly temperature, precipitation and pan evaporation in Jiangsu province.

Figure 3 shows monthly mean temperature, precipitation and pan evaporation in Jiangsu province. The changing processes of three meteorological variables within a year were similar to each other. Precipitation had obvious seasonal distribution characteristics that it is mainly concentrated in summer (from June to August) during which total precipitation was up to 512.1 mm, almost accounted for half of annual precipitation. Conversely, the amount of winter precipitation (from December to next February) is 95.4 mm, only accounted for one tenth of annual precipitation and it is smaller than any other season. Meanwhile, precipitation in spring (from March to May) and autumn (from September to November) accounted for about 20% of annual precipitation respectively.

3.2 *Variation and trend analysis*

3.2.1 *Inter-annual variation and trend*

Figure 4 shows interannual variation hydrographs of average temperature, precipitation and pan evaporation in Jiangsu province. Figure 4(a) is the one for average temperature with the period from 1956 to 2011. We could see that the average annual temperature fluctuated between 14.3 to 16.9°C in recent 56 years experiencing a changing process from low to high. In addition to the effects of global warming background, the rapidly increase trend may also have a certain relationship with the development of the second industry and tertiary industry. Meanwhile, agriculture land reduction, thermal power generation and traffic development all will make further efforts on gas emissions increase and exacerbate temperature rise.

The change of annual precipitation in Jiangsu Province is shown in Figure 4(b). The average annual precipitation of the 56 years was 1003.7 mm, but there was a wide variation in inter-annual precipitation, and the difference between the least (537.9 mm) in 1978 and the most (1407.1 mm) in 1991 was nearly 870 mm. The trend line indicated that the change of annual precipitation in the province was not significant. And value of MK-test in precipitation was 0.389, not passing the significance test at the significant level of 95%. Figure 4(c) shows the annual pan evaporation in the recent 32 years in Jiangsu Province. Pan evaporation was between 759.3 to 968.6 mm during

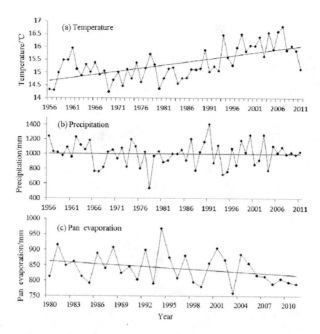

Figure 4. Long-term changes of temperature, precipitation and pan evaporation in Jiangsu province.

1980–2011. The figure indicates that the pan evaporation was decreasing significantly in recent 32 years with a decreasing amplitude of 14.9 mm per decade. The value of MK-test in pan evaporation was −3.049, passing the significance test at the significant level of 99%, confirmed the truth that the pan evaporation is conspicuously decreased in the 32 years in the province.

The research shows that the average annual temperature was increasing while the precipitation was decreasing in the past 56 years, and the pan evaporation was in the stable decreasing trend in the recent 32 years in Jiangsu Province, which may result from the decreasing total solar radiation (Zeng et al., 2007).

3.2.2 Inner-annual variation and trend

Values of MK-test of monthly temperature, precipitation and pan evaporation in Jiangsu Province are shown in table 1. From the table, temperature in Jiangsu Province increases obviously, and temperature variation in summer was not obvious while that in spring was the most significant. Also, the temperature in March to May all passed the significance test at the significant level of 99%. We also find that the variation in autumn and winter was significant but inferior to that in spring. Meanwhile, the precipitation variation in February and September was notable. Both of the variations passed the significance test ($\alpha = 0.05$), and that in February showed rising trend while that in September was contrary.

The trends of monthly pan evaporation in Jiangsu Province in the years during 1980–2011 were decreasing except that in March. And values of MK-test in July and August passed the significance test ($\alpha = 0.05$), indicating a rapid decreasing trend.

3.2.3 Spatial variation of change trend

The MK statistical values of hydrological and meteorological variables of these 13 districts in Jiangsu province are shown in table 2.

As show in table 2, temperature of 13 districts was significantly increased from 1956 to 2011. The MK statistical value of Zhenjiang was as high as 4.99, and the rising trend of Nanjing is slightly smaller than others. The MK statistical values of these 13 districts had indicated an obvious variation of spatial distribution of precipitation tendency. The precipitation of those 5 districts in Northern Jiangsu showed a slight decreasing trend, while those 8 districts in Southern Jiangsu was tending to

Table 1. MK statistics of the monthly average temperature, precipitation and pan evaporation in Jiangsu.

Month	Temperature	Precipitation	Pan evaporation
Jan	2.52*	1.38	−1.69
Feb	2.75**	2.11*	−0.91
Mar	2.61**	1.11	1.17
Apr	3.24**	−1.32	−0.58
May	4.63**	−0.4	−0.91
Jun	1.8	0.45	−1.14
Jul	−0.11	0.4	−2.11*
Aug	0.04	1.04	−2.47*
Sep	3.12**	−2.37*	−1.95
Oct	4.02**	−0.93	−0.62
Nov	2.42*	−0.26	−0.32
Dec	2.07*	1.48	−0.78

*means that the statistics get though the test at significant level 95%,
**means getting though the test at significant level 99%

Table 2. MK statistics of temperature, precipitation and pan evaporation of 13 districts in Jiangsu.

Districts	Temperature	Precipitation	Pan evaporation
Xuzhou	4.42**	−1.42	−0.44
Lianyungang	4.88**	−0.13	−1.9
Suqian	4.06**	−0.69	−1.12
Huaian	4.16**	−0.19	−1.44
Yancheng	4.38**	−0.7	−2.42*
Yangzhou	4.60**	1.39	−2.64**
Taizhou	4.93**	1.11	−4.59**
Nantong	4.45**	0.83	−4.01**
Suzhou	4.67**	1.46	−3.81**
Wuxi	4.95**	2.01*	−4.3**
Changzhou	4.94**	2.13*	−3.45**
Zhenjiang	4.99**	2.16*	−2.84**
Nanjing	4.05**	1.76	−1.99*

*means that the statistics get though the test at significant level 95%,
**means getting though the test at significant level 99%.

increase, of which the MK statistics values in Wuxi, Changzhou and Zhenjiang passed the significance test ($\alpha = 0.05$) indicating that precipitation increasd a lot. During the period of 1980 to 2011, the pan evaporation decreased around all the districts of Jiangsu province, also along with an obvious variation of spatial distribution. Huaian, Suqian, Lianyungang and Xuzhou, which all lies to the north of Yancheng, showed no obvious decline trend, while districts lying to the south of Yancheng, including Yancheng, showed an obvious decline trend. Besides, the pan evaporation of Taizhou decreased the most significantlty among recent 32 years in Jiangsu province, followed by Nantong.

Table 2 shows a significant trend of meteorological variables of Wuxi, Changzhou and Zhenjiang, this may have an important impact on the economic and social development in these three districts.

4 CONCLUSIONS

(1) In space, annual mean temperature increased from high latitudes to low latitudes and precipitation decreased from southeast to northwest in Jiangsu Province. And in time, precipitation mainly concentrated in summer, which accounts for 50% of the annual precipitation amout.

(2) From 1956 to 2011, the annual mean temperature of Jiangsu Province showed an obvious increasing trend and inner-annual variation. The change of temperature was not significant in summer, while in spring, autumn and winter, it showed an obvious rising trend. Besides, the annual precipitation increased slowly. However, the pan evaporation decreased significantly with a decreasing amplitude of 14.9 mm/10a over the past 32 years.
(3) In recent years, the temperature over all 13 districts of Jiangsu Province increased obviously, with a significant downward trend of pan evaporation. However there are differences in spatial distribution of precipitation. Precipitation in northern Jiangsu showed a downward trend, but an upward trend in the central and southern regions of Jiangsu Province where the pan evaporation decreased significantly.

ACKNOWLEDGEMENTS

This work is supported by the Natural Science Foundation of Jiangsu Province of China (Grant no. BK20131368), the National Natural Science Foundation of China (grants No. 41001012), the Foundation for the Author of National Excellent Doctoral Dissertation of PR China (Grant no. 201161), the Special Public Sector Research Program of Ministry of Water Resources (Grants No. 201301040), the Qing-Lan Project and Program for New Century Excellent Talents in University (Grant no. NCET-12-0842).

REFERENCES

Chen, H. S., Zhu, W. J., Wang, Z. W. & Ni, D. H. (2003) Summer Anomalous Precipitation in Jiangsu Province and Its Possible Mechanism (in Chinese). *Journal of Nanjing Institute of Meteorology*, 721–732.
Deng, Z. W., Zhou, X. L. & Chen, H. S. (2004) Spatial Variation of Precipitation Trend And Interdecadal Change In Jiangsu (in Chinese). *Journal of Applied Meteorological Science*, 696–705.
Hu, A. Y., Guo, S. L., Chen, H. & Guo, H. J. (2006) Multiple Time Scale Analysis of the Han River Runoff Based on Wavelet Analysis (in Chinese). *Yangtze River*, 61–62+89.
Min, L., Yanjun, S., Yan, Z. & Changming, L. (2010) Trend in pan evaporation and its attribution over the past 50 years in China. *Journal of Geographical Sciences*.
Qin, C. Y. (2012) Precipitation Anomalies and the Relationship between Large-scale Atmospheric Circulation Field In Jiangsu (in Chinese). *Strengthening science and technology foundation for promoting the modernization of weather – the 29th annual meeting of China meteorological society.* Shenyang, Liaoning, China.
Qiu, X. F., Zhang, X. L., Zeng, Y. & Wu, J. G. (2008) Precipitation Variation Trend in Jiangsu Province From 1961 to 2005 (in Chinese). *Meteorological Monthly*, 82–88.
Sun, Y., Zhang, X. L., Tang, H. S. & Yin, D. P. (2009) Spatia-l temporal variations of summer temperature In Jiangsu (in Chinese). *Scientia Meteorologica Sinica*, 133–137.
Sun, Y., Zhang, X. L., Zhao, X. Y., Zhang, J. & Liu, M. (2007) Spatial and Temporal Variations of Winter Temperature Anomalies in Jiangsu Province (in Chinese). *The Chinese meteorological society annual conference 2007.* Guangzhou, Guangdong, China.
Wan, X. L. & Ma, Q. (2005) Characteristics Analysis Of Evaporation Capacity In Jiangsu Province (in Chinese). *Jiangsu Water Resources*, 29–30.
Wang, B. M. (1992) The Change of Rainfall During The Last 40 Years In Jiangsu Province (in Chinese). *Scientia Meteorologica Sinica*, 295–302.
Wang, W. S., Ding, J. & Xiang, H. L. (2002) Application and prospect of wavelet analysis in hydrology (in Chinese). *Advances In Water Science*, 515–520.
Wu, Z. W., Cao, N. H. & Zhu, X. Y. (2003) Variation of 1961-1998 Annual Temperature Background Features. *Scientia Meteorologica Sinica*, 332–338.
Zeng, Y., Qiu, X. F., Liu, C. M., Pan, A. D. & Gao, P. (2007) Changes of pan evaporation in China in 1960–2000 (in Chinese). *Advances In Water Science*, 311–318.
Zhou, F. (2005) Comparative Study in Hydrological Sequence Trend Analysis (in Chinese). *Pearl River*, 35–37.
Zhou, X. L., Gao, Q. J., Deng, Z. W. & Chen, H. S. (2006) Long-Term Temperature Trends and Spatial Patterns of the Inter-Decadal Variations in Jiangsu (in Chinese). *Journal of Nanjing Institute of Meteorology*, 196–202.
Zhu, Z., Tao, F. L. & Lou, Y. S. (2013) Temperature Change Characteristics and High Temperature Hot against Rice in Jiangsu Province from 1980 to 2009 (in Chinese). *Jiangsu Agricultural Sciences*, 311–315.

Trichloroethylene biodegradation under sulfate reduction conditions

Ying Guo & Kang-ping Cui
School of Resources and Environmental Engineering, Hefei University of Technology, Hefei, China

ABSTRACT: Simulated experiments were carried out to study the effects of sulfate reduction on the TCE biodegradation with a series of sulfate and TCE concentration ratios (1, 5, 10, 15 and 20). The results showed that: sulfate reduction can strengthen the TCE degradation. Within the scope of this experimental study, TCE degradation performance enhances, with the increase of the mass concentration ratio. Optimal dosing ratio is 20. There is a certain role between sulfate reduction and TCE degradation in promoting mutual.

1 INTRODUCTION

Volatile chlorinated hydrocarbons are important organic solvents and chemical raw materials, and widely used in refrigerants, pesticides, rubber industry, degreasing and detergent manufacturing, etc. (Zuo, 2003). Because of improper handling of waste as well as some storage tank leaks in the production and application process, volatile chlorinated hydrocarbons have become the most widespread groundwater organic pollutants (Liu, 2006; Lu, 2002; Meng, 2005), of which the most common contaminant is trichloroethylene (TCE). Moreover, TCE and its degradation products chloride (DCE) and vinyl chloride (VC) are carcinogenic. In particular VC has caused great human health threats (He, 2002; Aulenta, 2002). In the natural environment, (including TCE) volatile chlorinated hydrocarbons can be degraded by anaerobic microorganisms. Therefore, the use of anaerobic degradation of these compounds is of important practical significance.

In the natural environment, TCE, whose decay rate is very slow, migrates through the groundwater, and contaminates the groundwater (Bruin, 1992; Shen, 1998). Groundwater system is usually anaerobic environment. Anaerobic biodegradation pathways primarily are produced by reductive dechlorination degradation under reducing conditions (Wu, 2007). For this reason, many researchers have proposed the use of microorganisms in the anaerobic environment degradation TCE. In the process of microbial degradation of TCE, adding some of metabolic substrates, such as sodium lactate, can improve the TCE degradation. At the same time, anaerobic degradation of organic matter also occurs in the sulfate-reducing conditions. In anaerobic environment, the two main types of microbes as methanogens and sulfate-reducing bacteria, the competition between them is a key factor affecting TCE degradation (Jia, 2003; Chen, 2004). Related studies have also shown that many persistent organic compounds can promote the degradation under sulfate-reducing conditions (Hood, 2008; Seungho, 2004). Research on sulfate-reducing bacteria and methanogens degradation of TCE has not been reported yet. In this paper, TCE was studied as the goal contamination, setting groups in accordance with sulfate and TCE concentration ratios to study the role of sulfate about reduction degradation of TCE in groundwater. So we could determine the most appropriate sulfate dosing ratio, designed to provide theoretical and practical basis for the bioremediation of contaminated groundwater.

2 MATERIALS AND METHODS

2.1 Materials and experimental apparatus

Reagents: TCE (HPLC grade), sodium lactate (AR), sodium sulfate (AR). Inorganic culture solution mass concentration of each component (g/L): KH_2PO_4, 0.48; Na_2CO_3, 0.8; NH_4Cl, 2.0; $CaCl_2$, 0.02. Trace element solution concentration of each component(g/L): H_3BO_3, 0.03; $ZnCl_2$, 0.10; $MgSO_4 \cdot 4H_2O$, 0.20; $NiCl_2 \cdot 6H_2O$, 0.75; $MnCl_2 \cdot 4H_2O$, 1.00; $CuCl_2 \cdot 2H_2O$, 0.10; $FeCl_2 \cdot 4H_2O$, 0.01; $CoCl_2 \cdot 6H_2O$, 1.50.

Agilent 7820A gas chromatograph, Hash IL550 TOC analyzer, constant temperature water bath, electronic balance. Agilent 7820A gas chromatograph instrument conditions: Inlet temperature, 200°C; Column flow, 1.0 ml/min; Furnace temperature, 70°C; Retention time, 10 min; Detector, FID; Temperature, 200°C. The detection limit of this method is 0.05 μg/L.

2.2 Experimental methods

Simulate groundwater contaminated by TCE, add TCE in water samples, make sure the TCE concentration was 10 mg/L and dispensed into six groups 100 ml brown bottles. Sodium sulfate was added, respectively, so that the initial concentrations of sulfate were 0 mg/L, 10 mg/L, 50 mg/L, 100 mg/L, 150 mg/L, 200 mg/L and the concentration ratios of sulfate and TCE were 0, 1, 5, 10, 15, 20. And set three control groups adding sulfate only, and ensure the initial concentrations of sulfate were 1 mg/L, 100 mg/L and 200 mg/L. These groups were added 1 g/L of sodium lactate as a metabolic substrate and acclimated anaerobic sludge quantitatively. Filled with nitrogen gas to maintain anaerobic conditions inside the reactor, the temperature was maintained at room temperature (28°C), adjusted to pH about 7.5, and sealed with rubber stopper and aluminum-plastic lid. Samples were sampled and analyzed every five days to determine the most appropriate sulfate and TCE dosing ratio. Sulfate measured by ion chromatography.

3 RESULTS AND DISCUSSION

3.1 TCE degradation results

In six groups, actual initial TCE concentrations were 10.216 mg/L, 10.386 mg/L, 10.192 mg/L, 10.17 mg/L, 10.081 mg/L, 10.121 mg/L. Changes of six groups TCE concentrations were shown in Figure 1. Figure 1 showed that, in each group added sulfate, the degradation of TCE were significantly better than the group of TCE alone, indicating the presence of sulfate could promote

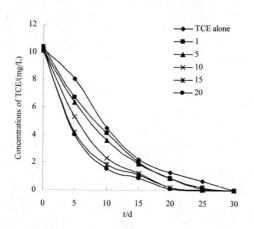

Figure 1. Degradation curve of TCE under different concentrations ratios.

the degradation of TCE. Moreover, the sulfate and TCE concentration ratio greater, the TCE degradation performance better. TCE removal rate at concentration ratio of 20 was 99.93% at 30 days, which was the best.

Previous literature has studied the impact of sulfate-reducing on degradation of toluene and xylene, and proved their degradation is due to the role of sulfate (Edwards, 1992; Ndon, 2000). In this experiment, setting different sulfate and TCE concentration ratio to create different sulfate reducing environment, results show increasing concentration of sulfate can enhance the degradation of TCE, and reactions are all particularly evident at the first 10 days.

3.2 Sulfate reduction results

Actual initial sulfate concentrations were 10.75 mg/L, 56.12 mg/L, 100.5 mg/L, 153.22 mg/L, 201.3 mg/L. Changes of sulfate concentration were measured at the same time. Results were shown in Figure 2. Figure 2 shows, while the TCE degradation, sulfate has also been reduced and the variation of sulfate concentration had significant correlation with the degradation of TCE. Sulfate reduced fastest in the group of the concentration ratio 20, and the first ten days it has reduced 89.73%. Further, since the first ten days of the experiment were the most obvious, select the first 10 days of data for analysis. Five groups of sulfate reduction rates were 72.05%, 77.34%, 82.31%, 89.85%, and 89.73%. The TCE removal rates were 60.36%, 64.64%, 77.09%, 81.45%, and 84.88%. Results of mutual comparison were shown in Figure 3.

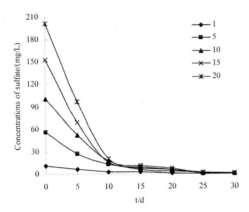

Figure 2. Sulfate curve under different concentrations ratios.

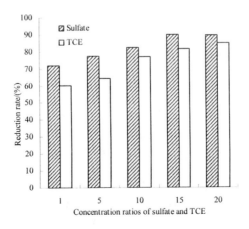

Figure 3. Sulfate and TCE reduction rate comparison chart.

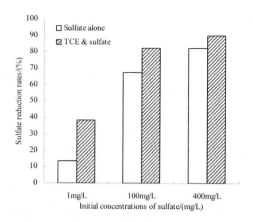

Figure 4. Sulfate reduction rate comparison chart.

As can be seen from Figure 3, in the first 10 days, removal efficiency of TCE is high, but the removal rate at the concentration ratio of 20 is the highest. Sulfate reduction rate is very different. But still at concentration ratio of 20 the reduction rate was the highest.

3.3 Relationships between sulfate change and TCE degradation

Compared three sulfate groups which were very representative (initial concentration of 1 mg/L, 100 mg/L, and 200 mg/L) with before three groups in which concentrations of TCE were 10 mg/L. Select data at the first 10 days when the reactions were most obvious to observe, sulfate reduction rate comparisons were shown in Figure 4.

As can be seen from Figure 4, the reduction rate of sulfates, when TCE and sulfate exist at the same time, were significantly higher than the groups of sulfate alone. Therefore, this comparative experiment results showed that TCE degradation could also enhance sulfate reduction effect. Because TCE degradation is mainly due to methanogenic bacteria, and sulfate reduction is mainly the role of sulfate-reducing bacteria, this experiment can show that the two strains have a certain role in mutual promoting.

3.4 Relationships between sulfate change and TCE degradation

Compared three sulfate groups which were very representative (initial concentration of 1 mg/L, 100 mg/L, and 200 mg/L) with before three groups in which concentrations of TCE were 10 mg/L. Select data at the first 10 days when the reactions were most obvious to observe, sulfate reduction rate comparisons were shown in Figure 4.

As can be seen from Figure 4, the reduction rate of sulfates, when TCE and sulfate exist at the same time, were significantly higher than the groups of sulfate alone. Therefore, this comparative experiment results showed that TCE degradation could also enhance sulfate reduction effect. Because TCE degradation is mainly due to methanogenic bacteria, and sulfate reduction is mainly the role of sulfate-reducing bacteria, this experiment can show that the two strains have a certain role in mutual promoting.

3.5 TCE degradation kinetics analysis

In order to fully study the dynamics of TCE degradation, to find the most suitable dynamic model of this process, the paper used the first-order and second-order kinetic equation to analysis. Then we can analyze the linear correlation coefficient R^2 to determine the two kinds of kinetic models suitability. According to the experimental data obtained we made $\ln(C/C_o) - t$ curve shown in Figure 5 and the $t/C_t - t$ curve shown in Figure 6.

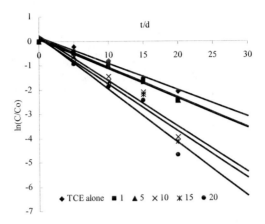

Figure 5. First-order reaction kinetics curve.

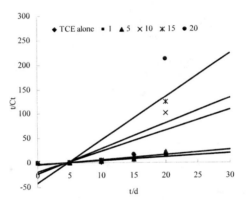

Figure 6. Second-order reaction kinetic curve.

Table 1. Kinetic parameters.

Concentration ratio	First-order kinetics		Second-order kinetics	
	k	R^2	k	R^2
0	0.1086	0.9783	0.7432	0.8340
1	0.1219	0.9815	1.0247	0.7670
5	0.1216	0.9910	1.0465	0.7788
10	0.1869	0.9456	4.2841	0.6000
15	0.1922	0.9456	5.2441	0.5910
20	0.2178	0.9450	8.8325	0.5687″

According to Figure 5, Figure 6 and Table 1, in this experiment, various concentrations of sulfate and TCE ratio, the first-order kinetic equation can describe the TCE degradation behavior well, and the linear correlation coefficient R^2 are all greater than 0.94. The second-order kinetic equation was relatively poor results, the correlation coefficient R^2 are less than 0.83. Therefore, it can determine the TCE degradation was in accordance with a first-order kinetic model under sulfate reducing conditions.

4 CONCLUSIONS

Sulfate reduction can strengthen the TCE degradation. Within the scope of this experimental study, as the concentration ratio of sulfate and TCE increases, performance of TCE degradation was enhanced. Optimal dosing ratio of sulfate and TCE was 20. There is a certain role between sulfate reduction and TCE degradation in mutual promoting. TCE degradation was in line with first-order kinetic model under sulfate-reducing conditions.

ACKNOWLEDGEMENT

This work was supported by the Key Special Program on the S&T for the Pollution Control and Treatment of Water Bodies (2012ZX07205-002) and the National Natural Science Foundation of China (No. 41072194).

REFERENCES

Aulenta F. et al. 2002. Complete dechlorinating of tetrachloroethene to ethene in presence of methanogenesis is and acetogenesis by an anaerobic sediment microcosm. Biodegradation 103(1):411–424.
Chen Cuibai et al. 2004. A Laboratory Study of Biodegradation and Adsorption and Desorption of Trichloroethylene to mixed bacteria. Hydrogeology & Engineering Geology 1(1):47–51.
Edwards EA. et al. 1992. Anaerobic degradation of toluene and xylene by aquifer microorganisms under sulfate-reducing conditions. Appl. Environ. Microbiol 58(3):794–800.
E.D. Hood et al. 2008. Demonstration of enhanced bioremediation in a TCE source area at launch complex 34, cape canaveral air force station. Groundwater Monitoring & Remediation 28(2):98–107.
He J. et al. 2002. Acetate versus hydrogen as direct electron donors to stimulate the microbial reductive dechlorination process at chloroethene-contaminated sites. Environ Sci. Technol 36(18):3945–3952.
Jia Xiaoshan & Li Shunyi. 2003. Kinetic Competition between Methanogenic and Sulfate-reducing Bacteria in Anareobic Mixed Cultures I: Model Development. Acta Scientiarum Naturalium Universitatis Sunyatseni 42(6):103–106.
Liu Mingzhu et al. 2006. Modeling of transformation and transportation of PCE and TCE by biodegradation in shallow groundwater. Earth Science Frontiers 13(1):155–159.
Lu Xiaoxia et al. 2002. Oxidative Degradation of Chlorinated Hydrocarbons under Anaerobic Conditions. Environmrntal Science 23(4):37–41.
Meng Fansheng & Wang Yeyao. 2005. Remediation of Groundwater and Soil Polluted by Trichloroethylene. China Water & Wastewater 21(4):34–36.
Ndon U.J. et al. 2000. Reductive dechlorination of tetrachloroethylene by soil sulfate-reducing microbes under various electron donor conditions. Environmental Monitoring and Assessment 60(3):329–336.
Shen Runnan & Li Shuben. 1998. The Biodegradation of Trichloroethethylene by A Methanotrophic Bacterium. Acta Microbiologica Sinica 38(1):63–69.
W.P. de. Bruin et al. 1992. Complete biological reductive transformation of tetrachloroethene to ethane. Appl.Environ.Microbiol 58(6):1996–2000.
Wu Yaoguo et al. 2007. Bio-translation of Aniline During Riverbank Filtration Process Under Sulfate-reducing Conditions. Journal of Agro-Environment Science 26(1):108–112.
Seungho Yu & Lewis Semprini. 2004. Kinetics and modeling of reductive dechlorination at high PCE and TCE concentrations. Biotechnology Bioengineering 88(4):451–464.
Zuo Jiane et al. 2003. A review on reductive dechlorination of chlorinated organic pollutants under anaerobic condition. *Techniques and Equipment for Environmental Pollution Control* 4(6):43–48.

Tidal range analysis of the central Jiangsu coast based on the measured tide level data

Yanyan Kang
College of Harbor, Coastal and Offshore Engineering, Hohai University, Nanjng, China

Xianrong Ding
College of Hydrology and Water Resources, Hohai University, Nanjng, China

ABSTRACT: Four new tide-gauge stations were established along the southern and northern flanks of Tiaozini, the core region of the central Jiangsu coast. Based on the measured tide level data, the quantitative analysis of the spatial distribution of the tidal range was carried out in this paper. The result showed that the maximum tidal range 9.39 m, a significant and reliable data, was observed in Xintiaoyugang station. It is higher than 9.28 m observed in Xiaoyangkou near Xintiaoyugang station in the 1980's. Moreover, the tidal range in the north was smaller than in the south. Thus, the tidal range presented an irregular convex shape. "Movable standing tidal wave" was the main reason of large tidal range in the central Jiangsu coast.

1 INTRODUCTION

Tidal range is the vertical difference in height between the consecutive high and low tides. Large tidal ranges often occurred in narrow harbors and estuaries, such as the Hangzhou Bay, Bay of Fundy, Amazon, and Red Mouth. However, the appearance of large tidal ranges in the central Jiangsu coast is unique in the entire coastal area of China in the Southern Yellow Sea. Radial sand ridges (RSR), a special continental shelf sand body is located off the central part of Jiangsu coast. RSR is fan-shaped with an apex at Tiaozini stretching seawards with a central angle of about 160° and covering an area of 20,000 km² (Figure 1). The two important characteristics of the RSR are the strong tidal flow and large tidal ranges. For decades, the studies on tidal range are often a qualitative description rather than a quantitative analysis because of the lack of long continued measured tide level data. Tidal range of 9.28 m observed in Xiaoyangkou in the 1980's is still in use. Moreover, according to the previous studies (Liu, Z.X., 1989; Zhang, D.S., 1996; Zhu, Y.R., 1997; Yan, Y.X., 1999; Zhang, C.K., 1999; Tao, J.F., 2005), numerical models were used to simulate the hydrodynamic environment of the sea area. However, the aforementioned studies only have some qualitative description on tidal range and rarely involved quantitative analysis due to the lack of measured tide level data. Besides, Wu (2008) developed the formula to calculate the maximum possible tidal range using the improved G. Godin's tidal level harmonic analysis and prediction procedures. However, the accurate value of the largest tidal range and precise position were not obtained by using this method. Further, the tidal range of 9.28 m has long been questioned. In this study, the tidal range of 9.39 m, observed in the Xintiaoyugang tide-gauge station is employed for the quantitative description of the spatial distribution of tidal range and for exploring the cause of large tidal range in the RSR sea area based on the measured tidal level data from six tidal stations.

2 MATERIALS

2.1 Tide-gauge stations

From October 2010 to October 2012, four new tide level automatic telemetry stations have been completed around Tiaozini. They are Dafeng, Dongdagang, Xintiaoyugang, and Yangkou stations

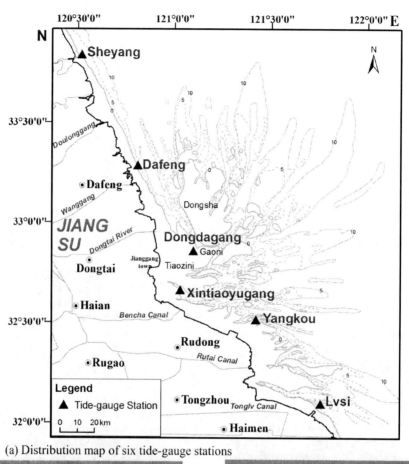

(a) Distribution map of six tide-gauge stations

(b) Photos of four new stations

Figure 1. Tide-gauge stations.

Table 1. Details of four newly built tide-gauge stations.

Name	Dafenggang	Dongdagang	Xintiaoyugang	Yangkougang
Build time	2010.10	2012.10	2012.10	2010.10
Latitude/°	33.3	32.9	32.7	32.5
Longitude/°	120.8	121.1	121.0	121.4
Station platform	Pier	Column steel tower		Pier
Distance from the shore/km	8	11.1	9.1	13
Low tide water depth/m	10	3	3	12
Limnimeter	Water level ranging radar			
Radar altitude/m	6.9	15.1	15.5	8.5
Observation accuracy	±1 mm	±1 mm	±1 mm	±1 mm
Transmission	Mobile network synchronous transmission			

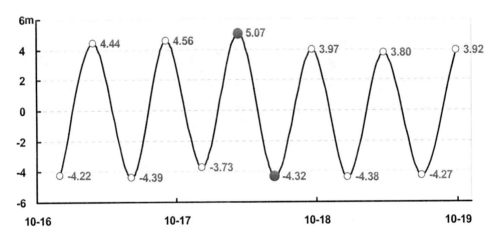

Figure 2. Tide curve of Xintiaoyugang station.

(Figure 1). A tide level automatic telemetry station consists of an outdoor sea monitoring equipment and an interior information collection system. The support platforms of the outdoor sea monitoring equipment are of two types: a Column Tower made of steel (Xintiaoyugang and Dongdagang) and the trestle of the pre-existing port (Dafeng and Yangkou). Moreover, mobile communications are used to achieve data synchronization between outdoor equipment and interior system (Table 1).

2.2 *The maximum value of tidal range*

The largest tidal range of 9.39 m was observed in Xintiaoyugang station. On October 17, 2012, the measured tidal level was 5.07 m (local mean sea level) at 13:03, and −4.32 m at 19:11. Therefore, the tidal range was 9.39 m (Figure 2), larger than the currently used tidal range of 9.28 m measured in the 1980's. The observed data is significant and reliable because of the following reasons: First, the observation platform is stable. The support platforms of outdoor equipment of Dafeng and Yangkou stations are set up in the trestle of the pre-existing port. And also, in Xintiaoyugang and Dongdagang stations, the platform consist of a Column Steel Tower fixed using 3–6 grouped wire cables. The underwater part of the wire cable is 13–15 m in length. Second, range laser radar can complete accurate data acquisition having a collection interval of 5 min. Each monitored data is the average of consecutive 3s radar pulse values rather than the instantaneous pulse value.

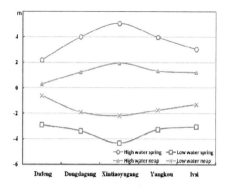

Figure 3. Rope skipping effect.

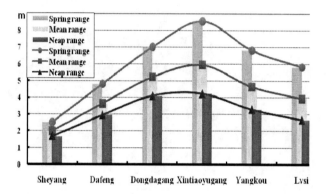

Figure 4. Tidal range along the coast.

3 RESULTS

The maximum tidal range 9.39 m, a significant and reliable data, was observed in Xintiaoyugang station. It is higher than 9.28 m observed in Xiaoyangkou near Xintiaoyugang station in the 1980's.

In addition, tide level connection curves obtained from six stations along the coast showed the morphological characteristics of skipping known as 'rope skipping effect'. Figure 3 shows four water-level curves: high water spring, low water spring, high water neap, and low water neap. During the period of high water, the tide level of the central area (Xintiaoyugang station) was higher than the northern and southern flanks. In contrast, during the period of low water, the central tidal level was lower than the flanks. Because of the continuous ebb and flow of tides, the tide level curves presented a rope skipping shape. This was one of the main features of the coast in RSR.

Figure 3 shows the characteristics of the tidal range along the coast. Based on the measured water level data of six stations, spring, neap, and mean tidal ranges were calculated as the statistical key figures. As shown in Figure 4, among the six stations, Xintiaoyugang has the largest tidal range of 9.39 m in the spring tide. The station observing the second largest tidal range is Dongdagang in the north of Tiaozini. On moving towards north and south, the tidal range gradually reduced. Along the coast, the shape of tidal range was convex. Moreover, the tidal range in the north was smaller than in the south. Thus, the tidal range presented an irregular convex shape and the maximum tidal range occurred in Xintiaoyugang station.

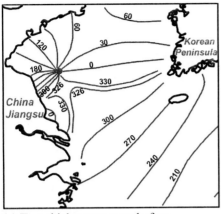 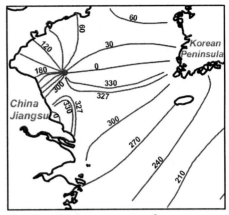

(a) Two tidal waves meet before   (b) Two tidal waves meet after

Figure 5. Co-phase lines of M_2 component in southern yellow sea (Zhang, D.S., 1996).

4 DISCUSSION

What's the reason for such a large tidal range in RSR? Tidal flow movement, broadly speaking, is determined by the tidal wave system. It belongs to semi-diurnal tide in the RSR area and is controlled by two tidal wave systems: the first one is the propagating tidal wave from the East China Sea and the second one is the rotational tidal wave system from the Southern Yellow Sea. The incoming wave from the north at the beginning travels towards the south and gradually turns to the shore of Jianggang (Tiaozini tidal flat), where it meets the wave from the south. Two tidal wave peak lines converge and forms a standing tidal wave in the RSR area. The standing tidal wave moves gradually towards the shore of Tiaozini area because of the friction of the seafloor. Hence, it is known as the movable standing tidal wave, resulting in the existence of the large tidal range in the entire RSR area and the largest tidal range in Tiaozini area. Figure 5 shows the change of co-phase lines of M_2 component in Southern Yellow sea when the two waves meet.

5 CONCLUSION

Based on the water levels data of six tide-gauge stations, along the coast, the shape of tidal range was convex. The vertex was Tiaozini area. The maximum tidal range of 9.39 m was observed in Xintiaoyugang station in the south of Tiaozini area. It not only confirms the reliability of the historical tidal range 9.28 m in Xiaoyangkou, but also refreshes the largest tidal range record of the Jiangsu coast. Moreover, the "movable standing tidal wave system" is the main hydrodynamic factor resulting in the large tidal range.

ACKNOWLEDGEMENTS

This work is financially supported by the National Basic Research Program of China (973 Program), No. 2010CB429001, National Key Technologies Research and Development Program of China (Grant no. 2012BAB03B01), Special Fund for Marine Scientific Research in the Public Interest (Grant no. 201005006-3), and Graduate Student Research and Innovation Project of Jiangsu General Higher Learning Institution, No. 2013B20814, No. CXLX12_0256.

REFERENCES

Liu, Z.X., et al. 1989. Tidal current ridges in the southwestern Yellow Sea. *Journal of Sedimentary Petrology* 59: 432–437.
Ren, M.E. (Ed.) 1986. *Comprehensive Investigation of the Coastal Zone and Tidal Land Resources of Jiangsu Province*. Beijing: Ocean Press.
Tao, J.F. et al. 2005. Numerical simulation of water environment for radial sandy ridge area of the Yellow Sea. *Journal of Hohai University: Natural Sciences* 33(4): 472–475.
Wu, D.A. et al. 2008. Calculation and analysis of possible maximum tidal difference in Jiangsu radial sand ridges sea area. *Journal of Dalian Maritime University* 34(4): 75–78.
Wang, Y. 2002. *Radiative sandy ridge field on continental shelf of the Yellow Sea*. Beijing: China Environmental Science Press.
Yan, Y.X. et al. 1999. Hydromechanics for the formation and development of radial sandbanks (I)—Plane characteristics of tidal flow. *Science in China (Ser. D)* 42(1): 13–21.
Zhang, C.K. 2013. *The basic status of marine environment and resources in Jiangsu province*. Beijing: Ocean Press.
Zhang, C.K. 2013. *Jiangsu offshore investigation and assessment report*. Beijing: Science Press.
Zhang, C.K. et al. 1999. Tidal current-induced formation-storm-induced change-tidal current-induced recovery-Interpretation of depositional dynamics of formation and evolution of radial sand ridges on the Yellow Sea seafloor. *Science in China (Ser. D)* 42(1), 1–12.
Zhang, D.S. et al. 1996. M_2 tidal wave in the Yellow Sea Radiate Region. *Journal of Hohai University: Natural Sciences* 24(5): 35–40.
Zhu, Y.R. et al. 1997. Explanation of the origin of radiant sand ridges in the southern Yellow Sea with numerical simulation results of tidal currents. *Journal of the Ocean University of Qingdao* 27 (2): 218–224.

Research of Linyi Municipal groundwater resources management information system based on GIS

Zhencai Cui, Jiyu Yu, Weiqun Cui, Xiaoli Gao & Yuzhi Li
Shandong Water Polytechnic, Rizhao, Shandong Province, China

Guangyun Ren
Linyi Municipal Water Conservancy Bureau, Linyi, Shandong Province, China

ABSTRACT: According to the demand of Linyi Municipal Water Conservancy Bureau, Linyi municipal groundwater resources management information system based on GIS is designed and implemented. An algorithm for auto-generating of groundwater isoline based on TIN and the same color weighted minimum distance is proposed, and the hydraulic characteristics of the contour lines produced by the algorithm are consistent with actual situation. The query operation between map and text mutually based on information mining technology which makes the spatial features integrate with the parameters of groundwater resources, is implemented. The management information system has been used in Linyi Water Conservancy Bureau and achieved good economic and social benefits.

1 INTRODUCTION

In Linyi city, the amount of average groundwater resources for years is 1.925 billion m^3, the factor of average groundwater resources for years is 112 thousand m^3/km, the gross amount of average years for groundwater resources is 5.536 billion m^3. The amount of water resources per capita is 537.6 m^3, so Linyi city has become one of the most water-short area.

The regional distribution of underground water resources is very uneven, the overall condition is clear – the plain regions are greater than the hilly regions and the karst mountainous regions are greater than the general hilly regions, the challenge of decaying shallow groundwater has become a more and more severe.

In Linyi city, the distribution acreage of shallow groundwater, whose water quality is class I or class IV and above, accounts for 55.3% of the total administrative region area, and covers an area of 44.6 of total groundwater resources. At the same time, the distribution acreage of shallow groundwater, whose water quality is class IV or class V, accounts for 31.9% or 12.8% of the total administrative region area, respectively, and covers an area of 39.1% or 16.3% of total groundwater resources, respectively.

The management echnologies of Linyi municipal groundwater resources are not adaptations to the modern water conservancy demands and the most strict water resources management. To some degree, the above situation has restricted the development of social economy. Therefore, It is significant for Linyi city to try to break the traditional extensive management model, and to build the water resources management information system which is an important part of water conservancy modernization management.

Our research is based on advanced technologies of GIS (Geographic Information System) and information mining. The research puts forward some improved algorithms, such as the algorithm for auto-generating of groundwater isoline based on TIN (Triangulated Irregular Network) and the same color weighted minimum distance, and the algorithm for query between map and text mutually based on information mining technology which makes the spatial features integrate with the parameters of groundwater resources.

Based on our research, we design and implement Linyi municipal groundwater resources management information system. This system can complete the drawing of groundwater level contour line, the drawing of groundwater level hydrograph, in a visual way. It can complete the statistics of groundwater resources (mainly includes maximum, minimum and extremes ratio), the analysis of the development and utilization of groundwater resources, the analysis of groundwater level regression between different wells, the calculation of groundwater resources in the plain regions, the calculation of groundwater resources in the hilly regions and the result can be queried, displayed or printed as the forms of thematic map, histogram, pie chart or data report form, etc.

The practical application in Linyi Municipal Water Conservancy Bureau shows that this system provides a stable and reliable and convenient solution, and experimental results show that this algorithm is effective and efficient,therefor, this system has a great popularization value and application prospect.

2 SYSTEM DESIGN AND CONSTRUCTION

Based on the ArcGIS Engine, Microsoft SQL Server and Visual Studio.NET framework, we established Linyi municipal groundwater resources management information system. The logic structure, developmental environment and platform, and functional framework of this system will be discussed summarily.

2.1 *System structure*

The logic structure of the system contains two layers of objects: Client and Server. These two objects will communication with each other via TCP/IP protocol. This structure enables Client to manipulate any data on Server side directly. Client is divided into two layers of functions: data request layer and GIS functional layer. The former layer is the realization of GIS data processing, while the GIS functional layer implements all the functions of GIS. This design style of hierarchical comes with many advantages. On the one hand, the system can make the most of the original GIS research results with single version.On the other hand, the development of GIS function layer and the development of data requests layer can be carried out at the same time, and the changes occurring in one layer will not affect the other layer as long as the interface standards unchanged.

2.2 *System structure*

The network environment of information management system is based on TCP/IP protocol Intranet, the server adopts the departmental server, connects the digitizers, color plotters and other peripheral devices. The system adopts C/S (Client/Server) mode. Aiming at the system's multi-source data, spatial characteristics and practical application need, the international leading ESRI corporation's ArcGIS, Microsoft's Visual Studio.NET and SQL Server as developing environment and database system are selected. Choose the stable Microsoft SQL server 2005 as database to store spatial data and attribute data and to manage them uniformly through the ESRI corporation's ArcSDE data engine. Choose the ESRI corporation's ArcGIS Engine and ArcObjects to do the secondary development of GIS. Choose the Microsoft Corporation's latest development platform Visual Studio.NET. C# or VB is a powerful visualization software development program language and it has powerful database access function. It can quickly access a large database such as SQL Server, Oracle, etc. So, C# and VB are very suitable for the development of this system.

2.3 *Developmental environment and developmental platform*

According to the requirements, the groundwater resources management subsystem based on the electronic map realizes many functions relative to groundwater resources in a visual way. The

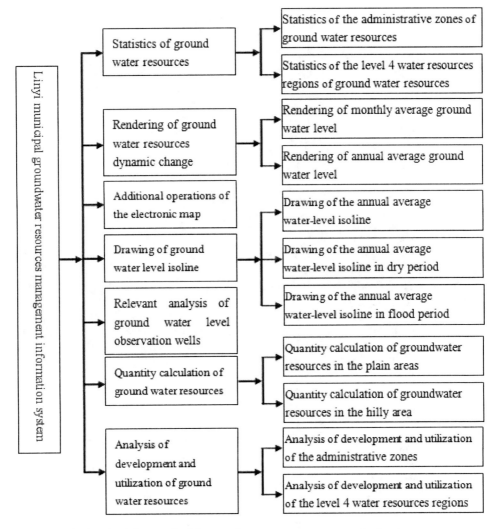

Figure 1. The functional framework of the groundwater resources management subsystem.

functions include the drawing of groundwater level contour line, the drawing of groundwater level hydrograph, the statistics of groundwater resources, the calculation of groundwater resources, and the rendering, displaying or printing of processing results. The functional architecture of the groundwater resources management subsystem as shown in figure 1.

3 CORE TECHNOLOGIES AND IMPLEMENTATIONS

The major technologies involved in this system platform will be discussed summarily.

3.1 *Auto-generating of groundwater isoline based on TIN and the same color weighted minimum distance*

Through contrast analysis, an algorithm for auto-generating of groundwater isoline based on TIN and the same color weighted minimum distance is proposed. According to the algorithm, we

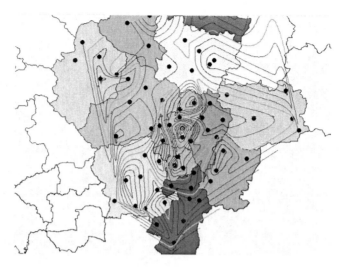

Figure 2. The example of groundwater level isoline.

implements the functionality of auto-generating of groundwater isoline with combinations of drawing ability of ArcObjects and MapObject. Practices show that the water isolines produced by the algorithm are accurate, beautiful and their hydraulic characteristics are consistent with actual situation.Results as shown in figure 2.

3.2 *Information mining based on spatial features and parameters of groundwater resources*

In order to realize the query functionality between map and text mutually, and to achieve the information mining technology based on spatial features and parameters of groundwater resources, The associated mechanism of spatial information data and attribute data has to be built. At the same time, to create the attribute tables through SQL statements, thus map data and text data can be accessed and manipulated mutually through the connected attribute.

At system initialization phases, store the names and the identifiers (ID number) of graphic elements in record sets and establish all indexes of them on a table. They must be bound together to the tree view list. when a certain graphic element is selected, its ID number is acquired and the graphic element associated with the ID number is retrieved in the electronic map. To render the graphic element to the view port, it will be flashing with a varying color. Click on the graphic element to query in the electronic map, the graphic element will be flashing. At the same time, Commit operation command and the view mode, the related query results will be displayed immediately (to save space, omit the illustrations).

3.3 *The organic integration of MIS and GIS*

MIS (Management Information System) has superiority in database management and dada model, but it has difficult in spatial analysis and operation, information display and representation. But GIS can combine the spatial data and attribute data organically and has powerful ability to analyze and manage spatial data.

Generally, the water resources system has spatial distribution characteristics. With its efficient spatial data and attribute data processing ability and powerful query function, GIS provides an effective and reliable technical support platform for the process of acquiring information and decision analysis of the water resources management information system.

According to the characteristic and demand of Linyi water resources management, Linyi municipal groundwater resources management information system selects Microsoft's Visual Studio.NET and ESRI Corporation's ArcGIS as developing environment, chooses Microsoft SQL server as database to store spatial data and attribute data and to manage them uniformly through the ArcSDE data engine. Through the above technical support, the information system realizes the organic integration of GIS and MIS, and implements the data sharing and exchange with other systems. Practices prove that the above solutions are effective and feasible.

The groundwater resources management information system is provided with functions of spatial data retrieval and analysis, attribute information query and statistics, map operation, map visualization, data and map management, etc. The query of management information is completed mainly by means of executing the menu command or toolbar command, and the query of map features is completed mainly through the map operations. MIS and GIS, the two systems are unified together through the establishment of electronic map platform and multi-level water resource administrative regions. Because of the organic integration of MIS and GIS operation, the groundwater resources management information system has many advantages, including rich functionalities, high performance, efficiency, and flexibility.

4 APPLICATIONS OF THE SYSTEM

In accordance with the functional requirements of Linyi municipal groundwater resources management information system, the attribute data are mainly taken from the historical data of hydrology and water resources from Linyi city, or from the data of "The water resources evaluation of Linyi city" or "The water resources overall planning of Linyi city", and so on. These data can be input and stored in the database through manual inputting or automatic importing, with the help of data management and maintenance subsystem.

5 CONCLUSION

The major technologies involved in this system platform will be discussed summarily.

In conclusion, the Linyi municipal groundwater resources management information system makes full use of the privilege of spatial analysis and spatial database management of ArcGIS Engine, data management of Microsoft SQL Server and the rapid, visualized programming of Visual Studio.NET. The system comprehensive utilizes information technology and GIS technology, provides an effective and reliable technical support platform for the process of acquiring, storing, managing, and analyzing spatial data information and decision analysis of the groundwater resources. The very system realizes the integrated management of hydrology, groundwater resources, ecological environment etc. and offers attribute query statistics, spatial retrieval and analysis. Also it can provide scientific basis for research, management and decision-making of regional groundwater resources, environmental and ecological.

The research makes use of drawing ability of ArcObjects and MapObjects, realizes the algorithm for auto-generating of groundwater isoline based on TIN and the same color weighted minimum distance via VB programming language. The result proves the water isoline built by this way is accurate, beautiful and it can fit the users'requirement.The research implements query operation between map and text mutually based on information mining technology which makes the spatial features integrate with the parameters of groundwater resources.

The Linyi municipal groundwater resources management information system has been used in Linyi Municipal Water Conservancy Bureau, and achieved good economic and social benefits.

In the future, we'll study The system further, improve its performance and add more functions to the system, such as real time interface to field sensor, more abundant map operation and display format etc., so that it can be used more conveniently.

REFERENCES

Tolga Elbir. 2004. A GIS Based Decision Support System for Estimation, Visualization and Analysis of Air Pollution for Large Turkish Cities. J. Atmospheric Environment. 38(2): 83–87.
Xue hui Cui, Honghan Chen, Chen Hongkun. 2006. Regional Hydrogeology GIS System and Accessorial Design for Engineering Based on ActiveX. J. Journal of Jilin University (Earth Science Edition). 36(1): 91–95.
Yang Xu, Suozhong Chen, Tan Tao. 2007. Application on Com GIS in Groundwater Resource Management System. J. Hydrogeology & Engineering Geology. 5(1): 46–49.
Yuzhi Li, Zhencai Cui. 2008. Algorithm Research of Generating Groundwater Isoline Based on TIN. J. Surveying and Mapping of Geology and Mineral Resources. 24 (4): 9–12.
Yuzhi Li, Zhencai Cui. 2009. Algorithm for Creating Contour of Underground Water Based on.
Zhaoming Deng, Jun Wang. 2006. Alication of GIS to the Information System of Regional Groundwater Resource. J. Hydro geology & Engineering Geology. 12(1): 106–108.

Evolution of water exchanging period in Lvdao lagoon induced by different water depth

Xinying Pan, Duo Li, Bingchen Liang & Peng Du
College of Engineering, Ocean University of China, Qingdao, China

ABSTRACT: Lvdao lagoon is located in top of Sanggou bay, China. The water depth of Lvdao lagoon will have to be changed as a result of building living area around the lagoon. However, it is well known that the man-made topography changes will have impacts on water exchanges. Therefore, the present work will focus on studying and demonstrating which scheme is the most reasonable by simulating the water exchanges under the two schemes provided by the designing company. The two schemes are specified by two water depths, which are 2 m and 3 m respectively. The simulating area is the whole Sanggou Bay, which includes the Lvdao lagoon. However, more attention will be paid to the Lvdao lagoon as the result of no changes occurring except for the lagoon. The moving boundary technology is applied since bathymetry of engineering area of Sanggou lagoon includes large area of tidal flat. Unstructured triangle grid system is adopted in modeling of current and water exchanges environment of Lvdao lagoon. Around location of and within the lagoon, high grid resolution is provided because the lagoon has significant length being 1500 m only. Open boundary conditions of tide level and tidal current are provided with outputting results of "father" simulation with lower grid resolution and larger modeling area. The present engineering simulating area has the highest resolution being 30 m around the lagoon. The present concrete jobs consist of: 1) studying the effects of the two schemes of layout of yacht on current; 2) calculating and predicting water exchanges induced by the changes of water depth for the two layouts. Generally, the most important purpose of this study is to find the more reasonable layout available, in other words, the one most attenuates the severe effects on environment. As the original case concerned, the values of time series of water level and current velocity provided by simulation for Lvdao Lagoon have a good agreement on observed data. Therefore, the model can be reliable to simulate the evolution of water exchanges induced by changes of water depth.

1 INTRODUCTION

In 1973, Leendertse (Leendertse J J, 1973) carry out three-dimensional flow simulation firstly. However, the development of the model was not smooth due to the complexity of problems and computational inability of computers at that time. By the end of 1970s', POM and ECOM (Mellor G L, 1998) were given by Princeton university; There are other famous models just like TRIM3D (Casulli V, 1994) and TRISULA of Delft (TRISULA, 1993) and so on (Zhu Y L, 2000) (Bai Y C, 2000), which makes three-dimensional numerical models get faster improvement.

This study first predicts the influence area on water exchange in the lagoon under different project schemes based on the natural environmental situation of the aiming project location via setting up a tidal and water exchange numerical model. Secondly, the study provides scientific basis for the feasibility study, the planning and construction of the project according to the analysis of different schemes.

2 MODEL DESCRIPTION

The governing equations are given as:

$$\frac{\partial \zeta}{\partial t} + \frac{\partial (Hu)}{\partial x} + \frac{\partial (Hv)}{\partial y} = 0 \tag{1}$$

$$\frac{\partial u}{\partial t} + u\frac{\partial u}{\partial x} + v\frac{\partial u}{\partial y} = -g\frac{\partial \zeta}{\partial x} + fv - \frac{g}{C^2}\frac{\sqrt{u^2+v^2}}{H}u + \frac{\tau_x}{\rho H} + \varepsilon(\frac{\partial^2 u}{\partial x^2} + \frac{\partial^2 u}{\partial y^2}) \tag{2}$$

$$\frac{\partial v}{\partial t} + u\frac{\partial v}{\partial x} + v\frac{\partial v}{\partial y} = -g\frac{\partial \zeta}{\partial y} - fu - \frac{g}{C^2}\frac{\sqrt{u^2+v^2}}{H}v + \frac{\tau_y}{\rho H} + \varepsilon(\frac{\partial^2 v}{\partial x^2} + \frac{\partial^2 v}{\partial y^2}) \tag{3}$$

where: ζ – water surface elevation from mean sea level; $H = \zeta + H_0$ – water depth (H_0 depth from mean sea level); f – Coriolis coefficient; $g = 9.81$ m/s² – gravitational constant; u, v – current velocities in x, y directions; t – time coordinate; ε – horizontal eddy viscosity coefficient; $\tau_x = r^2 \rho_a W^2 \cos\theta$, $\tau_y = r^2 \rho_a W^2 \sin\theta$ are wind stress of surface level (W wind velocity, ρ_a air density, θ wind direction, r^2 wind stress coefficient, the value of it is 0.0026); ρ – water density; C – Chezy coefficient, which is dependent on sea bottom roughness and water depth.

The transport and diffusion of pollutant in the sea can be expressed by the two-dimensional convection-diffusion equation:

$$\frac{\partial (HP)}{\partial t} + \frac{\partial (HPU)}{\partial x} + \frac{\partial (HPV)}{\partial y} = \frac{\partial}{\partial x}(HK_x \frac{\partial P}{\partial x}) + \frac{\partial}{\partial y}(HK_y \frac{\partial P}{\partial y}) + S$$

where: $H = h + \zeta$, h is the depth from mean sea level; ζ water surface elevation from mean sea level: U, V: current velocities in x, y directions; P: density of the pollutant; K_x, K_y horizontal turbulent diffusion coefficient; S: source strength per unit area.

3 MODEL SETTING

The computational area is the whole Sanggou Bay (fig. 1). At the same time the highest resolution is 20 m and the minimum resolution is 400 m.

4 SIMULATION RESULTS

4.1 Verification of tidal model simulation

According to the observed tide data in Lijiang Jetty (fig. 1 point C), as well as the tide level curve shown in fig. 2a, we can see that the calculated tide level can be identical with observation data. According to the data measured by Ocean University of China, in top layer of Sanggou Bay when it's spring tide (fig. 1a and b), to verify the measurement and simulation result. As shown in Fig. 2b and Fig. 2c, there exist deviation between the measurement and simulation result while at the moment of maximum flux, the main flow direction can give a reasonable agreement with the measurement. Therefore, the tidal current measurement of this area can be replaced by the tidal current simulation result, and consequently can be the hydrodynamic field of water exchange simulation.

4.2 Analysis of numerical experiment cases

According to the plan of layout schemes, three schemes (Fig. 3): *Case 1*: close the sluice; *Case 2*: open the sluice, dredging depth 2 m; *Case 3*: open the sluice, dredging depth 3 m.

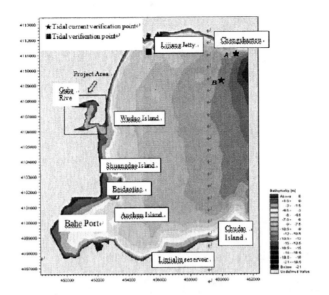

Figure 1. The big area and tidal current, tide level verify points.

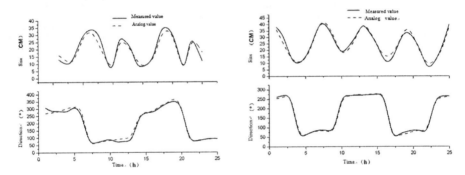

Figure 2(a, b, c). Tide level and tidal current verification.

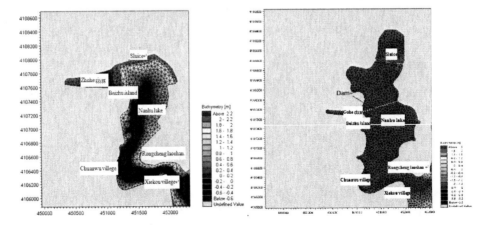

Figure 3. Computational grid around area before and after construction (Left Panel is case before construction and right one is case after construction).

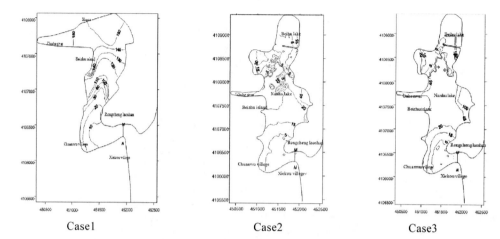

Figure 4(a, b, c). Distribution of water exchanging half period.

4.3 *Effects of project on water exchanging ability*

On the basis of establishing a 2-D material transport model, by using the density of dissolved formed conservative substances as a tracer to set up a water exchange numerical model of this sea area, applying MN section as exchange boundary, the initial density field in north of the section is 1.0 and given a density of 0 when flow into the section. Take note of the time when density of every point is 0.5, which is the water exchange period of this sea area. Water exchanging half period distribution of case 1, case 2 and case 3 are given in Fig. 4(a, b, c).

Analysis of water exchange period (WEP) prediction before the construction
According to Fig. 8a, before the construction, water flow is weak due to the impact of sedimentation, the water semi-exchange ability is consequently weak. The whole WEP of Lvdao Lake is 1 to 150 days before the construction. Water exchange ability is weak near the inlet of Guhe River, the WEP is over 140 days. South from the inlet to MN section, water semi-exchange period (WSEP) reduces linearly down to 10 days at the sea inlet.

(1) Project 1 (dredging depth 2 m) Analysis of WEP prediction
After dredging and opening the north gate, water flow of the Green Island Lake increased, accordingly water exchange ability got increased too. From Fig. 8b, the completed WEP of Green Island lake decreased to 1 to 100 days. Except the area nearby the lagoon inlet of Guhe River and part of east section of South lake, WEP of other areas is less than 20 days.

(2) Project 2 (dredging depth 3 m) Analysis of WEP prediction
Compared to Project 1, water flow of the Green Island Lake became smaller, and accordingly water exchange ability was weakened. From Fig. 8c, the complete WEP of Green Island lake is 1 to 120 days. Except the area nearby the lake inlet of Guhe and part of east section of South lake, WEP of other areas is less than 30 days.

5 CONCLUSIONS

(1) Through verifying calculation of the tidal current simulation under shoreline shows that the sea water flows into the bay from medium and north of the mouth to southwest when tide rising up, and at the same time flows out of the bay from south by east. To the contrary, the sea water flows out of the bay from medium and north of the mouth to northeast, and flows into the bay

from south to west by north. The flow velocity in south mouth of the bay is much bigger than it in north area. And the velocity in the bay mouth is larger than it inside bay, and the biggest velocity is 65 cm/s.

(2) In the area near the project before the construction, the sea water flows along the south-lagoon shoreline from the estuary to south and into the Lvdao Lagoon when tide rising up. And influenced by the bathymetry, the flow velocity in north area of Xiekou Village is much bigger whose biggest velocity is 16 cm/s.

(3) Case 2 and case 1: in the case of opening the north sluice, compared with it before the construction.

Before the construction, water flow is weak due to the impact of sedimentation, thus the water semi-exchange ability is consequently weak. The whole WEP of Lvdao Lake is 1 to 150 days before construction. Water exchange ability is weak near the inlet of Guhe River, WEP is over 140 days. South from the inlet to MN section, water semi-exchange period (WSEP) reduces linearly down to 10 days at the sea inlet.

(3) Analysis of WEP prediction before the construction
After dredging and opening the north gate, water flow of the Green Island Lake increased; which also result in the increase in water exchange. From Fig. 8b, the completed WEP of Green Island Lake decreased to 1 to 100 days. The WEP of other areas is less than 20 days, with the exception of the area nearby the Lake Inlet of Guhe and part of east section of South Lake.

(4) Scheme 2 (dredging depth 3m) Analysis of WEP prediction after the construction
Compared to Project 1, water flow of the Lvdao Lagoon become smaller, and accordingly water exchange ability was weakened. The completed WEP of Lvdao Lagoon is 1 to 120 days. Except the area nearby the Lake Inlet of Guhe and part of east section of South Lake, WEP of other areas is less than 30 days.

ACKNOWLEDGMENTS

The authors would like to acknowledge the support of the National Science Fund (Grant No. 51179178), the Program for New Century Excellent Talents in University of China (Grant No. NCET-11-0471) and Doctor Science Fund of Shandong Province (Grant No. BS2011HZ016).

REFERENCES

Bai Y C, Sheng H T, Hu S X. 2000. Three dimensional mathematical model of sediment transport in estuarine region [J]. *Int. J. Sediment Res.* 15(4):410–423.
Casulli V, Cattani E. Stability, accuracy and efficiency of a semi-implicit method for three-dimensional shallow water flow [J]. Comput. Math. Appl., 1994, 27(4):99–112.
Leendertse J J, Alexander R C, Liu S K. 1973. A three dimensional model for estuaries and coastal seas [R]. Rand Corp. R-1417-OWRR.
Mellor G L. 1998. Users guide for a three-dimensional, primitive equation, numerical ocean model [R]. Princeton University, Princeton, NJ 08544-0710.
TRISULA: A program for the computation of non-steady flow and transport phenomena on curvilinear coordinates in 2 or 3 dimensions [R], Delft Hydraulics, Delft, 1993.
Zhu Y L, *et al* 2000. Three dimensional nonlinear numerical model with inclined pressure for salt in intrusion at the Yangtze River estuary [J]. *J. Hydrodynamic*, Ser. B.1:57–66.

Mechanics analysis of arch braced aqueduct

Xiang Lu, Cong Li, Yong Ye & Yu Zhao
College of Hydraulic & Environmental Engineering, Three Gorges University, Yichang, Hubei, China

ABSTRACT: Aqueduct is a kind of bridge type overlapping buildings surmounting other constructions as canal, marsh land, path and rail road, as well as a group of water conveyance system formed by bridges, tunnels canals and so on. Based on its massive utilization in hydraulic engineering and the diversion works in China, the article, through Solver and static load experiment, analyses the mechanical characteristics and calculates the strength of the arch braced aqueduct model with U-shaped section. The result shows that the real deformation stays almost the same with the experimental deformation and the load range of normal use is 15.5~37.7 kg.

1 INTRODUCTION

1.1 Selection of the supporting structure

According to the supporting structure pattern, the aqueduct may be divided into the beam plate, the arch bound, the truss type, the combined type as well as the suspension or the cable-strayed type. This sorting technique can better reflect its unique feature, force situation, load transfer mode and structure computational method (Zhu, 2009).

The arch is a kind of with fold line or curve shaped axial line, whose skewback generates horizontal propelling force under the vertical load on the precondition of horizontal restraint. On the arch is the frame bent, whose upside lays the body and lower segment is solidified to the main arch ring.

The beam plate aqueduct body is on the top of the pier, the frame receives pressure on the lower part and tension on the upper part and have bending strain under the vertical load while the supporting point only has the vertical reaction. And according to the difference of the pivot number and the position, the beam plate aqueduct can be divided into the simple support, the twin cantilevers, the single bracket and continuous beam. After the repeatedly comparative analysis, finally we decided on the arch braced aqueduct as the model structural style. The reason are as follows (Luo, 2005):

(1) Arch braced aqueduct spans bigger compared with the beam plate aqueduct, and is suitable to construct in the rough terrain and matches more with our research goal (the mechanical analysis of mountain massif aqueduct).
(2) Arch braced aqueduct is characterized by the main arch ring withstanding axial compressive force, If properly designed, it may let each point on arch only withstands the thrust, fully using the mechanic characteristics of the arches (Ma, 1984).
(3) Arch braced aqueduct saves more materials than the beam plate aqueduct, so it is more economical in the model manufacture and the actual construction.

1.2 Selection of aqueduct body

For cross section pattern of the aqueduct body, there are the rectangle, the U shape, the circular and the parabolic, among which commonly used is the rectangle and the U shape. The U shape aqueduct

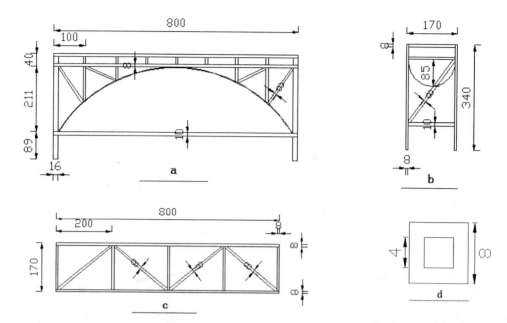

Figure 1. Schematic diagram of model size (a: front view, b: side view, c: top view, d: sectional drawing).

has good durability, practical artistic, big water volume and convenient construction; therefore we select the U shaped aqueduct (Cui, 2010).

2 MODEL DESIGN

2.1 Overall model design drawings

The model is a single span. The span length is 800 mm, beam width 170 mm, beam height 340. The circular arc radius of the U-shaped aqueduct is 85 mm (The size is shown in Fig. 1).

2.2 Model manufacturing

The bar is cut to the required length and cross-sectional forms, then stick two poles together with emulsion. To prevent the lever split up or crack at the node, the node is filled with a paper napkin to enhance the structural integrity. Because the white card paper can't be bent after rolled into the bar, we can only paste together layers of equal-sized notes to make it curve a radian that we need.

3 MODEL EXPERIMENT

3.1 Measurement of elastic modulus

The operating process is as follows:

(1) Choose different types and lengths of the paper bar and number them.
(2) Draw some equidistant lines in the middle of the paper bar and measure its length.
(3) Fill in the two pole of the paper bar with corresponding real solid wood plugs, and wrap the external with a certain thickness of the tape. To prevent the bar ends from being damaged during the experiment due to stress concentration, then resulting to too large experimental error.

Table 1. Water loading and deformation.

load/kg	12.0	13.0	14.0	15.0	16.0
deformation/mm	0.8	1.5	2.0	3.0	3.5

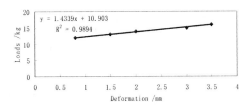

Figure 2. Analysis of water loading.

Table 2. Sand loading and deformation.

load/kg	23.8	26.7	30.1	37.7
deformation/mm	5.0	6.2	7.4	9.0

(4) Use well-processed bar to make experiment by universal testing machine. Choose the experimental data of the bars that are destroyed in the middle part to calculate the Elastic Modulus and average them. Fetch $E = \frac{\sigma}{\varepsilon} = 1400$ MPa finally.

3.2 Water static loading experiment

In order to facilitate water storage, we seal the ends of the aqueduct model, at the same time, use plastic wrap to waterproof model to prevent water intrusion in the loading process that damages model structure and reduces its carrying capacity. After the model to dry, add water slowly into the aqueduct from one end, 1 kg each time. Measure the distance of the vault from the ground after the water added stabilized. The whole water process is divided into 16 times, a total of 16 kg water is loaded, It was not until the 12th time, the load is 12 kg that a slight deformation of 0.75 mm was measured. When it reaches 16 kg, the aqueduct was filled with water, no further loading. Detailed in table 1.

Removed after being loaded, vault distance recovered again. Analyze the data dependency through EXCEL and the results are shown in Fig. 2.

R^2 is the coefficient of determination of the EXCEL trend line, referring to the degree of fitting between the relation curve and the data provided. The closer to 1, the better they fit. Linear regression analysis of the experiment data shows the coefficient of determination at 0.9894 provides better guarantee of accuracy. In other words, in hydraulic test, when load and deformation fit the equation of linear regression $y = 1.4339x + 10.903$, the model is in the state of elastic deformation.

3.3 Sand static loading experiment

In order to obtain the safe load range the model, besides water loading experiment, we carry out the sand loading experiment according to the fact that the sand density is bigger than water density and with an equal volume; the sand loading is greater than the water load. The loading test data is in table 2.

When added to 37.7 kg, the model has a serious deformation, to stop loading. Analysis the data dependency through EXCEL and the results are shown in Fig. 3. Because it is no longer linear

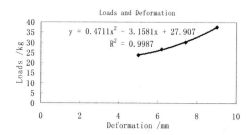

Figure 3. Analysis of sand loading.

relation between load and deformation, but the relation of y = 0.4711x² − 3.1581x + 27.907, the model has already had plastic deformation. Analyzes through the water load experiment and the sand load experiment, the model can withstand load approximately in 15.5~37.7 kg.

4 MECHANICAL CALCULATION

4.1 *Plane simplification computation and load analysis*

The structure net length is 80 cm, net width 17 cm, rectangle height 4 cm, semi height 8.5 cm. Based on volume and water density for $\rho = 1.0 \times 10^3$ kg/m³, the maximum carrying capacity of the structure is:

$$G = 0.95 \times 1.0 \times 10^3 \text{ kg/m}^3 \times [0.5 * 3.14 * (0.085)^2 + 0.04 * 0.17] * 0.8 * 9.8 = 135.142 \text{ kg}$$

1. Simplifying the strength recalculation of the U-shaped aqueduct

After checking the data, the paper bar parameters are as follows: $E = 1400$ MPa, $\nu = 0.3$, $\rho = 900$ kg/mm³, $\sigma_t = 20$ N/mm², $\sigma_c = 7$ N/mm².

When calculated, the water force on the U-shaped aqueduct body can be replaced by a continuous uniform load. When the aqueduct is filled with water, under the action of gravity, the water generates pressure to both the U-shaped body and the rectangle body on its upside. Both kinds can be represented by linear load. Through converting, for U-shaped body, the middle is the maximum, $q_1 = 6.894$ N/mm; both ends are the minimum 0 (symmetrical on both sides); for the rectangle side pole, the lower end is the maximum, $q_2 = 3.316$ N/mm; the upper end is the minimum 0. By Mechanical Solver: tape largest axial force $N = 29.80$ N, the axial force of the bar $N = 29.33$ N.

According to the computed result, the force distribution of U-shaped aqueduct can be simplified to side montant on both sides. The bending moment and shearing force are very small, and therefore can be neglected. For each section, the force can be converted into pressure of the two side bar. So in the graphic, each side of the rectangle can be seen as withstanding uniform load as axial force of 29.23. The following will be the cross-section recalculation of security and stability.

(1) Side montant: Cross-sectional area $A = 15$ mm², Section moment of inertia $I = 141.25$ mm⁴, known by the cross-sectional force diagram $M = 0.00082178$ N · cm, $N = 29.33$ N;

$$\sigma_t = \frac{N}{A} = \frac{29.33}{15} = 1.95 \text{ N/mm}^2 < 20 \text{ N/mm}^2, \text{ it meets the requirements of strength;}$$

(2) U-shaped tape: Cross-sectional area $A = 4$ mm², according to the force diagram, the maximum axial force is 29.80 N; maximum bending moment of 118.1 N · mm. For the u-shaped tape, the bending moment and shear force can be considered as infinite and therefore only need to consider axial force can be.

$$\sigma_t = \frac{N}{A} = \frac{29.80}{4} = 7.45 \text{ N/mm}^2 < 20 \text{ N/mm}^2, \text{ it meets the requirements of strength;}$$

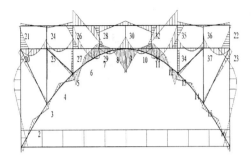

Figure 4. Bending moment diagram.

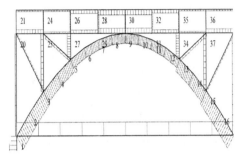

Figure 5. Axial force diagram.

Contrast tape and vertical rod the force, at the same loads, the tape is reached first vertical rod than the limit equilibrium state, that is $\sigma_t = 20$ N/mm, then N = 80 N. By a proportional relationship to be at this time, the load $M = 37.84$ kg, and the tape will be destroyed.

2. The overall simplifying and internal force calculation
(1) androgenic force computation

According to the symmetry, choose the calculation sketch in the whole structure lever. Through calculating (Long, 2002), get the structural internal force diagram. They are shown in Fig. 4 and 5.

By the structural internal force diagram for the stability of the structure, we find for the checking point unit 9 on the left, unit 1 and tape, the solution is:

For unit 9: N = −30.4 N, M = 33.1 N · cm

$$\sigma_t = \frac{N}{A} + \frac{M}{W} = -\frac{30.4}{56} + \frac{330 \times 6}{8 \times 7^2} = 4.51 \text{ N/mm}^2 < 20 \text{ N/mm}^2,$$ so the tensile stress meets the strength requirement;

$$\sigma_c = \frac{N}{A} - \frac{M}{W} = \frac{30.4}{56} + \frac{330 \times 6}{8 \times 7^2} = 5.59 \text{ N/mm}^2 < 7 \text{ N/mm}^2,$$ so the compressive stress meets the strength requirement;

For unit 1: N = −41.9 N, M = 4.79 N · cm

$$\sigma_t = \frac{N}{A} + \frac{M}{W} = -\frac{41.9}{56} + \frac{47.9 \times 6}{8 \times 7^2} = -0.015 \text{ N/mm}^2 < 0 \text{ N/mm}^2,$$ so the compressive stress meets the strength requirement;

$$\sigma_c = \frac{N}{A} - \frac{M}{W} = \frac{41.9}{56} + \frac{47.9 \times 6}{8 \times 7^2} = 1.48 \text{ N/mm}^2 < 7 \text{ N/mm}^2,$$ so the compressive stress meets the strength requirement;

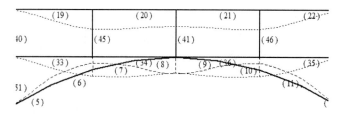

Figure 6. The displacement map.

Table 3. The deformation.

Pole or node	20	21	41	45	46	9
deformation/mm	−1.755	−1.487	−1.487	−1.755	−1.755	−1.487

For the tape: The tape is mainly for the tensile stress.

$$\sigma_t = \frac{N}{A} = \frac{33.24}{8 \times 0.5} = 8.31 \text{ N/mm}^2 < 20 \text{ N/mm}^2,$$ So the tensile stress meets the strength requirement.

In a word, the structure as a whole meets the strength requirements.

(2) Calculation of displacement

By Solver structures available, maximum deformation in the structure is 1.755 mm, located approximately in the middle arch, as shown in Fig. 6, the deformation data is in table 3:

But the maximum deformation is 3 mm obtained by water experiment, experimental analysis and solver data have error, but the combination of the experimental conditions and the middle of the uncertainty modeling, which is within the range of allowable error. But also by experimental deformation data are available; the structure of the most unfavorable position is not a point, but the highest point in the arch structure around a certain range.

5 CONCLUSIONS

(1) Through theoretical calculation and real loading experiment, the limit load of u-shaped aqueduct is 37.7 kg and the destroying point is the highest point of the arch braced structure. The maximum displacement at this time is 9 mm.
(2) There is a slight error between theory and real condition of the model, but in the experimental allowable error range, the result is the same.
(3) In model manufacturing process, both the size of the paper bar and the cohesion of the emulsion should be well handled to prevent big error.

REFERENCES

Cui Na. The U arch bound aqueduct optimized design studies (D). Shaanxi Yang to insult: Northwest farming and forestry scientific and technical university, 2010
Long Yuqiu, Chief Editor Bao Shihua. Structure Mechanics (M) Volume one. Second edition, 2002
Luo Yehui. The large-scale priestesses concrete aqueduct trough body finite element analysis research and applies (the D). Jiangsu Nanjing: River Sea University, 2005
Ma Zhen. Steel bar concrete arch truss most superior design. East China Water conservation Institute journal, 1984(4)
Zhu Huizhu, Chen Deliang, Manages maple tree year writing. Aqueduct. Chinese Water conservation Water and electricity Publishing house, 2009

Spatial coupling relationship between settlement and land and water resources – based on irrigation scale – A case study of Zhangye Oasis

L.C. Wang, R.W. Wu & J. Gao
College of Geography and Environmental Science, Northwest Normal University, Lanzhou, China

ABSTRACT: Heihe River Basin is the second largest inland river basin in the arid northwest China. In arid inland basin, the core problem is water resources, while the core area is an oasis. It is a complex interdependence of Irrigation District, through spatial integration after the formation of the modern oasis. Under the strong control and guidance of the artificial canal system, water supply quantity determines Irrigation District, and irrigated land determines population carrying capacity, so water-soil-human systems exist a high degree of interdependence and symbiosis. All of irrigation units in Zhangye, the average of coupling degree about settlements and water, soil resource is 0.714 and the spatial differences are significant. The spatial distribution characteristics of the coordination degree is essentially similar coupling degree. Accordingly, the 27 irrigation districts can be divided into the water-soil-human resources harmonious area; water-soil-human resources run-in areas; water-soil-human resources antagonistic area; water-soil-human resources low coupling region.

1 INTRODUCTION

In the arid inland area of northwest China, Heihe River Basin is the second largest inland river basin. At present, Heihe River Basin has been a hot region of study. Extensive research shows that the core problem are water resources, while the core area are oasis. As the family is the basic unit of the society, irrigation district is the basic cell of the whole oasis. The modern oasis is the product of artificial canal directional control. In other words, "irrigation district" is the actual users of water resources, rather than the whole oasis. You can also say that irrigation district which has the complex interdependent link with each other lead the formation of the modern oasis through spatial integration. The area of Irrigation District determines the size of the oasis, the distribution and spatial organization of irrigation district determine the distribution patterns and range of oasis. To deep understanding of modern oasis, it is necessary to transfer from macroscale scale to microscopic scale.

Surface water and groundwater exists repeated conversion in the oasis, which lead to the water consumption is greater than the water supply, but generally speaking, the key factor of controlling soil is water. In theory, in the same natural potential cases, the size of the irrigation area determines the output capacity of the land, which determines the human supplies ability, and then affect the size of the settlement and the distribution pattern!

In the latest 20 years, foreign study on residential environment in arid areas mainly include the oasis settlement archaeology, spatial pattern of settlement in the oasis, and the relationship of water conservancy, agriculture, land use and oasis settlement distribution and so on. Since the 1990's, research on irrigation area of Heihe River basin, mainly in water supply-demand balance and the optimal allocation of water resources and water-saving irrigation technology, water productivity, and exploring the relationship between oasis and settlement: yuanmei Jiao found that residents of oasis who have a high concentration link to the land and irrigation channels closely.

Most of these studies focused on the river basin macro-scale or oasis meso-scale, from the perspective of water resource, ecological, economic, demographic and urban research, come to a fruitful and enlightening conclusion. But unfortunately, they did not go deep into the basic units of the oasis–irrigation district, which made certain conclusions are often "unreal", then its operation is subject to a certain degree of damage.

So we must discuss the formation and developmental law of settlement system according to the coupling relationship of micro-process, micro-mechanism and microscopic structure, analyze the spatial distribution pattern of the settlement system in different types of irrigation districts and explore the coupling relationship between the particle system, corridor system and domain surface system.

There are 29 irrigation districts in Zhangye city. Because natural water system which those irrigation districts rely on are different, it can be divided into three regions: Heihe River irrigation districts, Liyuan River irrigation districts and the hillside irrigation districts. Generally speaking, the hillside irrigation districts, relying on a relatively independent river, relatively few water resources, its size are relatively small, structure are relatively simple. However, in the Heihe irrigation districts which relying on Heihe River mainstream, there are a complex series of channels, between irrigation districts exists the complex series and parallel relationship, so irrigation districts have a more complex and profound impact on the distribution pattern of land, population and settlements.

2 DATE SOURCES AND RESEARCH METHODS

2.1 *The data source*

The land use data come from the Zhangye city second land survey database. Then we extract the residential land and irrigated land area, respectively represent the "settlement" and "land". Irrigated district data come from the "management report" database, from which we retrieve attribute data of canal system and water resources represent the "water resources", other economic data come from the Zhangye city Statistical Year book.

2.2 *The research methods*

According to the theoretical and practical significance of the indicator, to the greatest extent possible reflect the features and benefits of the system, we select representative indices (table 1), respectively reflecting the actual status and level of settlement space, water, soil and the scale of resources subsystem. We select 29 irrigation district units in Zhangye city as the object of study.

2.3 *Effect function*

In oasis irrigation district, people-water-soil coupled system are consists of settlements, land resources and water resources subsystems. Then each subsystem are consists of a number of

Table 1. The comprehensive evaluation index system.

System	Index	Explain
Settlement	Residential area	Reflect settlement scale
Water Resources	Canal system length	Reflecting the irrigation channel developed
	Channel utilization ratio	Reflect conveyance power of water£» Irrigation water use
	diverted water volume	Reflect the diversion capacity of canal
	water consumption	Reflect the water resources consumption in irrigation district
Land Resources	Area of irrigated land	Reflecting the main direction of water use in irrigation district
Economic System	production of food-crops	Reflect the planting structure in the irrigation area and the output efficiency of water resource

indicators. We can assume that the subsystem i is made up of n indices, respectively are the x_1, $x_2,\ldots, x_n$. When the larger the value of x_{ij}, the better the system function is, in other words, the value of x_{ij} which contributes to the effectiveness of the system is positive, then it is known as a positive indicator. When the smaller the value of x_{ij}, the better the system function is, it is known as a negative indicator. In people-water-soil coupled systems efficacy coefficient calculation formula of different indicators is:

$$d_{ij} = (x_{ij} - x_{ijmin}) / (x_{ijmax} - x_{ijmin}) \text{ the positive indicator}$$

$$d_{ij} = (x_{ijmax} - x_{ij}) / (x_{ijmax} - x_{ijmin}) \text{ the negative indicator} \quad (1)$$

where d_{ij} as the function number of the index j of system i; x_{ijmax} as the maximum of index j of the system i; x_{ijmin} as the smallest of the index j of system i; x_{ij} as the value of the index j of system i. D_{ij} reflects the satisfaction achievement of the goals, $0 \le d_{ij} \le 1$. When d_{ij} is 0, for the most dissatisfied; and when the d_{ij} is 1, for the most satisfaction.

2.3.1 The subsystem function

The comprehensive efficacy of settlements and soil and water resources subsystem is the combination of all indexes in the system's contribution to the subsystem and it can be done by integrated method. Its computation formula is:

$$U_i = \Sigma W_{ij} \times d_{ij} \text{ among them, } W_{ij} \ge 0, \Sigma W_{ij} = 1, j = 1, 2, \ldots n \quad (2)$$

where W_{ij} as weights of the index j of subsystem I. Index weight is one of the important information of comprehensive evaluation. It should be based on the relative importance of indexes, namely it should be determined by the contribution of comprehensive evaluation from indexes. As the main component analysis method not only considers the data under a single index distribution, also considers the informational overlapping and mutual interference among indexes. This article uses principal component analysis, in the premise of selecting all indicators as the main component, use the variance of each principal component as the weight, and process a comprehensive evaluation to every subsystem, including settlement and water and soil resources scale subsystem.

2.3.2 Evaluation of the coupling

With the help of the capacitance coupling concept and capacity of the coupling coefficient model in physics, we can promote the interaction coupling model of multiple system.

$$C = \left\{ \frac{U_1 \times U_2 \times L U_n}{\prod (U_i + U_j)} \right\}^{1/n} \quad (3)$$

where, C = the degree of coupling (interdependence, mutual influence); the U_i = the comprehensive effect of each subsystem. The size of the coupling is determined by the size of each subsystem U_i. This article measured the settlement space, the coupling model of water and soil resources subsystem composition, therefore n = 3. Since U_i is a value between 0~1, the degree of coupling C values are also bounded between 0~1. When C = 0, the minimal degree of coupling among systems or among elements within the system is independent, then the system will provide a disorderly development. When the C = 1, the coupling is the largest. According to many scholars on the classification of the coupling now, using the median segment method, when $0 < C \le 0.3$, it indicate that the system is in a low coupling stage; when $0.3 < C \le 0.5$, it indicate that the system is in an antagonistic stage; when $0.5 < C \le 0.8$, it indicate that the system enters run-in phase; when $0.8 < C \le 1$, the system enters a high level coupling phase.

2.3.3 Coordination degree

Coordination degree model can judge the coordination degree of each subsystem better. Its computation formula is:

$$T = \alpha U_1 \times \beta U_2 \times \lambda U_3 \quad (4)$$

$$D = C \times T \quad (5)$$

where D = coordination degree; C = the degree of coupling; T is the index of subsystem comprehensive coordination, it reflect the synergy effect or contribution of people-water-soil system; α, β and λ for the undetermined coefficient, due to the regional economy and environment are equally important, so take the $\alpha = \beta = \lambda = 0.33$; U_1, U_2, U_3 for comprehensive efficacy of subsystem.

According to the distribution of the U value, degree of coordination D value is between 0~1.

The higher coupling values and coordination degree, the higher comprehensive efficacy is and the people, water and soil subsystem to the system, harmonious degree is also higher and vice versa. So we use the middle score method, coordinate degree to divide into four types: $0 < D \leq 0.3$ for the low coordinated coupling; $0.3 < D \leq 0.5$ for the moderately coordinated coupling; $0.5 < D \leq 0.8$ for the highly coordinated coupling; $0.8 < D \leq 1$ is extremely coordinated coupling.

3 THE RESULTS OF ANALYSIS

3.1 Space pattern of the degree of coupling

There are 27 irrigation districts in Zhangye oasis, the average coupling degree about settlements-water-soil resources is 0.174, belonging to the high level of coupling state, but the spatial differences are significant, the coupling value between 0.20 to 0.96. The highest is Yingke irrigation district (0.96) belongs to the typical oasis irrigation district and the lowest is Minghua irrigation district (0.20) belongs to the typical independent irrigation district.

Irrigation districts located in the oasis hinterland, as well as irrigation districts along the Heihe River, its coupling degree are obviously higher than irrigation districts located in the oasis fringe; joint irrigation districts are higher than independent irrigation districts (Fig. 1).

Highly coupled irrigation districts (coupling degree above 0.8–0.96) are mainly distributed in some typical oasis hinterland, along the heihe river, including Yingke, Liyuanhe, Xijun, Daman,

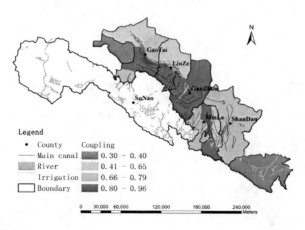

Figure 1. The distribution map coupling degree.

Shangsan, Huazhai, Youlian, Hongshuihe, Shahe, and Luquan irrigation districts. In those irrigation districts, the soil is very fertile, artificial canal system are well-developed and canal system utilization rate is high, relatively dense population distribution. Run-in phase's irrigation districts (coupling degree: 0.66–0.79) mainly distributed in hillside irrigation districts, as well as irrigation districts located in the oasis fringe, including Anyang, Dazhuma, Hongyazi, Laojunhe, Liuba, Sigouhe, Suyoukou, Xinba, Youlian, Yanuan and Tongziba irrigation districts. There are rivers almost in every irrigation district, such as Daduma River, Sigou River, Suyoukou River, Tongziba River, etc, river runoff is relatively larger (table 2), water resources relatively abundant.

Antagonistic phase's irrigation districts (coupling degree: 0.41–0.65) are mostly near the outlying areas of the lower reaches of Heihe River, mainly including Luocheng, Banqiao, Pingchuan irrigation districts. Affected by upstream "deprivation effect", their water resource are relative short and low levels of water use. For example, Luocheng Irrigation District, average diverted water volume is only 5306.06×10^4 m^3 in 1998–2002, the effective irrigation area is 0.46×10^4 hm^2. In the Banqiao Irrigation District, average diverted water volume is only 1378.4×10^4 m^3 in 1998–2002, the effective irrigation area is 0.6×10^4 hm^2. At the same time, these irrigation districts spread along roads and railways, and gathered a large number of the population, so the relationship among people-water-soil system is relatively tight. Low degree coupling irrigation districts (coupling degree: ≤0.3) distribute in the Huangcheng Irrigation Districts and the Minghua Irrigation Districts in Sunan County, belonging to more independent hillside irrigation districts. The geographic conditions is very inclement, and the protection level of water resources is very poor. Among them, there is no surface water in Minghua Irrigation District and agricultural water mainly relies on groundwater. In the 1950s and 1980s, people built two dams (Hejiadun Dam and Nangou Dam), its water source are spring water, later because the springs dried up, it cannot be used. So it is in the state of low intensity coupling.

3.2 Space pattern of the degree of coordination

(1) Among the 27 irrigation districts, the average coordination degree about settlements-water-soil resource is from 0.12 to 0.643. The highest value in Liyuanhe Irrigation Districts and Xijun Irrigation Districts, the lowest value in Daquangou Irrigation Districts (Fig. 2).

(2) Highly coordinated irrigation districts (coordination degree: 0.75–0.89) mainly distributed in along the Heihe River mainstream, those irrigation districts are typical oasis-irrigated farmland, including Liyuanhe, Banqiao, Hongyazi, Yanuan, Youlian, Xijun and Daduma irrigation districts. Moderate coordinated irrigation districts (coordinate degree: 0.55–0.74) distribution in the edge of the oasis irrigation districts and partial hillside irrigation districts, including Anyang, Daman, Hongshuihe, Huazhai, Luquan, Pingchuan and Xinba irrigation districts; Low coordinated irrigation districts (coordinate degree: 0.31–0.54) are mainly distributed in Shandan country, including Laojunhe, Liuba, Mayinghe, Shangsa, and Suyoukou irrigation districts. Uncoordinated irrigation districts (coordinate degree: 0.1–0.3) are typical independent irrigation district, including Minghua, Huangcheng in the Sunan, and Sigou in Sandan.

Table 2. Runoff of major tributaries of Heihe River Basin.

River name	Station Name	Basin area (km^2)	Run of volume (10^4 m^3)
Maying River	Lijiaqiao	1143	0.740
Hongshiu River	Shuangshusi	578	1.260
Daduma River	Wafangcheng	217	0.890
Xiaoduma River	Xintian	101	0.190
Tongziba Daduma River	Biandukou	331	0.644
BaiLang Daduma River	Gaotaixindi	211	0.400

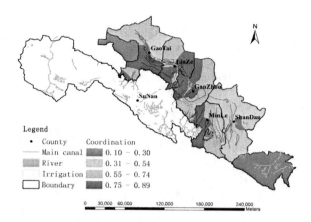

Figure 2. The distribution map coordination degree.

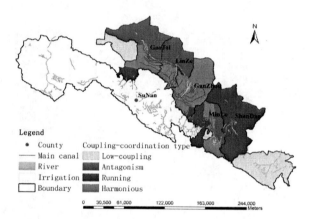

Figure 3. Type of coupling-coordination.

3.3 Space – type of coupling–coordination

According to the coupling degree and coordination degree of each irrigation district, the 27 irrigation districts can be divided into 4 types (Fig. 3).

(1) Harmonious type, namely high coupling-high coordination areas. They mainly distribute in the riverine oasis and typical oasis. It Means that those irrigation districts have already been from high-water-soil resources highly coupled to the harmony.
(2) Run-in phase's type, namely high coupling-low harmonious areas. They are mainly in the basin of Dazhuma River, Shandan River and Liyuanhe River, which have bigger runoff and abundant water resources. The coupling degree is above 0.8 and the coordination degree is between 0.3 and 0.5.
(3) Antagonistic type, namely middle coupling-low harmonious region. They mainly distribute in irrigation districts located in the oasis fringe, in the downstream of Heihe River. Coupling degree is above 0.8, coordination degree is below 0.3, because of the limitation of topography and soil infertile, most of the area cannot conduct agricultural production.
(4) Low coupling type, namely low coupling-low harmonious region. The geographic conditions are very poor and water resource is very limited, so the relationship among people-water-soil system is relatively tight. Both coupling and coordination degree are below 0.3.

ACKNOWLEDGMENTS

This research is funded by the Natural Science Foundation of China (40235053) and the Natural Societal Science Foundation of China (05XSH010). Northwest Normal University Backbone of the project (SKQNGG10029). and 2010 Gansu College tutor Fund (1001-22).

REFERENCES

Andrew, W. & Kandel. & Nicholas J. & Conard. 2012. Settlement patterns during the Earlier and Middle Stone Age around Langebaan Lagoon, Western Cape (South Africa). Quaternary International, 270(23):15–29.

Gideon, G. 1982. Selecting sites for new settlements in arid lands: Negev case study. Energy and Buildings, 4, (1): 23–41.

Grace, M. & Leonie, S. & Clive, M. A. & Chris, J. & Bronwyn P. & Greg B. 2011. Land cover change under unplanned human settlements: A study of the Chyulu Hills squatters, Kenya. Landscape and Urban Planning, 99(2):, 154–165.

Guangjin, T. & Zhi, Q. & Yaoqi, Z. 2012. The investigation of relationship between rural settlement density, size, spatial distribution and its geophysical parameters of China using Landsat TM images. Ecological Modelling, 231(24): 25–36.

Hadas. 2012. Ancient agricultural irrigation systems in the oasis of Ein Gedi, Dead Sea, Israel. Journal of Arid Environments, 86:75–81.

Hill, M. 2003. Rural settlement and the Urban Impact on the Countryside. Hodder & Stoughton. 20–36; 58–72.

Hu, G. L. & Zhao, W. Z. 2009. Change of water productivity of wheat at different scales in oasis irrigation districts. Transaction of the Chinese sociaty of agricultural emgomeering. 25(2):24–30.

Jessica, G. 2009. The evolution of settlement patterns in the eastern Oman from the Neolithic to the Early Bronze Age (6000–2000 BC). Comptes Rendus Geoscience, 341(8–9):739–749.

Ji, X. B. & Kang, E. & Zhao W. Z. 2005. Analysis on Supply and Demand of Water Resources and Evaluation of the Security of Water Resources in Irrigation Region of the Middle Reaches of Heihe River, Northwest China. Scientia Agricultura Sinica, 38(5):974–982.

Jiao, Y. M. & Xiao, D. N. & Ma, M. G. 2003. Spatial pattern in residential area and influencing factors in oasis landscape Acta ecologica sinica, 23(10):2092–2100.

Jiao, Y. M. & Ma, M. G. & Xiao, D. N. 2003. Research on the Landscape Pattern of Zhangye Oasis in the Middle Reaches of Heihe River. Journal of glactioligy and geocryilogy. 25(1):94:99.

Local Chronicles Compilation Committee of Gansu Province, Gansu Province, water conservancy codification of the leading group. Gansu Province, 23 volumes, Water conservancy Lanzhou: Gansu Culture Press, 1998:166 of-168.

Luo, Y. L. & Li, Q. J. & Zhang, X. 2005. The Water-saving transformation effect analysis in middle reach of the Heihe River Irrigation District. Water saving irrigation. (6):40–42.

Marjanne, S. & Marc, A. Settlement models, land use and visibility in rural landscapes: Two case studies in Greece. Landscape and Urban Planning, 2007, 80(4):362–374.

Shi, M. J. & Wang, L. & Wang X. J. 2011. A Study on Changes and Driving Factors of Agricultural Water Supply and Demand in Zhangye after Water Reallocation of the Heihe River. Resources Science, 33(8).

Siebert, J. & Häser, M. & Nagieb, L. & Korn, A. & Buerkert. 2005. Agricultural, architectural and archaeological evidence for the role and ecological adaptation of a scattered mountain oasis in Oman. Journal of Arid Environments, 62(1): 177–197. G.

Stefan, S. & Maher, N. & Andreas, B. 2007. Climate and irrigation water use of a mountain oasis in northern Oman. Agricultural Water Management, 89(1–2): 1–14.

HadasSaaroni. & Arieh, B. &Eyal, B. D. & Noa, F. 2004. The mixed results concerning the 'oasis effect' in a rural settlement in the Negev Desert, Israel. Journal of Arid Environments, 58(2): 235–248.

Sylvi Haldorsen. & Holger, T. 2012. Palaeo groundwater dynamics and their importance for past human settlements and today's water management. Quaternary International, (257): 1–3.

Wang, T. X. 2002. The analysis on water-saving irrigation mode application in Zhangye. China rural water and hydropower. 4: 25–26.

ns
Numerical simulation of tidal current near the artificial island sea area

Zegao Yin, Zhenlu Wang, Tong Xu & Ningning Ma
Engineering College, Ocean University of China, Qingdao, Shandong, China

Wanqing Chi
First Institute of Oceanography, SOA, Qingdao, Shandong, China

ABSTRACT: Based on Mike21 software, a two-dimensional hydrodynamic mathematical model was established. The mathematical model was verified by using the measured data. With the validated model, the simulation of the tidal current near the artificial island sea area was completed. It is found that the current velocity variation and local scour occur between two islands, and the similar phenomenon also occurs between the island and the coastline. It is necessary to consider the influence of project at program selection and construction period.

1 INTRODUCTION

An artificial island is constructed by people rather than formed by nature. It is often created by expanding existing islets, construction on existing reefs, or amalgamating several natural islets into a bigger island. And now most of the islands were built by reclaiming land from sea. The construction of the artificial islands which was mainly used for military, fishery, etc. has started from ancient times. With the continuous development of economy, human demand for marine space utilization is increasing, and so are the artificial island projects. In recent years, the island has been widely used in maritime platform, aquatic products processing base, deep sea port, tourism projects and so on (Lin & Wu, 2008).

The construction of artificial island will change the hydrodynamic and coastal dynamic balance of sediment transport to a certain extent. Numerical simulation technology is an effective method in the prediction and assessment of artificial island on the impact of adjacent water. A systematic summarization has been carried through the method of numerical simulation in the tidal field of coastal estuary area (Li & Cao, 1999; Yuan & Yu, 2004). The mathematical models with irregular grid of current and suspended sediment in Huludao sea area were established based on TK-2D software (Li, 2008). A 2-D mathematical model with multiple nested grids of current and sediment in estuary was adopted near the artificial island in Hebei province (Wei et al., 2012). Numerical simulation technology has also been applied in the analysis of the influence of reef arrangements on artificial reef flow fields (Su et al., 2009; Liu & Su, 2013).

In this paper, the mathematical model of tidal current is established and validated by the measured data. The artificial island influence on the current variation and bed deformation is discussed. It is helpful to choose the artificial island design and construction.

2 MATHEMATICAL MODEL OF 2-D TIDAL CURRENT

2.1 *Control equation*

Continuity equation

$$\frac{\partial \zeta}{\partial t} + \frac{\partial p}{\partial x} + \frac{\partial q}{\partial y} = 0 \qquad (1)$$

Figure 1. Location of the stations of tidal level, tidal current velocity.

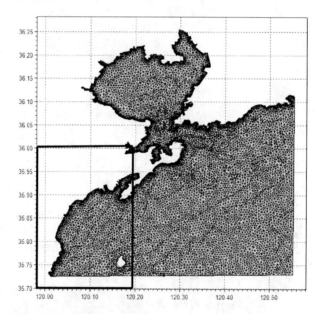

Figure 2. Mess generation of the nested mesh model.

Momentum equation

$$\frac{\partial p}{\partial t}+\frac{\partial}{\partial x}\left(\frac{p^2}{h}\right)+\frac{\partial}{\partial y}\left(\frac{pq}{h}\right)+gh\frac{\partial \zeta}{\partial x}+\frac{gp\sqrt{p^2+q^2}}{C^2h^2}$$
$$-\frac{1}{\rho_w}\left[\frac{\partial}{\partial x}(h\tau_{xx})+\frac{\partial}{\partial y}(h\tau_{xy})\right]-\Omega q-fVV_x+\frac{h}{\rho_w}\frac{\partial}{\partial x}(p_a)=0 \quad (2)$$

$$\frac{\partial p}{\partial t}+\frac{\partial}{\partial x}\left(\frac{p^2}{h}\right)+\frac{\partial}{\partial y}\left(\frac{pq}{h}\right)+gh\frac{\partial \zeta}{\partial x}+\frac{gp\sqrt{p^2+q^2}}{C^2h^2}$$
$$-\frac{1}{\rho_w}\left[\frac{\partial}{\partial x}(h\tau_{xx})+\frac{\partial}{\partial y}(h\tau_{xy})\right]-\Omega q-fVV_x+\frac{h}{\rho_w}\frac{\partial}{\partial x}(p_a)=0 \quad (3)$$

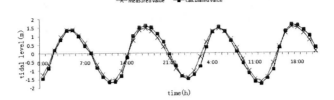

(a) Tidal level history at T station

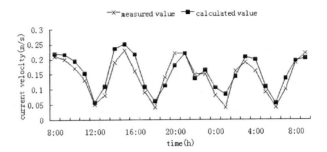

(b) Current velocity history at D station

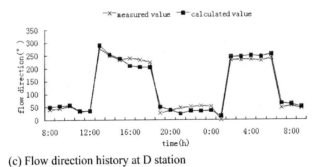

(c) Flow direction history at D station

Figure 3. The validation between mathematical results and measured data.

where ζ = tidal level; p, q = the discharge per unit width of x, y direction; $C(x,y)$ = Chezy Coefficient; $f(V)$ = wind resistance factor; $V, V_x, V_y(x,y,t)$ = wind speed and component of x, y direction; $\Omega(x,y)$ = Coriolis force parameter; $\tau_{xx}, \tau_{xy}, \tau_{yy}$ = the effective shear Stress Element.

2.2 *Initial and boundary conditions*

2.2.1 *Initial condition*
Cold start when calculating begins:

$$\zeta(x,y,t) = 0; \quad h(x,y,t)_{t=0} = h_0(x,y); \quad u(x,y,t)_{t=0} = 0; \quad v(x,y,t)_{t=0} = 0.$$

2.2.2 *Boundary condition*
Tangential velocity and normal velocity are both zero in the close boundary. Input the tidal level in the open boundary:

$$\zeta = \sum_{i=1} \{f_i H_i \cos[\sigma_i t + (V_{0i} + V_i) - G_I]\} \tag{5}$$

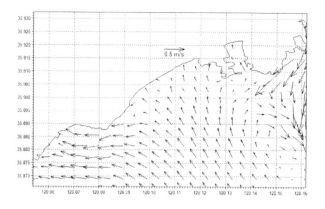

(a) The moment during spring tide

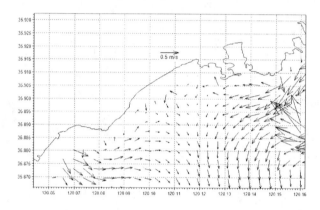

(b) The moment during neap tide

Figure 4. The calculated current velocity in the status quo.

where σ_i = angel velocity of the i-th component tide (taking a total of four component tides: M2, S2, O1, K1); f_i, θ_i = nodal factor & epoch correction of the i-th partial tide; H_i = harmonic constant (amplitude of component tide); G_i = harmonic constant (epoch of component tide); V_i = hour angle of component tide.

2.3 Validation of mathematical model

T tidal level station and D current station were set as Figure 1, and the related data were measured from 8:00, October 12, 2010 to 9:00, October 13, 2010.

In order to simulate water and sediment transport behavior, the nested model was used in the numerical simulation, including 2 scopes of mesh. The calculation results of large area provide the boundary conditions for small area model. The large area domain is 35°40′08″-36°14′01″N, 120°06′46″-120°36′25″E, while the small area domain is 35°40′22″-36°00′46″N, 120°06′46″-120°18′15″E, and the time step is 10 seconds (Figure 2).

The tidal current of the research area was simulated by using the mathematical model, and the results were compared with the measured data (Figure 3).

The figure 3 shows that the calculation of tidal level, tidal velocity, flow direction are in good agreement with the measured data, and the mathematical model can be used for the further prediction.

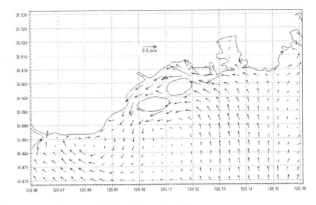

(a) The moment during spring tide

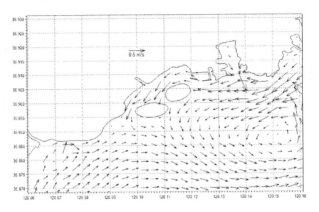

(b) The moment during neap tide

Figure 5. The calculated current velocity after the artificial.

3 WATER CHARACTERISTICS BEFORE AND AFTER THE ARTIFICIAL ISLAND CONSTRUCTION

3.1 *Project profile*

The project is located in the northwest coast of the Yellow sea; two independent islands are arranged about 600 m away from the coastline. There is about 200 m distance between them, the shape of them is almost the oval, and the area is about 400,000 square meters each.

3.2 *Tidal current field*

Figure 4 shows the tidal current field during spring tide and neap tide of the status quo. Figure 4 shows that the maximum velocity of the spring tide is 0.3 m/s and the flow direction is northwest while the maximum velocity of the neap tide is 0.35 m/s and the flow direction is southwest.

Figure 5 shows the tidal current field after the artificial island construction. As can be seen from figure 5, the maximum velocity of the spring tide between the two islands is about 0.4 m/s, while it is 0.6 m/s between the islands and the coast, and the tidal direction is southwest. The maximum velocity is about 0.2 m/s between the two islands, while it is 0.5 m/s between the islands and the coast which tidal direction is southwest. A possible reason is that the tidal current direction is northwest during spring tide after the artificial islands construction. So the velocity of the northeast waterway will increase due to the narrower waterway than the status quo. This waterway is much wider than other waterways, so the unit discharge is larger. Alongshore current will be formed in southwest direction when water flows along the coastline, and the flow velocity will increase compared with the status quo. The tidal current will be jacked between the two islands by the alongshore current, which will cause small increase of the flow velocity between the two islands. The tidal direction is southwest during neap tide, while the waterway direction between two islands is northwest, so the flow velocity will decrease.

4 ISLANDS CONSTRUCTION CONCLUSION

A two-dimensional hydrodynamic mathematical model was established based on the Mike 21 software. Tidal level, tidal current velocity, flow direction and concentration were verified by using the observation data. And they are both in good agreement which suggests that the mathematical model aforementioned was reasonable and reliable.

The calculation and analysis of tidal current before and after the artificial islands construction was completed. The results show that the current velocity variation and local scour occurs between two islands, and the similar phenomenon also occurs between the island and the coastline after the artificial islands construction. The optimization study of artificial island layout is an interesting work to be further investigated in future.

ACKNOWLEDGEMENT

This study is financed by Qingdao Science and Technology Development Plan (11-2-4-1-(7)-jch).

REFERENCES

Li Mengguo & Cao zude, 1999. Studies on the numerical simulation of the coast estuarine tidal current. *Acta Oceanologica Sinica* 21(1): 111–125. (In Chinese)

Li Wendan, 2008. Numerical simulation of tidal current and sediment surrounding the Huludao sea area. *Journal of Waterway and Harbor* 29(2): 94–98. (In Chinese)

Lin Yuanjun & Wu Jiaming, 2008. Discussion of the numerical analysis method on the influence of the artificial island engineering construction to the marine environment. *Guangdong Ship Building* 26(4): 35–37. (In Chinese)

Liu Tsung-Lung & Su Dong-Taur, 2013. Numerical analysis of the influence of reef arrangements on artificial reef flow fields. *Ocean Engineering* 74: 81–89.

Su, D.T., 2009. Numerical simulation investigation into submerged artificial reef flows. *National Taiwan Ocean University, Taiwan, ROC*. PhD thesis.

Wei Long et al., 2012. Numerical simulation of the tidal current and sediment near the artificial island of south fort in eastern Hebei province. *Port & Waterway Engineering* (6): 43–47.

Yuan Hang & Yu Dingyong, 2004. Review and studies on the numerical simulation of tidal current and sediment. *Advances in Marine Science* 22(1): 97–106. (In Chinese)

Using ANN and SFLA for the optimal selection of allocation and irrigation modes

J.X. Xu & Y.J. Yan
Farmland Irrigation Research Institute, Chinese Academy of Agricultural Sciences, Xinxiang, Henan, China;
School of Water Conservancy, North China University of Water Resources and Electric Power, Zhengzhou, Henan, China

Y. Zhao
College of hydrology and water resources, Hohai University, Nanjing, Jiangsu, China

H.Y. Zhang & Y. Gao
School of Water Conservancy, North China University of Water Resources and Electric Power, Zhengzhou, Henan, China

ABSTRACT: Total 12 different allocation and irrigation modes were used for the optimal selection in the typical irrigation area. An evaluation index system with three layers was established and evaluating by the method of Artificial Neural Network (ANN) based on Shuffled Frog-Leaping Algorithm (SFLA). The results show that the mode with ground hose irrigation and high irrigation quantity under the present water supply is the best allocation and irrigation mode. Under the present water supply condition, the modes with ground hose irrigation are superior to that with sprinkler and small border irrigation. But if there are new sources of water supply, the modes with sprinkler irrigation will be superior to the others.

1 INTRODUCTION

Under the level of regional economic development and utilization of water resources, the optimal selection of allocation and irrigation modes is important to water conservation irrigation decision and regional agriculture sustainable development. Most of current researches focus on the mode selection with irrigation benefit and irrigation schedule (Xu et al. 2002, Feng & Xu 2009). However, there are few studies on combining irrigation mode and allocation mode as a whole.

Generally, the methods of analytic hierarchy process, fuzzy comprehensive evaluation, grey correlation analysis, etc. are used to evaluate given modes (Xu et al. 2002, Li et al. 2006). But, these methods still have some limitations in the process of optimal selection. For example, the results of fuzzy comprehensive evaluation could be affected by the correlation among the evaluation indices.

Artificial neural networks (ANN) are computational models inspired by animal central nervous systems (in particular the brain) that are capable of machine learning and pattern recognition. Due to the flexibility and effective utilizing existent data of ANN, it has been applied in the field of water resource widely (Sahoo et al. 2005, Singh et al. 2009). However, ANN is easy to fall into local optima. Shuffled Frog Leaping Algorithm (SFLA) is a recent memetic meta-heuristic algorithm proposed by Eusuff and Lansey in 2003 (Eusuff & Lansey 2003). This algorithm is used to calculate the global optima of many problems and proves to be a very efficient algorithm.

In this paper, based on the coupling relationship between irrigation mode and allocation mode, an evaluation index system is established for the typical irrigation area. The method of ANN and based on SFLA is then used to evaluate different allocation and irrigation modes.

Table 1. The 12 different allocation and irrigation modes in the typical irrigation area.

Water supply scheme	Irrigation methods	Low irrigation quantity	High irrigation quantity
present water supply	U type channel lining and small border irrigation	M_1	M_4
	low pressure pipe of water conveyance and sprinkler irrigation	M_2	M_5
	ground hose irrigation	M_3	M_6
new water supply	U type channel lining and small border irrigation	M_7	M_{10}
	low pressure pipe of water conveyance and sprinkler irrigation	M_8	M_{11}
	ground hose irrigation	M_9	M_{12}

2 METHODOLOGY

2.1 Index evaluation system

Crop water use includes water allocation and output water two different courses. The process of water allocation is relevant to the determination of allocation mode, and the other one is relevant to the selection of irrigation mode. The allocation mode and irrigation mode are interacted between water resources development and irrigation scheme. For example, the different allocation modes can make different strategies of water supply and have different impacts on the irrigation water use efficiency. Therefore, there should be corresponding irrigation modes to the different allocation modes.

Based on the regional economic development and utilization of water resources in the typical irrigation area, irrigation mode focused on the impacts of field water-saving irrigation methods and irrigation system for the optimal allocation of water resources. For the irrigation methods, U type channel lining and small border irrigation, low pressure pipe of water conveyance and sprinkler irrigation, ground hose irrigation three irrigation methods were considered. As for the irrigation system, $11100\,m^3\,hm^{-2}$ and $14100\,m^3\,hm^{-2}$ were used as low irrigation quantity and high irrigation quantity respectively according to the main crops irrigation water. The multi-source allocation mode mainly considered the present and new sources of water supply. Total 12 different allocation and irrigation modes were established (Table 1).

In this study, we evaluated different allocation and irrigation modes by establishing an evaluation index system with three layers (Figure 1). 13 indicators derived from economy, technology, resources and society were chosen, which have more influences and less dependence to other indicators. The characteristic values of evaluation indexes are shown in Table 2.

2.2 ANN based on SFLA

An ANN usually contains three layers, namely input layer, hidden layer and output layer. Each layer has several nodes. There are weight W_{ij} and threshold θ_j between the input layer and hidden layer. Besides, the weight and threshold between the hidden layer and output layer are V_j, and γ. The values of the above four parameters compose the position of the individuals in SFLA. The output value can be calculated:

$$y_0 = f\left(\sum_{j=1}^{h} V_j x_j' - \gamma\right) \quad (1)$$

$$x_j' = f\left(\sum_{i=1}^{d} W_{ij} x_i - \theta_j\right) \quad (2)$$

where: f is using the function expression $f(x) = (1 + e^{-x})^{-1}$; d is the number of nodes in input layer; h is the number of nodes in hidden layer.

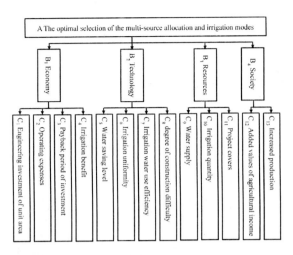

Figure 1. Framework structure of the evaluation index system.

Table 2. Values of the indicators for the 12 allocation and irrigation modes.

Indicator	Allocation and irrigation modes											
	M_1	M_2	M_3	M_4	M_5	M_6	M_7	M_8	M_9	M_{10}	M_{11}	M_{12}
C_1 (Yuan hm^{-2})	5295	4687	2895	5295	4687	2895	6000	5311	3281	6000	5311	3281
C_2 (Yuan·hm^{-2}·a)	300	956	504	300	956	504	340	1083	571	340	1083	571
C_3 (a)	1.55	1.88	1.12	1.20	1.37	0.84	1.55	1.88	1.12	1.20	1.37	0.84
C_4 (Yuan·hm^{-2})	3716	3446	3100	4720	4377	3938	4211	3905	3513	5349	4960	4462
C_5 (%)	34	40	37	34	40	37	34	40	37	34	40	37
C_6 (%)	75.0	85.2	83.2	75.0	85.2	83.2	75.0	85.2	83.2	75.0	85.2	83.2
C_7 (%)	65	85	80	65	85	80	65	85	80	65	85	80
C_8	0.8	0.4	0.6	0.8	0.4	0.6	0.8	0.4	0.6	0.8	0.4	0.6
C_9 (Billion m^3)	3.98	3.98	3.98	3.98	3.98	3.98	4.51	4.51	4.51	4.51	4.51	4.51
C_{10} (Million m^3·hm^{-2})	1.11	1.11	1.11	1.41	1.41	1.41	1.11	1.11	1.11	1.41	1.41	1.41
C_{11} (%)	10.0	4.5	4.0	10.0	4.5	4.0	11.3	5.1	4.5	11.3	5.1	4.5
C_{12} (Million yuan·hm^{-2})	0.51	0.48	0.43	0.65	0.61	0.54	0.58	0.54	0.49	0.74	0.69	0.62
C_{13} (kg·hm^{-2})	2570	2383	2140	3265	3027	2718	2912	2700	2425	3699	3430	3080

The algorithm steps of ANN based on SFLA are described below:

Step 1: N-dimensional population initialization is randomly given, and each individual element is recorded as weights and thresholds of each layer.
Step 2: Each individual adaptation is calculated in the current state.
Step 3: All the individual's fitness are sorted by descending, and then divided into the sub populations in turn.
Step 4: The worst individual of each current sub population should be updated. Then, the process is repeated until the iterative of sub populations reaches the maximum number of iterations.
Step 5: When all sub populations complete the update operation, steps 3–5 are repeated until it reaches a maximum number of iterations.

3 RESULTS AND CONCLUSIONS

Because the dimension of each indicator is different, we firstly standardized the original data.

Table 3. Assessment results of the allocation and irrigation modes.

modes	M_1	M_2	M_3	M_4	M_5	M_6	M_7	M_8	M_9	M_{10}	M_{11}	M_{12}
value	4.04	4.97	4.98	4.89	4.00	1.09	4.98	4.98	3.53	4.98	1.79	2.30

With the bigger the better indicators:

$$r_{ij} = \frac{0.0001 + 0.9999 \times (x_{ij} - \min_j(x_{ij}))}{\max_j(x_{ij}) - \min_j(x_{ij})} \quad (3)$$

With the smaller the better indicators:

$$r_{ij} = \frac{0.0001 + 0.9999 \times (\max_j(x_{ij}) - x_{ij})}{\max_j(x_{ij}) - \min_j(x_{ij})} \quad (4)$$

The numbers of nodes in input layer and hidden layer were determined by testing and comparing. In this study, 12 nodes were in input layer and 5 nodes were in hidden layer. During the optimization with SFLA, after the processes of trial and comparison, the number of memeplexes was chosen as 20. But the number of frogs in each memeplex was limited to 10. The lower limit and upper limit of the parameters value in each frog were −6 and 6. The maximum step size was 6. The position of the frog contained the weight and threshold between input layer and hidden layer W_{ij} and θ_j, the weight and threshold between hidden layer and output layer V_j and γ. In this study, at the optimizing stage of SFLA, the above four parameters were optimized. After 596 iterations optimizing, the converged error is less than 0.001.

Finally, the threshold θ_j is: $\theta_j = [2.616\ 5.821\ 5.020\ 1.047\ 4.154]$, the weight V_j is: $V_j = [-5.947\ -5.125\ 4.331\ -5.532\ -5.306]$, and the threshold γ is 3.690.

The assessment results of the allocation and irrigation modes are shown in Table 3.

The sorting for the advantages of the modes is:

$$M_6 > M_{11} > M_{12} > M_9 > M_5 > M_1 > M_4 > M_2 > M_7 > M_3 > M_{10} > M_8$$

The results show that the mode with ground hose irrigation and high irrigation quantity under the present water supply is the best allocation and irrigation mode. The modes of higher irrigation quantity are superior to that of the lower quantity with the non-sufficient irrigation condition. Under the present water supply condition, the modes with ground hose irrigation are superior to that with sprinkler and small border irrigation. But if there are new sources of water supply, the modes with sprinkler irrigation will be superior to the others. These results are consistent with the actual situation of regional agriculture, and would provide basis for irrigation when water supply scheme changes.

REFERENCES

Eusuff, M.M. & Lansey, K.E. 2003. Optimization of water distribution network design using the shuffled frog leaping algorithm. *Journal of Water Resources Planning Management* 129(3): 210–225.

Feng, F. & Xu, S.G. 2009. Improved fuzzy-optimized multi-level evaluation for comprehensive benefit of water resources in irrigation area. *Transactions of the Chinese society of agricultural engineering* 25(7): 56–61.

Li, H.L., Wang, X.G. & Cui, Y.L. 2006. Comprehensive evaluation methods for irrigation district. *Advances in Water Science*, 17(4): 543–548.

Sahoo, G.B., Ray, C. & Wade, H.F. 2005. Pesticide prediction in ground water in North Carolina domestic wells using artificial neural networks. *Ecological Modelling* 183(1):29–46.

Singh, K.P., Basant, A., Malik, A. & Jain, G. 2009. Artificial neural network modeling of the river water quality-A case study. *Ecological Modelling* 220(6):888–895.

Xu, J.X., Wang, P., Shen J. & Zhang M.A. 2002. Study on theories of optimal selection of irrigation module and its application. *Water Saving Irrigation* (6): 53–55.

Influences of cement temperature on the performance of concrete mixed with superplasticizer

Shu-yin Wang, Wen-kang Guo & Zhi-yang Gao
Changjiang River Scientific Research Institute, Wuhan, Hubei, China
Research Center of National Dam Safety Engineering Technology, Wuhan, Hubei, China

ABSTRACT: In order to investigate the influence of temperature of low heat Portland cement on the performance of concrete mixed with polycaboxylate superplasticizer, this paper selected two kinds of polycaboxylate superplasticizers, and explored the effect of cement temperature changes on the mixes properties and early age mechanical properties of concrete mixed with polycaboxylate superplasticizer. The results showed that 1) temperature of low heat Portland cement has greater impact on the workability of concrete, when the cement temperature was higher than 60°C, the concrete occurred slight bleeding phenomenon; 2) the setting times of concrete and cement temperature were following inverse relationship; 3) the slump loss rate and air content loss rate of concrete was proportional to cement temperature, the higher cement temperature, and the greater loss rate with time; 4) the early ages compressive strength of concrete were increasing with the changing of cement temperature, the same age's compressive strength was greatly impacted by the changing of cement temperature; 5) since the quality performance of polycaboxylate superplasticizers were different, the sensitivity of two kinds of superplasticizers to cement temperature changes were different too.

1 INTRODUCTION

As we all know, the temperature significantly influencing the performance of cement concrete, for example, the temperature rise could accelerate the hydration rate of cement particles, improve the early strength, in addition, the higher temperature could even lead to a false setting of cement concrete, etc. However, the sensitivity of superplasticizers to cement temperature changes and the influence law of temperature of low heat Portland cement on the performance of concrete mixed with polycaboxylate superplasticizer is still unclear. As polycarboxylate superplasticizer have significant advantages (such as lower dosage, higher water reducing rate, better slump performance, lower shrinkage, molecular structure can be adjusted and environmental friendliness etc.) could overcome many defects of traditional water reducing agent, generally agreed that it represents the current direction of the development of superplasticizer (Collepardi M, 2006; Hu Hong-mei, 2009; Sun Zhen-ping, 2009; Sakai E, 2006), so that, more and more used in engineering practice. There are many factors could affect the performance of polycarboxylate superplasticizer, in addition to the characteristics of superplasticizer and cement themselves, it's also influenced by the external environment, such as temperature, humidity, etc. Zhou Dong-liang, Ran Qian-ping etc. (Zhou Dong-liang, 2011) investigated the influence of temperature on performance (especially for the superplasticizer's adsorption and dispersion properties) of polycaboxylate superplasticizer, the results showed that, the initial dispersion capacity gradually increased with temperature, but the dispersion-retaining capacity was reduced with increasing of temperature, and it was showed a strong temperature dependence. Therefore, it is necessary to system study on the sensitivity problems of polycarboxylate superplasticizer to low heat cement temperature changes.

There is a large-scale hydropower diversion tunnel project in southwestern China, it require low heat cement concrete with polycaboxylate superplasticizer about one million cubic meters. However, as the increasing of concrete pouring scales and the impacted by summer high temperatures, the cement supply pressure has increased day by day, leading to the mixing floor cement temperatures as high as 85°C to 120°C and causing concrete slump, air content, 1 hour slump loss rate and air content loss rate in the outlet presented a larger fluctuation. So that, it's urgent to address the sensitivity of superplasticizers to cement temperature changes and the impact of cement temperature on concrete performance issues. Furthermore, the related research results about these problems at home and abroad is relatively small. Therefore, this paper combined with engineering actual selected two kinds of polycaboxylate superplasticizers, through fresh concrete mixes properties and harden concrete early age mechanical properties comparative tests explored the sensitivity of superplasticizers to cement temperature changes and the influence law of temperature of low heat Portland cement on the performance of concrete mixed with polycaboxylate superplasticize.

2 EXPERIMENTAL DETAILS

2.1 Materials

In this study, all constituent materials used in the program were tested to comply with the relevant Chinese Standards. A kind of P·LH42.5 type low heat Portland cement conforming to GB 200-2003 standard (GB200-2003, 2003) was used, the chemical and mineral composition of cement were given in Table 1. Clean and saturated surface dry artificial basalt sand was used in concrete mixture, with fineness modulus of 2.75. Artificial coarse aggregate employed in this study was basalt. GK-9A air entraining agent and two different types of polycarboxylic superplasticizer admixtures (SP-1 and SP-2, water reduction rate are 30.1% and 34.9, respectively) were used. A low-calcium types of F and Class I fly ash conforming to GB/T 1596-2005 standard (GB/T 1596-2005, 2005) were used as the cementitious materials.

2.2 Mixture proportions

In this study, selected a $C_{90}30$ pumping concrete mixtures which used for a diversion tunnel project in western China (incorporating two different types of superplasticizers) with the same water/cement ratio of 0.47 were prepared. The slump and air content flow of the mixtures was kept constant at 170 ± 10 mm and $5.0 \pm 0.5\%$, respectively. The corrected mix proportions of pumping concrete mixtures were listed in Table 2.

Table 1. Chemical and Mineral composition (by mass) of cement.

Chemical composition (%)					Mineral composition (%)				
CaO	SiO_2	Al_2O_3	Fe_2O_3	f-CaO	MgO	C_3S	C_2S	C_3A	C_4AF
60.80	22.80	4.00	5.27	0.32	3.62	33.88	40.0	1.66	16.02

Table 2. Proportioning of pumping concrete mixtures.

W/C	Fly ash/%	sand-ratio/%	Slumps/mm	Materials/(kg/m³)					Water reducing agent	Air entraining agent
				Water	Cement	Fly ash	Sand	Aggregate		
0.47	25	42	140–160	144	230	77	820	1160	0.246	0.0154

2.3 Test methods

According to the pumping concrete mixtures, selected 5 kinds of different cement temperatures (20°C, 40°C, 60°C, 80°C and 100°C) and two different types of polycarboxylic superplasticizer (SP-1 and SP-2), and conducted a series of fresh mixtures properties and harden concrete early age mechanical properties comparative tests, studied 1) the sensitivity of superplasticizers to cement temperature changes and 2) the influence law of temperature of low heat Portland cement on the performance of fresh mixtures (such as workability, slumps and air content loss with different times) and physical & mechanical properties (setting times, compressive strength of each ages) of concrete mixed with polycaboxylate superplasticize.

3 TEST RESULTS AND DISCUSSION

3.1 Influences of temperature of low heat Portland cement on the performance of concrete mixed with polycarboxylate superplasticizer

3.1.1 The workability of fresh mixtures

According to the fresh properties of the mixtures, the cohesiveness of fresh mixtures which mixed with two kinds of superplasticizer and different temperatures low heat cement was good, however, when the cement temperature rose to 60°C~100°C, the fresh concrete with two kinds of superplasticizer all occurred slight bleeding phenomenon, and the fresh mixtures with PS-2 superplasticizer even appeared slightly aggregate and slurry separation and concrete compaction phenomenon. Therefore, the cement temperature rises would impact the fresh properties of the mixtures. The main reason for this phenomenon as follows: to some extent, the temperature of fresh mixtures system was rising with the temperature rose of cement, under the effect of higher temperatures, the molecular movement of water and superplasticizer significantly accelerated, so that, the superplasticizer's adsorption-dispersion action rapidly enhanced, and the water reduction function was excited in a short time, then produced more and more free water, eventually resulted in a certain amount of bleeding.

3.1.2 The setting times of concrete

It can be seen from Fig. 1, with the temperature rising of cement, the setting times of concrete which mixed with PS-1 and PS-2 superplasticizers showed a decreasing trend, the higher the cement temperature was, and the shorter concrete setting time, in which the initial setting time's decreasing trend was more evident. Compared with 20°C cement, in the condition of cement temperature were 100°C, 80°C, 60°C and 40°C, the initial setting times of concrete (mixed with

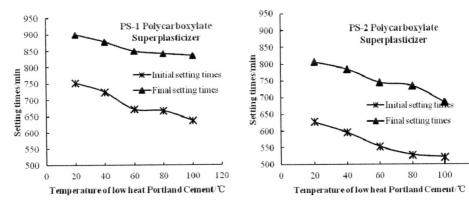

Figure 1. Influence of low heat Portland cement temperature on setting times of concrete.

Table 3. Influence of low heat Portland cement temperature on slump loss rate and air content loss rate of concrete.

Water reducing agent	Temperature of cement/°C	Slump loss rate/%					Air content loss rate/%		
		0 min	30 min	60 min	90 min	120 min	0 min	30 min	60 min
PS-1	100	0	8.6	42.9	77.1	94.3	0	47.3	72.2
	80	0	4.1	34.3	70.3	94.2	0	39.1	56.5
	60	0	4.4	27.8	60.6	86.1	0	10.0	32.0
	40	0	2.9	20.0	47.4	82.9	0	25.5	36.2
	20	0	2.9	17.1	40.0	77.1	0	11.5	23.1
PS-2	100	0	17.1	37.1	51.4	97.1	0	66.7	77.8
	80	0	9.1	30.3	45.5	93.9	0	45.7	56.5
	60	0	4.7	14.7	41.2	93.5	0	33.3	40.0
	40	0	0	8.6	40.0	88.6	0	25.5	40.4
	20	0	0	8.6	37.1	82.9	0	24.0	36.0

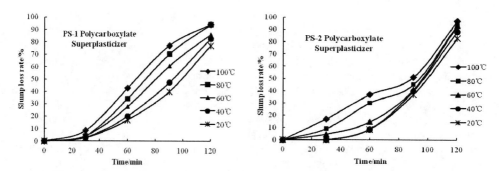

Figure 2. Influence of low heat Portland cement temperature on slump loss rate of concrete.

PS-1 superplasticizers) were shortened 115 min, 86 min, 81 min and 28 min, respectively; final setting times were shortened 64 min, 58 min, 50 min and 22 min, respectively. Compared PS-1 and PS-2 two kinds of superplasticizers, the former's retarding effect was better.

In the superplasticizer concrete, the temperature of fresh concrete system was rising with the temperature rose of cement, under the effect of higher temperatures, the rate of cement hydration was significantly accelerated, so that, the water reduction function was excited immediately, then produced more and more free water to participate in cement hydration reaction. Finally, accelerating concrete setting and hardening rate directly, leading to the setting time of concrete decrease with increasing of the cement temperature. Compared with two kinds of superplasticizer's setting time differences, the retarding effect of concrete was mainly influenced by the property of superplasticizer.

3.1.3 *The slump loss rate of concrete*

Better slump performance is one of the notable features of polycaboxylate superplasticize. In order to investigate the influence of temperature of low heat Portland cement on the performance of the slump loss of concrete mixed with polycaboxylate superplasticizer, the paper tested the slump loss rate values of the mixtures at 0 min, 30 min, 60 min, 90 min and 120 min five time nodes. The test results were shown in Table 3 and Fig. 2.

Test results showed that, at the same time node, the slump loss rate values of the mixtures were proportional to the temperature of cement, and this law was more obvious within 60 min. When slump loss time period were 90 min~120 min, 1) under the condition of different cement

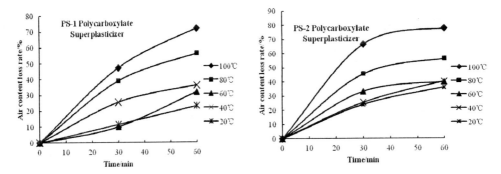

Figure 3. Influence of low heat Portland cement temperature on air content loss rate of concrete.

temperatures the slump loss rate law of the concrete mixed with different superplasticizer were almost the same, but slump loss rate of the concrete mixed with PS-1 was higher than that mixed with PS-2; 2) the slump loss rate law of the concrete has decreased, but at this time the slump retention values was less than 20%, it was unable to meet the performance requirements of pumping concrete construction.

At the same slump loss rate, the lower the cement temperature was, the longer the time to achieve the same slump loss rate. For example, when the cement temperature was 100°C, the concrete(mixed with PS-1) 40 min slump loss rate was about 20%, while the cement temperature was 20°C the slump loss rate reach 20%, should take about 65 min. Thus, the cement temperature change has seriously influenced the slump loss of concrete mixed with superplasticizer, the lower cement temperature was beneficial to reduce slump loss of concrete.

3.1.4 *The air content loss rate of concrete*

When concrete mixed with a small amount of air-entraining agent, could uniformly dispersed a large number of tiny bubbles into fresh mixtures, could effectively improving the durability of concrete, especially frost resistance of concrete (Yan Hua-quan, 2005). Thus, air content has a greater impact on the durability of concrete. In order to investigate the influence of temperature of low heat Portland cement on the performance of the air content loss of concrete mixed with polycaboxylate superplasticizer, the paper tested air content loss rate values of the mixtures at 0 min, 30 min and 60 min, three time nodes. The test results are shown in Table 3 and Fig. 3.

Test results showed that, cement temperature change has a greater impact on the air content loss rate of concrete with mixed with PS-1 and PS-2 polycaboxylate superplasticizer, the change relationship between cement temperature and air content loss rate of concrete was similar to the slump loss of concrete. That was, except when the cement temperature was 60°C, the air content loss rate of concrete(mixed with PS-1) was lower than the cement temperature was 40°C and 20°C, others basically conformed to the relationship, the air content loss rate of concrete was increased with the temperature rising of cement. When the cement temperature was higher than 80°C, the air content loss rate of concrete has dramatically increased, significantly higher than other temperatures.

3.1.5 *Early age compressive strength of concrete*

Because of the temperature has a greater impact on the early age compressive strength of concrete, this paper selected PS-1 and PS-2 polycaboxylate superplasticizer and tested 1d, 2d, 3d and 7d early ages compressive strength of concrete, aimed to investigate the influence of temperature of low heat Portland cement on the early age compressive strength of concrete mixed with polycaboxylate superplasticizer. The test results are shown in Table 4 and Fig. 3.

Figure 4 shows the early age compressive strength of concrete (mixed with polycaboxylate superplasticizer) was proportional to cement temperature, the strength increased with the rising of

Table 4. Influence of low heat Portland cement temperature on early age compressive strength of concrete.

Water reducing agent	Temperature of cement/°C	Compressive strength/MPa			
		1d	2d	3d	7d
PS-1	100	4.7	7.1	9.6	14.3
	80	4.4	6.2	8.7	13.8
	60	4	4.9	6.9	13.5
	40	3.5	4.2	6.5	12
	20	3.2	4	6.3	10.1
PS-2	100	6.3	8.0	10.4	15.1
	80	5.2	7.8	9.4	14.6
	60	5.4	7.2	9.8	14.0
	40	4.7	6.9	9.0	13.3
	20	4.0	6.3	8.9	12.4

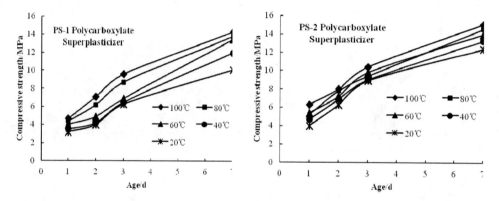

Figure 4. Influence of low heat Portland cement temperature on early age compressive strength of concrete.

cement temperature. When the cement temperature between 20°C and 100°C, 1d and 7d compressive strength of concrete (mixed with PS-1) were 3.2~4.7 MPa and 10.1~14.3 MPa, the average values were 4.0 MPa and 12.7 MPa, 1d and 7d compressive strength of 20°C cement were lower than 100°C cement 1.5 MPa and 4.2 MPa, respectively; 1d and 7d compressive strength of concrete (mixed with PS-2) were 4.0~6.3 MPa and 12.4~15.1 MPa, the average values were 5.1 MPa and 13.9 MPa, 1d and 7d compressive strength of 20°C cement were lower than 100°C cement 2.3 MPa and 2.7 MPa, respectively.

Leading to the early age compressive strength of concrete proportion to cement temperature maybe have the following two reasons, 1) compressive strength of concrete is mainly controlled by the rate of cement hydration reaction, higher cement temperature could promote early hydration of cement and accelerate the setting and hardening of cement & concrete, then increasing the early strength cement concrete; 2) the temperature of fresh mixtures system was rising with the temperature rising of cement, under the effect of higher temperatures, the water reduction function of superplasticizer was excited quickly, then producing much more free water to participate in cement hydration reaction and generating more hydration products, ultimately improving the strength of concrete.

3.2 *The sensitivity of superplasticizers to the changes of cement temperature*

The test result shows that under the condition of the same cement temperature, most part of variations of fresh mixtures and mechanical properties of concrete, which mixed with PS-1 and PS-2, were almost the same, but still have some differences.

In terms of concrete workability, when the cement temperature rose to 60°C~100°C, the fresh mixtures with two kinds of superplasticizer all occurred slight bleeding phenomenon, and the fresh mixture with PS-2 superplasticizer even appeared slightly aggregate and slurry separation and concrete compaction phenomenon. This was mainly attributed to the water reduction rate of PS-2 was relatively higher than PS-1, under the effect of higher temperatures, the water reduction function of PS-2 was excited immediately, then produced much more free water to participate in cement hydration reaction, however, the water retention property of PS-2 was poor, so that, the excess moisture precipitated from the aggregate surface leading to slightly bleeding and aggregate & slurry separation and concrete compaction phenomenon.

The regularity between two kinds of superplasticizer's setting times, slumps and air content loss rate of concrete and cement temperature was similar. At the same cement temperature, the retarding effect of PS-1 was better than PS-2 superplasticizer.

In terms of early age compressive strength of concrete mixed with two kinds of superplasticizer, each age's compressive strength of concrete mixed with PS-1 were lower than that mixed with PS-2 superplasticizer. This was mainly attributed to the own properties of PS-2, such as the following reasons: the water reduction rate of PS-2 was higher than PS-1 and the setting time was shorter than PS-1, could fully display dispersion and adsorption function of superplasticizers, so that, the cement hydration reaction rate was accelerated and the setting time of concrete was shortened, that was conducive to the growth of strength at early ages.

On the whole, PS-2 superplasticizer was more sensitive to cement temperature changes than PS-1. Therefore, the fresh mixture and mechanical properties of concrete mixed with PS-2 superplasticizer were greatly influenced by the cement temperature changes.

4 CONCLUSIONS AND SUGGESTIONS

In this experimental study, the effect of low heat Portland cement temperature change on the fresh mixtures properties and early age mechanical properties of concrete mixed with two kinds of polycaboxylate superplasticizer were investigated. The following conclusions and suggestions were drawn.

The temperature changes of low heat Portland cement has greater impact on the workability of concrete, when the cement temperature was higher than 60°C, the concrete occurred slight bleeding and even appeared slightly aggregate and slurry separation and concrete compaction phenomenon. Suggestion from our test goes that before adding into concrete mixing system the cement temperature should be controlled within 60°C.

The setting times of concrete and cement temperature were following inverse relationship; the slump loss rate and air content loss rate of concrete was proportional to cement temperature, the higher cement temperature was, the greater loss rate with time.

The early age compressive strength of concrete (mixed with polycaboxylate superplasticizer) was proportional to cement temperature, the strength increased with the rising of cement temperature. Because of the cement hydration reaction rate and dispersion and adsorption function of superplasticizers was significantly affected by cement temperature changes, the same age compressive strength changing with cement temperature changes.

PS-2 superplasticizer was more sensitive to cement temperature changes than PS-1, therefore, the fresh mixture and mechanical properties of concrete mixed with PS-2 superplasticizer were greatly influenced by the cement temperature changes.

REFERENCES

Collepardi M, Valente M. Recent developments in superplasticizers[C]//8th CANMET/ACI International Conference on Superplasticizers and other Chemical Admixtures in Concrete. Michigan: American Concrete Institute, SP-239, 2006: 1–14.

Hu Hong-mei, Yao Zhi-xiong, Huang Chun-quan. The Performance Comparison of Home and Abroad Polycarboxylate-type Water Reducer[J]. Journal of Xiamen University (Natural Science) 2009, 48(4): 538–542.

Sun Zhen-ping, Zhao Lei. Study of synthesis of polycarboxylate based superplasticizer[J]. Journal of Building Materials, 2009, 12(2): 127–131. (in Chinese)

Sakai E, Ishida A, Ohta A. New trends in the development of chemical admixtures in Japan[J]. Journal of Advanced Concrete Technology, 2006, 4(2): 1–13.

Zhou Dong-liang, Ran Qian-ping, Jiang Jiang. Influence of Temperature on Properties of Ester-type Acrylate Polycarboxylate Superplasticizer[J]. Journal of Building Materials, 2011, 14(6): 757–760.

Zhou Dong-liang, Yang Yong, Ran Qian-ping. Temperature on the Adsorption and the Dispersion Properties of Polycarboxylate Superplasticizer[J]. New Materials, 2011, 1:54–56. (in Chinese)

GB200-2003. Moderate heat Portland cement, Low heat Portland cement and Low heat Portland slag cement. Beijing: Chinese standard; 2003. (in Chinese)

GB/T1596-2005. Fly ash used for cement and concrete. Beijing: Chinese standard; 2005. (in Chinese)

Yan Hua-quan, Li Wen-wei. Hydraulic Concrete Research and Application. Beijing: China Water Power Press, 2005:98~99. (in Chinese)

Application of ceramic filter in seawater desalination pretreatment

Yuzheng Lv & Jiafu Yang
Building Design and Research Academy of Canbao, Beijing, China

Jie Shi & Hengguo Liang
Logistical Engineering University of PLA, Chongqing, China

ABSTRACT: Performance of ceramic filter pretreatment unit was studied with the experimental object of seawater. Test results showed that: Bearing performance of ceramic filter pretreatment unit was good and could be used as desalination osmosis of seawater with high pressure. The turbidity of effluent water would be less than 1NTU when the feed turbidity was 460NTU, and the turbidity can satisfy the reverse osmosis influent water quality requirements. The permeate flux recovery rate maintained over 97% after the cleaning of ceramic filter.

1 INTRODUCTION

Conventional pretreatment technology has been running smoothly in desalination by reverse osmosis for many years, which covers a large area, and runs more complexly in the design, operation and maintenance, and also makes reverse osmosis membrane prone to fouling. Ceramic filter as desalination pretreatment, without adding flocculating agent, fungicide, residual chlorine removal agent and other chemicals (Nanping Xu, 2002; Yuzhong Zhang, 2004; Ebrahimi M, 2010), can eliminate the need for security filters and effectively prolong the life cycle of reverse osmosis membrane and meet the requirements of field water, because of the substantial increase in the influent water quality.

2 MECHANISM ONCERAMIC FILTER PRETREATMENT UNIT

The basic mechanism of Ceramic filter in the liquid phase filtering is as follows: under the pressure driven, through the membrane pores' sieving effect, the molecules bigger than the membrane pore size are retained, while small molecules are flowing through the membrane layer, thereby producing the desired product (Xiangfeng Li, 2006; Zengji Cai, 2009). In the actual filtering process, the filtering mechanism is not just the simple sieving effect, particles smaller than the membrane pore size will also be retained and removed, the ceramic filter even can remove partial dissolution impurities, for the reason that they can be attached to the other macromolecular impurity particles (Jegatheesan V, 2009; Teplyakova V, 2006).

In desalination pretreatment process, the ceramic filter with high pressure-bearing capability can meet the water pressure requirements of reverse osmosis membrane on the premise of structural integrity; the recovery rate of contaminated ceramic filter after cleaning is high. In other words, its repeated utilization rate is high, which can decrease the cost of reverse osmosis technology; the main function of the reverse osmosis membrane is desalination with low affordability of the contaminants, so the life cycle is highly affected by the inlet water quality, the reverse osmosis membrane pollutants in seawater basically have suspended particles, such as clay ($>1\,\mu m$), colloidal particles (about $0.2\,\mu m \sim 1.0\,\mu m$), microorganisms, dissolved organic matter and total dissolved solids, etc. The porous ceramic filter has a relatively neat, uniform porous structure with pore size of $0.05\,\mu m \sim 10\,\mu m$, which can remove these contaminants and well protect the reverse osmosis membrane elements, using its better screening and adsorption.

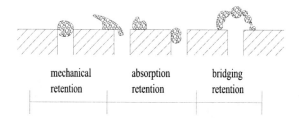

Figure 1. Different retention role of the ceramic filter.

3 EXPERIMENTAL DESIGN

3.1 Test process

Technological process used in the test was as follows:

Seawater → pump → ceramic filter → the effluent water of pretreatment

In test reports, different turbidity and salinity of sea water samples were prepared with the "tap water + sea salt + clay", and pre-treated with the ceramic filter unit, then the operating pressure and the yielding water flux was tested, wherein the average pore diameter of the ceramic filter was 1 μm, and the new filter element to a set pressure value was taken as the first filter cycle, and was tested again after cleaning, in proper order to enter the second and third filtration cycle.

3.2 The primary evaluation project

Turbidity was selected as the main water quality technical indicator, which was mainly caused by the suspended solids and colloids. The basic purpose of desalination pretreatment was to reduce the turbidity and examine the filtration efficiency of the filter element, including the flux of water cycle process and the permeate flux recovery degree after cleaning.

4 RESULTS AND DISCUSSIONS

4.1 Pressure performance test of pretreatment unit

The normal working pressure of the reverse osmosis unit was 4 MPa~6 MPa, therefore, it was necessary to carry out pressure performance test of ceramic filter pretreatment unit, which was tested in the continuous operation under high pressure conditions. The test results were showed in table 1.

From the Table 1, under the premise of the flow rate no less than 30 mL/second, ceramic filter pressure performance was good, which could satisfy the feed water pressure requirements of the reverse osmosis unit, and could be used as the pretreatment process for reverse osmosis unit.

4.2 Water flow test of pretreatment unit

4.2.1 Normal pressure performance test

Tap water, sea salt and clay were mixed into 20NTU and 40NTU of sea water, with pH value 8 and salinity 35000 mg/L. Seawater was filtrated by ceramic filter under ordinary pressure, and the test results were shown in figure 2.

Figure 2 showed that, when the influent turbidity was 20NTU, the ceramic filter water production cycle was growing with the increase of temperature, when the influent turbidity was 40NTU, the ceramic filter water production cycle basically was the same under different water temperatures. When the temperature was kept constant, high water turbidity produced slightly less water production cycle, which was due to high water turbidity caused the impurities deposited on the filter surface quickly, and increased the impact on the production of the water cycle.

Table 1. Pressure maintaining performance test of ceramic filter under high pressure.

Pressure (MPa)	Test project			Pressure-bearing capability	Average pressure difference (MPa)
	flux (mL/times)				
	①	②	③		
4	30.5	32.8	32.4	good	0.4
5	31.6	30.2	31.8	good	0.4
6	32.1	30.4	30.9	good	0.5

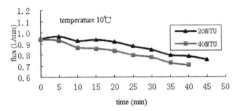

(1) the water temperature is 10℃

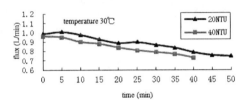

(2) the water temperature is 30℃

Figure 2. Effect of turbidity on water flow under normal pressure.

4.2.2 High pressure performance test

To test the ceramic filter performance characteristic as pretreatment unit of desalination process, the high pressure performance test of the ceramic filter has been carried out, which was shown in figure 3.

Figure 3 showed that, when the operating pressure was 4 MPa, the yielding water flux of the ceramic filter decreased with time, in the beginning of filtration, the water flow decreases faster with time, and then the rate of decrease tends to slow down stably. Comparison with Figure 1 showed that, when the water turbidity was 40NTU, the water flow increased with the increase of pressure and diminishing of water production cycle, thereby affecting water flow per cycle, so in actual operation, ceramic filter should be cleaned promptly.

4.3 Pretreatment unit water flow and pressure test under per cycle

For the ceramic filter pretreatment unit in a filtration cycle, operating pressure, and the processing flow over time were tested. According to the theory of fluid mechanics, the fluid resistance P, resistance coefficient S and flux Q has the following relations[5]:

$$P = S \times Q^2 \qquad (1)$$

where: the fluid resistance P is the outlet pressure of pretreatment unit.

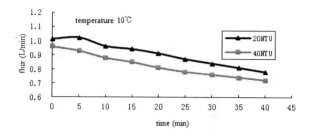

(1) the water temperature is 10℃

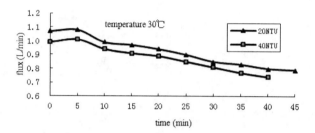

(2) the water temperature is 30℃

Figure 3. Effect of turbidity on water flow under 4 MPa pressure.

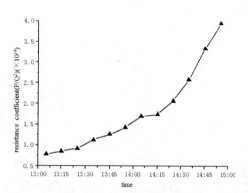 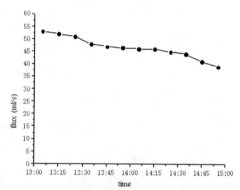

Figure 4. Regulation of ceramic filter preprocessing unit resistance coefficient variation.

Figure 5. Regulation of ceramic filter preprocessing unit flow variation.

Based on the experimental data and formula (1), it will be able to get resistance coefficient of ceramic filter pretreatment unit, and pretreatment unit resistance coefficient values were shown in figure 4, figure 5 shown the flow change rule of ceramic filter pretreatment unit.

As can be seen from figure 4 and 5, the resistance coefficient of the ceramic filter unit gradually increased with the filtration time increase, while the yielding water flux gradually decreased with the increase of filtration time, which was mainly due to the resistance coefficient increases; the pollution layer of the membrane surface become density by penetrating drag force, lower porosity and higher penetration resistance lead to the water flow falling out. But pretreatment unit running two hours later, the water flow was still higher than 35 ml/s, thus, ceramic filter had a good filtering performance.

Table 2. Properties of ceramic filter preprocessing unit treated with high turbidity water.

1st life cycle (new ceramic filter)			2nd life cycle (1st cleaning of water-sediment)			3rd life cycle (2nd cleaning of water-sediment)		
time	flux (mL/min)	yielding water turbidity (NTU)	time	flux (mL/min)	yielding water turbidity (NTU)	time	flux (mL/min)	yielding water turbidity (NTU)
16:33	1600	0.4	17:04	1590	0.2	17:36	1520	0.6
16:37	864	0.2	17:09	780	0.1	17:40	850	0.45
16:41	670	0.2	17:13	630	0.05	17:44	670	0.2
16:45	550	0.2	17:17	540	0.2	17:48	560	0.2
16:49	490	0.1	17:21	470	0.1	17:52	490	0.2
16:53	450	0.0	17:25	440	0.1	17:56	450	0.18
16:57	380	0.0	17:29	380	0.1	18:00	410	0.1

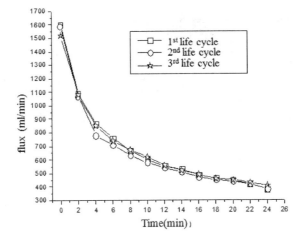

Figure 6. Pretreatment unit processing flow varying with time treated with high turbidity seawater.

4.4 Water flow test of pretreatment unit under different cycle

With water turbidity being 460NTU, the water flow stability of the ceramic filter pretreatment unit and the change of filtration cycle were tested and the results were shown in table 2 and figure 6, when ceramic filter pretreatment unit deals with high turbidity water, the yielding water turbidity was less than 1NTU, meeting the turbidity requirements of *drinking water health standards* (GB5749-2006). The treated water flow of pretreatment unit had not changed significantly with the filtration cycle, which was due to the uniform pore distribution of the ceramic filter, and the retaining mostly occur in membrane surface, then the original capacity can be restored by cleaning. The flow and water quality did not change, and the preprocessing unit outer water own stable performance.

4.5 Ceramic filter cleaning test

The cleaning of ceramic filter occurs at the end of each filtering process. The cleaning test operating pressure was 5 MPa, with water turbidity 30NTU, the temperature 20°C, and two filters are numbered $1^{\#}$ and $2^{\#}$, respectively.

After $1^{\#}$ filter completing a filtration cycle, detaching the filter housings, and carefully removing the filter, cleaning the filter cake from the surface layer with sponge gently under the running water,

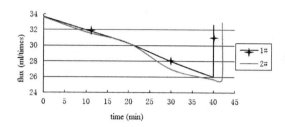

Figure 7. Permeate flux recovery of different filter.

when filter surface has been cleaned, wiping filter surface gently with sand paper until yellow ceramic filter grows slightly white, then stop to wiping the filter, clean the fines of surface with a brush under running water, put the filter cartridge into the filter housing and tighten, then test the water flow under the same conditions. Using the same method with $2^{\#}$ filter, the results were shown in Figure 7.

The figure 7 showed that the permeate flux recovery rates of $1^{\#}$ and $2^{\#}$ filter was higher. When the ceramic filter flow was down to 75% (26/34 ≈ 75%) of the initial water, cleaning experiment was carried out, the filter element permeate flux recovery rate remains above 97%, this is because the filter pollution mainly occur in the shallow part of the surface and near surface of the filter, filter internal basically hasn't been blocked and polluted.

5 CONCLUSIONS

(1) The pressure performance of ceramic filter pretreatment unit is good, and the turbidity of outer water is less than 1NTU, satisfying the high pressure and water quality requirements of reverse osmosis desalination.
(2) The raw water turbidity has greater impact on water production cycle of pretreatment units, and the cycle of water production with raw water turbidity increased substantially slow down after the fast trend, in actual applications a lower turbidity seawater as the desalination pretreatment unit source water should be chosen.
(3) When the pretreatment unit permeation flux decreased to 75% of the initial permeation flux, it's time for the filter cleaning process. After cleaning the contaminant with sandpaper on the filter surface and subsurface, the permeate flux recovery rate remained at 97% above.

REFERENCES

Ebrahimi M, Willershausen D, Ashaghi S. 2010. Investigations on the use of different ceramic membranes for efficient oil field produced water treatment. *Desalination* 250 (3):991–996.
Jegatheesan V, Phong D, Shu L. 2009. Performance of ceramic micro-and ultrafiltration membranes treating limed and partially clarified sugar cane juice. *J. Membrane Sci* 327: 69–77.
Nanping Xu, Weihong, Xing, Yijiang Zhao. 2002. *The Separation Technology and Application of Ceramic Membranes*. Beijing: Chemical Industry Press.
Teplyakova V, Tsodikova M, Magsumova M. 2006. Intensification of gas phase catalytic processes in nano-channels of ceramic catalytic membranes. *Desalination* 199:161–163.
Xiangfeng Li, Changhong Dai, Haisheng Sun. 2006. Applications of ceramic membrane separation Technology in water treatments. *Filtration and separation* 16 (1): 8–10.
Yuzhong Zhang, Lingying Zheng, Congjie Gao. 2004. *The Separation Technology and Application of Liquid*. Beijing: Chemical Industry Press.
Zengji Cai, Tianyu Long. 2009. *Fluid Mechanics of pump and fan*. Beijing: China Building Industry Press.

Simulation of sedimentation in the channel of Lianyungang Harbor by using 3D sediment transport model of FVCOM

Bing Yan & Hua Yang
Tianjin Research Institute for Water Transport Engineering, M.O.T., China

Qinghe Zhang & YuRu Wu
Tianjin University, China

ABSTRACT: Lianyungang Harbor is located on the western coast of Yellow Sea, China. Since the 1960s, a series of research including field observations and laboratory experiments, have been carried out to investigate tidal current motions, wave propagations and sediment properties around the harbor. It is important for manager, engineer and designer to understand the channel siltation along the channel during storm events. For prediction of channel siltation, a 3D numerical model of cohesive fine-grained sediment transport based on the coupling of modified FVCOM model, WRF model and SWAN model is developed to investigate the sediment transport. It is shown from simulated results that the sedimentation in the channel was well simulated in typhoon events, and the high sediment concentration near sea bed for muddy coast during storm process can also be basically reflected by the model.

1 INTRODUCTION

Lianyungang Harbor is located on a muddy coast. The sediments on a muddy coast are usually fine and cohesive, and their transport is driven by the combined actions of tidal currents and waves. Sedimentation physics is extremely complicated including e.g. flocculation, settling, consolidation, erosion and deposition. Extensive researches have been carried out to explore the mechanism of sediment transport and channel siltation of Lianyungang Harbor. Rodriguez and Ashish (2000) used a longshore transport formula of fine-grained sediment due to waves to estimate the suspended sediment concentration and discharge of Lian Island near Lianyun Port, which is a part of Lianyungang Harbor. Xie et al. (2010) set up a 3D numerical model to discuss the characteristics of suspended sediment concentration (SSC) distribution around Lianyungang Harbor, based on the annual mean calibration and verification approaches. Zhang et al. (2011) predicted the channel siltation of Xuwei Port, which is another part of Lianyungang Harbor, during Typhoon Wipha in 2007 with MOHID model and SWAN model.

As we know, there are several important factors for simulation of sediment transport and sedimentation on the muddy coast. First of all, the sediment is mainly suspended and transported by wind wave and current for typical mud coasts interested here, so the combined wave-current action is very important. Furthermore, the wave breaking also plays important roles in the vertical concentration of suspended sediment in shallow water. Then lutoclines and fluid mud are two important characteristics of high concentration environment on the muddy coast. The development of lutoclines and fluid mud layers can have a significant effect on the flowing suspension fluid, which in turn affects the lutocline or fluid mud layer. Yang et al. (2013) developed a one-dimensional model to simulate the fluid mud flowing along the navigation channel. In this paper, the one dimensional model is incorporated with FVCOM model to calculate the fluid mud transport in channel. In addition, the flocculation, hindered settling and consolidation are also important for prediction of navigation channel siltation and considered in this paper.

A 3D numerical model with fluid mud based on the coupling of modified FVCOM model, WRF model and SWAN model is developed and applied in the present paper to investigate sediment transport and the channel siltation of LianYun Harbor during Typhoon Wipha in 2007.

2 NUMERICAL MODELS

2.1 Modified FVCOM model

FVCOM is a prognostic, unstructured-grid, Finite-Volume, free-surface, three-dimensional (3-D) primitive equations Community Ocean Model developed originally by Chen et al. (2003). It is fully coupled ice-ocean-wave-sediment-ecosystem model system with options of various turbulence mixing parameterization, generalized terrain-following coordinates, data assimilation schemes, and wet/dry treatments with inclusion of dike and groyne structures under hydrostatic or non-hydrostatic approximation. However it doesn't take the wave breaking effect on suspended sediment, flocculation, hindered settling and consolidation into consideration.

The governing equations for the FVCOM hydrodynamic model equations are given below.

The momentum equation is:

$$\frac{\partial u}{\partial t}+u\frac{\partial u}{\partial x}+v\frac{\partial u}{\partial y}+w\frac{\partial u}{\partial z}-f_v=-\frac{1}{\rho_0}\frac{\partial(p_a+p_H)}{\partial x}+\frac{\partial}{\partial z}\left(K_m\frac{\partial u}{\partial z}\right)+F_u \quad (1)$$

$$\frac{\partial v}{\partial t}+u\frac{\partial v}{\partial x}+v\frac{\partial v}{\partial y}+w\frac{\partial v}{\partial z}+f_u=-\frac{1}{\rho_0}\frac{\partial(p_a+p_H)}{\partial y}+\frac{\partial}{\partial z}\left(K_m\frac{\partial v}{\partial z}\right)+F_v \quad (2)$$

$$\frac{\partial w}{\partial t}+u\frac{\partial w}{\partial x}+v\frac{\partial w}{\partial y}+w\frac{\partial w}{\partial z}=\frac{\partial}{\partial z}\left(K_m\frac{\partial w}{\partial z}\right)+F_w \quad (3)$$

$$\frac{\partial P}{\partial z}=-\rho g \quad (4)$$

The continuity equation is:

$$\frac{\partial u}{\partial x}+\frac{\partial v}{\partial y}+\frac{\partial w}{\partial z}=0 \quad (5)$$

where x, y, and z are the east, north, and vertical axes in the Cartesian coordinate system; u, v, and w are the x, y, z velocity components; ρ is the density; p_a is the air pressure at sea surface; p_H is the hydrostatic pressure; f is the Coriolis parameter; g is the gravitational acceleration; K_m is the vertical eddy viscosity coefficient; F_u and F_v represent the horizontal momentum diffusion terms; F_w is the vertical momentum diffusion term.

We choice the Smagorinsky eddy parameterization method to calculate the horizontal diffusion coefficient and Mellor and Yamada level 2.5 (MY-2.5) turbulent closure model for the parameterization of the vertical eddy viscosity (K_m).

The cohesive sediment transport is controlled by a 3D equation of convective diffusion in which sediment setting is considered in vertical movement, which is

$$\frac{\partial(C)}{\partial t}+\frac{\partial(uC)}{\partial x}+\frac{\partial(vC)}{\partial y}+\frac{\partial((w-w_s)C)}{\partial z}=\frac{\partial}{\partial x}\left(\varepsilon_h\frac{\partial C}{\partial x}\right)+\frac{\partial}{\partial y}\left(\varepsilon_h\frac{\partial C}{\partial y}\right)+\frac{\partial}{\partial z}\left(\varepsilon_v\frac{\partial C}{\partial z}\right) \quad (6)$$

where C is the concentration of sediment, ε_h is the horizontal diffusivity and ε_v is the vertical diffusivity, w_s is the settling velocity.

In nearshore zones, the wave breaking plays an important role on vertical concentration distribution and sediment transport, so we modified the FVCOM model through incorporating the wave breaking effect by using the sediment diffusivity due to wave breaking presented by van Rijn (2007). The vertical diffusion coefficient of sediment suspension considering the wave breaking effect can be seen in Eq. (7)–(12). The diffusivity for current and waves is represented by a non-linear combination of the current-related and the wave-related diffusion coefficients (Eq. (7)). Here we assume that the current-related diffusivity ε_c equal to the vertical eddy viscosity coefficient (K_m) rather than is calculated by van Rijn's equation. In addition, stratification effects due to the presence of sediment particles is represented by Eq. (12).

$$\varepsilon_v = \phi\left[\varepsilon_c^2 + \varepsilon_w^2\right]^{0.5} \quad (7)$$

$$\varepsilon_w = \begin{cases} \varepsilon_{w,bed} & z \leq \delta \\ \varepsilon_{w,bed} + (\varepsilon_{w,max} - \varepsilon_{w,bed})\left(\dfrac{z-\delta}{0.5h-\delta}\right) & \delta < z < 0.5h \\ \varepsilon_{w,max} & z \geq 0.5h \end{cases} \quad (8)$$

$$\varepsilon_{w,bed} = 0.018 \beta \delta u_w \quad (9)$$

$$\varepsilon_{w,max} = 0.035 \gamma_{br} h H_s / T_p \quad (10)$$

$$\gamma_{br} = \begin{cases} 1 + \left(\dfrac{H}{h} - 0.4\right)^{0.5} & \dfrac{H}{h} > 0.4 \\ 1 & \dfrac{H}{h} \leq 0.4 \end{cases} \quad (11)$$

$$\phi = 1 + \left(\dfrac{C_s}{C_s^0}\right)^{0.8} - 2\left(\dfrac{C_s}{C_s^0}\right)^{0.4} \quad (12)$$

The settling velocity equation of Winterwerp (2002) is adopted to substitute the original equations of FVCOM for the effects of flocculation and hindered settling. A one-dimensional model of fluid mud transport (Yang, 2013) and a consolidation model (Sanford, 2008) are also incorporated with FVCOM model to the effect of fluid mud transport on the sedimentation in the channel.

2.2 WRF model

The Weather Research and Forecasting (WRF) Model is a mesoscale numerical weather prediction system designed to serve both atmospheric research and operational forecasting needs. It features two dynamical cores, a data assimilation system, and a software architecture allowing for parallel computation and system extensibility. There are two dynamical core versions of WRF, i.e., Advanced Research WRF (ARW) and Nonhydrostatic Mesoscale Model (NMM). ARW is used to calculate the wind field of Typhoon Wipha in this paper. The detailed WRF model can be seen in its manual.

2.3 Swan model

SWAN model is a wind-wave model based on the action density balance equation. The wave process during storm was simulated by SWAN model successively. The detailed SWAN model can be seen in Booij (1999) and would not be given here. Wave in sea area around Lianyungang Harbor is mainly wind wave. So SWAN model is adopted to simulated wave field under the action of typhoon Wipha with wind field calculated by WRF.

2.4 Coupling between FVCOM, WRF and SWAN models

The FVCOM, WRF and SWAN models are coupled through data exchange in regional and coastal domains respectively, in which WRF calculates the wind field for SWAN model, FVCOM provides SWAN with time series of water elevation and current velocity, and SWAN supplies FVCOM with wave field and arrays of radiation stress (see Fig. 1).

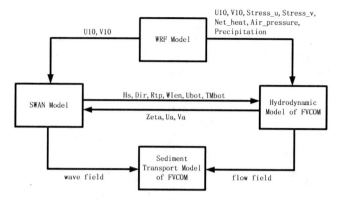

Figure 1. Computational procedures of wind, wave, current, sediment transport and channel siltation.

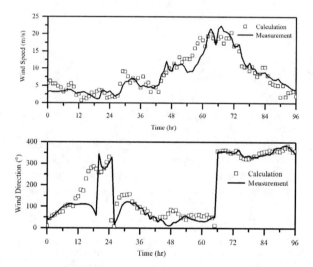

Figure 2. Comparison of computed and measured wind speed and direction.

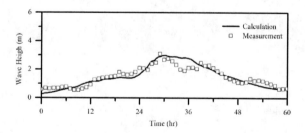

Figure 3. Comparison of computed and measured wave height.

2.5 Verification

The models have been calibrated with field observations and the results would be given here. Wind data at Daxishan ocean station, tide, current and sediment concentration data from a temporary observation station at −5 m are used to verified and validate the present model. By comparison, the numerical calculation results are in good agreement with the measured data (see Fig. 2–5).

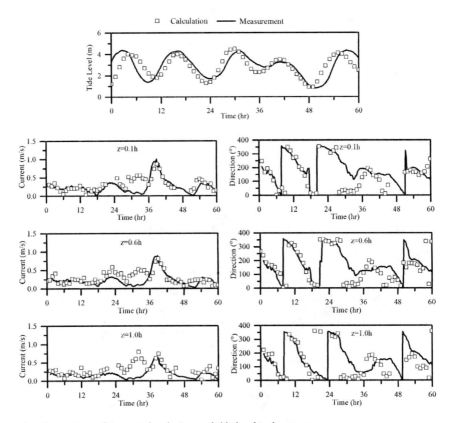

Figure 4. Comparison of computed and measured tide level and current.

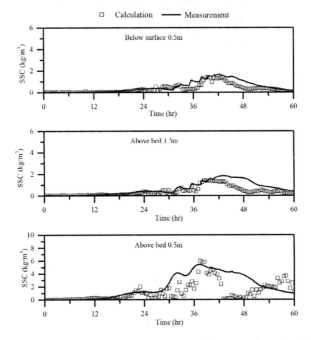

Figure 5. Comparison of computed and measured suspended sediment concentration (SSC).

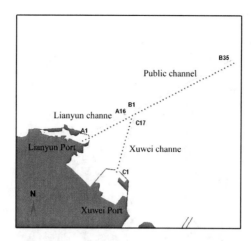

Figure 6. The layout of the Lianyungang Harbor.

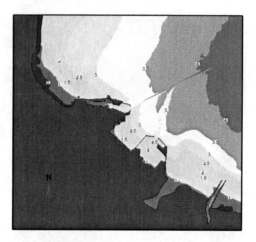

Figure 7. The wave field at 0:0 September 20, 2007.

3 MODEL APPLICATION

The Lianyungang harbor is composed of two ports (Lianyun Port and Xuwei Port). The navigation channel is planned to have three parts, i.e. public channel (B1–B35), Lianyun channel (A1–A16) and Xuwei channel (C1–C17) (see, Fig. 6), which is designed to have an effective width of 350 m and bottom elevation of $-22.5 \sim -23$ m for 300 000 dwt ships. The model introduced above is used to estimate channel siltation of Lianyungang Harbor during Typhoon Wipha in 2007.

Figure 7 shows the wave field of Lianyungang Harbor at 00:00 September, 20, 2007, when the wave reaches the maximum. The wave height decreases gradually towards land and reaches around 3 m at entrances of Lianyun port and Xuwei port. The maximum of suspended sediment concentration is later than maximum wave, which is shown in Fig. 8. It could be noticed that (1) the SSC values are higher near the shoreline, and gradually reduce to the offshore; (2) the SSC values around Guan River estuary are relatively higher, where the values are around 6–7 kg/m^3 near bed and 1.2–2 kg/m^3 in the surface layer; (3) the SSC values become lower from Guan River estuary towards North; and (4) The SSC values in the bed layer is higher than in the surface layer, which there is 3–4 times of differentiation between.

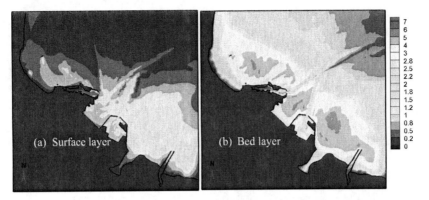

Figure 8. The Distribution of suspend load of Lianyungang sea area at 7:00 September 20, 2007.

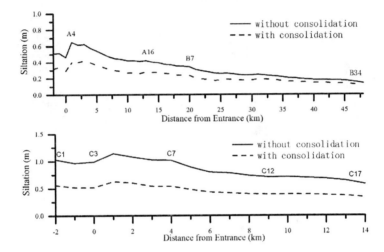

Figure 9. The Distribution of deposition thickness along the channels.

Figure 9 shows the distribution of deposition thickness along the channels. There are two values with and without consolidation process. We can see that the siltation without consolidation is larger than with consolidation. This implies that there is a process of fluid mud deposition. The value with consolidation is a real and effective sedimentation of channel. The largest deposition thickness of Xuwei channel is 0.5 m in the area 4.0 km away the entrance. The largest deposition thickness of Lianyun channel is 0.4 m in the area 2.0 km away the entrance.

4 CONCLUSION

(1) The 3D FVCOM model was modified by incorporating the wave breaking effects in nearshore zones, fluid mud, flocculation, hindered settling and consolidation. It is able to predict the sediment transport nearshore reasonable.
(2) Wind waves were simulated by SWAN model and WRF model.
(3) The simulated wind speed, wave height, water level, current, suspended sediment concentration basically agreed with the measured data.
(4) The model is used to simulate the sedimentation in the channel under the effect of "Wipha" typhoon. The maximum of suspended sediment concentration is later than maximum wave. The maximum depositions occur outside the entrance.

ACKNOWLEDGMENTS

The research was supported by the National Natural Science Foundation of China under grant no. 51209111.

REFERENCES

Booij, N.C., Ris R.C. & Holthuijsen L.H. 1999. A third-generation wave model for coastal regions:1. Model description and validation. *Journal of Geophysical Research*, 104: 7649–7666.

Chen, C., Liu, R. C. & Beardsley, R. C. 2003. An unstructured, finite-volume, three-dimensional, primitive equation ocean model: application to coastal ocean and estuaries. *Journal of Atmospheric and Oceanic Technology*, 20: 159–186.

Rodriguez, H.N. & Ashish J.M. 2000. Longshore tansport of fine-grained sediment. *Continental Shelf Research*: 1419–1432.

Sanford, L.P. 2008. Modeling a dynamically varying mixed sediment bed with erosion, deposition, bioturbation, consolidation, and armoring. Computers & Geosciences, 34: 1263–1283.

Skamarock, W.C., Klemp, J.B., et al. 2008. A description of the advanced research WRF Version3. *National Center for Atmospheric Research*, Boulder, Colorado, USA.

van Rijn L.C. 2007. Unified view of sediment transport by currents and waves: suspended transport. *Journal of Hydraulic Engineering*, 133(6): 668–689.

Winterwerp, J.C. 2002. On the flocculation and settling velocity of estuarine mud. *Continental Shelf Research*, 22: 1339–1360.

Xie, M.X., Zhang, W. & Guo, W.J. 2010. A validation concept for cohesive sediment transport model and application on Lianyungang Harbor, China. *Coastal Engineering*, 57: 585–596.

Yang, X.C., Zhang, Q.H., Zhao, H.B. & Zhang, N. 2013. Numerical simulation of fluid mud flow along Navigation channel. *Journal of Sediment Research*, (4): 14–21.

Zhang, N., Yang, H., Yan, B. & Zhao, H.B. 2011. Three-dimensional numerical simulation of cohesive sediment transport due to a typhoon. Proceedings of the Sixth International Conference on Asian and Pacific Coasts.

Assessment based on Monte Carlo for sample rotation under stratified cluster sampling

Y. Fu, G. Gao & S.X. Liu
School of Public Health, Soochow University, Suzhou, Jiangsu, China

ABSTRACT: The purpose of this research is to evaluate the reliability and validity of successive survey using sampling rotation under stratified cluster sampling and its formulae based on Monte Carlo simulation. Physical indicators (e g. vital capacity) of some students in a city were taken as simulative population parameters. Monte Carlo simulation was conducted with SAS programming when sampling ratios were 10% and 40% respectively. 100 samples in each sampling ratio were simulated and 95% CIs (confidence intervals) could be acquired. Almost all the CIs of the 100 estimated population means at each sampling ratio include the corresponding simulative population mean. Successive survey using sample rotation under stratified cluster sampling and its formulae are of high reliability and validity which can be applied to large-scale investigation.

1 INTRODUCTION

In order to reflect the overall levels and changes at different times, many successive surveys were applied in our statistical system, such as sampling survey of earnings (NBS survey office in Shenzhen, 2013), sampling survey of nutritional status (Zhu & D, 2013) and sentinel surveillance of chronic diseases (Tang & Sun, 2013). Reasonable methods of sample rotation and correct formulae are needed, because of frequent sample loss in epidemiological cohort study (Sen, 1972). In this paper, successive survey using sample rotation under stratified cluster sampling was applied to investigate physical indicators (e.g. vital capacity) of primary and middle school students in a city. Monte Carlo simulation was carried out to simulate 100 samples at two sampling ratios (10% and 40%) respectively while survey results were taken as analog overall parameters.

2 METHODS

2.1 *Successive survey using sample rotation under stratified cluster sampling*

Suppose population is divided into L layers. The numbers of groups from each layer are $N_1, N_2, \ldots, N_L$ and $\sum_{k=1}^{L} N_k = N$. The ith group from layer k contains M_{ki} secondary units. Sample rotation under cluster sampling is conducted independently, and $\sum_{k=1}^{L} n_k = n \cdot m_{hk}$ stands for number of retained groups in layer k at h time. u_{hk} stands for number of rotational groups in layer k at hth sample rotation. For any h, $m_{hk} + u_{hk} = n_k$ and $k = 1, 2 \ldots, L$. $\overline{\overline{Y}}_{hk}$ stands for overall mean of secondary units from layer k at h time. $\overline{\overline{y}}_{hku}$, $\overline{\overline{Y}}_{hku}$ respectively represent sample mean and overall mean of the sample rotation parts which are secondary units from layer k at h time. $\overline{\overline{y}}_{hkm}$, $\overline{\overline{Y}}_{hkm}$ respectively represent sample mean and overall mean of retained sample which are secondary units from layer k at h time. $\overline{\overline{y}}_{h-1,km}$ stands for sample mean which are secondary units in layer k at h time and retained from $h-1$ time. b_{hk} represents regression coefficient for the same objects on the same indicators. $\overline{\overline{y}}_{hk}$, s_{hk}^2 respectively stand for sample mean of secondary units and sample variance in layer k at h time. ρ_{hk} denotes correlation coefficient for the same groups on the same indicators that are present in layer k both at h time and $h-1$ time.

2.2 Statistical formulae

The estimator of population mean at h time is:

$$\hat{\bar{Y}}_h = \sum_{k=1}^{L} W_k \hat{\bar{Y}}_{hk} \tag{1}$$

$$W_k = \frac{N_k}{N} \tag{2}$$

$$\hat{\bar{Y}}_{hk} = \Phi_{hk} \hat{\bar{Y}}_{hku} + (1-\Phi_{hk}) \hat{\bar{Y}}_{hkm}$$
$$= \Phi_{hk} \bar{y}_{hku} + (1-\Phi_{hk})[\bar{y}_{hkm} + b_{hk}(\hat{\bar{Y}}_{h-1,k} - \bar{y}_{h-1,km})] \tag{3}$$

The overall mean variance estimator of secondary units at h time is:

$$V(\hat{\bar{Y}}_h) = \sum_{k=1}^{L} W_k^2 V(\hat{\bar{Y}}_{hk}) \tag{4}$$

$$V(\hat{\bar{Y}}_{hk}) = \Phi_{hk}^2 V(\hat{\bar{Y}}_{hku}) + (1-\Phi_{hk})^2 V(\hat{\bar{Y}}_{hkm})$$
$$= \Phi_{hk}^2 \frac{S_{hk}^2}{u_{hk}} + (1-\Phi_{hk})^2 [\frac{S_{hk}^2(1-\rho_{hk}^2)}{m_{hk}} + \frac{\rho_{hk}^2 S_{hk}^2}{n_k}] \tag{5}$$

Suppose sample rotation under cluster sampling is conducted in layer k. Regarding groups as observation units, the optimum weight of layer k at h time is shown to be:

$$\Phi_{hk} = V(\hat{\bar{Y}}_{hkm}) / [V(\hat{\bar{Y}}_{hku}) + V(\hat{\bar{Y}}_{hkm})]$$
$$= \frac{(1-\rho_{hk}^2)/m_{hk} + \rho_{hk}^2/n_k}{(1-\rho_{hk}^2)/m_{hk} + \rho_{hk}^2/n_k + 1/u_{hk}} \tag{6}$$

Optimum ratio of sample rotation in layer k at h time is:

$$\frac{u_{hk}}{n_k} = [1 + (\frac{1-\sqrt{1-\rho_{hk}^2}}{1+\sqrt{1-\rho_{hk}^2}})^{h-1}]/2 \tag{7}$$

2.3 Application example

A survey to reveal characteristics of physical indicators was conducted twice separately on October, 2010 and June, 2011. All the students at a middle school consisted of 24 classes and a primary school consisted of 18 classes were brought into the survey. Middle and primary school were as two layers and classes were as groups. With an eye to representativeness of samples, human and material resources, first time ($h=1$) in 2010 we randomly chose 9 primary classes and 12 secondary classes setting sampling ratio was 50%. All the students in these classes were brought into the study. Second time ($h=2$) in 2011 we held a pre-survey to estimate correlation coefficient (ρ) and regression coefficient (b). Then formal investigation with sample rotation began. Kept a portion of old classes, replaced a portion of new classes, altogether still maintained 9 primary classes and 12 secondary classes unchanged. There were altogether 1971 persons to be investigated. Survey indicators included juvenile body composition, aerobic endurance, strength and endurance, and flexibility evaluation index. Taken vital capacity for instance, population mean was 1907 ml for the first survey in 2010. The estimators of population mean, variance, 95%CI were 2220 ml, 6174.4 and (2065, 2374) ml by equations (1)~(7).

3 MONTE CARLO SIMULATION

3.1 *Simulated population*

Hierarchical structure was easily set while the results of two surveys above were as simulative parameters and vital capacity as analog indicator. For the first survey, simulative overall mean were 989 and 2596 respectively, and simulative standard deviation were 273.3 and 403.3 respectively. For the second survey, simulative overall means were 1315 and 2903 respectively, and simulative standard deviations were 138.5 and 89.9 respectively. Overall population was 100000. The two layers, whose weights were 0.43 and 0.57, contained 1000 clusters. Part of SAS program codes were as follows:

```
data population2;
    do id=1 to 1E5*0.43*0.7;h1=989+273.3*rannor(100);
        h2=(1315/989)*h1+138.5*rannor(100);cluster=ceil(ranuni(100)*430); strata=1;
        output;end;
    do id=1E5*0.43*0.7+1 to 1E5*0.43;
        h1=989+273.3*rannor(100); h2=1315+138.5*rannor(100);
        cluster=ceil(ranuni(100)*430); strata=1;output;end;
    do id=1E5*0.43+1 to 1E5*0.7;
        h1=2596+403.3*rannor(100); h2=(2903/2596)*h1+89.9*rannor(100);
        cluster=ceil(ranuni(100)*570); strata=2; output;end;
    do id=1E5*0.7+1 to 1E5;
        h1=2596+403.3*rannor(100); h2=2903+89.9*rannor(100);
        cluster=ceil(ranuni(100)*570);strata=2; output;end;
run;
```

3.2 *Simulating sample rotation*

For the first survey, we extracted some groups from the total. For the second survey, kept the sampling ratio in the layers, retained some old groups, and rotated some new groups to simulate 100 samples respectively at two sampling ratios. Then various statistics, 200 overall means at two ratios and their 95% CIs could be acquired. Part of the corresponding SAS program codes were as follows:

```
%macro stratify_cluster; %do seed=1001 %to 1100;
proc sql;
    create table total1 as select distinct strata, cluster, mean(h1) as h1, mean(h2) as h2
        from population2 group by strata, cluster; quit;
data total2; set total1;if strata=1 then wi=0.43; if strata=2 then wi=0.57; run;
proc sort data=total2 out=total3; by strata; run;
PROC SURVEYSELECT DATA = total3 OUT =sample1 (drop=selectionprob
    samplingweight) SAMPRATE =(0.1 0.1) SEED = &seed; STRATA strata; RUN;
proc reg data=total3 outest=reg; by strata; model h2=h1/Rsquare; quit;
proc sql; create table sample1_2 as select a.*, sqrt(b._rsq_) as row2i, b.h1 as b2i
    from sample1 as a left join reg as b on a.strata=b.strata; quit;
proc sql; create table sample1_3 as
    select *, count(cluster) as smalln from sample1_2 group by strata; quit;
data sample1_final;set sample1_3;
    rate2i=(1+((1-sqrt(1-row2i**2))/(1+sqrt(1-row2i**2))))**1)/2;
    u2i=round (smalln*rate2i, 1); m2i=smalln-u2i;
    fai2i=((1-row2i**2)/m2i+row2i**2/smalln)/((1-row2i**2)/m2i+row2i**2/smalln+1/u2i);
run;
proc sql; create table param as
    select distinct strata, smalln, row2i, b2i, rate2i, u2i, m2i,fai2i, wi from sample1_final;
```

```sas
quit;
proc sql; create table total_final as
      select a.*, b.smalln, row2i, b2i, rate2i, u2i, m2i, fai2i from total3 as a left join param as b
      on a.strata=b.strata; quit;
data rotate1;set sample1_final;by strata; samplesize=m2i;run;
PROC SURVEYSELECT DATA =rotate1 OUT =rotate2 SAMPSIZE =rotate1
      SEED = &seed; STRATA strata; RUN;
data rotate_old; set rotate2;flag='old'; run;
proc sql; create table rotate3 as select *, u2i as samplesize from total_final
      where put(strata, 8.)||put(cluster, 8.) not in (select put(strata, 8.)||put(cluster, 8.) from
      sample1_final);quit;
PROC SURVEYSELECT DATA =rotate3 OUT =rotate4 SAMPSIZE =rotate3
      SEED = &seed; STRATA strata; RUN;
data rotate_new; set rotate4;flag='new'; run;
data sample2_final;
          set rotate_old rotate_new; drop samplesize SelectionProb SamplingWeight; run;
proc sql; create table mean1 as
      select strata, round(mean(h2), 1) as mean_new from sample2_final
          where flag eq 'new' group by strata;
      create table mean2 as select strata, round(mean(h2), 1) as mean_old from sample2_final
          where flag eq 'old' group by strata;
      create table mean3 as select strata, round(mean(h1), 1) as mean_old_last
          from sample2_final where flag eq 'old' group by strata;
      create table mean4 as
          select strata, round(mean(h1), 1) as mean_last from sample1_final group by strata;
quit;
proc sql;
      create table mean5 as
          select a.*, b.mean_old from mean1 as a left join mean2 as b on a.strata=b.strata;
      create table mean6 as
          select a.*, b.mean_old_last from mean5 as a left join mean3 as b on a.strata=b.strata;
      create table mean7 as
          select a.*, b.mean_last from mean6 as a left join mean4 as b on a.strata=b.strata;
      create table mean8 as
          select a.*, b.fai2i, b2i, wi from mean7 as a left join param as b on a.strata=b.strata;
quit;
data mean9;
    set mean8; yi=fai2i*mean_new+(1-fai2i)*(mean_old+b2i*(mean_last-mean_old_last));
run;
proc sql; create table mean_final as select round(sum(yi*wi), 1) as y from mean9;quit;
proc sql;
      create table var1 as
          select strata, round(var(h2), 0.1) as s2i from sample2_final group by strata;
quit;
proc sql;
      create table var2 as
      select a.*, b.fai2i, row2i, wi, u2i, m2i, smalln from var1 as a left join param as b
      on a.strata=b.strata;
quit;
data var3;set var2;
      vari=fai2i**2*s2i/u2i+(1-fai2i)**2*(s2i*(1-row2i**2)/m2i+row2i**2*s2i/smalln);
run;
proc sql; create table var_final as select sum(vari*wi**2) as var from var3;quit;
```

Table 1. Computer simulation results (sampling ratio was 10%).

Sample	Mean	95% confidence interval		Sample	Mean	95% confidence interval	
		Upper limit	Lower limit			Upper limit	Lower limit
1	2220	2214	2226	51	2222	2216	2228
2	2220	2214	2226	52	2220	2215	2225
3	2223	2218	2228	53	2224	2218	2230
		...				...	
21	2229*	2224	2234	71	2220	2214	2226
29	2229*	2224	2234	79	2224	2218	2230
		...				...	
48	2220	2214	2226	98	2222	2216	2228
49	2221	2215	2227	99	2227	2221	2233
50	2223	2218	2228	100	2220	2214	2226

*95% CI didn't include simulative population mean.

Table 2. Computer simulation results (sampling ratio was 40%).

Sample	Mean	95% confidence interval		Sample	Mean	95% confidence interval	
		Upper limit	Lower limit			Upper limit	Lower limit
1	2221	2218	2224	51	2222	2219	2225
2	2222	2219	2225	52	2222	2219	2225
3	2224	2221	2227	53	2223	2220	2226
		...				...	
48	2223	2220	2226	98	2222	2219	2225
49	2224	2221	2227	99	2221	2218	2224
50	2221	2218	2224	100	2222	2219	2225

*95% CI didn't include simulative population mean.

```
proc sql;create table ci01 as select a.y, b.var from mean_final as a, var_final as b;quit;
data ci&seed;
    set ci01;cilower=round(y-1.96*sqrt(var), 1); ciupper=round(y+1.96*sqrt(var), 1);
run; %end;
data ci_all; set ci1001-ci1100; run; %mend stratify_cluster; %stratify_cluster;
```

3.3 Reliability and validity assessment

Simulative population mean was 2222. Made a simulation on sample rotation under stratified cluster sampling 200 times at two sampling ratios (10% and 40%) respectively. The analog results were shown as table 1 and table 2.

Table 1 and table 2 showed that only two 95% CIs didn't include the simulative population mean at the sampling ration of 10%, while all the simulative population were included at the sampling ratio of 40%. No matter how large or small the sample fraction was, 100 samples at each ratio were almost close to the overall mean, which indicated high reliability and validity of the sampling method and its formulae.

4 DISCUSSION

Sample rotation is almost applicable to all long-term successive survey since society is developing rapidly (Xu, & W, 2011). Sample rotation is an important means to reduce or control non-sampling errors, which can achieve a balance between sampling fees and precision (Du, & Liu, 2002).

Successive survey is of wide promising application, however, theoretical research is relatively immature, which largely restricts the application of the survey (Chen, 2012). Most of studies focus on simple random sample rotation (U.K. Office of Population Censuses and Surveys, 1984), which leads to estimated formulae just applicable to simple random sample. However, in reality we are facing more complex large-scale sampling (Kish, 1999; Lu, & Chen, 2005). Monte Carlo simulation was used in this paper to assess the reliability and validity on complex sample rotation survey (stratified cluster sampling). The results of simulation revealed high reliability and validity of the survey and its formulae. The survey in this paper was applied to the estimators on physical indicators and was of great practical significance.

By using random numbers generator, Monte Carlo simulation can extract data meeting the requirements of input variables (Jin, & Hu, 2013). For the statistical questions hard to prove in theory, Monte Carlo simulation can verify through a mass of simulations using infinite approximation. Its algorithm complexity is only associated with simulation times (Du, & Gao, 2013). Described above, Monte Carlo is worth to be popularized.

5 CONCLUSIONS

In this paper, Monte Carlo simulation was conducted with SAS programming to assess the reliability and validity on complex sample rotation survey (stratified cluster sampling). Almost all the CIs of the 100 estimated population mean at each sampling ratio include the simulative population mean. Successive survey using sample rotation under stratified cluster sampling and its formulae are of high reliability and validity which can be applied to large-scale investigation.

ACKNOWLEDGEMENT

The study was supported by National Natural Science Foundation of China (No. 81273188.). We are grateful to professor Ge Gao for his invaluable help in field investigations and calculations.

REFERENCES

Chen, G.H. 2012. The review of theoretical research on successive sampling survey. *Application of statistics and management*, 31(3): 426–433.
Du, Q.Q., & Gao, G. 2013. Reliability and validity evaluation based on Monte Carlo simulations in two-stage cluster sampling on multinomial sensitive question survey. *Chinese Journal of Health Statistics*, 30(2), 217–220.
Du, Z.F., & Liu, A.Q. 2002. Actual effect analysis of sample rotation. *Statistical research*, (6): 38–40.
Jin, S., & Hu, F.X. 2013. Mnining investment risk analysis based on Monte Carlo simulation. *Morden Mining*, (5): 148–151.
Kish L. 1999. Data collection for details over space and time. *Statistical Methods and the Improvement of Data Quality*, (34).
Lu, Z. H., & Chen, R.E. 2005. On sample rotation in social economics survey. *Journal of radio & TV university (Philosophy & Social Sciences)*, (1): 82–85.
NBS survey office in Shenzhen. 2013. A special investigation on high and middle gross household income in Shenzhen. *China statistics*, (1):11–13.
Sen, A.R. 1972. Successive Sampling with p Auxiliary Variables. *The Annals of Mathematical Statistics*, 43: 2031–2034.
Tang, Y., & Sun, Q. 2013. Sentinel surveillance of HIV, hepatitis C and syphilis infections. *J Diagn Concepts Pract*, 12(3): 352–354.
U.K. Office of Population Censuses and Surveys. 1984.General Household Survey 1982. London: HMSO.
Xu, G.X., & W,F. 2011. Application of Sample Rotation in Repeated Surveys and Its Real Diagnosis Study. *Statistical Research*, 28(5): 89–96.
Zhu, Y.P., & D,F. 2013. Nutritional risk screen and nutritional interventions in the elderly inpatients. *Chinese Journal of Gerontology*, 33(11): 2609–2611.

Reliability and validity evaluation based on Monte Carlo simulations by computer in three-stage cluster sampling on multinomial sensitive question survey

Shaochun Yang, Kejin Chen, Xiangyu Chen, Jiachen Shi, Wei Li, Qiaoqiao Du, Zongda Jin & Ge Gao
School of Public Health, Soochow University, P.R. China

ABSTRACT: In this paper, Randomized Response Technique (RRT) model through using computer was presented in application to investigate multiple classifications of sensitive issues by three-stage stratified sampling. Monte Carlo simulation by computer was successfully feasible. Almost 95% confidence intervals always include overall parameters. So the sensitive question survey method in this study has good reliability and validity.

1 INTRODUCTION

Sampling survey has been working as one of the important methods in the research and health survey and an important mean to obtain statistics. In recent years, by the increasingly attention of the government departments, enterprises, the academia and the public, the width and depth of its application get rapidly development, which has become a main sampling method [Jin & Gao 2000]. In some sampling investigation, researchers will often encounter sorts of sensitive problems. The so-called sensitive issues refer to the highly personal confidentiality or most people think that it is the inconvenient in public places [Cochran & Wu 1985]. For example, there are in particular public health research fields, such as: drug abuse, prostitution, whoring, male behavior, sexually transmitted disease, AIDS and so on [Fan 2008].

After years of development, a variety of random response technology is widely used in abortion rate, women drinking, tax evasion, other kinds of sensitive problems of the investigation process, and obtain better results [Liu & Chow 1976]. However, at home and abroad about random response technology research mainly focuses on the randomized design and improvement of the randomization device such as theoretical research [Zhang & Yan 2009]. Randomized Response Technique (RRT) was the clever use of a random device in the case of the protection of personal privacy by calculating the estimated probability character of people with sensitive issues to help to obtain sensitive issues for the true probability.

2 METHOD

2.1 The RRT model for multinomial sensitive question

A RRT model for three-stage cluster sampling on multinomial sensitive question survey was adopted in the investigation. Supposing sensitive questions were divided into many kinds of incompatible categories. In the model there is random device which includes some small balls written respectively $0, 1, 2 \ldots, K$. And the proportion of each kind of ball is $P0, P1, P2, \ldots, Pk(P0 + P1 + P2 + \cdots + Pk = 1)$. Every respondent was instructed to pick out a ball from the bag randomly, If the ball is signed 0, the answer was the corresponding sensitive question they belong to. However, the number was picked out.

2.2 Important part of program code

```
%macro sample(popsubjnum = ,
N1M = ,
N2M = ,
n1_m= ,
n2_m= ,
rate= ,
times= );
```

**Step2: simulating three-stage sampling from the population;
**calculate num of subjects per site, per district and per strata;
proc sql noprint;
create table pop_site as
select distinct strata, district, site, count(id) as SubjPerSite
from all02
group by 1,2,3;
quit;

**calculating total numbers of districts and sites in the population;
proc sql noprint;
create table alldist as
select distinct district
from all02;
quit;

**calculating population parameter;
proc sql noprint;
create table pop as
select distinct ques3, count(id)/&popsubjnum as pop_prop
from all02
group by ques3;
quit;

**fist stage, sampling district;
proc surveyselect data=alldist noprint
method=srs n=&n1_m
seed=&seed out=Sampledist;
run;

data _null_;
set sampledist;
call symput('dist'||left(_n_), strip(district));
run;

**Second stage, sampling site;
%do i=1 %to &n1_m;

%let seed1=%eval(&seed+&i);

proc surveyselect data=allsite noprint
method=srs n=&n2_m
seed=&seed1 out=Samplesite;
run;

proc sql noprint;
select distinct site into :site separated by ' '

```
from samplesite;
quit;
%put &site;
```

**Third stage, sampling subjects;
```
data sub01;
set all02;
where district=&&dist&i and site in (&site);
run;

proc sort;
by strata site;
run;

proc surveyselect data=sub01    noprint
method=srs rate=&rate
seed=&seed1 out=Sample01;
strata strata site;
run;

data sample02;
%if &i=1 %then %do; set sample01; %end;
%else %do; set sample02 sample01;   %end;
run;
%end;

data sample03;
set sample02;
rand1=ranbin(&seed,1,0.7);
if rand1=1 then t3=ques3;
else if rand1=0 then do;
rand2=ranbin(&seed,1,0.1/(1-0.7) );
if rand2=1 then t3=1;
else if rand2=0 then do;
rand3=ranbin(&seed,1,0.5 );
if rand3=1 then t3=2;
else if rand3=0 then t3=3;
end;
end;
run;
```

**calculate num of subjects per site, per district and per strata in samples;
```
proc sql noprint;
create table pop_site_s as
select distinct strata, district, site, count(id) as SubjPerSite_s
from sample03
group by 1,2,3;
quit;
```

**for each site;
```
proc sql noprint;
create table sample03a as
select a.*, b.subjpersite
from sample03 as a left join pop_site as b
on a.strata=b.strata and a.district=b.district and a.site=b.site;
```

```
create table sample04 as
select a.*, b.subjpersite_s
from sample03a as a left join pop_site_s as b
on a.strata=b.strata and a.district=b.district and a.site=b.site;
quit;

**calculate pct per district;
proc sql noprint;
create table sample05 as
select a.*, b.subjperdist
from sample04b as a left join pop_dist as b
on a.strata=b.strata and a.district=b.district ;
quit;

**calculate pct per strata(strata);
proc sql noprint;
create table sample07 as
select a.*, b.subjpergrp
from sample06 as a left join pop_strata as b
on a.strata=b.strata;
quit;

data out;
%if &seed=101 %then %do; set sample10; %end;
%else %do; set out sample10; %end;
run;

%end;
%mend;
```

Fill in the corresponding macro parameters, and implement the macro program, we can let computer calculate 100 times by using multiple sensitive issues under a single sample random response model of RRT survey to get its estimation variance, CL, P value of the chi-square test of 100 samples. Macro arguments are as follows:

%sample(popsubjnum = 67750,
N1M = 16,
N2M = 15,
n1_m = 3,
n2_m = 5,
rate = 0.6,
times = 100);

3 MONTE CARLO SIMULATION

3.1 *Simulated population*

According to Jianfeng Wang, Ge Gao [Wang & Gao 2009], sample size obtained the relevant literature. The three-stage stratified sampling survey investigated the gay men in Beijing aged from 15 to 29 and 30 to 49 years old. Their proportion was W1 (15 to 29 years old) = 58.24%, W2 (30 to 49 years old) = 41.76%. The 16 districts and counties of Beijing was as primary units, and let gay bath, clubs, bars, parks and so on as the second units, MSMs as tertiary units where people are a total of 67,750. On average, at 15 to 29 and 30 to 49 years old, every two counties were about the number of gay men $\overline{N}_{21} = 2466, \overline{N}_{22} = 1768$.

The first stage is that three counties ($n_{11}=n_{12}=3$) were randomly selected from 16 counties; the second phase of the extracted counties randomly selected five survey points ($n_{21}=n_{22}=5$); At last, the third stage was that 15 survey sites randomly selected 2,062 people. In the inner layer of the survey drawn equally from each point was taken approximately $\bar{n}_{31}=80, \bar{n}_{32}=57$. Then RRT model for multinomial sensitive question is presented for application to ask MSMs about the sexual behavior they have with male partners lately (achieve ejaculation orgasm). Besides, sexual behavior pattern of k: k = 1 (anal sex), k = 2 (oral sex), k = 3 (otherwise). The formula (1)–(4) to calculate the proportion of MSMs for anal sex, oral sex and other forms were 65.83%, 18.70%, 15.47%.

3.2 Simulated investigation results

To get both acceptable accuracy and reliability in parameter estimates, we could calculate sample size of simulated sampling from the relevant research [Wang & Gao 2011]. At the first stage, 3 primary units were randomly selected from 16 districts and counties from the simulated population. At the second stage, 5 secondary units were randomly drawn from the selected primary unit. The third stage is to choose MSMs simulated respondents. All the results were given in Tables.

Monte Carlo simulation sampling was repeated 100 times under SAS program. Each tertiary unit has produced a response value. Then we could get the estimators of the simulated population proportion and the simulated population proportion variance of that sensitive question according to formulas.

For single sample model applied for multichotomous sensitive questions investigation, there are individually 96, 99, and 97 CIs for the proportion of population out of 100 computer simulation results including the real population proportion for each category, which shows high validity and reliability of the sampling survey method and the formulae we deduced for improved model.

4 DISCUSSION

Monte Carlo Simulation method (Monte Carlo Simulation, MCS) was also known as a statistical simulation method because of the development of science, technology and the invention of computer. Moreover MCS is presented as a guide to the theory of probability and statistics and a very important numerical method to use random numbers (or more commonly pseudo-random numbers) to solve many computer problems [Xu 1985].

Monte Carlo method solved the problem based on the probability of random events or the expected value of a random variable through some kind of experimental approaches to estimate the frequency of some events happened or the probability of the randomly events. Besides, make getting some character of this random numerical variable as the solution to the problem. The basic idea of the Monte Carlo method by people was discovered and utilized in the 1940s of the emergence of computer, especially, in recent years, the emergence of high-speed computer make computer use mathematical methods popularly. At present, this method has been widely applied to mathematics, physics, management, genetic engineering, social science and other fields and has already showed special lots of advantages.

With the development of society, many problems such as homosexuality, prostitution, drug abuse, multiple sexual partners, sexually transmitted diseases and AIDS have increasingly become unavoidable and neglected public health and social problem. But for such a sensitive issue, it is easy to produce the serious bias of using traditional methods investigation. However, the use of RRT model can obtain accurate and reliable data. In recent years, sampling method RRT combination with various sensitive issues of statistical survey methods has become an important research topic of statistical theory as well as methodology. Now, stratified multi-stage sampling RRT model survey method in evaluating reliability and validity of sensitive issues has become popular trends in sampling survey researches.

5 CONCLUSIONS

In this paper, a large number of SAS programs for computer simulated sampling were making samples repetition of sensitive issues under a single sample model three-stage stratified sampling method and statistical formulas for the evaluation of the validity and reliability to obtain the high degree of reliability and efficient evaluation findings [Wang, Wang & Zhang 2006]. And in the reliability and validity of the computer simulation sampling survey as well as the simulation results of interval estimation methods about the overall parameter are innovative. Evaluation results showed that sensitive issues through multi-classification model RRT three-stage stratified sampling method, sensitive issues can get real survey data. Besides, the relevant departments will take better measures to prevent HIV/AIDS and other social and public health policy issues by providing a scientific and efficient method.

ACKNOWLEDGEMENT

The study was supported by National Natural Science Foundation of China, No. 81273188.

We are grateful to Ge Gao, Zongda Jin, Qiaoqiao Du, for their invaluable help in field investigations and calculations.

REFERENCES

Cochran, W. G. & Wu, H. 1985. Sampling techniques [M] China Statistics Press, 87.

Jin, P. H. & Gao, G. 2000. Research and Application of a three-stage stratified sampling sample size. Chinese Health Statistics, 17 (6):325–327

Fan, Y. B. 2008. Sensitive issues Simmons model (hierarchical) cluster sampling study. Chinese Health Statistics, 25 (6):562–565.

Liu, P. T. & Chow L. P. 1976. The efficiency of the multiple trial randomized response technique [J]. Biometrics, 32(3):607–618.

Wang, J. F. & Gao, G. 2009. The estimation of sampling size in multi-stage sampling and its application in medical survey [J]. Applied Mathematics and Computation; 178(2006): 239–249.

Wang, L. & Gao, G. 2011. Other sensitive problem has nothing to do double model of stratified two-stage cluster sampling statistics methods and applications. China health statistics, 2011, 28 (1): 37–39.

Wang, C. P. Wang, Z. F. Zhang, G. C. et al. 2006. Attributes sensitivity problem of design, analysis and evaluation [J]. China health statistics, 23 (1): 60–62.

Xu, Z. J. 1985. Monte Carlo Method [M]. Shanghai: Shanghai science and technology publishing house.

Zhang, Q. & Yan, Z. Z. 2009. Under the binomial sample randomized survey than estimation model [J]. Journal of Inner Mongolia university of technology, 28 (2): 81–85.

… Advanced Engineering and Technology – Xie & Huang (Eds)
© 2014 Taylor & Francis Group, London, ISBN 978-1-138-02636-0

Parameter estimation for sample rotation of successive survey in stratified sampling and its application

S. Liu, Y. Fan & G. Gao
School of Public Health, Soochow University, P.R. China

ABSTRACT: In this paper, methods and theorems of mathematical statistics and probability were applied to deduce the formulae for estimating the composite population mean as well as the variance in stratified sampling for sample rotation of successive survey. These formulae were applied in the successive survey of staffs' body mass index (BMI) of one of the nuclear power stations in China.

1 INTRODUCTION

Sample rotation in successive survey is highly recommended when investigating population's value and its change in continuous time. By sample rotation, parts of the old sample units were retained and the others were replaced by new units in next survey, so it assembles the advantages of both totally new sample and fixed sample, which can not only avoid sample fatigue but also can attain the balance among sample representative, estimation precision and cost.

However, at present, both theory and application of sample rotation are seldom researched. Formulae are only for simple random sampling, but not for complex samplings. To deal with this problem, in this paper, we deduced the formulae for composite estimate of population mean as well as the variance for stratified sampling for sample rotation of successive survey, moreover, we successfully employed them in practical surveys.

2 FORMULA DEDUCTION

2.1 Concepts, notations and statistics

Suppose the population (N) is made up of L strata, the size of each stratum is denoted respectively with $N_1, N_2, \ldots, N_L$. $\sum_{i=1}^{L} N_i = N$. Sample rotation is carried out in each stratum independently. That is to say, on occasion $h-1$, n_i sample units are drawn from the ith stratum, and on occasion h, n_i sample units are still drawn from the ith stratum, among which some units were replaced by new ones. The retained units (matched portion) and new units (unmatched portion) are denoted respectively with m_i, u_i. $m_i + u_i = n_i, i = 1, 2, \ldots, L$. The total sample size is denoted with n, $\sum_{i=1}^{L} n_i = n$. Population parameters are sized up according to the weighted sums of each stratum. Assume that the population variance and correlation coefficient of each survey is constant, besides, ignoring the population correction coefficient.

The following notations will be used:

h the occasion of survey,
ρ_{hi} the correlation coefficient of the same sample units between the survey on occasion h and $h-1$ for the identical variable of the ith stratum,
b_{hi} the regression coefficient of the same sample units between the survey on occasion h and $h-1$ for the identical variable of the ith stratum,
Φ_{hi} the weight coefficient of the ith stratum on occasion h,
λ_{hi} the optimum sample rotation rate of the ith stratum on occasion h,

y_{hij} the values obtained for the jth ($j = 1, 2, \ldots n_i$) unit in ith ($i = 1, 2, \ldots L$) stratum on occasion h of the sample,

W_i the weight of the ith stratum, $W_i = \frac{N_i}{N}$,

$\hat{\bar{Y}}_{hi}$ the estimator of the population mean of the ith stratum on occasion h,

$\bar{y}_{hi}$ the sample mean of the ith stratum on occasion h,

$\bar{y}_{hui}$ the mean of unmatched portion of the ith stratum on occasion h,

$\bar{y}_{hmi}$ the mean of matched portion of the ith stratum on occasion h,

S_{hi}^2 the population variance of the ith stratum on occasion h, $S_{hi}^2 = \frac{1}{N_i}\sum_{j=1}^{N_i}(Y_{hij} - \bar{Y}_{hi})^2$,

s_{hi}^2 the sample variance of the ith stratum on occasion h, $s_{hi}^2 = \frac{1}{n_i-1}\sum_{j=1}^{n_i}(y_{hij} - \bar{y}_{hi})^2$,

$\bar{Y}_i$ the population mean of the ith stratum,

$V(\hat{\bar{Y}}_h)$ the variance of estimate mean of the population on occasion h,

$V(\hat{\bar{Y}}_{hi})$ the variance of estimate mean of the ith stratum on occasion h.

2.2 Parameter estimation

2.2.1 Population mean

In stratified sampling, estimate of population mean is gained by the weighted mean of $\hat{\bar{Y}}_{hi}$ ($i = 1, 2, \ldots, L$) as per W_i. That is

$$\hat{\bar{Y}}_h = \sum_{i=1}^{L} W_i \hat{\bar{Y}}_{hi}. \tag{1}$$

For stratified simple random sampling of sample rotation, in each stratum, $\hat{\bar{Y}}_{hi}$ is calculated according to the computing method in simple random sampling of sample rotation. By Cochran W.G., we can get

$$\begin{aligned}\hat{\bar{Y}}_{hi} &= \Phi_{hi}\hat{\bar{Y}}_{hui} + (1-\Phi_{hi})\hat{\bar{Y}}_{hmi} \\ &= \Phi_{hi}\bar{y}_{hui} + (1-\Phi_{hi})[\bar{y}_{hmi} + b_{hi}(\hat{\bar{Y}}_{h-1,i} - \bar{y}_{h-1,mi})]\end{aligned} \tag{2}$$

From the formulae (1) and (2), we gain the estimator of population mean on occasion h:

$$\hat{\bar{Y}}_h = \sum_{i=1}^{L} W_i \{\Phi_{hi}\bar{y}_{hui} + (1-\Phi_{hi})[\bar{y}_{hmi} + b_{hi}(\hat{\bar{Y}}_{h-1,i} - \bar{y}_{h-1,mi})]\}. \tag{3}$$

As to $\hat{\bar{Y}}_{hi}$, as long as the sample rotation rate isn't zero, this estimator has the characteristic of minimum linear unbiased estimate. To stratified sampling, if $\hat{\bar{Y}}_{hi}$ is unbiased estimate of the population mean of the ith stratum, $\hat{\bar{Y}}_h$ is the unbiased estimate of the population mean $\bar{Y}_h$.

2.2.2 Variance

Because the sampling is independent in each stratum, $\hat{\bar{Y}}_{hi}$ is independent. We can get

$$V(\hat{\bar{Y}}_h) = \sum_{i=1}^{L} W_i^2 V(\hat{\bar{Y}}_{hi}). \tag{4}$$

For stratified simple random sampling of sample rotation, $V(\hat{\bar{Y}}_{hi})$ is calculated according to the computing method in simple random sampling of sample rotation. By Cochran W.G., we can get

$$V(\hat{\bar{Y}}_{hi}) = \Phi_{hi}^2 V(\hat{\bar{Y}}_{hui}) + (1-\Phi_{hi})^2 V(\hat{\bar{Y}}_{hmi}), \tag{5}$$

where, population variance is substituted with sample variance, that is to say

$$V(\hat{\bar{Y}}_{hui}) \approx v(\hat{\bar{Y}}_{hui}) = \frac{s_{hi}^2}{u_{hi}}, \qquad (6)$$

$$V(\hat{\bar{Y}}_{hmi}) \approx v(\hat{\bar{Y}}_{hmi}) = s_{hi}^2 (\frac{1-\rho_{hi}^2}{m_{hi}} + \frac{\rho_{hi}^2}{n_{hi}}). \qquad (7)$$

So formula (5) can be written as

$$V(\hat{\bar{Y}}_{hi}) = \Phi_{hi}^2 \frac{s_{hi}^2}{u_{hi}} + (1-\Phi_{hi})^2 [\frac{s_{hi}^2(1-\rho_{hi}^2)}{m_i} + \rho_{hi}^2 \frac{s_{hi}^2}{n_{hi}}]. \qquad (8)$$

From the formulae (4) and (8), we may gain the variance of the estimator of population mean on occasion h:

$$V(\hat{\bar{Y}}_h) = \sum_{i=1}^L W_i^2 \{\Phi_{hi}^2 \frac{s_{hi}^2}{u_{hi}} + (1-\Phi_{hi})^2 [\frac{s_{hi}^2(1-\rho_{hi}^2)}{m_{hi}} + \rho_{hi}^2 \frac{s_{hi}^2}{n_{hi}}]\}. \qquad (9)$$

2.2.3 Optimum weight and optimum sample rotation rate
Because stratified sampling of successive survey in each stratum is simple random sampling of successive survey, by Cochran W.G., the optimum weight and optimum sample rotation rate of the *i*th stratum on occasion *h* is as follow respectively:

$$\Phi_{hi} = V(\hat{\bar{Y}}_{hmi}) \big/ [V(\hat{\bar{Y}}_{hui}) + V(\hat{\bar{Y}}_{hmi})], \qquad (10)$$

$$\lambda_{hi} = \frac{u_{hi}}{n_i} = [1 + (\frac{1-\sqrt{1-\rho_{hi}^2}}{1+\sqrt{1-\rho_{hi}^2}})^{h-1}]/2. \qquad (11)$$

From the formulae (6), (7) and (10), we get

$$\begin{aligned}\Phi_{hi} &= V(\hat{\bar{Y}}_{hmi}) \big/ [V(\hat{\bar{Y}}_{hui}) + V(\hat{\bar{Y}}_{hmi})] \\ &= \frac{S_{hi}^2(1-\rho_{hi}^2)/m_{hi} + S_{hi}^2 \times \rho_{hi}^2/n_i}{S_{hi}^2(1-\rho_{hi}^2)/m_{hi} + S_{hi}^2 \times \rho_{hi}^2/n_i + S_{hi}^2/u_{hi}} \\ &= \frac{(1-\rho_{hi}^2)/m_{hi} + \rho_{hi}^2/n_i}{(1-\rho_{hi}^2)/m_{hi} + \rho_{hi}^2/n_i + 1/u_{hi}}\end{aligned} \qquad (12)$$

3 APPLICATION

Stratified sampling for sample rotation of successive survey was employed to survey the staff's body mass index (BMI) of one of the nuclear power stations in China, which has four departments. The production dept. and service dept. which are chiefly physical labor were regarded as the first stratum (named as production and service stratum). The technic dept. and quality assurance dept. which are chiefly mental labor were regarded as the second stratum (named as technic and quality stratum). The population size of the two strata is 1102 and 378 respectively.
According to the formula $W_i = \frac{N_i}{N}$, we could get

$$W_1 = \frac{1102}{1102+378} = 0.74468, \quad W_2 = \frac{378}{1102+378} = 0.25532.$$

According to the results of investigation materials that had been got before, we could get the variance of the two strata respectively:

$$S_1^2 = 7.06976, S_2^2 = 6.91686.$$

3.1 Survey in the first year ($h=1$)

The upper limit of the variance is $V = 0.01021$, according to proportion sampling, the sample size is almost 470.

According to the weight, we could get the sample size of the two strata respectively of the survey in the first year ($h=1$):

$$n_1 = 470 \times 0.74468 \approx 350, \quad n_2 = 470 \times 0.25532 \approx 120.$$

The estimators and variances which were computed by survey data are as follows:

the estimator of the population mean (arithmetic mean) is $\hat{\bar{Y}}_1 = \bar{y}_1 = 23.52888$,

the variance of the estimator of the population mean is $V(\hat{\bar{Y}}_1) = v(\hat{\bar{Y}}_1) = 7.02540$.

3.2 Survey in the second year ($h=2$)

Firstly, we compute the estimator of the population mean of the first stratum (the production and service stratum).

For the item of BMI, according to the force-survey, we could estimate the regression coefficient and correlation coefficient of the first stratum between the survey in the first and second year:

$$b_{21} = 0.94860, \quad \rho_{21} = 0.95387.$$

From the formula (11), we could get the optimum sample rotation rate in the first stratum:

$$\lambda_{21} = \frac{u_{21}}{n_1} = [1 + (\frac{1-\sqrt{1-0.95387^2}}{1+\sqrt{1-0.95387^2}})^{2-1}]/2 = 0.769.$$

Because the sample size is constant on each occasion, we get

$$u_{21} = n_1 \times 0.769 = 350 \times 0.769 \approx 269, \quad m_{21} = 350 - 269 = 81.$$

From the formula (12), we could get the optimum weight:

$$\Phi_{21} = \frac{(1-0.95387^2)/81 + 0.95387^2/350}{(1-0.95387^2)/81 + 0.95387^2/350 + 1/269} \approx 0.5$$

For the first stratum, according to the survey data, we could get

$$\bar{y}_{2u1} = 23.56613, \qquad \bar{y}_{2m1} = 23.62222,$$

$$\hat{\bar{Y}}_{11} = \bar{y}_{11} = 23.50927, \qquad \bar{y}_{1m1} = 23.59012.$$

From the formula (2), we could get the estimator of the population mean of the first stratum:

$$\hat{\bar{Y}}_{21} = \Phi_{21}\bar{y}_{2u1} + (1-\Phi_{21})[\bar{y}_{2m1} + b_{21}(\hat{\bar{Y}}_{11} - \bar{y}_{1m1})]$$
$$= 0.5 \times 23.56613 + (1-0.5) \times [23.6222 + 0.9486 \times (23.50927 - 23.59012)]$$
$$= 23.55583.$$

Then we compute the variance of the estimator of the population mean of the first stratum (the production and service stratum).

From the formula $s_{hi}^2 = \frac{1}{n_i - 1} \sum_{j=1}^{n_i} (y_{hij} - \bar{y}_{hi})^2$, we could get $s_{21}^2 = 7.06976$.

Then from the formula (8), we get

$$V(\hat{\bar{Y}}_{21}) = \Phi_{21}^2 \frac{s_{21}^2}{u_{21}} + (1-\Phi_{21})^2 [\frac{s_{21}^2(1-\rho_{21}^2)}{m_{21}} + \rho_{21}^2 \frac{s_{21}^2}{n_1}]$$

$$= 0.5^2 \frac{7.06976}{269} + (1-0.5)^2 [\frac{7.06976(1-0.95387^2)}{81} + \frac{0.95387^2 \times 7.06976}{350}]$$

$$= 0.01313.$$

In the same way, we could work out the population mean and its variance of the second stratum (the technic and quality stratum):

$$\hat{\bar{Y}}_{22} = 23.32273, \ V(\hat{\bar{Y}}_{22}) = 0.03762.$$

So from the formula (1), we could get the estimator of the population mean of the survey in the second year:

$$\hat{\bar{Y}}_2 = W_1 \times \hat{\bar{Y}}_{21} + W_2 \times \hat{\bar{Y}}_{22}$$

$$= 0.74468 \times 23.55583 + 0.25532 \times 23.32273$$

$$= 23.49631.$$

From the formula (4), we could get the variance of the estimator of the population mean of the survey in the second year:

$$V(\hat{\bar{Y}}_2) = W_1^2 \times V(\hat{\bar{Y}}_{21}) + W_2^2 \times V(\hat{\bar{Y}}_{22})$$

$$= 0.74468^2 \times 0.01313 + 0.25532^2 \times 0.03762$$

$$= 0.00973.$$

With the purpose of comparison with that sample rotation is not employed, we assume that the data surveyed in the second year are from fixed sample or totally new sample. According to the calculation method of stratified sampling (without sample rotation), we get $V(\hat{\bar{Y}}_2') = 4.08788$.

It can be seen that the $V(\hat{\bar{Y}}_2)$ is much less than the $V(\hat{\bar{Y}}_2')$. That is to say, the estimation precision for that sample rotation is employed is better than that sample rotation is not employed.

4 DISCUSSION

The formulae for the estimation of population mean and its variance in stratified sampling for sample rotation of successive survey were deduced for the first time in this paper, filling a vacancy of parameter estimation for sample rotation of successive survey in complex sampling. With the formulae deduced in this paper, we scientifically estimated the mean of the staff's BMI of one of the nuclear power stations in China.

The method of sample rotation assembles the advantages of both the fixed sample and fresh sample, and overcomes their shortcomings. It employs composite regression estimation, of which the estimation precision is better than that of common estimation. Because of the better precision, the estimate for the population is more reliable.

This study is not only the demand of the development of statistics research, but also providing scientific method of parameter estimation for practical successive survey. It can raise the precision and efficiency of statistic survey and provide reliable scientific basis for some planning and policy, having important practical value for the development of social economy.

ACKNOWLEDGEMENT

National Natural Science Foundation of China (No. 30972548).

REFERENCES

Brint A.T. 2006. Sampling on successive occasions to re-estimate future asset management expenditure, *European Journal of Operational Research, Stochastics and Statistics*, 175: 1210–1223.
Cochran W.G. 1977. *Sampling Techniques*. New York: Wiley.
Du Z. 2005. *Sampling Techniques and Practices*. Beijing: Tsinghua University Press.
Lu Z. 1998. *System Study of Sampling Method*. Beijing: China Statistics Press.
Qi S. 2004. The study on sample rotation methods of sampling survey. *Statistical Research*, 20–24.
Ruedaa M., Muza J.F., Gonzalezb S., Arcosa A. 2006. Estimating quantiles under sampling on two occasions with arbitrary sample designs. *Computational Statistics & Data Analysis*, 25: 111–120.

Earthquake influence field dynamic assessment system

Baitao Sun, Xiangzhao Chen, Hongfu Chen & Yingzi Zhong
Institute of Engineering Mechanics & Key Laboratory of Earthquake Engineering and Engineering Vibration of China Earthquake Administration, Harbin, Heilongjiang, China

ABSTRACT: To solve the key problems in earthquake influence field assessment systems, the author developed a WebGIS based dynamic earthquake influence field assessment system, implemented data sharing and dynamic assessment of earthquake influence field from different platform and sources. This paper firstly describes the current status of earthquake influence field research, then puts forward solutions for the existing problems and functional structure design for the system. According to the working process of earthquake intensity evaluation on site and the formation process of earthquake intensity map, the author develops the dynamic WebGIS based assessment model and realizes its function. By comprehensively considering earthquake influence field data format in various platforms, the author develops online data sharing function of earthquake influence field vector layer. Finally, this paper describes the function of this system in earthquake emergency and earthquake prediction work, and expounds its referable meaning to disaster prevention and mitigation planning job.

1 INTRODUCTION

As one of the most serious natural disasters, earthquakes have caused heavy casualties and economic losses. Ever since the 20th century, there were more than 1200 earthquakes over magnitude 7 around the world. Some strong earthquakes have broken out in recent years, such as the Wenchuan earthquake, Yushu earthquake and Lushan earthquake. The forthcoming "The Ground Motion Parameter Zoning Map" in China shows that more than 75% of the cities in China are facing with the potential severe earthquake disaster threat. The earthquake industry is still in serious challenge.

Accurately and efficiently determining the earthquake impact area is very important for the earthquake emergency rescue and damage prediction (Chen H.F. et al. 2013). Earthquake emergency system needs to promptly determine the scope of affected area and the most serious area after the shock in order to arrange the emergency rescue. The earthquake damage prediction system generates setting earthquakes from earthquake influence fields. The U.S. Geological Survey inversely gives out earthquake influence fields based on simulating the rupture process of faults every time after the earthquake shakes and generates various kinds of file formats for users to download (Earle P.S. et al. 2009). So far, China has completed earthquake damage prediction in more than 30 cities and large industrial and mining enterprises, and has established the corresponding information systems. During the ninth five-year period, China have established the "two cities (Beijing, Tianjin) and one province (Hebei)" earthquake damage prevention and mitigation command center. During the period of "eleventh five-year", the integration of earthquake emergency command technology system network covering mainland China has been basically constructed relying on "China Digital Earthquake Observation Network" project. In recent years, various provinces and cities have established their earthquake emergency command systems and earthquake damage prediction systems (Fan L.C. et al. 2011). All the above systems contain earthquake influence field modules based on C/S or B/S, but these systems are built independently and their data cannot be shared mutually. The earthquake influence field is hardly to be on-line modified after generated.

Therefore, the author develops the WebGIS based earthquake influence field dynamic assessment system (hereinafter referred as EIFDAS) to realize data sharing and dynamic assessment between different platforms and sources.

2 THE OVERALL DESIGN OF SYSTEM

The system is developed based on B/S architecture, users could input earthquake three elements (time, epicenter, magnitude) through a browser, the background calculates intensities and the long and short axis of each according to the intensity attenuation relationships after receives the parameters. Then the background would draw the earthquake influence field vector surface layer based on ArcEngine and feedback the affected range by calculating each intensity layer's area. After the calculation is completed, release the earthquake influence field vector layer based on ArcServer and returned the assessment result to the front end users.

Earthquake intensity assessment on site is a dynamic process. The system firstly produces a preliminary estimation earthquake influence field according to attenuation models. Earthquake field commanders could identify severely afflicted areas and properly distribute on-site earthquake intensity assessment staffs referring to the preliminary estimation influence field and approximate estimated affected areas. As on-site intensity assessment staffs constantly send feedback like assessment result data and aftershock distribution, etc. (Yuan Y.F. & Tian Q.W. 2012), on-site commanders could correct the preliminary earthquake influence field directly through a browser without professional GIS software, and finally accomplish the intensity distribution map.

At present, both domestic and abroad has researched a lot on earthquake influence field, developed a large number of statistical models based on instrument data or empirical statistical data and established many assessment systems. During previous earthquakes, lots of earthquake intensity distribution maps which could reflect actual earthquake impact range were obtained through field investigation. The earthquake influence fields from different sources mentioned above have very important referential meaning on co-earthquake emergency action and pre-earthquake disaster prevention. Therefore, this system develops earthquake influence field opened sharing system based on WebGIS, to implement the data standardization of different formats and online data import and export function. The overall function logic architecture is shown as Figure 1.

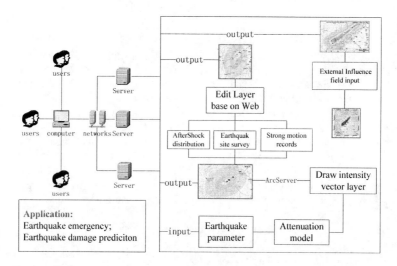

Figure 1. The function logic architecture of system.

3 MODELS, ALGORITHMS AND TECHNICAL ROUTE

The EIFDAS adopts oval empirical intensity attenuation models (Zhou K.S. & Fu L.L. 1995). The long and minor axis could be calculated according to attenuation models and regression parameters. Then the intensity vector layer would be drawn based on ArcEngine by long and minor axis, the calculate workflow is shown by Figure 2.

$$I_a = C_{1a} + C_{2a}M - C_{3a}\ln(R_a + R_{0a}) + \varepsilon_{la} \qquad (1)$$

$$I_b = C_{1b} + C_{2b}M - C_{3b}\ln(R_b + R_{0b}) + \varepsilon_{lb}$$

The meaning of I_a and I_b are average intensity along long and minor axis direction, R_a, R_b is long and minor axis radius, C_{1a}, C_{2a}, C_{3a} is regression parameters along long axis direction, C_{1b}, C_{2b}, C_{3b} is regression parameters along minor axis direction, R_{0a}, R_{0b} is saturation factors along long and minor axis direction, ε_{la}, ε_{lb} is modification values along long and minor axis. All the above parameters could be calculated by iterative regression method.

The rapture direction of fault is an important factor to determine the long axis direction. The system would identify the nearest fault from the epicenter, and acquire the fault direction as the long axis of intensity rings accordingly. However, the nearest fault may not be the seismogenic one in some circumstances, users could estimate and manually input the rapture direction instead.

The EIFDAS adopts service oriented software development method, using the business process to integrate software services. The system includes five layers: hardware and network platform, operating system platform, infrastructure platform (application server), the general business support platform and earthquake disaster risk assessment system software platform. This layered architecture could reduce the correlation between application analysis and underlying data, and also make development of different business analysis processing systems based on the same underlying data available. In addition, this architecture could realize the integration of loose couplings among different parts and reduce the system complexity. The data structure would follow the data exchange standard and regulations of earthquake industry, and be deployed and accessed through the business process. The technical architecture is shown by Figure 2.

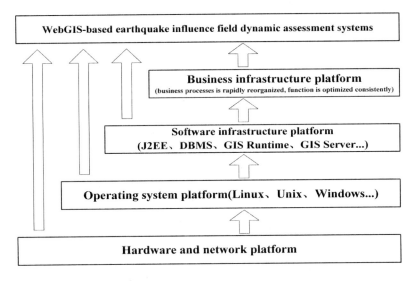

Figure 2. Technical architecture of system.

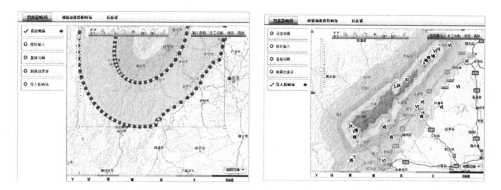

Figure 3. Modification and import of earthquake influence field.

Based on former experiences, the earthquake influence fields generated by attenuation models are normally idealized oval. However, the actual shapes of earthquake influence field are irregular, influenced by site conditions and other factors. Therefore, the earthquake influence fields generated by empirical models need to be constantly modified according to the feedback information from the field investigations and other sources. This system could carry out online editing for intensity circle layer based on the browser, including changing intensity range, and sketching intensity anomaly area, etc.

To enhance the openness, this system has realized the earthquake influence field vector data sharing function, including online export and external data import. The final intensity map formed by earthquake field work could be online imported into the system in a browser, and used as accurate input for earthquake disaster loss estimation. In addition, other sources such as ShakeMap figures form the USGS could be imported into the system for comparatively analysis and reference. As the formats of external earthquake influence fields are not unified, required data format is given out for users to edit while importing, and the system would check the geometric regularity and data validity of the vector layer afterwards. The function is specifically shown in Figure 3.

The EIFDAS system gave out prompt estimation result of earthquake influence field right after the Lushan Ms 7.0 strong earthquake, and modified the result immediately according to the feedback information on site and aftershock distribution. The results provided necessary support for local government to carry out emergency rescue and disaster mitigation work. Moreover, the system could be applied in urban earthquake damage prediction and disaster prevention and mitigation planning.

4 CONCLUSION

In order to solve the problems of online sharing and dynamic evaluation at present, the EIFDAS system was developed. This paper describes the research status of earthquake influence field, as well as logic structure, algorithm models, programming realization and application of the system. This system mainly has the following innovations: (1) Data standardization for different formats based on WebGIS is researched, and the result sharing of different platforms is realized. (2) Layer editing based on WebGIS and online dynamic modification of earthquake influence field are achieved; (3) In order to implement the standardization of the empirical attenuation formula, the formula of different formats are stored in the database as parameters. There remain some functions to be improved. For instance, the earthquake influence field could be generated based on the projective coordinate system. However, deviation might appear due to the inconsistent spatial reference during online map show. In addition, the earthquake influence field models would be more meaningful for actual assessment with other factors, such as site effects, being considered.

ACKNOWLEDGEMENT

This paper was supported by International Technical Cooperation Project of China-U.S. (Grant No. 2011DFA71100), Basic Scientific Research Special Project of Institute of Engineering Mechanics, CEA (Grant No. 2009A01), and National Natural Science Foundation (Grant No. 51308511).

REFERENCES

Chen H.F., Sun B.T., et al. 2013. HAZ-China earthquake disaster loss estimation system. *China Civil Engineering Journal* 46(S2): 294–300.
Earle P.S., Wald D.J., Jaiswal K.S., et al. 2009. *Prompt assessment of global earthquakes for response (PAGER): A system for rapidly determining the impact of global earthquakes worldwide.* U.S. Geological Survey Open-File Report 1131.
Fan L.C., Hu B., Li Y.R. 2011. The design and realization of Xian earthquake emergency command technical system. *Technology for Earthquake Disaster Prevention* 6(2):190–198.
Yuan Y.F. & Tian Q.W. 2012. *Engineering seismology*, Beijing: Earthquake Press.
Zhou K.S. & Fu L.L. 1995. Symmetrical transform at ion on attenuation laws of seismic ground motion. *South China Journal of Seismology* 15 (4): 1–9.

Analysis of safety-accident on residential building construction

Hong-Bo Zhou, Xi-Xing He, Xin Lu & Mao-Sen Zhou
Shanghai Jianke Engineering Consulting Co., Ltd. Shanghai, China

Li Feng
Shanghai Jiao Tong University, Shanghai, China

ABSTRACT: According to the statistical analysis of the accident cases collected from residential engineering, four common types of the accident are showed in this paper. By conducting statistical analysis of the time of the accident, the cause of the accident and the casualties caused by different types the accident; it can be seen that the security situation of residential construction is still serious recently, the confusion of construction management, the lack of supervision and the poor quality of construction workers are the main cause of accidents of residential construction. Taking Tower crane as an example, this paper proposed a method for safety risks control based on BIM, it has certain reference value for controlling the safety risks of residential engineering.

1 INTRODUCTION

With the rapid development of China's economic construction, the urbanization develops rapidly, more and more residential projects are under constructing. Consequently, the security issues have been revealed. According to relevant statistics, during 2012, 487 engineering safety accidents happened in China and 624 persons lost their lives in those accidents. Relatively serious accidents and more serious accidents happened 29 times in total and 121 persons died, which was 4 times more and 11 persons more compared with last year respectively, increased by 16% and 10% respectively. Residential engineering has always been high construction density and high building height, followed by high-risk project such as deep foundation pit, high crane, construction elevator, high formwork. The accidents happen during the construction will cause great loss of property, serious casualties and negative social influence undoubtedly. For example: on December 21, 2012, the ninth building's foundation pit collapsed during construction in Yunmeng County, Hubei Province, which led to 3 persons being killed; on November 15, 2012, a crane collapsed in Fushan, Shandong Province, three persons were killed and three were badly hurt in the accident; on September 14, 2012, the construction elevator felled from 30th floor in a residential project called "east lake garden" in Wuhan, 19 persons lost their lives in the accident; on July 14, 2012, template support system collapsed suddenly during casting concrete in the main hall roof in "Mansion Town accompanied by Hok", Zouping County, Binzhou City, Shandong Province, 4 persons died and 8 persons hurt in the accident.

This article is based on the accident cases collected from residential engineering (Liu Hui et al. 2010, Wang Jun et al. 2011), statistical analysis was conducted about the cause of the accident, trying to reveal the occurrence rules and characteristics of the accident and offer basic data for residential engineering safety risks control.

2 PROMINENT ISSUES

The accident cases of residential engineering were collected by using public media such as publicly available papers, openly published Journal, the relevant reports from mass media and so on.

Table 1. Annual distribution of the accidents.

Year	Before 2005	2005	2006	2007	2008
Quantity	20	7	13	10	12
Year	2009	2010	2011	2012	2013
Quantity	25	38	49	21	16

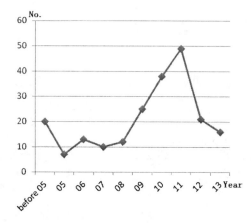

Figure 1. Annual distribution of the accidents.

By conducting statistical analysis, the accidents are concentrated on foundation pit, crane, formwork and construction elevator.

2.1 *Data distribution*

According to accident cases collected, the quantity of the accident happened each year had been added up to show the tendency of the accident's occurrence. The annual distribution of the quantity of accidents can be seen in Table 1 and Figure 1.

As is showed in Table 1 and Figure 1, with the development of residential engineering, the quantity of the accident is becoming larger and larger. Before the year of 2008, the amount of the accident is much smaller, while after the year of 2008, owing to the implementation of large-scale economic stimulus program, lots of residential engineering are under construction, the amount of accidents increased rapidly year by year, it reached its peak in year of 2011, after 2011, the amount of accidents began to decrease, but the amount is still much larger than that of the year before 2008, the security situation of residential engineering is still very severe.

2.2 *Severity and amount of loss*

According to the Reporting and Investigation Disposal Ordinance of Production Safety Accident, the accident grade is classified based on the casualties. From both the accident type and the casualties the statistical analysis is conducted.

It is showed in Table 2 and Figure 2 that most accidents are ordinary accidents, which accounted for 77% of the total, besides, relatively serious accidents accounted for 21% of the total, which caused the greatest casualty loss. Extraordinarily serious accidents and serious accidents were fewer, while in case they happen, they must be vicious accidents that cause great casualties.

Depending on the different types of the accidents, the formwork accidents are consisted of 59 ordinary accidents, 35 relatively serious accidents and 3 serious accidents. The probability of the relatively serious accidents and the serious accidents is apparently higher than the others. When it

Table 2. Statistics at the grade of accidents.

Accident grade	Accident type							
	Tower crane		Formwork		Elevator		Foundation pit	
Amount of Accident								
							8	
Relatively serious	13		35		1		2	
Serious	/		3		2		/	
Extraordinarily serious	/		/		/		/	
Total amount	100		97		36		10	
Type	Death	Injury	Death	Injury	Death	Injury	Death	Injury
Casualties								
Ordinary	76	136	54	176	3	7	7	/
Relatively serious	63	25	191	346	0	19	11	/
Serious	/	/	54	50	31	/	/	/
Extraordinarily serious	/	/	/	/	/	/	/	/
Total casualties	139	161	299	572	34	26	18	0

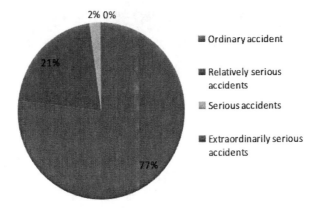

Figure 2. The statistic distribution of the accident grade.

comes to the casualties, 299 persons were killed and 572 persons were injured by the formwork accidents, which is also greater than the others.

2.3 Accident type

From four aspects of foundation pit, tower crane, formwork and construction elevator, the situation of the residential engineering is as follows:

2.3.1 Foundation pit accidents

During the construction of residential engineering, the excavation and maintenance of the foundation pit are essential. The unreasonable excavation procedures, the insufficient dewatering and maintenance always caused the foundation pit collapsing. Based on the occurrence reason, the foundation pit accidents can be divided into seepage damage, supporting instability, landslide, skirting damage and Inrushing damage.

2.3.2 *Tower crane accidents*

Tower crane is the most widely used lifting equipment in construction site, which is always used to lift rebar, wood, scaffold tube and so on. Tower crane has been the essential construction equipment in the construction of the multi-storey and high-rise residential engineering. The accidents caused by non-standard installation, the poor skills and safety awareness of the workers, the mistake of the management personnel are becoming a very common accident type among the residential engineering accidents. Tower crane accidents are mainly consisted with collapse, fracture, falling and wire rope breaking.

2.3.3 *Elevator accidents*

In the construction industry, the elevator always plays a very important role in vertical transportation, which has been widely used in residential engineering construction. Elevator accidents can be mainly divided into construction elevator accidents and residential elevator accidents. As the construction elevator is usually accompanied with simple facilities and insufficient security measures, once the elevator falls, great casualties will be caused.

2.3.4 *Formwork and scaffold accidents*

As the reinforced concrete structure is the most common structure in residential engineering in our country, a considerable application of formwork and scaffolding is essential. The collapse of formwork and scaffolding has become one of the most common residential engineering accidents. Formwork collapse mainly refers to instability collapse of formwork support system collapse caused by the irregularities installation which can't meet the requirements of the bearing capacity. The collapse of the scaffold mainly refers to the collapse of the scaffold system bearing external loads caused by the unreasonable installation, the insufficient support and so on.

3 CAUSES AND CONTROL

3.1 *Causes analysis*

According to the statistical analysis of the causes of the accident cases, it can be found that the causes can mainly divided into several types as follows: the human factor, the unstandardized installation and disassembly, the unqualified qualifications, the majeure force, the quality defects of the construction equipment, the construction management and the lack of supervision (Wu Sheng-ying et al. 2011).

a) the human factor: it refers to the operational errors of the worker, the lack of the safety awareness and the illegal operation.
b) the unstandardized installation and disassembly: it means that the installation and disassembly do not meet the specifications, the capacity isn't calculated before the installation and disassembly and the special construction program isn't formulated before implementing.
c) the unqualified qualifications: the contractor contracted projects beyond the qualifications or without the qualifications and the special operations personnel have no special operations qualification are contained.
d) the majeure force: the natural phenomena of storm and earthquake are included, which can cause construction accidents.
e) the quality defects of the construction equipment: it mainly refers to the quality defects of the construction equipment and the insufficient connection strength.
f) the construction management: it means that the construction unit doesn't prepare construction organization, the safe production responsibility system is not perfect, the management of the construction site is chaotic and the necessary maintenance of the construction machinery isn't carried out.

Table 3. Statistical analysis of the cause.

Causes	a)	b)	c)	d)	e)	f)	g)
Account	22	33	10	2	8	32	26

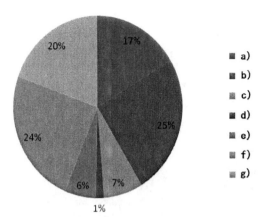

Figure 3. The probability of the accident.

g) the lack of supervision: including the inadequate supervision of the supervision units, the supervisors' being absent, being not timely of detecting and solving the risk, besides, the lack of supervision from the related government departments is also included.

The statistical analysis of the causes of the collected accident cases is conducted based on the seven aspects above; the consequence is showed in Table 3 and Figure 3.

As is showed in Table 3 and Figure 3, the unstandardized installation and disassembly and the construction management appear 33 and 32 times respectively, which account for 25% and 24% of the total respectively. They cause the most accidents compared with the others. Besides, the human factor and the lack of supervision appear 22 and 26 times respectively, accounting for 17% and 20% of the total, which also cause relatively more accidents. Finally, the unqualified qualifications and the quality defects of the construction equipment appear 10 and 8 times, accounting for 7% and 6% of the total. As the engineering accidents always cause great casualties and property damage, both of causes should be also paid attention to avoid construction accidents.

3.2 Preventive action

Based on the above analysis, the main factors of the accident includes human, management, physical, and regulatory, etc., so the following recommendations are made to ensure safe construction and reduce security incidents

3.2.1 Improving the workers' technical level and safety awareness

At present, the safety awareness and operational level of the operations personnel at construction site are very poor, the construction enterprises should strengthen the skills and safety awareness of the construction workers. the critical measures should be taken are as follows:

1) Technical level training should be carried out timely to improve the workers' occupational skill levels
2) Safety education should be carried out regularly to enhance the staffs' safety awareness.

Table 4. Tower Crane working risk list (based on accidents collection).

Accident Type	Reason & Risk	Key factors
Tower Crane broken	Overweight Lifting	Overloading
	Operating mistake, integral slipped and back clasp fracture	Operation
	Gale blew down	Windy
	Serious violation erection procedure	Violation operation
	Without qualification or no work license	No qualification or license
	Mechanical Failure	Mechanical Failure
	Personnel operating irregularities	Personnel
	Improper Maintenance	Maintenance
Steel wire fracture	Replace wire rope is not timely	Accessories Maintenance
	The unqualified wire rope	Material Quality
Tower Crane Collapse	The installation is uncensored, the competent authorities is not informed before installation.	Absence of Supervision
	Operation of non-standard installation and faults, defects	Operation mistakes
	The bolt screw and nut to fix the tower crane is not standard, which cannot bear the axial force and overturning of tower crane	Quality Defects
	During the climbing installation, the construction personnel violated the operation procedure	Violated the Operation Procedure
	The unqualified welding	Welding Fault
	Construction enterprises ignored management of the tower crane.	
	The wind pressure.	Wind
	The nonstandard installation and the illegal operation	Nonstandard or Illegal
	The site operations command violates the rules and the regulations.	Violate the commands.
	Tower crane operations, collision led to the loss of balance and fell.	Loss of Balance.
	Tower bottom bearing capacity is poor.	Poor Bearing.
	Demolishing the tower crane without the hook up to the tower crane beam, leading to the collapse of losing balance.	Losing Balance.
	Lack of supervision, inspection and safety supervision is not in place.	Without effective supervision
Tower Crane Fall	The quality problems of tower crane	Quality
	The crane wire rope fracture, the tower arm fall.	Wire rope fracture
	No safety and technical clarification before maintenance.	Without clarification
	Tower crane construction teams operate illegally.	Operate Illegally.
	Construction behavior against rules is not found in time.	Behavior Against Rules.
	High-altitude operations, without safety belts fasten.	Lack of Safety Protection.
Other	Hanging overload	Over Load.
	Jib direction and tower crane is not vertical, results the force unbalance.	Force Unbalance.
	The crane arm collides with voltage and results fire.	Collision and Fire.

3.2.2 *Strengthening site construction management level*

Management deficiencies can cause equipment failure or human mistakes, the contractor should organize the labor rationally, improve safety procedures, strengthen safety inspection and guidance and routine maintenance of construction machinery, enhance education and training efforts, implement the accident prevention measures earnestly.

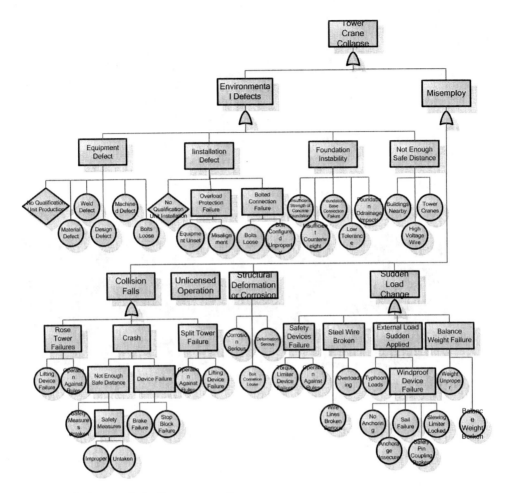

Figure 4. Fault tree analysis of tower crane collapse.

3.2.3 *Strengthening supervision*

First, we should strengthen supervision on the subcontractor and strict identification of illegal subcontracting and ultra-qualification; Again, supervisor units should strengthen the supervision of the work site, including the review of constructor qualification, construction design, key nodes place oversight and daily supervision; Finally, to strengthen the administrative supervision of production safety, strengthen basic construction procedures and formalities seriousness, departments at all levels should strictly not allowed without formalities project started, vigilant and prevent bidding, corruption in the bidding process.

4 RISK ASSESSMENT BASED ON BIM (EX. TOWER CRANE)

4.1 *Tower crane accident*

Accident generally results great loose or casualties, such as Zhejiang Wenzhou, a construction site crane collapse accident in July 29th 2011, two workers was hurt, and the crane collapsed hit on a 5-storey residential buildings, causing the building body parts damage. Here, taken the tower crane as an example to introduce the application of BIM (Building Information Model) and risk assessment.

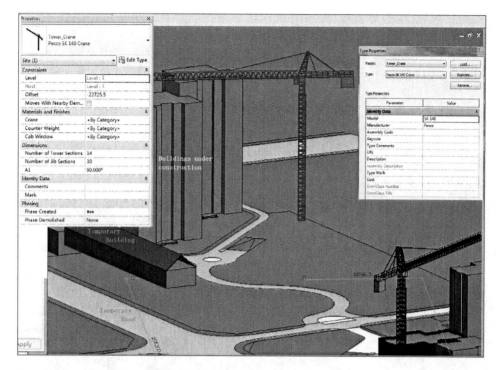

Figure 5. Tower crane layout and operation.

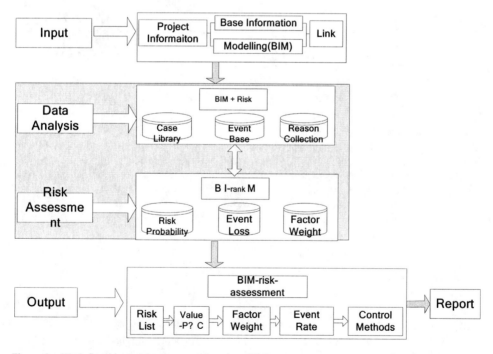

Figure 6. Work flow chart-risk assessment based on BIM.

4.2 *List the risk and key factors*

Based on the collected examples of tower crane accidents, the risk event and key factors are recognized as follows.

Use the method of Fault Tree Analysis (Wang Yu-ping et al. 2007, Zhou Hong-bo et al. 2009), and based on analysis of the accident, to create the Tower Crane Collapse Fault Tree, see Figure 4.

4.3 *BIM and risk assessment*

From the analysis of the collected crane accident, risk events and key factors can been listed, and the fault tree of tower crane collapse also is created. How to use these valuable information or data? BIM (Building Information Model) plus Risk assessment may be a better way to control the Crane Collapse accident under construction. By using BIM technology, tower crane models are created before or under construction, and risk information are combined in its. From the example model, see Figure 5, the tower crane working environment and condition can be planning and simulating, so risk can be early recognized, and then better tendency will be selected and bad situation will be controlled.

BIM plus Risk Assessment system also been designed, the main work flow is shown as follow Figure 6. Main Steps include: a) Input Data or Information; b) Data Analysis; c) Risk Assessment; d) Result Output or Create Report.

5 CONCLUSION

(1) It can be seen from the accident occurrence tendency that after the year of 2007, with the rapidly development of the urbanization of our country, the safety accident of residential project increase very fast, since 2010 the number of accidents has dropped, but the security situation overall residential construction is still grim;
(2) according to the statistical analysis of the causes of the safety accidents, mounting and dismounting construction management problems, the lack of supervision, illegal operation and equipment not standard and so on are the main cause of security, strengthening construction management and supervision and improving the quality of the staff are the most important task to ensure the safety accidents not occurring;
(3) from the casualties caused by different types of accidents, formwork and scaffolding accidents is obviously higher than that of other types of security accidents, it should be paid sufficient attention to prevent the accidents happening in residential construction;
(4) Taking the tower project as an example, the construction safety risk assessment based on BIM is introduced, by accident analysis of tower crane, the engineering risk list is obtained, the safety accident control can be realized by studying the risk BIM model combined with pre assessment, dynamic tracking and control measures.

ACKNOWLEDGEMENT

The research was supported by the National Technological Support Program for the 12th-Five-Year Plan of China (2012BAJ03B07).

REFERENCES

Liu Hui, Zhang Zhi-chao, Wang Lin-juan. Statistical Analysis of Tunnel Construction Accidents In China 2004-2008. Journal Of China Safety Science [J]. 2010. 1.
Wang Jun, Zheng Rong-hu, Lv Bao-he. The Construction Death Accident Statistical Analysis and Countermeasures of A City Area, Journal of Safety Science and Technology [J]. 2011, 3.

Wang Yu-ping, Bai Jie. Fault Tree Analysis on the Accident Reasons of Tower Crane Things Falling to Person and the Control Measures of the Accident, Ceramic research and occupation education [J]. 2007. 3.

Wu Sheng-ying. Discussions about Construction Safety Risk Analysis And Control, Building Safety [J]. 2011. 8.

Zhou Hong-bo, Cai Lai-bing, Gao Wen-jie. Statistical Analysis Of The Accidents At Foundation Pit Of The Urban Mass Rail Transit Station, Hydrogeology and Engineering Geology [J]. 2009. 6(2).

Study on the harmony of emergency management to urban major accidents

Dianjian Huang
National Research Center of Work Safety Administration, Beijing, China

ABSTRACT: The biology origin of emergency management to urban major accidents is discussed. It is point out that the crux of emergency management is cooperation. The mechanism of urban emergency cooperation system is studied. The constitution of urban emergency system is studied. It is composed of inspection and alarming system, social controlling system, public response system, emergency rescuing system and resource guarantee system. The corporation relation model of urban emergency system is studied. The harmony model is built on the basis of self-organization theory. It provide basis to quantify harmony of emergency management and is provided with theory significance.

1 INTRODUCTION

From the point of biology, everybody has the appetence of stress reaction. It is to say that the apparatus of body can be adjust automatically when one body suddenly suffering from severe stimulation or great damage of environment, and make them harmony and keep the best state to confront the attack from environment. This is the appetence of self-protection. Stress reaction is all the strained states of non-different, which arose by bad effect to animals due to stressor. The concept was brought forward by H. Selye in 1936 on the basis of the reaction of body in the case of cold condition. Stress reaction sometimes is local, sometimes is entire, however, it is general the uniform reaction of economy in despite of which reaction of both. But stress reaction also has destruction reaction to animals, which is called adaptable disease. When animals are in the stress state, the system activities of the back of thalamencephalon, pituitary and adrenal gland cortex boost up, thus hormone excretion of these apparatuses increases. In addition, it also affects immunity system, then the recovery mechanism of economy also change either good or bad. Many people will put up abnormal capability and do the thing that they never do in normal time when facing accidents suddenly. But there is another condition of stress reaction. When accidents happen, it restricts people's capability of replying and they can't do the thing that they can do easily in normal time. The reason is that the apparatuses of body can be cooperated in special state or not. And governments and enterprises also have the capability of the stress action. In normal condition, every department and institute work normally in order, society develops safely and steadily, living and working of people are in peace and contentment. However, the level and capability of stress reaction of governments and enterprises will be test once accident happens, which threaten the safety of life and property.

2 THE MECHANISM OF EMERGENCY COOPERATION TO URBAN MAJOR ACCIDENTS

The key of emergency to urban major accidents is cooperation, so the study of mechanism of emergency cooperation to urban major accidents has important significance. There are many understandings of cooperation mechanism for many social and economical scholars in our country.

The mechanism of emergency cooperation to urban major accidents can be understood from the two points as following (F. Fiedrich, 2000).

1) In emergency to urban major accidents, cooperation depends on the being of emergency main body, which has cooperation capability and brings cooperation behavior. Emergency cooperation has intention which is the behavior aim to cooperate main body. Cooperation behavior has uniformity, which emphasize that the reasonable relations are kept among idiographic cooperation behavior of main body.
2) In emergency to urban major accidents, cooperation is that every compositive part of system is in balance state at the volley, change for corresponding information, material, energy among each other, to submit the uniformity of the whole system in space and time, keep the whole system in order, and keep homeostasis state with the environment in which the system place.

Though the two understandings are different, both of them contain the understanding of main condition to cooperation. Emergency system to urban major accidents is as a complex unit which is composed of diversiform factors. The development of emergency system depends on establishing and achieving the relation of interaction, inter-effect and control ability among these factors.

Sum up, the mechanism of emergency cooperation to urban major accidents determines the necessary to study of emergency system to urban major accidents through using cooperation theory in self-organization theory. So it can achieve better effects through the study of cooperation of emergency to urban major accidents with using nonlinear method of cooperation subject.

3 COOPERATION THEORY OF EMERGENCY SYSTEM TO URBAN MAJOR ACCIDENTS

Self-organization phenomenon is that the macroscopical phenomenon are in sequence formed spontaneously in nature. A mass of the phenomenon is existent in nature. The main content of self-organization theory includes cooperation theory and so on, which attempt to communicate with physics, biology, and even social science. It shows the phenomenon of biological and social field in order at certain degree. Cooperation theory thinks that different systems having dissimilar attributes, but in the whole environment, there are relations of inter-effect and inter-cooperation among every system. It also includes usual social phenomenon, such as inter-cooperation among different unit, collaboration of relations among department, inter-competition among enterprise, and inter-interference and inter-restriction in system and so on (Huang Dianjian, 2003).

The emergency system to urban major accidents is a complex system, which is composed of inspection and alarming system, social controlling system, public response system, emergency rescuing system and resource guarantee system. These five subsystems compose the organic integer of emergency system to urban major accidents. Inspection and alarming subsystem is the precondition of the whole emergency process. Only having real-time inspection and exact alarming to all kinds of accidents, we can reduce disaster efficiently and make stable basis for successful emergency. Social controlling subsystem is the pivotal factor to achieve the aim of emergency management, playing the functions of mobilizing social power and collocating social resource in emergency management. Public is not only the object of emergency management, but also is the main body to which emergency management exerts functions. Local emergency rescuing subsystem plays important function in the whole emergency process, which is the emphasis in the process. Complete and enough emergency resource is the basis and important tache of disaster emergency working. The function of emergency system to urban major accidents depends on emergency hardware and emergency software to urban major accidents and the inter-relations, which not only come down to direction system with uniformity, efficiency and authority, but also come down to

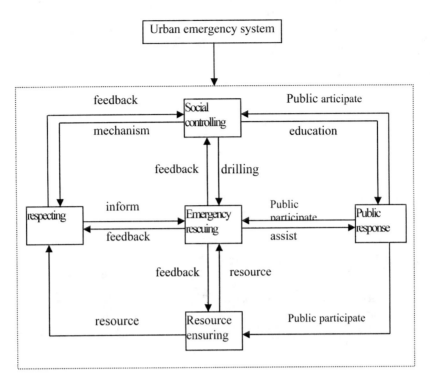

Figure 1. The model to cooperation relation.

the status and level of multifold emergency resource, like credible and sensitive communication net, firm and practical equipment and establishment for preventing disaster and rescuing disaster, rapid and capable emergency rescuing troop, and experts troop for consultation with extract techno-knowledge and so on. The model to cooperation relation among every subsystem of emergency system to urban major accidents is described fig. 1.

4 THE COOPERATION MODEL OF EMERGENCY SYSTEM TO URBAN GRAVE ACCIDENTS

The cooperation activities of the emergency system to urban grave accidents include the cooperation in subsystems and that among subsystems. The models being built from the two aspects are as follows (Wu Zongzhi and Liu Mao, 2003).

The Cooperation Model of Subsystems

We here divide the emergency system to urban grave accidents into m subsystems. Let n_i denote the ith subsystem's index number, O_{ij} denote the ith subsystem's jth sub-target value, where $i = 1, 2, \ldots, m$ and $j = 1, 2, \ldots, n_i$.

The target value, O_{ij}, can be described as follows:

In some cases, we hope that the value of the corresponding index gets as big as possible, which is the bigger-expect index, then O_{ij} is the upper limit of the index. When we hope that the value of the corresponding index gets as small as possible, which is the smaller-expect index, O_{ij} is the lower limit of the index. In other situations we may hope that the value of the index should be close to some mesial magnitude, which is the better-expect index, then O_{ij} is the best target value.

X_{ij} is used to represent the value of the ith subsystem's jth index variable. Here we firstly define U_{ij} as the efficacy coefficient of the ith subsystem's jth index, which is described as the following formula:

$$U_{ij} = \begin{cases} \dfrac{1-e^{k_1(O_{ij}-X_{ij})}}{1+e^{k_1(O_{ij}-X_{ij})}} & \text{(when } X_{ij} \text{ denotes bigger-expect index)} \\ \dfrac{1-e^{k_2(O_{ij}-X_{ij})}}{1+e^{k_2(O_{ij}-X_{ij})}} & \text{(when } X_{ij} \text{ denotes smaller-expect index)} \\ \dfrac{1-e^{k_3(O_{ij}-X_{ij})^2}}{1+e^{k_3(O_{ij}-X_{ij})^2}} & \text{(when } X_{ij} \text{ denotes better-expect index)} \end{cases} \quad (1)$$

In above formula, k_1, k_2 and k_3 are coefficients, which value could be constants above zero. The function of the three coefficients is to adjust the sensitivity of data's calculation. The bigger k is, the better the result's sensitivity is.

Note that U_{ij} represents the measurement of the ith subsystem's jth index's contribution to the efficacy of the emergency system to urban grave accidents.

The characteristics of the efficacy coefficient which is constructed according to above formula are as follows: U_{ij} reflects the degree of satisfaction of the index's reaching target in each subsystem. $U_{ij} = -1$ means it is the most unsatisfying, while $U_{ij} = +1$ means the most satisfying. When $X_{ij} = O_{ij}$, both the bigger-expect index and the smaller-expect index have $U_{ij} = 0$, which represent they have reached the basic requirements. When $X_{ij} = O_{ij}$, the better-expect index has $U_{ij} = 1$, which means it has satisfactorily reached the target requirement. The bigger U_{ij} is, the bigger the efficacy is, and the more satisfying it is, where

$$U_{ij} \in (-1, +1).$$

Efficacy coefficient shows the capability or efficiency which the emergency system to urban grave accidents possesses when achieving emergency goals. And the efficacy coefficient may be used to reflect the cooperatively developing situation in subsystems. The inner cooperatively developing coefficient of subsystem is:

$$U_i = \sum_{j=1}^{n_i} \lambda_{ij} U_{ij} \quad (i=1, 2, \ldots, m) \quad (2)$$

where λ_{ij} is index weighing, and

$$\sum_{j=1}^{n_i} \lambda_{ij} = 1$$

5 THE COOPERATION MODEL AMONG SUBSYSTEMS

If we want to know the cooperatively developing situation among subsystems of the emergency system to urban grave accidents, we could firstly analyze the interactions of those subsystems' indexes, then calculate the interaction among those subsystems, and finally review the cooperation status among subsystems. Therefore, we analyze the interaction among indexes of subsystems.

Let α_{ij}^{pq} denote the effect coefficient of the ith subsystem's jth index being effected by the pth subsystem's qth index, where $i, p = 1, 2, \ldots, m$ and $j = 1, 2, \ldots, n_p$. α_{ij}^{pq} can be acquired by experts' qualitative analysis or associative analysis, where $\alpha_{ij}^{pq} \in (-1, +1)$. Let $\alpha_{ij}^{pq} > 0$ represents interpromotion, $\alpha_{ij}^{pq} < 0$ represents mutual inhibition and $\alpha_{ij}^{pq} = 0$ represents no effect. Specially, $\alpha_{ij}^{pq} = 1$ and $\alpha_{ij}^{pq} \neq \alpha_{pq}^{ij}$, when $i \neq p$ and $p \neq j$.

Calculate whole effect of each subsystem's index being effected by other subsystems

Let α_{ij}^p represents the effect of all the pth subsystem's indexes acting on the ith subsystem's jth index. It is:

$$\alpha_{ij}^p = \sum_{q=1}^{n_p} \alpha_{ij}^{pq} X_{pq} \tag{3}$$

While the sum of the other $m - 1$ subsystems' effect on the ith subsystem's jth index can be described as follows:

$$\alpha_{ij}^p = \sum_{p=1}^{m}\sum_{q=1}^{n_p} \alpha_{ij}^{pq} X_{pq} \quad (i=1,2,\ldots,m;\; j=1,2,\ldots n_i) \tag{4}$$

The Combined Effect of the Subsystem's Index Being Effected by Other Subsystems, the Cooperation Index among Subsystems' Indexes.

The combined effect of the subsystem's index being affected by other subsystems, the cooperation index among subsystems' indexes, C_{ij} can be given by the following formula:

$$C_{ij} = \alpha_{ij} \Big/ \sum_{p=1}^{m}\sum_{q=1}^{n_p} \alpha_{pq} \quad (i=1, 2,\ldots, m;\; j=1,2,\ldots n_i) \tag{5}$$

The Cooperation Index among Subsystems

On the base of having known C_{ij}, we could calculate the cooperation index among subsystems, C_i:

$$C_i = \sum_{j=1}^{n_i} C_{ij} X_{ij} \quad (i=1, 2\ldots m) \tag{6}$$

The Whole Cooperation Coefficient of Emergency System to Urban Grave Accidents

We can review the cooperation status of emergency system to urban grave accidents according to both the system's inner cooperation coefficient and the cooperation index among subsystems. Let C denote the cooperation index of the emergency system to urban grave accidents:

$$C = \sum_{i=1}^{m} \beta_i (u_{i1} U_i + u_{i2} C_i) \tag{7}$$

where the weighing meets:

$$\sum_{i=1}^{m} \beta_i = 1, \quad u_{i1} + u_{i2} = 1 \tag{8}$$

6 CONCLUSIONS

In safety field, some questions which are difficult to solve temporarily, usually are looked after the resolving methods from the origin or biology. So the author discuss the biology origin of emergency management to urban major accidents, and point out the same between emergency system to urban major accidents and stress reaction of body, which the key of both is cooperation, in term of this thought. This article discuss the mechanism of emergency cooperation to major accidents, and build the model to cooperation relation of emergency system to urban major accidents, then we have profound understanding to cooperation of emergency to major accidents. Finally, the article build the model to cooperation of emergency to urban major accidents, which provide basis to quantify harmony of emergency management and is proved with theory significance.

REFERENCES

F. Fiedrich, (2000). Optimized resource allocation for emergency response after earthquake disasters. Safety Science 35:41–47
Huang Dianjian, (2003). SARS globalization and the study on urban emergency mechanism development [J]. Urban Development Study, 2003, (3):16–20
Huang Dianjian, (2003). UN major hazards controlling system [J]. Modern Safety Health, 2003(6):28–32
Wu Zongzhi and Liu Mao, (2003). Major Incident Emergency Rescuing System and Planning Guideline [M]. Beijing: Metallurgy Industry Press, 2003, 1–70
Xiu Jigang, (2001). Quakeproof, disaster reducing and community building [J]. City and Disaster Reducing, 2001 (4):16–18

Advanced Engineering and Technology – Xie & Huang (Eds)
© 2014 Taylor & Francis Group, London, ISBN 978-1-138-02636-0

The prediction of razor clams shelf-life based on kinetics model

Qi He & Xiang-dong Lin
Food Science College, Hainan University, Haikou, China

Le Wang
Sea College, Hainan University, Haikou, China

Lin-na Feng & Ming Xiong
Food Science College, Hainan University, Haikou, China

ABSTRACT: This paper focuses on shelf-life prediction of razor clams in frozen storage by water activity method. The changes of texture, pH, conductivity, water activity and TVB-N of razor clams stored at −5°C, −20°C and −40°C along with storage time were measured. According to their correlation, water activity was chosen as the optimal indicator, the reaction activation energy Ea was 53.53 kJ/mol, pre-exponential factor k_0 was 215.64 and water activity up to 0.20 was the shelf-life end. Based on the experimental results, the predicted shelf-life values were close to the real measured values. This experiment results showed that water activity could be used to predict the shelf-life of razor clams accurately.

1 MATERIALS AND METHODS

1.1 Material

Razor clams: Fresh, purchased from San-xi agricultural products trade market in Haikou city. Length: 4.54–5.68 cm; width: 1.62–2.01 cm, without damage. Washed them by clean water and stored them at −5°C, −20°C and −40°C (Le, W. 2013).

1.2 Methods

1.2.1 Quality measured

PH value was measured by precision pH meter method (Jing, X, et al. 2011). Samples were thawed, weighed 5 g sample and mixed with 50 mL distilled water.

Water activity was measured by water activity portable instrument method (Qiang, Z, et al. 2011). Samples were thawed, weighed 2–3 g sample into measurement chamber.

Electrical conductivity was measured by conductivity meter method (Li-rong, S, et al. 2011). Samples were thawed before the measured.

The preservation of TVB-N was measured by semi micro Kjeldahl method (Wei, Z, et al. 2011; Xiao-mian, C, et al. 2011). Samples were thawed, weighed 10 g sample and mixed with 50 mL distilled water.

Sensory evaluation method was used to estimate quality of frozen sample (Herbert, S, et al. 2004). The assessment team composed of 10 members and sensory value took the average. The aspects of sensory description changed between 0 to 5, 5 point represented the best quality and 0 points represented total corruption.

Texture was tested by texture analyzer (Bourne, M. C. 2002). Text conditions were as followed. Text method: TPA (Texture Profile Analysis); probe type: TA41; target value: 2.0 mm; the trigger point load: 5 g; the speed: 2.00 mm/s.

Texture analyzer was connected with the computer through Texture Pro CT software, and ONEView1.0 software was used to analysis the data. Selected parameters included the first cycle hardness (the maximum deformation place in sample compression process); springiness (the ratio between two compression peak height) and cohesion (two peak area ratio, selection the results with instrument error correction).

1.2.2 Correlation analysis

Pearson computing correlation coefficients between quality changes and each index was analyzed by SPSS13.0 software (SPSS Company, USA). The bigger correlation coefficient was, the higher correlation can be considered (Li-ping, Z, et al. 2007).

1.2.3 Shelf life prediction

One of parameter was chosen as the key quality factor, its curves changed by the storage time t can be used to calculate constant change rate k, so the Arrhenius relationship between k and storage temperature T can be established. So the sensory end of key factor and the corresponding shelf-life in the corresponding temperature can be obtained as followed.

$$lnk = (-E_a/RT) + lnk_0 \qquad (1)$$

From equation (1), E_a and k_0 can be measured, so the rate constants k at different storage temperature can be calculated (Xiao-qin, Y, et al. 2007).

2 RESULTS AND ANALYSIS

2.1 Changes of quality parameters

As Figure 1 (1), at $-5°C$, initial pH was around 6.9, then it increased to 7.3 after 20 days. Followed, the trend began to go down, after 40 days storage, it went down to 7.1. The reason of this phenomenon was that amino acids in the samples would decompose and produced alkaline volatilenitrogen

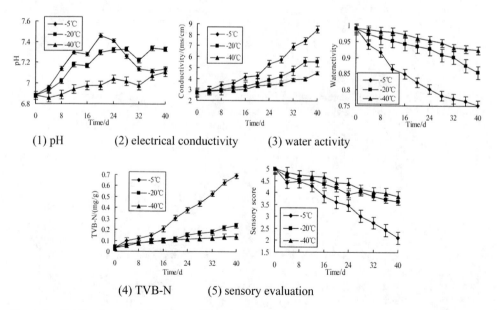

Figure 1. Quality change trends of razor clams.

in preservation period, so the pH value rose at the beginning. With the passage of time, glycollysis in clams broke down and produced lactic acid, ATP and creatine produced phosphate and other substances produced phosphoric acid, those acids produced made pH value decreased. While at −20°C and −40°C, the pH value remained in upward trend because experimental time was short.

Electrical conductivity was closely related to the storage temperature because of the material electrolyte ionization degree and ion migration velocity. We can see the trend from Figure 1(2).

As Figure 1(3), when the samples preserved at different temperatures, their water activity (thawed samples) decreased slowly with storage time increased, when storage temperature was higher, the changes were more obvious.

We can see the trend from Figure 1(4), TVB-N value increased with the storage time increased, and the rate of change was rapid. The reason was that TVB-N value was associated with the propagation of bacteria and decomposition of protein during the samples storage (Cobb, B. F, et al. 1973).

In the freezing process, the sensory evaluation scores were decreased over time gradually. Figure 1(5) shows the results. When razor clams preserved at −40°C, the change speed of its sensory quality was slower than −20°C and −5°C significantly.

2.2 Change of texture

As Figure 2(1), the first cycle hardness amplitude stress in the range of 25–70 g. (When hardness value was 60 g stress, the corresponding standard peak pressure was 11768 dyn/cm^2) The trend was irregular.

As Figure 2(2), elasticity decreased with the time prolonged. At −5°C, the trend was significantly higher than that of −20°C and −40°C.

As Figure 2(3), the trend of cohesion change was similar to that of springiness.

2.3 The Pearson correlation of quality parameters

As shown in Table 1, Pearson correlation coefficients of hardness and pH is relatively low (most of them less than 0.9), so we excluded them on the later. According to the experimental results, water activity has best Pearson correlation (most of r > 0.95), so we choose it as the key quality parameters to predict quality changes and shelf-life.

2.4 The shelf life prediction

2.4.1 Dynamic model of quality

Based on the relationship between $-\ln k$ and $-1/T$, linear equation $y = -6.4258x + 6.2459$ ($R^2 = 0.9047$) can be found out (as Table 2), so we can find that the activation energy

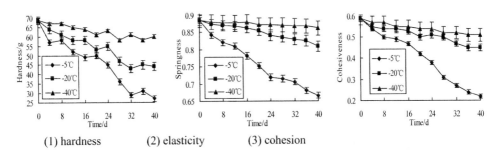

(1) hardness (2) elasticity (3) cohesion

Figure 2. Texture change trends of razor clams.

Table 1. Pearson correlation coefficients between the parameters at −5°C, −20°C, −40°C temperature.

	Hardness (value: −5°C/ −20°C/ −40°C)	Springiness (value: −5°C/ −20°C/ −40°C)	Cohesiveness (value: −5°C/ −20°C/ −40°C)	pH (value: −5°C/ −20°C/ −40°C)	Conductivity (value: −5°C/ −20°C/ −40°C)	Water activity (value: −5°C/ −20°C/ −40°C)	TVB-N (value: −5°C/ −20°C/ −40°C)	Sensory evaluation (value: −5°C/ −20°C/ −40°C)
Hardness Springiness	0.958 0.949 0.863							
Cohesiveness	0.980 0.961 0.958	0.976 0.959 0.928						
pH	−0.266 −0.857 −0.954	−0.436 −0.834 −0.865	−0.262 −0.840 −0.896					
Conductivity	−0.951 −0.920 −0.821	−0.944 −0.965 −0.961	−0.979 −0.954 −0.912	0.150 0.762 0.882				
Water activity	0.957 0.945 0.897	0.982 0.983 0.946	0.948 0.970 0.955	−0.536 −0.837 −0.891	−0.897 −0.976 −0.971			
TVB-N	−0.961 −0.949 −0.922	−0.968 −0.992 −0.920	−0.990 −0.962 −0.976	0.219 0.857 0.889	0.991 0.976 0.875	−0.925 −0.990 −0.915		
Sensory evaluation	0.980 0.946 0.919	0.978 0.975 0.965	0.989 0.972 0.973	−0.274 −0.886 −0.887	−0.978 −0.929 −0.967	0.950 0.957 0.992	−0.991 −0.974 −0.938	

Table 2. Regressive equation of water activity of razor clams at different temperature.

Temperature/°C	Regression equation	Regression coefficient R^2	k
−5	$y = 0.0068x + 0.0411$	0.955	0.0068
−20	$y = 0.0033x + 0.0082$	0.965	0.0033
−40	$y = 0.0019x + 0.0028$	0.975	0.0019

$E_a = 53.424$ kJ/mol, the pre-exponential factor $k_0 = 515.893$, so we can obtain the Arrhenius equation between cohesion rate constant k and storage temperature T:

$$k = 515.893 \exp(-53424/RT) \qquad (2)$$

When sample reach the corruption point, the corresponding water activity is 0.78–0.83, so we choose 0.80 as the quality change point of water activity in our study.

2.4.2 Prediction of shelf-life based on the kinetic model and validation

According to the dynamic model, the shelf-life of razor clams stored at −5, −20, −40°C or other temperatures can be forecasted and compared with the actual value, as the results in Table 3

Table 3. Predicted and observed shelf-life of razor clams during storage at different temperature.

Temperature/°C	Predicted value by Arrhenius equation/d	Measured value/d	The error between predicted values and measured value/%
0	6.53	7	7.20
−5	26.05	25	−4.03
−10	54.14	57	5.28
−20	112.13	108	−3.68
−40	965.70	−	−

(the value of −40°C has not got yet because it is too large). The most of relative errors between Arrhenius forecast value and actual measured value were less than 5 present.

The experiment data prove water activity can be used to predict the shelf-life of razor clams under frozen conditions.

3 CONCLUSION

According to quality analysis of razor clams, water activity can used to evaluate quality changes and predict shelf-life. As the experiment result: water activity 0.80 was the corruption end point of shelf life, related activation energy E_a was 53.424 kJ/mol, pre-exponential factor k_0 was 515.893. Based on the establishment of Arrhenius equation between the change rate constant of water activity k and storage temperature T, the predicted values of razor clams shelf-life were calculated and they were closed to measured values. This experimental results showed water activity could be as an optimal indicator to predict razor clams shelf-life accurately.

REFERENCES

Alasalvar, C. Taylorkd, A. Oksuz, A. et al. 2001. Freshness assessment of cultured sea bream (Sparus aurata) by chemical, physical and sensory methods. *Food Chemistry* 72(1):33–40.

Bourne, M.C. 2002. Food texture and viscosity. *Second Edition, Academic Press*, New York.

Cobb, B.F. Alaniz, I. Thompson, C.A. 1973. Biochemical and microbial studies on shrimp: volatile nitrogen and amino nitrogen analysis [J]. *Food Sci.* 38:43–46.

David, W. Aldridge. Barry, S. et al. 1995.Oxygen consumption, nitrogenous excretion, and filtration rates of *Dreissena polymorpha* at acclimation temperatures between 20 and 32°C. *Can J Fish Aquat Sci* 52: 1761–1767.

Herbert, S. Joel, L.S. 2004. Sensory evaluation practices. *Third Edition: Elsevier Pte Ltd. Press.*, New York.

Jing, X. Sheng-ping, Y. 2011. Effects of biopreservative combined with modified atmosphere packaging on shelf-life of trichiutus haumela. *Transactions of the Chinese Society of Agricultural Engineering* (01).

Johnson, W.A. Nicholson, F.J. et al. 1994. Freezing and refrigerated storage in fisheries. *Food and Agriculture Organization of the United Nations* (1): Rome.

Labuza, T.P. Schmidl, M.K. 1985. Accelerated shelf-life testing of foods. *Food Technology* 39(9): 64–67.

Le, W. 2013. Effects of different frozen storage processing methods for *Sinonovacula constricta (Lamarck)* qualities. *Hainan university*.

Liang, W. Ming-Yong, Z. Shi-Yuan, D. et al. 2011. Evaluation quality of shrimp (peneaus vannmei) Under uontrolled freezing-point storage [J]. *Periodical of Ocean University of China* (21).

Li-ping, Z. Xiao-qin, Y. Hua-rong, T. 2007. Application of Kinetic model to predict dry salted duck Shelf-life. *Food Science* (11).

Li-rong, S. Jin-feng, P. Yang, W. et al. 2010. Research of shelf-life with partial freezing tilapia fillets during storage. *Food and Machinery* 11, 26(6): 25–29.

Li-rong, S. Shu-xiang, C. Xiang-dong, L. et al. 2011. Research on determination of freezing point of food materials. *Food Science* (51).

Nourian, F. Ramaswamy, H.S. Kushalappaa, C. 2003. Kinetics of quality change associated with potatoes stored at different temperatures. *Lebensm-Wissu-Technol* 36(1): 49–65.

Palmer, M. E. 1980. Bechavioral and rhythmic aspects of filtration. *Asgopecten irradians concentricus* (Say), and the oyster, *Crassastrea virginica* (Gmelin). *J Exp Mar Biol Ecol* 45:273–295.

Qiang Z. Yuan-yuan, L. Xiang-dong, L. 2011. Fresh-keeping technology for tilapia fillets by vacuum packaging followed by partial freezing. *Food Science* (04).

Sivertsvik, M. Rosnes, J. T. Kleiberg, G. H. 2003. Effect of modified atmosphere packaging and super chilled storage on the microbial and sensory quality of Atlantic salmon (*Salmo salar*) fillet. *Food Microbiology and Safety* 68(4): 1467–1472.

Wei, Z. Ai-min, W. Guo-yuan, X. 2011. Effect of compound natural preservative on chilled chicken. *Food Science* (06).

Xiang-dong, L. Qi, Z. Shi-cong, J. et al. 2006. Study on the lotus root freezing point by mensurating electrical conductivity. *Beverage & Fast Frozen Food Industry* (04).

Xiao-qin, Y. Xiao-yan, C. Li-ping, Z. et al. 2007. Predicting shelf-live of food -a review. *Food Research and Development* (03).

Xiao-mian, C. Xiao-ping, W. Chu-jin, D. et al. 2011. Comparison of preservative effects of chitosan and tea polyphenols on cold storage of tilapia. *Modern Food Science and Technology* (03).

DEA model evaluating the supply efficiency of rural productive public goods in Hubei Province, China

Fang Luo, Yewang Zhou & Caihong Sun
School of Commerce, Huanggang Normal University, Huanggang, Hubei, China

Lijun Chen
School of Tourism Culture and Geographical Science, Huanggang Normal University, Huanggang, Hubei, China

ABSTRACT: As China entering the period of industry nurturing agriculture, financial funds for agriculture has grown substantially, and then improving funds service efficiency is important as well as increasing fiscal input. As a big agricultural province, Hubei Province evaluating the supply efficiency of rural productive public goods is the premise of improving service efficiency of fiscal input. Based on the non-parameter DEA model, the supply efficiency of rural productive public goods in Hubei Province in the past twenty years is evaluated and the relative efficiency to the other fellow provinces is estimated. It is concluded that Hubei Province has the trend of supply efficiency of rural productive public goods decreasing and relative efficiency is low among the 13 granary provinces.

1 INTRODUCTION

Rural economy development and farmers' well-off influence the national progress fundamentally. Recently, Chinese Government pays high attention to the problems in agriculture, rural areas and farmers. With the fiscal agriculture expenditure increasing, how to promote the use efficiency of the fiscal fund becomes a major topic (Yu, 2012). As a big agricultural province and main commodity grain base, Hubei Province raised its economic aggregate and fiscal revenue fastly, and the agricultural productive capacity upgrades significant in recent years (Mei, Yao and Fan, et al, 2013). However, comparing with the goal of building agricultural strong province and agricultural modernization, the supply of rural productive public goods is insufficient and that of structure is incomplete. Hence, it directly impacts social harmony that how to improve the efficiency of productive fiscal agricultural expenditure under current conditions.

Academics critically focused on the relationship between fiscal expenditure and rural economic development, and financial rewards, etc. More empirical researches concerned on the situation of financial support in agriculture. The research group of Development Research Center of the State Council (2003) proposed questions of aggregate amount insufficiency, structure unreasonable and management system incomplete in Chinese financial support in agriculture. Li, Tang and Zhang (2007) raised that the absolute quantity of poverty alleviation fund of Finance Ministry rose yearly, but its share of the national GDP fluctuated acutely, and even showed a significant downtrend in the period of 1986–2003. Chen and Chen (2007) calculated the service efficiency of financial support in agriculture in Fujian Province during the period of 1995–2005, and it was concluded that this efficiency kept 40 or so before 1997, and then it decreased although the funds for supporting agriculture was promoted. About the government actions of financial support in agriculture, Zhao (2001) raised that the source and scale of agricultural R&D investment were hungry in China, and the government didn't increase investing the higher-yielding agricultural scientific research which weren't arranged with priority because it competed with the other potent plans, high investing risk and externality. Jiang (2004) put forward that besides minority developed area, the financial support capacity of county and township in agriculture was weak, which couldn't provide matching funds to the special fund for agriculture by the central, province and city government, and even withholding and appropriating happened occasionally, which root was the fiscal difficulties of county and township generated by system and policy imperfect.

Based on the Data Envelopment Analysis (DEA) model of the rural productive public goods, supply efficiency of Hubei Province is estimated from both its time series and cross-compared with other fellow provinces, what creative point is that the previous researches of DEA efficiency assessment payed great attention to comparison of cross-section units, however, this paper adds dynamic efficiency comparative study of time units which helps to understand the issue of rural productive public goods.

2 VARIABLE SELECTION AND DATA SOURCES

Rural productive public goods is that of directly making for the agricultural production, which covering the infrastructure of irrigation, electric power, disaster reduction and prevention, and environmental protection, etc. The input indexes incorporate reservoir capacity, x_1 (hund. mil. cub. met.), controlling area of waterlogging, x_2 (ten thous. hm^2), controlling area of soil erosion, x_3 (ten thous. hm^2), capacity of rural generating electricity, x_4 (hund. mil. kw. hr.), and planting area, x_5 (thous. hm^2). The output indexes conclude gross output value of agriculture, y_1 (hund. mil. yuan), grain yield per unit area, y_2 (kg/hm^2) and gross grain yield, y_3 (ten thous. ton). To eliminate the influence of price fluctuation, gross output value of agriculture is calculated at base period, 1992, price. The data comes from Statistical Yearbook of Hubei and that of China 1993–2012, etc.

Based on input-oriented, the rural productive public goods efficiency evaluation is modeling by DEA method, and the slacks are calculated with multi-stage approach. To understand the trend of supply efficiency of rural productive public goods in Hubei Province in the period of 1992–2011 and the relative efficiency compared with other similar provinces in China, both DEA efficiency of time unit and cross-section unit are evaluated.

3 PEARSON CORRELATION ANALYSIS

To satisfy the hypothesis of isotonicity, that is, the increase of input doesn't result in the decrease of output, the Pearson correlation between input and output variables during the period of 1992–2011 is analyzed. The statistic software of EViews 6.0 is adopted and the result is shown in Table 1. Among output variables, the correlation coefficient of gross grain yield (y_3) only with controlling area of soil erosion (x_3) is significant, but the sign is negative, which doesn't meet the isotonicity prerequisite. Hence, the variable y_3 should be taken out. Among input variables, although the correlation coefficient of planting area (x_5) with gross output value of agriculture (y_1) and grain yield per unit area (y_2) are significant, separately, the signs are negative, which doesn't meet the isotonicity prerequisite. Hence, the variable x_5 should be taken out. Lastly, the model incorporates four input variables and two output variables.

4 DYNAMIC ANALYSIS BASED ON TIME DMU

4.1 Efficiency evaluating

The DEA software of DEAP 2.1 is used to evaluate the supply efficiency of Hubei Province rural productive public goods in the period of 1992–2011, and the result is shown in Table 2. The DMUs, i.e. years which technical efficiency (TE) equal 1 are 1992, 1994, 1995, 1997, 1999, 2000, 2004, 2006, 2008 and 2011. Due to TE being pure technical efficiency (PTE) multiplied by scale

Table 1. The isotonicity test.

Variables	y_1	y_2	y_3
x_1	0.907***	0.754***	−0.097
x_2	0.411*	0.175	0.312
x_3	0.837***	0.903***	−0.514**
x_4	0.655***	0.696***	−0.094
x_5	−0.623***	−0.793***	0.208

***, ** and * represent significantly at level of 1%, 5% and 10%, respectively.

Table 2. The supply efficiency of rural productive public goods in Hubei Province.

DMU	TE	PTE	SE	returns to scale
1992	1.000	1.000	1.000	cons.
1993	0.976	1.000	0.976	ins.
1994	1.000	1.000	1.000	cons.
1995	1.000	1.000	1.000	cons.
1996	0.975	1.000	0.975	ins.
1997	1.000	1.000	1.000	cons.
1998	0.978	0.997	0.981	ins.
1999	1.000	1.000	1.000	cons.
2000	1.000	1.000	1.000	cons.
2001	0.978	0.979	0.999	ins.
2002	0.988	0.989	0.999	ins.
2003	0.961	1.000	0.961	ins.
2004	1.000	1.000	1.000	cons.
2005	0.997	0.999	0.998	ins.
2006	1.000	1.000	1.000	cons.
2007	0.952	0.969	0.982	ins.
2008	1.000	1.000	1.000	cons.
2009	0.999	1.000	0.999	drs.
2010	0.977	0.983	0.994	ins.
2011	1.000	1.000	1.000	cons.

efficiency (SE), these units' PTE and SE equal 1, separately, and they are DEA efficient. The DMUs, 1993, 1996, 2003 and 2009, which PTE = 1 and SE < 1 are weak DEA efficient. The DMUs, 1998, 2001, 2002, 2005, 2007 and 2010, which PTE < 1 and SE < 1 are not DEA efficient.

In recent twenty years, the supply efficiency of rural productive public goods in Hubei Province shows a deteriorative trend, which is inconsistent with the mainstream viewpoint brought forward by the government and academia in China who insist that it is general effective and improved yearly (Huang and Ye, 2009). The possible reason may be that the absolute scale of financial support agriculture rises with the low level of relative scale, the supply structure of public goods are seriously unbalanced. Additionally, as the modern agriculture development, the ecological environment deteriorates, water and soil erode, soil fertility declines, and industrialization occupies lots of arable land.

4.2 Efficiency improvement

Related to PTE and SE, TE improves through two ways. On the one hand, the scale rises. In Table 2, there are ten years which SE < 1, and they should be improved by adjusting the scales. The years, 1993, 1996, 1998, 2001–2003, 2005, 2007 and 2010 are increasing returns to scale could raise supply scale. However, the year, 2009 is decreasing returns to scale could lower supply scale. On the

Table 3. Efficiency improvement.

DMU	Radial movement: input				Slacks movement					
					Input				Output	
	x_1	x_2	x_3	x_4	x_1	x_2	x_3	x_4	y_1	y_2
1998	−1.48	−0.35	−1.08	−0.19	0.00	0.00	−16.43	0.00	0.00	49.24
2001	−11.74	−2.58	−8.47	−1.17	0.00	0.00	−3.32	0.00	197.72	0.00
2002	−5.99	−1.32	−4.55	−0.56	0.00	−0.71	−25.02	0.00	239.54	0.00
2005	−0.70	−0.15	−0.52	−0.08	0.00	−3.49	−1.07	0.00	15.95	0.00
2007	−29.04	−3.66	−12.91	−2.22	−262.88	0.00	0.00	−3.38	0.00	103.77
2010	−16.58	−2.04	−7.80	−1.37	−95.89	0.00	0.00	−8.57	0.00	41.47

The radial movements of output index omitted are zero.

other hand, PTE rises. Not PTE effective DMUs may be improved by radial and slacks movement as shown in Table 3. Due to the model selecting input-oriented mode, the radial movements of output index omitted are zero.

The radial and slacks movements of input of six DMUs are negative which shows the existing of input surplus, namely reducing input value listed in Table 3 and the output keeping constant. The output slacks are positive which shows the existing of output deficit, namely at the current input scale, the higher output could be obtained as shown in Table 3 (Wu, 2009). For example, in 1998, not PTE effectiveness could be improved by reducing reservoir capacity (x_1) 1.48 hund. mil. cub. met., controlling area of waterlogging (x_2) 0.35 ten thous. hm^2, controlling area of soil erosion (x_3) 17.51 (1.08 + 16.43) ten thous. hm^2, and capacity of rural generating electricity (x_4) 0.19 hund. mil. kw. hr., or grain yield per unit area (y_2) rises 49.24 kg/hm^2.

In brief, in recent twenty years, there are six years which rural productive public goods supply are not PTE effective, and ten years that are not SE effective. In the same year maybe neither PTE effective and nor SE effective. It's easy to find that the degree of not PTE effective (account for 30% of the sample) is lighter than that of not SE effective (account for 50% of the sample), in which not PTE effective units could be improved by adjusting supply structure of public goods, and not SE effective units could be improved by adjusting supply scale. Among the ten not SE effective years, there are nine that are increasing returns to scale which need increase input, and one that is decreasing returns to scale which need decrease input. As a whole, inadequate investment in rural productive public goods plays a primary role. Therefore, the first issue is promoting input scale, and the second is adjusting supply structure.

5 COMPARATIVE STATIC ANALYSIS

To learn the position of rural productive public goods of Hubei Province in China, the comparative static analysis between Hubei and other fellow provinces is necessary. The thirteen granary provinces, Henan, Hunan, Hebei, etc. are selected as DMUs, the input and output variables are the same as above, namely four input variables and two output variables, which data adopt the mean value of each province during the period of 2009–2011, and the estimated result is shown in Table 4. Nine provinces, Henan, Hunan, Sichuan, etc. are DEA efficient which's PTE = 1 and SE = 1. Hebei Province is weak DEA efficient which's PTE = 1 and SE < 1. Anhui, Heilongjiang and Hubei Province are not DEA efficient which's PTE < 1 and SE < 1, and the improving of not PTE effective is shown in Table 5. For example, not PTE effectiveness of Hubei Province could be improved by reducing x_1 668.98 hund. mil. cub. met., x_2 22.24 ten thous. hm^2, x_3 84.88 ten thous. hm^2, and x_4 13.71 hund. mil. kw. hr., or y_2 rises 97.05 kg/hm^2. Hubei Province is increasing returns to scale, thus not SE effectiveness (SE < 1) could be improved by raising public goods input.

Table 4. Supply efficiency of rural productive public goods in Chinese granary provinces.

DMUs	TE	PTE	SE	Returns to scale
Henan	1.000	1.000	1.000	cons.
Hunan	1.000	1.000	1.000	cons.
Hebei	0.961	1.000	0.961	irs.
Anhui	0.715	0.913	0.783	irs.
Sichuan	1.000	1.000	1.000	cons.
Neimenggu	1.000	1.000	1.000	cons.
Liaoning	1.000	1.000	1.000	cons.
Jilin	1.000	1.000	1.000	cons.
Heilongjiang	0.698	0.878	0.795	irs.
Jiangsu	1.000	1.000	1.000	cons.
Shandong	1.000	1.000	1.000	cons.
Jiangxi	1.000	1.000	1.000	cons.
Hubei	**0.804**	**0.817**	**0.984**	**irs.**

Table 5. DMUs' (PTE < 1) efficiency improvement.

	Radial movement: input				Slacks movement					
					Input				Output	
DMUs	x_1	x_2	x_3	x_4	x_1	x_2	x_3	x_4	y_1	y_2
Anhui	−25.89	−19.73	−18.61	−1.71	−24.13	0.00	0.00	0.00	695.76	1497.86
Hlj*	−21.64	−40.65	−57.42	−0.69	0.00	−91.64	0.00	−4.00	975.40	1123.45
Hubei	**−182.09**	**−22.24**	**−84.88**	**−13.71**	**−486.89**	**0.00**	**0.00**	**0.00**	**0.00**	**97.05**

*Representing Heilongjiang. The radial movements of output index omitted are zero.

Among Chinese thirteen granary provinces, the rural productive public goods supply of nine provinces is DEA efficient, that of one province is weak DEA efficient, and that of three provinces concluding Hubei Province is not DEA efficient. That is to say Hubei Province is a granary province, but not an advancing one, which is inappropriate for the government strategy of "Rise of Central China" and the blueprint of "loins-strengthening Engineering". The main reasons are that each granary province generally benefits from the preferential policies for the farmers taken by the government which results in both rural public goods supply and agricultural production level rising, and Hubei Province supply efficiency declines in recent twenty years as mentioned above.

6 SUMMARY

Based on the DEA efficiency evaluating, it is concluded that since the input scale of rural productive public goods is insufficient and the PTE is not efficient in some years, the supply of it has trended down over the twenty years in Hubei Province, and compared with other granary provinces, the supply efficiency of Hubei Province is low. Because the first issue is scale deficiency and, secondly, the pure technical efficiency is low, both deepening the reform of financial budget management system and strengthening management and integration of funds for agriculture are suitable for resolving the scale questions, and institutional innovation is the right resources for low technical efficiency.

REFERENCES

Chen, J.H. & Chen, W. 2007. Empirical analysis on the performance of financial support in agriculture and its policy choice in Fujian Province. *Fujian Forum: Humanities and Social Sciences* (9): 115–118.

Huang, L.H. & Ye, H. 2009. Evaluating the supply efficiency of rural productive public goods in China. *Statistics and Decision* (12): 51–53.

Jiang, C.Y. 2004. County and township fiscal difficulties and its effect on the capacity of financial support in agriculture. *Management World* (7): 61–68.

Li, X.Y., Tang, L.X. & Zhang, X.M. 2007. Analysis on the mechanism of financial poverty alleviation funds investing. *Issues in Agricultural Economy* (10): 77–82.

Mei, Z.S., Yao, W. & Fan, X.W., et al. 2013. The problems and solutions of modern agriculture support system in Hubei Province. *Study Monthly: the Second Half* (5): 32–33.

Research Group of Development Research Center of the State Council. 2003. The Chinese government funds for agriculture use and management. *Report of Research Group of Development Research Center of the State Council.*

Wu, D. 2009. Research on the harmonious development of integrated traffic system evaluation based on DEA. *Master Dissertation.* School of Economics and Management, Beijing Jiaotong University.

Yu, L. 2012. An empirical analysis on the contribution of financial support to development in agriculture. *Journal of Agrotechnical Economics* (9): 60–65.

Zhao, Y. 2001. Theoretical analysis on the input for agriculture scientific research and its policy suggestion. *China Rural Survey* (1): 2–8.

The experimental study of the compressive mechanical behaviors of the H-shaped grouted masonry dry-stacked without mortar

Ni Lou
Tianjin University, Tianjin, China

Guohui Yi & Lanying Zhang
China National Engineering Research Center for Human Settlements, Beijing, China

ABSTRACT: In this paper, the compressive mechanical behaviors of the H-shaped grouted masonry dry-stacked without mortar were studied experimentally through 6 groups of compressive strength elements. The failure modes of that grouted masonry were studied. The design methods of the average compressive strength and bearing capacity were proposed in order to provide a reference for its application and spread.

1 INTRODUCTION

In the 1980s, the research and application had been carried out in foreign countries, and several representative assembly-block building systems have been built up to now, such as the haenar block in Canada, R-Thallon block in American, Mecano block in Peru and the interlocking block in Malaysia, etc. However, the research of the industrialized masonry system has been falling behind in China, although the promotion has been carrying on by local standards or enterprise standards. The construction experiences and the experimental data are still fully in need. In order to provide a reference for its application and spread, a large number of experiments were carried out by our research group to study its mechanical behaviors. This article is part of the results.

2 EXPERIMENT OVERVIEW

In this experiment, the H-shaped dry-stacked concrete block is selected as shown in Fig. 1. The strength grade of the main type of this block with the hole ratio of 55% is marked by MU10 and the section size is 400 mm × 200 mm × 200 mm.

The compressive mechanical behavior have been studied through 6 groups of 36 compressive strength elements according to *the Uniform technical code for wall materials used in buildings (GB50574-2010)*, and the experiment overview is shown in table 1.

Figure 1. The main type of the dry-stacked block.

Table 1. The overview of the experiment.

Groups	The design Strength grade of the block	The design Strength grade of the concrete	Length (mm)	Width (mm)	Height (mm)	Grout rate (%)	Slenderness ratio	Numbers per group	Remark
bky1	MU10	C20	600	200	1000	100	5	6	Carried out
bky2	MU10	C20	600	200	1000	100	5	6	according to
bky3	MU10	C20	600	200	1000	100	5	6	GB/T50129-2011
bky4	MU10	C30	600	200	1000	100	5	6	
bky5	MU10	C30	600	200	1000	100	5	6	
bky6	MU10	C30	600	200	1000	100	5	6	

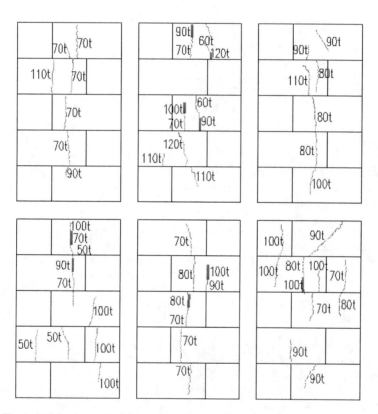

Figure 2. The typical development of the cracks on the experimental members.

3 THE EXPERIMENTAL PHENOMENA AND ANALYSIS

Fig. 2 shows the typical development of the cracks on the experimental members.

This type of grouted masonry is dry-stacked without mortar, so the compactness between the adjacent blocks cannot be guaranteed. Normally, there are four stages in the compressive process.

1) The first stage—the adjacent blocks become more compact. In the beginning of the compression process, the gap between the vertical adjacent blocks become smaller and smaller gradually. The concrete core columns is the main bearing members, and the lateral deformation of the concrete results in the moment and rises some tension at the far end of the cantilever half hole of the

Figure 3. The typical failure mode of the experimental members.

H-shaped block. In this period, both the concrete core columns and the blocks are elastic and no cracks appear.

2) The second stage—the cracks appear. With the increase of vertical pressure, the lateral deformation of the concrete core columns becomes larger. So does the tension at the far end of the cantilever half hole of the H-shaped block. The blocks become more and more compact and begin to take some vertical pressure. In this period, the first crack appears. And in most cases, it appears at the far end of the cantilever half hole.

3) The third stage—the cracks develop. The cracks go upwardly or downwardly and the numbers of cracks increase along with the increase of vertical pressure. The vertical load goes up slowly. When the vertical load stops increasing, the splitting sounds can be heard and the cracks develop automatically. In this period, both the concrete core columns and the blocks are in plastic stage.

4) The failure stage—When the load exceeds the ultimate bearing capacity, the deformation continues to increase, but the stress goes down quickly. Finally, the rib of the block break, and the block burst and fall down. The pressure on the blocks transfer to the concrete core column instantly, and then the core column lost its capacity and the whole member breaks.

The typically broken photos are shown in Fig. 3. From these pictures, many slim cracks in the concrete core columns and a few thick cracks on the blocks can be seen. And the cracks on the blocks appear mostly at the far end of the cantilever half hole. So the damage happens characterized by the crushing of the concrete and the bursting of the blocks.

4 THE EXPERIMENTAL RESULTS AND ANALYSIS

The calculation formulas are shown as following according to the *Code for design of masonry structures (GB50003-2011)*.

$$f_m = k_1 f_1^\alpha (1 + 0.07 f_2) k_2 \quad (1)$$

$$f_{gm,0} = f_m + 0.63 \frac{A_c}{A} f_{cu,m} \quad (2)$$

where k_1 = parameter related to the type of the block, here it is equal to 0.46; α = parameter related to the height of the block, here it is equal to 0.9; f_1 = the average axial compressive strength of block; f_2 = the average axial compressive strength of mortar; k_2 = parameter related to the strength grade of mortar, when f_2 is equal to 0, k_2 should be 0.8; $f_{gm,0}$ = the average axial compressive strength of the grouted masonry; f_m = the average axial compressive strength of the hollow masonry; A_c = the

Table 2. Comparations between the test results and calculated value according to GB 50003-2011.

Groups	The average compressive strength of the blocks got in tests f_1(MPa)	The average axial compressive strength of concrete cubes got in tests $f_{cu,m}$(MPa)	Calculated value of the grouted masonry according to GB50003-2011 $f_{gm,0}$(MPa)	The average compressive strength of the grouted masonry got in tests f'_{gm}(MPa)	Test results/ calculated value in criterion $f'_{gm}/f_{gm,0}$
bky1	11.33	26.45	12.44	12.07	0.97
bky2	11.33	28.68	13.21	12.47	0.94
bky3	11.33	26.84	12.57	13.13	1.04
bky4	12.71	28.44	13.48	12.11	0.90
bky5	12.71	30.18	14.08	12.14	0.86
bky6	12.71	31.89	14.68	11.78	0.80
Average value					0.92
Coefficient of variation					0.09

section area of the concrete core column, here it is equal to $0.55A$; A = the section area of the masonry and $f_{cu,m}$ = the average axial compressive strength of concrete cubes.

The compare of the average axial compressive strength of this grouted masonry between the test results and calculated value according to *GB50003-2011* is shown as table 2.

The results of the tests which are carried out according to *GB/T50129-2011* are lower than the calculated value in criterion as we can see in the table 2. And we think there are two reasons as following:

1) The dry-stacked blocks whose compactness cannot be guaranteed result in worse stability. So the effects of the slenderness ratio of the masonry should be considered.
2) The formulas of the code are based on the coordinate relationship of the strain between the masonry and the concrete core column. When the load arrives to peak load, the peak strain of the concrete should be 0.002, and the peak strain of the masonry should be 0.0015. The strain-stress relationship of the concrete is shown as following:

$$\sigma = \left[2\left(\frac{\varepsilon}{\varepsilon_0}\right) - \left(\frac{\varepsilon}{\varepsilon_0}\right)^2\right] f_{c,m} \quad (3)$$

And we can get the peak stress of the concrete from formula (3).

$$\sigma = \left[2\left(\frac{0.0015}{0.002}\right) - \left(\frac{0.0015}{0.002}\right)^2\right] f_{c,m} = 0.94 f_{c,m} = 0.63 f_{cu,m} \quad (4)$$

This grouted masonry is dry-stacked, so the compactness is not good enough. When the load arrives to peak load, the strain of the block cannot arrive to its peak. The damage happens characterized by the crushing of the concrete and the bursting of the blocks, so the formulas in code cannot be applied to this grouted masonry.

The effects of the slenderness should be considered when we calculate the compressive bearing capacity of this grouted masonry. And the factor of the slenderness can be calculated according to Appendix D of the code *(GB50003-2011)*. So the formula of the compressive bearing capacity for this grouted masonry can be written as following:

$$N_{gm} = \varphi f_{gm} A \quad (5)$$

Table 3. The calculation of the adjusted factor of the compressive strength for the dry-stacked masonry.

Groups	The average compressive strength of the blocks got in tests (MPa)	The average axial compressive strength of concrete cubes got in tests (MPa)	The factor of the slenderness φ	The average axial compressive capacity of the grouted masonry. N'_{gm} (KN)	The average axial compressive strength based on formula (5). f_{gm} (MPa)	The max value of the k calculated by formula (5), formula (6), formula (7) and formula (8)
bky1	11.33	26.45	0.82	1461.00	14.91	1.26
bky2	11.33	28.68	0.82	1528.00	15.60	1.23
bky3	11.33	26.84	0.82	1574.00	16.07	1.51
bky4	12.71	28.44	0.82	1456.00	14.86	0.97
bky5	12.71	30.18	0.82	1418.00	14.48	0.74
bky6	12.71	31.89	0.82	1421.00	14.51	0.61
min						0.61

$$\beta \leq 3: \quad \varphi = 1$$
$$\beta > 3: \quad \varphi = \frac{1}{1 + 0.009\beta^2} \tag{6}$$

where N_{gm} = the average axial compressive bearing capacity of the grouted masonry; f_{gm} = the average axial compressive strength of the grouted masonry; φ = factor of the slenderness

The compressive strength of this grouted masonry should be constituted by two parts, one comes from the hollow block masonry, and the other comes from the concrete core column. The compressive strength of the hollow block masonry can be calculated according to the formula (1), but the parameter related to the strength grade of mortar is suggested to modify. Since the cracks appear at the far end of the cantilever half hole of the H-shaped block beforehand, the block cannot provide the effective horizontal restraint to the concrete core column. So the compressive strength of the concrete can be the axial compressive strength directly. And the formula of the compressive strength for this grouted masonry can be written as following.

$$f_{gm} = f_m + \delta\rho f_{c,m} = f_m + 0.67\delta\rho f_{cu,m} \tag{7}$$

$$f_m = 0.46 k f_1^{0.9} \tag{8}$$

where δ = hole ratio; ρ = grouted ratio; k = adjusted factor of the compressive strength for the dry-stacked masonry.

The adjusted factor of the compressive strength for the dry-stacked masonry can be calculated by formula (5), formula (6), formula (7) and formula (8) according to the average compressive capacity get in the tests. And it is shown in table 3.

So the adjusted factor of the compressive strength for the dry-stacked masonry is suggested to be 0.6 according to table 3. And the average axial compressive strength of this grouted masonry is suggested to calculate according to formula (9). The average axial compressive capacity is suggested to calculate according to formula (5).

$$f_{gm} = 0.276 f_1^{0.9} + 0.67\delta\rho f_{cu,m} \tag{9}$$

The average axial compressive strength and capacity calculated according to formula (9) and formula (5) are compared with the experiments, as shown in table 4.

Table 4. Calculations compared with the experiments.

Groups	The average axial compressive strength based on formula (9) f_{gm}(MPa)	The average axial compressive capacity based on formula (5) N_{gm}(KN)	The average axial compressive strength got in test when considering the factor of the slenderness f'_{gm} (MPa)	The average axial compressive capacity got in test N'_{gm} (KN)	f'_{gm}/f_{gm} & N'_{gm}/N_{gm}
bky1	12.20	1194.65	14.91	1461	1.22
bky2	13.02	1275.12	15.60	1528	1.20
bky3	12.34	1208.72	16.07	1574	1.30
bky4	13.20	1292.53	14.86	1456	1.13
bky5	13.84	1355.31	14.48	1418	1.05
bky6	14.47	1417.01	14.51	1421	1.00
The average value					1.15
Coefficient of variation					0.10

5 CONCLUSION AND SUGGESTION

1) When the H-shaped grouted masonry dry-stacked without mortar is built based on the test method provided in *GB/T50129-2011* to test its compressive strength, the factor of slenderness should be considered, and it is suggested to calculate based on the appendix D of *GB50003-2011*.
2) The formulas of the compressive strength of grouted masonry in *GB50003-2011* cannot be applied to this H-shaped grouted masonry dry-stacked without mortar, the average axial compressive strength and capacity of this grouted masonry are suggested to calculate according to formula (9) and formula (5).

REFERENCES

Guohui Yi, Ni Lou and others. 2013. Preliminary study for the test method of the compressive strength of H-shaped hollow concrete block. Building block & block building. 13–15. Beijing.

Lianyu Gao, Jian Xu and others. 2011. Code for design of masonry structures (GB 50003-2011). China architecture & building press. Beijing.

Lianyu Gao, Jianguo Liang and others. 2010. The Uniform technical code for wall materials used in buildings (GB 50574-2010). China architecture & building press. Beijing.

Ti Wu, Rui Lin and others. 2011. Standard for test method of basic mechanics properties of masonry (GB/T 50129-2011). China architecture & building press. Beijing.

Evaluation based on Monte Carlo simulation under stratified three-stage sampling

K.J. Chen, Y.B. Fan & G. Gao
School of Public Health, Soochow University, Suzhou, Jiangsu, China

ABSTRACT: In this paper, Randomized Response Technique (RRT) model was presented in application to investigating quantitative sensitive questions under stratified three-stage sampling. Using SAS programming to build a simulation in general and do simulate stratified three-stage sampling and random response process 100 times with statistic value of actual survey as population parameters. Monte Carlo simulation was successfully performed in the assessment of reliability and validity of stratified three-stage sampling investigation on quantitative sensitive questions. The results show that the stratified three-stage sampling method and corresponding formulae were feasible.

1 INTRODUCTION

Sampling survey is a statistical analysis method extracting part of the actual data from a population survey which is based on the principle of random and calculating overall indicators according to the sample data by using probabilistic estimation methods. As a research method of sampling, it has mainly been used in medical research and health work in recent years. A great deal of development has also taken place in government departments, enterprises and academia (Wang, 2012, Lu & Chen, 2007, Sun & Xu, 2010). Sensitive questions such as drug and alcohol abuse, sexually transmitted diseases, domestic violence, tax evasion (Li & Gao, 2011), could be frequently encountered in a sample survey (Sun & Zhao, 2012). For sensitive questions, if applying in traditional direct questions and answers, respondents often deliberately false answer and the sample information will be difficult to control, resulting in findings generated information bias (Zhang & Huan, 2012).

In this paper, we not only focus on statistical method of additive RRT model of stratified three-stage sampling on quantitative sensitive question investigation, but also employ Monte Carlo simulation to examine the reliability and validity of that statistical method. Meanwhile, it provides a scientific and reliable method for the sensitive issues investigation in large and complex sampling (He & Gao, 2010).

2 METHODS

2.1 *RRT model for quantitative sensitive question*

Design a random device like this: Place 10 balls in a small bag which have same size, weight and texture. These balls were affixed with $0, 1, 2, \ldots, 9$ digital label. Every respondent (tertiary unit) was instructed to picked out a ball with replacement from the device randomly. In the absence of others, fill questionnaire with result Z, resulting from one's own characteristic number of the sensitive issue plus the number of small selected ball (random variable Y).

2.2 The concept of three-stage stratified sampling

Suppose the population is divided into L levels, the hth level is composed of N_{1k} primary units (h = 1, 2, ..., L), and the ith primary unit of the hth level contains N_{i2h} second-stage units (i = 1, 2, 3, ..., N_{1h}). On average, each primary unit of the hth level contains $\overline{N}_{2h}$ second-stage units. The jth second-stage unit of the ith primary unit of the hth level contains N_{ij3h} third-stage units (i = 1, 2, 3, ..., N_{1h}, j = 1, 2, 3, ..., N_{i2h}). On average, each second-stage unit containster $\overline{N}_{3h}$ tiary units. The hth level includes N_h tertiary units as well as the population totally includes N tertiary units.

At the first stage, n_{1h} primary units were randomly selected from the hth level (h = 1, 2, ..., L). At the second stage, n_{i2h} second-stage units were randomly drawn from the ith selected primary unit of the hth level (i = 1, 2, 3, ..., N_{1h}). On average, $\overline{n}_{2h}$ secondary units were drawn from per chosen primary unit of the hth level. At the third stage, n_{ij3h} third-stage units were randomly opted for from the jth drawn second-stage unit of the ith selected primary unit of the hth level (j = 1, 2, 3, ..., N_{i2h}). On average, $\overline{n}_{3h}$ tertiary units were drawn from per chosen secondary unit of per selected primary unit of the hth level. Every selected tertiary units were investigated with additive RRT model on quantitative sensitive questions.

2.3 Statistical formulae

Let $(\hat{\mu})$ and $V(\hat{\mu})$ stand for the estimator of the population mean (μ) and its variance respectively. Furthermore, μ_{ih} and μ_{ijh} denote the estimator of the population mean (μ_{ih}) of ith chosen primary unit of the hth level and the estimator of the population mean (μ_{ijh}) of jth chosen second-stage unit of the ith selected primary unit of the hth level. Then, y_{ijkh} is defined as index value of the kth tertiary unit of the jth chosen second-stage unit of the ith selected primary unit of the hth level. According to the formulae given by GAO Ge, JIN Pihuan(Gao & Jin, 2000), the estimators are shown to be

$$\hat{\mu} = \sum_{h=1}^{L} \frac{N_h \hat{\mu}_h}{N} = \sum_{h=1}^{L} W_h \hat{\mu}_h \qquad (W_h = \frac{N_h}{N}) \qquad (1)$$

$$V(\hat{\mu}) \approx \sum_{h=1}^{L} W_h^2 \left[\frac{\sigma_{1h}^2}{n_{1h}} \left(1 - \frac{n_{1h}}{N_{1h}}\right) + \frac{\sigma_{2h}^2}{n_{1h}\overline{n}_{2h}} \left(1 - \frac{\overline{n}_{2h}}{N_{2h}}\right) + \frac{\sigma_{3h}^2}{n_{1h}\overline{n}_{2h}\overline{n}_{3h}} \left(1 - \frac{\overline{n}_{3h}}{N_{3h}}\right) \right] \qquad (2)$$

3 MONTE CARLO SIMULATION

3.1 Simulated population

In the study, sample size was calculated by the formula of the relevant literatures on three-stage stratified sampling (Wang & Gao, 2006). This sampling was implemented to investigate behavioral features of MSMs in Beijing, in August 2012.

As population of this simulation, the number of homosexual men aged 15–49 years who accessing to Beijing's gay venues is 67750. The overall simulations were divided into 15–29 and 30–49 layers by age, each proportion was $W_1 = 58.24\%$, $W_2 = 41.76\%$. In the first stage, 3 districts/counties (primary units) were randomly selected from 16 districts/counties of simulated primary units. In the second stage, 5 chambers of MSMs (survey sites) were randomly sampled from the primary units selected in the first stage as simulated secondary units (survey sites), such as bathhouses, nightclubs, bars, etc. MSMs were randomly sampled by setting the sampling ratio as simulated tertiary units.

For the quantitative sensitive issue, "ages of the first occurrence of MSM", use SAS to write macros % sample to simulate 100 survey samples of quantitative additive RRT model on that question by means of three-stage stratified sample (three-stage sampling within each layer), each sample contains 2533 simulated respondents on average. According to formula (1) (2), we could get 100 sample means, variance estimations and the population means of 95% confidence interval respectively. Compare sample statistics with population parameters using the method of 95% confidence interval, and resluts are equivalent to hypothesis testing of $\alpha = 0.05$. If 100 95% confidence intervals of the proceeds almost contain the population mean (true value), indicate that the survey methodology and formulas have good validity; and also shows good reliability due to that 100 sample means are all close to the same number. Macro code of simulating three-stage stratified sampling and using quantitative additive RRT model to investigate sensitive issues could be written as follows.Fill in the appropriate macro parameters and execute written macro procedures, we could get 100 sample means, variance estimations and the population means of 95% confidence interval respectively, resulting from using additive RRT model on quantitative questions to investigate specimen of analog sampling 100 times.

3.2 Part of program code

```
%sample(popsubjnum = 67750,N1M = 16,N2M =15, n1_m=3,n2_m= 5 ,rate= 0.6, times=100);
**Step2: simulating three-stage sampling from the population;
**calculate num of subjects per site, per district and per strata;
proc sql noprint;
create table pop_site as select distinct agegrp, district, site, count(id)
as SubjPerSite from all02group by 1,2,3;quit;
proc sql noprint;
create table pop_dist as select distinct agegrp, district, count(id)
as SubjPerdist from all02 group by 1,2;quit;
proc sql noprint;create table pop_agegrp as select distinct agegrp, count(id)
as SubjPerGrp from all02 group by 1;quit;
proc sql noprint;
create table pop as select distinct mean(ques2) as pop_mean from all02; quit;
data _null_;set pop;call symput('pop_mean', left(pop_mean));run;
    %put &pop_mean ;
proc sql noprint;create table num_mean as select distinct agegrp, count(id)/&N1M/&N2M
as N3M, (calculated N3M)*&rate as n3_m,
count(id)/&popsubjnum as W from all02 group by agegrp; quit;
proc sql noprint;create table alldist as select distinct district from all02;quit;
proc sql noprint; create table allsite as select distinct site from all02;quit;
%do seed=101 %to 200;
**fist stage, sampling district;proc surveyselect data=alldist noprint
method=srs n=&n1_m seed=&seed out=Sampledist;run;
data _null_;set sampledist;call symput('dist'||left(_n_), strip(district));run;
**Second stage, sampling site;%do i=1 %to &n1_m;%let seed1=%eval(&seed+&i);
proc surveyselect data=allsite noprint method=srs n=&n2_m
seed=&seed1 out=Samplesite;run;
proc sql noprint;select distinct site into :site separated by ' ' from samplesite;quit;
%put &site;
**Third stage, sampling subjects;data sub01;set all02;
where district=&&dist&i and site in (&site);run;
proc sort;by agegrp site;run;
proc surveyselect data=sub01 noprint method=srs rate=&rate
seed=&seed1 out=Sample01;strata agegrp site;run;
data sample02;%if &i=1 %then %do; set sample01; %end;
%else %do; set sample02 sample01; %end;run;%end;
```

3.3 Simulated investigation results

Monte Carlo simulation sampling was repeated 100 times under SAS program. Each tertiary unit has produced a response value. For the quantitative sensitive issue, "ages of the first occurrence of MSM", the population mean calculated from the simulated population was 21.96 years old. For analog population, simulate 100 stratified three-stage sampling specimens, and for all simulated respondents of each specimen, use quantitative sensitive additive model to simulate the investigation, all the results were given in Table 1.

3.4 Reliability and validity assessment

The principles of reliability and validity are fundamental cornerstones of the scientific method (Liu & Gao, 2011, Gao & Fan, 2008). Reliability is defined as the extent to which a measurement is repeated under identical conditions. Validity refers to the degree to which a test measures what it purports to measure (He & Gao, 2009). In this paper, Monte Carlo simulation was used to evaluate the reliability and validity. Establishing good quality studies need both high reliability and high validity (Kim & Warde, 2004). According to Table 1, we could know that all of the population mean of 95% confidence interval obtained from 100 analog samples contains population mean. We could consider that the difference between 100 sample means and population mean (simulated true value) was not statistically significant, illustrating that the survey methodology and formulas of additive RRT model on quantitative sensitive question with stratified three-stage sample studied in this paper have good validity. Due to that 100 sample means are all close to the same number (simulated population mean), the survey methodology and formulas are also deemed to have good reliability.

4 DISSCUSSION

Monte Carlo Simulation method (Monte Carlo Simulation MCS) is a kind of numerical calculation method which is based on statistical sampling theory and studying random variable through computer (Li & Gao, 2009). Recently, with the rapid development of computer simulation technology, Monte Carlo method has developed into an important research tool (Lau & Lin, 2011). In the our research, Monte Carlo sampling simulation, employed under three-staged complex survey designs for sensitive question, had certain innovation and high-applied value. Most of the literatures on theory of RRT are restricted to simple random sampling, especially for the research on sensitive questions (Du & Gao, 2012). Moreover, objects of investigation on sensitive questions confined to a small range are selected by a simple random design. However, evaluations of reliability and validity on sensitive questions in survey using the RRT have been seldom reported (Fido & Al Kazemi). In our present research, these weaknesses have been successfully overcome.

The method and formulae for additive RRT model of stratified three-stage sampling on quantitative sensitive question investigation show higher reliability and validity. Thus, stratified three-stage sampling appeared in this paper is presented as an effective method for obtaining real data of sensitive questions in a wide range of area. This would be expected to not only allow local policy makers to better formulate public health policy and guide efficient allocation of resources, but also provide the scientific basis for effective prevention and control of HIV/AIDS among high risk group (Gage & Ali, 2004). At the same time, our research results will fill the research blank for the statistical survey method and calculation formulae.

5 CONCLUSIONS

In this paper, Monte Carlo simulation was conducted with SAS programming to assess the reliability and validity under stratified three-stage sampling. All the CIs of the 100 estimated population mean

Table 1. Computer simulation results.

Sample number	Sample mean $\hat{\mu}$	Sample size	95% confidence interval		Sample number	Sample mean $\hat{\mu}$	Sample size	95% confidence interval	
			Lower limit	Upper limit				Lower limit	Upper limit
1	22.49	2612	21.49	23.49	51	22.22	2580	21.32	23.12
2	22.15	2582	21.16	23.14	52	21.95	2545	20.78	23.11
3	21.97	2553	21.09	22.85	53	22.50	2615	21.44	23.57
4	22.05	2568	21.16	22.94	54	22.88	2655	21.67	24.08
5	22.00	2558	21.13	22.88	55	22.78	2651	21.63	23.93
6	21.56	2504	20.53	22.60	56	21.62	2507	20.68	22.57
7	21.37	2475	20.25	22.48	57	22.38	2603	21.38	23.38
8	21.94	2562	21.06	22.81	58	22.12	2572	21.23	23.01
9	21.89	2545	20.99	22.79	59	22.08	2564	21.05	23.10
10	21.70	2525	20.78	22.63	60	21.37	2494	20.37	22.38
11	21.97	2561	20.98	22.95	61	21.91	2541	21.04	22.77
12	21.54	2506	20.60	22.47	62	22.22	2562	21.06	23.38
13	22.53	2603	21.57	23.49	63	21.31	2468	20.18	22.43
14	22.01	2559	21.12	22.89	64	22.42	2602	21.31	23.53
15	21.69	2528	20.80	22.58	65	22.03	2565	21.15	22.91
16	22.14	2578	21.20	23.08	66	22.64	2629	21.52	23.75
17	21.61	2505	20.63	22.59	67	22.02	2556	21.14	22.91
18	21.71	2521	20.73	22.69	68	21.42	2503	20.46	22.39
19	22.61	2626	21.54	23.67	69	21.71	2530	20.83	22.59
20	21.94	2563	20.88	23.01	70	22.11	2580	21.14	23.08
21	21.99	2560	21.12	22.86	71	22.08	2560	21.19	22.97
22	22.19	2579	21.25	23.12	72	21.73	2515	20.81	22.64
23	22.21	2572	21.33	23.09	73	22.31	2601	21.36	23.25
24	21.91	2553	21.03	22.78	74	21.69	2525	20.76	22.62
25	21.63	2516	20.73	22.53	75	21.89	2548	21.02	22.76
26	22.30	2603	21.35	23.25	76	22.36	2590	21.36	23.36
27	21.85	2545	20.89	22.81	77	21.98	2553	21.09	22.87
28	22.48	2614	21.34	23.63	78	22.08	2564	21.18	22.98
29	22.35	2594	21.33	23.37	79	21.66	2513	20.70	22.61
30	21.55	2507	20.53	22.57	80	22.40	2610	21.41	23.38
31	22.01	2548	21.04	22.97	81	22.04	2559	21.15	22.93
32	21.30	2479	20.27	22.32	82	22.15	2563	21.20	23.10
33	21.77	2534	20.86	22.69	83	21.49	2509	20.52	22.46
34	22.19	2582	21.28	23.10	84	22.22	2582	21.32	23.12
35	21.66	2532	20.65	22.68	85	21.75	2529	20.87	22.63
36	21.93	2553	21.07	22.80	86	22.35	2607	21.40	23.30
37	22.08	2564	21.20	22.95	87	22.26	2577	21.22	23.30
38	22.38	2612	21.37	23.39	88	21.76	2541	20.79	22.74
39	22.57	2616	21.53	23.62	89	22.42	2595	21.44	23.40
40	22.37	2606	21.42	23.32	90	21.68	2513	20.77	22.59
41	21.74	2531	20.82	22.67	91	22.34	2601	21.39	23.28
42	21.65	2511	20.63	22.67	92	21.82	2540	20.95	22.70
43	22.02	2569	21.11	22.92	93	21.58	2514	20.68	22.48
44	21.86	2530	20.89	22.84	94	21.90	2550	21.03	22.78
45	21.81	2525	20.88	22.74	95	22.47	2592	21.53	23.41
46	22.18	2586	21.13	23.24	96	22.25	2571	21.37	23.13
47	21.90	2535	21.03	22.76	97	21.65	2513	20.70	22.60
48	22.20	2577	21.32	23.08	98	21.37	2494	20.23	22.52
49	22.33	2595	21.18	23.48	99	21.70	2521	20.79	22.60
50	21.72	2527	20.74	22.71	100	22.33	2599	21.37	23.29

include population mean. Successive survey using additive RRT model under stratified three-stage sampling and its formulae are of high reliability and validity which can be applied to large-scale investigation.

ACKNOWLEDGEMENT

The study was supported by National Natural Science Foundation of China (No. 81273188).
We are grateful to professor Ge Gao for his invaluable help in field investigations and calculations.

REFERENCES

Du, Q.Q. & Gao, G. et al. 2012. Application of monte carlo simulation in reliability and validity evaluation of two-stage cluster sampling on multinomial sensitive question. *Information Computing and Applications*: 261–268.
Fido, A. & Al Kazemi, R. 2002. Survey of HIV/AIDS knowledge and attitudes of Kuwaiti family physicians. *Family Practice*: 682–684.
Gage, A.J. & Ali, D. 2004. Factors associated with self-reported HIV testing among men in Uganda. *AIDS Care*: 153–165.
Gao, G. & Fan, Y.B. 2008. Stratified cluster sampling and its application on the Simmons RRT model for sensitive question survey. *Chinese Journal of Health Statistics*: 562–565.
Gao, G. & Jin, P.H. et al. 2000.The method to estimate the sample sizes for stratified three-stage sampling and its application. *Chinese Journal of Health Statistics* (06):325–327.
He, Z.L. & Gao, G. 2009. Multiple choices sensitive questions survey in two-stage sampling and its application. *Recent Advance in Statistics Application*: 1160–1164.
He, Z.L. & Gao, G. et al. 2010. Multiplication models of quantitative sensitive questions survey in two-stage sampling and its application. *Data Processing and Quantitative Economy Modeling*: 6–10.
Kim, J.M. & Warde, W.D. 2004. A stratified Warner's randomized response model. *Journal of Statistical Planning and Inference*: 155–168.
Lau, J.T.F. & Lin, C. et al. 2011. Public health challenges of the emerging HIV epidemic among men who have sex with men in China. *Public Health*: 260–265.
Li, W. & Gao, G. et al. 2011. Statistical methods of two-stage sampling on simmons model for sensitive question survey with and its application. *Studies in Mathematical Sciences*: 46–51.
Li, X.D. & Gao, G. et al. 2009. Stratified random sampling on the randomized response technique for multi-class sensitive question survey. *Recent Advance in Statistics Application and Related Areas*: 800–803.
Liu, P. & Gao, G. et al. 2011. Two-stage sampling on additive model for quantitative sensitive question survey and its application. *Progress in Applied Mathematics*: 67–72.
Lu, S.Z. & Chen, F. 2007. *Medical Statistics*. Beijing: China Statistics Press.
Sun, P. & Zhao, D.F. 2012. Design of Warner model and card parameter P in questionnaire about sensitive issues. *Statistics and Decision* (09): 72–73.
Sun, Z.Q. & Xu, Y.Y. 2010. *Medical Statistics*. Beijing: People's Medical Publishing House.
Wang, J. & Gao, G. et al. 2006. The estimation of sampling size in multi-stage sampling and its application in medical survey. *Applied Mathematics and Computation*: 239–249.
Wang, Y.F. 2012. Application and inspiration of sampling techniques in the statistics of the Hong Kong government. *Statistics and Management* (03): 88–89.
Zhang, Q.Q. & Huan, X.P. et al. 2012. Prevalence of sexually transmitted disease and risk factors among female sex workers in Jiangsu province. *Acta Universitatis Medicinalis Nanjing(Natural Science)* (04): 473–478.

Nonlinear stability analysis for composite laminated shallow spherical shells in hydrothermal environment

Shang-yang Jiang, Yao-shun Deng & Jia-chu Xu
Department of Mechanics and Civil Engineering, Jinan University, Guangzhou, China

ABSTRACT: Based on the first-order shearing deformation theory, considering hydrothermal effect, the nonlinear stability of laminated composite shallow spherical shells under uniform load is investigated. An analytic solution for rigidly clamped edges is obtained by using modified iteration method. The extremum buckling principle is employed to determine the critical buckling load. The influences of geometric parameters, temperature and humidity on buckling behavior are discussed as well.

1 INTRODUCTION

With good performances of light weight, high strength as well as high fatigue and severe environment resistance, advanced composite materials have been widely used in aviation, aerospace, machinery, shipbuilding, civil engineering and other areas. When the composite structure is used in the atmosphere, the effect of humidity and temperature will significantly reduce the mechanical property of composite materials. Thus, investigating the mechanical behaviors of composite materials under hydrothermal environment is of highly importance in conducting safety evaluation. In recent years, a great deal of research have been devoted to the bending and stability of composite laminate plate and cylindrical shell under hydrothermal environment. Based on the previous works (Hui-shen, 2001, J. M. Whitney, 1971, P. C. Upadhyay, 2000, Yang Jia-ming 2005, K. S. Sai Ram, 1992, S. Y. Lee, 1992, Z. M. Li, 2008, ZHU Yong-an, 2008), considering the effect of temperature and humidity, nonlinear stability of composite laminate shallow spherical shell is investigated in this paper.

2 FUNDAMENTAL EQUATIONS

Consider a thin symmetrical laminated cylindrical orthotropic composite shallow spherical shell, in the temperature and humidity environment, under the action of uniform loading q shown in Fig. 1. Assume that $Q_{ij}^{(k)}, \alpha_r^{(k)}, \alpha_\theta^{(k)}, \eta_r^{(k)}$ and $\eta_\theta^{(k)}$ are the reduced stiffness, thermal coefficient of expansion and wet coefficient of expansion of the kth layer respectively. ΔT is the difference between the usage temperature and the initial temperature. ΔC is the content of water absorption.

Assume that w, ψ, N_r are the deflection of the middle surface, the rotation of a normal to the middle surface and the radial membrane stress respectively. Based on the first-order shearing deformation and von-Karman large deflection theory, considering hydrothermal effect, the governing equation for nonlinear stability of laminated composite shallow spherical shells under uniform load is derived as follows

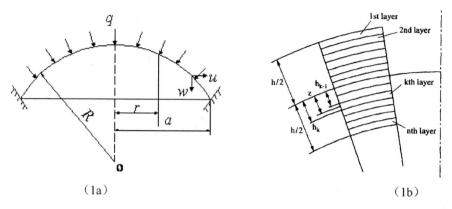

(1a)　　　　　　　　　　　　　　　　(1b)

Figure 1.　Geometry of laminated shallow spherical shell.

$$A_3\frac{d}{dr}[r\frac{d}{dr}(rN_r)]-A_1N_r+\frac{1}{2}(\frac{dw}{dr})^2+\frac{r}{R}\frac{dw}{dr}$$
$$=[(A_1+A_2)\overline{\alpha_r}-(A_2+A_3)\overline{\alpha_\theta}]\Delta T+[(A_1+A_2)\overline{\eta_r}-(A_2+A_3)\overline{\eta_\theta}]\Delta C,$$
$$D_{11}\frac{d}{dr}(r\frac{d\psi}{dr})-(Gr+\frac{D_{22}}{r})\psi-Gr\frac{dw}{dr}=0, \qquad (1)$$
$$rG(\frac{dw}{dr}+\psi)+rN_r(\frac{r}{R}+\frac{dw}{dr})+\frac{1}{2}qr^2=0$$

where

$$A_1=\frac{A_{22}}{A_{11}A_{22}-A_{12}^2},\quad A_2=\frac{A_{12}}{A_{11}A_{22}-A_{12}^2},\quad A_3=\frac{A_{11}}{A_{11}A_{22}-A_{12}^2}$$

where A_{ij} and D_{ij} are the extensional rigidities and the flexural rigidities respectively, G is the shear rigidity. $A_{ij}, D_{ij}, G, \overline{\alpha_r}, \overline{\alpha_\theta}, \overline{\eta_r}$ and $\overline{\eta_\theta}$ are given by

$$A_{ij}=\sum_{k=1}^{n}Q_{ij}^{(k)}(h_k-h_{k-1}),\quad D_{ij}=\frac{1}{3}\sum_{k=1}^{n}Q_{ij}^{(k)}(h_k^3-h_{k-1}^3)\quad (i,j=1,2)$$

$$G=\sum_{k=1}^{n}G_{rz}^{(k)}G=\frac{4h^2}{9\sum_{k=1}^{n}\frac{1}{G_{rz}^{(k)}}[h_k-h_{k-1}-\frac{8}{3h^2}(h_k^3-h_{k-1}^3)+\frac{16}{5h^4}(h_k^5-h_{k-1}^5)]}(h_k-h_{k-1})$$

$$\overline{\alpha_r}=\sum_{k=1}^{n}(Q_{11}^{(k)}\alpha_r^{(k)}+Q_{12}^{(k)}\alpha_\theta^{(k)})(h_k-k_{k-1}),\quad \overline{\alpha_\theta}=\sum_{k=1}^{n}(Q_{12}^{(k)}\alpha_r^{(k)}+Q_{22}^{(k)}\alpha_\theta^{(k)})(h_k-k_{k-1}),$$

$$\overline{\eta_r}=\sum_{k=1}^{n}(Q_{11}^{(k)}\eta_r^{(k)}+Q_{12}^{(k)}\eta_\theta^{(k)})(h_k-k_{k-1}),\quad \overline{\eta_\theta}=\sum_{k=1}^{n}(Q_{12}^{(k)}\eta_r^{(k)}+Q_{22}^{(k)}\eta_\theta^{(k)})(h_k-k_{k-1})$$

Consider the edge of the shell is rigidly clamped, the boundary conditions are given by

$$w=0,\ \psi=0,\ u=0 \quad \text{at}\ r=a \qquad (2)$$

$$\psi=0,\ N_r\ \text{finite} \quad \text{at}\ r=0 \qquad (3)$$

where

$$u=r\left[A_3\frac{d(rN_r)}{dr}-A_2N_r\right] \qquad (4)$$

Introducing the following dimensionless parameters

$$y = \frac{r}{a}, \; W = \frac{w}{h}, \; \Phi = ky + \frac{\partial W}{\partial y}, \; \Psi = \frac{a}{h}\psi, \; S = \frac{a}{D_{11}}rN_r, \; P = \frac{a^4}{2D_{11}h}q, \; k = \frac{a^2}{Rh},$$

$$\lambda = \frac{a^2}{D_{11}}G, \; \beta_1^2 = \frac{D_{22}}{D_{11}}, \; \beta_2^2 = \frac{A_1}{A_3}, \; \beta_3 = \frac{h^2}{A_3 D_{11}}, \; \beta_4 = \frac{A_2}{A_3},$$

$$T = \frac{a^2}{A_3 D_{11}}[(A_1 + A_2)\overline{\alpha}_r - (A_2 + A_3)\overline{\alpha}_\theta]\Delta T,$$

$$C = \frac{a^2}{A_3 D_{11}}[(A_1 + A_2)\overline{\eta}_r - (A_2 + A_3)\overline{\eta}_\theta]\Delta C$$

By substitution, the governing equations (1) and the boundary conditions (2), (3) are transformed to the dimensionless form

$$\lambda y \Phi = -\left[S\Phi + \lambda y \Psi + (P - k\lambda) y^2\right] \quad (5)$$

$$y^{\beta_1} \frac{d}{dy} y^{1-2\beta_1} \frac{d}{dy}(y^{\beta_1}\Psi) = \lambda y(\Psi + \Phi - ky) \quad (6)$$

$$y^{\beta_2} \frac{d}{dy} y^{1-2\beta_2} \frac{d}{dy}(y^{\beta_2} S) = -\frac{1}{2}\beta_3(\Phi^2 - k^2 y^2) + T + C \quad (7)$$

$$W = 0, \; \Psi = 0, \; \frac{dS}{dy} - \beta_4 \frac{S}{y} = 0 \quad \text{at} \quad y=1 \quad (8)$$

$$\Psi = 0, \; S = 0 \quad \text{at} \quad y=0 \quad (9)$$

3 SOLUTION OF NONLINEAR BOUNDARY VALUE PROBLEM

By using modified iteration method to solve the dimensionless nonlinear boundary value problem (5)–(9). The dimensionless center deflection W_0 of the shell is taken as an iteration parameter

$$W_0 = -\int_0^1 (\Phi - ky) dy \quad (10)$$

In the first approximation, the liner boundary value problems are as follows

$$\lambda y \Phi_1 = -\left[\lambda y \Psi_1 + (P - k\lambda) y^2\right] \quad (11)$$

$$y^{\beta_1} \frac{d}{dy} y^{1-2\beta_1} \frac{d}{dy}(y^{\beta_1}\Psi_1) = \lambda y(\Psi_1 + \Phi_1 - ky) \quad (12)$$

$$y^{\beta_2} \frac{d}{dy} y^{1-2\beta_2} \frac{d}{dy}(y^{\beta_2} S_1) = -\frac{1}{2}\beta_3(\Phi_1^2 - k^2 y^2) + T + C \quad (13)$$

$$\Psi_1 = 0, \; \frac{dS_1}{dy} - \beta_4 \frac{S_1}{y} = 0 \quad \text{at} \quad y=1 \quad (14)$$

$$\Psi_1 = 0, \; S_1 = 0 \quad \text{at} \quad y=0 \quad (15)$$

Using expression (10), the solutions of boundary value problem (11)–(15) are

$$\Phi_1 = \alpha_0 W_0 \left(a_1 y^{\beta_1} + a_2 y^3 + a_3 y\right) + ky \quad (16)$$

$$S_1 = \alpha_0^2 W_0^2 \sum_{i=1}^{7} b_i y^{t_i} + k\alpha_0 W_0 \sum_{i=8}^{11} b_i y^{t_i} + \sum_{i=12}^{13} b_i y^{t_i} \quad (17)$$

Limited by the length of paper, the coefficients α_0, a_i and b_i are not given. For the second approximation, we have the following linear boundary value problem

$$\lambda y \Phi_2 = -[S_1 \Phi_1 + \lambda y \Psi_2 + (P - k\lambda) y^2] \tag{18}$$

$$y^{\beta_1} \frac{d}{dy} y^{1-2\beta_1} \frac{d}{dy} (y^{\beta_1} \Psi_2) = \lambda y (\Psi_2 + \Phi_2 - ky) \tag{19}$$

$$\Psi_2 = 0 \quad \text{at} \quad y=0 \tag{20}$$

$$\Psi_2 = 0 \quad \text{at} \quad y=1 \tag{21}$$

Solving boundary value problem (18)–(21), we can obtain the solutions for the second approximation Φ_2 and Ψ_2. Using expression (8), we have the nonlinear characteristic relation of the symmetrically laminated cylindrically orthotropic shallow spherical shell in hydrothermal environment

$$P = c_1 W_0^3 + c_2 W_0^2 + c_3 W_0 + c_4 \tag{22}$$

Using the extremal condition

$$\frac{dP}{dW_0} = 0 \tag{23}$$

We can obtain the dimensionless critical center deflection of the shell when buckling occurs

$$W_0^* = -\frac{c_2 \pm \sqrt{c_2^2 - 3c_1 c_3}}{3c_1} \tag{24}$$

Substituting (24) into (22), we obtain the dimensionless critical buckling pressure

$$P = c_1 W_0^{*3} + c_2 W_0^{*2} + c_3 W_0^* + c_4 \tag{25}$$

Limited by the length of paper, the coefficients c_i are not given.

4 NUMERICAL EXAMPLE AND DISCUSSION

To simplify calculations, without loss of generality, assume that the different layers of the shells have the same thickness and elastic constants, and

$$\frac{E_\theta}{E_r} = 1.5, \quad v_{r\theta} = 0.2, \quad \frac{\alpha_\theta}{\alpha_r} = \frac{1}{20}, \quad \frac{\eta_\theta}{\eta_r} = \frac{1}{20}$$

In such a case we have

$$\beta_1^2 = \beta_2^2 = 1.5, \quad \beta_3 = 16.92, \quad \beta_4 = 0.3$$

Numerical results are given in Fig. 2–Fig. 4.

Fig. 2 shows the characteristic curves for several value k. It is seen from Fig. 2 that when geometric parameter k is smaller, there is no buckling take place for the shell. When the geometric parameter k is larger, the curves become crooked which mean buckling occurs. Fig. 2 also shows that the critical buckling load increases with the increasing of k.

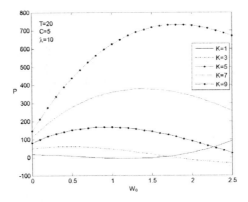

Figure 2. Characteristic curves for k.

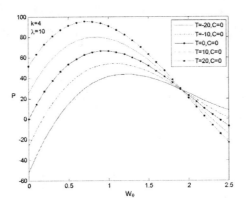

Figure 3. The effect of temperature on load-deflection.

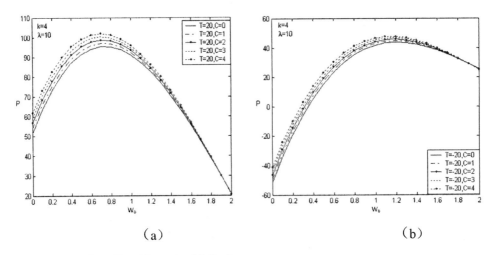

(a)   (b)

Figure 4. The effect of humidity on load-deflection.

Fig. 3 shows the effect of temperature on buckling behavior. It is clear that the critical buckling load decreases with the increasing of temperature difference T.

Fig. 4 (a) and (b) show the effects of humidity on buckling behavior when heating and cooling respectively. It is seen from Fig. 4 that the critical buckling load decreases with the increasing of humidity C. Fig. 4 also shows that the effects of humidity on buckling behavior when heating is larger than that when cooling.

ACKNOWLEDGEMENTS

The authors gratefully acknowledge the financial support from the National Natural Science Foundation of China (No.11032005).

REFERENCES

Hui-Shen, S. 2001. Buckling and postbuckling of laminated thin cylindrical shells under hydrothermal environments[J]. *Applied Mathematics and Mechanics*, 22(003), 228–238.
J. M. Whitney, J. E. A. 1971. Effect of environment on the elastic response of layered composite plates[J]. *AIAA Journal*, 9(9), 1708–1713.
K. S. Sai ram, P. K. S. 1992. Hydrothermal effects on the buckling of laminated composite plates[J]. *Composite Structures*, 21(4), 233–247.
P. C. Upadhyay, J. S. L. 2000. Effect of hydrothermal environment on the bending of PMC laminates under large deflection[J]. *Journal of Reinforced Plastics and Composites*, 19(6), 465–491.
S. Y. Lee, C. J. C., J. L. Jang, J. S. Lin. 1992. Hydrothermal effects on the linear and nonlinear analysis of symmetric angle-ply laminated plates[J]. *Composite Structures*, 21(1), 41–48.
Yang Jia-Ming, S. L.-X., Wu Li-Juan, Zhang Yi-Long. 2005. Geometrically nonlinear analysis of laminated composite plates under hydrothermal environments[J]. *Engineering Mechanics*, 22(005), 59–63.
Z. M. Li, H. S. S. 2008. Postbuckling analysis of three-dimensional textile composite cylindrical shells under axial compression in thermal environments[J]. *Composites Science and Technology*, 68(3–4), 872–879.
Zhu Yong-An, W. F., Liu Ren-Huai. 2008. Thermal buckling of axisymmetrically laminated cylindrically orthotropic shallow spherical shells including transverse shear[J]. *Applied Mathematics and Mechanics*, 29(003), 263–271.

Evidential uncertainty quantification in fatigue damage prognosis

H. Tang
State Key Laboratory for Disaster Reduction in Civil Engineering, Tongji University, Shanghai, China

J. Li
Research Institute of Structural Engineering and Disaster Reduction, Tongji University, Shanghai, China

L. Deng
China Aerospace Construction Group Co. Ltd, Beijing, China

ABSTRACT: Various sources of uncertainty exist in fatigue crack growth analysis, such as variability in loading conditions, material parameters, experimental data, model uncertainty and unclear information in the modeling. The object of this paper is to present a methodology for fatigue damage prognosis under epistemic uncertainty. The parameters in fatigue crack growth model are obtained by fitting the available sparse experimental data and then the uncertainty in these parameters is taken into account. Evidence theory and differential evolution algorithm are proposed to characterize and propagate the epistemic uncertainty. The overall procedure is demonstrated using experimental data of Ti-6Al-4V aluminum alloy specimens. With comparison of probability theory and interval method, the computational efficiency and accuracy of this approach are also investigated.

1 INTRODUCTION

Mechanical members are often subjected to cyclic loads which will lead to fatigue and progressive crack growth, so it is necessary to predict the performance of such members to facilitate risk assessment and management. Fatigue crack growth is a stochastic process and there are different kinds of uncertainty, such as physical variability, data uncertainty and modeling errors associated with it. Natural variability in many input variable also introduces uncertainty in model output.

Numerous studies (Keith & Graeme, 2008; Sankararaman & Ling & Shantz & Mahadevan, 2009; Surace & Worden, 2011) have carried on about crack growth and life prediction, but focused mainly on natural variability in loading, geometry and material properties, and epistemic uncertainty due to lack of knowledge and unclear information is not included. Uncertainty quantification is the most important part in uncertainty analysis. In previous researches about the natural variability, probability theory is the most common way of uncertainty quantification, in which loading conditions and geometric and material properties are treated as random variables. However, probability theory is not an ideal method when there is insufficient data or the system model is not accurate enough, because the epistemic uncertainty caused by unclear information which cannot be described by probability theory.

As a result, numerous promising uncertainty theories were proposed to represent epistemic uncertainty, including fuzzy set theory, rough set theory, evidence theory (Bae & Grandhi & Canfield, 2004; Bao & Li & An, 2012; Helton & Johnson & Oberkampf & Storlie, 2007; Salehghaffari & Rais-Rohani & Marin & Bammann, 2012) and interval analysis. Among these theories, evidence theory has great potential in uncertainty quantification, which can handle both aleatory uncertainty and epistemic uncertainty. Evidence theory is applied in this paper for presentation of epistemic uncertainty.

As uncertain variable is represented by many discontinuous set instead of smooth and continuous explicit function, onerous cost is inevitable in uncertainty quantification with evidence theory. Uncertainty propagation based on evidence theory essentially is to find the maximum and minimum values of the system response in certain intervals and can be solved through optimization methods. Among numerous optimization methods, differential evolution (DE) algorithm is selected in this paper, which is a fast and robust algorithm (Storn & Price, 1997).

The main subject of this paper is to propose an evidential uncertainty quantification method for the fatigue damage prognosis under epistemic uncertainty. Paris law is selected as damage propagation model and experimental data provided in literature (Morten & Stuart, 2001) is used.

2 COMPUTATIONAL MODEL

2.1 Fatigue-crack growth

In common with most polycrystalline alloys, crack growth rates for Ti-6Al-4V are given by the Paris expression:

$$\frac{da}{dN} = C(\Delta K)^m \tag{1}$$

where C and m are constants of the material, and $\Delta K = K_{max} - K_{min}$, the stress range. The stress intensity range for a central mode 1 crack can be expressed by

$$\Delta K_1 = (\Delta \sigma)\sqrt{\pi a \sec\left(\frac{\pi a}{W}\right)} \tag{2}$$

Under conditions of cyclic loading the crack length will grow monotonically. The point at which the crack-tip SIF exceeds the fracture toughness will occur at a critical crack length a_C, which can be calculated as

$$a_c = \frac{(1-R)^2 K_{1C}^2}{\pi (\Delta \sigma)^2} \tag{3}$$

2.2 Design of specimen

In this article, the structure under consideration is a thin rectangular plate (Ti-6Al-4V aluminum alloy) under cyclic loading ($P_{max} = 200$ kN, $P_{min} = 100$ kN), with width W = 500 mm and thickness T = 5 mm as illustrated in Fig. 1. The damage is assumed to be a central mode 1 through crack for which the stress intensity factor range can be calculated as Eq.(2).

Experimental data provided in literature (Morten & Stuart, 2001) is used in this paper, which gives different values of C and m of Ti-6Al-4V under different environment, as shown in Table 1, and it can be seen that the material parameters of the damage propagation law are uncertain. The paper is to investigate the effects of the parametric uncertainty on the fatigue lifetime prediction of the specimen.

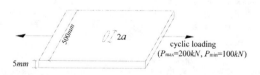

Figure 1. The finite plate under cyclic loading.

Table 1. Values of material constants C and m of Ti-6Al-4V.

No.	1	2	3	4	5	6	7	8
log C	−13.9	−12.7	−15.0	−11.6	−12.5	−13.3	−12.6	−12.9
m	5.4	5.0	6.2	4.0	4.5	5.5	4.7	4.1
No.	9	10	11	12	13	14	15	16
log C	−11.9	−13.4	−12.8	−12.4	−12.8	−11.7	−14.6	−13.2
m	3.8	4.9	4.4	4.6	4.3	3.7	6.1	4.4

3 EVIDENTIAL UNCERTAINTY QUANTIFICATION OF THE FATIGUE DAMAGE PROGNOSIS

Before applying evidence theory for uncertainty modeling of crack growth model, a brief introduction is provided in this section. The universe of evidence theory (Shafer 1976) is the frame of discernment Θ which concludes all the possible answers about the investigated problem, and all the elements in Θ mutual exclusion between each other, and evidence theory is a mapping from $2^{\Theta} \to [0, 1]$. Define mass function m a mapping from $2^{\Theta} \to [0, 1]$, and A is a subset of frame of discernment Θ, denoted by $A \subseteq \Theta$, then, this mass function is given as:

$$m(\Phi) = 0, \quad \sum_{A \subseteq \Theta} m(A) = 1, m(A) > 0 \qquad (4)$$

In fact, m(A) is also called basic probability assignment (BPA), and it represent confident degree in event A. BPA is estimated by the obtained data, or given by experience. For event A, the lower and upper bounds of uncertainty interval are respectively called the belief function Bel(A) and the plausibility function Pl(A). Bel(A) and Pl(A) are given as:

$$Bel(A) = \sum_{B \subseteq A} m(B) \qquad (5)$$

$$Pl(A) = \sum_{B \cap A \neq \Phi} m(B) \qquad (6)$$

The interval $[Bel(A), Pl(A)]$ represents the belief degree of proposition A. Evidence from different sources can be aggregated by Dempster's combinational rule.

Fatigue crack propagation life can be obtained through the integration

$$N_f = \int_{a_0}^{a_c} \frac{dN}{da} da \qquad (7)$$

Uncertainty quantification (UQ) of the fatigue damage prognosis is essentially to obtain the uncertainty of N_f when uncertain parameters are given. The quantification of uncertainty framework involves three necessary steps: uncertainty representation, propagation and measurement. For uncertainty representation, the UQ framework uses all possible obtained values of material constants provided by experimental data in constructing separate belief structures for material constants. Then, differential evolution (DE) global optimization method is used for propagation of the represented uncertainty through fatigue crack growth model. Finally, observed evidence on simulation responses is used in determination of target propositions to estimate uncertainty measures, i.e. cumulative belief function (CBF) and cumulative plausibility function (CPF). Detailed explanation of uncertainty quantification procedure with corresponding results is provided below.

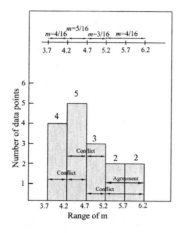
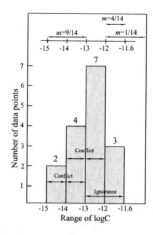

Figure 2. Data distribution and the corresponding belief structure for constants C and m.

Table 2. Belief structures of constants C and m for Ti-6Al-4V.

Interval No.	lg C Range	BPA	lg C Range	BPA
1	[−15.0, −14.0]	0.125	[3.7, 4.2]	0.250
2	[−14.0, −13.0]	0.250	[4.2, 4.7]	0.313
3	[−13.0, −12.0]	0.438	[4.7, 5.2]	0.187
4	[−13.0, −11.6]	0.187	[5.2, 6.2]	0.250

3.1 Uncertainty representation

For the purpose of uncertainty representation of material constants using evidence theory, separate belief structures for each uncertain parameter should be constructed. Salehghaffari and Rais-Rohani (Salehghaffari & Rais-Rohani & Marin & Bammann, 2012) developed a general methodology that can extract the necessary information from available data and express them in the mathematical framework of evidence theory. The methodology involves two principal steps (1) representation of uncertain parameters in interval form using all available data; and (2) categorization of different types of relationship between all adjacent intervals. Two adjacent intervals can be identified as having ignorance, conflict, or agreement relationship. The distinction depends on the number of data points in each interval.

Following the methodology, the data distribution and the corresponding belief structure for uncertain material constants C and m are constructed for Ti-6Al-4V aluminum alloy using the data provided in Table 1 as shown in Fig. 2 and Table 2.

A commonly used description of the relation between C and m is $\lg C = \alpha + \beta m$. Taking a statistical analysis on the data in Table 1, it is obtained that the correlation coefficient $R = 0.9062$. It indicated that lg C and m are highly linear correlated. Further consider the standard deviation, the correlated expression is obtained by least square method as Eq. (8) and Fig. 3.

$$\lg C = -7.5775 - 1.1384m \pm 0.414 \qquad (8)$$

3.2 Uncertainty propagation

As the focal element of uncertain variable represented by evidence theory is usually a series of intervals, uncertainty propagation is to find the maximum and minimum values of the system response. There are two main approaches to finding the maximum and minimum of the response

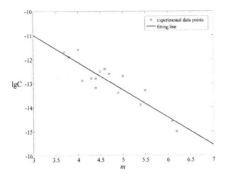

Figure 3. Fitting line of lg C and m.

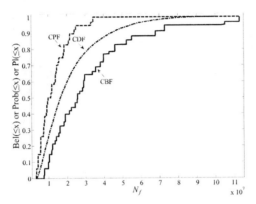

Figure 4. Cumulative distribution of fatigue lifetime N_f by evidence theory and probability theory.

value: sampling and optimization. The accuracy of sampling approach is highly dependent on the number of samples, and the process is very costly. On the contrary, optimization methods have the potential to dramatically reduce the computational work.

Differential evolution used in this paper is a novel evolutionary computation technique, which resembles the structure of an evolutionary algorithm (EA), but differs from traditional EA in its generation of new candidate solutions and by its use of 'greedy' selection scheme. The characteristics make DE a fast and robust algorithm as an alternative to EA. The DE algorithm is a population based algorithm like genetic algorithms using the similar operators: crossover, mutation, crossover and selection.

In this paper, m is chosen as the uncertain parameter, and C changes with m as Eq. (8). The propagation terminated at a critical crack size of 280 mm. This is determined from Eq. (3) using a fracture toughness value of 75 MPa·m$^{1/2}$. Basing on the belief structure of m given in Table 2, the results of propagation using evidence theory (DE algorithm) and probability theory (MC sampling) are shown in Fig. 4. For comparison with probability theory, the approximate PDFs of uncertain variables are obtained by the assumption that probability mass (BBA) in each interval is distributed uniformly.

3.3 Uncertainty measurement

In the context of evidence theory, uncertainty quantification requires assessment of uncertainty measures (belief, plausibility) for a defined target proposition set using the obtained propagated belief structure. Table 3 shows some information of interest from the results calculated by probability theory, interval method and evidence theory. The result of interval calculation is $[3.28, 105] \times 10^6$, which does not give any probabilistic information. The lower and upper bounds on

Table 3. Prognosis results by three methods.

Results \ Methods	Evidence theory	Probability	Interval analysis
Results with 95% guarantee probability	$[3.31, 7.59] \times 10^6$	4.72×10^6	—
Entire range	$[3.21, 113] \times 10^6$	—	$[3.21, 113] \times 10^6$

the estimated lifetime based on evidence theory are 3.21×10^6 and 113×10^6 respectively, and the range is larger than $[3.28, 105] \times 10^6$. This shows that the uncertainty propagation process based on DE algorithm does better in searching the maximum and minimum values in focal element intervals.

The accuracy of probability (MC) results could be improved by increasing the number of samples. 10000 runs are carried out for the result in Fig. 5, and the execution time is over an hour compared with 50 seconds for the interval analysis and several minutes for DE based evidence theory, and the CDF curve obtained by MC is totally in the region enclosed by the belief function and plausibility function curves. For a 95% guarantee probability, evidence theory gives the interval as $[3.31, 7.59] \times 10^6$, according to which the conservative estimation of lifetime can be taken as the minimum 3.31×10^6, while the result of probability theory is 4.72×10^6.

4 CONCLUSIONS

As for the uncertainty induced by the lack of knowledge or incomplete, inaccurate, unclear information, simulation and measurement in fatigue damage prognosis, the method based on evidence theory is proposed as an alternative to the classical probability theory to handle uncertainty. In order to alleviate the computational difficulties in the evidence theory based UQ analysis, a differential evolution (DE) based interval optimization for computing bounds method is developed.

In this study, the computational efficiency and accuracy of the proposed method are demonstrated by experimental data of Ti-6Al-4V aluminum alloy specimens. With comparison of probability theory and interval method, the proposed method can quantify epistemic uncertainty successfully and has significant advantages in saving time and providing more information.

REFERENCES

Bae, H.R., Grandhi, R.V. & Canfield, R.A. 2004. An approximation approach for uncertainty quantification using evidence theory. *Reliability Engineering and System Safety* 86:215–225.

Bao, Y., Li, H. & An, Y. 2012. Dempster- Shafer evidence theory approach to structural damage detection. *Structural Health Monitoring* 11:13–26.

Helton, J.C., Johnson, J.D., Oberkampf, W.L. & Storlie, C.B. 2007. A sampling-based computational strategy for the representation of epistemic uncertainty in model predictions with evidence theory. *Computer Methods in Applied Mechanics and Engineering* 196:80–98.

Keith, W. & Graeme,M. 2008. Prognosis under uncertainty-An idealized computational case study. *Shock and Vibration* 15:231–243.

Morten, A. & Stuart, R. 2001. Fatigue-crack growth in Ti-6Al-4V-0.1Ru in air and seawater. *Metallurgical and materials transactions* 32A:2297–2314.

Salehghaffari, S., Rais-Rohani, M., Marin, E.B. & Bammann, D.J. 2012. A new approach for determination of material constants of internal state variable based plasticity models and their uncertainty quantification. *Computational Materials Science* 55:237–244.

Sankararaman, S., Ling, Y., Shantz, C. & Mahadevan, S. 2009. Uncertainty quantification in fatigue damage prognosis. *Annual Conference of the Prognostics and Health Management Society*:1–13.

Surace, C. & Worden, K. 2011. Extended analysis of a damage prognosis approach based on interval arithmetic. *Strain* 47:544–554.

Storn, R. & Price, K. 1997. Differential evolution-A simple and efficient adaptive scheme for global optimization over continuous spaces. *Journal of Global Optimization* 11:341–359.

Side force proportional control with vortex generators fixed on tip of slender

J. Zhai, W.W. Zhang & C.Q. Gao
Northwest Polytechnical University, Xian, Shanxi, China

Y.H. Zhang & L. Liu
Air Force Engineering University of PLA, Xian, Shanxi, China

ABSTRACT: There is a large side force on slender at high angle of attack. In this thesis, vortex generators (a micro delta wing) are fixed on the tip of slender model, which have 10° semi-apex angle, rotate the vortex generators along the model axis, the side force change approximate linear proportional at high angle of attack (AOA). Moreover, the control effect of vortex generators' half-wingspan length, sweepback of lead edge, AOA and Reynold number had been researched. Base on this research, the vortex generators with 6 mm half-wingspan and 45° sweepback had appear better control effect, it can control the side force approximate linear proportional at a series of Re and AOA. This device is so simple that it may be a useful appliance in the future.

1 INTRODUCTION

The leeward of slender appear complex multi-vortex flow at high AOA, this phenomenon bring a side force as large as lift, which induce yaw and roll moment. In addition, the magnitude and direction of this side force is uncertain, so this phenomenon is called "Phantom yaw". At the same time, the normal aerodynamics rudders in the leeward cannot provide necessary landscape orientation control force, so control the aircraft become difficultly.

With deep research of side force at high AOA, researchers had got that micro stir on slender tip change the asymmetric vortex obviously. When put some passive or active stir on slender tip, the asymmetric flow can be controlled. Early research is mainly focused on weaken or even eliminate the asymmetric force, such as: change the blunt or coarseness of slender tip, fix strake or nose-boom at slender tip, all of above techniques don't need import energy, so they are called passive control methods.

Recently, more active control methods are used to control the asymmetric flow at high AOA. Various deployable strakes and jet blowing mechanisms have been studied and reviewed by Malcolm (Malcom, G. 1991 & 1993) and Williams (Williams, D. 1997). However, most of these active control techniques are based on steady methods, that is, control actuation is a static or steady excitation.

Williams and Bernhardt (Williams, D. 1990 & Bernhardt, J. 1998) used an unsteady jet blowing technique near the forebody tip to control the asymmetric flow. The vortex position can continuously vary when the control input is at $\alpha = 45°$, which is consistent with a convective type of instability, thus proportional control of the side force can be achieved. However, this process is not feasible under the condition $\alpha = 55°$, in which bistable modes dominate. From this study, many researchers realized that the side force may continuously respond to dynamic alternating excitations. Hanff et al. (Hanff, E. 1999) used alternate blowing from two forward-facing nozzles near the model tip to switch vortices between their bistable states with given duty cycles at high enough frequencies. Liu et al. (Liu, F. 2008) used a pair of SDBD plasma actuators as alternating excitations to replace blowing nozzles. Gu and Ming (Gu, Y. 2003 & 2004) used an oscillating microtip strake to control asymmetric vortices. Asymmetric side force will be thoroughly eliminated when the microtip strake vibrates with an appropriate frequency and amplitude driven by a power unit, and the effective angle of attack ranges from 30° to 80°. The flow and the side force will continuously respond by varying the mean setting angles of the oscillation of the strake.

The aforementioned studies demonstrate the feasibility of using dynamic alternating excitations not only to suppress asymmetric aerodynamic force, but also to provide lateral control of aircraft at high angles of attack. In those studies, however, the bistable vortices can only be controlled when the active alternating excitations achieve enough intensity, otherwise, the excitation will be whelmed by the strong free stream. The other shortcoming of these techniques is the complexity of the excitation facility, which is not convenient to set in the small space of the slender tip.

Vortex generator is an ingenious flow control device, which is widely used to control flow separate in boundary layer, airfoil, increase-lift device, and engine inlet. The local AOA of above vortex generators are smaller than its vortex break angle, in fact their trailing vortex bring energy to downstream flow, then control the flow separate. In this thesis, vortex generators (a micro delta wing) are installed on a slender tip, which have 10° semi-apex angle, rotate the vortex generators along the model axis, the side force is approximate proportional change. Comparing with the theory of controlling flow separate of normal vortex generator, the vortex generated by vortex generator has broken in this thesis's angle range; in fact the trailing flow is a high-frequency pulse flow. This research just uses the pulse flow induced by the broken vortex to control the asymmetry force of slender at high AOA.

2 MODEL AND EXPERIMENT DEVICE

2.1 Wind tunnel and slender model

This experiment was taken at low-speed tunnel in AFEU, experiment section of tunnel is rectangle, high 1.2 m, width 1 m, length 1.5 m, velocity range 5 m/s~75 m/s. Experiment model is a conical-cylinder, refer to reference paper (Modi, V.J. 1983). The length of model is 600 mm, diameter of bottom is 100 mm, the length of conical section is 300 mm, as sketched in Fig. 1. The head conical section can rotate with the model axis, and the bottom circle of head conical section was marked, 5° interval, the leeward of symmetry plane of transition section was marked benchmark. The model is support through trail, as sketched in Fig. 2.

The experiment Re of reference paper is 9.45×10^4, the relation between Re and velocity in this research is as Table 1.

2.2 Vortex generator

Vortex generators (VG) used in this experiment are micro delta wing indeed which is made by aluminum (thick 0.18 mm). Vortex generators' parameters: half wingspan length – b, sweepback of lead edge – α, dihedral angle – β, as sketched in Fig. 3. Branch of Vortex generators is made by steel cylinder whose diameter is 1 mm. There are 8 different Vortex generators in this experiment, the corresponding parameters b, α, β are VG-1 (8 mm, 37°, 15°), VG-2 (4 mm, 37°, 15°), VG-3 (8 mm, 37°, 60°), VG-4 (4 mm, 37°, 60°), VG-5 (6 mm, 37°, 0°), VG-6 (6 mm, 45°, 0°), VG-7 (6 mm, 70°, 0°), VG-8 (6 mm, 80°, 0°). In Fig. 4 VG are fixed on model.

Table 1. Relation between Re and wind velocity.

Velocity m/s	8	10	12	14	15	16	18	20
Re/10^4	5.04	6.3	7.56	8.82	9.45	10.08	11.34	12.6

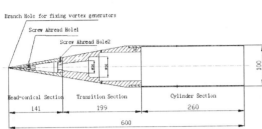

Figure 1. Sketch of model.

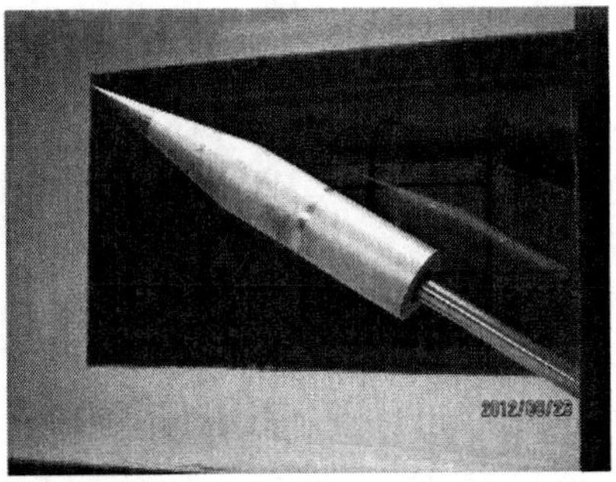

Figure 2. The fixed sketch of slender model.

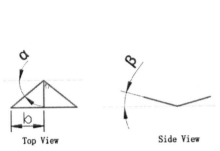

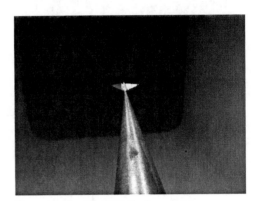

Figure 3. Sketch of VG.

Figure 4. Sketch of VG fixed on model.

3 EXPERIMENT RESULT AND ANALYSIS

3.1 *Side force experiment characteristic of bareheaded model*

The side force result with AOA change of bareheaded model at $Re = 9.45 \times 10^4$ is showed in Fig. 5, this result basic match with the result of reference paper which is at the same experiment Re. In Fig. 5, we can see the results in this experiment are smaller than that in reference paper's experiment at 55° AOA, and the begin angle of side force is at 30° in this experiment, which is at 25° in reference paper.

For analysis the difference between this experiment and the reference paper's experiment, there is the side force change with the roll angle at 27.5°, as sketched in Fig. 6. In this Fig. the coefficient of side force change remarkable, form −0.78∼1.5, but the trend of change is smoothly, there is no jumping phenomenon. This result shows that the flow has been asymmetry, but flow continuum depend on boundary change, it is a convection instability flow indeed.

In Fig. 5, we can see that at 45° AOA, the side force of model used in this experiment and reference paper's experiment reach the max. For study the flow instability property of above condition, there is side force change of bareheaded model with rotate the conical section at 45° AOA in Fig. 7. From the Fig. 7, we can see the experiment model have "double-stability" property at 45°. Consequently, 45° AOA is the base AOA used to validate the side force control device in this experiment.

3.2 *Side force experiment characteristic of model with head-vortex generators*

In Fig. 8, two VGs with different half-wingspan are fixed on the model at AOA = 45°. From the Fig., VG-2 can change the magnitude of side force smoothly, but it can't change the force direction

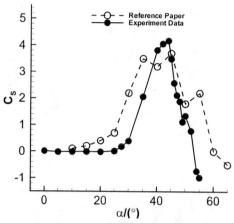

Figure 5. Side force with AOA change.

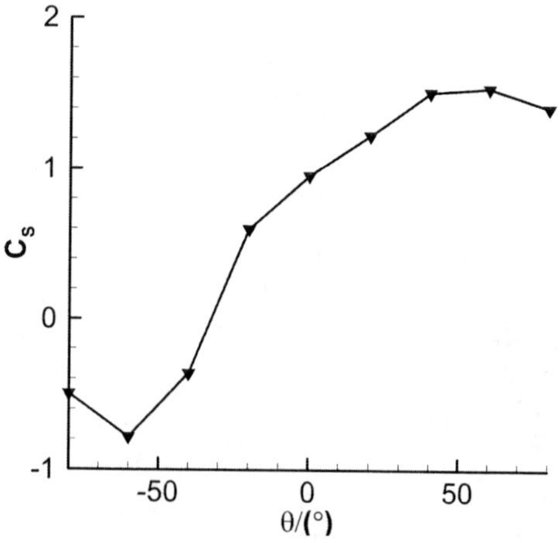

Figure 6. Side force with roll angle change at AOA = 27.5°.

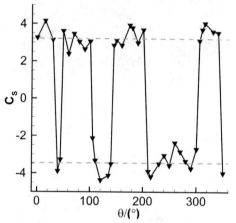

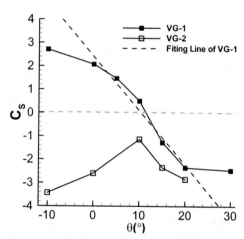

Figure 7. Side force of bareheaded slender with roll angle change at AOA = 45°.

Figure 8. Side force with roll angle change.

between −10° and 30° roll angle. In the same roll angle range, VG-1 can not only change the magnitude of side force but also the direction, the breadth of side force coefficient is −2.7∼2.5. This indicates that only the vortex generators are big enough, their pulse separate flow can control the asymmetry lee-vortex. Also, VG-1 can change the side force with roll angle approximate linear, this linear control relationship is a convenience for design the control law.

From the Fig. 9, all the four VGs with different sweepback can control the side force direction availability, but only the one with 45° sweepback (VG-6) can change the magnitude of side force continuously. Others VGs change the side force from −3 to 3 only in 5° range, smaller roll angle interval maybe can change the side force continuously, but that is out the limit of this experiment.

From the Fig. 9, we can see VG-6 is the best at control the side force, so the control effect of it at different AOA have been taken on. In Fig. 10, the side force with roll angle change at 40°, 45°, 50°, 55° AOA are showed. It can change the side force approximately linear from 40° to 55°.

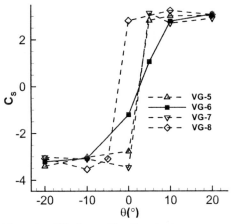 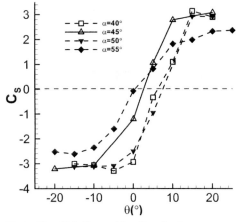

Figure 9. Side force with roll angle change.   Figure 10. Side force with roll angle change.

4 CONCLUSION

It develops a new control side force method in this thesis. By fixed vortex generators on the head of slender, rotate the head-conical section to change the roll angle, it could change the side force approximately linear. The feasibility is validated by wind experiment. The influence of the vortex generators magnitude and sweepback of vortex generators in the side force control are researched. Vortex generators whose parameters are (6 mm, 45°, 0°) has better control effect. It can change the side force approximately linear with rotate roll angle at 40°, 45°, 50°, 55°. This experiment is underway, more data about flow display and relative CFD may show the mechanism more convincingly.

REFERENCES

Bernhardt, J. E. & Williams, D. R. 1998. Proportional control of asymmetric forebody vortices. *AIAA Journal* 36(11): 2087–2093.

Gu, Y. & MING, X. 2003. Forebody vortices control using a fast-swinging micro-tip-strake at high angles of attack, ACTA Aeronautica Et Astronautica Sinica 24(2): 102–106.

Gu, Y. 2004. Control of forebody flow asymmetry in high-angle flows, PHD thesis, Nanjing University of Aeronautics and Astronautics.

Hanff, E., Lee, R. & Kind, R. J. 1999. Investigations on a dynamic forebody flow control system. *Proceedings of the 18th International Congress on Instrumentation in Aerospace Simulation Facilities, Inst. of Electrical and Electronics Engineers, Piscataway, NJ*, 1–28 September 1999.

Liu, F., Luo, S. & Gao, C. 2008. Flow control over a conical forebody using duty-cycled plasma actuators. *AIAA Journal* 46(11): 2969–2973.

Modi, V. J. & Ries, T. 1983. Aerodynamics of pointed forebodies at high angles of attack. AIAA Paper 83-2117.

Malcolm, G. 1991. Forebody vortex control. *Progress in Aerospace Sciences* 28(3): 171–234.

Malcolm, G. 1993. Forebody vortex control: a progress review. AIAA Paper 93-3540.

Williams, D. & Bernhardt, J. 1990. Proportional control of asymmetric forebody vortices with the unsteady bleed technique. AIAA Paper 1990-1629-273.

Williams, D. 1997. A review of forebody vortex control scenarios. AIAA Paper 97-1967.

Characteristic of unsteady aerodynamic loads with a synthetic jet at airfoil trailing edge

C.Q. Gao, W.W. Zhang & J. Zhai
Northwestern Polytechnical University, Xi'an, China

X.B. Liu
The 8th Institute of Shanghai Academy of Spaceflight Technology, Shanghai, China

ABSTRACT: In order to study the characteristics of the synthetic jet near the trailing edge of an airfoil, an experimental project in wind tunnel is designed for a NACA0015 airfoil. It contains the design of the experimental model, the design of low frequency high power synthetic jet actuator and the method of measurement. The results show that the interaction of the synthetic jet with the upstream can modify the aerodynamic loads obviously. The amplitudes of the lift and momentum coefficients show a good proportional relationship to the amplitude of the jet velocity, and the amplitudes of the loads will decrease with the increase of the reduced frequency. When the stroke of the synthetic jet is fixed, changing the rotate speed of the servo-motor, there is a linearly relation between the amplitude of the loads coefficients and the rotate speed.

1 INTRODUCTION

In recent years, many new aerodynamic control methods have been developed to improve the traditional control device actuation mechanism. The Gurney flap (Liebeck 1978) is a vertical tab added to the trailing edge on the pressure side of a wing. Several researches carried by Jeffrey et al. (2000) and Zerihan (2001) show that the Gurney flap can increase lift and moment significantly. Inspired by the Gurney flap, Bieniawski & Kroo (2003) design the Micro-Trailing Edge Effectors(MiTEs). Lee & Kroo (2002) applied the MiTEs in the research of active control of flutter. As another aerodynamic control method, Jet flap is an airstream ejected from the wing's trailing edge with certain direction and momentum coefficients. The study of jet flap begins around 1950s, and a recent study has developed a hinge-less control method based on the jet flap method, proposed by Tranb & Lund (2006, 2008).

Synthetic jet is widely concerned in recent studies (Smith & Amitay 1998, Roeland & Mostafa 2007), for its application potentials in engineering with many advantages, such as structure simplicity, light weight and no-external-gas-source-quality. It extends a new field of research by developing the hinge-less control (Traub & Miller 2004, Patrick & Douglas 2009) and aeroelastic active control (Preetham et al. 2000) based on the technology of synthetic jet. In this field, Donnel & Marzocca (2007) and Schober & Marzocca (2006) have carried out a series of active flutter suppression study.

The key to conduct such an application is to design a synthetic jet actuator with higher momentum coefficients. In studying the aircraft hinge-less control, Traub & Miller (2004) designed a compact high-power synthetic jet actuator, which can speed the jet up to 100 m/s. For more general and complex cases, such as the wing aeroelastic control, the optimum jet position is in the vicinity of the trailing edge. Therefore, it is necessary to design a kind of trailing edge of low-frequency high-power synthetic jet actuator. However, the ongoing study of this is still rare. This paper proposed a set of wind tunnel experimental model. The objectives of these experiments are to investigate the effect of the synthetic jet at airfoil trailing edge on the aerodynamic loads.

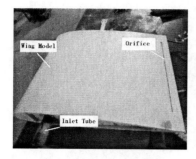 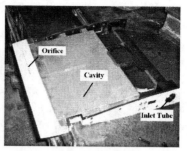

a. The outward of the model   b. The internal structure

Figure 1. Schematic of the wing model.

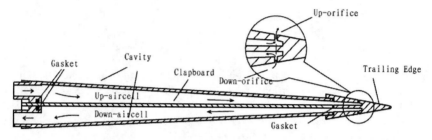

Figure 2. Schematic of the cavity body.

2 EXPERIMENT SETUP

The experiment was conducted in low-turbulence wind tunnel of Northwestern Polytechnical University (LTWT). The size of the 2-dimensional test section is 0.4 m × 1.0 m. Tunnel's free stream turbulence is less than 0.3%. The model was a rectangular wing section to simulate the two-dimensional flow situation. A NACA 0015 profile was used. Geometric details of the model are 500 mm in chord and 398 mm in span. The orifice is located at 95% of the chord from the wing's leading edge. It is 1 mm in width and 300 mm in length, as shown in Figures 1–2.

The Figure 3 shows the assembly diagram of the synthetic jet. Driven by the servo-motor, the turntable rotates at uniform speed, which will lead cylinder piston to start a reciprocating movement by the linkage mechanism. Then this reciprocating movement will lead to the expansion and compression of spaces within the cylinder, and further lead to the periodical "blowing/sucking" on the trailing edge. The two cylinders are symmetrically installed. When one side is in the "blowing" process, the other side will be precisely in the "sucking" process. The model installs in the wind tunnel as shown in Figure 4.

To facilitate the study of the test parameters in the experiment, the turntable uses a special design, which can achieve 12 kinds of the cylinder stroke. Experimental data are collected by the dynamic data acquisition system, which is composited by DH-3846 strain amplifier and LMS dynamic data acquisition devices with the sampling frequency of 400 Hz, the sample spectrum line numbers of 4096 and the sampling time of $4096/400 = 10.24$ s.

3 CHARACTERISTICS OF TYPICAL DYNAMIC LOADS

3.1 *The effect of the synthetic jet*

Results in Figure 5 show that the two comparative results in the upstream velocity of $U_\infty = 0$ m/s and $U_\infty = 20$ m/s, including the lift, the moment, and the strain response of the orifice, when the

Figure 3. Final assembly of the synthetic jet.

Figure 4. Schematic of the model installation.

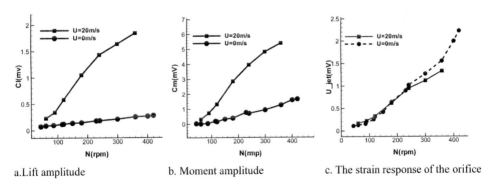

a. Lift amplitude   b. Moment amplitude   c. The strain response of the orifice

Figure 5. The comparison of the amplitude for different upstream velocity, $U_\infty = 0$ m/s and $U_\infty = 20$ m/s.

cylinder stroke is fixed at L = 100 mm. The amplitude of the loads response(lift and moment) in the upstream velocity of $U_\infty = 20$ m/s is significantly larger than that in steady flow, even though that two strain responses of the orifice are the same(see Fig. 5(c)). Thus, the interaction of the jet and the upstream can indeed modify the loads distribution over the wing.

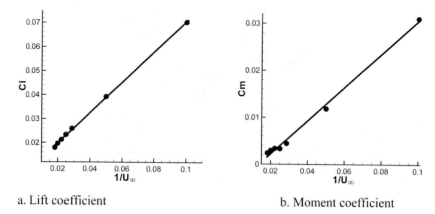

a. Lift coefficient　　　　　　　　　　　　b. Moment coefficient

Figure 6. Effect of upstream velocity on the loads coefficients at L = 10 mm, N = 720 rpm.

3.2 Research on synthetic jet parameters

The jet momentum coefficient is an important parameter of the synthetic jet, which is defined as follow:

$$C_\mu = \frac{2\rho_{jet} U_{jet}^2 d}{\rho_\infty U_\infty^2 c} \quad (1)$$

where ρ_{jet} and U_{jet} are the density and velocity of the jet, d is the orifice characteristic length, ρ_∞ and U_∞ are the density and velocity of the upstream, and c is the chord of the airfoil.

Williams et al. (1961) noted that the aerodynamic loads acting on the airfoil and the square root of the momentum coefficient have an approximate linear relationship in his study in 1961, namely:

$$C_l \propto \sqrt{C_\mu}, \quad C_m \propto \sqrt{C_\mu} \quad (2)$$

However, in actual experimental operation, we just change the velocity of the jet U_{jet} and the velocity of the upstream U_∞ to achieve a variable jet momentum coefficient, confined by the experimental conditions, especially the constraint of the fixed model.

At low-power (L = 10 mm) and medium-frequency (N = 720 rpm) synthetic jet working condition, the relationship of the lift coefficient and moment coefficient with the reciprocal upstream velocity($1/U_\infty$) is shown in Figure 6. The upstream velocities U_∞ are 10 m/s, 20 m/s, 35 m/s, 40 m/s, 45 m/s, 50 m/s and 55 m/s, respectively. As revealed in Figure 6, the aerodynamic loads, including the lift coefficient and the moment coefficient, and the reciprocal of the speed of the upstream showed an approximate linear relationship.

Analogously, under the fixed upstream velocity $U_\infty = 20$ m/s and a certain frequency (N = 229 rpm) condition, we got a set of cylinder strokes, which is the power source of the jet, by adjusting the position of the hinge shaft on the turntable. The investigated cylinder strokes are 10 mm, 25 mm, 40 mm, 55 mm, 70 mm, 85 mm, and 100 mm. The results are shown in Figure 7. Except in two data points (10 mm and 25 mm), the amplitude of the lift and moment coefficient exhibit fine linear characteristics with the cylinder stroke (Figs. 7a, b). Such relationship is also satisfied in the jet velocity and the cylinder stroke case. Thus the amplitude of the lift and moment coefficients shows a fine proportional relationship with the amplitude of the jet velocity, namely:

$$C_l \propto U_{jet}, \quad C_m \propto U_{jet} \quad (3)$$

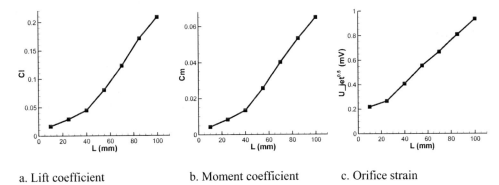

a. Lift coefficient   b. Moment coefficient   c. Orifice strain

Figure 7. Effect of the jet velocity on loads coefficients at $U\infty = 20$ m/s and $N = 229$ rpm.

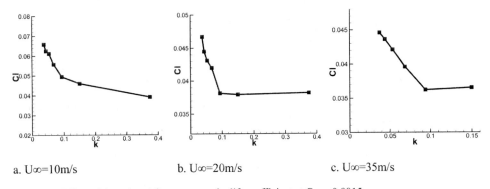

a. $U\infty=10$m/s   b. $U\infty=20$m/s   c. $U\infty=35$m/s

Figure 8. Effect of the reduced frequency on the lift coefficient at $C\mu = 0.0015$.

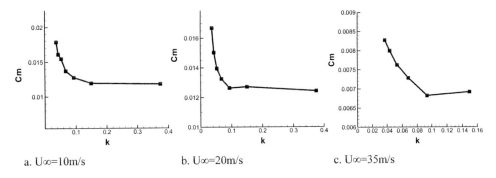

a. $U\infty=10$m/s   b. $U\infty=20$m/s   c. $U\infty=35$m/s

Figure 9. Effect of the reduced frequency on the moment coefficient at $C\mu = 0.0015$.

Another important parameter is the jet's reduced frequency k, when the synthetic jet is working at unsteady jet condition. k is defined as $k = b\omega/U_\infty$, where $b = c/2$ is the reference length, c is the chord of the airfoil and U_∞ is the velocity of the upstream.

When the jet momentum coefficient is fixed at the theoretical value ($C_\mu = 0.0015$), we change the reduced frequency k through a method combination of adjusting the cylinder stroke and the rotating speed of the servo-motor. The velocities of the upstream are 10 m/s, 20 m/s and 35 m/s, respectively. In Figures 8–9, the curves show that the effect of the reduced frequency k on the loads coefficients (lift and moment), that the amplitudes of the loads will decrease with the increasing reduced frequency. When the reduced frequency k is small, generally less than 0.1, the amplitudes of the loads change quickly with a changing k, but when the reduced frequency k is greater than 0.1, the above changes will become gentle.

4 CONCLUSIONS

A study of the experimental results obtained with synthetic jet near the trailing edge of NACA0015 airfoil has shown that:

1. The interaction of the synthetic jet and the upstream can modify aerodynamic loads. The effect of such interaction is larger than the jet's recoil effect on the model.
2. The amplitudes of the lift and moment coefficients show a good proportional relationship to the amplitude of the jet velocity.
3. The amplitude of the loads will decrease as the reduced frequency increases. When the reduced frequency is small, the amplitudes of the loads change quickly, but as the reduced frequency become greater, the corresponding changes of loads will become gentle.

REFERENCES

Bieniawski, S. & Krool, M. 2003. Flutter suppression using micro-trailing edge effectors. *AIAA Paper*, 2003–1914.
Donnel, K. & Marzocca, P. 2007. Design of a wind tunnel apparatus to assist flow and aeroelastic via net zero mass flow actuators. *AIAA Paper*, 2007–1711.
Lee, H.T. & Kroo, I.M. 2002. Flutter suppression for high aspect ratio flexible wings using micro flaps. *AIAA Paper*, 2002–1717.
Jeffrey, D. & Zhang, X. & Hurst, D.W. 2000. Aerodynamics of gurney flaps on a single-element high-lift wing. *Journal of Aircraft* 37(2): 295–301.
Liebeck, R. H. 1978. Design of subsonic airfoils for high lift. *Journal of Aircraft* 15(9): 547–561.
Patrick, R.S. & Douglas R.S. 2009. Aerodynamic control of a rectangular wing using gurney flaps and synthetic jets. *AIAA Paper*, 2009–886.
Preetham, P. R. & Thomas W. S. & Othon K. R. 2000. Control of aeroelastic response via synthetic jet actuators. *AIAA Paper*, 2000–1415.
Roeland, D.B. & Mostafa, A. 2007. Aeroelastic control and load alleviation using optimally distributed synthetic jet actuators. *AIAA Paper*, 2007–2134.
Smith, D. R. & Amitay, M. 1998. Modification of lifting body aerodynamic using synthetic jet actuators. *AIAA Paper*, 98–0209.
Schober, S. & Marzocca, P. 2006. Reduced order models for synthdtic jet actuators on a lifting surface for optimization and control. *AIAA Paper*, 2006–1906.
Traub, L.W. & Miller, A. 2004. Distributed hinge-less flow control and rotary synthetic jet actuation. *AIAA Paper*, 2004–224.
Traub, L.W. & Lund, D. 2006. Preliminary flight test of a hing-less roll control effecter. *Journal of Aircraft* 43(4): 1244–1246.
Traub, L.W. & Agarwal, G. 2008. Aerodynamic characteristic of a furney/jet flap at low reynolds numbers. *Journal of Aircraft* 45(2): 425–429.
Williams, J. & Butler, S.F. & Wood, M.N. 1961. The aerodynamics of jet flaps. *R.&M* 3304.
Zerihan, J. & Zhang, X. 2001. Aerodynamics of gurney flaps on a wing in ground effect. *AIAA Journal* 39(5):772–780.

Simulation analysis on surface subsidence near tunnel influenced by lining partially flooded in seepage field

Zhenhua Jiang
Jiangsu Traffic Planning and Design Institute Co., Nanjing, Jiangsu Province, China

Fei Ma
Changjiang Rive Adminstration of Navigational Affairs. MOT, Wuhan, Hubei, China

Xupeng Dong & Xiaochun Zhang
Intelligent Transportation System Engineering Research Center, Southeast University, Nanjing, Jiangsu Province, China

ABSTRACT: The long-term effect of groundwater seepage and seepage tunnel lining partially flooded on upper surface construction and road is an unavoidable problem for urban underground space development. By establishing the Finite Element Model (FEM) for tunnel lining partially flooded in seepage field, surface subsidence is analyzed in four situations, namely different lining parts, different permeability ratio, different tunnel depth with the same underground water level and different underground water level with the same tunnel depth. The laws of surface subsidence and stability time are gained, which give guide to construction and post-surface subsidence monitoring in practice.

1 INTRODUCTION

With rapid urbanization, it has become an inevitable choice for urban modernization to develop rationally underground space, aiming to address the growing demand of transport. However, all kinds of problems have come into being along with the construction of subway in city. For example, in southern China with rich groundwater, the seepage field around the tunnel always change due to the effect of tunnel excavation and long-term cyclic loading, which has caused accidents. Some data indicate that quite a few geotechnical accidents in domestic are caused by large area of rock (soil) instability due to water seepage in soil. A direct consequence caused by the loss of groundwater is soil consolidation and the percentage of consolidation subsidence in total surface subsidence can exceed 60%.

The surface subsidence caused by city tunnel seepage has become a hot and difficult issue in the field of tunnel construction. A series of studies have been launched by domestic and international scholars and engineers. These studies include the variation of pore water caused by shield tunnel construction (Kerry et al. 1993; AbuFarsakh et al. 1999), the long-term characteristics of groundwater movement after shield tunnel construction (JH Shin et al. 2002), the surface subsidence and settling time (Mair, 1997; Taylor, 1994; Shirlaw, 1995; O'Reilly, 1991) and the impact analysis of lining partially flooded on surface subsidence, sedimentation tanks and ground loss (Dongmei et al. 2005; Shaorui et al. 2005; Fei, 2008). Thus, the long-term effect of groundwater seepage and tunnel lining on the upper surface construction and road cannot be.

This paper select a city sedimentary soil parameters and use the numerical software of FLAC3D to simulate the surface subsidence near tunnel in for different situations, namely different lining parts, different permeability ratio, different tunnel depth with the same underground water level and different underground water level with the some tunnel depth.

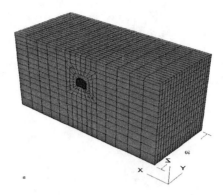

Figure 1. Three-dimensional numerical calculation model.

Table 1. Geological parameter table in a region.

Water ratio W/%	Wet density ρ g/cm^3	Moisture %	Soil weight γ kN/m^3	Void ratio e	Plasticity index I_P	Liquidity index I_L
32.6	1.82	37.7	18.3	1.055	14.8	1.07

Compressibility α MPa	Compress modulus E MPa	cohesive force kPa	Internal friction angle $\varphi/°$	Poisson ratio μ	Coefficient of vertical permeability K_v 10^{-7} cm/s	Coefficient of horizontal permeability K_b 10^{-7} cm/s
0.40	3.812	18.6	20.6	0.34	5.31	16.1

2 NUMERICAL SIMULATION ANALYSIS MODEL

2.1 Model establishment

In this section, the following assumptions are made firstly:

1) The soil is isotropic elastic-plastic body;
2) The soil is saturated soil;
3) The fluid is isotropic and one-way flow.

The three-dimensional numerical model (unit: m) with horseshoe-shaped cross-section is shown in Figure 1. The unit division takes locally encrypted at tunnel section and then spread out.

The model boundary conditions are fully constrained (ie, displacement constrains) at the bottom of the model. Four sides respectively take the displacement constrained with the perpendicular direction to side.

Seepage is treated as head boundary. Bottom aquitard is treated as undrained boundary. Lining seepage is regarded as permeable boundary. The next content analyzes different situations in terms of permeable boundary.

Mohr-Columb elastoplastic constitutive model is used and soil parameters in a region are shown in table 1. Linear elastic constitutive model is used for tunnel lining. Material parameters include elastic modulus and Poisson ratio. Fluid model is isotropic model (ie. The fl_iso model in the software of FLAC3D).

Based on this model, the surface subsidence near tunnel in four situations are analyzed in this paper, which are different lining parts, different permeability ratio, different tunnel depth with the same underground water level and different underground water level with the some tunnel depth.

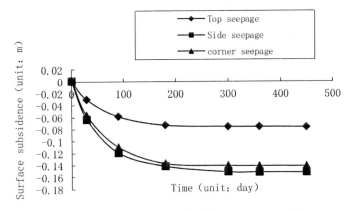

Figure 2. The surface subsidence versus time curve with different seepage position.

3 NUMERICAL ANALYSIS ON SURFACE SUBSIDENCE NEAR TUNNEL DUO TO PARTIALLY FLOODED IN SEEPAGE FIELD

3.1 Surface subsidence with different seepage position

According to the survey about tunnel seepage position, this paper analyzes the following three different lining seepage situations. That is the top arc parts, side parts and corner parts of tunnel. The tunnel depth is 20 m. The distance between underground water level and surface is 2 m. Permeability ratio is 1 (ie, soil and seepage lining parts share the same permeability). The computation time of seepage is 450 days.

The surface subsidence of middle point above the tunnel versus time curve in three different lining positions is shown in figure 2. According to the figure, it shows that for the top arc and corner seepage case, the time when the subsidence becomes stable is 200 days while for the side seepage that is 300 days. Besides, the surface subsidence in side and corner seepage case is almost the same (about 15 cm) and is twice as big as that in top seepage case.

3.2 Surface subsidence with different permeability

Considering the partial permeability of tunnel lining, this paper adopts the concept of limited permeability proposed by O'Reilly. The permeability of lining is denoted by the relative permeability. Then, the permeability of lining can be discussed under the following four circumstances:

(1) Undrained: Tunnel lining is undrained boundary;
(2) Partly drainage: Tunnel lining is drainage boundary;
(3) Partly drainage: Tunnel lining is drainage boundary;
(4) Completely drainage: Tunnel lining is drainage boundary.

According to the calculation results in section 3.1, the case of tunnel corner seepage is used. The surface subsidence with three different permeation ratios (1, 0.1 and 0.01) is analyzed. The upper surface subsidence variation with time under the condition of three different permeation ratios is described in Figure 3.

As is shown in Figure 3, the surface subsidence with permeation ratio of 1 (ie, fully seepage case) is largest and after 200 days, the subsidence becomes stable with the number of 14 cm. However, those in the other two cases have no trend to be stable. Moreover, the slop of subsidence-time curve with permeation ratio of 0.1 is approximate ten times as big as that with permeation ratio of 0.01.

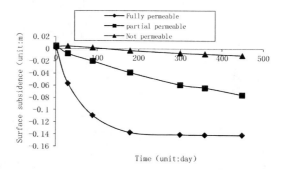

Figure 3. The surface subsidence with different permeation ratio.

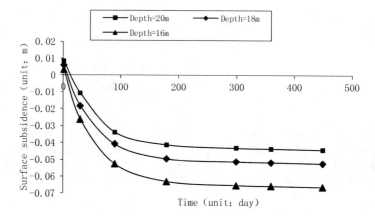

Figure 4. The subsidence versus time curve with different tunnel depth.

3.3 *Surface subsidence impacted by different tunnel depths with the same groundwater level*

Lining corner seepage is also used in this section and the distance between groundwater level and the tunnel top is 8 m. Tunnel depth is 20 m, 18 m and 16 m. Designated time is 450 days.

Figure 4 depicts the subsidence of ground middle point versus time curve in the condition of different tunnel depth with the same groundwater level. It can be seen that the tunnel has the largest subsidence (about 7 cm) when the tunnel depth is 16 m and has the smallest subsidence (about 4 cm) when the depth is 20. It can be got that with the same time and groundwater level, surface subsidence decreases as the depth increases. The figure also shows that the time when subsidence becomes stable with three different conditions is almost the same. So, the tunnel depth has no significant influence on the stable time.

3.4 *Surface subsidence impacted by different groundwater level with the same tunnel depth*

Select corner seepage in this section and tunnel depth is 20 m (ie, h is 20 m in Figure 3-2). The groundwater level is 8 m, 10 m, 12 m and 14 m. Time is 450 m for calculation.

Fig. 5 is the subsidence of ground middle point versus time curve in the condition of different groundwater levels with the same tunnel depth. It can be seen that the tunnel has the largest subsidence (about 14 cm) with the groundwater level of 16 m and has the smallest subsidence (about 4 cm) the groundwater level of 8 m. It can be analyzed that with the same time and tunnel depth, surface subsidence increases as the groundwater level increases. The figure also shows that the time when subsidence becomes stable with four different conditions is almost the same. So, the groundwater change is not significant to the stable time.

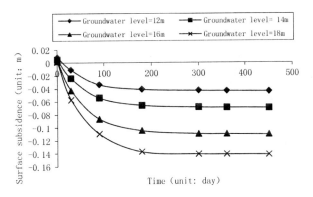

Figure 5. The subsidence of ground middle point versus time curve with different groundwater.

4 CONCLUSION

This paper simulates the surface subsidence caused by lining partially flooded in some groundwater-rich area with the help of the Finite Element Method. Surface subsidence is analyzed in four situations, which different lining parts, different permeability ratio, different tunnel depth with the same underground water level and different underground water level with the some tunnel depth, respectively. Based on numerical analyses, it can be concluded that:

(1) In terms of lining seepage position, surface subsidence in the top of lining is smallest and that at corner and lining middle position is almost the same.
(2) With the same seepage condition and seepage time, the surface subsidence decreases non-linearly with the decrease of k_t/k_v while the seepage time increases with the decrease of k_t/k_v.
(3) With the same groundwater level and seepage time, surface subsidence decreases as the tunnel depth increases. With the same tunnel depth and seepage time, that increases with the increase of groundwater level. However, there is no influence between seepage stable time and groundwater level. That is to say, stable time is only relative to permeation ratio.

REFERENCES

Dongmei, Zhang and Hongwei, Zhang etc. Study about long-term surface subsidence effected by lining partial drainage tunnel [J]. Geotechnical Engineering, 2005, 27 (12):1430–1436.

Fei, Ma. Numerical simulation study about surface subsidence affected by seepage around on the subway train loads [D]: [Master thesis]. Jiangsu: Southeast University Road and Railway Engineering, 2007.

Shin, J.H., Addenbrooke T.I., Potts D.M. A numerical study the effect of groundwater movement on long-term tunnel behaviors. Geotechnique, 2002, 52 (6):391–403.

Jinzhou, Xi. Numerical Simulation Study of surface subsidence caused by underground tunnel in water-rich stratum [D]: [Master thesis]. Sichuan: Southwest Jiaotong University Bridge and Tunnel Engineering, 2005.

Murad Y. Abu-Farsakh, George Z. Voyiadjis. Computational model for the simulation of the shield tunneling process in cohesive soils. Int. J. Anal. Mech. Geomech, 1999, 23:23–24.

Mair R.J., Taylor R.N. Theme lecture: Bored tunneling in the urban environment [A]. Proceedings of the Fourteenth International Conference on Soil Mechanics and Foundation Engineering [C]. 1997. 2353–2385.

O'Reilly M.P., Mair R.J., Alderman G.H. Long-term settlements over tunnels: an eleven-year study at Grimsby [A]. Proceedings of Conference Tunneling [C]. London: Institution of Mining and Metallurgy, 1991: 55–64.

Shirlaw J.N., Subsidence owing to tunneling II Evaluation of a prediction techniques: Discussion. Canadian Geotechnical Journal, 1994, 31:463–466.

Shirlaw J.N. Observed and calculated pore pressures and deformations induced by an earth balance shield: Discussion. Canadian Geotechnical Journal, 1995, 32:181–189.

Shizhong, Lu. Status and Prospects of Mine Slope in China [C]. Fourth National Engineering Geology Conference Proceedings, 1992.

Shaorui, Sun and Jimin Wu, etc. Characteristics of seepage field of large-span double-arch tunnel [J]. Exploration Science and Technology, 2005 (6):14–19.

Xucheng, Bian. Dynamic response analysis of subbase and tunnel under the high-speed train moving loads [D]: [PhD thesis]. Zhejiang: Geotechnical Engineering, Zhejiang University, 2005.

Yi. X., Kerry Rowe R., Lee K.M. Observed and calculated pore pressures and deformations induced by an earth balance shield. Can. Geotech. J. 1993, 30:476–490.

Numerical simulation and dynamic response analysis of wedge water impact

Y.Q. Zhang, F. Xu & S.Y. Jin
School of Aeronautics, Northwestern Ploytechnical University, Xi'an, Shaanxi, China

ABSTRACT: At the beginning of the water impact, the aircraft or capsule structure suffers from serious impact load, which has an important effect on the safety of the structure and the people inside, such as aircraft ditching and capsule landing on water. Taking the wedge as an analysis object, the process of the wedge impact water is simulated by LS-DYNA and the pressures on the wedge surface are got. The numerical results are greed well with the test results of El-Mahdi Yettou, so the correctness of the simulation method is proved. The maximum values of the pressure with the respect of the wedge drop height, deadrise angle and mass are discussed. A valuable reference for the design of the aircraft and the capsule are obtained.

1 INTRODUCTION

The water impact problem such as aircraft ditching and space capsule sea landing is a very complex fluid structure interaction problem that occurs between structural components and the water. In the ditching case, the impact loads are transferred to the aircraft via loads that act on the skin panels. If skin panels are not designed to withstand out of plane impulsive loads, and therefore rupture easily during a water impact. And large impact forces are experienced by the passengers, generally with fatal consequences. There is a forced landing accident of An-24 aircraft in Siberia in July 2011. There were 7 peoples dead and the aircraft broke up. Maximum impact loads occur during the initial stages of impact over a few milliseconds, and are a crucial parameter for structural design. The study of hydrodynamic impact is of great importance to structural designers and engineers.

Von Karman (1929) introduced significant work on this subject. He developed an analytical formula which allows estimation of the maximum pressure on seaplane floats during water landing. Wagner (1936) modified the von Karman solution by taking into account the effect of water splash on the body. Base on their works, Zhao et al. (1993) introduced a complement to Wagner's studies, with linear approximation of the free-surface boundary condition for the two-dimensional impact problem. Mei et al. (1999) proposed a purely analytical method of resolution for the global two-dimensional impact problem of arbitrary bodies. With the theory research continuing, the water impact test is also done all the time. Recently water impact tests of wedge and shell structure have been done by Carcaterra A (2004), El-Mahdi Yettou(2006, 2007), Tveitnes T (2008) and Tanvir Mehedi Sayeed (2010) and Panciroli R (2012). Many parameters such as impact velocity, wedge deadrise angle and mass are analyzed for wedge surface pressure. Because of little limit on numerical calculative model such as irregular geometry and nonlinear boundary, numerical research becomes more popular. A review describing water entry studies related to aerospace structures between 1929 and 2003 was recently published by Seddon C.M (2006). This review pointed out that finite element techniques for analyzing both fluid and structural response in a single model were only recently available. It also stated that the aircraft industry requires a validated numerical tool in which the interaction between the fluid and the structure can be modeled accurately. One final suggestion of the paper is that numerical modeling techniques such as hybrid analysis, finite element modeling and smoothed particle hydrodynamics will certainly be utilized for future work. However a large amount of work needs be done to fully validate these numerical modeling techniques. Explicit nonlinear dynamic finite element codes with an Arbitrary Lagrangian Eulerian (ALE) solver for modeling water landing impact are currently available in commercial codes such as LS-DYNA, which can produce useful results and provide valuable insight into water impact problems. However, these studies also reveal the shortcomings such as the lack of good correlations between test data and numerical prediction, and the computationally intensive issues related to the explicit method and the ALE solver.

This paper provides the results of a preliminary study using LS-DYNA for simulating a wedge impacting on water. The objective of this preliminary study is to demonstrate that the numerical results obtained are reasonable and can be used to adequately assess design configurations. Section one is the introduction of the numerical model of a wedge with the surrounding air and water. Section two is numerical results. The pressures on the wedge surface are discussed and influencing factors, wedge drop height, deadrise angle and mass, are analyzed. Section three is concluding remarks given to summarize the findings from this study.

2 NUMERICAL MODEL

The model shown in Figure 1 includes wedge, water and air. The wedge is simulated by Lagrange element. The water and the air are simulated by Euler elements. When the model meshed, the nodes of water and air element must be merged. The grid of wedge surface must be uniform along the symmetry plane, or the wedge will be oblique after impacting water because of error accumulation. The nodes of model edges are fixed. Nonreflecting boundary condition is used. The wedge is assumed as rigid. *MAT_NULL is used for water and air model. The pressure is got by *EOS_GRUNEISEN. The pressure gradient of water is produced by the gravity with *LOAD_BODY_OPTION and *DEFINE_CURVE.

3 NUMERICAL RESULT

3.1 Dynamic response analysis

The pressure is an important parameter that the researchers are interested in water impact problem. If the pressure is big in local area, the deformation of structure will be large, even structure broken. Base on the test of Mahdi Yettou, the pressure of wedge impact water is simulated and analyzed.

The drop height of wedge is 1.3 m. The deadrise angle is 25°, and the mass is 94 kg. There are 12 pressure transducers from the wedge apex on the side (Fig. 2).

Figure 3 is the test and numerical pressures on the No. 2, 5, 8 positions on the wedge side. The curve of the numerical pressure is almost same as the test and the maximum value is also much closed on all pressure test positions (Tab. 1). So the correctness of simulating pressure is proved.

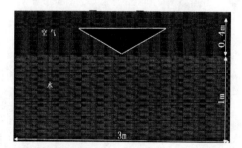

Figure 1. Model.

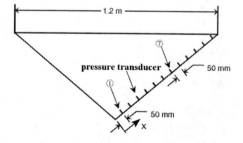

Figure 2. Position of pressure transducer.

From the curve of pressure, when the water reaches the pressure transducer, the pressure increases very fast until the maximum value, and then decreases. Figure 4 is the impact situation of wedge when the pressure reaches the maximum value on the No. 2, 5, 8 positions. When the pressure reaches the maximum value, the pressure transducer is at the root of water splash and the splash water has passed transducer. This is to say that the splash water has little effect on the wedge pressure.

With the wedge impact water, maximum pressure is 124 KPa at 5.5 ms on No. 2 transducer and it is 86 KPa on No.1 transducer. There is no pressure on others because of not submerging water. At 18.5 ms, the water just passes the No.5 transducer and the pressure reaches the maximum value 71.8 KPa. The pressure on No.2 has decreased to 32 KPa. The No.8 pressure is also zero. At 31ms, the No.8 pressure reaches the maximum value 38.1 KPa (Fig. 5). From the pressure analysis, each transducer on the wedge surface will go through its maximum pressure value when the wedge impact water (Fig. 6). A distribution of the maximum pressures along the wedge side can be obtained. The fitting curve is $P = 381x^2 - 463x + 163$, x is the distance away from the wedge apex along the side (Fig. 2). But the maximum value decreases gradually because of impact energy losing.

3.2 Influence factor

(1) Drop height

Three different drop heights simulated are 1 m, 1.3 m and 2 m.

With the drop height increasing, the maximum value of pressure and acceleration is bigger. The difference of maximum pressure on the transducer near the wedge apex is obvious because the impact energy is big in the beginning. With the wedge impact, the water is splashed upward along

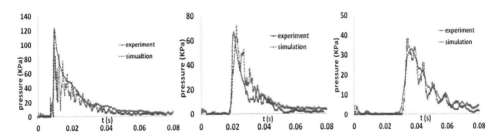

Figure 3. Pressure on the No. 2, 5, 8 positions.

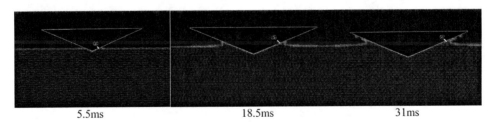

Figure 4. Process of wedge impact water.

Table 1. Maximum value of pressure.

Pressure transducer	1	2	3	4	5	6	7	8	9	10	11	12
Test (KPa)	132.6	125	102.6	87.5	67.6	51	42.1	33.4	26.2	20	16.9	13.2
simulation (KPa)	137	124	101	88.2	71.8	54	43.2	38.1	32.8	27.8	24.1	20.1

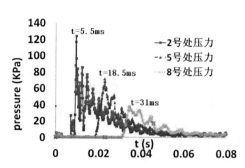

Figure 5. Pressure on No. 2, 5, 8 positions.

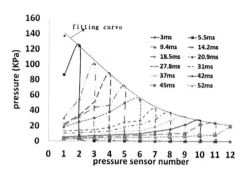

Figure 6. Pressure on the transducer in different time.

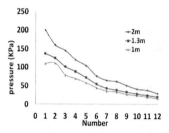

Figure 7. Maximum pressure.

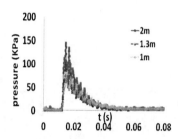

Figure 8. Pressure on No.3 transducer.

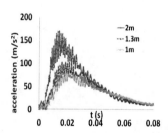

Figure 9. Wedge acceleration.

the wedge surface. The drag force on the wedge becomes small, so the difference of maximum pressure is also small at the end of wedge surface. From the pressure and acceleration curve, the drop height is main effect on the maximum pressure. The trend of pressure changing is almost same.

(2) Deadrise angle

Three different deadrise angles simulated are 15°, 25° and 35°.

With the deadrise angle decreasing, the maximum value of pressure and acceleration is bigger. If the deadrise angle is big, the water will be easier to splash and the force on the wedge will be small, so the maximum pressure will be also small. The changing of maximum pressure on different transducers is same as the drop height effect. From the pressure and acceleration curve, deadrise angle has obvious effect on the maximum value and the response time. For a bigger deadrise angle, the impact energy will decrease slowly because of smaller drag force, so the response time will be longer. In the later stage of impact, the acceleration of 35° wedge is obvious bigger. The reason is that the water splash goes back to the top of the wedge and a force will act on the wedge (Fig. 13). The normal water splash of 15° wedge is show in Figure 14.

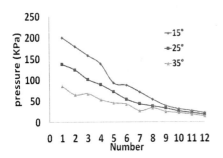

Figure 10. Maximum pressure.

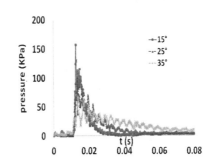

Figure 11. Pressure on No.3 transducer.

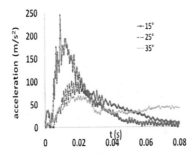

Figure 12. Wedge acceleration.

Figure 13. Water splash of 35° wedge.

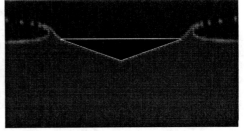

Figure 14. Water splash of 15° wedge.

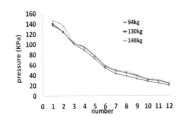

Figure 15. Maximum pressure.

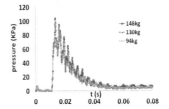

Figure 16. Pressure on No.3 transducer.

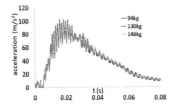

Figure 17. Wedge acceleration.

(3) Mass

Three different masses simulated are 94 kg, 103 kg and 148 kg.

The maximum pressure and acceleration are almost same in different wedge masses. From the curve, the trend of pressure and acceleration changing is also same. So the mass of wedge has a little influence on the pressure and acceleration.

4 CONCLUSION

The simulation results are agreed well with the test data on the pressure and the acceleration. This preliminary study demonstrates that the ALE method in LS-DYNA is suitable to analyze the water impact problem.

The water is separated by the wedge apex when the wedge impact water. The pressure of wedge will increase rapidly until the maximum value. Then the water splashes upward along the wedge surface and the value decreases gradually. So the maximum value of pressure occurs in the beginning of impact. The distribution of the maximum pressures along the wedge side is the same as a quadratic curve.

Three parameters that influence the acceleration and the pressure of the wedge are analyzed.

(1) The maximum value of pressure and acceleration is bigger and the response time is shorter when the drop height of wedge is higher.
(2) The maximum value of pressure and acceleration is bigger and the response time is shorter when the deadrise angle of wedge is smaller.
(3) The mass has a little effect on the pressure and the acceleration.

These analysis results can give a valuable reference for the design of water impact test and the water entry structures.

REFERENCES

Carcaterraa, A. & Ciappi, E. 2004. Hydrodynamic shock of elastic structures impacting on the water: theory and experiments. *Journal of Sound and Vibration* 271(1): 411–439

Karman, V.T. 1929. The impact of seaplane floats during landing. *National Advisory Committee for Aeronautics, NACA Technical Notes 321.*

Mei, X., Lui, Y. & Yue, D.K.P. 1999. On the water impact of general two-dimensional sections. *Applied Ocean Research* 21(1): 115

Panciroli, R., Abrate, S., Minak, G. & Zucchelli, A. 2012. Hydroelasticity in water-entry problems: Comparison between experimental and SPH results. *Composite Structures* 94(2): 532–539

Sayeed, T.M. Peng, H. & Veitch, B. 2010. Experimental investigation of slamming loads on a wedge. *The International Conference on Marine Technology.* Dhaka: Bangladesh

Seddon, C.M. & Moatamedi, M. 2006. Review of water entry with applications to aerospace structures. *International Journal of Impact Engineering* 32(7): 1045–1067

Tveitnes, T. Fairlie-Clarke, A.C. & Varyani, K. 2008. An experimental investigation into the constant velocity water entry of wedge-shaped sections *Ocean Engineering* 35(14): 1463–1478

Wagner, V.H. 1947. Phenomena associated with impacts and sliding on liquid surfaces *NACA107421075*

Yettou, E.M. Desrochers, A. & Champoux, Y. 2006. Experimental study on the water impact of a symmetrical wedge. *Fluid Dynamics Research* 38(1): 47–66

Yettou, E.M., Desrochers, A. & Champoux, Y. 2007. A new analytical model for pressure estimation of symmetrical water impact of a rigid wedge at variable velocities. *Fluid Dynamics Research* 23(3): 501–522

Zhao, R. Faltinsen, O.M. & Aarsnes, J. 1996. Water entry of arbitrary two-dimensional sections with and without flow separation *In: Proceedings of the 21st Symposium on Naval Hydrodynamics,* Trondheim: Norway, National Academy Press, Washington, DC, USA

Aeroelastic study on a flexible delta wing

J.G. Quan, Z.Y. Ye & W.W. Zhang
Northwestern Polytechnical University, Xi'an, P.R. China

ABSTRACT: Numerical simulation of aeroelastic analyses for a flexible delta wing at high angle of attack is a hard research. In this work, the aeroelastic analysis of a 70-degree cropped delta wing is studied using a computational aeroelastic solver. The aeroelastic solver couples a Navier-Stokes code with structural motion equations. It's shown that, with the angle of attack increasing, the leading-edge vortices breakdown, and the location moves forward. When vortex breakdown is present over the wing, the unsteady fluctuation characteristics of flow become highly obvious. The external excitation loads exerted on the wing will induce a buffet phenomenon, and has a significant effect on the aeroelastic characteristics of the wing. Before the vortex breakdown, the aeroelastic instability presents itself as just a simple flutter. However, after the vortex breakdown, the aeroelastic phenomenon presents neither a simple flutter, nor a simple buffet, but a complex problem including both flutter and buffet.

1 INTRODUCTION

With the requirements of high maneuverability of the aircrafts in modern air combat, the new unmanned combat air vehicle (UCAV) and missile usually install high-sweepback delta wing, and fly at high angle of attack. For the high-sweepback delta wing, the flow always separate at the leading edge (Wang & Ye 2008). And with the angle of attack increasing to a sufficiently high value, the vortices undergo a sudden expansion known as vortex breakdown. The micro-scale vortex flow dominated by the viscosity shows obviously unsteady fluctuation characteristics, which causes high nonlinear effect on the aerodynamics of the wing.

In recent several decades, the vertical flow and vortex breakdown over a high-sweepback delta wing at high angle of attack have been studied (Ye & Zhao 1994, Liang & Ye 2001, Lu et al. 2004, Yang & Ye 2002, Zhang & Ye 2006) extensively at home and abroad, but the corresponding research on aeroelastic problem is few. Ye & Zhao (1994) used the nonlinear lifting line method to analyze the wing's flutter characteristics at a low speed and high angle of attack. Gordnier & Visbal 2003 and Attar 2006 coupled a well-validated Euler code with a nonlinear finite element plate model to perform relevant aeroelastic analysis for a flexible delta wing at high angles of attack. Li et al. (2007) studied the buffeting characteristics of the twin vertical fin on the aircraft of strake-wing layout based on the Euler equations. Zhang & Ye 2009 has used the unsteady aerodynamic reduced-order model based on the Euler equations to study the flutter characteristics of a 70° cropped delta wing at high angles of attack. However, this work just focused on the influence of the angle of attack on the flutter characteristics before the main vortex breakdown on the delta wing, never involved the aeroelastic characteristics subsequently.

After vortex breakdown, the aeroelastic problem of the delta wing contains strong nonlinear aerodynamics, whose essence is likely neither a simple flutter, nor a simple buffet, but a complex problem including both. As a continuation of an earlier study (Zhang & Ye 2009), this paper performs aeroelastic study on a flexible cropped delta wing at high angle of attack, by using Navier-Stokes equations to compute unsteady aerodynamic forces, and coupling with the structural motion equations.

2 THEORY

The Reynolds-Averaged Navier-Stokes (RANS) equations are used to simulate the vertical flow and vortex breakdown of cropped delta wing at high angle of attack, which is shown in Eq. (1):

$$\frac{\partial}{\partial t}\iiint_{\Omega} Q dV + \iint_{\partial\Omega} F(Q) \cdot n dS = \iint_{\partial\Omega} G(Q) \cdot n dS \qquad (1)$$

where, Ω is the control volume, $\partial\Omega$ is the boundary of the control volume, Q is the conservation vector, $F(Q)$ is the inviscid flux vector, $G(Q)$ is the viscous flux vector.

Applying the Lagrange's Equations, the structural equations of motion in matrix notation can be shown as follows:

$$M \cdot \ddot{\xi} + G \cdot \dot{\xi} + K \cdot \xi = Q \qquad (2)$$

where, ξ is the structural generalized displacement vector, M is the generalized mass matrix, G is the generalized damping matrix (Generally, G is measured by experiments. Here, $G = 0$), K is the generalized stiffness matrix, and Q is the generalized aerodynamic forces matrix provided by the unsteady flow solving codes.

By defining the structural state-vector $x = [\xi_1, \xi_2 \cdots \dot{\xi}_1, \dot{\xi}_2 \cdots]^T$, the structural equations in state-space form are as follows:

$$\dot{x} = F(x,t) = Ax + BQ(x,t) \qquad (3)$$

where

$$A = \begin{bmatrix} 0 & I \\ -M^{-1}K & -M^{-1}G \end{bmatrix}, B = \begin{bmatrix} 0 \\ M^{-1} \end{bmatrix}, 0 \text{ is the zero square matrix, } I \text{ is the identity matrix.}$$

According to the traditional mode vibration concept, the wing deflection can be defined as follows:

$$\omega(x,y,z,t) = \sum_{i}^{N} \Phi_i(x,y,z) \cdot \xi_i(t) \qquad (4)$$

where, N is the number of the natural mode, w is the wing deflection, t is the time, $\Phi_i(x,y,z)$ is the i-order natural mode.

To analyze the aeroelastic characteristics of the flexible delta wing at high angle of attack, the aerodynamic and structural equation must be coupled.

3 RESULTS

A flexible cropped delta wing with 70-degree sweep angle (Zhang & Ye 2009) is used to be the computational model, which is semi-model. The Mach number is 0.3. Fig. 1 shows the first three mode shapes of the wing. Fig. 2 depicts the mesh distribution of the wing used for this study.

3.1 Flow characteristic at high angle of attack

Firstly, the unsteady vertical flow of cropped delta wing at different angle of attack ($\alpha = 20°$, $25°$, $27°$) is computed separately.

The iso-surface of total pressure represents the vortex core of the vortex. Figure 3 shows the slices of total pressure at different locations, from which we can see the change of the vortex core before and after vortex breakdown clearly. At $\alpha = 20°$, the vortex core is uniform, and has no expansion, which means that the vortex has not broken. At $\alpha = 25°$, the vortex core experiences a

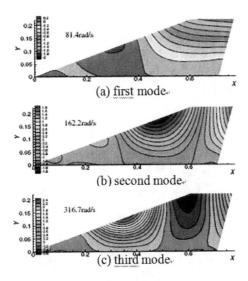

Figure 1. First three mode shapes of cropped delta wing.

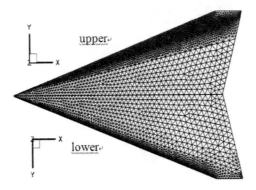

Figure 2. Mesh distribution of the wing.

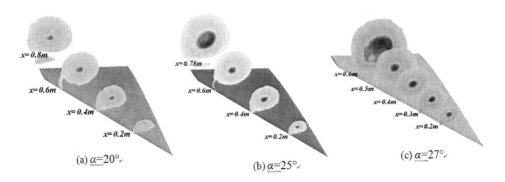

Figure 3. Slices of total pressure at different locations at different angles of attack.

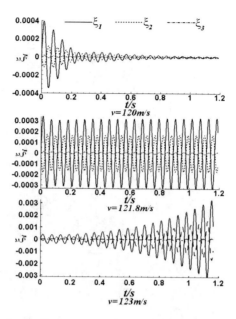

Figure 4. Time responses of generalized displacements at various velocity at $\alpha = 20°$.

dramatic expansion at the location of $x = 0.78$ m, which is at the tailing edge. The computational location of vortex breakdown is in region of $x = 0.74 \sim 0.78$ m. At $\alpha = 27°$, the vortex core becomes large suddenly after the location of $x = 0.5$ m. At this time, the location of vortex breakdown is between $x = 0.5$ m and $x = 0.6$ m. From this figure, we know that the location of vortex breakdown moves forward with the increasing angle of attack.

3.2 Aeroelastic characteristic at high angle of attack

Following, the aeroelastic characteristics of the cropped delta wing at different angles of attack will be analyzed.

Fig. 4 shows the time responses of generalized displacements at different velocities at $\alpha = 20°$. The computed flutter velocity is 121.8 m/s and the flutter frequency is 113 rad/s. Both of them agree well with the result by the ROM analysis method in reference Zhang & Ye 2009.

Fig. 5 shows the time responses of generalized displacements at different velocities at $\alpha = 25°$. The computed flutter velocity is about 124.3 m/s, and the flutter frequency is 102.6 rad/s. As is evident in the figure, although the curve of the time responses at $v = 120$ m/s is convergent, it still has a local high frequency fluctuation. That is because, after vortex breakdown, the frequency of the unsteady loads is much higher than the natural frequency of the structure, and the external excitation loads on the wing induce buffet. In this condition, the aeroelastic problem is no more a simple flutter, but a complex problem including flutter and buffet. However, the effect of buffet is a little slight, because the location of the vortex breakdown is at the tailing-edge.

Fig. 6 shows the time responses of generalized displacements at different velocities at $\alpha = 27°$. Compared to the results at $\alpha = 25°$, the high frequency irregular fluctuation becomes more pronounced. It's because that as the angle of attack increases, the location of vortex breakdown moves forward and the external excitation loads produced by the breakdown vertical flow rises, so the effect of buffet becomes much stronger. In Fig. 6(a), the time responses are neither convergent nor divergent, but present a forced vibration form. That is because when the velocity is 100 m/s, the system is far away from the critical flutter state, so the responses owing to self-excited vibration diverge quickly, and then the system is soon in buffet state. At this time, even the total computational

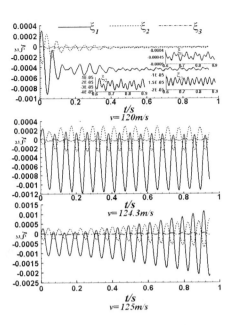

Figure 5. Time responses of generalized displacements at various velocity at $\alpha = 25°$.

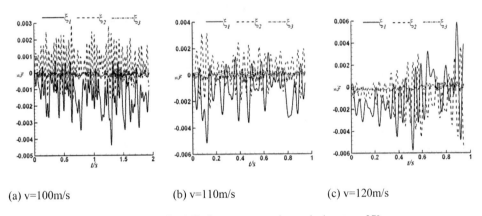

(a) v=100m/s (b) v=110m/s (c) v=120m/s

Figure 6. Time responses of generalized displacements at various velocity at $\alpha = 27°$.

time increases to two seconds, the responses still don't have obvious convergent trend or divergent trend. In Fig. 6(b) the time responses show obvious convergent trend at $v = 110$ m/s. In Fig. 6(c) the time responses show obvious divergent trend at $v = 120$ m/s. It means that there is still a chance for the wing to experience a vibration divergence due to flutter, even the vortex breakdown occurs in large area and the effect of buffet becomes much pronounced. Thus, when the velocity is large, close to the wing's flutter velocity before vortex breakdown, the aerodynamic characteristics at 27-degree angle of attack present an obvious complex phenomenon including flutter associated with buffet.

Furthermore, in order to study carefully the aeroelastic characteristics before and after vortex breakdown at high angle of attack, this paper analyzes the power spectra density (PSD) of generalized displacements and forces for each mode at different angles of attack.

Fig. 7 shows the distribution curve of power spectra density of generalized displacements and forces at $\alpha = 20°$, $v = 121.8$ m/s. It can be seen that the generalized displacements and forces both

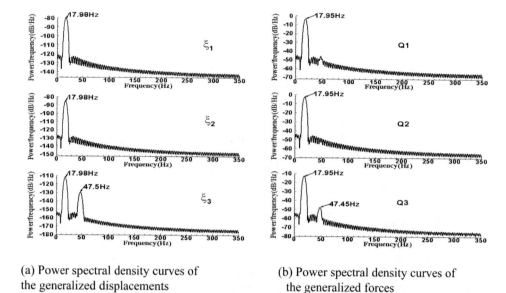

(a) Power spectral density curves of the generalized displacements

(b) Power spectral density curves of the generalized forces

Figure 7. The distribution of power spectra density at $\alpha = 20°$, $v = 121.8$ m/s.

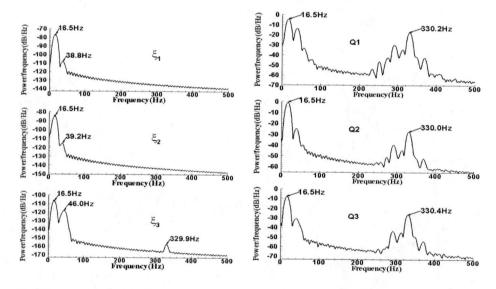

(a) Power spectral density curves of the generalized displacements

(b) Power spectral density curves of the generalized forces

Figure 8. The distribution of power spectra density at $\alpha = 25°$, $v = 124.3$ m/s.

have a peak at 17.95 Hz, which is consistent with the calculated flutter frequency 113 rad/s, and is between the first and the second mode natural frequency. It indicates that the occurrence of the flutter at 20-degree angle of attack is caused by the coupling of the first mode and the second mode. The generalized displacements and forces of the third mode both have a lower peak at 47.5 Hz, which is close to the third mode natural frequency of 50 Hz.

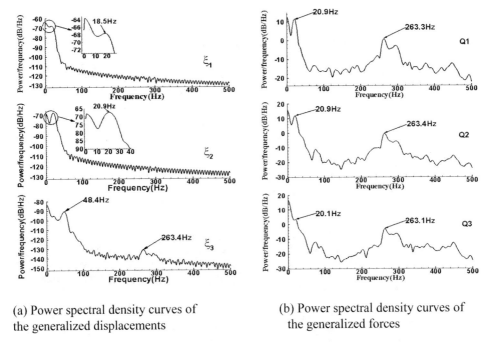

(a) Power spectral density curves of the generalized displacements

(b) Power spectral density curves of the generalized forces

Figure 9. The distribution of power spectra density at $\alpha = 27°$, $v = 120$ m/s.

Figure 8 shows the distribution curve of power spectra density of generalized displacements and forces at $\alpha = 25°$, $v = 124.3$ m/s. It can be seen that both of the generalized displacements and forces have a large peak at 16.5 Hz. This main peak frequency is consistent with the calculated flutter frequency 102.6 rad/s, and between the first and the second mode natural frequency. It also indicates that the coupling of the first mode and the second mode causes the occurrence of the flutter. In Figure 8(b), the generalized forces also have a peak around 330 Hz, which is the frequency of the vortex breakdown.

Figure 9 shows the distribution curve of power spectra density of generalized displacements and forces at $\alpha = 27°$, $v = 120$ m/s. Differently, the main peak frequency of generalized displacements for each mode is different, that is because the fluctuation loads coupled with the structural vibration. In Figure 9(b), the power spectra density distribution curve of generalized forces has two peaks. One is in the region of less than 21 Hz, the other is near 263 Hz, corresponding separately to the structural vibration frequency and fluctuation loads frequency.

4 CONCLUSION

By coupling the aerodynamic and structural equations, the aeroelastic study on a flexible cropped delta wing at high angle of attack is carried out in this paper, and the following conclusions are gained:

1) $\alpha = 20°$, the leading-edge separation vortex doesn't break down, so the aeroelastic problem is just a simple flutter.
2) $\alpha = 25°$, the leading-edge separation vortex breaks down at the tailing edge of the wing. The external excitation loads on the wing produced by the vortex breakdown induce buffet, so the time responses show local irregular fluctuation. The aeroelastic characteristics present a complex phenomena mixed with flutter and buffet, but the effect of buffet is relatively weak.

3) $\alpha = 27°$, the location of the vortex breakdown moves forward, and the effect of buffet enhances, so the time responses present pronounced high frequency irregular fluctuation. At a relatively high velocity, the aeroelastic characteristics present an obvious complex phenomena mixed with flutter and buffet.
4) After the vortex breakdown, the fluctuating loads on the wing induced by vortex breakdown rise as the angle of attack increases, the influence of buffet thus increases, and the aeroelastic phenomenon presents a complex problem involving flutter and buffet.

REFERENCES

Attar, P.J. & Gordnier, R.E. 2006. Numerical simulation of the buffet of a full span delta wing at high angle of attack. *AIAA Paper* 2006–2075.

Gordnier, R.E. & Visbal, M. 2003. Computation of the aeroelastic response of a flexible delta wing at high angle of attack. *AIAA Paper* 2003–1728.

Li, J.J., Yang, Q. & Yang Y.N. 2007. The numerical investigation of twin-vertical tail buffet of strake-wing configuration. *Acta Aerodynamics Sinica* 25(2): 205–210.

Liang, Q. & Ye Z.Y. 2001. Analysis of airfoil flutter characteristics. *Journal of Northwestern Polytechnical University* 19(3): 341–344.

Lu, Z.L., Guo, T.Q. & Guan, D. 2004. A study of calculation method for transonic flutter. *Acta Aeronautica et Astronautica Sinica* 25 (3): 214–217.

Wang, G. & Ye, Z.Y. 2008. Study of the unsteady flow around a delta wing at high incidence using detached eddy simulation. *Journal of Northwestern Polytechnical University* 26(4): 418–426.

Ye, Z.Y. & Zhao, L.C. 1994. Nonlinear flutter analysis of wings at high angle of attack. *Journal of Aircraft* 31(4): 973–974.

Yang, Y.N & Ye, Z.Y. 2002. An investigation of airfoil flutter characteristics with structure nonlinearity at transonic speed. *Chinese Journal of Computational Physics* 19(2): 152–162.

Zhang, W.W & Ye, Z.Y. 2006. Numerical simulati-on of aeroelasticity basing on identification technology of unsteady aerodynamic loads. *Acta Aeronautical et Astronautica Sinica*, 27(4): 579–583.

Zhang, W.W. & Ye, Z.Y. 2007. Transonic aeroelastic analysis basing on reduced order aerodynam-ic models. *Chinese Journal of Computational Mechanics* 24(6): 768–772.

Zhang, W.W. & Ye, Z.Y. 2009. Effects of leading-edge vortex on flutter characteristics of high sweep angle wing. *Acta Aeronautica et Astronautica Sinica* 30(11): 1–6.

Zhang, W.W. & Ye, Z.Y. 2009. Analysis of supersonic aeroelastic problem based on local piston-theory method. *AIAA Journal* 47(10): 2321–2328.

Advanced Engineering and Technology – Xie & Huang (Eds)
© *2014 Taylor & Francis Group, London, ISBN 978-1-138-02636-0*

Efficient numerical simulation based on computational fluid dynamics for the dynamic stability analysis of flexible aircraft

W.G. Liu, R. Mei, W.W. Zhang & Z.Y. Ye
Northwestern Polytechnical University, Xi'an, China

ABSTRACT: A unified approach of numerical simulation is presented to analyze the interaction of flight dynamics and aeroelasticity of flexible aircraft. Using an unsteady aerodynamic reduced-order model (ROM) based on computational fluid dynamics (CFD), an efficient numerical simulation is proposed to simulate the dynamic response of flexible aircraft and determine flight dynamic stability characteristics from dynamic responses in the time domain. The numerical results show that strong interacting between the short-period modes of flight dynamics and the structural elastic modes of low natural frequency will impair the stability of flexible planes with high aspect-ratio wings and flexible missiles with high slenderness-ratio bodies. Two numerical simulations confirm the effectiveness and efficiency of ROM in relation to the full CFD method. The simulation yields satisfactory numerical results in the engineering field and improves the computational efficiency up to two orders of magnitude.

1 INTRODUCTION

The rapid development of modern aircraft has significantly increased the elasticity of structure to satisfy different performance requirements increase evidently. In terms of structural design, increasing structural size inevitably reduces natural frequencies. The traditional stability problems of flight dynamics interact with the traditional aeroelastic stability problem, and several new stability problems will occur if elastic frequencies is similar to modal frequencies of flight dynamics. A strong interaction between the oscillation of flight attitude and the vibration of elastic structure was observed in a UAV video to cause unstable vibration and and disintegration of the structure.

Chen & Chen (1985), Liu & Yang (1988) and other scholars first studied the coupling problem between flight dynamics and aeroelasticity at Northwestern Polytechnical University in the early 1980s. Traditional aerodynamic computation was also based on the linear potential theory and quasi-steady assumption at that time. With the rapid development of aircraft, this interaction problem has gradually threatened aircraft design and safety. Many studies have investigated the interaction between the aeroelasticity and flight dynamics of flying wing aircraft (Zhang & Xiang 2011, Patil & Hodges 2006 and Chang & Hodges 2008). However, most aerodynamic models are based on two-dimensional quasi-steady theory about the layout characteristics of the flying wing, which cannot solve several complex coupling problems, such as transonic flow. To establish a method to solve the complex interaction problem between flight dynamics and aeroelastic dynamics, Mei & Ye (2011) develops a numerical simulation technique based on CFD. However, its computational efficiency is low.

To study the interaction between flight dynamics and aeroelasticity efficiently and effectively, an efficient algorithm based on CFD (Quan & Ye 2011, Zhang & Ye 2009, Song & Lu 2009, 2011 and Zhang 2006) for the interaction problem is developed. Results show that the proposed method has the same accuracy as complete CFD technology, and that its computational efficiency is improved by two orders of magnitude.

2 THEORETICAL MODEL

To achieve a unified form of both flight dynamics and structural motion equations, a centroid-based rotating coordinate system is created, which translates with the aircraft but cannot rotate. The aircraft motion equations based on the coordinate system are expressed as follows:

$$m \cdot \left(\begin{bmatrix} \dfrac{dV_{fx}}{dt} \\ \dfrac{dV_{fy}}{dt} \\ \dfrac{dV_{fz}}{dt} \end{bmatrix} + \begin{bmatrix} \dfrac{dV_{fgx}}{dt} \\ \dfrac{dV_{fgy}}{dt} \\ \dfrac{dV_{fgz}}{dt} \end{bmatrix} \right) = T_{gb} \cdot \begin{bmatrix} T \cdot \cos\sigma \\ 0 \\ -T \cdot \sin\sigma \end{bmatrix} + T_{gb} \cdot T_{ba} \cdot \begin{bmatrix} -D \\ C \\ -L \end{bmatrix} + m \cdot \begin{bmatrix} 0 \\ 0 \\ g \end{bmatrix} \quad (1)$$

$$\begin{bmatrix} I_x & & \\ & I_y & \\ & & I_z \end{bmatrix} \cdot \begin{bmatrix} \dfrac{d\omega_{fx}}{dt} \\ \dfrac{d\omega_{fy}}{dt} \\ \dfrac{d\omega_{fz}}{dt} \end{bmatrix} = \begin{bmatrix} L_f \\ M_f \\ N_f \end{bmatrix} \quad (2)$$

where V_{fx}, V_{fy}, V_{fz} are three components of the velocity vector in thng coordinate system; V_{fgx}, V_{fgy}, V_{fgz} are three components of the velocity vector in the coordinate e rotatisystem relative to the inertial coordinate system on the ground; and T, D, C, L are the thrust, drag, lateral force and lift; T_{gb}, T_{ba} are the transformation matrices converted from the corresponding coordinate system to the inertial coordinate system (Xiao & Jin 1993); I_x, I_y, I_z are the rotational inertia of the three directions; ω_{fx}, ω_{fy}, ω_{fz} are the angular velocity of the three directions; and L_f, $M_f N_f$ are the total torque of the three directions.

Because the main purpose is to study the stability, the aircraft flies straight and horizontally. The above body coordinate system is the inertial coordinate system. Thus, equation1 (1) and (2) are uniformed as the following matrix

$$\tilde{M}\ddot{\xi} = Q \quad (3)$$

where $\tilde{M}$ can be regarded as the modal mass matrix. Similarly, in the coordinate system, the aeroelastic kinetic equation can be expressed as follows [4]

$$M\ddot{\xi} + C\dot{\xi} + K\xi = Q \quad (4)$$

where M is the modal generalized mass matrix; C is the structural damping matrix; K is the modal generalized stiffness matrix; Q is the modal generalized aerodynamic matrix; and ξ is the modal displacement. Equation (3) is a special form of the equation (4), where matrices C and K are both zero. Therefore, these equations have a unified form, and this paper solves equations by using a multistep method with the fourth-order accuracy. The results show that a high-precision numerical result can be obtained with this method. In each calculation step, the generalized aerodynamic force can be directly applied to the CFD results, however, the number of operations is large. To meet the efficiency demand of practical engineering problems, we present an unsteady aerodynamic model.

The ARX model using reduced-order (ROM) technology, which is based on the unsteady aerodynamic ROM of system identification method is an auto-regressive moving average exogenous

(ARMAX) model that ignores the moving average (MA) item of the noise and thus becomes a simplified auto-regressive model. For a multiple-input multiple-output system, the mathematical form of the model is as follows

$$Y(k) = \sum_{i=1}^{na} A_i Y(k-i) + \sum_{i=0}^{nb-1} B_i U(k-i) + E(k) \qquad (5)$$

where $Y(k)$ is the k-th observation value of the output; $U(k-i)$ is the $(k-i)$-th observation value of the input, $E(k)$ is a zero-mean random noise, A_i and B_i are the coefficient matrices not yet identified; na and nb are the delay orders of the output and input. In this paper, the modal displacement of each movement is the input, and the generalized aerodynamic coefficients corresponding to modal displacement are the outputs. The tests on the specific method are described in [12].

3 NUMERICAL RESULTS AND ANALYSIS

3.1 The plane model with high aspect-ratio wings

The first example is a model similar to the Global Hawk UAV aerodynamic layout. The form layout and surface CFD grid of this model are shown in Fig. 1. The body length is 13.5 m, and half the wingspan is 17.5 m, and the centroid is located at 6.0 m away from the nose. The wing material is assumed to be steel, the natural frequencies of the first three bend modes are 1.2978 Hz, 4.9418 Hz and 11.8400 Hz.

Fig. 2 shows the responses of the aircraft according to the ROM unsteady aerodynamics modeling method and full CFD method. The Mach number is 0.5, and the dynamic pressure is 17.7 Kpa. The elastic vibration response of the wing is shown in Fig. 2a, and the modal response of the rigid-body motion of the aircraft (three typical short-period movements of flight dynamics) is shown in Fig. 2b. The result of the calculation based on ROM is consistent with that of the calculation based on full CFD method. Calculation efficiency is improved by two orders of magnitude.

Fig. 3 shows the responses of the short-period motion of the rigid body. These responses are simulated by full CFD methods at the same mach number and dynamic pressure. The responses of the pitching motion are shown in Fig. 3a, whereas the responses of yawing and rolling modes of rigid-body motion are shown in Fig. 3b. The decline of the longitudinal pitching modes of the rigid-body motion slows when the elastic modes of vibration are considered, and the lateral

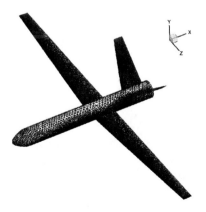

Figure 1. Aircraft model with high aspect-ratio wing.

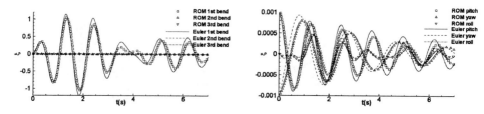

Figure 2. Calculated responses between full CFD and ROM.

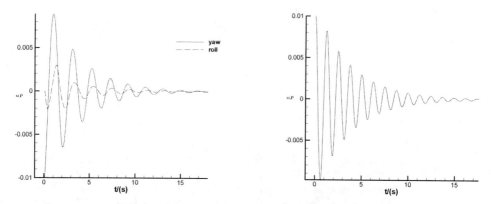

Figure 3. Responses of longitudinal pitch and lateral movement with full CFD.

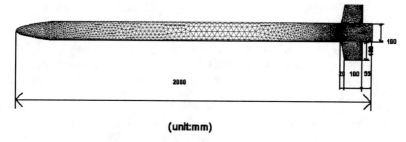

(unit:mm)

Figure 4. Missile model with high slenderness-ratio body.

yawing modes remain nearly unchanged. With regard to aerodynamic layout characteristics, this phenomenon is the result of the interaction between the longitudinal pitching motion of the rigid body and the bending elastic modes of the wing. The lateral modes are less significantly affected by the elastic deformation of the wing. Therefore, the aerodynamic damping is close to the decline of the rigid body. The qualitative result agrees with the result of [3].

3.2 The missile model with high slenderness-ratio body

In this paper, we create a missile model with a slenderness ratio of 20, and an aerodynamic shape similar to that of the AIM-9X missile without the front wing. The shape characteristics and surface mesh of the model are shown in Fig. 4. The centroid is located at 1.2 m away from the warhead, and it has a total length of 2 m and a design mach of 2.0. The first four modes are considered. The first- and second-order bending modal natural frequencies of the missile body are 12.430 Hz and 33.957 Hz, respectively; the natural frequency of the first-order bending mode and first-order torsional mode of the rudder are 228.36 and 538.26 Hz, respectively.

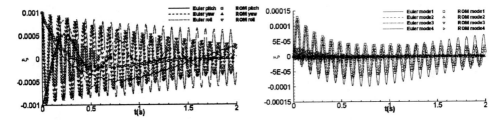

Figure 5. Responses with full CFD and efficient aerodynamic modeling (fighting dynamic pressure = 10 KPa).

Fig. 5 shows the responses of modal displacement at a Mach of 2.0, and a dynamic pressure of 10 kPa. The elastic modes are the above four modes and the short period modes of the aircraft are the longitudinal pitching, yawing and rolling modes.

The figures show the results of full CFD and efficient aerodynamic modeling. The result of efficient aerodynamic modeling method is consistent with that of full CFD at the initial stage of response. However, the deviation increases with the time and amplitude of vibration because a small-amplitude vibration trains the unsteady aerodynamics model according to analysis, and the deviation is reduced further if the training method of the unsteady aerodynamic simulation is improved. However, trends of the vibration remain consistent although there is some error.

3.3 *Calculation efficiency*

The above results show only the dynamic responses. However, the above method can also predict the critical flight parameters of the stability of elastic aircraft, such as critical dynamic pressure and parameter changes. For example, the effect of changes in the center of gravity on stability can be further investigated. In practical engineering efficient unsteady aerodynamic modeling can save a large amount of computation time. To quantify the efficiency of the method, comparative experiments are conducted with the same computer resources. The results show that 150 hours are required to obtain the computing results by using the fully CFD method, while just 2 hours are required by using the efficient unsteady aerodynamic modeling. From the above comparison, it can be found that the efficient algorithm will be able to improve the efficiency about two orders of magnitude.

4 CONCLUSION

In conclusion, we present a unified analysis of the interaction between flight dynamics and aeroelastic mechanics of elastic aircraft. This analysis uses unsteady aerodynamic modeling method based on ROM technology. Numerical simulation results show that in planes with high aspect-ratio wings and missiles with high slenderness-ratio bodies, a strong interaction may occur between traditional aeroelasticity and flight dynamics (main short-period modes of rigid-body motion). The natural vibration frequency of the structure decrease such that the structural frequency is similar to that of the short-period modes of rigid-body motion when a ratio of aspect or slenderness ratio increases to a certain extent. The interaction forms a new stability problem. The interaction occurs and impairs aircraft stability. The numerical results also show that the results of calculation is suitable for engineering application when performed through unsteady aerodynamic modeling. The calculation accuracy is close to that of the full CFD. Computational efficiency is improved to two orders of magnitude. Therefore, the proposed method is an efficient and practical algorithm for the abovementioned engineering problems.

REFERENCES

Chang, C. S. & Hodges, D. H. 2008. Flight dynamics of highly flexible aircraft. *Journal of Aircraft* 45(2): 538–545.

Chen, S. L., Chen, X. J. & Yan, H. Y. 1985. Longitudinal stability of elastic aircraft. *Acta Aeronautica et Astronautica Sinica* 6(4):321–328.

Liu, Q. G. & Yang, Y. N. 1988. A united method for analyzing the stability and buffeting characteristic of Elastic aircraft. *Acta Aeronautica et Astronautica Sinica* 9(9):35–43.

Mei, R. & Ye, Z. Y. 2011. Study on interaction between rigid-body movement and elastic movement of aircraft. *Aeronautical Computing Technique* 41(5):52–55.

Patil, M. J. & Hodges, D. H. 2006. Flight dynamics of highly flexible flying wings. *Journal of Aircraft* 43(6):1790–1798.

Quan, J. G., Ye, Z. Y. & Zhang, W. W. 2011. Study on aeroelastic characteristics of a cropped delta wing before and after vortex breakdown. *Acta Aeronautica et Astronautica Sinica* 32(3):379–389.

Song, S. F., Lu, Z. Z. & Zhang, W. W. 2011. Uncertainty importance measure by fast fourier transform for wing transonic flutter, *Journal of Aircraft* 48(2):449–455.

Song, S. F., Lu, Z. Z. & Zhang, W. W. 2009. Reliability and sensitivity analysis of transonic flutter using improved line sampling technique, *Chinese Journal of Aeronautics* 22(5):513–519.

Xiao, Y. L. & Jin, C. J. 1993. Flight principle in atmospheric disturbance. *National Defense Industry*.

Zhang, J. & Xiang, J. W. 2011. Static and dynamic characteristics of coupled nonlinear aeroelasticity and flight dynamics of flexible aircraft. *Acta Aeronautica et Astronautica Sinica* 32(9):1569–1582.

Zhang, W. W., Ye, Z. Y. & Zhang, C. N. 2009. Aeroservoelastic analysis for transonic missile based on computational fluid dynamics. *Journal of Aircraft* 46(6):2178–2183.

Zhang, W. W. 2006. Study on an efficient method of aeroelasticity based on the CFD technology. *Doctoral thesis of Northwestern Polytechnical University*.

Hydro-mechanical analysis on the fluid circulation in a coaxial cylinder rheometer

Jun Xing, Shiqiang Ding & Jirun Xu
Dalian University, College of Environment & Chemical Engineering, Dalian, Liaoning, China

ABSTRACT: The sedimentation of flocs or solid particles has to be taken into account when a coaxial cylinder rheometer is used to deal with the flocculated suspensions or pulps. An important approach to the problem is to make the test fluid circulate from the bottom to the top of the rheometer through an external channel. In order to examine the influence of fluid circulation on the rheometer operation, a careful hydro mechanical analysis about the shear stress on the piddle surface is done in the paper, and some important results founded. The investigation proves that if the circulated fluid does not introduce any additional velocities in the tangential and radial directions within the operation space of the rheometer, the basic principles of the rheometer will not be changed although an axial movement is attached to the fluid because of the circulation. In addition, the circulated fluid needs to be evenly distributed on the whole cross-section of the operation space, otherwise the correction of the measurements is necessary.

1 INTRODUCTION

The coaxial cylinder rheometer has been widely used to measure the rheological parameters of non-Newtonian fluid at present. However, when this kind of rheometer deals with a flocculated suspension or a pulp, the sedimentation of flocs or particles due to gravity will result in the accurate measurement impossible (Assael & Dalaouti, 2000). Alternatively, the fluid circulation has been presented as an approach to the problem by different researchers (Kawatra et al., 1996; Xu et al., 2010).

Naturally, the circulated fluid is required not to interfere the rheometer operation in both theory and practice, therefore a suitable circulation method should be carefully designed to qualify the requirement. In this paper, authors will develop a series of conclusions about the fluid circulation on the basis of a systematical hydro-mechanical analysis on the stress acting on the spindle surface.

2 STRESS DISTRIBUTION ON SPIDDLE SURFACE

Taking the coaxial cylinder rheometer with internal cylinder as spindle as example, a column coordinate system is set up as Figure 1 where the spindle rotates at angular speed ω along the direction of θ coordinate. Then the stresses of the rotating spindle toward the fluid on its surface can be listed respectively as follows. In θ-direction,

$$\tau_\theta = \tau_{r\theta} + \tau_{\theta\theta} + \tau_{z\theta} \tag{1}$$

In r-direction,

$$\tau_r = \tau_{rr} + \tau_{\theta r} + \tau_{zr} \tag{2}$$

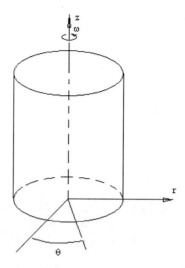

Figure 1. Adopted column coordinate system.

and in z-direction

$$\tau_z = \tau_{rz} + \tau_{\theta z} + \tau_{zz} \quad (3)$$

In these equations, the first subscript of the stress denotes the normal direction of the stress action surface and the second one denotes the stress direction.

The nine fluid stress divisions consist of a stress tensor:

$$[\tau] = \begin{pmatrix} \tau_{rr} & \tau_{r\theta} & \tau_{rz} \\ \tau_{\theta r} & \tau_{\theta\theta} & \tau_{\theta z} \\ \tau_{zr} & \tau_{z\theta} & \tau_{zz} \end{pmatrix} \quad (4)$$

According to the General Newtonian Law, stress tensor $[\tau]$ is related with deformation rate tensor $[\varepsilon]$ by (Zhao & Liao, 1983)

$$[\tau] = 2\eta[\varepsilon] \quad (5)$$

where η = the fluid apparent viscosity. The deformation rate tensor $[\varepsilon]$ has also nine divisions:

$$[\varepsilon] = \begin{pmatrix} \varepsilon_{rr} & \varepsilon_{r\theta} & \varepsilon_{rz} \\ \varepsilon_{\theta r} & \varepsilon_{\theta\theta} & \varepsilon_{\theta z} \\ \varepsilon_{zr} & \varepsilon_{z\theta} & \varepsilon_{zz} \end{pmatrix} \quad (6)$$

For the incompressible fluid the divisions of deformation rate tensor can be expressed by three divisions of velocity (u_r, u_θ, u_z) in column coordinate system if the pressure is neglected:

$$\begin{cases} \varepsilon_{rr} = \dfrac{\partial u_r}{\partial r} \\ \varepsilon_{\theta\theta} = \dfrac{\partial u_\theta}{r\partial \theta} + \dfrac{u_r}{r} \\ \varepsilon_{zz} = \dfrac{\partial u_z}{\partial z} \\ \varepsilon_{r\theta} = \varepsilon_{\theta r} = \dfrac{1}{2}\left(\dfrac{\partial u_r}{\partial \theta} + \dfrac{\partial u_\theta}{\partial r} - \dfrac{u_\theta}{r}\right) \\ \varepsilon_{z\theta} = \varepsilon_{\theta z} = \dfrac{1}{2}\left(\dfrac{\partial u_z}{r\partial \theta} + \dfrac{\partial u_\theta}{\partial z}\right) \\ \varepsilon_{rz} = \varepsilon_{zr} = \dfrac{1}{2}\left(\dfrac{\partial u_z}{\partial r} + \dfrac{\partial u_r}{\partial z}\right) \end{cases} \quad (7)$$

Within the operation space i.e. the column-ring area between spindle and external cylinder in a traditional rheometer, there are no axial and radial flow, meanwhile the tangential velocity is symmetrical to the rotating axis and is independent to the axial coordinate. In other words, $u_r = u_z = 0$, $\dfrac{\partial u_\theta}{\partial \theta} = 0$ and $\dfrac{\partial u_\theta}{\partial z} = 0$. These yield that from equation (7):

$$\begin{cases} \varepsilon_{rr} = \varepsilon_{\theta\theta} = \varepsilon_{zz} = 0 \\ \varepsilon_{r\theta} = \varepsilon_{\theta r} = \dfrac{1}{2}\left(\dfrac{\partial u_\theta}{\partial r} - \dfrac{u_\theta}{r}\right) \\ \varepsilon_{z\theta} = \varepsilon_{\theta z} = \varepsilon_{rz} = \varepsilon_{zr} = 0 \end{cases} \quad (8)$$

Therefore the divisions of stress tensor becomes

$$\begin{cases} \tau_{rr} = \tau_{\theta\theta} = \tau_{zz} = 0 \\ \tau_{r\theta} = \tau_{\theta r} = \eta\left(\dfrac{\partial u_\theta}{\partial r} - \dfrac{u_\theta}{r}\right) \\ \tau_{z\theta} = \tau_{\theta z} = \tau_{rz} = \tau_{zr} = 0 \end{cases} \quad (9)$$

Then the stress acted by fluid on the spindle surface in the θ-direction can be written as

$$\tau_w = \tau_\theta + k\tau_r = \tau_{r\theta} + k\tau_{\theta r} = (1+k)\eta\left(\dfrac{\partial u_\theta}{\partial r} - \dfrac{u_\theta}{r}\right) \quad (10)$$

with $k =$ the friction coefficient of fluid on the spindle surface. If the friction is neglected, equation (10) becomes

$$\tau_w = \eta\left(\dfrac{\partial u_\theta}{\partial r} - \dfrac{u_\theta}{r}\right) \quad (11)$$

where $\left(\dfrac{\partial u_\theta}{\partial r} - \dfrac{u_\theta}{r}\right)$ is the fluid shear rate on the spindle surface and noted as D_w. The apparent viscosity η can be obtained by measuring the shear stress and shear rate on the spindle surface:

$$\eta = \dfrac{\tau_w}{D_w} \quad (12)$$

The analysis above is the operation basis of a coaxial cylinder rheometer.

3 REQUIREMENTS FOR FLUID CIRCULATION

In order that the operation principles mentioned above can be still effective when the tested fluid is circulated from the bottom to the top of a coaxial cylinder rheometer, some requirements have to be satisfied. These conditions can be presented as follows. Firstly, the radial flow has to be avoided, i.e. $u_r = 0$; Secondly, the circulated fluid has no division in θ-direction so that the fluid tangential velocity in operation space is just resulted from the spindle rotation and no external shear force to be resisted by the spindle; Thirdly, the circulated fluid should be distributed evenly on the whole ring cross section between spindle and external cylinder to qualify that $\frac{\partial u_z}{\partial \theta} = 0$; In addition, the circulated fluid should no impaction on the upper section of the spindle so that no additional loading exacted to the spindle. In fact all these requirements can be summarized as that the circulated fluid has only the axial velocity without other two divisions within the operation space of the remoter. If this can be qualified, the equations (7), (8) and (11) are still effective and the rheometer is able to work according to the traditional theories. It should be pointed out that if the circulated fluid goes into the operation space from some special position and does not evenly distribute on the whole cross-section, a correction to the measurements will be necessary (Xu et al., 2010).

On the basis of hydro-mechanical analysis mentioned above, a circulation fluid distributor has been designed by authors. In the distributor, the circulated fluid is distributed evenly on the cross-section, has no tangential and radial velocity within the operation area and only is the axial velocity kept, but the details will be reported elsewhere because of the limited space.

4 CONCLUSIONS

When a coaxial cylinder rheometer is applied to test the rheological parameters of flocculated suspensions or pulps containing solid particles, the sedimentation of flocs and particles due to gravity will interfere seriously the measurements, and an alternative approach to the problem is make the tested fluid circulate within the operation area between spindle and external cylinder of the rheometer. A systematic hydro-mechanical analysis on the circulation fluid shows that if the fluid circulation attaches only axial velocity to the measured suspension or pulps and yields no external tangential and radial movement, the traditional basic principles of the rheometer will kept effective and do not need any modification. Then the sedimentation problems of flocs or particles in the operation space of the rheometer can be resolved successfully.

ACKNOWLEDGEMENT

Authors present their thanks to NSFC (21276032) for its support to the research.

REFERENCES

Assael M. J., Dalaouti N. K. 2000. Prediction of the Viscosity of Liquid Mixtures: An Improved Approach. Rheologica Acta, 21(2), 356–357.

Kawatra, S. K., Bakshi, A. K. and Miller Jr., T. E. 1996. Rheological characterization of mineral suspensions using a vibrating sphere and a rotational viscometer. Int. J. Miner. Process., 44–45, 155–165.

Xu Jirun, Xing Jun and Ding Shiqiang. 2010. The circulation system of coaxial cylinder rheometer. Innovation Patient of China, No. 201010107268.7.

Zhao Xueduan, Liao Qidian. 1983. Viscous Fluid Mechanics (in Chinese). Beijing, Mechanical Industry Press.

Analysis of adhesively bonded pipe joints under torsion by finite element method

J.Y. Wu, H. Yuan & J. Han
MOE Key Laboratory of Disaster Forecast and Control in Engineering, Institute of Applied Mechanics, Jinan University, Guangzhou, China

ABSTRACT: Adhesively bonded pipe are extensively used in pipe system and pipeline failure under torsional loading is one of the most common failure modes. Owing to the complication of theoretical analysis, development of numerical method for optimized design and stress analysis of a strong joint remains a pressing issue. In this paper, a realistic bilinear local bond-slip law is employed for analyzing the behavior of adhesively bonded pipe joints under torsion, based on a parameters study by ANSYS finite element software. Several parameters (i.e. the shear modulus of pipe coupler) are considered for assessing shear stress distribution in the adhesive and load-displacement relationship of the interface. Finally, the factors influencing the interfacial behavior is analyzed.

1 INTRODUCTION

Pipe structures are a very important structural form for energy industry and construction industry. With the development of materials science and manufacturing, the mechanical properties of pipe itself has been dramatically improved (Ouyang and Li, 2009b, Ouyang and Li, 2009a). Among all the possible structural failure, pipeline failure under torsional loading is one of the most common failure modes. Few studies were conducted to investigate the interfacial behavior of the adhesively bonded pipe joints under torsion load (Hosseinzadeh and Taheri, 2009, Chen and Cheng, 1992, Zhao and Pang, 1995). The failure analysis of pipe joints under torsion have been presented numerically (Hosseinzadeh and Taheri, 2009). A relatively simpler method is provided to predict the interfacial debonding and failure of bonded joints by using fracture mechanics (Hutchinson and Evans, 2000, Rizzi et al., 2000). In the present work, efforts have been made to illustrate how the pipe coupler material and thickness of joint affect the interfacial shear stress and debonding process. Therefore, it is emphasized that the proposed methodology used in piping applications can be also further extended to handle fiber-reinforced composites pipes.

2 MODEL DESCRIPTION

The model mainly consist of three parts: the main pipe, the adhesive layer and the pipe coupler; and all the geometrical parameters are given, as shown in Figure 1. The adhesive layer is assumed no thickness here. In order to consider the influence of material properties, five kinds of materials of pipe coupler assumed to be isotropic and are considered in this study. Additionally, all the material of main pipe and pipe coupler are assumed to be isotropic, elastic and no damage mechanism is considered. The behavior of the adhesive is chosen as bilinear in calculation. Only right part of main pipe is discussed for the reason of symmetry in the following.

Various bond–slip models have been considered in previous work (Teng et al., 2002). In the paper, the bilinear local bond-slip model is apply to model the interface debonding. A typical

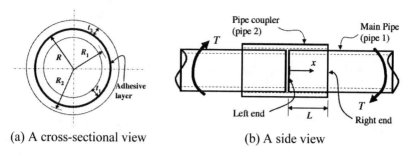

(a) A cross-sectional view   (b) A side view

Figure 1. Adhesively bonded pipe joint.

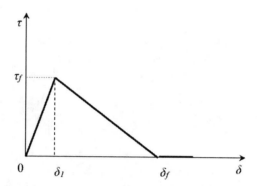

Figure 2. Local bond-slip model.

bilinear local bond-slip model consists of a linear elastic branch and a linear softening branch, as illustrated in Figure 2. According to this model, the bond shear stress increases linearly with the interfacial slip until it reaches the peak stress τ_f at which the value of the slip is denoted by δ_1. After the occurrence of an interfacial microcrack (softening), the local bond-slip relation is linearly descending with a range from δ_1 to δ_f. The value of shear stress is reduced to zero and an interfacial macrocrack (debonding) occurs when the value of slip exceeds τ_f (Wu et al., 2002, Yuan et al., 2004).

3 NUMERICAL SIMULATIONS

To gain a clear understanding of the interfacial shear stress distribution and crack propagation along main pipe and pipe coupler, finite element simulations were conducted to simulate the behavior of adhesively bonded pipe joints under torsion. The material properties and geometry parameters in the numerical analysis are selected as follows: $t_1 = 20$ mm, $R_1 = 150$ mm, $t_2 = 2$ mm, $R_2 = 161$ mm, $G_1 = 20$ GPa. It should be particularly noted how to select G_2. The shear modulus and the polar moment of inertia of main pipe and pipe coupler are G_1, G_2 and J_1, J_2, respectively. $\beta = (G_2 J_2)/(G_1 J_1)$, so it can be obtained that $G_2 = 165.52$ GPa when $\beta = 1$. In order to discuss the interfacial debonding behavior and shear stress propagation, five kinds of materials with different shear modulus of pipe coupler will be considered. When β equals 0.5, 0.75, 1.5 and 2.5, G_2 can be calculated to be equal to 82.76 GPa, 124.14 GPa, 248.28 GPa and 413.80 GPa, respectively. And the interfacial characteristic parameters are selected as $\delta_1 = 0.034$ mm, $\tau_f = 7.2$ Mpa, $\delta_f = 0.16$ mm.

The parametric study was conducted by ANSYS finite element software. Main pipe and pipe coupler are modeled with SOLID65 element and SHELL181, respectively; both are connected by zero-thickness COMBIN39 element, which are used to simulate the bond interface. A torsion T

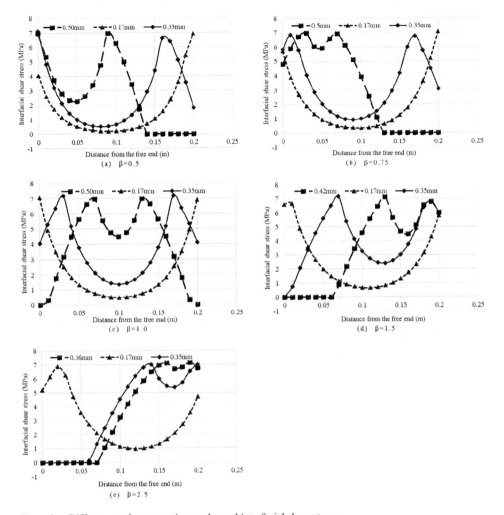

Figure 3. Different crack propagation modes and interfacial shear stress.

applied at the left end of pipe coupler, and the right end of the main pipe is fixed. Finite element simulations are performed by the displacement control method.

As shown in Figure 3, at small loads there is no interfacial debonding or softening along the interface as long as the interfacial shear stress at both ends is less than τ_f. After shear stress attains the value $\tau_f(\delta=\delta_1)$ at any end, softening appears, and the whole interface is in a combined elastic-softening stage. The load increases with increase of the length of the softening zone, and the maximum transferable load appears at this stage. Debonding is initiated when the length of the softening zone attains $\delta=\delta_f$ and $\tau=0$ at any ends. After this crack propagation happens, the whole main pipe and pipe coupler interface is in a combined elastic-softening-debonding stage, and the shear stress peak τ_f moves toward the middle of the bonded joint.

When $\beta < 1$, softening and debonding appear at the right end first, and when $\beta > 1$, softening and debonding appear at the left end first. The larger $|1-\beta|$, the earlier softening and bebonding appear.

As seen from Figure 4, when $\beta=1$, the ultimate load is the largest. Both $\beta > 1$ and $\beta > 1$, the ultimate load will decreases with the $|1-\beta|$ increasing.

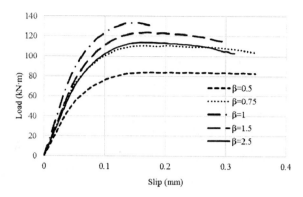

Figure 4. Load-slip relationship at the loaded end.

4 CONCLUSIONS

In this paper, the influence of parameters involved in design of an adhesively bonded joint under torsion was studied by using the ANSYS finite element software. Five kinds of different shear modulus of pipe coupler were considered to examine the influence of various material parameters on the joint shear stress distribution and strength. The following conclusion were drawn.

1. When $\beta = 1$, softening and debonding appear at both ends at the same time. The larger $|1 - \beta|$, the earlier softening and bebonding appear.
2. When $\beta = 1$, the ultimate load is the largest. Both $\beta > 1$ and $\beta > 1$, the ultimate load will decreases with the $|1 - \beta|$ increasing.

ACKNOWLEDGEMENT

The authors gratefully acknowledge the financial support provided by the Science and Technology Scheme of the Guangdong Province (2012A030200003) and the Natural Science Foundation of China (National Key Project No. 11032005).

REFERENCES

Chen, D. & Cheng, S. 1992. Torsional stress in tubular lap joints. *International Journal of Solids and Structures*, 29, 845–853.
Hosseinzadeh, R. & Taheri, F. 2009. Non-linear investigation of overlap length effect on torsional capacity of tubular adhesively bonded joints. *Composites Structures*, 186–195.
Hutchinson, J. W. & Evans, A. G. 2000. Mechanics of materials: topdown approaches to fracture. *Acta Materialia*, 48, 125–135.
Ouyang, Z. & Li, G. 2009a. Cohesive zone model based analytical solutions for adhesively bonded pipe joints under torsional loading. *International Journal of Solids and Structures*, 46, 1205–1217.
Ouyang, Z. & Li, G. 2009b. Interfacial debonding of pipe joints under torsion loads: a model for arbitrary nonlinear cohesive laws. *International Journal Fractures*, 155, 19–31.
Teng, J., Chen, J., Smith, S. & Lam, L. 2002. *FRP strengthened RC structures*, UK, John Wiley and Sons.
Wu, Z. S., Yuan, H. & Niu, H. D. 2002. Stress transfer and fracture propagation in different kinds of adhesive joints. *Journal of Engineering Mechanics-Asce*, 128, 562–573.
Yuan, H., Teng, J. G., Seracino, R., Wu, Z. S. & Yao, J. 2004. Full-range behavior of FRP-to-concrete bonded joints. *Engineering Structures*, 26, 553–565.
Zhao, Y. & Pang, S. S. 1995. Stress–strain and failure analyses of pipe under torsion. 117, 273–278.

Plastic deformation of Honeycomb sandwich plate under impact loading in ansys

Qiang Huan & Yongqiang Li
College of Science, Northeastern University, Shenyang, Liaoning, China

ABSTRACT: Honeycomb sandwich plate will produce huge plastic deformation under impact loading. The impact energy is converted to the energy of plastic deformation, so it has a good effect on withstanding impact. The finite element analysis software ansys/ls-dyna is used to simulate the process of honeycomb sandwich plate impacted by a bullet in this article. According to the different design parameters of honeycomb sandwich plate, such as thickness and the thickness of its sandwich layer, compare the stress and strain of honeycomb sandwich plate. Research the dynamic response of honeycomb sandwich plate under load of impact in medium speed.

1 INTRODUCTION

In recent years, ultra light porous metals, such as aluminum honeycomb and metal foam, are concerned by engineering increasingly because of its superior physical and chemical properties. Honeycomb materials are widely used in automobile, shipbuilding, aerospace and other fields because the structure is high specific strength, high specific modulus and light. Honeycomb sandwich plate absorbs the huge energy caused by impact by plastic deformation .The scholars at home and abroad study on honeycomb sandwich plate deeper and deeper as time goes by in recent year. Sun Jia has studied the nonlinear dynamics of honeycomb sandwich plate (Sun Jia, 2008). Zhao Jinsen has studied the equivalent model of honeycomb sandwich plate's mechanical property (Zhao Jinsen, 2006). Empirical formula on perforation diameter of aluminium plate which is impacted by a bullet in a high speed is given by Sawle D R (Sawle D R, 1969), Maiden C J (Maiden C J, 1964) and Nysmith C R (Nysmith C R, 1969) personally. Because the plastic dynamic response process of honeycomb sandwich plate is complex, it is difficult to study it directly. The researchers found that it is an accurate way to study the plastic dynamic response process of honeycomb sandwich plate by using a bullet to impact it. The dynamic response of honeycomb sandwich plate under the impact loading has been an important direction to the academic circles. The research in this field is not deep, so it still needs further studies on the specific process of deformation, efficacy-losing and energy absorption mechanisms about honeycomb sandwich plate.

The article uses finite element analysis ansys/ls-dyna to simulate the process of honeycomb sandwich plate impacted by a bullet in a medium speed. In view of the different design parameters of honeycomb sandwich plate, research the change of stress and strain on honeycomb sandwich plate under the different design parameters and compare the results, and find the specific effects caused by different design parameters on honeycomb sandwich plate.

2 EMPIRICAL PROCESS

Figure 1 is modal experiment pattern. The length of honeycomb sandwich plate is a, the width of it is b, the thickness of it is h, the thickness of its sandwich layer is hc, and the speed of bullet

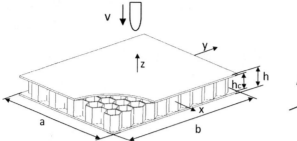
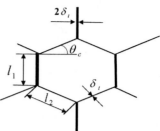

Figure 1. Modal experiment pattern.

Figure 2. Structure of cell in regular hexagon.

Table 1. Specific design parameters of honeycomb sandwich plate.

a/m	b/m	hc/h	h/a	l/m	θ	v/m/s
0.15	0.15	0.6	0.01–0.1	0.004	Π/6	13
0.15	0.15	0.85	0.01–0.1	0.004	Π/6	13
0.15	0.15	0.95	0.01–0.1	0.004	Π/6	13

is v. Figure 2 is structure of cell in regular hexagon. It is the main structure of sandwich layer. The length of it is l, the thickness of it is δ and 2δ, and the angle of it is θ.

The impact process is analyzed by using ansys/ls-dyna in this experiment. The impact process is direct impact, the speed of a bullet is 13 m/s, the experimental temperature is room temperature and the fixed form is fixed bearing. The ansys/ls-dyna contact model is set to be Eroding, contact time is 2 seconds and contact step is 20. Aimed at the design parameters changes of honeycomb sandwich plate such as thickness and the thickness of its sandwich layer, find the result of stress and strain on honeycomb sandwich plate under different design parameters after it impacted by a bullet.

Because the structure of honeycomb sandwich plate is complex, microsoft visual c++ is used to write command stream to establish the experimental model in ansys/ls-dyna. Then use macro-variable to change the design parameters of honeycomb sandwich plate and get different kinds of honeycomb sandwich plate to run research experiment. See the specific design parameters of honeycomb sandwich plate in table 1.

3 RESULTS AND ANALYSIS

The main design parameters of honeycomb sandwich plate comprises plate thickness, sandwich layer thickness. The change of those design parameters has great influence on impact resistance performance and energy absorption mechanism of honeycomb sandwich plate. In order to make the changes of stress and strain that resulted from changes of design parameters clear, need figure out the specific data of stress and strain on honeycomb sandwich plate under different parameters. Compare and summarize the experimental data, find the specific influence about stress and strain on honeycomb sandwich plate under different design parameters. Optimize the design parameters and make honeycomb sandwich plate have a good performance on withstanding impact.

In order to judge the differences between energy absorption mechanisms and withstanding impact which are resulted from different design parameters on honeycomb sandwich plate, control variable method is used in this experiment. The variables are sandwich layer thickness to plate thickness ratio (hc/h), plate thickness to plate length ratio (h/a).

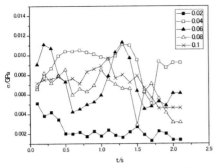
Figure 3-1. hc/h is 0.6, h/a is from 0.01 to 0.1, von mises stress variation curves.

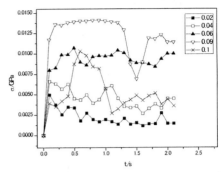
Figure 3-2. hc/h is 0.85, h/a is from 0.01 to 0.1, von mises stress variation curves.

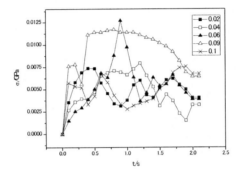
Figure 3-3. hc/h is 0.95, h/a is from 0.01 to 0.1, von mises stress variation curves.

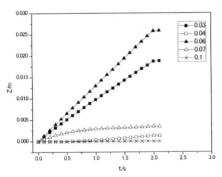
Figure 4-1. hc/h is 0.6, h/a is from 0.01 to 0.1, Z displacement.

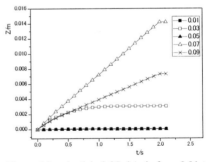
Figure 4-2. hc/h is 0.85, h/a is from 0.01 to 0.1, Z displacement.

(1) The results of changing h/a

When hc/h and l are constant, study the influence about energy absorption mechanisms and withstanding impact on honeycomb sandwich plate which are resulted from h/a. The hc/h in this experiment are 0.6, 0.85, 0.95. The l is 0.004 m. Compare stress and strain variation curves when hc/a is from 0.01 to 0.1. Figures from 3-1 to 3-3 are its von mises stress variation curves. Figures from 4-1 to 4-3 are its strain variation curves in Z axis. Because part of the data is too close, only select half the data to draw.

hc/h is 0.6, when h/a is less than 0.04 or more than 0.08, the maximal value of von mises stress on honeycomb sandwich plate is lesser; when h/a is from 0.04 to 0.08, the maximal value of von mises stress on honeycomb sandwich plate is larger.

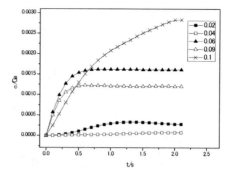

Figure 4-3. hc/h is 0.95, h/a is from 0.01 to 0.1, Z displacement.

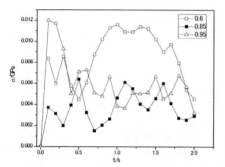

Figure 5-1. h/a is 0.01, von mises stress variation curves.

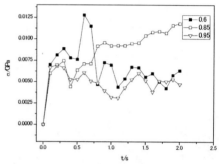

Figure 5-2. h/a is 0.03, von mises stress variation curves.

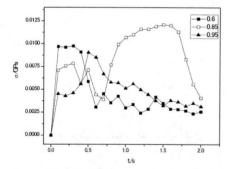

Figure 5-3. h/a is 0.05, von mises stress variation curves.

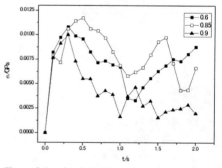

Figure 5-4. h/a is 0.07, von mises stress variation curves.

hc/h is 0.85, when h/a is less than 0.04 or more than 0.09, the maximal value of von mises stress on honeycomb sandwich plate is lesser; when h/a is from 0.04 to 0.09, the maximal value of von mises stress on honeycomb sandwich plate is larger.

hc/h is 0.95, when h/a is less than 0.06 or more than 0.09, the maximal value of von mises stress on honeycomb sandwich plate is lesser; when h/a is from 0.06 to 0.09, the maximal value of von mises stress on honeycomb sandwich plate is larger.

hc/h is 0.6, the Z displacement of honeycomb sandwich plate increases linearly with time, Z displacement is lesser when h/a is more than 0.07 or nearby to 0.04.

hc/h is 0.85, the Z displacement of honeycomb sandwich plate increases linearly with time firstly and then be almost constant, Z displacement is lesser when h/a is less than 0.05.

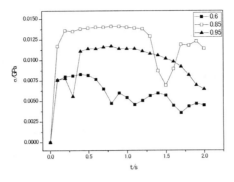

Figure 5-5. h/a is 0.09, von mises stressvariation curves.

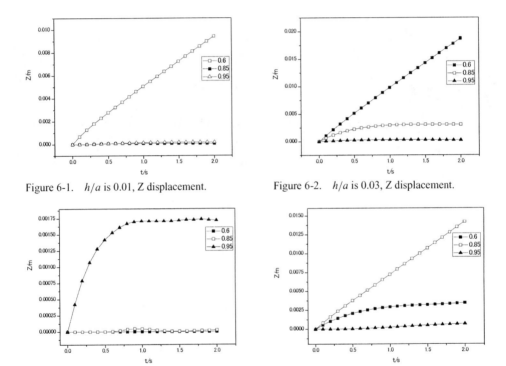

Figure 6-1. h/a is 0.01, Z displacement.   Figure 6-2. h/a is 0.03, Z displacement.

Figure 6-3. h/a is 0.05, Z displacement.   Figure 6-4. h/a is 0.07, Z displacement.

hc/h is 0.95, the Z displacement of honeycomb sandwich plate increases linearly with time firstly and then be almost constant, Z displacement is lesser when h/a is less than 0.04.

(2) The results of changing hc/h

When h/a and l are constant, study the influence about energy absorption mechanisms and withstanding impact on honeycomb sandwich plate which are resulted from hc/h. The h/a in this experiment is from 0.01–0.1. The l is 0.004 m. Compare stress and strain variation curves when hc/h are 0.6, 0.85, 0.95. Figures from 5-1 to 5-5 are its von mises stress variation curves. Figures from 6-1 to 6-5 are its strain variation curves in Z axis.

h/a is less than 0.02, the maximal value of von mises stress on honeycomb sandwich plate is minimum when hc/h is 0.85. h/a is from 0.02 to 0.08, the maximal value of von mises stress on honeycomb sandwich plate is minimum when hc/h is 0.95. h/a is more than 0.08,

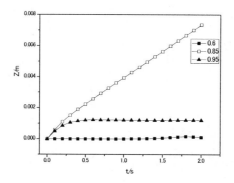

Figure 6-5.　h/a is 0.09, Z displacement.

the maximal value of von mises stress on honeycomb sandwich plate is minimum when hc/h is 0.6.

h/a is less than 0.02, Z displacement of honeycomb sandwich plate increases linearly with time when hc/h is 0.6 and has small change when hc/h is 0.85 or 0.95. When hc/h is 0.85, Z displacement is minimum.

h/a is from 0.02 to 0.08, Z displacement of honeycomb sandwich plate increases linearly with time when hc/h is 0.6 and has small change when hc/h is 0.85 or 0.95. When hc/h is 0.95, Z displacement is minimum. All of above except hc/h is 0.05.

h/a is more than 0.08, Z displacement of honeycomb sandwich plate increases linearly with time when hc/h is 0.85 and has small change when hc/h is 0.6 or 0.95. When hc/h is 0.6, Z displacement is minimum.

4　CONCLUSIONS

This experiment uses ansys/ls-dyna to proceed with modal experiment, and analyzes the honeycomb sandwich plate impacted by a bullet in medium speed. Find the dynamic response of the honeycomb sandwich plate. We have Conclusions as following:

(1) When hc/h, l are constant, h/a is less than 0.04 or more than 0.09, the stress and strain of honeycomb sandwich plate is lesser, the effect of withstanding impact is well. The energy absorption effect of honeycomb sandwich plate is best when h/a is 0.02.
(2) When l is constant, the energy absorption effect of honeycomb sandwich plate is best and the stress and strain of its is least when h/a is lee than 0.02 and hc/h is 0.85; The energy absorption effect of honeycomb sandwich plate is best and the stress and strain of its is least when h/a is from 0.02 to 0.08 and hc/h is 0.95; The energy absorption effect of honeycomb sandwich plate is best and the stress and strain of its is least when h/a is more than 0.08 and hc/h is 0.6.

REFERENCES

Maiden C J, McMillan A. An investigation of the protection afforded a spacecraft by a thin shield [J]. American Institute of Aeronautics and Astronautics, 1964, 2(11):1992–1998.
Nysmith C R, Denardo B P. Experimental investigations of momentum transfer associated with impact into thin aluminum targets [R]. NASA TN D-5429, October, 1969.
Sawle D R. Hypervelocity impact in thin sheets semi-infinite targets at 15 km/s [C]. AIAA Paper No. 69-378, AIAA Hypervelocity Impact Conference, Cincinnati, OH, 1969.
Sun Jia, Nonlinear dynamics of honeycomb sandwich plate, Beijing University of Technology, 2008-04-01.
Zhao Jinsen, The mechanical properties of the equivalent model on aluminum honeycomb sandwich plates, Nanjing University of Aeronautics & Astronautics, 2006-12-01.

ced Engineering and Technology – Xie & Huang (Eds)

The responses of self-anchored suspension bridges to aerostatic wind

Tian-Fei Li
Institute of Bridge Engineering, Dalian University of Technology, Dalian, China

Chang-Huan Kou
Department of Architecture and Urban Planning, Chunghua University, Hsinchu, Taiwan

Chin-Sheng Kao
Department of Civil Engineering, Tamkang University, Taipei, Taiwan

Je Jang
Institute of Bridge Engineering, Dalian University of Technology, Dalian, China

Pei-Yu Lin
Department of Civil Engineering, Chunghua University, Hsinchu, Taiwan

ABSTRACT: This paper uses an improved incremental iteration nonlinear method to analyze the nonlinear response to aerostatic wind of self-anchored suspension bridges. By subjecting the Jinzhou Straight Bridge to aerostatic loads, the bridge was analyzed to investigate the effect of factors such as initial wind attack angle, the cable sectioning, added wind attack angle and the wind speed cross-section parameter on the aerostatic displacement of the tower.
  Results of the analysis show that the interaction between the pile foundation and the soil is not considered, the vertical and horizontal displacements of tower due to the wind load are underestimated. Whether the cables are divided into sections or not has only a relatively small effect on the displacements of the tower. On the other hand, the wind load on the cable system has significant influence on the lateral displacement of the tower.

1 INTRODUCTION

Early aerostatic stability analysis made use of the lateral buckling method and the structural torsion collapse method (Cheng Jin 2000). These two methods belong to linear analysis. A subsequent method was to combine the eigenvalue with the iterations. Using this approach, Ming-Shan Fang (1997) combined the incremental method with the iteration method and applied it to calculate the second kind aerostatic instability problems. Zhi-Tian Zhang (2004) introduced a low relaxation factor which increased the speed of convergence. Starting with responses to aerostatic wind and vibration, Xiao-Lun Hu (2006) proposed a method of analysis with one main rule and three secondary rules and investigated the effect of aerostatic wind on the long spanned cable-stayed bridges. Using the aforementioned aerostatic stability analysis to calculate iteratively the nonlinear added attack angle requires a parameter to end the convergent calculation. It requires a great deal of experience to set the magnitude of this parameter correctly and sometimes it is set by the practical environment. If the magnitude of the parameter is too large, then even though the number of iterations is reduced, the error in the calculated result is large.

2 AEROSTATIC RESPONSE AND STABILITY ANALYSIS

2.1 *Procedures of analysis*

The analysis procedure for the improved incremental nonlinear iteration method used in this paper is given below:

(1) Assume a wind speed and an aerostatic initial wind attack angle. From these, calculate the aerostatic loads based on the three component coefficients of static force in each bridge component. Apply these loads to the finite element model.
(2) Perform a nonlinear static analysis of the bridge structure to calculate the twisting angle at the nodes on the girder. Recalculate the wind loads acting on the girder based on the girder's three component coefficients of static force corresponding to various aerostatic attack angles (initial attack angle and added attack angle).
(3) Continue the iteration process until a nominated number of iterations are reached or until two successive iterations in displacement or twisting angle responses have differences smaller than a nominal value. At this point, the structural displacement for the considered wind speed can be determined and we can proceed to perform the eigenvalue analysis and to calculate the model's frequency of vibration.
(4) Increase the wind speed to the next strength and repeat steps (1)–(3). If the nonlinear analysis is convergent, then the displacement response of the structure at that wind speed is confirmed.
(5) If the nonlinear analysis is not convergent, then reduce the wind speed and repeat the analysis. The interpolation method was used to calculate aerostatic instability wind speed.

2.2 *Calculation of aerostatic loads for structural components*

The girder's three component coefficients of static force used in this calculation are shown in Figure 1.

Following the code JTG/T D60-01-2004(2004), the transverse wind load applied to the tower, hanger, and main cable can be calculated from the following equation

$$F_H = \frac{1}{2}\rho V_g^2 C_H A_n \tag{1}$$

where V_g is the static wind speed; C_H is drag coefficient due to structural components of side span, and A_n is the projected area of structural component along the wind direction (m^2).

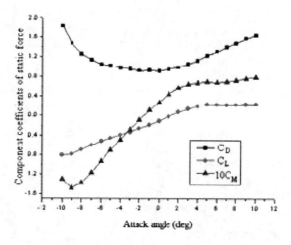

Figure 1. Distribution curves of the three component coefficients of static force.

Based on the code JTG/T D60-01-2004(2004), the drag coefficient of tower is chosen as 1.55 and that for cable system is taken as 0.7 for this paper.

2.3 Basic information for analysis

The main bridge length of the Jinzhou Straight Bridge is 660 m. Its central span is 400 m, and at each side spans a self-anchored 130 m suspension bridge with a width of 23.5 m. The main girder is a steel box girder. The steel material is Q345C. The main cable and hangers are made of galvanized high strength cables with a strength of 1670 MPa. The main cable is made of 37 bundles. The tower has a reinforced concrete frame structure with a total height of 96.35 m. The superstructure is 76.35 m high. The configuration of the bridge is shown in Figure 2. and the cross-section of the steel box girder is shown in Figure 3.

In order to analyze the effect of pile foundations on the aerostatic response of the structure, two separate boundary conditions are considered, with one case where there is interaction between the pile foundation and the soil and another case without this interaction. To analyze the effect of dividing the hanger into sections, two models for the hanger are considered, with one model where the hanger is divided into sections and another without divisions. To test the above conditions into the models used for the aerostatic wind load calculation are Model A, Model B, and Model C, shown in Table 1.

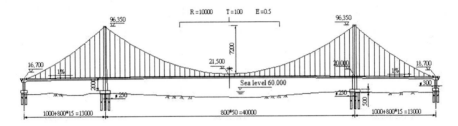

Figure 2. Side View of the Jing-Zhou Bay Bridge. (Unit:cm).

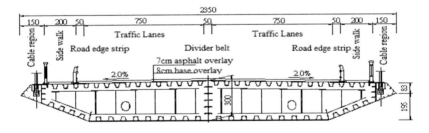

Figure 3. The cross-section of the steel boxgirder.

Table 1. Calculation model of Jinzhou Strait Bridge under static wind load.

Models	Pile foundation		Hangers	
	Interaction with soil	No interaction with soil	Divided	Not divided
A	V		V	
B		V	V	
C		V		V

3 RESULTS OF ANALYSIS AND THEIR COMPARISON

3.1 *Effect of initial wind attack angle on tower displacements due to static wind*

Figures 4.1–4.3 show the Model B results obtained for tower top's longitudinal and lateral displacements and rotation angle about vertical direction when the initial attack angle is set at $-3°$, $0°$ and 3.

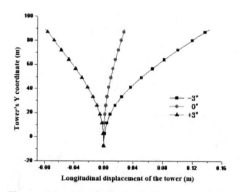

Figure 4-1. Longitudinal displacement of the tower as initial attack angle varies.

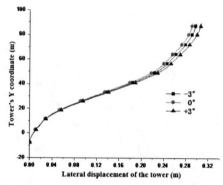

Figure 4-2. Lateral displacement of the tower as initial attack angle varies.

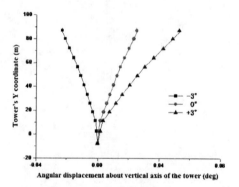

Figure 4-3. Angular displacement of the tower about vertical axis as initial attack angle varies.

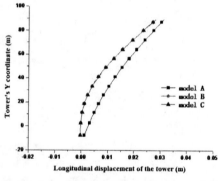

Figure 4-4. Longitudinal displacement of the tower.

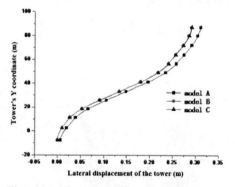

Figure 4-5. Lateral displacement of the tower.

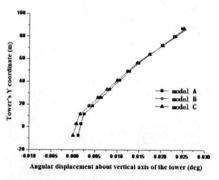

Figure 4-6. Angular displacement of the tower about vertical axis.

In addition, the top of left tower has a positive displacement in the longitudinal direction of the bridge indicating an inclination of the tower toward the mid-span. It is because, when the initial attack angle is at 0°, the wind lift coefficient has a negative value of −0.1066.

3.2 *The effect of pile foundation and sectioning of cable system on aerostatic displacement*

When the initial attack angle is 0°, the wind speed cross-section parameter is taken to be 0.12, and the analysis considers the static wind load on the hangers. We investigated the effect of pile foundations on the static response by comparing the Models A and B. We also investigated the effect of sectioning of hangers on the static response by comparing Models B and C. The results are shown in Figures 4.4–4.6.

The figures show that, if the pile foundation effect is disregarded, the aerostatic longitudinal and lateral displacements of tower are under-estimated. It is shown that whether or not the cable system is divided into sections has relatively little influence on tower displacements. The pile foundation has a relatively small influence on the tower's twisting angle about the vertical axis. The pile's effect on the tower's bottom is larger than its effect on the upper regipart.

4 CONCLUSION

This paper uses an improved nonlinear incremental iteration method to analyze the nonlinear aerostatic response of self-anchored suspension bridge to aerostatic wind. The Jinzhou Straight Bridge was analyzed under aerostatic load to investigate the effect of factors such as initial attack angle, pile foundation, sectioning of cables and wind speed cross-section parameters on the displacements.

Also we know, if the pile foundation effect is neglected, the longitudinal and lateral displacements in both the girder and the tower will be under-estimated. The sectioning of the cable system has a relatively small effect on the displacements in the girder and the tower.

REFERENCES

Fang, Ming-Shan, 1997. *Large-span cable supported bridge nonlinear aero-static stabiLity theory*. Shanghai: Tongji University.
Hu, Xiao-Lun, 2006. *Stability analysis of large span cable-stayed bridge shaking vibration response and calm wind*. Shanghai: Tongji University.
Jin, Cheng, 2000. *Study on Nonlinear Aerostatic Stability of Cable-Supported Bridges*. Shanghai: Tongji University.
JTG/T D60-01-2004, 2004. *Highway Bridge Wind Load Design Specification*. Beijing: China Communications Press.
Zhang, Zhi-Tian, 2004. *Nonlinear Buffeting Vibration and It's Effects on Aerostatic and Aerodynamic Stability of Long-Span Bridges*. Shanghai: Tongji University.

Distribution features and protection for heritage in the metropolitan area

Yufang Zhang & Xuesong Xi
College of Water Resources & Civil Engineering, China Agricultural University, Beijing, P.R. China

Xuemei Wang
Chaoyang Division of Beijing Municipal Commission of Urban Planning, Beijing, P.R. China

ABSTRACT: Cultural heritage protection campaign has been rapidly developed since the 1970s, and the protection of cultural heritage has become a common view for all people. However, under China's current context of rapid urbanization, the number of urban cultural heritage has sharply decreased, and the "constructive destroy" towards the heritage is very common. The research chooses Beijing Chaoyang District as the study object that has the largest district area and population and possesses a long history and rich cultural sediments. Through spatial analysis in GIS on the correlation between the heritage and the systems of road, water, and park greens, the research finds the spatial distribution features of the heritage. Through the analysis on least cost distance, the research constructs a conservation network and proposes an integrated protection strategy for the historic and cultural heritage in Chaoyang. The research involves the issue of heritage protection in the context of urbanization.

1 INTRODUCTION

Cultural heritage protection campaign has been rapidly developed since the 1970s, and the protection of historic and cultural legacy has become a common view for all people. However, in China's current context of rapid urbanization, cultural heritage protection in cities has been trapped into a dilemma of damage by development and conservation (Zhang 2005, Wei 2000). On the one hand, the complete reconstruction at old districts and flourishing new development result in damages at various degrees to the cultural heritage in the area (Yang & Yu 2004), severing the historic line of the city and blurring its cultural identity (Ruan 2004). On the other hand, the lack of measures for the integrated protection and sustainable utilization (Ruan 1998), weak public participation (Ruan & Ding 2007, Wu 2011, Jiao 2002), and the lagging behind of protective legislation and regulations (Wang 2000, Shao & Ruan 2002), all make the activities of heritage conservation and sustainable utilization hard to be implemented and promoted (Zhang 2005). Therefore, the study on effective protective measures for the historic and cultural heritage in the process of urban development has become a key issue in urban and rural planning and in relic protection research (Chen 2002, Yu & Li 2003).

The research chooses Beijing Chaoyang District as a case study. Based on the analysis about the spatial relationship between the historic and cultural heritage in this district and the elements of roads, water system and park greens, it proposes the strategy for the integrated protection of the cultural heritage (Liu 2010, Jiang 2005), trying to find a balance point between the development of Beijing's central area and the conservation of cultural heritage and a foundation for the strategy planning of Chaoyang's cultural development.

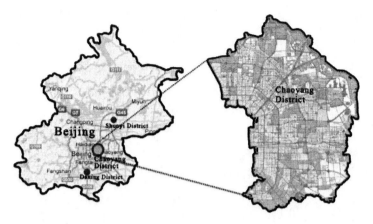

Figure 1. Study Area.

2 AREA UNDER THE RESEARCH

The area under the research is Beijing Chaoyang District. This district is located at the east and northeast of Beijing central downtown, 39°48′ to 40°09′ degrees north latitude, and 116°21′ to 116°42′ east longitude. It neighbors Tongzhou District in the east, Haidian District, Xicheng District and Dongcheng District in the west, Fengtai District and Daxing District in the south, and Shunyi District and Changping District in the north. Chaoyang District has the largest area, and the largest population. With a north-south length of 28 kilometers and an east-west width of 17 kilometers, the total area of this district is 470.8 square kilometers.

Chaoyang District has a long history. The district was first established in 1925, named Beijing Dongjiao (east suburb) District at the time. In 1928, it was renamed Beiping Dongjiao District, which was changed as Suburb One District and Suburb Two District in 1945, and later the 13th District and the 14th District in the same year. In 1949, the two districts merged into one as the 13th District. In 1952, its name was changed back to Dongjiao District, and modified as Chaoyang District under the approval of the State Council in 1958 (Chen 2006). As of 2012, this district has got 24 streets and 19 townships.

3 METHODOLOGY

The research uses a methodology combining GIS spatial analysis and least cost distance analysis.

3.1 *GIS spatial analysis*

GIS spatial analysis refers to a method of using geographical information system to collect and transfer the spatial features and their correlation according to the position and morphological characteristics of the geographical subject, for spatial decision making. The research inputs the geographical information data of Bejing Chaoyang District into GIS system and constructs a spatial information platform, analyzing the correlation between the historic and cultural relic sites and the elements of road hierarchy, water system and park greens, computing the number of heritage distributed along levels of road and water system, and in park greens, together with the distribution features, and thereby generating the spatial distribution features of the historic and cultural heritage in Chaoyang District.

3.2 *Least cost distance analysis*

For various dispersive elements, the connecting route is decided by degrees of attraction, spatial relationship, distance, and surrounding barriers among these elements. Considered from the

Table 1. Historical and cultural heritage constituted system of Chaoyang District.

Protection level	National	City	Distract	Registration items	Others	Total
Quantity	4	5	8	89	5	111

perspective of relic protection, the journey from one relic site to another is a process of overcoming spatial relationship, distance and surrounding barriers. The stronger the resistance is, the more unsuitable it is to conduct heritage protection, and therefore more unsuitable to build a protection network. On the contrary, the area with weaker resistance has higher suitability, meaning more suitable for relic protection network. In GIS, this is reflected as the resistance coefficient value under different land surface features. In reality, it stands for the value of comprehensive evaluation over a series of suitable or unsuitable elements, such as land use type, road hierarchy, and accessibility. In GIS, these elements are reflected as the cost distance resistance value under different earth surface features. Based on the above analysis, the Least-Cost Distance model is used to simulate and establish the resistance surface, with the following formula (Yu 1995):

$$MCR = f \min \sum_{j=n}^{i=n} (D_{ij} \times R_i)$$

of which D_{ij} stands for the distance from one element to another, and R_i the resistance of the element. Based on this formula, after the resistance coefficient for different earth surface features is determined, the optimal path can be computed and simulated with the cost distance tool frequently used in GIS. The method is also widely applied in ecological protection and cultural heritage conservation (Yu et al. 2005).

4 SPATIAL DISTRIBUTION FEATURES OF THE HISTORIC AND CULTURAL HERITAGE IN CHAOYANG DISTRICT

4.1 Overview of the historic and cultural heritage in Chaoyang District

Data from the 3rd National Survey on Cultural Heritage (Chaoyang District section) in 2010 shows that there are 106 immovable cultural heritage distributed in 36 streets in this district. In the first list of Beijing Excellent Modern Structures to Be Protected, released by Beijing municipality in 2007, 8 structures are located in Chaoyang District. The 798 Workshop, Beijing Workers' Stadium, and Beijing Workers' Gymnasium have been listed as cultural heritage under conservation, and the other 5 structures are not. In total, 111 historic and cultural heritages, 106 historic and cultural sites under protection and 5 excellent modern structures will be studies in this research.

The number of permanent residents in Chaoyang District, 3.545 million, surpasses those of any other districts in Beijing. GIS analysis shows that the distribution of the historic and cultural heritage is positively correlated to population density, i.e. the higher density of the population, the more historic and cultural sites, while the lower density, the less sites.

The analysis on the 108 historic and cultural sites (the construction periods of the other 3 are unknown) in Chaoyang District shows that, apart from 2 heritage built in Han Dynasty, the others form a continual chronosequence that covers from Yuan Dynasty to modern periods. The types of heritage are also rich, including 5 categories: ancient cultural heritage, ancient tombs, ancient architecture, stone inscriptions, and modern historic sites and memorial buildings.

Before analyzing the spatial distribution features of these cultural heritage in Chaoyan District, the research first withdraws these heritage from their complicated surrounding environment with modern urban structures, and simplifies the space into a basic system formed by cultural sites and common landscape elements (roads, water system, and vegetated areas); and in turn based on this system, the spatial distribution features of the cultural heritage in Chaoyang District are to be analyzed.

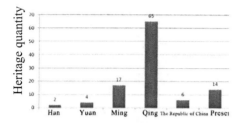

Figure 2. Age constitution of heritage.

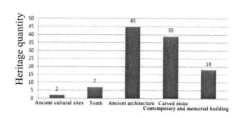

Figure 3. Type constitution of heritage.

Figure 4. Distribution of cultural heritage.

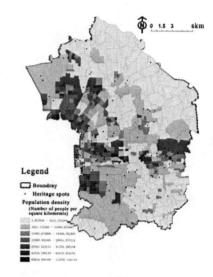

Figure 5. Heritage and population density.

4.2 The spatial relationship between cultural heritage and road system

Chaoyang District has formed a "radial-circle shaped" road network in a framework consisting urban expressways, arterials, inner ring road, medium ring road and outer ring road. At present, the district contains more than 1000 kilometers of road, and the density of road network is 2.77 li/square kilometers.

The overlapping of district road system and the historic and cultural sites through GIS system indicates that, as shown in the diagram, the curved line continues stretching within 100 meters, and tends to be flat after 100 meters. At the 200-meter points, the number of cultural sites reaches 92, accounting for over 80% of the total heritage.

For cultural heritage within 200 meters from the roads, further analysis is conducted based on the 25–50 m interval and 50 m interval.

The number of cultural heritage within 25 meters from the road has the highest percentage, 27.93%; 52 heritage within 50 meters, accounting for 46.85% of the total; 77 within 100 meters, accounting for 69.37%; 92 within 200 meters, accounting for 82.88%.

The above data and diagram show that the historic and cultural sites in Chaoyang District are closely related to the integrated road network. Over 80% of the heritage are distributed within 200 meters from the road system.

Next, the relic sites are overlapped with a hierarchy of roads (highway and expressway, major arterials, minor arterials, connectors) for analysis. Over 80% of the relic sites, 90 in all, are distributed within 1000 meters from the highway and expressway, nearly 70% of the relic sites, 77

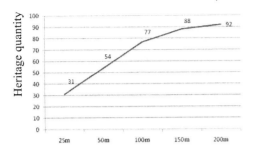 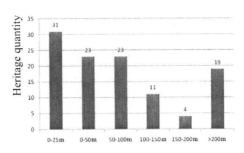

Figure 6. Heritage quantity in different distance between heritage and road.

Figure 7. Relationship between heritage quantity and road spatial distance.

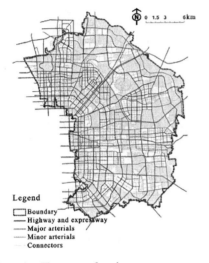 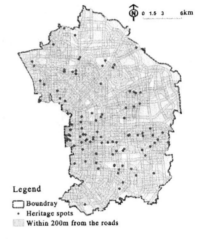

Figure 8. The status of road system.

Figure 9. Heritage distribution within 200 m of whole road system.

Figure 10. Heritage distribution in highway, expressway and major arterials.

in all, are within 500 meters from the major arterials; over 80% of the relic sites, 92 in all, are within 500 meters from the minor arterials; nearly 70% of the relic sites, 77 in all, are within 200 meters from the connector. These figures indicate that the distribution of the cultural heritage in Chaoyang District is relatively concentrated, and they are evenly distributed around each level of road.

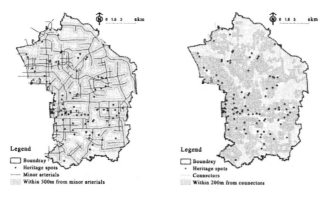

Figure 11. Heritage distribution in minor arterials and connectors.

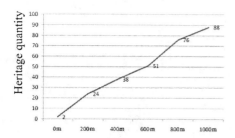

Figure 12. Heritage quantity in different distance between heritage and river.

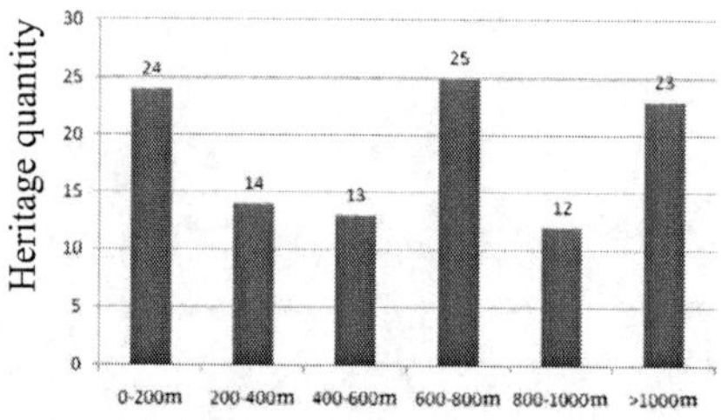

Figure 13. Relationship between heritage quantity and river spatial distance.

4.3 The spatial relationship between cultural heritage and water system

Chaoyang District is at the tail of Beijing's drainage system. The district has many rivers and lakes. Sections of Liangshui River, Xiaotaihou River and Tonghui irrigation cannel run through the southern part of Chaoyang District. The total length of rivers in this area is 151 kilometers, in addition to 110 middle and small sized ditches, coming up to 320 kilometers. There are lakes like Chaoyang Park lake, Yaowa Lake, Honglingjin Lake, and Gaobeidian Lake, together with fish ponds, pools and swales, totaling around 70 in number, and covering an area of 980 hectares.

The overlapping between the river and lake system in Chaoyang District and the historic and cultural heritage through GIS shows that the curved line continues stretching within 1000 meters. The number of relic sites reaches 88 at 1000-meter, approaching to 80% of the total.

Taken 1000-meter as the spatial distance boundary, the number of heritage within 1000 meters is further analyzed with 200-meter as an interval.

Within the 200-meter intervals, the distribution of heritage at each interval is relatively even. The percentages of sites distributed within 200 meters and in 600–800 meters are the highest, both of which exceed 20% of the total. 51 heritage are within 600 meters around the water system, accounting for 45.94%; 76 heritage are within 800 meters, accounting for 68.47%; 88 are within 1000 meters, accounting for 79.28%. When the research zooms in to Tonghui River section and analyzes the relic sites distribution here, it is found that there are 88 heritage within 1000 meters around the water, and about 1/3 are located along the river. The above data and diagram proves that the historic and cultural heritage in Chaoyang District are closely related to the water system of rivers and lakes.

4.4 The spatial relationship between cultural heritage and green area

There are 16 forests existing or under construction in Chaoyang District, each of whose area covers over 10,000 mu (1 mu equals 666.67 square meters), 13 parks with various peculiarities, and 27

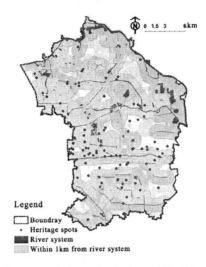

Figure 14. The status of river system.

Figure 15. Heritage distribution within 1 km of whole river system.

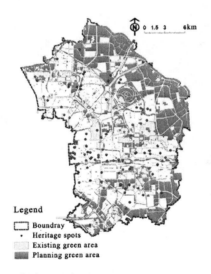

Figure 16. Relationship between heritage and green area.

community parks. The green coverage ratio in the district is 30.6%. Of the 82-kilometer long and 50-100-kilometer wide greenbelt around Beijing city, 55 kilometers are located in Chaoyang District.

GIS overlapping shows that 11 cultural heritage are in existing large-scale greens, and 38 heritage are in greens planned. As they are further developed in Chaoyang District, greens in the future will cover around 40% of the cultural heritage.

5 INTEGRATED PROTECTION STRATEGIES FOR THE HISTORIC AND CULTURAL HERITAGE IN CHAOYANG

First, to construct a protection network for these cultural heritage. In reference of elements such as land use type, road hierarchy, and park greens, and the spatial relationship among elements, as well as degrees of people's willingness, such as their preference to different levels of road

Table 2. Construction of related factors and relative resistance value.

Types	Related factors	Effects	Relative resistance values
Land Types	Residential land	Assignment	200
	Commercial and financial land		600
	Administrative land		500
	Multifunctional land		800
	Educational land		600
	Other land		1000
Road Types	Highway and expressway		20
	Major arterials		10
	Minor arterials		40
	Connectors		30
Others	Green area		30
	River	Correction	0

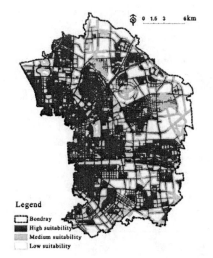

Figure 17. Suitability analysis of network system.

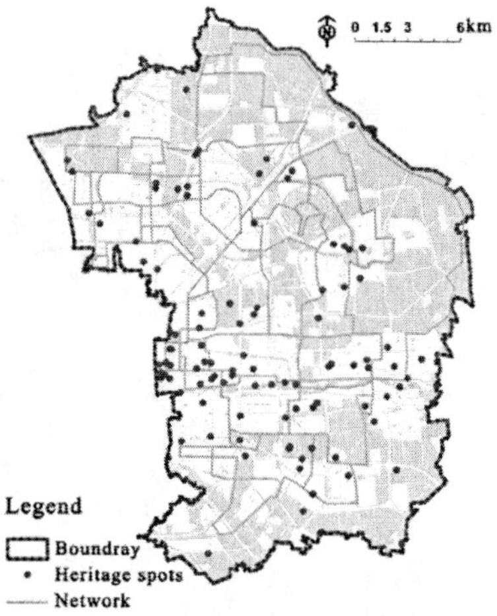

Figure 18. Relationship between heritage and network system.

(connectors>minor arterials>major arterials>expressways), the resistance surface is established for the computation of least cost distance. The specific resistance values are as follows.

In computing, the warehouse land, village public land, land for further research, industrial land, square parking land, city greens land, agricultural land, agricultural facilities land, municipal facilities land, sports land, railway land, foreign affairs land, medical and health land, land for religion and social welfare are listed as miscellaneous land type which cannot be passed through.

On the basis of the above study and preparation, the research computes cost distance, establishes resistance surface and generates the suitability analysis as shown in the diagram, from which the relic protection network is evolved.

By evaluating the designed relic protection network, we see that the network covers each section of public greens, connects the major rivers and lakes in this district, and links a majority of the historic and cultural heritage. The conservation network, constructed with elements of roads, water system and park greens, is able to cover 80% of the relic sites, and forms an integrated roadmap for the relic resources in Chaoyang District.

Second, to improve the illustration and presentation system for the heritage. Based on the designed relic protection network, an illustration and presentation system shall be built up, with

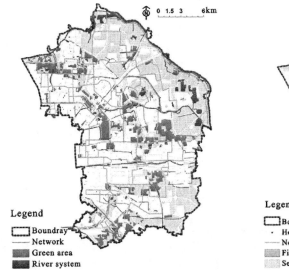
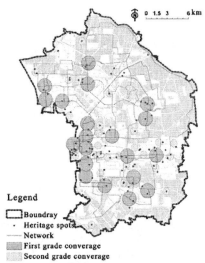

Figure 19. Relationship between network system and river, network system and green area.

Figure 20. Distribution of heritage interpretation dots.

elements such as road hierarchy and the density of relic resources distribution taken into consideration. According to people's travelling habits and requirements, the illustration system is designed to set 1 km as the radius of range for first-tier service centers, and 0.5 km for the second-tier centers, as supplements for the first-tier ones.

Third, to retain the authenticity and integrity of the heritage. For the protection of historic and cultural heritage, the authentic styles and features shall be retained, and their surrounding environment shall also be fully referred to. All the historic information in those heritages must be conserved, and state intact, for the authenticity. And for those that have blemishes, the rehabilitation work shall be done according to the original status in order to remain the integrity of the heritage.

Fourth, to raise public awareness, and attract public participation. Public participation to protect the historic and cultural heritage can be motivated by improving public cognition towards the heritage, raising their participation awareness, and supplying various participating access and necessary technical support, so that the cultural heritages and their space are able to vividly enter into public vision, and the public can have more opportunities to understand the historic and cultural heritage and prolong their memories and thinking on the heritages.

6 CONCLUSION AND DISCUSSION

The research overlaps the historic and cultural heritage sites in Chaoyang District with resources of roads, water system and green area by using GIS spatial analysis, and generates the spatial distribution features of those sites. Based on that, the research, through least cost distance analysis, further assigns resistance values to each spatial element and constructs a network for the relic protection, and thus discusses an effective method for the integrated protection over the historic and cultural heritage in cities.

According to the research, there are 106 historic and cultural heritage (referring to tangible cultural heritage) listed as protected sites in Beijing Chaoyang District, and 5 structures as the first batch of modern excellent buildings to be protected, totaling 111 heritage sites in the district. In terms of space, the heritage sites are closely related to roads. Over 80% of the heritage sites are distributed within 200 meters around the roads, and the sites have a close relationship with each level of the road hierarchy. The heritage sites are also tightly related to the water system. Over 80% of the heritage are distributed within 1000 meters around the water system of rivers and lakes. The relationship of the heritage sites with park greens is loose. Only 10% of the heritage are located in existing greens.

The research analyzes the spatial distribution features of the cultural heritage in Chaoyang District, assigns values to elements of different land types, road hierarchy and green area, and simulates a heritage protection network with the cost distance model. The finalized heritage protection network connects each section of public greens, important rivers and lakes, and most of the historic and cultural heritage sites which account for 80% of the total in this district. Meanwhile, the research proposes protective strategies of improving the illustration and presentation system, retaining the authenticity and integrity of the heritage, and motivating public participation.

ACKNOWLEDGEMENTS

The thesis is sponsored by the Chaoyang Division of Beijing Municipal Commission of Urban Planning under the Research of Resource survey and mapping of cultural heritage in Beijing Chaoyang District.

REFERENCES

Chen, Gang 2006. History of Chaoyang distract. Beijing: Beijing Press.
Chen, Junyu 2002. Revisit the earth Landscaping and Urban Landscaping. Journal of Chinese Landscape Architecture 18 (3):3–6.
Jiang, Bin 2005. Research of Cultural Heritage in Jining City and Discussion on the Integrated Conservation Strategy. Beijing: Peking University.
Jiao, Yixue 2002. Non-governmental organization in historic conservation in UK. Planners 18(5): 79–83.
Liu, Qing 2010. Integrated protection theory of urban heritage. Urban Problems (2): 13–17.
Ruan, Yisan 1998. The course of historical and cultural heritage protection of the world and China. Journal of Tongji University (social science section) 9(1):1–8.
Ruan, Yisan 2004. Retain our roots: Urban development and urban heritage protection. Urban and rural development (7):8–11.
Ruan, Yisan & Ding, Feng 2007. The growth of civil power of urban heritage protection in China. Construction science and technology (17): 54–55.
Shao, Yong & Ruan, Yisan 2002. Legal Construction of historical and cultural heritage protection: the Inspiration of French historical and cultural heritage protection system development. Urban planning forum (3): 57–80.
Wang, Lin 2000. The comparative study on the protection system for the historical heritage in China and Abroad. City Planning Review 24 (8):49–61.
Wei, Yaping 2000. Dilemma of confusion: Perception of contemporary urban heritage conservation. Planners 16 (1):118–119.
Wu, Zuquan 2011. Public participation in urban heritage protection: a case study of Henning District Reconstruction of Guangzhou. Transformation and Reconstruction. 2011 China Urban Planning Annual Conference Proceedings: 3852–3862.
Yu, Kongjian 1995. Ecological security patterns in landscapes and GIS application. Annals of GIS1 (2):88–102.
Yu, Kongjian & Li Dihua 2003. Landscape Architecture: The Profession and Education. Beijing: Building Press: 70–92.
Yu, Kongjian et al. 2005. Suitability analysis of heritage corridor in rapidly urbanizing region: a case study of Taizhou City. Geographical Research 24 (01):69–77.
Yang, Lixia & Yu, Xuecai 2004. A Summary of China's Researches on the Protection and Utilization of Cultural Heritage. Tourism Tribune 19:85–91.
Zhang, Fan 2005. Historical and cultural protection measures in urban Development. Nanjing: Southeast University.
Zhang, Song 2005. Keeping the trace of times, maintaining the spirit of city: frontier issues in urban conservation and revitalization. Urban planning forum (3):31–35.

Landscape planning for city parks with rainwater control and utilization

Yun Yang & Xuesong Xi
College of Water Resources & Civil Engineering, China Agricultural University, Beijing, China

ABSTRACT: City parks are carriers for urban forestation and ecosystem improvement. In park design and planning by cities in China, rainwater control and utilization is often missing. As a result, city parks do not mitigate water-logging, and their water consumptions are acutely increasing. The thesis takes Tuanjiehu Park as a case study. By analyzing the current rainwater utilization status and existing problems, combines rainwater utilization and landscape ecological design, and constructs a rainwater utilization system with pervious pavement, sunken lawn, and detention tanks, in order to improve the efficiency of rainwater utilization, alleviate urban drainage burden, and tackle water crisis in the park.

1 INTRODUCTION

In recent years, water-logging and water scarcity have become a new urban illness pestering many Chinese cities. On one hand, during the fast urbanization process, a large amount of forestry and farmland with good permeability is replaced by impermeable surface. As a result, the natural water cycle is affected, hydrological ecosystem seriously deteriorates (Grimm 2008), urban surface runoff increases and concentration time is shortened. These changes form a strenuous pressure on rainwater control (Hollis 1977, Zhao & Li 2011, Jennings & Jarnagin 2002). Many big cities, such as Beijing, Xi'an, Changsha, Guangzhou and Chongqing, have been trapped in the predicament of water-logging whenever it rains. On the other hand, cities are facing severe water scarcity. China is identified by the United Nations as a country in short of water resources. There are more than 400 cities that do not have enough water supply, and 110 of them are in serious shortage (Song & Yu 2007, Jiang 2008). One of the key reasons for the coexistence of water-logging and water scarcity is that we simply treat "rainwater", a key element in water cycle, as a kind of "waste" and just "discharge" it (Che et al. 2009). Obviously, this conventional "water-proof" concept cannot meet the rainwater control requirement in modern cities; therefore, to solve flood disaster and water crisis by utilizing rainwater as a resource has become a key research topic in urban planning (Cao & Che 2002, Barraud & Gautier 1992, Nancy & Grimm 2008, Wang & Cui 2003).

City parks, as an important component of city system, are urban public green space combining functions of recreation, leisure and entertainment, open spatial places for urban residents to have public activities (Cai 2000, Gao & Ni 2005), and carriers for urban forestation and ecosystem improvement (Meng et al. 2005). In recent years, many Chinese cities have entered into a faster city park construction period. However, as rainwater control design does not match with park planning, city parks, instead of mitigating urban water-logging through their lakes, water bodies, greens and squares, causes the acute increase of water demand. When China is faced by water shortage, the conventional way of feeding city parks with tap water and ground water has become a huge form of extravagancy. Therefore, under the context of building ecological parks advocated by Chinese government, it is an inevitable trend to add rainwater control and utilization into landscape designing for city parks. By introducing the concept of rainwater management, the thesis takes Tuanjiehu Park at Beijing Chaoyang District as an instance, and based on the current level of rainwater utilization in the park, analyzes the feasibility of rainwater utilization and explores a landscape planning model with rainwater control and utilization.

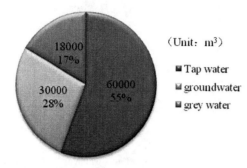
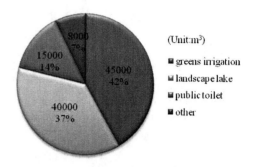

Figure 1. The water source proportion in the park.

Figure 2. The ratio of the different uses of the water in the park.

2 OVERVIEW OF THE AREA UNDER RESEARCH

Beijing is a typical city where flooding and drought coexist. The annual precipitation is around 585 mm, few in total amount and distribution uneven within a year. Most rainfall concentrates in flood season in form of rainstorm, which results in frequent and severe urban water-logging. The gap of water supply in Beijing every year is up to 1.1 billion cubic meters, while a huge amount of rainwater is directly discharged through drainage system, 66% of which runs off without full utilization (Li et al. 2001).

The Tuanjiehu Park under research is located at North Road of the East 3rd Ring Road of Beijing Chaoyang District. It was built in 1986, 8 kilometers away from the center of Beijing. As a city park featured with ancient garden characteristics in South of Yangtze River, it covers an area of 123,000 m^2, among which water surface 54,100 m^2, green area 42,800 m^2, road pavement 17,200 m^2, square 8,900 m^2 and structure area 400 m^2. As the park was constructed many years ago, the design did not take rainwater utilization into consideration. Although the park has undergone several rehabilitations later, the improvements were focused on forestation. The research finds that the park is situated in a low-lying area, and the drainage facilities can just meet a low level of design standards, so there is always water-logging in rainstorm days, while its water consumption and water cost are climbing up year by year.

3 CURRENT WATER UTILIZATION IN THE PARK AND PROBLEMS

3.1 Water consumption status quo and problems

Water is consumed in the park for daily life (hand washing, drinking), toilet flushing, road cleaning, park greens irrigation, and landscape lake water recharge. Preliminary statistics show that the annual water consumption in the park is about 10,800 m^3, among which grey water 18,000 m^3, ground water 30,000 m^3, and tap water 60,000 m^3 (Fig. 1). Water for park greens irrigation is replenished by grey water and tap water, with an annual amount of about 45,000 m^3; water for landscape lake is from grey water and ground water, with an annual amount of 40,000 m^3; water for public toilet 15,000 m^3, recharged by grey water and tap water; other usages 8,000 m^3 (Fig. 2). The park greens irrigation and landscape water recharge take up 80%, a major chunk of the total water consumption in the park. The replenishing sources are mainly tap water and ground water, and the water usage structure is irrational because too much quality water is put into inferior usage.

3.2 Rainwater utilization status quo and problems

Rainwater catchment area in the park consists of road, square, park greens and water body. As the permeability of the pavement materials for road and square is weak, the large area of impermeable pavement becomes the major site where rainwater runoff is generated and drained through municipal pipe network, not been utilized at all. Moreover, the edge of park greens adjacent to the road and square is usually lined by stone curbs which are higher than the road surface, and the park greens is elevated even higher than the road surface, which makes it impossible for the park greens to harvest and collect the rainwater runoff. When rainfall is strong or the park greens soil is saturated, rainwater runoff also emerges from the park greens and is then drained to the road surface in natural flow, exerting additional burden to the municipal drainage network that has already got a large amount of rainwater and forming a severe loss of rainwater (Chen et al. 2003).

4 FEASIBILITY ANALYSIS ON RAINWATER UTILIZATION IN THE STUDIED AREA

4.1 Target and principle of water balance analysis

To calculate water volumes in each part of the park and do water balance analysis is a fundamental work to decide on rainwater utilization scheme and the utilization system design, as well as the structures. An accurate water balance analysis is a crucial basis for scheme development and optimization (Li & Che 2002). Water balance analysis refers to the calculation of water input, output and the difference between them in a certain period, under the condition of normal park operation, i.e. maintaining normal water supply as designed. The difference between water input and output will help to determine if the water volume is in deficit or surplus under current condition, if water demand can be satisfied, and if efflux or recharge is needed, together with the recharge cost. Water balance equation goes as:

$$\sum Input - \sum output = \Delta difference$$

where $\sum Input$ = rainwater catchment + recharge; $\sum output$ = evaporation + consumption + loss; when Δ is positive, water volume is stored in water body till reaching overflow level; when Δ is negative, water volume is in deficit, and water level drops below the constant level.

4.2 Water balance calculation

According to experience, as temperature in January and February is very low and the lake is frozen, lake water recharge will be out of calculation. Summer rainfall is abundant but its time distribution is not even. The precipitation of one rainstorm could take up 70% of the whole volume in that month, and the excessive rainwater is just drained away. Therefore, a large amount of water recharge in summer is quite possible. Calculation shows that the annual water recharge for the lake is about 40,000 m^3. The lawn area in the park is 36,970 m^2, plants and trees 8,676, park greens total area 42,800 m^2. Pursuant to Beijing Water Quota for Park Greens, an amount of 1.1 m^3/m^2 for comprehensive water usage can be applied to Tuanjiehu Park. Taken the lawn area and plant and tree area into consideration, it is computed that the water consumption for park greens irrigation every year is around 45,000 m^3. Road area is 17,200 m^2, 0.0015 m^3/m^2 of water is used for sprinkling. From March to December (low temperature in the other two months, and sprinkling would result in icy road surface), one time per month, or 10 times in all for road washing every year will consume 200 m^3 of water (Table 1).

The total annual precipitation in the studied area is calculated using the following formula:

$$W = H \times A$$

where W = the total annual precipitation, m^3; H = the annual average depth of rainfall, m (value is 0.58m according to the above analysis); A = the catchment area of the studied area, m^2 (value is 123,000m^2, according to data available). The volume of rainwater utilizable at Tuanjiehu Park is calculated as shown in Table 2 below.

The above analysis shows that the utilizable rainwater in the park can reach 15,000m^3 (Table 3) every year. If rainwater can be reused within the park by constructing rainwater utilization system and using facilities such as infiltration and detention tanks, ground water will be effectively replenished, and city rainwater drainage burden is mitigated. Moreover, part of the rainwater can

Table 1. The statistics of water consumption of lake filling, greens irrigation road cleaning, toilet flushing and other use each month (Unit: m^3).

Month	Lake filling water	Greens irrigation water	Road cleaning water	Toilet flushing water	Other water	Total
1	0	1000	0	1250	700	2950
2	0	1000	0	1250	700	2950
3	3000	5000	20	1250	700	9970
4	5000	5000	20	1250	700	11970
5	6000	5000	20	1250	700	12970
6	8000	5000	20	1250	700	14970
7	3000	5000	20	1250	700	9970
8	3000	5000	20	1250	700	9970
9	4000	5000	20	1250	700	10970
10	4000	4000	20	1250	700	9970
11	2000	2000	20	1250	700	5970
12	2000	2000	20	1250	700	5970
Total	40000	45000	200	15000	8400	108600

*Rainfall data from the national weather service site platform (54511).

Table 2. The amount of available rainwater in Tuanjiehu Park each month (Unit: m^3).

Mouth	Annual precipitation (mm)	Road pavement Runoff coefficient 0.85	Greens Runoff coefficient 0.15	Square Runoff coefficient 0.87	Total
1	2.3	30	14	17	61
2	3.5	4	23	25	52
3	7.5	110	48	53	211
4	18.2	266	118	130	514
5	40.4	584	261	288	1133
6	71.4	1037	461	508	2006
7	202.9	2949	1311	1445	5705
8	141.4	2059	913	1007	3979
9	41	584	26	292	902
10	26.3	380	170	187	737
11	10.9	147	70	78	295
12	2.6	32	17	18	67
Total	568.4	8182	3432	4048	15662

*The underlying surface runoff coefficient data from Che Wu (urban rainwater to use technology and management).

be harvested and detained as grey water to replace tap water or ground water which is currently used for park greens irrigation, landscape lake recharge, and toilet flushing. Therefore, to utilize rainwater as a resource is an economical, promotable, simple, fast and effective way to tackle existing problems, and the trend of exploring and utilizing rainwater in city parks is inevitable.

5 PLANNING STRATEGIES FOR RAINWATER UTILIZATION IN STUDIED AREA

5.1 *Principles of utilizing rainwater*

(1) Infiltration as a major approach, supplemented by harvesting. The portion of park greens and water surface is relatively high in the park. Through park greens, gardens and pervious pavement,

Table 3. Potential Analysis of available Rainwater Utilization (Unit: m^3/year).

Items	Total amount of available rainwater	Road cleaning water	Toilet flushing water	Greens irrigation water	Lake filling water
consumption	15662	200	15000	45000	40000

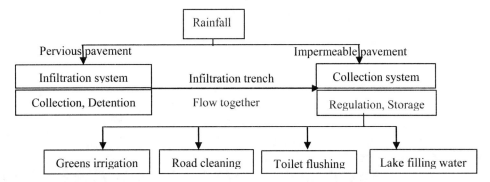

Figure 3. Rainwater utilization system schematic diagram.

rainwater will naturally infiltrate into and replenish ground water. In addition, due to high level of management by the park and good quality of water, the collection pipe and rainwater facilities can be fully utilized to harvest rainwater from roads and the square and in turn supply water to the park greens, water body and toilets, so that there will be no rainwater efflux in the park.

(2) Proximity reduces the cost. Rainwater is harvested through the nearby collection system to the nearby detention tanks, and after purification, used to nearly park greens, toilets, or diverted to nearby water body. This would reduce the complexity of the system and the difficulty in maintenance, and rainwater resources can be reused.

5.2 Rainwater utilization system

All landscape elements are aimed to be dynamically integrated, and the design is targeted at the zero drainage of rainwater within the park by various technical measures in order to maximize the utilization of rainwater resources (Fig. 3). In general, the rainwater utilization system consists of two sub-systems: rainwater infiltration system and rainwater collection and reuse system. These two sub-systems interact with each other and jointly complete the task of rainwater utilization.

5.2.1 Rainwater infiltration system

The rainwater infiltration system is composed of pervious pavement, infiltration trench and sunken lawn. As rainwater goes through multi-layers of permeable surface, permeable underlying surface and soil matrix, infiltrated and purified, a large amount of pollutants will be removed. Meanwhile, the vegetated surface, the underlying surface and the edge grooves are able to retain a large amount of infiltrated rainwater and prolong the time of infiltration. The system not only controls the aggregated runoff generated on the surface, but also recharges underground water, retains water resource, and improves the ecological environment (Wang 2007). Therefore, multi-point dispersed infiltration system with facilities like pervious pavement and sunken lawn will improve the permeability for rainwater, effectively detain water, and reduce runoff from the road, square and parking lot. The technology is simple and easy for maintenance.

5.2.2 Rainwater collection and reuse system

By facilities like infiltration ditches, perforated pipes and detention tanks, rainwater collection and reuse system will effectively detain surface runoff, and adjust water balance between rainy

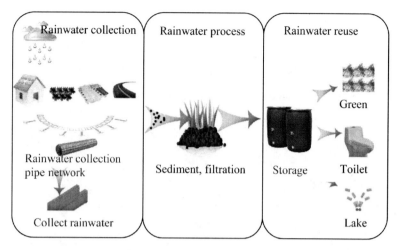

Figure 4. Rainwater collection and reuse system.

season and non-raining periods, converting the discontinuous rainwater resource in different time and space into a system with stable water supply. The rainwater, after being harvested, purified and detained, can function as grey water and replace part of the tap water or ground water for park greens irrigation, road sprinkling, toilet flushing and landscape water body recharge in the park (Fig. 4).

(1) Rainwater harvesting system for road

The system is designed to maximize the use pervious pavement materials in a sensible way, and to increase the slope of road bed, so that rainwater runoff can rapidly flow to the pretreated infiltration ditch. When there is heavy precipitation, stormwater will infiltrate to the perforated pipes pretreated at both sides of the road and under the sidewalk. This would replenish the ground water in time, and prevent the road surface from water-logging. The extra surface water, after preliminary infiltration through the road surface, then concentrates to the rainwater collection system.

(2) Rainwater harvesting system for herbaceous field

The efficiency of rainwater runoff concentration is relative low. When the precipitation is little, the amount of rainwater concentrated is almost zero. But as the catchment area is large in the herbaceous field, the stormwater volume would be significant (including accepting runoff from buildings and roads adjacent to the field). Moreover, the quality of rainwater concentrated is relatively good and easier to be processed. Taking into consideration the landscape, vegetation distribution, and the vegetation roots depth, together with the road slope degrees, the system is designed to construct sunken lawn, rain detention pond, and rain garden, as well as collecting pipes and perforated pipes under the field, draining the infiltrated rainwater into retention facilities for further reuse.

(3) Rainwater harvesting system for the square

The spacious paved square is also an effective site to collect rainwater. Segments of catchment area can be divided according to elevation, and infiltration trenches are fixed along the edges of the segments. The surface rainwater runoff will infiltrate through the permeable wall of the trenches into the infiltration underlying surface or soil matrix. The captured rainwater will then be piped into the detention tanks for further treatment.

5.3 Benefits assessment for the planning scheme

5.3.1 Benefits assessment for the rainwater infiltration system

The planned rainwater infiltration system is mainly composed of sunken lawn and pervious pavement. The sunken lawn offers abundant sunken space for rainwater detention and remarkably

Table 4. The amount of infiltration rainwater under the different proportion of sunken lawn.

Catchment area (m²)	Green area (m²)	Sunken lawn rate	Sunken lawn area (m²)	Sunken lawn total flow (m³)	Infiltration of rainwater (m³)
68900	42800	25%	10700	4845	3906
		30%	12840	4945	4690
		35%	14980	5049	5473
		40%	17120	5149	6256
		45%	19260	5249	7035
		50%	21400	5354	7818

*Analysis method reference: Mo Lin & Yu kongjian (Structure the Urban Green Sponge: Study on Planning an Ecological Stormwater Regulation System), Zhang jingling, & Zhangzhiming Study on Rain Storage and Infiltration of Sunken Lawn and Its Influencing Factors).

prolongs the time of infiltration. The infiltration volume is calculated through the following formula:

$$S = s * (t_2 - t_1) * J * 60$$

where S = rainwater volume infiltrated through the field, m³; s = infiltration capacity, m³/s; J = hydraulic gradient, for vertical infiltration, $J = 1$; t_1 = time when the volume of stormwater runoff equals the infiltration capacity as storm intensity increases in the duration of rainfall from the initial stage, min; t_2 = time when the volume of stormwater runoff equals the infiltration capacity as storm intensity decreases in the duration of rainfall from the medium and late stage, min. In this formula,

$$s = k * F'$$

where k = permeability coefficient of soil, or steady infiltration rate, m/s; F' = field area, m².

When rainwater infiltration volume is calculated, the portions of sunken lawn are also taken into consideration (Table 4). The design sets the portion of sunken lawn at 30%, and lawn depth 100 mm (Zhang et al. 2008), and therefore the rainwater volume infiltrated is 4690 m³.

Pervious pavement is mainly applied to the square and sidewalks, with the normal function of which unaffected. The area is around 10,000 m², with runoff coefficient 0.15 after pavement (Zhao & Li 2011), and the rainwater volume infiltrated through the pavement is 5,100 m³.

In total, about 9,700 m³ of rainwater is infiltrated through sunken lawn and pervious pavement to recharge the ground water. This effectively alleviates the municipal drainage network burden and reduces the occurrence of urban water-logging at the same time.

5.3.2 *Benefits assessment for the rainwater collection and reuse system*

Rainwater detention tank is the primary facility in rainwater collection and reuse system. According to the terrain and water demand in Tuanjiehu Park, the design categorizes the park into two clusters. Cluster A consists of outer ring of greenbelt, primary roads and the lake, and two toilets included. Cluster B consists of central greenbelt and the square at the center of the park, and one toilet included. Based on the principle of rainwater utilization, the rainwater concentration direction and location is concluded according to the terrain in the park (Fig. 5), by which 6 water collection points are determined. Rainwater collected at these points will be taken in by the detention tanks.

The range of rainwater collection by the detention tanks covers rainwater runoff from roads, park greens and mountain surface. As the size of the tanks is fixed, the following formula will be used to calculate volume of the 6 tanks.

$$V = \alpha \psi H A 10^{-3}$$

● collection point
→ main direction

Figure 5. Rainwater collection of main direction and position after planning.

Table 5. The collection amount of rainwater in Cluster A.

Volume designed of detention tanks	Full number of storage	Runoff coefficient	Area for runoff collection	Reduction factor	Volume of detention tanks	Collection amount of the rain water per year (m^3)
1	61	0.85	3000	1.0	3.0	183
5	23	0.85	3000	1.0	15	345
10	14	0.85	3000	1.0	30	420
15	9	0.85	3000	1.0	42	378
25	5	0.85	3000	1.0	69	345
30	4	0.85	3000	1.0	84	336
40	2	0.85	3000	1.0	120	240

Table 6. The collection amount of rainwater in Cluster B.

Volume designed of detention tanks	Full number of storage	Runoff coefficient	Area for runoff collection	Reduction factor	Volume of detention tanks	Collection amount of the rain water per year (m^3)
1	61	0.9	6450	1.0	6.45	393
5	23	0.9	6450	1.0	29	667
10	14	0.9	6450	1.0	58	812
15	9	0.9	6450	1.0	87	783
25	5	0.9	6450	1.0	145	725
30	4	0.9	6450	1.0	174	696
40	2	0.9	6450	1.0	232	464

where V = the volume of detention tanks, m^3; α = the reduction factor for initial rainwater removal; ψ = runoff coefficient; A = the area for runoff collection; H = the precipitation volume designed in the detention tanks. Based on the above formula, the rainwater volume collected by detention tanks at the 4 collection points in Cluster A (Table 5) and at the 2 points in Cluster B (Table 6) are calculated respectively.

After the efficiency of runoff concentration and investment cost are taken into consideration, it is determined that the volume of the 4 detention tanks in Cluster A is 30 m^3, and that of the 2 detention tanks in Cluster B is 60 m^3 (Fig. 6–7). The 6 rainwater detention tanks are able to collect 3,200 m^3 per year, which can function as grey water and be corresponding substitutes for part of ground water or tap water to irrigate park greens, flush toilets, or recharge the landscape lake.

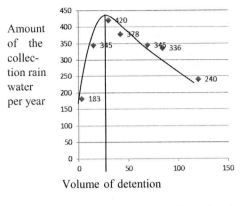

Figure 6. Relationship between detention tanks volume and amount of collection rain water in cluster A.

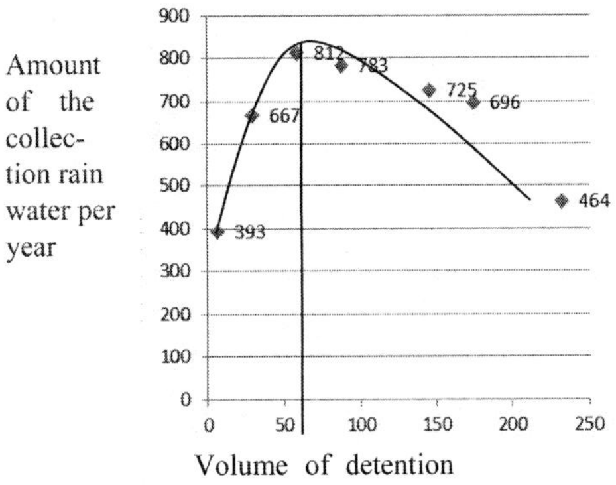

Figure 7. Relationship between detention tanks volume and amount of collection rain water in cluster B.

Table 7. Before and after the planning rainwater utilization efficiency analysis.

Before			After				
Catchment area	Rain runoff (m^3)	Digestion of rainwater	Measures	Main function	Collection rainwater (m^3)	Infiltration rainwater (m^3)	Digestion of rainwater
Green	3432	Mainly drained through municipal pipe network, not been utilized at all	Sunken lawn	Infiltration	—	4690	Through infiltration and collection, it can reduce 83% of the discharged rainwater in the region
Road	8182		Pervious pavement	Infiltration	—	5100	
Square	4048		Detention tanks	Storage	3200	—	
Total	15662		Total		12990		

5.3.3 Comprehensive benefits assessment

The above analysis proves that the rainwater utilization system that is composed by sunken lawn, pervious pavement and detention tank can absorb 13,000 m^3 of rainwater in the park, among which 10,000 m^3 of rainwater is for ground water recharge through the infiltration system, and 3,000 m^3 of it is stored in the collection and reuse system to substitute part of the tap water and ground water consumption in the park. As a result, the planned rainwater utilization system can reduce 83% of the discharged rainwater in the region, contributing to the mitigation of urban rainwater drainage burden and water shortage in the park (Table 7).

6 CONCLUSION

The combination of rainwater control and utilization with landscape design for city parks has changed the conventional mindset for rainwater management that is focused on "drainage". The designed rainwater utilization system for city parks, with facilities like sunken lawn, pervious pavement and detention tanks, can reduce 83% of rainwater efflux in the park (about 13,000 m^3). In the current situation where urban water-logging and water scarcity coexist, the designed system, by rainwater harvesting and utilization, will not only contribute to the release urban rainwater drainage burden, but also to the alleviation of water crisis faced by city parks, so it has good application prospects.

ACKNOWLEDGEMENTS

The thesis is sponsored by the Chaoyang Division of Beijing Municipal Commission of Urban Planning under the Research of Rainwater Control and Utilization in Beijing Chaoyang District. We would like to express our gratitude to all those who helped during the writing of this thesis, especially, Zhang Danming, who has offered valuable suggestion in the academic studies.

REFERENCES

Barraud S. & Gautier A. 1992. The impact of intentional stormwater infiltration on soil and groundwater. Water Science and Technology 2: 185–192.
Cao Xiuqin & Che Wu 2002. City roof rainwater harvesting system design analysis. Water & Wastewater Engineering 1: 13–15.
Cai Yanwei 2000. Urban Space. Beijing: Science Press.
Che Wu et al. 2009. Typical Stormwater and Flood Management Systems in Developed Countries and Their Inspiration. China Water & Wastewater 20(29): 12–18.
Chen Jiangang & Ding, Yueyuan et al. 2003. Beijing urban rainwater utilization engineering measures. Beijing Water Resources 6: 12–14.
Gao Hei & Ni Qi 2005. Preliminary Discussion about the Ecological and Tactics in the Contemporary Landscape Design. Huazhong Architecture 4: 127–130.
Grimm, N. B. & Faeth, S.H, et al. 2008. Global Change and the Ecology of Cities. Science 319: 756–760.
Hollis, G.E. 1977. Water yield changes after the urbanization of the Canon's Brook catchment, Harlow, England. Hydrological Sciences Bulletin 22: 61–75.
Jennings, D. B. & Jarnagin, S. T. 2002. Changes in anthropogenic impervious surfaces, precipitation and daily streamflow discharge: A historical perspective in amid-Atlantic subwatershed. Landscape Ecology 17: 471–489.
Jiang Wen 2008. A Discussion on Relationship between Water Resource and Sustainable Development. Environmental Science & Technology 31(2): 155–157.
Li Junqi & Che Wu, et al. 2001. Urban Design and rainwater utilization technical and economic analysis. Water & Wastewater Engineering 27(12): 25–28.
Li Junqi & Che Wu 2002. Review of Rainwater Utilization Technology in Germany Cities. Urban Environment & Urban Ecology 15(1): 47–49.
Meng Gang & Li Lan, et al. 2005. Urban Park Design. Shanghai. Tongji University Press.
Nancy B. & Grimm 2008. Global Change and the Ecology of Cities. Science (319): 756–760.
Song Yun & Yu Kongjian 2007. Construction of urban stormwater management systems landscape planning approach – for example in Weihai. Urban Problems 8: 64–70.
Wang Bo & Cui Ling 2003. From the "resource perspective" of urban rainwater utilization. Urban Problems 3: 50–53.
Wang Yanmei 2007. Rainwater utilization of domestic and overseas. Journal of Anhui Agricultural Sciences 35(8): 2384–2385.
Zhao Jing & Li Dihua 2011. Urbanization context stormwater management approach – based on the perspective of the development of low-impact. Urban Problems 9: 95–101.
Zhang Shuhan & Chen Jiangang, et al. 2008. Analysis on Technical Indexes and Design Methods of Permeable Brick Pavement. China Water & Wastewater 22(11): 15–17.
Zhang Wei & Che Wu 2008. Application of Graphical Approach for Low Elevation Greenbelt Design. China Water & Wastewater 24(20): 35–39.

The study on the evolution of wind-sand flow with non-spherical particle

S. Ren, Y. Luo & X. Hu
Chengdu University of Technology, College of Environment and Civil Engineering

ABSTRACT: Saltation is one of the important transport modes in the movement of wind-blown grains. Here, based on the theoretical model of sand saltation evolution with consideration of two feedback mechanisms which are the wind-sand mechanism and the particle mid-air mechanism, the entire process of wind-sand flow with non-spherical particle is simulated quantitatively. The results showed that the sand shape would effect on the time of the entire process to reach a steady state and the transport flux at the steady state.

1 INTRODUCTION

As an important physical process of wind erosion and sand/dust storms, sand movement under wind flow has severely influenced the natural environment and human activity. Saltation is the primary means by which sand-sized particles travel in most realistic winds (Bagnold, 1941), so the researchers have done many works on the numerical simulation, and they have discussed the features of sand flows on the different conditions (Willetts, 1986, Duran, 2006, Zheng, 2006, Huang, 2007, Hang, 2008, Kok, 2009, Ren, 2010).

But, in many research, considereds and particle as sphere. In fact, through the studies on the dune sand shape, researchers suggested the particle shape or psephicity had definite relation with sand grain size. According to the statistical results, edge angle sand account for more than 80% in the original sand while the hypo-edge angle sand account for more than 50% (Wu, 1987). Furthermore, some researchers showed that the quartz grains are with pockmark pits, dished pits and ridges on their surface by SEM study.

In this paper, based on the theoretical model of sand saltation evolution with consideration of two feedback mechanisms (Ren, 2010) which are the wind-sand mechanism and the particle mid-air mechanism, the entire process of sand saltation with non-spherical particle issimulated quantitatively.

2 ANALYSIS OF SAND SHAPE

By using the Beckman-Coulter, Blott S. J. and Pye K. (Blott, 2006) suggested: the average sphericity of natural sand (Ψ) is between 0.82 to 0.86, and could be expressed as $\Psi = A_v/A_p = \pi d_v^2/A_p$, in which A_v is the superficial area of sphere which volume is same as particle, A_p is the superficial area of particle, d_v is the equivalent diameter of sphere which volume is same as particle. Here, considered two shapes (ellipsoid, cylinder) of sand which their sphericity is between 0.82 to 0.86.

1) Ellipsoid: use ellipticity $\lambda = \frac{\sqrt{ab}}{c}$ to describe ellipsoid sand, in which a, b, c mean the size of major axis, medial axis and minor axis respectively. Thus the equivalent diameter $d_v = (abc)^{1/3}$. In the paper, take the revolution ellipsoid in the simulation model: $S_1(l, l, \frac{2}{5}l)$, $d_v = l(\frac{2}{5})^{1/3}$, $\Psi_1 = 0.848$.
2) Cylinder: consider another shape, cylinder ($S_5(l, 3l/5)$, $d_v = 3l(\frac{1}{50})^{1/3}$, $\Psi_5 = 0.850$), in the simulation model, which basal plane diameter is $3l/5$, height is l.

3 BASIC EQUATION

3.1 Force analysis

For sand particle, only the force due to gravity and aerodynamic drag are considered: $W = \frac{1}{6}\pi D_p^3(\rho_p - \rho)g$, $F_D = \frac{1}{2}C_D\rho A V_r^2$, in which A means the maximum area of sand section with different value for different shape; V_r is relative velocity between sand and airflow; μ is aerodynamic viscosity coefficient; C_D is drag coefficient.

For non-spherical particle, the empirical formulae are given by researchers (Loth, 2008):
for round cross-section particle:

$$C_D^* = \frac{24}{R_{ep}^*}[1 + 0.15(R_{ep}^*)^{0.687}] + \frac{0.42}{1 + 42500/(R_{ep}^*)^{1.16}} \qquad (1)$$

for non-round cross-section particle:

$$C_D^* = \frac{24}{R_{ep}^*}[1 + 0.035(R_{ep}^*)^{0.74}] + \frac{0.42}{1 + 33/(R_{ep}^*)^{0.5}} \qquad (2)$$

in which R_{ep}^*, C_{shape}^1, C_{shape}^2 is expressed in reference (Loth, 2008).

3.2 Sand grains lifted off from the sand bed per unit time and unit area

At moment t, there are $N(t)$ grains lift-off from the sand-bed per unit time and unit area. So, $N(t)$ can be expressed as (Ren, 2010):

$$\begin{aligned} N(t) &= N_a + \bar{N}_j + \bar{N}_r \\ &= N_0(\tau_a - \tau_c) + 0.95\int_{v_{0\min}}^{v_{0\max}}\int_{v_{im\min}}^{v_{im\max}} \exp[-\frac{(v_0 - 0.56v_{im})}{(0.2v_{im})^2}]dv_0 dv_{im} \\ &+ 1.75\int_{v_{0\min}}^{v_{0\max}}\int_{v_{im\min}}^{v_{im\max}} v_{im}\exp[-\frac{v_0}{0.25v_{im}^{0.3}}]dv_0 dv_{im} \end{aligned} \qquad (3)$$

3.3 Wind field equation

Here, considering the reaction of sand and wind, the wind field can be expressed as in the X-direction (Ren, 2010):

$$\rho\frac{\partial u}{\partial t} = \frac{\partial}{\partial y}(\rho k^2 y^2 \left|\frac{\partial u}{\partial y}\right|\frac{\partial u}{\partial y}) + \int_{v_{0\min}}^{v_{0\max}} N(t)m_p f(v_0,t) \begin{bmatrix} p_\uparrow(v_0,y,t)\frac{a_{x\uparrow}}{|\dot{y}_\uparrow|} + p_\downarrow(v_0,y,t)\frac{a_{x\downarrow}}{|\dot{y}_\downarrow|} + \\ p_\uparrow(v_0,y,t)\left(\frac{a_{x\uparrow}}{|\dot{y}_\uparrow|}\right)_p + p_\downarrow(v_0,y,t)\left(\frac{a_{x\downarrow}}{|\dot{y}_\downarrow|}\right)_p \end{bmatrix} dv_0 \qquad (4)$$

where the subscripts "↑" and "↓" mean respectively the sand in ascending stage and descending stage; $(\cdots)_p$ represents the physical quantity with consideration of the mid-air collision; $p_\uparrow(v_0,y)$ and $p_\downarrow(v_0,y)$ are respectively the probabilities that the saltating particle with lift-off speed v_0 collides with another particle at the height y in ascending stage and in descending stage.

3.4 Sand mass flux

When the mid-air collision is considered, the sand transport rate at moment t is given by Ren(Ren, 2010):

$$\begin{aligned} Q_{mc}(t) &= \int_0^y \int_{v_{0\min}}^{v_{0\max}} N(t)m_p f(v_0,t)\left[p_\uparrow(v_0,y,t)\frac{\dot{x}_\uparrow(v_0)}{\dot{y}_\uparrow(v_0)} - p_\downarrow(v_0,y,t)\frac{\dot{x}_\downarrow(v_0)}{\dot{y}_\downarrow(v_0)}\right]dv_0 \\ &+ \int_0^y \int_{v_{0\min}}^{v_{0\max}} N(t)m_p f(v_0,t)\left(p_\uparrow(v_0,y,t)\left(\frac{\dot{x}_\uparrow(v_0)}{\dot{y}_\uparrow(v_0)}\right)_p - p_\downarrow(v_0,y,t)\left(\frac{\dot{x}_\downarrow(v_0)}{\dot{y}_\downarrow(v_0)}\right)_p\right)dv_0 \end{aligned} \qquad (5)$$

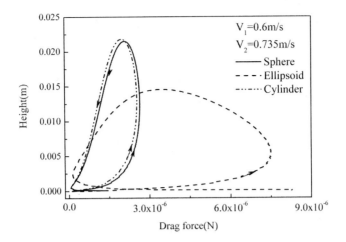

Figure 1. Drag of different sand particle.

4 COLLISION MODEL

Here, use Huang (Huang, 2007) et. al's model to calculate the probability of the mid-air inter-particle collisions. The event that a saltating particle with random lift-off velocity collides with another particle (Huang, 2007):

$$p(C) = \sum p(C_i|L_i)p(L_i) = \sum p(C_i)p(L_i) \qquad (6)$$

L_i represents the event that a sand particle takes off with lift-off velocity v_{0i}, C_i denotes the event that a sand particle with lift-off velocity v_{0i} collides with another particle. Velocity of sand particle after mid-air collision can be calculated by hard ball mode (Huang, 2008).

5 CALCULATING STEP

The calculating procedures are taken as follows:

1) Taking the initial lift-off speeds of saltating grains as $v_{0\min} = \sqrt{2gD}$, $v_{0\max} = 5u_*$.
2) Dividing the duration time into several moments, that is, $t_j = t_0 + j \times \Delta t$ ($t_0 = 0, j = 1, 2, \ldots$), in which the time step $\Delta t = \sqrt{2D/g}$.
3) Having respectively taken the initial iteration value of the wind profile at t_j moment as $u^{0j}(y) = u^{j-1}(y)$ and $N^{0j}(k) = N_0(v^{j-1}(y) - \tau_c)$, use the equations (1)-(4), (6) to obtain $N^{ij}(k)(i = 1, 2, \ldots)$ and $u^{ij}(y)$. The iteration repeats until the expression $|u^{i+1,j}(y) - u^{i,j}(y)| < \varepsilon_1$ is attained. In the meantime, we can get the sand mass flux by using the equations (5).

Repeat the above procedures until the expression $|u^{j+1}(y) - u^j(y)| < \varepsilon_2$ is satisfied, which means that the saltating population of windblown and grains reaches an equilibrium.

6 CALCULATION RESULTS AND DISCUSSION

Figure 1 shows the drag of different sand particles vary with height at friction velocity of 0.49 m/s, equivalent diameter of 0.35 mm, horizontal lift-off velocity (v_1) of 0.6 m/s and vertical lift-off velocity (v_2) of 0.735 m/s (solid line is for sphere sand, dashed line is for ellipsoid sand, dashed-dotted line is for cylinder). From figure 1, it is shown that variation law of drag for different shapes

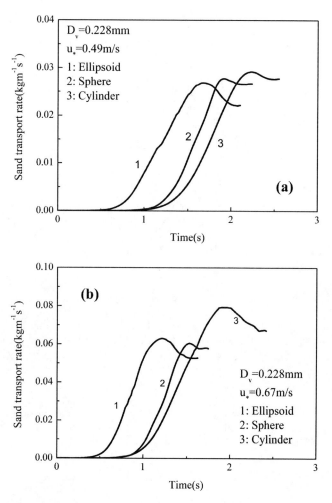

Figure 2. The sand transport rate of grains against time for different shape sand (D = 0.228 mm).

with height are similar. Namely, at ascending stage increase at first, then decrease with height, while at descending stage decrease at first, then increase with the decrease of height. Moreover, the values of drag for different shapes at same height are different, in which drag for cylinder sand is little less than drag for sphere sand. Furthermore, we can see that the larger the maximum drag value, the less the maximum height of sand. For example, for sphere sand, the maximum height and drag is respectively 2.15 cm and 2.65×10^{-6} N, and for ellipsoid sand, the corresponding value is respectively 1.45 cm and 7.5×10^{-6} N. The maximum drag ratio and the maximum height ratio of sphere sand to ellipsoid sand is respectively 1.48 and 0.35. The higher maximum height that a sand can get, the more energy the sand can obtained from the wind field, the larger velocity when the sand return back the bed, the more particle number that can be ejected from the bed. So the shape of sand particle would effect on the wind-sand flow.

Figure 2 represents the transport rate of grains vary with time at friction wind velocity of 0.49 m/s and 0.67 m/s. It can be seen that the trends of different sand particle with time are similar, namely, gradually increase at the beginning of the evolution process, then decrease slightly, finally reach the steady state. But, the duration time and the transport flux at steady stage have great relationship withthe different shapes. The minimum duration time and transport flux is for ellipsoid, next is for sphere, the maximum is cylinder. For instance, at friction wind velocity of 0.67 m/s (figure 2-b), the duration time and the transport flux of the ellipsoid, cylinder and sphere particles is respectively 1.48 and 0.053 kg/(ms), 2.37s and 0.067 kg/(ms), 1.62s and 0.058 kg/(ms). The simulated transport fluxes are in the same order with the experimental value which is 0.0497 kg/(ms) (Zheng, 2006). That's because the maximum height of ellipsoid is less than the sphere's which is less than the cylinder's. On the one hand, the less height that sand particle can get, the less time that one saltation process takes, thus, the sand return back earlier to eject the other sands on the bed. On the other hand, the higher height that sand particle can get, the more energy that the sand obtained, the more sand particles would be ejected. That's why the transport flux of ellipsoid is higher at early stage of the evolution process.

7 CONCLUSION

In this paper, the entire process of sand saltation with non-spherical particle is simulated quantitatively, based on the theoretical model of sand saltation evolution with consideration of two feedback mechanisms which are the wind-sand mechanism and the particle mid-air mechanism. The results showed trend of the drag force that increases at first, then decrease with height at ascending stage, while at descending stage decrease at first, then increase with the decrease of height. And, the drag forces are different with sand shape. Moreover, the sand shape would effect on the time of the entire process to reach a steady state and the sand transport flux at the steady state.

ACKNOWLEDGEMENT

This research work is supported fromScientific Research Project of Chengdu University of Technology KR1112. The authors sincerely appreciate this support.

REFERENCES

Bagnold R. A., 1941, The Physics of Blown Sand and Desert Dunes, London: Methuen, 105–106.
Blott S. J., Pye K., 2006, Particle size distribution analysis of sand-sized particles by laser diffraction: an experimental investigation of instrument sensitivity and the effects of particle shape, Sedimentologym, 53 (3): 671–685.
Duran O., Hermann H., 2006, Modeling of saturated sand flux, Journal of Statistical Mechanics: Theory and Experiment, P07l1, doi:10.1088/1742-5468/2006/07/P07011.
Huang N., Zhang Y. L. and Adamo R. D., 2007, A model of the trajectories and mid-air collision probabilities of sand particles in a steady-state saltation cloud, J. Geophys. Res., 112: D08206.
Huang N., Ren S., Zheng X. J., 2008, Effects of the Mid-air Collision on Sand Saltation, Science in China (Series G), 38(3): 260–269.
Kok, J. F., Renno N. O., 2009, A comprehensive numerical model of steady-state saltation, J. Geophys. Res., 114, D17204, doi:10.1029/2009JD011702.
Loth E., 2008, Drag of non-spherical solid particles of regular and irregular shape, Power Technology, 182: 342–353.
Ren, S., Huang, N., 2010, A numerical model of the evolution of sand saltation with consideration of two feedback mechanisms, Eur. Phys. J. E., 33: 351–358.
Willetts, B.B., Rice, M.A., 1986, Collisions in aeolian saltation, Acta Mechanica 63: 255–265.
Wu, Z., 1987, The aeolian geomorphology [M], Beijing: Science press.
Zheng, X. J., Huang, N., Zhou, Y. Y., 2006, The effect of electrostatic force on the evolution of sand saltation cloud [J], The European Physical Journal E, 19, 129–138.

Comprehensive ecosystem assessment based on the integrated coastal zone management in Jiaozhou Bay area

Cui Wang, Juanjuan Dai, Zhouhua Guo & Qingsheng Li
Third Institute of Oceanography, State Oceanic Administration, Xiamen, Fujian, China

ABSTRACT: Based on the Integrated Coastal Zone Management, the index system of coastal ecological assessment was constructed, which were divided into four subunits: coastal land subsystem, coastal waters subsystem, socio-economic subsystem and management regulation subsystem, and the Ecological Index in the coastal zone of Jiaozhou Bay from 1988 to 2007 was calculated. The results showed that the ecological quality in the coastal zone on Jiaozhou bay showed obvious fluctuation and descending trend form 1988 to 2005, and the ecosystem quality is not optimistic. It is very important theoretically and practically for ICM and sustainable development in coastal zone to study comprehensive ecosystem assessment based on the Integrated Coastal Zone Management, which can effectively improve the ability of the pollution treatment and management.

1 INTRODUCTION

With the rich natural resources and unique geographic position, the coastal zone has become the heartland of human activities. Along with the rapid economic development and rapidly increasing population, the coastal zone ecosystem has been bearing greater pressure, with more and more serious resource exhaustion, ecological degradation, environmental pollution and problems (Lakshmi et al., 2000). As an efficient tool and means to solve the above problems (Christie et al., 2003; Pomeroys et al., 2003), the study and practice of ICZM (integrated coastal zone management) have been widely applied (Saeda et al., 2005; Stesda et al., 2006) internationally. The traditional coastal zone management mode cannot efficiently solve the prominent problems in the current coastal zone. As the leading edge and hotspot of the coastal zone management at home and abroad, ecosystem approach has provided one new method and thought for the integrated coastal zone management (Chen et al., 2005).

Having an integrated coastal zone management based on the ecosystem approach to realize the harmonious development of the ecological environment and social economy in the coastal zone lies in the assessments of decision makers on the status and development of social economy and ecosystem in the coastal zone. Therefore, the establishing one assessment model to quantize the quality of the coastal zone ecosystem and measuring the effects of integrated coastal zone management measures can provide the scientific basis for the formulation of the management policy and at the same time call out the public to participate in the integrated coastal zone management.

2 RESEARCH PROGRESS OF ECOSYSTEM ASSESSMENT

Along with the rapid development of social economy, the increasing population and numerous ecological environment problems, the ecosystem assessment is developed to analyze the status quo, changes and influences of the ecosystem. The ecosystem assessment started from the mid-stage of 1960s (Wang et al., 2008) and the initial ecosystem assessments mostly limited to certain region and objects, ignoring the influences on the overall ecosystem, especially the potential ecological influences of the region future development. Furthermore, with the profound attention of the structure, function and bearing capacity on the ecosystem in recent years, the integrated ecosystem assessment has gradually be a new research focus (Qiu et al., 2008; Tian and Yue, 2003), for example, millennium ecosystem assessment (MA) launched in 2001 (Fu et al., 2001; MA, 2003).

MA aims at predicting the future trend by aid of the assessment of the present situation and proposing the corresponding measures to improve the ecosystem management status, including ecosystem and its service function, human welfare and poverty elimination, the driving force of ecosystem and its service function changes, the interaction between ecosystems of different scales and assessments, the value and assessment of ecosystem (Zhao and Zhang, 1997) and so on. China also launched Western China ecosystem assessment research program in 2001. It adopted system simulation and geosciences information scientific method system to assess the current situation and change trend of the various ecosystems and their service functions and propose the policy suggestions of ecosystem protection and repair (Yang and Lu, 2003).

3 RESEARCH METHOD OF INTEGRATED ECOSYSTEM ASSESSMENT

3.1 Selection of assessment index

The index system is a set of closely-linked specific indexes which can reflect the whole or part of the assessment object. The selection of assessment index shall follow the following fundamental principles, scientificity, operability, simplicity, data availability, strong representativeness and so on. Referring to the relevant literatures, this study proposed the index system of the integrated coastal zone ecosystem assessment (Table 1).

3.2 Standardization of assessment index

Due to the different properties of each index, their dimensions vary, so they cannot be compared directly. The original data shall be standardized to uniform non-dimensional numerical value. This study adopted the difference value method for standardization.

When the index value has a positive correlation (positive effect) with the assessment object,

$$X'_{ij} = \frac{X_{ij} - X_{min}}{X_{max} - X_{min}}$$

When the index value has a negative correlation (negative effect) with the assessment object,

$$X'_{ij} = \frac{X_{max} - X_{ij}}{X_{max} - X_{min}}$$

where, X'_{ij} represents the value of one index after standardization; X_{ij} represents the original value of one index; X_{min} represents the minimum value of each index in original data; and X_{max} represents the maximum value of each index in original data.

3.3 Integrated ecosystem assessment method

This paper proposed to use the integrated ecosystem assessment (IEA) indexes of the coastal zone ecosystem to measure the quality of the coastal zone ecosystem. The calculation formula of IEA is as below.

$$IEA = \sum_{i}^{m} W_i C_i$$

where IEA represents the comprehensive assessment index of the coastal zone ecosystem; C_i represents the normalization value of Index i, $0 \leq C_i \leq 1$; W_i represents the weight of Index i, which

Table 1. The index of coastal comprehensive ecosystem assessment corresponding to the management objectives.

Object Layer A	Mandatory Layer B	Index Layer C	Unit
Indexes of the integrated coastal zone ecosystem assessment (IEA)	Terrestrial ecosystem (B1)	forest coverage rate (C_1)	%
		cultivated land area per capita (C_2)	hm^2/person
		index of landscape fragmentation (C_3)	—
		River water qualification rate (C_4)	%
		Biological abundance index (C_5)	—
	Marine ecosystem (B2)	Accumulative reclamation area (C_6)	km^2
		Clarification time (C_7)	d
		Target rate of offshore area functional zone (C_8)	%
		Marine comprehensive index of water quality (C_9)	—
		Risk index of heavy metal pollution in sediment (C_{10})	—
		Wetland area variation factor (C_{11})	%
		Assessment index of biological residual hazard (C_{12})	—
		Allowable catch of marine fishery (C_{13})	10,000 t
		$E_x(C_{14})$	—
		$E_{xsr}(C_{15})$	—
		Phytoplankton biomass (C_{16})	t/km^2
	Socioeconomic system (B3)	Per Capita GDP (C_{17})	Yuan
		Population density (C_{18})	person/km^2
		Engel coefficient (C_{19})	%
		Tertiary industry ratio in national GDP (C_{20})	%
	Management control system (B4)	Satisfaction degree water environmental capacity (C_{21})	%
		Treatment rate of urban domestic sewage (C_{22})	%
		Proportion of the protected area in the national land (C_{23})	%
		Proportion of environmental protection investment (C_{24})	%
		Reduction rate precedence-controlled pollutant (C_{25})	%

can be obtained by aid of analytic hierarchy process (Pan, 2004); and n represents the number of the index. The calculation results of the comprehensive assessment index of the coastal zone ecosystem can be divided into 5 scales, respectively excellent ($0.8 \leq IEA < 1.0$); good ($0.6 \leq IEA < 0.8$); normal ($0.4 \leq IEA < 0.6$); bad ($0.2 \leq IEA < 0.4$); and worse ($IEA < 0.2$).

4 INTEGRATED ECOSYSTEM ASSESSMENT OF JIAOZHOU BAY COASTAL ZONE

4.1 Assessment results

According to the index system in Table 1, this paper utilized Delphi method to score through pairwise comparison and got the judgment matrix. Then it calculated the single hierarchical arrangement results and carried out the consistency check on the computed results. After determining the weight and standardization value of each index, we can assess the quality of the assessment ecosystem only by aid of the calculation of comprehensive indexes. This paper made use of the weighting sum of the multi-index comprehensive assessment model to obtain the value of each index, shown as Figure 1.

4.2 Comprehensive analysis

From the four assessment sub-system in Figure 1, it can be known that the change trend of each system was different. During 1988–2009, the qualities of the marine ecosystem and terrestrial ecosystem were both in a downward trend while the states of socioeconomic system and the

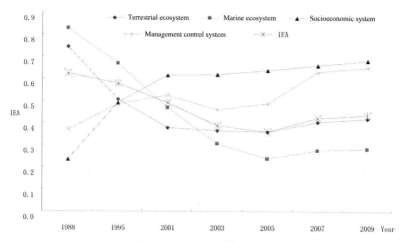

Figure 1. Integrated assessment indices of ecosystem in coastal zone of Jiaozhou Bay.

management control system were in a upward state, which indicated that along with the rapid development of social economy, the stress effect of the coastal zone ecosystem on human became more and more intense and the coastal zone ecosystem is bearing the pressure from various aspects.

In view of the comprehensive assessment index of Jiaozhou Bay coastal zone ecosystem, the ecosystem state in Jiaozhou Bay was good in 1988; while during 1988–2001, namely 1990s, the quality of Jiaozhou Bay coastal zone ecosystem decreased dramatically and the scale turned to normal; during 2001–2005, the quality continued decreasing and the decreasing trend gradually tended to slow down and reached the maximum, 0.36 in 2005, with poor quality; along with the continuous progress of human society and increasing management control ability, the ecosystem state during 2005–2009 tended to be better (0.43) while it still approached the critical value of the bad scale ($0.2 < IEA < 0.4$), which indicated that the ecosystem in Jiaozhou Bay is still not optimistic.

5 CONCLUSION

As the leading edge and hotspot of the integrated coastal zone management at home and abroad, ecosystem approach has provided one new method and thought for the integrated coastal zone management. Scholars at home and abroad have made many studies on the integrated coastal zone management based on ecosystem approaches while there has no specific index system of the integrated coastal zone ecosystem assessment. This paper established the index system and scale method of integrated coastal zone ecosystem assessment in view of terrestrial ecosystem, marine ecosystem, socioeconomic system management control system according to the requirements of the integrated coastal zone management and applied it to the comprehensive assessment studies of Jiaozhou Bay coastal zone management. Analytic hierarchy process was adopted to assess the ecosystem states of Jiaozhou Bay coastal zone in 1988, 1995, 2001, 2003, 2005, 2007 and 2009 and the results showed that the quality of Jiaozhou Bay coastal zone has been in a decreasing trend. Though in recent years, the deteriorating condition has been basically restrained, the Jiaozhou Bay basin ecology state is still serious and needs efficient management measures to improve the coastal zone ecological quality. Through the comprehensive assessment study on the coastal zone ecosystem, we can efficiently identify the sustainable development constrains of the coastal zone, provide scientific basis for the scientific establishment of the integrated coastal zone management based on ecosystem and the reasonable implementation of the integrated coastal zone management planning, and give technical supports and guarantee to the sustainability and sound development of the social economy and environment in coastal zone.

REFERENCES

Chen B H, Yang S Y, Zhou Q L. 2005. Implying ecosystem management approach in integrated coastal zone management [J]. Journal of Oceanography in Taiwan Strait, 24(1):122–130.

Christie P, White A T, Stockwell B, et al. 2003. Links between environmental condition and integrated costal management sustainability [J]. Silliman Journal, 44(1):285–323.

Fu B J, Liu S L, Ma K M. 2001. The contents and methods of integrated ecosystem assessment. Acta ecologica sinica [J], 21(11): 1885–1892.

Lakshmi A, Rajagopalan R. 2000. Sicio-economic implications of coastal zone degradation and their mitigation: a case study from coastal villages in India [J]. Ocean and Coastal Management, 43:746–762.

MA (Millennium Ecosystem Assessment). 2003. Ecosystems and human well-being a framework for assessment[M]. Washington DC: Island Press.

Pan Z H. 2004.The study of sustainable development index system of town in China [D]. Chong Qin: Chong Qin univisty: 35–52.

Pomeroys S, Oracion E G, Caballesed A, et al. 2003. Economic benefits and integrated coastal management sustainability [J]. Silliman Journal, 44(1):75–94.

Qiu J, Zhao J Z, Deng H B, et al. 2008. Ecosystem-based marine management principles, practices and suggestions [J]. Marine Environmental science, 27(1):74–78.

Saeda R, Avlac C, Mora J. 2005. A methodological approach to be used in integrated coastal zone management process: the case of the Catalan Coast (Catalonia, Spain) [J]. Estuarine, Coastal and Shelf Science, 62:427–439.

Stesda S M, Mcglashan D J. 2006. A Coastal and Marine national Park for Scotland in partnership with Integrated Coastal Zone Management [J]. Ocean & Coastal Management, 49:22–41.

Tian Y Z, Yue T X. 2003. Discuss of several issues on ecosystem assessment. China population, resources and environment [J], 13(2):17–22.

Wang M, Bi J G, Duan Z X. 2008. Study on marine management model of ecosystem [J]. Marine Environmental science, 27(4):378–382.

Yang H X, Lu Q. 2003. Review on ecological region assessment in north America and U.N. millennium ecosystem assessment [J]. China population, resources and environment, 13(1):92–97.

Zhao S D, Zhang Y M. 1997. Concepts, contents and challenges of ecosystem assessment [J]. Advances in earth science, 14(5):650–657.

Spatial variation and source apportionment of coastal water pollution in Xiamen Bay, Southeast China

Qingsheng Li, Cui Wang, Yurong Ouyang & Juanjuan Dai
Third Institute of Oceangraphy, State Oceanic Administration, Xiamen, Fujian, China

ABSTRACT: Multivariate statistical techniques, such as cluster analysis (CA) and principal component analysis (PCA) were applied to investigate the spatial variation and source of coastal water pollution in Xiamen Bay. According to CA, the sampling sites were significantly classified into two clusters, Inner bay (1st cluster) and outer bay (2nd cluster). The 1st cluster was distributed in Jiulong River Estuary (JRE), Eastern Sea (ES) and Dadeng Sea (DS), and the other one was distributed in Western Sea (WS) and Tongan Bay (TB). Nutrients of 1st cluster were higher than the other. based on PCA, 88.65% and 90.97% of the total variance can be explained for 1st and 2nd cluster respectively. Three and two Varifactors (VFs) were intentified for the two clusters respectively. VFs of 1st cluster represent river input and agricultural wastewater, livestock and poultry and organic pollutant of domestic sewage, while VFs of 2nd cluster can interpreted by industrial wastewater, sewage and agricultural wastewater discharge. The focus of coastal pollution control in Xiamen Bay was land-based pollution, and how to make the watershed comprehensive management.

1 INTRODUCTION

Coastal zone is cited the high productive area of the world (Costanza, 1999), But coastal waters are also affected by anthropogenic activities, are seriously impacted by land- and coast- based pollution, such as wastewater, soil erosion, agricultural activities and livestock poultry. In recent years, land- and coast-based pollution caused red tide frequently occur and eutrophication level rise in Xiamen Bay (Lin and Zhang, 2008), so it is imperative to study the spatialtemporal characters and to control the pollution of coastal water in Xiamen Bay.

Multivariate statistical techniques, such as cluster analysis (CA), principle component analysis (PCA) and factor analysis (FA) have been applied to a variety of environmental application, such as assessment the spatialtemporal variation, source apportionment of surface water (Singh et al., 2004, Zhou et al., 2007a, Li and Zhang, 2009, Bu et al., 2010, Li et al., 2010, Wu et al., 2010, Huang et al., 2012), however, applications of different Multivariate statistical methods to the analysis of coastal water are still uncommon. The research objectives of this study are: (1) to find the spatial variability characteristics of coastal waters; (2) to identify the underlying source of main pollution in Xiamen Bay; (3) to provide scientific support of coastal pollution control.

2 MATERIALS AND METHODS

2.1 Study area

Xiamen Bay, a semi-enclosed bay, located in southeast of Fujian Province, surrounded by Golden Gate islands, Dadan island and Erdan island, which form a natural barrier for Xiamen Bay. Xiamen Bay consists of five seaareas: Tongan Bay, Dadeng Sea, Southeast Sea, Jiulong River Estuary and the Western Sea, etc. (Figure 1).

Figure 1. Sampling sites in the study bay.

2.2 Data source

In order to explore coastal water environmental conditions in Xiamen Bay, 30 stations water sampling in may, august and october of 2010 were performed (Figure 1). Sampling and analysis methods follow standard methods. Water quality parameters consist of pH, salinity (S), permanganate index (COD_{Mn}), nitrate (NO_3-N), nitrite (NO_2-N), ammonia (NH_4-N), phosphate activity salt (PO_4-P), total nitrogen (TN), total phosphorus (TP), suspended matter (SPM), total organic carbon (TOC), reactive silicate (SiO_3-Si), oil (Oil), chlorophyll a (Chl$a$).

2.3 Data analysis

2.3.1 Data pretreatment

In this study, missing data were estimated as average values from corresponding datasats, Kurtosis and Skewness method was used to test whether the data is normally distributed prior to analysis (Wunderlin et al., 2001, Simeonov et al., 2003). When the data is not normally distributed, logarithmic transformation was performed to reduce the data amplitude and became homogeneous, nearly normal. Finally, the mean and cariance were set to zero and one, which can minimize the effects of difference in measurement units.

2.3.2 Statistical analysis

Hierarchical cluster analysis, a widely used method of exploratory, can assemble objects based on the characteristics they possess (Wunderlin et al., 2001, Simeonov et al., 2003). This study used Ward's method and Euclidean distances for spatial similarity analysis. Principal component analysis is a widely applied technique that can explain the variance of alarge set of intercorrelated variables by transforming them into a smaller set of independent variables (Simeonov et al., 2003, Li and Zhang, 2009, Bu et al., 2010). Varimax rotation method was used in this research.

3 RESULT AND DISCUSSION

3.1 Spatial similarity of coastal water pollution

Based on the monitoring data, spatial cluster analysis was perform to generated a dendrogram grouping the 30 sites into two clusters, and the difference between the cluster was significant, the results shown in Figure 2. Cluster 1 (1st Sea Area, 1st SA) consisted of site 8–10, 12, 14–17, 19, 22–26, 28 and 30, cluster 2 (2nd Sea Area, 2nd SA) comprised site 1–7, 11, 13, 18, 20–21, 27 and

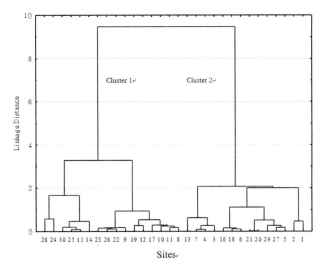

Figure 2. Dendrogram of cluster analysis.

29. Sites of Cluster 1 are mainly distributed in the Jiulong Estuary, southeast sea and Dadeng sea and that of Cluster 2 is mainly distributed in western sea and Tongan Bay.

The hydrodynamic conditions of 1st Sea Area are better than that of 2nd Sea Area, most human activities, such as fish culture, agricultural activities, reclamation, disposal of treated dffluent from major public sewage treatment works of 2nd Sea Area are more intensive than 1st Sea Area, so coastal waters in the two sea areas are significantly different.

3.2 *Spatial variation of coastal water pollution*

Spatial variation of water pollutant is large (Figure 3). The average concentrations of TN, TP, DIN and PO_4-P in 2nd SA were higher than that of 1st SA, and DIN and PO_4-P concentrations of 2nd SA exceed the fourth class water quality standards, DIN of 1st also the same, indicating that Nitrogen and phosphorus are the main water pollutants in Xiamen Bay especially in 2nd SA. The TOC of 2nd SA is greater than 1st, while SPM is the opposite, which may be related with the injection of river, brought a large number of SPM.

3.3 *Source identification*

Varimax rotated principal component analysis can explained by 88.65% and 90.97% of the variance in Cluster 1 and 2 respectively (Table 1). Three components can be extracted from Cluster 1, the first varifactor (VF1) explains 45.72% of the total variance, had strong positive loadings on NO_3-N, SiO_3-Si, NO_2-N and TN. VF1 represents the river pollutant inputs and agricultural non-point source pollution (Singh et al., 2005, Zhou et al., 2007a, Wang et al., 2008, Chen et al., 2009), some rivers such as Jiulong River inject 1st SA, in which nitrate is a major nitrogen pollutant (Chen et al., 2009), and related research also shows that the river input is the most important source of marine Si (Wang et al., 2008); VF2 explains explained 23.05% of the total variance , had strong positive loadings on TP, NH_4-N and PO_4-P, which might be interpreted as non-point source pollution from watershed (Chen et al., 2009), in particular pollution caused by livestock poultry (Huang et al., 2012). Non-point source pollution is a major source of phosphorus pollution in Jiulong River watershed, while livestock poultry is a major source of non-point source phosphorus pollution (Huang et al., 2012); VF3 had strong positive loadings on TOC, which likely represents organic matter from sewage discharge (Singh et al., 2005, Zhou et al., 2007b).

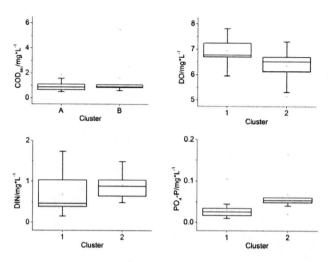

Figure 3. Spatial variation of main coastal water quality parameters in Xiamen Bay.

Table 1. Explained total variance and principal component loading matrix.

Parameters	Cluster 1			Cluster 2	
	VF1	VF 2	VF 3	VF 1	VF 2
NO_3-N	0.96	0.24	0.12	−0.10	−0.93
SiO_3-Si	0.94	0.09	0.24	0.45	0.50
NO_2-N	0.97	0.17	0.23	0.88	0.38
pH	−0.81	−0.27	−0.10	−0.91	−0.07
TN	0.88	0.24	0.29	0.86	0.30
S	−0.90	−0.41	−0.09	−0.97	−0.09
COD_{Mn}	0.63	0.24	0.49	0.94	0.35
DO	−0.71	−0.63	0.11	−0.70	−0.29
PO_4-P	0.51	0.88	0.27	0.90	0.32
TP	0.31	0.75	0.27	0.95	0.27
NH_4-N	0.32	0.80	0.47	0.83	0.31
SPM	−0.22	−0.63	0.21	−0.89	0.19
TOC	0.16	0.13	0.91	0.79	0.60
Chla	0.22	−0.12	0.73	0.83	0.37
Oil	0.20	0.45	0.43	0.89	0.21
Total variance	6.55	3.16	3.07	12.18	2.81
Variance explained/%	45.72	23.05	19.88	73.65	17.32
Cumulative variance explained/%	45.72	68.77	88.65	73.65	90.97

For cluster 2, two VFs can be extracted, VF1 had strong positively loadings on TN, NH4-N, NO2-N, Oil, PO4-P and TP, VF2 was strong positively correlated with TOC, representing land-based sewage discharges and agricultural non-point source pollution respectively (Zhou et al., 2007a, Zhou et al., 2007b). Most sewage outfalls of Xiamen Bay are distributed in Western Sea and Tongan Bay, so land-based outfall is important sources of pollution in Western sea and Tongan Bay. Fertilizer use per unit area in Xiamen is high, and livestock poultry are common in the catchment around Western sea and Tongan Bay, which are contribute to pollution in Western Sea and Tongan Bay.

3.4 *Management implications*

The main source of coastal environment pollution in Xiamen Bay is land-based pollution and River input. In order to control coastal water pollution, there are two methods maybe effective: First, to strengthen management of land-based sewage outfall to ensure the normal operation of sewage treatment facilities and sewage discharge; moreover, to strengthen comprehensive improvement of river watershed around Xiamen Bay, reducing watershed agricultural nonpoint source pollution, enhancing comprehensive management of livestock and poultry and sewage discharge. It is also important to reduce interference of human activities on coastal ecosystems, and to maintain health coastal ecosystem.

4 CONCLUSION

(1) Based on the average of monitoring data of 30 monitoring sites, spatial cluster analysis was perform to generated a dendrogram grouping the 30 sites into two clusters. The average concentrations of TN, TP, DIN and PO_4-P in 2nd SA were higher than that of 1st SA, and DIN and PO_4-P concentrations of 2nd SA exceed the fourth class water quality standards, DIN of 1st also the same. The SPM of 1st SA is greater than 2nd
(2) Varimax rotated principal component analysis can explained by 88.65% and 90.97% of the variance in Cluster 1 and 2 respectively. Three and two components can be extracted from Cluster 1 and 2. For Cluster 1, VF1 represents the river pollutant inputs and agricultural non-point source pollution, VF2 might be interpreted as non-point source pollution from watershed while VF3 is likely representing organic matter from sewage discharge; the two VFs of Cluster 2 represented land-based sewage discharges and agricultural non-point source pollution respectively.
(3) In order to control coastal water pollution, two methods maybe effective: First, to strengthen management of land-based sewage outfall to ensure the normal operation of sewage treatment facilities and sewage discharge; moreover, to strengthen comprehensive improvement of river watershed around Xiamen Bay.

REFERENCES

Bu, H. M. Tan, X. Li, S. Y. 2010. Water quality assessment of the Jinshui River (China) using multivariate statistical techniques, *Environmental Earth Sciences* 60(8):1631–1639.
Chen, N. W. Hong, H. S. Zhang, L. P. 2009. Preliminary results concerning the spatio-temporal pattern and mechanism of nitrogen sources and exports in the Jiulong River watershed, *Acta Scientiae Circumstantiae* 29 (4): 830–839. (in Chinese with English abstract)
Costanza, R. 1999. The ecological, economic and social importance of the oceans. *Ecological Economics* 31: 199–213.
Huang, J. L. Huang, Y. L. Li, Q. S. et al. 2012. Preliminary analysis of spatialtemporal Variation of water quality and its influencing factors in the Jiulong River Watershed, *Environmental Science* 33(4): 1098–1107. (in Chinese with English abstract)
Huang, J. L. Lin, J. Zhang, Y. et al. 2012. Analysis of phosphorus concentration in a subtropical river basin in southeast China: Implications for management, *Ocean and Coastal Management* DOI: 10.1016/j.ocecoaman.2012.09.016
Li, S. Y. Li J. Zhang, Q. F. 2011. Water quality assessment in the rivers along the water conveyance system of the Middle Route of the South to North Water Transfer Project (China) using multivariate statistical techniques and receptor modeling, *Journal of Hazardous Materials* 195: 306–317.
Li, S. Y. Zhang, Q. F. 2009. Geochemistry of the upper Han River basin, China2: Seasonal variations in major ion compositions and contribution of precipitation chemistry to the dissolved load, *Journal of Hazardous Materials* 170(2-3): 605–611.
Lin, H. Zhang, Y. B. 2008. A study on the changes of seawater eutrophication in Xiamen Bay, *Journal of Oceangraphy in Taiwan Strait* 27(3): 347–355. (in Chinese with English abstract)
Simeonov, V. Stratis, J. A. Samara, C. et al. 2003. Assessment of the surface water quality in Northern Greece, *Water Research* 37: 4119–4124.

Singh, K. P. Malik, A. Mohan, D. et al. 2004. Multivariate statistical techniques for the evaluation of spatial and temporal variations in water quality of Gomti River (India) a case study, *Water Research* 38(18): 3980–3992.

Singh, K. P. Malik, A. Singh, V. K. et al. 2005. Chemometric analysis of groundwater quality data of alluvial aquifer of Gangetic Plain, North India, *Analytica Chimica Acta* 550(1-2): 82–91.

Wang, L. J. Ji, H. B. Ding, H. J. et al. 2008. Advances of the research on the biogeochemical cycle of Silicon, *Bulletin of Mineralogy, Petrology and Geochemistry* 27(2): 188–194. (in Chinese with English abstract)

Wu, M. L. Wang, Y. S. Sun, C. C. et al. 2010. Identification of coastal water quality by statistical analysis methods in Daya Bay, South China Sea, *Marine Pollution Bulletin* 60: 852–860.

Wunderlin, D. A. Diaz, M. P. Ame, M. V. et al. 2001. Pattern recognition techniques for the evaluation of spatial and temporal variations in water quality. A case study: Suquia river basin (Cordoba, Argentina), *Water Research* 35: 2881–2894.

Zhou, F. Huang, G. H. Guo, H. C. et al. 2007a. Spatio-temporal patterns and source apportionment of coastal water pollution in eastern Hong Kong, *Water Research* 41: 3429–3439.

Zhou, F. Guo, H. C. Liu, Y. et al. 2007b. Chemometrics data analysis of marine water quality and source identification in Southern Hong Kong, *Marine Pollution Bulletin* 54(6): 745–756.

Isolation and identification and characterization of one phthalate-degrading strain from the active sludge of sewage treatment plant

Bo Li
School of Public Health, Zhengzhou University, Zhengzhou, China
School of Nursing, Henan University, Kaifeng, China

Jun Cao
Zhengzhou Wastewater Purification Co. Ltd, Zhengzhou, China

Chuanhong Xing
School of Water Conservancy & Environmental Engineering, Zhengzhou, China
Research Institute of Nanjing University at Lianyungang, Lianyungang, China

Zhijin Wang & Liuxin Cui
School of Public Health, Zhengzhou University, Zhengzhou, China

ABSTRACT: In this study, one phthalate-degrading strain was isolated from the active sludge of a sewage treatment plant. The 16S rDNA sequence of the isolate was determined and referred to the Genebank to identify the biological species. The isolate was belonged to Pseudomonas sp. and was a new strain which was not reported before. Orthogonal experiment was designed to explore the optimistic conditions for the degradation of the bacterium, and the results showed that: pH 8.0, temperature 23°C. Under the optimistic conditions, PAEs stimulate the isolate grow, the phthalate-biodegradation rate of it sped up with the incipient concentration of PAEs rising and the half-life time would be descending.

1 INTRODUCTION

Phthalate acid esters (PAEs) are anthropogenic compounds widely-used in plastics, clothing, building materials, home furnishings, transportation components, cosmetics, surface lubricants, and medical product packaging (Vamsee-Krishna et al. 2006). These compounds easily leach out, and they have been found to be ubiquitously distributed in the environment (Andrady 2011; Thompson et al. 2009). Even at very low concentration, PAEs can affect endocrine system, especially reproduction of humans and wildlife (Lottrup et al. 2006). Therefore, the United States Environmental Protection Agency (USEPA) have classified diethylhexyl phthalate (DEHP), butyl benzyl phthalate (BBP), dimethyl phthalate (DMP), diethyl phthalate (DEP), dibutyl phthalate (DBP) and dioctyl phthalate (DOP), as priority environmental pollutants (USEPA, 1992).

Though PAEs in sewage all through the world are often lower than effluent standard (Clara 2012), long-term and massive discharging and bioconcentrating will enhance their concentrations in animal body or plants. Sewage, a complex mixture of organic and inorganic chemicals, is considered to be a major source of environmental pollution. PAEs ubiquity in the environment and tendency to bioconcentrate in animal fat are well known (Jobling et al. 1995). Hence, many technologies that eliminate PAEs from sewage are used widely. One of the major routes for environment degradation of PAEs is microbial degradation, while the rate of abiotic degradation is very low (Scholz et al. 1997; Staples et al. 1997). By now, some of phthalate-degrading strains have been reported, such as *Pseudomonads* (Nomura et al. 1992), *Pseudomonas testosterone* (Ono et al, 1970),

Pseudomonas fluorescens (Wang et al. 1997; Zeng et al. 2004), *Agrobacterium, Alcaligens, Bacillus, Micrococu* (Nomura et al. 1989), *Sphigomonas* sp. and *Corynebacterium* sp. (Chang et al. 2004), and *Comamonas acidovoran* (Wang et al. 2003).

In this study, a new specy of *Pseudomonads* that can degrade PAEs was isolated from the active sludge of one WWTP. And it was found that Its phthalate-biodegradation kinetics was obviously different from other phthalate-degrading bacteria.

2 MATERIALS AND METHODS

2.1 Chemicals

Dimethyl phthalate (DMP) (99% purity), dibutyl phthalate (DBP) (99% purity), diethyl phthalate (DEP) (99% purity) and di-n-Propyl Phthalate (DPrP) (98% purity) were purchased from A Johnson Matthey Company (Alfa Aesar, USA). In these experiments, the word, PAEs, was defined as the mixture of DEP, DPrP, DBP and DMP with same weight. All other chemical reagents were of analytical grade and all solvents were of HPLC grade.

2.2 Enrichment and isolation of bacteria

The mineral salts medium (MSM: K_2HPO_4 $1.0\,g\,L^{-1}$, NaCl $1.0\,g\,L^{-1}$, NH_4Cl $0.5\,g\,L^{-1}$, $MgSO_4$ $0.4\,g\,L^{-1}$) was supplemented with the four kinds of phthalate ester as the sole source of carbon and energy to compose the enrichment culture with increasing concentration of PAEs ($1\,mg\,L^{-1}$, $10\,mg\,L^{-1}$, $100\,mg\,L^{-1}$ and $1000\,mg\,L^{-1}$).

A gram active sludge, collected from a sewage treatment plant, was inoculated into the enrichment culture (PAEs $1\,mg\,L^{-1}$) in shaking bed with 30°C and 150 rpm. After 7 days, 1 ml of culture was inoculated into the culture (PAEs $10\,mg\,L^{-1}$), which was been putting into the same conditions 7 days. After another two subcultures, PAEs $100\,mg\,L^{-1}$ and PAEs $1000\,mg\,L^{-1}$, the some survival strain(s) was inoculated into the nutrient agar plate by streaking and remained in the worming box with 37°C for 48 hours, then some pure colonies would appear on the surface of plate. One colony was picked out to inoculate into the culture (PAEs $1000\,mg\,L^{-1}$) again, which was put into shaking bed with 30°C and 150 rpm. After 7 days, the culture was spraided on the plate by streaking again. The same steps were circulated 3 times, and the pure strain was isolated and inoculated into the enrichment medium.

2.3 Identification and characterization of bacteria

The pure strain was inoculated into nutrient agar plate. After 48 hours, the sole colony was picked up to extract the genomic DNA for 16S rDNA gene amplification by PCR (TaKaRa Biotechnology (Dalian) Co. Ltd). 16S rDNA gene sequence was determined and referred to GenBank. The sequence was compared with the bacterial gene sequences online to characterizate the strain (http://www.nebi.nlm.nih.govzblast).

2.4 Degradation optimal conditions

The pure strain was inoculated into the Nutrient Agar medium, and was put into the shaken bed with 37°C and 120rpm for 24 hours. Cells were collected (5000rpm for 10 mins) and washed twice with phosphate buffered saline (PBS), then were mixed with PBS. After diluted by PBS, the cells were incorporated into nutrient agar plate and cultivated with 37°C for 48 hours. Count the member of colonies and calculate the concentration of cell suspension, which helped us to make sure inoculate 108 cells of strain into one MSM supplemented with PAEs as the sole source of carbon and energy.

Degradation of DMP, DBP, DEP and DPrP in MSM was carried by using 150 ml culture in 250 ml flask shaken at 120 rpm for 5 days. The degradation rates were measured by high performance liquid chromatography (HPLC). The orthogonal experiment, with pH (5, 6, 7, 8), temperature (16, 23, 30, 37°C) and PAEs concentration (total 200, 400, 600, $800\,mg\,L^{-1}$) as effect factors, was designed to determine the optimal degradation conditions. All the experiments were carried out 3 times and the data was the mean. K1, K2, K3 and K4 were the mean degradation rate of every levels of each factor, and R was the range of them. The importance of each factor was dependent R value. K value suggested under which level of one factor the phthalate-biodegrading strain worked more efficiently.

2.5 Degradation

The same number of cells was inoculated into the MSMs with PAEs (total concentration, 200, 400, 600, 800 mg L^{-1}) and put into shaken bed with the optimal conditions and 120 rpm. The culture was tested every day from the second to the sixth to measure the residual substrates by HPLC.

The degradation tendency of 4 kinds of PAEs by the strain was assumed to fit the Monod first-order kinetic equation as follows:

$$LnC = -kt + b \qquad (1)$$

$$t_{1/2} = Ln_2 k^{-1} = 0.693 \cdot k^{-1} \qquad (2)$$

where C is PAEs concentration at t time, k is the first order rate constant and b is a constant, $t_{1/2}$ is the half-life value.

2.6 Analysis of residual substrates by HPLC

The aliquots, collected from the culture, were filtered through a 0.22 μm membrane filter before injecting 20 μl into the HPLC system (Agilent 1100 series) equipped with a chromatography column (Hypersil ODS, 25 μm × 4.6 mm × 250 mm), and were detected by 224 nm ultraviolet at normal temperature. The mobile phase was methanol/water (3/1, v/v), and the flow rate was 1 ml/min.

3 RESULTS AND DISCUSSION

After the steps of isolation experiment, one strain that could utilize PAEs as sole source of carbon and energy was obtained from the active sludge. It was named as CL001 temporarily. It formed white and round colonies on the agar plates, and it was Gram-negative.

The 16S rDNA sequence was detected by PCR, and was referred to Genebank (query ID: lcl|32413). Based on BLAST analysis, we found that CL001 belonged to the genus Pseudomonas, and the Fig. 1 illustrates the phylogenetic relationship to its close relatives, the similarity between CL001 and its relatives was beyond 99%. The present studies showed that many Pseudomonas species could degrade organics, include PAEs, efficiently (Nomura et al. 1992; One et al. 1970; Wang et al. 1997; Zeng et al. 2004).

The concentration of PAEs in the nominal samples could be illustrated by the area of peaks in the HPLC maps that were listed in *Table 1*. After statistic analyses, we found that the concentration and the area obeyed linear regression, and the equations were listed in *Table 2*. The values of R^2 in the table 2 were approximate to or larger than 0.99, which illustrated that the fitting result of regression equation was good enough to certify that the test method was reliable.

In this study, the orthogonal experiment was applied to explore the main factors that affected the PAEs biodegradation. As the results showed in the tables from 3 to 6, temperature, pH and the incipient concentration of PAEs were the important factors, and the last one was the most important, temperature was the second and pH was inferior again. The optimum temperature was 23°C and pH was 8.0, which were different from some previous studies about *Pseudomonas*. For example, *Pseudomonas fluorescens* FS1, the highest biodegradation rate of DMP, DEP, DnBP, DIBP, DnOP, and DEHP was achieved at the ranges from pH 6.5 to 8.0 at 30°C (Zeng et al. 2004). The most interesting and strange was that the optimum incipient concentration of PAEs was 800 mg L^{-1}. In theory, PAEs is toxic to the bacteria, therefore, higher incipient concentration causes stronger toxicity and the rate of degradation should be lower. The unusual phenomenon will be mentioned in this article again.

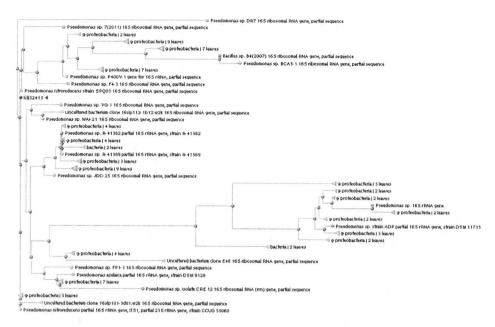

Figure 1. The phylogenetic relationship of CL001.

Table 1. The HPLC Peak Area of PAEs in the Nominal Samples.

PAEs	The peak area of various concentration				
	10 mg l^{-1}	20 mg l^{-1}	40 mg l^{-1}	80 mg l^{-1}	100 mg l^{-1}
DEP	623.089	1053.22	2171.53	4374.73	4782.33
DPrP	443.101	766.754	1572.36	3170.42	3921.31
DBP	350.389	623.055	1197.5	2407.38	3425.66
DMP	285.872	513.636	1133.64	2121.36	3151.2

Table 2. The Linear Regression Equation of PAEs.

PAEs	Linear regression equation	R^2
DEP	y = 49.546x + 103.065	0.990
DPrP	y = 39.218x + 11.582	1.000
DBP	y = 32.848x − 34.673	0.991
DMP	y = 30.070x − 51.977	0.987

Comparing the degradation rate of four kinds of PAE, we concluded that CL001 degraded DEP, DBP and DMP more efficiently than DPrP under same conditions. The present studies manifested that the phthalates with shorter ester chains were readily biodegraded, whereas those with longer ester chains were only marginally degraded, entirely consistent with previous studies (Chang et al, 2004; Zeng et al, 2004). This study showed that CL001 degraded DEP, DBP and DMP following the above rule except for DPrP, the cause should be explored in future.

The first-order kinetic equations and the half-life time that explain the degraded rules of PAEs by CL001 were listed in the table 7. Under the optimal conditions, the degradation rules of PAEs with diverse incipient concentration by CL001 were described by the first-order kinetics equations

Table 3. The Orthogonal Experiment Program and Result of DEP Degraded by CL001.

No.	A Temp. (°C)	B pH	C concentration of substrate (mg l^{-1})	peak area of HPLC	vestigial concentration of substrate (mg l^{-1})	rate of degradation (%)
1	16	5	50	11596.6	15.46	69.07
2	16	6	100	15850.4	21.19	78.81
3	16	7	150	19986.5	26.75	82.16
4	16	8	200	12145.1	16.20	91.90
5	23	5	100	11924.7	15.91	84.09
6	23	6	50	12335	16.46	67.08
7	23	7	200	7864.63	10.44	94.78
8	23	8	150	8195.35	10.89	92.74
9	30	5	150	13415.9	17.91	88.06
10	30	6	200	17978.7	24.05	87.97
11	30	7	50	15119.3	20.20	59.59
12	30	8	100	3946.34	5.17	94.83
13	37	5	200	5367.74	7.08	96.46
14	37	6	150	16851.1	22.53	84.98
15	37	7	100	11009.7	14.67	85.33
16	37	8	50	9074.52	12.07	75.86
K1	80.49	84.42	67.90			
K2	84.67	79.71	85.77			
K3	82.61	80.47	86.99			
K4	85.65	88.83	92.78			
R	5.16	9.12	24.88			

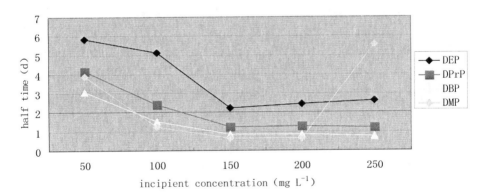

Figure 2. The half time of PAEs degraded by CL001.

(table 3). After analyzing the data of the substrates utilization testing, we could draw a conclusion that the half-life time would be descending with the incipient concentration of PAEs rising (table 3 and Fig. 2), though the data of several groups didn't present significant difference. The phenomenon was consistent with the result of orthogonal experiment, which meant the biodegradation rate of PAEs by CL001 was sped up with the incipient concentration of PAEs rising, which was consistent with the orthogonal experiment. The fact demonstrated that the phenomenon was not because of the error or fault. This seemed inexplicable, because the previous studies presented PAEs were somewhat toxic to bacterium and its activity should be restrained more and more strongly with the concentration of PAEs rising, the half-life time should be prolonged, not be shorten (Chang et al, 2004). In previous study, one *Pseudomonas* strain, named FS1, was isolated and used to

Table 4. The Orthogonal Experiment Program and Result of DPrP Degraded by CL001.

No.	A Temp. (°C)	B pH	C concentration of substrate (mg l^{-1})	peak area of HPLC	vestigial concentration of substrate (mg l^{-1})	rate of degradation (%)
1	16	5	50	17964.9	30.52	38.96
2	16	6	100	21632.4	36.75	63.25
3	16	7	150	25425.4	43.20	71.20
4	16	8	200	13662.1	23.20	88.40
5	23	5	100	13014.7	22.10	77.90
6	23	6	50	6638.52	11.27	77.47
7	23	7	200	3330.81	5.64	97.18
8	23	8	150	5134.42	8.71	94.19
9	30	5	150	12479.3	21.19	85.87
10	30	6	200	18738.7	31.83	84.08
11	30	7	50	14100.9	23.95	52.10
12	30	8	100	6108.36	10.36	89.64
13	37	5	200	3260.77	5.52	97.24
14	37	6	150	23781.9	40.41	73.06
15	37	7	100	14638.8	24.86	75.14
16	37	8	50	8723.93	14.81	70.38
K1	65.45	74.99	59.73			
K2	86.68	74.47	76.48			
K3	77.92	73.90	81.08			
K4	78.95	85.65	91.72			
R	21.23	11.75	31.99			

Table 5. The Orthogonal Experiment Program and Result of DBP Degraded by CL001.

No.	A Temp. (°C)	B pH	C concentration of substrate (mg l^{-1})	peak area of HPLC	vestigial concentration of substrate (mg l^{-1})	rate of degradation (%)
1	16	5	50	8283.12	16.88	66.24
2	16	6	100	11242.5	22.89	77.11
3	16	7	150	12091.2	24.61	83.59
4	16	8	200	4527.03	9.26	95.37
5	23	5	100	5203.21	10.63	89.37
6	23	6	50	1112.17	2.33	95.34
7	23	7	200	537.714	1.16	99.42
8	23	8	150	707.165	1.51	99.00
9	30	5	150	4200.62	8.60	94.27
10	30	6	200	13818.3	28.12	85.94
11	30	7	50	9050.66	18.44	63.12
12	30	8	100	5796.92	11.84	88.16
13	37	5	200	923.145	1.94	99.03
14	37	6	150	23462	47.69	68.21
15	37	7	100	10763.2	21.91	78.09
16	37	8	50	4644.11	9.50	81.01
K1	80.58	87.23	76.43			
K2	95.78	81.65	83.18			
K3	82.87	81.05	86.27			
K4	81.58	90.89	94.94			
R	15.20	9.84	18.51			

Table 6. The Orthogonal Experiment Program and Result of DMP Degraded by CL001.

No.	A Temp. (°C)	B pH	C concentration of substrate (mg l^{-1})	peak area of HPLC	vestigial concentration of substrate (mg l^{-1})	rate of degradation (%)
1	16	5	50	4370.4	9.80	80.39
2	16	6	100	8392.07	18.72	81.28
3	16	7	150	6846.4	15.29	89.80
4	16	8	200	2187.4	4.96	97.52
5	23	5	100	2526.53	5.72	94.28
6	23	6	50	206.921	0.57	98.85
7	23	7	200	75.619	0.28	99.86
8	23	8	150	76.851	0.29	99.81
9	30	5	150	2112.52	4.80	96.80
10	30	6	200	12398.1	27.60	86.20
11	30	7	50	7263.3	16.22	67.56
12	30	8	100	5607.17	12.55	87.45
13	37	5	200	268.29	0.71	99.64
14	37	6	150	24083.7	53.51	64.33
15	37	7	100	8942.09	19.94	80.06
16	37	8	50	3649.7	8.21	83.59
K1	87.25	92.78	82.60			
K2	98.20	82.66	85.77			
K3	84.50	84.32	87.69			
K4	81.90	92.09	95.80			
R	16.30	9.43	13.20			

Table 7. The Kinetic Equations and the Half-life Time of PAEs Degraded by CL001.

PAEs	incipient concentration (mg l^{-1})	kinetic equation	R^2	F	P	The half-life time (d)
DEP*	50	ln C = −0.119t + 3.922	0.712	7.407	0.072	5.82
DPrP	50	ln C = −0.166t + 4.183	0.873	20.644	0.020	4.17
DBP	50	ln C = −0.225t + 3.902	0.796	11.709	0.042	3.08
DMP*	50	ln C = −0.178t + 3.422	0.529	3.365	0.164	3.89
DEP*	100	ln C = −0.135t + 4.500	0.572	4.017	0.139	5.13
DPrP*	100	ln C = −0.290t + 3.819	0.704	7.121	0.076	2.39
DBP*	100	ln C = −0.453t + 3.294	0.696	6.852	0.079	1.53
DMP*	100	ln C = −0.565t + 2.952	0.579	4.129	0.135	1.23
DEP	150	ln C = −0.314t + 5.277	0.846	16.464	0.027	2.21
DPrP	150	ln C = −0.569t + 5.802	0.809	12.670	0.038	1.22
DBP	150	ln C = −0.803t + 5.831	0.842	15.988	0.028	0.86
DMP	150	ln C = −1.091t + 6.156	0.837	15.435	0.029	0.64
DEP*	200	ln C = −0.283t + 5.193	0.722	7.808	0.068	2.45
DPrP	200	ln C = −0.536t + 5.544	0.822	13.866	0.034	1.29
DBP	200	ln C = −0.813t + 5.516	0.867	19.551	0.021	0.85
DMP	200	ln C = −1.092t + 5.614	0.860	18.448	0.023	0.63
DEP	250	ln C = −0.264t + 5.343	0.803	12.256	0.039	2.63
DPrP	250	ln C = −0.571t + 5.925	0.785	10.975	0.045	1.21
DBP	250	ln C = −0.955t + 6.372	0.796	11.730	0.042	0.73
DMP	250	ln C = −1.282t + 6.692	0.792	11.427	0.043	0.54

*P > 0.05, there was not significant difference between in the numbers of vestigial concentration.

degrade PAEs, and the half-life time was prolonged from 6 h to 10 h with increasing the incipient concentration (Zeng et al, 2004). In this study, PAEs seemed stimulate the *Pseudomonas* species grow. The more tests should be designed and carried out in the following time to explore whether the specificity of the biodegradation rule is fact or just because of test design and operation faults.

4 CONCLUSION

The phthalate-degrading strain isolated from the active sludge of a sewage treatment plant was belonged to Pseudomonas sp. and was a new strain which was not reported before.

ACKNOWLEDGMENTS

This work was supported by a grant from the State Key Project "water pollution control and treatment" (No.: 2009ZX07210-001-001).

REFERENCES

Andrady, A.L. Microplastics in the marine environment. *Marine Pollution Bulletin*, 2011. 62(8): p. 1596–1605.
Chang B, Yang C, Cheng C, Yuan S. *Biodegradation of phthalate esters by two bacteria strains*. Chemosphere, 2004. 55(4): p. 533–538.
Clara M., Windhofer G., Weilgony P., Gans O, Denner M., Chovanec A., Zessner M. Identification of relevant micropollutants in Austrian municipal wastewater and their behaviour during wastewater treatment. *Chemosphere*, 2012. 87: p. 1265–1272
Lottrup G., Andersson A, Leffers H, Mortensen G, Toppari J, Skakkebaek N, Main K. Possible impact of phthalates on infant reproductive health. *International Journal of Andrology*, 2006. 29(1): p. 172–180.
Jobling S., Reynolds T., White R., Parker M. G., Sumpter J. P. A variety of environmentally persistent chemicals, including some phthalate plasticizers, are weakly estrogenic. *Environmental health perspectives* 1995. 103(6), 582–7.
Nomura Y., Takada N., Oshima Y. Isolation and identification of phthalate-utilizing bacteria. *Journal of Fermentation and Bioengineering*, 1989. 67(4): p. 297–299.
Nomura Y., Nakagawa M., Ogawa N., Harashima S., Oshima Y. Genes in PHT plasmid encoding the initial degradation pathway of phthalate in *Pseudomonas putida*. *Journal of Fermentation and Bioengineering*, 1992. 74(6): p. 333–344.
Ono K., Nozaki M., and Hayaishi O. Purification and some properties of protocatechuate 4,5-dioxygenase. *Biochimica et biophysica acta*, 1970. 220(2): p. 224–38.
Scholz, N., Diefenbach R., Rademacher I., Linneman D. Biodegradation of DEHP, DBP, and DINP: poorly water soluble and widely used phthalate plasticizers. *Bulletin of environmental contamination and toxicology*, 1997. 58(4): p. 527–34.
Stales, C.A., Peterson D.R., Parkerton T.F., Adams W.J. The environmental fate of phthalate esters: A literature review. *Chemosphere*, 1997. 35(4): p. 667–749.
Thompson, R.C., Moore C.J., Vom Saal F.S., Swan S.H. Plastics, the environment and human health: current consensus and future trends. *Philosophical Transactions of the Royal Society B-Biological Sciences*, 2009. 364(1526): p. 2153–2166.
USEPA, 2012. Code of Federal Regulation, Appendix A to 40 CFR Part 423.
Vamsee-Krishna, C., Y. Mohan, and P.S. Phale. Biodegradation of phthalate isomers by *Pseudomonas aeruginosa* PP4, *Pseudomonas* sp PPD and Acinetobacter lwoffii ISP4. *Applied Microbiology and Biotechnology*, 2006. 72(6): p. 1263–1269.
Wang J, Liu P, Qian Y. Biodegradation of phthalic acid esters by immobilized microbial cells. *Environment International*, 1997. 23(6): p. 775–782.
Wang Y, Fan Y, and Gu J. Aerobic degradation of phthalic acid by Comamonas acidovoran Fy-1 and dimethyl phthalate ester by two reconstituted consortia from sewage sludge at high concentrations. *World Journal of Microbiology & Biotechnology*, 2003. 19(8): p. 811–815.
Zeng F, Cui K, Li X, Fu J, Sheng G. Biodegradation kinetics of phthalate esters by *Pseudomonas fluoresences* FS1. *Process Biochemistry*, 2004. 39(9): p. 1125–1129.

Seedling emergence and growth of Ricinus communis L. grown in soil contaminated by Lead/Zinc tailing

Xinyu Yi, Lijuan Jiang, Qiang Liu, Mingliang Luo & Yunzhu Chen
Central South University of Forestry and Technology, Changsha, Hunan, China

ABSTRACT: A greenhouse study was conducted to observe seedling emergence and growth of *Ricinus communis* L. grown in substrate contaminated with lead-zinc tailings. The concentrations of heavy metals including Pb, Zn, Cu and Cd in seedling tissues were determined to know the bioaccumulation of heavy metals and tolerance. The results demonstrated that emergence percentage, emergence viability and survival rate of substrate E (60% lead-zinc tailings + 20% peat moss + 20% soil without pollution) was higher than others, while seedlings were restrained intensively by substrate C (80% lead-zinc tailings + 10% peat moss + 10% soil without pollution + manure) ($p < 0.05$). For the translocation ability of the four heavy metals in plant, the order of average was Cu > Zn > Pb > Cd and transfer coefficient was lower than 1. *Ricinus communist* L. was proved to possess good metal-tolerance ability and certain application prospects in remediation of lead-zinc contaminated soils.

1 INTRODUCTION

Environmental pollution by heavy metals (HM) is a serious problem in mining area (Liang et al., 2011). Mining areas suffer from considerable pollution with high HM concentrations, which causes great influence to the plant growth (Zhou et al., 2002). Phytoremediation is a relatively new approach to remove HM from soils by plants (Chaney et al., 1997) and has become a hot research field in last decades as it is safe and potentially cheap compared to traditional remediation techniques (Salt, 1998; Mitch, 2002; Glick, 2002; Pulford, 2003).

Ricinus communis L., a member of Euphorbiaceae, is a perennial with extensive adaptability to harsh conditions. It is an ecologically and economically important plant because castor oil can act as alternative materials for petroleum derivatives. Zheng et al. (2005) showed that castor has been applied to copper with a good agreement with plant uptake. Castor plants can accumulate 9686.8 mg/kg zinc in roots, 3748 mg/kg in shoots and 2042.5 mg/kg in leaves (Lu et al., 2005). Qu et al. (2006) also mentioned that the lead content in castor roots was greater than other tissues when they were treated at 100 mg/L lead for 72 hours. Castor was an oil plant with maximum oil content up to 70%, which draw attraction as non-food oil seed tree for biodiesel production. Growing oil plants in abandoned, marginal or infertile lands such as many metal-polluted mining areas are a most promising way to restore the vegetation cover and provide renewable resources for bioenergy. In this study, we investigated the influence of different substrates on seedling emergence and growth of castor plant. The concentrations of Pb, Zn, Cu and Cd were also quantified to understand the bioaccumulation and biotransformation of HM through different plant tissues.

2 MATERIALS AND METHODS

2.1 Pot experiment

Seeds of *Ricinus communis* L. 'Xiangbi No. 1' were obtained from Energy Research Institute of Hunan Academy of Forestry (Changsha, China).

Table 1. The composition of the substrates used in the study.

Substrate	Volume ratio
A	100% soil contaminated with lead and zinc
B	80% soil + 10% peat moss + 10% soil without pollution
C	80% soil + 10% peat moss + 10% soil without pollution + manure
D	80% soil + 20% soil without pollution
E	60% soil + 20% peat moss + 20% soil without pollution
F	60% soil + 40% soil without pollution
G (control)	50% soil without pollution + 50% peat moss

Soil contaminated with lead and zinc were collected from Zixing county (Hunan, China), about 342 km away from Changsha, Hunan, and then mixed with manure and peat moss at ratios described in Table 1. Pots of 1.5 L were filled up with 2 kg fresh substrates. They were watered to container capacity. Castor seeds were sown into the substrates at 1 cm deep. They were germinated and emerged in a greenhouse with a temperature range of 15~25°C and humidity of at least 70%.

2.2 Sample analysis

The substrates were sampled and oven dried at 75°C for 72 h. Soil samples were grounded and passed through a 0.8-mm nylon mesh sieve. The plants were harvested 40 days after sown (DAS). Morphological indexes were measured. Plants were separated into root and shoot and washed with deionized water and oven-dried at 70°C for 48 hours. 300 mg plant samples and substrates were digested in an agent containing HNO_3 (6 mL), HCl (2 mL) and H_2O_2 (0.5 mL) in a microwave oven (Tank, Hanon Technologies). Soil pH was investigated by pH meter (BPH-200B). Total N was determined using Kjeldhal method, total P was extracted by Antispetrophotography method, while total K by flame photometer (Bao Shidan, 2000).

The total heavy metal contents of soil and plant sample were determined by Inductively Coupled Plasma Emission Spectroscopy (Thermo iCAP6000, ICP-AES). Three samples were collected for each type of materials and repeating the measurements if the standard deviation exceeded 10%.

2.3 Seedling emergence

Emergence percentage (EP), Emergence viability (EV) and Survival percentage (SP) were calculated using the following formula:

$$EP(\%) = (N_{20}/N_{total}) \times 100\% \quad (1)$$

$$EV(\%) = (N_{15}/N_{total}) \times 100\% \quad (2)$$

$$SP(\%) = (N_{25}/N_{total}) \times 100\% \quad (3)$$

where N_t = number of survived seedlings after t days and N_{total} = total number of seeds.

2.4 Statistical analysis

The experiment followed a completely randomized design with three replications. Some data were analyzed by one-way analysis of variance (ANOVA). Means separation was conducted using LSD multiple comparison at $P < 0.05$. All statistical analyses were performed using SPSS 16.0 software (IBM Corporation).

Table 2. The nutrients and HM concentrations of substrates.

Substrate	pH	Nutrients (g·kg⁻¹)			HM concentrations (mg·kg⁻¹)			
		Total N	Total P	Total K	Pb	Zn	Cu	Cd
A	7.75	0.61	0.26	5.11	2339.6	2143.75	59.45	68.87
B	7.47	0.78	0.45	5.08	1364.17	1832.25	48.79	41.55
C	6.45	0.76	0.47	5.07	1298.06	2121	57.1	48.03
D	7.55	0.81	0.38	5.06	1236.66	1666.5	49.52	29.77
E	7.01	0.77	0.43	5.07	1255.37	1352.75	48.74	39.19
F	7.27	0.89	0.44	5.07	1064.73	1428	45.84	32.72
G	4.63	0.68	0.14	5.07	55.14	52.83	11.15	10.09

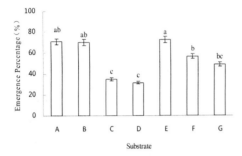

Figure 1. Emergence percentage of *castor grown* in substrates.

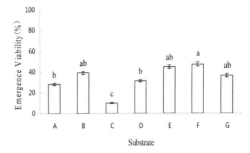

Figure 2. Emergence viability of *castor* grown in substrates.

3 RESULTS

3.1 The nutrients and HM concentrations of substrate

The nutrients and HM concentrations of substrate of all samples were presented in Table 2. Substrate pH ranged from 6.45 to 7.75 and was mostly circum-neutral. Total N and P in samples from substrate A to F were similar but significantly different from those in substrate G. Total K ranged from 5.06 to 5.11, while the highest concentrations and total HM concentrations found in the substrate A.

Of all the samples examined in substrate from A to G was mainly contaminated with Pb and Zn, which exceed the Grade III of National Soil Environmental Standard (GB15618-1995). Total Pb and Zn concentrations in samples collected from substrate were variable from 1064.73 to 2339.6 g·kg⁻¹ and 1428 to 2143.75 g·kg⁻¹, respectively. The range of concentrations of Cd (10.09~68.87 g·kg⁻¹) was lower, followed by Cd (11.15~59.45 g·kg⁻¹).

3.2 Seedling emergence

Seeds appeared not to differ in morphology, the seedling emergence of castor seeds according to the result was affected by the metal and substrates concerned but not at all by the concentrations. Seeds began to germinate after ten days and stopped three weeks later, and there were significantly difference in EP ($P < 0.05$, Fig. 1). Minimum EP were found in substrates C and D of 0.35 and 0.32, the EP of A, B and E got to 0.71, 0.7 and 0.73, respectively increased by 44%, 42.4% and 47.4% compared with the substrate G.

The EV data (Fig. 2) showed that substrate E (EV=45%) performed well in comparison with other substrates, whereas the EV of substrate C was lower than other groups where no significant difference existed compared with substrate G. The differentiation change between EV and SP were approximately similar, and the peak of SP appeared in the substrates B and E.

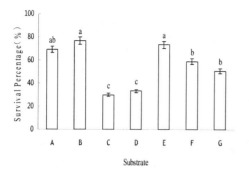
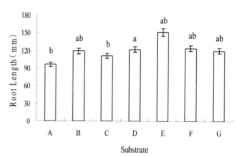

Figure 3. Survival rate of *castor* grown in substrates.

Figure 4. Effect of different heavy metal concentrations on Root length.

Means with same letters were not significantly different among substrates by LSD multiple comparison at P < 0.05. As could be seen from results, changes in emergence percentage, emergence viability and survival rate were observed between the control and treatment group in the range of 1064~2339 mg·kg^{-1} for Pb, 1352~2143 mg·kg^{-1} for Zn, 45~54 mg·kg^{-1} for Cu and 29~68.87 mg·kg^{-1} for Cd. Most of substrates with plants showed an increase in germination compared with the control, except on the substrate C and D. The results may thus reflect the greater tolerance to accumulate HM. As in the substrate C with manure, the values in all experimental results were lower than others, which means that the fertilizer may have contributed to its inhibition. This agreed well with other researchers, for example, Ci E (2007) thought that enzyme in manure was much suitable for increase of enhanced toxicity, and inhibition of enzymes was also important factors that control the protease activity related to emergence.

3.3 Seedling growth

3.3.1 Root growth

Plant root was more sensitive to metals, and metals being able to influence the plant growth due to the root had access to substrate at first. The data of root growth was showed (Fig. 4) that substrates D, E and F was higher than the substrate G of 2.2%, 26.4% and 3.7%, respectively, and there was difference among root length, this agreed with Wang L B (2010), who held that the quantities of metals was tended to accelerate root growth. However, the toxic soil contaminated with HM was not suitable for root growth in substrate A, as its length was fairly low compared with other substrates. The average root length was 95.97 mm and the minimum value was 80.76 mm, decreased 46.635% and 32.51% compared to substrates E and G.

3.3.2 Seedling growth

As shown in the Fig. 5, the significant influence of biomass was not necessarily dependent on the HM concentrations and the average value was 0.25 and 0.26 g per plant samples in substrate D and E. On the other hand, the negative effect was observed in the case of seedling emergence with the substrate C combined with dry biomass and height (Fig. 6) and the value of group D and E increased in the range of 30~35 mm. Besides of the above mentioned indexes, the basal diameter was used as an indicator for plant growth. The fact that none of substrates led to significantly lower basal diameter, in the contrary, the increasing extent reached 0.20–0.45 mm, so there were no basal diameter picture presented.

The ANVOA indicated that (Table 3) there was no significant difference in these morphological indexes. Castor seedling had the tolerance ability to Pb and Zn, it may be concluded that low concentration of HM could stimulate the activity of physiological metabolism in the plant. Brown et al. (1994) also held that the lower concentrations of HM was beneficial to the seedling growth.

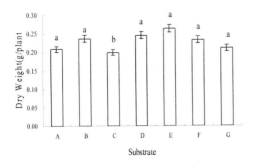
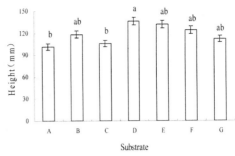

Figure 5. Effect of different heavy metal concentrations on dry weight.

Figure 6. Effect of different heavy metal concentrations on Height.

*Values within colums followed by different letters are significantly different at $P < 0.05$.

Table 3. Variance analysis of morphological indexes of castor.

	SS	df	MS	F	P-value
Root length/(mm)					
Factor	4976.21	6	829.37	1.7	0.19
Error	6819.62	14	487.12		
Total	11795.83	20			
Height/(mm)					
Factor	3031.09	6	505.18	1.74	0.18
Error	4066.01	14	290.43		
Total	7097.1	20			
Dry matter weigh (g)					
Factor	0.02	6	0.003	1.16	0.354
Error	0.06	14	0.002		
Total	0.08	20			
Basal diameter/(mm)					
Factor	0.55	6	0.09	0.61	0.719
Error	2.12	14	0.15		
Total	2.67	20			

But this positive effect may alter to negative damage for the plants under high concentration with lead substrate. Castor seedling grown in high concentration HM was poisoned as follows: the seedling dwarfed, the green leaves turned yellow, the bloom postponed.

3.4 Concentrations of HM in plants

The HM concentration data showed (Table 4) that with the decrease of metals in substrate, the concentrations in seedling tissue was decreased significantly ($p < 0.01$). In general, irrespective of group type, HM concentrations in underground part were, as was typical, higher than aboveground. Maximum Pb concentrations (1400.89 mg/kg) in roots was found in substrate A and the root to soil concentration ratio for Pb reaching 0.6. Castor was not a hyperaccumulator, concentrations of HM in leaves were always smaller than roots, contrary to its accumulating characteristics of the reported plant species which have been described as hyperaccumulator such as *Amaranthus tricolor* and *Rumex L.* (Liu Xiumei et al, 2002), *Vetiver grass* (Antiochia R et al, 2007) and *Chenopodium ambrosioides L.* (Wu Shuangtao et al, 2004). On the substrate C, concentrations in aboveground part were, for Pb, higher than substrate A of 72.74% and the translocation factor was over 3~9

Table 4. Heavy metal concentrations in shoot and root tissues following different types of substrate.

Elements	Substrate	Aboveground part	Underground part	Translocation Factor	Ratio of root and substrate
Lead (mg/kg)	A	121.15 ± 1.86b	1400.89 ± 13.7a	0.09 ± 0	0.6 ± 0.25
	B	88.88 ± 0.36c	922.39 ± 5.64b	0.1 ± 0.15	0.68 ± 0.2
	C	428.72 ± 0.88a	524.56 ± 1.93f	0.82 ± 0.35	0.4 ± 0.3
	D	61.41 ± 1.03e	542.33 ± 1.77e	0.11 ± 0.5	0.44 ± 0.09
	E	84.48 ± 0.62d	668.89 ± 2.65c	0.13 ± 0.02	0.53 ± 0.05
	F	58.68 ± 0.85f	578.06 ± 5.76d	0.1 ± 0.03	0.54 ± 0.07
	G	3.85 ± 0.03g	14.22 ± 0.02g	0.27 ± 0.01	0.26 ± 0.15
Zinc (mg/kg)	A	116.88 ± 1.48a	536.11 ± 4.34a	0.22 ± 0.02	0.25 ± 0.15
	B	68.52 ± 0.22c	319.39 ± 0.54b	0.21 ± 0.03	0.17 ± 0.09
	C	97.87 ± 0.24b	200.5 ± 1.02d	0.49 ± 0.06	0.09 ± 0.06
	D	57.61 ± 0.33e	175.67 ± 0.6e	0.33 ± 0.03	0.11 ± 0.08
	E	69.59 ± 0.24c	261.06 ± 0.51c	0.27 ± 0.01	0.19 ± 0.08
	F	60.02 ± 0.42d	170.83 ± 0.29f	0.35 ± 0.02	0.12 ± 0.07
	G	4.26 ± 0.02f	11.46 ± 0.03g	0.37 ± 0	0.22 ± 0.1
Copper (mg/kg)	A	12.58 ± 0.06b	26.84 ± 0.22a	0.47 ± 0.04	0.45 ± 0.25
	B	10.74 ± 0.15d	20.11 ± 0.13b	0.53 ± 0.06	0.41 ± 0.25
	C	12.74 ± 0.10b	16.99 ± 0.23d	0.75 ± 0.2	0.3 ± 0.2
	D	11.12 ± 0.05c	19.83 ± 0.13b	0.56 ± 0.06	0.4 ± 0.3
	E	14.07 ± 0.10a	18.26 ± 0.10c	0.77 ± 0.09	0.37 ± 0.2
	F	10.03 ± 0.14e	19.58 ± 0.95b	0.51 ± 0.06	0.43 ± 0.3
	G	0.96 ± 0.02f	1.59 ± 0.03e	0.6 ± 0.06	0.14 ± 0
Cadmium (mg/kg)	A	3.97 ± 0.09a	59.25 ± 0.37a	0.07 ± 0	0.86 ± 0.5
	B	2.46 ± 0.03b	38.06 ± 0.28b	0.06 ± 0	0.92 ± 0.6
	C	2.18 ± 0.02b	25.17 ± 0.07d	0.09 ± 0.01	0.52 ± 0.5
	D	2.06 ± 0.53b	23.03 ± 0.14e	0.09 ± 0.01	0.77 ± 0.45
	E	2.37 ± 0.04b	27.16 ± 0.07c	0.09 ± 0.02	0.69 ± 0.1
	F	1.32 ± 0.02c	20.81 ± 0.09f	0.06 ± 0.01	0.64 ± 0.2
	G	0.39 ± 0.10d	9.28 ± 0.005g	0.04 ± 0	0.92 ± 0.6

*The same letter are not significantly different at $P < 0.05$ level, according to LSD Test.

times. Bioavailable form of metal was necessary for uptaking by plants from root to shoot and leaf translocation. One of the most influential factors, controlling the conversion of metals from immobile solid-phase forms to bioavailability solution-phase, was the manure. It was expected that at low pH values, low trace metal availability should be found. The pH affected not only metal bioavailability, but also the process of metal uptake into roots (Brown et al, 1995). Enzymatic action that was another hypothesis for translocating these metals from underground part to aboveground part. As already showed by tested results, the influence of plant species on uptake of HM in this substrate was mostly evident which increased the toxic effect from the stage of germination and seedling growth.

Some plants appeared to be good candidates for stabilization, which may reduce metal mobilization and hazard in the ecosystem. The narrow TF range, the lower metal translocation and the more restrictions to metal translocation, comparison with each group was obvious, Pb concentrations in roots were higher than in aboveground because of Pb mostly existed in the form of phosphate and carbonate (Dahmani et al, 2000). Zn was essential trace element for plant, that was one of the reason for translocating Zn from root to shoot and leaf. The contents of total Zn in substrate were more than Pb, however, no such accordance between Zn and Pb measured in the plant and seedlings which seemed able to uptake more Pb from substrate.

In the case of Cu, as an essential element in plant growth, large quantities could cause irreversible damage to physiological metabolism (Sun Dan et al, 2007). Comparing total Cu and Cd uptake showed that in similar contents of same substrate, seedling took up more Cd in roots than Cu, increased by 54.7%, 47.2%, 32.5%, 14%, 32.8%, 0.9% and 82.9% of every substrate. The TF results indicated the capacity of seedling to translocate metals, in the order of Cu > Zn > Pb > Cd and in all cases where TFs were less than 1. Mechanism tended to be investigated deeply combined with hereditary, biological features and the contents of available metals.

4 DISCUSSION

In general, the HM greatly exceeded the ranges was considered to have a toxic effect on emergence and seedling growth of normal plant. Kabata-Pendias et al. (1984) studied that toxic effects on plant growing in a contaminated soil was found to be in the range of 100~400 mg/kg for Pb, 70~400 mg/kg for Zn and 60~125 mg/kg for Cu. Some traits made castor a promising-plant to use for phytoremediation in degraded areas. Some quantities of the HM containing in substrate have been involved in promotion to seedling emergence and growth, and especially the substrate A (100% soil contaminated with lead and zinc) was superior in aspects of emergence to that others except E. This finding was in accordance with other observation in domestic research reported by Zeng Xiaolong (2010), Duo Li'an et al. (2006) and Ran Xiaoming et al. (2004). It was recognized the role of large quantities of metals in the process of metabolism such as photosynthesis, respiration, chloroplast structure and protein synthesis (Shi Xiang et al., 2011), to plant for overcoming contaminant-induced stress response and influencing plant growth may have contributed to the weakening of plant growth in substrate A at late of growing stage.

Changes of metal bioavailability could be explained by the reaction between fertilizer and soil colloid. On the other hand, the reason why the plant showed a better accumulating power to HM may lie in the fact that fertilizer improved the biomass, in which plant could complemented nutrition (Wang Hongxin et al, 2012). The high translocation factors of castor in substrate C were related with the decrease of pH value and increase of metals availability and it was clear that this could lead to disadvantage of enhanced toxicity of HM for plants throughout the growing stage.

Most reviews focus on the screening for hyperaccumulator, particularly the exploration for accumulation and tolerance ability mechanism, which is the area of major scientific and technological progress in past years. We propose a combination of soil remediation and oil production with bioenergy value for metal-polluted site, taking into account low metal concentration in harvestable parts by screening of rhizosphere microorganism and interrupting the ion channels. Castor may represent more economical and ecological advantages of cropping this plant on polluted site because it is robust and fast growing oil yield and does not compete for land required for food production, more research will be addressed in the future.

REFERENCES

Antiochia R, Can panella L, Ghezzi P, M ovassaghi K. 2007. The use of vetiver for remediation of heavy metal soil contamination. *Analytical and Bioanalytical chemistry* (338), 947–956.

Shidan Bao. 2000. *Soil Agro-chemistrical Analysis*. Beijing: China Agricultures Press.

Brown S L, Chaney R L, Angle J S, et al. 1994. Phytoremediation potential of *Thlaspi caerulescens* and *Bladder Campion* for zinc- and cadmium-contaminated soil. *J Environ Qual* (23): 1151–1157.

Brown, S L., Chaney, R L., Angle, J S., Baker, A J M 1995. Zinc and cadmium uptake by hyperaccumulator *Thlaspi caerulescens* and metal tolerant *silene vulgaris* grown on sludge amended soils. *Environ. Sci. Technol* (29):1581–1585.

Chaney R L, Malik M, Li YM, et al. 1997. Phytoremendiation of soil metals. *Current Opinion in Biotechnology* 8(3):279–284.

En Ci, Ming Gao, Zifang Wang, Yun Wu, Jiancheng Qin. 2007. Effect of cadmium on seed germination and growth of alfalfa. *Chinese Journal of Eco-Agriculture* 15(1):96–98.

Dahmani, Muller H, Valorta F. et al. 2000. Strategies of heavy metal uptake by three plant species growing near a metal smelter. *Environmental Pollution* (109):231–238.

Li'an Duo, Yubao Gao, Shulan Zhao. 2006. Effects on initial growth of *Lolium perenne* under heavy metal progressive stress. *Bulletin of Botanical Research* 26(1):117–122.

Glick BR. 2002. Phytoremediation: synergistic use of plants and bacteria to clean up the environment. *Biotechnol Adv* (21):383–93.

Kabata-Pendias A, Pendias H. 1984. Trace Elements in Soils and Plants. Florida: CRC Press.

Guilian Liang, Jianping Qian. 2011. Features and Remediation technologies of Pollution in Lead-Zinc Mining Areas of China. *Mining Reaearch and Development* 31(04):84–95.

Xiumei Liu, Qingren Wang, Junhua Nie. 2002. Research on lead uptake and Tolerance in Six Plants. *Acta Phytoecologica Sinica* (5):533–537.

Xiaoyi Lu, Chiquan He. 2005. Environmental behavior of zinc in *Ricinus communis*. *Environmental Pollution& Control* 27(06):414–419.

Xiaoyi Lu, Chiquan He. 2005. Tolerance, Uptake and Accumulation of Cadmium by *Ricinus communis* L. *Journal of Agro-Environment science* 24(4):674–677.

Mitch M. 2002. Phytoremediation of toxic metals: a review of biological mechanism. *J Environ Qual* (31): 109–20.

Ministry of Environmental Protection of the People's Republic of China. 1995. *GB15618-1995 Environmental Quality Standard for Soils*. Beijing: China Environmental Science Press.

Pulford ID. Watson C. 2003. Phytoremediation of heavy metal-contamineated land by tree-a view. *Environ Int* (29):529–40.

Rongling Qu, Desen Li, Rongqian Du, Mingyao Ji. 2006. Phytoremediation for Heavy Metal Pollution in water II: The blastofiltration of Pb from water. *Agro-environmental Protection* 21(6):499–501.

Xiaoming Ran, Jinzhu He, Qingsong Miao. 2004. Progress in the research of phytoremediation technology of polluted soil. *Chinese Journal of Eco-Agriculture* 12(3):131–133.

Salt DE, Smith RD, Raskin I. 1998. Phytoremediation. *Annu Rev Plant Physiol Plant Mol Biol* (49):643–68.

Xiang Shi, Yitai Chen, Shufeng Wang, Xiaolei Zhang, Yuan Yuan. 2011. Growth and metal uptake of three woody species in lead-zinc and copper mine tailing. *Acta-Ecological Sinica* 31(7):1818–1826.

Dan Sun, Peifang Cong, Zili Cong. 2007. Accumulationas of Heavy Metal in Wheat Plant and the Growth Status Under Pot Experiment with Different Contents of Tailings. *Journal of Agro-Environment Science* 26(2):683–687.

Hongxin Wang, Shaoyi Guo, Feng Hu, Yangyang Jiang, Qunxin Li. 2012. Effects of Chelating Agents on Growth and Lead-Zinc Accumulation of Castor Seedlings Growth in Amended Substrate of Lead-Zinc Tailings. *Acta Pedologica Sinica* 49(3):491–497.

Libao Wang, Ninghua Zhu, Jianhua E. 2010. Effects of heavy metals lead, zinc and copper on young seedling growth of *Cinnamomum camphora* and *Loelreuteria paniculata*. *Journal of Central South University of Forestry & Technology*, 30(2):44–47.

Shuangtao Wu, Xiaofu Wu, Yueli Hu, Shaojing Chen, Jingzhao Hu, Yifei Chen, Ningzi Xie. 2004. Study on soil pollution around Pb-Zn smelting factory and heavy metals hyperaccumulators. *Ecology and Environment*(2):156–157, 160.

Jin Zheng, Wei Kang. 2009. Study on Heavy Metals in *Ricinus Communis* L. from Daye Tonglvshan Copper Mine. *Journal of Huangshi Institute of Technology* 25(1):36–40.

Dongmei Zhou, yujun Wang, Xiuzhen Hao, Huaiman Chen. 2002. Primary Study of Distribution of Heavy Metals in Copper Mines. *Agro-environmental Protection* 21(3):225–227.

Xiaolong Zeng. Research Progress for Resistance and Tolerance Mechanism of Castor. *Chinese Agriculture Science Bulletin* 26(4):123–125.

The acute toxicity of the organic extracts from two effluents of different wastewater treatments

Huijie Zhao, Zhihong Huang & Bo Li
School of nursing, Henan University, Kaifeng, Henan, China

ABSTRACT: The aims of our experiment were to evaluate the removal rate of the Trace Organic Pollutants (TOPs) in the wastewater by using new treatment process through comparing the acute toxicity of the organic extracts from effluents treated by different wastewater treatments. The Luminescent Bacteria Toxicity test (LBT) was used to detect the acute toxicity. Phenol equivalent was used to indicate the acute toxicity of the water samples. The results showed the toxicity decreased after wastewater treatment, which demonstrated the toxicity of the TOPs in the effluents was decreased significantly by using the new process. And it was safer by using the new process to treat the municipal water than the present.

1 INTRODUCTION

Under the serious circumstance of the global shortage of water, one of the most effective approaches to resolve the severe issue was to reuse the water. Today the advanced treatment and reclamation of the municipal wastewaters have already been the applied engineering works worldwide (Bixio et al. 2006). However studies have shown that municipal wastewater contain a variety of poisonous and harmful substances, such as heavy metals, pesticide, plasticizer etc., which can still be detected in the effluent even though after the advanced treatment (Al-Rifai et al. 2007, Singh et al. 2010, Zhang et al. 2011). Among them, the trace organic pollutants (TOPs) have the especially harmful influence on the environment because of the properties of hard to volatilize and degrade (MEPC 2011).

There was an increasing need to minimize the TOPs in reclaimed water through process modification. Oxidation Ditch (OD), the most commonly employed wastewater treatment process, was used by certain wastewater treatment plant (WWTP) in Zhengzhou in the past. For resolving the issues of reusing the wastewater, new process were used to treat the municipal wastewater by the WWTP in Zhengzhou now. The new process include Membrane-enhanced Oxidation Ditch (MEOD) and strengthened phosphorus and nitrogen removal based on recycling the phosphorus at the beginning.

One of the biological testing methods used in toxicology was using luminescent bacteria to detect the toxicity (MEPC 1995). The luminescent bacteria toxicity test (LBT), often being chosen as the first screening method in wastewater quality analysis, is a useful method supplementing chemical analysis and is widely employed in the field of wastewater treatment because of its simple operation, rapid determination, high sensitivity, low lost and certain correlations between chemical analysis and biological toxicity tests (Rosal et al. 2010, Gartiser et al. 2010, Ji et al. 2013). LBT was used to detect the acute toxicity in our experiment.

The aims of our experiment were to evaluate the removal rate of the TOPs in the wastewater by using new treatment process through comparing the acute toxicity of the organic extracts from two effluents treated by different wastewater treatments. This experiment could supply theoretical support on applying the new treatment process.

2 MATERIALS AND METHODS

2.1 *Materials*

2.1.1 *Main instrumentation*

NIKON microscope (YS100, Japan Nikon Co.), high-pressure sterilizer Hirayama (HVE-50, Japan Hirayama Co.), SARTORIUS AG scale (BS224S, German Sartorius Co.), high velocity low-temperature centrifuge (Sigma 3k30, American sigma Co.), spectrophotometer Hitachi (type 2000, Japan Nikon Co.), shaking incubator (Shanghai Yiheng Tech. Co. Ltd.).

2.1.2 *Main chemicals*

Culture medium of luminescent bacterium: $MgSO_4$ 2.47 g, $MgCO_3$ 0.79 g, $MgBr_2$ 0.09 g, $MgCl_2$ 0.09 g, $Mg(HCO_3)_2$ 0.50 g, $CaCO_3$ 0.03 g, KCl 10.22 g, NaCl 8.29 g, yeast extract 5 g, glycerin 3 g, tryptone 5 g, were dissolved into 1000 ml distilled water.

1000 mg/L phenol mother liquors: weigh out crystal phenol 100 mg quickly with analytical balance, instantly put them into 100 ml beaker with 30–50 ml 0.8% NaCl solution£¬and then put into 100 ml volumetric flask after fully dissolved, get constant volume with 0.8% NaCl solution, preserve at 4°C, storage time was 2d.

Phenol standard concentration: suction from the 1000 mg/L phenol mother liquors 0.5 ml, 1.0 ml, 2.0 ml, 4.0 ml, 5.0 ml, 6.0 ml, 8.0 ml, 10.0 ml, 12.5 ml, 15 ml, put them into 50 ml volumetric flask separately, then get constant volume with 0.8% NaCl solution, get 10 mg/L, 20 mg/L, 40 mg/L, 80 mg/L, 100 mg/L, 120 mg/L, 160 mg/L, 200 mg/L, 250 mg/L, 300 mg/L, ten concentration gradients.

Trace organic extractive concentration: suction 1ml trace organic extractive oil solution of the primary water, secondary water, tertiary water and the water using the new process and dissolve them into 4 ml DMSO, diluted 10 times with fresh water, that was DMSO : raw water = 1 ml : 100 ml.

2.2 *Methods*

2.2.1 *Collection of water sample*

The municipal wastewater (40 L), secondary effluent (200 L), tertiary effluent (200 L) and effluent using new process (200 L) were collected from a WWTP in Zhengzhou. Samples were put in the brown glass bottles, preserve below 4°C, dark area. All samples must be treated within 7d, tested and analyzed within 40d (Lin et al, 2004).

2.2.2 *Experimental methods*

Revive and cultivate freeze drying powder of luminescent bacterium: For the acute toxicity detect, *Vibrio-qinhaiensis sp.-Q67* was purchased from Beijing Hammatsu Photon Techniques Inc. The bacteria strains from a stock culture medium which was maintained at 4°C were inoculated to a liquid medium. The bacteria were grown in the liquid medium up to the logarithmic growth stage after 24h under 28°C while shaking at 2000 rpm.

Detect the toxicity: The analytic methods were chosen according to the literature (Liu et al, 2006). Put 100 μl samples into the wells of a sterile 96-well plate. The first line 12 wells added into 0.8% NaCl was designed as control group. Then, the other wells were added into 100 μl culture with bacteria. Put the plate into shaker for 20~35 s, stand it under ordinary temperature for 15 min and move into microplate spectrophotometer reader to record the Light-emitting Intensity. Every well measured thrice, it was the average that was the value of measurement. Each sample was determined 3 times. Light-emitting Inhibition Rate was calculated on Formula (1)

$$A = \frac{B-C}{B} \times 100\% \tag{1}$$

where A was Sample Light-emitting Inhibition Rate, B was Control Intensity of Light, C was Sample Intensity of Light.

Detects estimate: In order to quantitatively compare the toxicity of different water samples, the concentration (times) corresponding to the inhibition value of 50%, namely effective concentration (EC50) was used. By definition, the higher the EC50 value, the lower the toxic effect. However, if the relative intensity of light of the raw water >50%, the toxicity can only use the corresponding phenol concentration to express because of not testing the EC50 value.

2.2.3 *Experimental groups*

DMSO was used as blank control, as the first group, DMSO dissolved into corn oil was the second group, effluent using new process was the third group, the tertiary effluent was the fourth group, the secondary effluent was the fifth group, the raw wastewater was the sixth group.

2.2.4 *Phenol standard curve*

When testing sample, same bacteria solutions were added into the phenol standard concentration solution and was determined following the step above. The EC50 was calculated out by equation of linear regression that was listed as Formula (2) below. This method requested $p < 0.90$ and that EC50 should vary from 100 mg/L to 200 mg/L.

$$I = a + bC \tag{2}$$

where I was Light-emitting Inhibition Rate, C was the phenol concentration.

2.2.5 *Statistical analysis*

Statistical analysis was performed by ANOVA (SPSS 12.0). Statistical tests with P less than 0.05 was considered significant.

3 RESULTS

Phenol concentrations were used to indicate the acute toxicity of the water samples, the results showed the toxicity decreased after wastewater treating, as showed in stable 1. That demonstrated the toxicity of the TOPs in the effluent by using the new process decreased, which implied it is safer by using the new process to treat the municipal water.

3.1 *Phenol standard curve Table 1*

Table 1. Light-emitting inhibition rate of Phenol standard solution.

Concentration gradient (mg/L)	Light-emitting Inhibition Rate $\bar{x} \pm S$ (%)
10	7.24 ± 5.58
20	12.73 ± 7.93
40	21.59 ± 6.48
80	28.87 ± 7.64
100	30.57 ± 8.49
120	34.32 ± 7.36
160	47.33 ± 6.42
200	53.78 ± 4.83
250	68.96 ± 7.51
300	79.64 ± 6.03
F	923.464
p	0.000
R^2	0.991
regression equation	$I = 7.880 + 0.239C$

3.2 Light-emitting inhibition rate of organic extracts of water samples Table 2

Table 2. Light-emitting inhibition rate of organic extracts of water samples.

Groups	n	Light-emitting Inhibition Rate $\bar{x} \pm S$ (%)	Corresponding phenol concentration (mg/L)
1	3	8.21 ± 0.081	1.38
2	3	7.97 ± 0.013*	0.38
3	3	9.56 ± 0.060*#	7.03
4	3	10.58 ± 0.085*#△	11.30
5	3	11.53 ± 0.060*#△•	15.27
6	3	14.02 ± 0.034*#△•◊	25.69

$F = 222.535$, $P < 0.001$, the difference between all groups was significant.
*The difference between No. 1 group and others.
#The difference between No. 2 group and others.
△The difference between No. 3 group and others.
•The difference between No. 4 group and others.
◊The difference between No. 5 group and others.

4 DISCUSSION

At present, the approaches of using luminescent bacterium to detect the toxicity were applied to the wastewater treatment widely (Li et al., 2009). T3 luminescent bacteria were used as toxic indicator to detect the removal of the TOPs by using the active sludge technology in WWTP. This study found it was not ideal to remove the TOPs by this way (Ma et al. 1999). Photobacterium phosphoreum was used to detect the acute toxicity in the reclaimed water of a city in Northern China. The study showed the acute toxicity was enhanced in the raw reclaimed water after sterilizing by adding chlorine and ozone (Liu et al. 2011). *Vibrio-qinhaiensis sp. -Q67* was used to detect the acute toxicity of the influent and effluent water of seven sewage treatment plants, and the results showed the toxicity reduced significantly after treating and the luminescent bacterium tests were the effective supplement to the chemical analysis (Li et al. 2010). The comprehensive toxicity of influent and effluent water in sewage treatment plants with three different treatment processes in Shanghai was tested. The results showed that the bioassay based on luminescent bacteria test method in testing the toxicity of municipal wastewater was better than the other two methods. The results also demonstrated that the toxicity of raw municipal wastewater could be reduced by traditional activated sludge process, anoxic/acrobic process and uni-tank process respectively. Gas Chromatography-Mass Spectrometry (GC-MS) analysis showed that the refractory organic substance might be one of the factors causing the toxicity of effluents (Huang et al. 2005).

Based on relevant literatures (Ma et al. 1999) and the results of the pre-experiments, we diluted the TOPs of different water samples to 1ml, which concentration was equal to 100 ml raw water. That means we concentrated the TOPs in raw water to 100, which can make sure the lighting intensity of the luminescent bacterium meet the testing demands. Make the comparisons of the lighting inhibition rate of the luminescent bacterium among different water samples. We found that the difference between all groups was significant ($F = 222.535$, $P < 0.001$). The lighting inhibition rates of the effluent of by using new process, secondary effluent and tertiary effluent were lower than the rate of the influent, which demonstrates the toxicity was reduced significantly after secondary and tertiary treatment. The results also showed the biodegradable substance contributed to the most toxicity in the raw water. The significant differences existed among the different effluents, which showed it can remove the TOPs better by using the new process. The lighting inhibition rate of the tertiary effluent was less than that of the secondary effluent, which was a little different with the research results by Jiafei Liu (Liu et al. 2011). The author analyzes the reasons were the research subjects were the raw water of the different effluents, the interval of the sampling was short, which can result in the disinfection by-products left more, the action of the inorganic elements still had. However the water samples we used in our experiment extracted by XAD2 resin, and we just studied the TOPs in the water samples, so the results were different.

Phenol equivalent was used to indicate the acute toxicity of the water samples, the results showed the toxicity decreased after wastewater treating. As showed in stable 1, the phenol equivalent of the secondary effluent was 15.27 mg/L, the phenol equivalent of the tertiary effluent was 11.30 mg/L, and that of the effluent by using new process was even more low 7.03 mg/L. We needed to point out that the corresponding phenol equivalent could not reflect the direct toxicity of the TOPs in the water samples. The TOPs were hardly soluble in water but soluble in corn oil. So the auxiliary solvent must be used to dissolve the TOPs. However, the auxiliary solvent itself had the toxic action to the luminescent bacterium and inhibited its lighting rate (Liu et al. 2007). DMSO was chosen as the auxiliary solvent based on the past study results, because which is less toxic to the luminescent bacteria (Liu et al. 2007, Dong 2006). The toxic action of DMSO was included in the phenol equivalent in the experiment, so the results only reflected the acute toxicity of the TOPs indirectly. The real toxicity was less than the experimental results.

The bioassay based on luminescent bacteria test method was reliable relatively. However, it was hard to make sure the results not affected by some factors, because the components of the surveyed objects were very complicated. In addition, the interference factors existed, such as the salinity, turbidity, pH and the differences of the components of the culture medium, which could affect the results (Zhao 2010, Yu 2005). The bioluminescent bacteria toxicity test could be used to assess the toxicity of untreated wastewater and treated wastewater, the toxicity of some specified compounds, and the toxicity removal efficiency of treatment and then provide important supporting information for the toxicity assessment, the impact of effluent on the environment, and the reconstruction of the treatment processes. Other experiments, such as multifaceted, multi-species experimental designs, should be done to evaluate the acute toxicity of the surveyed objects comprehensively. Meanwhile, the application and development of bioluminescent bacterial in the field of wastewater treatment in the future are prospected.

5 CONCLUSION

The toxicity of the TOPs was decreased significantly by using the new process to treat municipal water in the WWTP and its efficiency was beyond the present.

ACKNOWLEDGEMENT

This work was supported by the Stage Key Project "water Pollution Control" (2009ZX07210-001-001).

REFERENCES

Al-Rifai J H, Gabelish C L, Schafer A I. 2007. Occurrence of pharmaceutically active and non-steroidal estrogenic compounds in three different wastewater recycling schemes in Australia. Chemosphere. 69: 803–815.

Bixio D, Thoeye C, De Koning J, et al. 2006. Wastewater reused in Europe. Desalination. 187: 89–101.

Dong Yuying, Lei Bingli, Ma Jing, et al. 2006. Influence of cosolvents of *Photobacterium phosphoreum* toxicity test. Journal of Chemical Industry and Engineering (China). 57(3): 636–639.

Gartiser S, Hafner J, Hercher C, et al. 2010. Whole effluent assessment of industrial wastewater for determination of BAT compliance. Part 2: metal surface treatment industry. Environ Sci Pollut Res. 17: 1149–1157.

Huang Manhong, Li Yongmei, Gu Guowei. 2005, Application of several bioassays on evaluating toxicity of municipal wastewater. Journal of Tongji University (natural science). 33(11): 1489–1493.

Ji Junyuan, Xing Yajuan, Ma Zitao, et al. 2013. Toxicity assessment of anaerobic digestion intermediates and antibiotics in pharmaceutical wastewater by luminescent bacterium. Journal of Hazardous Materials. 246-247: 319–323.

Li Xuemei, Ke Zhenshan, Du Qing, et al. 2009. Application of bioluminescent bacteria in wastewater treatment. China Water & Wastewater. 25(16): 1–5.

Li Xuemei, Ke Zhenshan, Du Qing, et al. 2010. Research on the toxicity of influent and effluent of a wastewater treat plant by using luminescent bacterium. 36(1): 130–134.

Lin Xingtao, Wang Xiayi, Chen Ming, et al. 2004. Analysis of phthalic acid esters of environmental hormone in water using solid phase extraction and high performance liquid chromatography. Research of Environmental Sciences. (6): 79–81.

Liu Baoqi, Ge Huilin, Liu Shushen. 2006. Microplate luminometry for toxicity bioassay of environmental pollutant on a new type of fresh water luminescent bacterium (*Vibrio-qinhaiensis sp.-Q67*). Asian Journal of Ecotoxicology. 1(2): 186–191.

Liu Jiafei, Zhang Yu, Yan Zhiming, et al. 2011. Toxicity evaluation of reclaimed wastewater in a city in North China by using luminescent bacterium and Daphnia magna. Chinese Journal of Environmental Engineering 5(5): 977–981.

Liu Shushen, Liu Fang, Liu Hailing. 2007. Toxicities of 20 kinds of water-soluble organic solvents to *vibrio-qinghaiensis sp, Q67*. China Environmental Science. 27(3): 371–376.

Ma Mei, Wang Yi, Wang Zijian. 1999. Changing ruls of concentration and toxicity of organic pollutant in the process of treating municipal wastewater. 19(6): 9–12.

Ministry of Environmental Protection of the People's Republic of China. 1995. Water quality-Determination of the acute toxicity-Luminescent bacteria test (GB/T 15441-1995).

Ministry of Environmental Protection of the People's Republic of China. 2011-01-18. 2010 Yearbook environmental statistics [EB/OL]. http://zls.mep.gov.cn/hjtj/nb/2010tjnb/201201/t20120118_222727.html.

Rosal R, Rodea-Palomweres I, Boltes K, et al. 2010. Ecotoxicity assessment of lipid regulators in water and biologically treated wastewater using three aquatic organisms. Environ Sci Pollut Res.17: 135–144.

Singh S P, Azua A, Chaudhary A, et al. 2010. Occurrence and distribution of steroids, hormones and selected pharmaceuticals in South Florida coastal environments. Ecotoxicology. 19: 338–350.

Yu Ruilian, Hu Gongren. 2005. Toxicity and QSAR research of luminescent bacterium at different PH. Environmental Science & Technology. 28(4): 20–22.

Zhang X, Zhao X, Zhang M, et al. 2011. Safety evaluation of an artificial groundwater recharge system for reclaimed water reuse based on bioassays. Desalination. 281: 185–189.

Zhao Yangyong, Hu Jianlin, Shao Lijun. 2010. Study on the affection of luminescent bacteria toxicity test. Modern Scientific Instruments. 3: 75–78.

Aerobic degradation of highly chlorinated biphenyl by *Pseudomonas putida* HP isolated from contaminated sediments

Chengyu Wang, Juntao Cui, Jihong Wang & Shumin Cui
College of Resources and Environmental Sciences, Jilin Agricultural University, Changchun, PR China

Yucheng Sun
Institute of Military Veterinary Medicine, Academy of Military Medical Sciences, Changchun, PR China

ABSTRACT: A polychlorinated biphenyls (PCBs) degrading bacterium HP was isolated using PCBs as sole carbon source and energy source from contaminated sites. Phenotypic features, physiological and chemotaxonomic characteristics, and phylogenetic analysis of 16s rDNA sequence revealed that the isolate belongs to the genus of *Pseudomonas*. The results of Detection of *bph*A1 gene of HP indicate that the isolate may express this gene. Strain HP could also degrade some PAHs and other environmental pollutants when provided as sole carbon and energy sources. Gas chromatography analysis of individual congeners in Aroclor 1260 following a 4-week incubation showed 78.3% degradation of PCBs with the need for yeast extraction as a substrate.

1 INTRODUCTION

Polychlorinated biphenyls (PCBs) have been shown to cause cancer in human and animals (Mayes et al. 1998), and also to cause a number of serious effects on the immune function, reproductive, nervous and endocrine systems, among others (ATSDR, 2000; Aoki, 2001; Faroon et al., 2001).

Several methods, including physical and chemical processs for the removal of PCBs in the environment, have been reported (Wirtz et al., 2000; Betterton et al., 2000; Jones et al., 2003). Microbial degradation of PCBs has been extensively studied in recent years and is regarded as one of the most effective procedures. Many PCB-degrading bacteria have been isolated (Sakai et al., 2005; Abramowicz, 1990). Lowly chlorinated biphenyls are usually aerobically degraded via co-metabolism, requiring biphenyl as a growth substrate to induce the requisite enzymes. However, biphenyl is also toxic and not easily dispersed in contaminated soils or sludges. Therefore, the isolation of bacteria growing on PCBs contaminated environment, without the need for biphenyl as a primary substrate, may solve this secondary pollution problem (Adebusoye et al., 2008)

Aerobic degradation of PCBs though have been reported to be limited to less chlorinated biphenyls (PCBs). We isolate a PCB-degrading strain, isolated from PCBs contaminated sites, named *Pseudomonas putida* HP. This bacteria that is able to efficiently utilize highly chlorinated PCBs as its sole source of carbon and energy in the absence of biphenyl.

2 MATERIALS AND METHODS

2.1 Isolation and identification of PCB-degrading bacteria

PCBs degrading bacterium, using Aroclor 1260 as sole carbon source and energy source, was isolated from activited sludge obtained from the wastewater treatment plant of PetroChina Petrochemical Company, China. Enrichment culture was evidenced by visual increase in turbidity and a color change in the medium. Physiological and chemotaxonomic characteristics were performed using the methods described in Bergey's Manual of Determinative Bacteriology (Holt et al., 1994) for the identification of strain HP. And phylogenetic analysis of 16s rDNA sequence were identified.

2.2 Detection of PCB degraders by Amplification of the gene bphA1 using specific primers

Aerobic PCB degraders were detected by biphenyl 2,3-dioygenase gene (*bph*A1) identification carried out by PCR method using specific primers (A1f: 5'-CTTGGGCACGAGAGTCATGTGC-3' and A1r: 5'-TCAGGGCTTGAGCGTGGCCCAGCT-3').

2.3 Growth and PCB degradation experiments

The influence of initial pH value, salt concentration and temperature on the PCBs degradation was evaluated by GC analyses. All experiments were performed in triplicates. Growth was evaluated via visual monitoring of turbidity, in conjunction with GC analyses, to measure the disappearance of test compounds.

2.4 Analytical methods

The PCB measurements were performed by gas chromatograph (Agilent Technologies 6890N). The temperature program was following: the initial temperature was 120°C for 0.5 min and then increasing by 15°C/min up to 200°C, then lasting for 2 min and then immediately increased to 280°C (rate 20°C/min) and then kept at this temperature for 10 min.

PCB-degrading activity was evaluated based on the reduction of the peak area of each congener in GC-ECD chromatograms of the culture extracts compared to the peak areas in the chromatograms of the control sample extracts (with autoclaved culture). Each value represents the mean of triplicate repeats with a standard deviation less then 15%.

3 RESULTS AND DISCUSSION

3.1 Visualization of the presence of biphenyldioxygenase in the isolated

There are many screening methods to detect the bacterial using PCBs, which enable to confirm the presence of biphenyldioxygenase, the first and one of the most important enzymes of the biodegradation pathway of PCBs. In this way, we detected bacteria producing dioxygenase, which indicates their potential to degrade PCBs. The enrichment medium turned yellow after cultivation. On the basis of the performed positive screening, we have continued in the isolation of the pure strain named HP, which was subsequently submitted to genetic analysis for the presence of the *bph*A1 gene encoding the enzyme biphenyldioxygenase.

3.2 PCR amplification of bphA1 gene

We have based our experiments on the fact that biphenyldioxygenase determines the specificity of PCB cleavage. The *bph*A1 of strain HP was submitted to the GenBank (accession number: JN559868). In Figure 1, the results indicate that the isolate may express this gene. Sequences of *bph*A1 previously isolated were also retrived from GenBank database and included in the phylogenetic analysis.

3.3 Identification of isolated PCB-degrading bacteria

After about six months of enrichment culture, a strain bacterium able to use PCBs as the sole carbon and energy sources was isolated by the appearance of intensive yellow colour. This strain was identified as a Gram-negative, catalase-positive, oxidase-positive, non-fermentative and soluble pigment, rod-shaped. The 16s rDNA of strain HP was submitted to the GenBank (accession

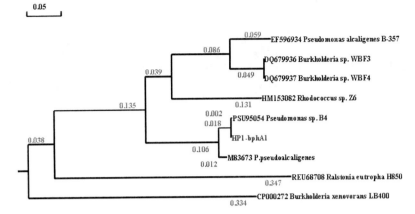

Figure 1. Phylogenetic relationship between strain HP and other bacteria in the same genera and previously isolated bacteria degrading PCBs based on *bph*A1 gene sequence analysis.

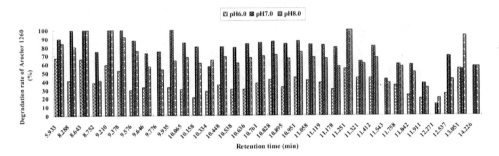

Figure 2. The influence of initial pH value on Aroclor 1260 degradation.

number JN559867). The sequence showed a 97% *Pseudomonas putida* SY45 (GenBank accession number FJ494693) identity. The nucleotide sequence was compared with those of known *bph*A1 gene sequences. The sequence showed a 77.43% *Pseudomonas* sp. B4 (GenBank accession number PSU95054) identity. Moreover, this analysis (16s rDNA) was in strong agreement with morphological identification.

From the above results, the PCBs-degrading bacterium was identified as Pseudomonas putida HP.

3.4 *Effect of pH on the biodegradation of PCBs*

Residue Aroclor 1260 was observed on week 4 when pH values ranged from 7–8, as shown in Figure 2. These data illustrate that strain HP is adaptable to pH7-8 culture fluids. The biodegradation of PCBs was almost inhibited in relatively acidic (pH5.0) and alkaline (pH 9.0) conditions.

3.5 *Effect of biphenyl and yeast extraction on the biodegradation of PCBs*

Analysis of the degradation of individual PCB congeners in Aroclor 1260, both in the presence and absence of biphenyl and yeast extraction (Figure 3). Percent reduction after 4 weeks in the presence and absence of biphenyl or yeast extraction for total PCB content were 34.5%, 68.8% and 78.3%, respectively. This results show that the isolate was able to grow using PCBs as sole sources of carbon and energy. Our result also show that the addition of yeast extract to increase the PCB degradation rate is more efficiently than the addition of biphenyl.

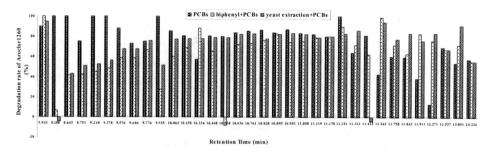

Figure 3. Analysis of the degradation rate of Aroclor 1260 components by the strain HP.

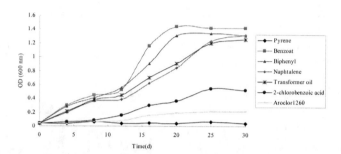

Figure 4. Growth of HP on different carbon sources.

3.6 Substrate diversity of the bacterial strain

Growth curve of the strain in LB and MSM with yeast extraction were compared with growth in seven different aromatic compounds (Figure 4). In the LB medium, strain HP entered the stationary phase of growth after 26 h with OD = 2.33 (data not shown). With a MSM and 100 mg benzoat, the maximum OD was 1.44. This strain entered the stationary phase of growth after 25 d, the OD was 1.31, 1.23 and 1.25 for media supplement with biphenyl, naphthalene and 45[#] transformer oil, respectively. After 28 d of growth, the OD was 0.52 and 0.21 for media supplemented with 2-chlorobenzoic acid and Aroclor 1260, respectively. However, none growth was discoved in a pyrene-supplemented medium. Therefore, the capability of the new isolate to grow on PCBs was found to be outstanding and unique.

The growth of this isolate using PCB congeners, PAHs and other environmental pollutants as sole carbon sources is analyzed. This strain utilized naphtalene, phenanthrene, anthrone, chlorobenzene and benzdine, but was not observed to grow on pyrene, anthrone, benzophenone and quintozene. It has been noted that PCB pathway enzymes may not only transform PCBs and their metabolites, but also several xenobiotics, including PAHs. In other words, strain HP is promising for the removal of various chlorinated compounds, including PCBs and PAHs, from contaminated sites. These results reinforce the conclusions of Arnett et al. (2000), Adebusoye et al. (2007) and Ashrafosadat et al. (2009).

3.7 Degradation of PCBs Aroclor 1260

Degradation of Aroclor 1260 was evaluated using washed, biphenyl-grown cells. No carbon source other than the commercial PCB mixture were provided. Growth on this mixture was evidenced by a significant reduction in PCB substrate concentration and a color change of the media culture to yellow. Analysis of the transformation of individual PCB congeners in Aroclor 1260 (Figure 5).

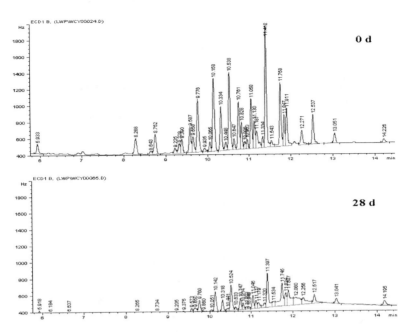

Figure 5. Gas chromatograms showing PCBs degradation by HP.

4 CONCLUSIONS

In this research, strain HP belongs to *Pseudomonas*. It can obviously degrade highly chlorinated biphenyl, at pH7 and it is a promising microorganism for PCBs and PAHs bioremediation.

REFERENCES

Abramowicz D A, 1990. Aerobic and anaerobic biodegradation of PCBs: a review. *CRC Critical Review Biotechnology*, 10:241–251.

Adebusoye S A, Picardal F W, Ilori M O, Amund O O, Fuqua C, Grindle N, 2007. Growth on dichlorobiphenyls with chlorine substitution on each ring by bacteria isolated from contaminated African soils. *Applied Microbiology and Biotechnology*, 74:484–492.

Aoki Y, 2001. Polychlorinated biphenyls, polychlorinated dibenzop-dioxins, and polychlorinated dibenzofurans as endocrine disrupters—what we have learned from Yusho disease. *Environmental Research*, 86:2–11.

Arnett C M, Parales J V, Haddock J D, 2000. Influence of chlorine substituents on the rates of oxidation of chlorinated biphenyls by the biphenyl dioxygenase of *Burkholderia* sp. strain LB400. *Applied Environmental Microbiology*, 66:2928–2933.

Ashrafosadat, 2009. Extensive biodegradation of highly chlorinated biphenly and Aroclor 1260 by *Pseudomonas aeruginosa* TMU56 isolated from contaminated soils. *International biodeterioration and biodegradation*, 63:788–794.

ATSDR, 2000. Toxicological profile for polychlorinated biphenyls (PCBs). Agency for Toxic Substances and Disease Registry, United States Department of Health and Human Services, Public Health Service, Atlanta, GA

Betterton E A, Hollan N, Arnold R G, Gogosha S, Mckim K, Liu Z J, 2000. Acetone-photosensitized reduction of carbon tetrachloride by 2-proponal in aqueous solution. *Environmental Science and Technology*, 34:1229–1233.

Faroon O, Jones D, de Rosa C, 2001. Effects of polychlorinated biphenyls on the nervous system. *Toxicology and Industrial Health*, 16:305–333.

Holt J G, Krieg N R, Sneath P H A, Stanley J T, Williams S T, 1994. Bergey's Manual of Determinative Bacteriology, ninth ed. Williams and Wilkins Co, Baltimore.

Jones C G, Silverman J, Al-sheikhly M, 2003. Dechlorination of polychlorinated biphenyls in industrial transformer oil by radiolytic and photolytic methods. *Environmental Science and Technology*, 37:5773–5777.

Mayes B, McConnell E, Neal B, Brunner M, Hamilton S, Sullivan T, Peters A, Ryan M, Toft J, Singer A, Brown J, Menton R, Moore J, 1998. Comparative carcinogenicity in Sprague-Dawley rats of the polychlorinated biphenyl mixtures aroclors 1016, 1242, 1254, and 1260. Toxicology Science, 41:62–76.

Miller S A, Dykes D D, Polesky H F. 1988. A simple salting out procedure for extracting DNA from Human nucleated cells. *Nucleic Acids Research*, 16:1215.

Sakai M, Ezaki S, Suzuki N, Kurane R, 2005. Isolation and characterization of a novel polychlorinated biphenyls-degrading bacterium, Paenibacillus sp. KBC101. *Applied Microbiology and Biotechnology*, 68:111–116.

Wirtz M, Klucik J, Rivera M, 2000. Ferredoxin-mediated electrocatalytic dehalogenation of haloalkanes by cytochrome p450cam. *Journal of Chemistry and Society*, 122:1047–1056.

Research on transport and diffusion of suspended sediment in reclamation project

Xiaofeng Guo, Chuhan Chen & Junjian Tang
Third Institute of Oceanography, SOA, Xiamen, Fujian, China

ABSTRACT: Suspended sediment in the sea increases rapidly as a result of reclamation projects, which in turn influence the marine eco-system. To investigate the possible ecological impact of suspended sediment produced during the construction of Fengwei reclamation project at Meizhou Bay, this research analyzes the suspended sediment extrusion by blast (5.17 t/s) and riprap (1.39 kg/s) respectively. Based on two-dimensional tidal current field model of the project area, the study analyzes each mesh point (40 m × 40 m) in the construction area using two-dimensional suspended load modal. The numerical simulation shows that the scope and pattern of suspended sediment diffusion are affected by the tidal current. The project construction can cause the suspended sediment whose density being more than 10 mg/L to form a maximum envelop area of 5.848 km^2 in tidal current. This impact mostly disappears 9 hours after the construction. The result provides a reference for the environmental management during the construction period.

1 INTRODUCTION

With the skyrocketing development of marine economy, offshore land resources become rarer. Reclamation projects bring about huge benefits while changing the original ecological environment and resulting in many problems for ecological environment. The construction process of reclamation projects leads to rapid increase of suspended sediment in the sea. The original water dynamic condition is changed, reducing the diffusion capability of pollutants and then influencing the surrounding marine ecological system. Therefore, the study on the transport and diffusion of suspended sediment of reclamation project holds great importance for regional ecological environment quality and comprehensive ecological influence assessment of project construction. Currently, for the suspended sediment diffusion during construction period, simplified two-dimensional diffusion equation is generally adopted to calculate its envelope area. This method cannot simulate the diffusion coverage figure of suspended sediment visually and its calculation accuracy is quite low. Huang Huiming (2008) and Lou Haifeng (2009) set representative points in project construction area, and adopt the method of numerical simulation to conduct initial exploration on suspended sediment diffusion by blast. Envelope figures of representative points in construction area are collected to gain the envelope coverage figure of suspended sediment for the constructed project. As construction coverage is a concept of area, this way in which the representative points are calculated still deviates from the actual construction process of project. Thus, the calculated result does not match the reality well. Through the calculation of source intensity of sediment extrusion by blast and riprap and based on the mathematic model of two-dimensional tidal current field in the project sea area, the paper adopts two-dimensional suspended sediment mathematic model to calculate the model mesh points in construction area point by point and conducts constructive exploration on the precise simulation of suspended sediment diffusion in construction.

Meizhou Bay is located along the coast in the middle of Fujian Province. It is a long and narrow bay penetrating into the land. The length from the north to south is about 35 km and the average width from east to west is more than 15 km. Fengwei reclamation project at Meizhou Bay is situated at Fengwei Town, Quangang District, which is on the southern bank of Meizhou Bay. The planned

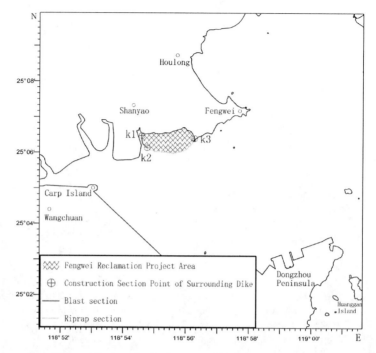

Figure 1. Location of Fengwei Reclamation Project at Meizhou Bay.

reclamation area covers 2.12 km². The total length of surrounding dike is 3,675 m, located in the south of the planned sea area and lying in a shape of arc from east to west. The construction of reclamation project is arranged in the sequence of first forming the surrounding dike and then filling with earth and stones. According to different thickness of sediment on sea bottom in different sections of construction, the extruded sediment in K1—K2 section of dike employs the construction process of blast while sediment extrusion by the weight of stones is adopted in K2—K3, seen in Fig. 1. Suspended sediments of project construction are general from sediment extrusion by blast and riprap.

2 RELATED MATHEMATIC MODELS

The numerical simulation of suspended sediment in construction includes two steps. First, the mathematic model of tidal current field in project sea area is established. Then the numerical simulation of suspended sediment diffusion follows.

2.1 *Mathematic model of tidal current field*

The water in Meizhou Bay is open. Tidal current field simulation selects two-dimensional shallow tidal wave equation:

$$\frac{\partial u}{\partial t}+u\frac{\partial u}{\partial x}+v\frac{\partial u}{\partial y}=-g\frac{\partial z}{\partial x}+f\cdot v-r\cdot u+A_x(\frac{\partial^2 u}{\partial x^2}+\frac{\partial^2 u}{\partial y^2}) \quad (1)$$

$$\frac{\partial v}{\partial t}+u\frac{\partial v}{\partial x}+v\frac{\partial v}{\partial y}=-g\frac{\partial z}{\partial y}-f\cdot u-r\cdot v+A_y(\frac{\partial^2 v}{\partial x^2}+\frac{\partial^2 v}{\partial y^2}) \quad (2)$$

$$\frac{\partial z}{\partial t}+\frac{\partial}{\partial x}[(d+z)u]+\frac{\partial}{\partial y}[(d+z)v]=0 \quad (3)$$

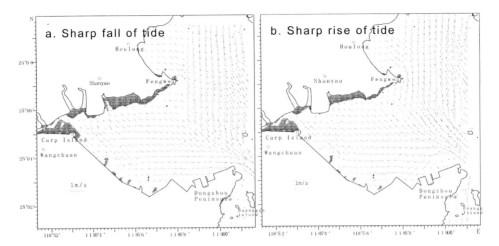

Figure 2. Typical tidal current field at Meizhou Bay.

In the equations (1–3): $r = g\sqrt{u^2+v^2}/c_n^2(H+H_1 \cdot e^{-P \cdot H})$, $H = d+z$, t is time (s); x and y are the plane rectangular coordinates; u and v are components of full flow along x and y direction respectively; H is water depth (m); d is the water depth (m) below average water level; z is instantaneous water level (m); f is Coriolis parameter; r is bottom friction factor; g is acceleration of gravity; c_n is Chezy coefficient; $c_n = H^{1/6}/n$, n is Roughness coefficient of sea bottom; A_x and A_y are viscosity coefficients of horizontal movement.

Semi-implicit finite difference method proposed by Vincenzo Cassulli is taken in the solutions of equations (1–3), calculated with 40 m × 40 m square mesh and 20 s time step. The verification of tidal currents for sharp rise and fall of spring tide at Meizhou Bay is shown in Fig. 2a and 2b. The shadow part represents the exposed beach. For sharp fall of tide, the falling tidal current flowing downward from the top of the bay in the north enters the main channel through Xiuyu Islet, and merges with the current from the hillside. Then it is divided into two branches after passing through Dongzhou Peninsula-Dongwu section. One flows out of Southern Entrance through the channel in the east of Dongzhou Peninsula; the other one with a higher velocity of flow has its outlet out of Eastern Entrance from Wenjia Entrance. For sharp rise of tide, current outside the Bay enters Meizhou Bay through Wenjia Entrance and Southern Entrance. Then, the trend of flood tidal current is generally opposite to that of falling tide.

2.2 *Mathematic model of suspended sediment transport and diffusion*

After entering the sea during construction process, the suspended sediment diffuses, transports, and settles down under the role of water dynamics to form concentration field of suspended sediment. While transporting and diffusing after entering the sea, the sediment settles downward until it is solidified on sea bottom. In consideration of the diffusion and sedimentation and the reciprocating action of tidal current, the physical process is represented by convection and diffusion equation. The sea area under construction is inside the Bay with shallow water whose depth is generally lower than 10 m in the diffusion coverage of sediment. As the vertical effect is not prominent, two-dimensional convection and diffusion equation can be used for simulation:

$$\frac{\partial (S \cdot D)}{\partial t} + \frac{\partial (S \cdot U \cdot D)}{\partial x} + \frac{\partial (S \cdot V \cdot D)}{\partial y} = \frac{\partial}{\partial x}(D \cdot K_x \cdot \frac{\partial S}{\partial x}) + \frac{\partial}{\partial y}(D \cdot K_y \cdot \frac{\partial S}{\partial y}) - \alpha \cdot \omega_g \cdot S + Q \quad (4)$$

In the equation, S is the average sediment content in the vertical line (kg/m³); D is water depth (m); Q is source intensity; ω_g is the average sedimentation velocity of suspended sediment (m/s); α is the sedimentation probability of suspended sediment; K_x and K_y are diffusion coefficients of sediment.

The dynamic factors related to suspended sediment diffusion in suspended sediment transport model include average sedimentation velocity ω_g, sedimentation probability α, and diffusion coefficients of sediment K_x and K_y. The selection of parameters related to dynamic factors is explained respectively in the following:

① Average sedimentation velocity of suspended sediment ω_g: there is no much difference for the average median particle diameters of suspended sediment in different places of hillside sea area under construction. It is fine sediment particles with a general value of about 0.01 mm under the conditions of coastal waters, when salinity is higher than 10–12, sediment flocculation sedimentation velocity approaches the normal value. Its average limit sedimentation velocity is within 0.015–0.06 cm/s. for the calculation, 0.0004 m/s is taken for the average sedimentation velocity of suspended sediment ω_g.

② Sedimentation probability of suspended sediment α: sedimentation probability of sediment, also called recovery saturation coefficient, is a parameter to reflect the recovery velocity where non-equilibrium sediment transport sediment content in the water approaches to the capability of carrying sediment. It is calculated with the following equation:

$$\alpha = 2\varphi(\frac{\gamma' \cdot \omega_s}{\sigma}) - 1 \tag{5}$$

In the equation, $\varphi(\frac{\gamma' \cdot \omega_s}{\sigma})$ is probability function; $\sigma = 0.033 u_*$ is the mean square deviation of vertical pulsation velocity; u_* is friction velocity.

③ Sediment diffusion coefficient K_x, K_y: it is to reflect the diffusion velocity of suspended sediment in water; the sediment content in the project sea area is lower. The sediment diffusion velocity is related to the local water flow velocity. It is typically calculated as follows:

$$K_x = \beta \cdot |u| \cdot d^{0.8333}, K_y = \beta \cdot |v| \cdot d^{0.8333}, \beta = 0.3$$

3 SIMULATION OF SUSPENDED SEDIMENT TRANSPORT AND DIFFUSION

3.1 Source intensity of sediment entering the sea

3.1.1 Source intensity of suspended solids produced by sediment extrusion by blast for surrounding dike

Sediment extrusion by blast is a method to remove sediment on sea bottom by blast. Meanwhile, the thrown stones take advantage of their own weight to move into the sediment to achieve the replacement of sediment with stones. Section K1—K2 of the surrounding dike adopts the construction process of sediment extrusion by blast. Due to the action of extrusion, only part of sediment enters the sea. In light of the blast experience of the Third Harbor Engineering Bureau, the ratio of sediment entering the sea is taken as 20% of the extruded sediment each time. Among the sediment entering the sea, particles of larger diameter will settle down on sea bottom rapidly. On the other hand, sediment of fine particles (mainly referring to particles with diameter lower than 0.063 mm) suspends in the water and moves with the tidal current to influence sea water quality. For the project sea area, sediment particles with diameter lower than 0.063 mm account for about 20%. The sediment ratio is taken as 1.8 t/m^3, and the average natural water content ratio is about 48.6%. The advanced dike length for each sediment extrusion by blast in the project is about 5–7 m. For the calculation, it is taken as 6 m. The amount of sediment extruded by blast per linear meter is 465.9 m^3. Then the source intensity of suspended sediment at the moment of entering the sea during the sediment extrusion by blast in this project is: 5.17 t/s.

3.1.2 Source intensity of suspended sediment generated by sediment extrusion by riprap for surrounding dike

Sediment extrusion by riprap is to throw a certain amount of large rocks into the bottom of soft foundation. It is to extrude the sediment from the foundation coverage in order to strengthen the foundation. Section K2—K3 of the surrounding dike employs the construction process to extrude sediment by the weight of riprap. On one hand, this method brings fine sediment particles into the water to increase the concentration of suspended sediment; on the other hand, suspended particles are also produced during the process of removing sediment by extrusion of riprap. For the former one, the construction of surrounding dike uses large block of rocks weighted 50–100 kg for throwing and compaction. There is few fine sediment particles. Thus, the sediment brought directly by riprap into the water is not counted in.

Due to extrusion, the amount of sediment entering the sea during sediment extrusion by riprap is 5% of the extruded sediment. The total amount of sediment extruded by project riprap is 53,4961 m^3. The construction lasts for 120 days with work of 8 hours per day. Thus, the intensity of sediment entering the sea is 0.0075 m^3/s in the sediment around the sea area; particles with diameter lower than 0.063 mm accounts for about 20%. Wet density of bottom is taken as 1.8 t/m^3. The natural water content ratio averages about 48.6%. Then the source intensity of suspended particles formed by riprap extrusion is: 1.39 kg/s.

3.2 Simulation of suspended sediment transport and diffusion

The simulation of sediment entering the sea during project construction is divided into Case one and two respectively; the former of which refers to the suspended sediment caused by the process of blast extrusion and the latter by riprap extrusion of surround dike. The simulation method is introduced as follows.

3.2.1 Suspended sediment entering the sea produced by blast extrusion

The constructed section area is K1—K2. Taking into account the fact that the transport and diffusion laws of suspended sediment under the water dynamics after blast extrusion in spring, middle and neap tide are similar and that the blasted section is close to land area in the west, ebb tide period of spring tide is taken to calculate the maximum influenced coverage. The calculation time for each blast point is 15 h. The pressure of shock wave produced by blast extrusion is quite large. Thus, for the research on large area of suspended sediment transport and diffusion process, the suspended sediment generated by blast extrusion can be considered similarly to conform to the instantaneous distribution of source. For the simulation, the blasting is conducted by following the mesh points covered by blast extrusion on the water front line of surrounding dike. A number of 20 mesh points are calculated in total to gain the concentration envelope of suspended sediment entering the sea by blast extrusion (it is mapped by the maximum instantaneous value of each mesh point in the calculated area. The same below). With the purpose of describing the process of the blasted sediment entering the sea, the single representative blasting point around K2 is taken as an example to provide the suspended sediment concentration distribution and envelope 1 h, 3 hs and 7 hs after blasting and then to give the envelope of blast extrusion constructed point by point.

3.2.2 Suspended sediment entering the sea by riprap extrusion

The constructed section area is K2—K3. For the simulation, the construction and calculation is done by following the grip points in the area of riprap extrusion on the water front line of surrounding dike. 70 mesh points are calculated in total. The construction time per day for each mesh point is taken as 12 hs for calculation (including the flood tide period and ebb tide period of spring tide). It is calculated by continuous and fixed sources to gain the concentration envelope of suspended sediment by riprap extrusion.

Table 1 indicates the envelope area (concentration increment which is higher than the allowed one) whose entering suspended sediment concentration increment of each case and the combined influence at spring tide is higher than 10 mg/dm^3 (it is the allowed concentration increment for both Level one and two sea water quality and fishing water quality standard). The maximum influenced envelope area at spring tide, whose suspended sediment concentration by combined influence of project construction is higher than 10 mg/dm^3, covers 5.848 km^2.

Table 1. Area of sediment flux influence envelope of blast and riprap silt extrusion combined.

Fengwei reclamation project at Meizhou Bay	Suspended sediment concentration/mg · dm^{-3}	Maximum envelope area influenced by spring tide/km^2
Sediment extrusion by blast	>10	4.259
Sediment extrusion by riprap	>10	4.389
Combined	>10	5.848

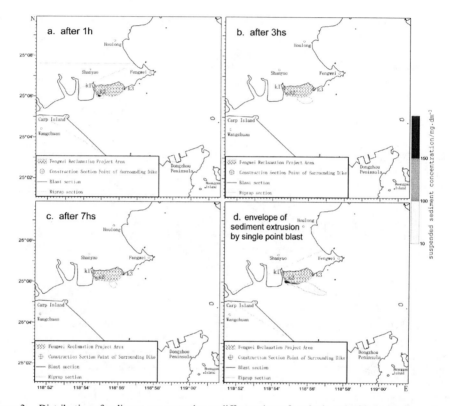

Figure 3. Distribution of sediment concentration at different time after single-point blast and riprap.

After blast extrusion, one part of the suspended sediment entering the water will settle down while the other part transports and diffuses in the water area around the blast extrusion point under the action of flood and ebb tidal current. With the passing of time, the concentration of suspended sediment generated by blast extrusion drops gradually. Fig. 3a–3c are the distribution diagrams of suspended sediment concentration 1 h, 3 hs and 7 hs after the single point blasting around k2. Fig. 3d shows the envelope of sediment entering the sea by blast extrusion at single representative point, i.e. maximum influenced area isopleths of suspended sediment diffusion along its movement. It is known from the figure that after blasting the suspended sediment diffuses outward with current to be far from the blasting point. The maximum value of suspended sediment drops gradually. The increment concentration of suspended sediment in the water takes on a changing trend that the farther it is from the blasting point, the lower the concentration increment is. After 9 hs, the influence of suspended sediment generally disappears.

Fig. 4 and 5 shows the envelopes of two cases and the combined influence for sediment entering the sea. It is seen from the figures that the suspended sediment in case one and two moves with the flood and ebb tidal current in the area within 2 km of the front line of surrounding dike. At the time

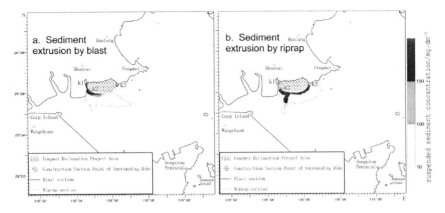

Figure 4. Sediment flux influence envelop of blast and riprap silt extrusion.

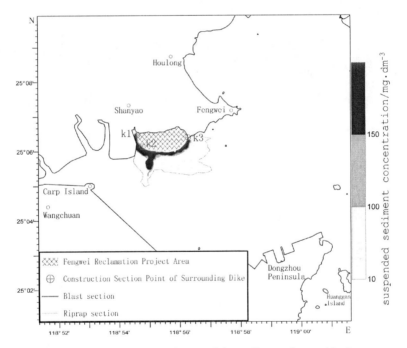

Figure 5. Sediment flux influence envelope of blast and riprap silt extrusion combined.

of flood tide, the current in front of the surrounding dike flows from the east to west. Suspended sediment diffuses toward the inside of Bay at hillside along with the current. At the time of ebb tide, the current in front of the surrounding dike flows from the west to east. Suspended sediment diffuses toward the east following the current. The diffusion coverage and shape of suspended sediment are mainly dominated by the tidal current.

4 CONCLUSION

This research is based on Fengwei reclamation project construction at Meizhou Bay. It adopts the method of numerical simulation to predict the transport and diffusion law of suspended sediment and to analyze comprehensively its influence on ecological environment. It is to provide reference for the environmental management during construction period of marine projects. The analysis above indicates the transport and diffusion of suspended sediment generated by sea wall construction is closely related to the movement of tidal current. The direction of tidal current and the dynamic strength of water to an extent determine the transport and diffusion direction and coverage of suspended sediment. The overall distribution of suspended sediment diffusion in construction is that the total concentration of suspended sediment is higher in the area close to project. And the farther the area is from the project, the closer to background concentration of sediment in natural status. To sum up, it is temporary for suspended sediment in construction to influence on marine ecology. The influence is to disappear with the completion of construction. There is no long-term adverse effect on marine ecology. A certain aquaculture area of kelp and laver exists around the project. The influenced aquaculture area is known by overlapping the spring-tide maximum envelope coverage figure of suspended sediment diffusion in construction with current surrounding aquaculture distribution figure. Measure of temporary requisition shall be taken for the aquaculture area during construction period of project to avoid influence of suspended sediment in construction of project on it. Meanwhile, the diffusion coverage of suspended sediment is calculated to offer scientific basis for quantitative calculation of biomass loss and ecological compensation resulted from suspended sediment.

This research calculates the model mesh points covered by the construction point by point, whose result match well with the actual construction of project. Future research shall verify the result of numerical simulation with the tracking and monitoring data in the actual construction process of project, and further determine the parameters of model.

REFERENCES

Casulli, V. & Cattani, E. 1994. Stability, accuracy and efficiency of a semi-implicit method for three-dimensional shallow water flow. *Computer Math Applic* 27(4): 99–112.

Casulli, V. & Cheng, R.T. 1992. Semi-implicit finite difference methods for three-dimensional shallow water flow. *International Journal for Numerical in Fluids* 15(6): 629–648.

Chen, X.H. & Tu, X.J. 2000. Prediction and analysis of influence of suspended solids of Yacht Project in Shenzhen Bay. *Marine Environmental Science* 19(1): 48–51.

Dou, G.R. 1999. Another study on starting speed of sediment. *Sediment Research* (6): 1–9.

Galappatti, G. & Vreugdenhil, C.B. 1985. A depth-integrated model for suspended sediment transport. *Journal of Hydraulic Research* 23(4): 359–377.

Huang, H.M., Wang, Y.G., Sun, S.Y. & Shang, J. 2008. Study on transport and diffusion simulation of suspended sediment generated by blast extrusion. *Ocean Engineering* 28(1): 70–75.

Lou, H.F. 2010. Research on transport and diffusion of suspended sediment generated by blast extrusion under Action of Tidal Current. *Zhejiang Hydrotechnics* 2010(5): 9–12.

Lu, R.H., Yu, D.S., Yang, J.Y. & Gu, J.Y. 2011. Initial research on accumulative influence of reclamation project on tidal current dynamics at Xiamen Bay. *Journal of Oceanography in Taiwan Strait* 30(2): 165–174.

Oliveira, A. & Baptista, A.M. 1995. A comparison of integration and interpolation Eulerian-Lagrangian methods. *International Journal of Numerical Methods in Fluids* 21(3): 183–204.

Peng, H. & Yuan J.X. 2012. Research on simulation of transport and diffusion of suspended sediment by blast extrusion for base of surrounding dike of Yinglianmen Reclamation Project at Daishan. *Port and Waterway Engineering* 8(8): 16–21.

Qiao, J.Y., Ding, Y. & Zheng, Z.M. 2004. Mechanism research on sediment extrusion by blast and riprap. *Journal of Geotechnical Engineering* 26(3): 350–352.

Tang, J.J. & Chen, C.H. 2011. Numerical simulation of tidal current of sea-crossing bridge at Quznhou Bay. *Proceedings of the 15th China Ocean (Coast) Engineering Academic Symposium*. Nanjing: Chinese Ocean Engineering Society.

Tang, J.J. & Wen, S.H. 2006. Two-dimensional numerical simulation of tidal current in Haitan Strait. *Journal of Oceanography in Taiwan Strait* 25(4): 533–540.

Zeng, X.M., Guan, W.B. & Pan, C. 2011. Accumulative effect of reclamation projects in years at Xiangshan Port on water dynamics. *Journal of Marine Science* 29(1): 73–83.

Zhang, Y.L., Baptista, A.M. & Myers, E.P. 2004. A cross-scale model for 3D baroclinic circulation in estuary–plume–shelf systems: I. Formulation and skill assessment. *Continental Shelf Research* 24(18): 2187–2214.

Isolation and characteristics of two novel crude oil-degrading bacterial strains

Yue Gao, Xun Gong, Hongmin Li, Min Ma, Jihong Wang, Shumin Cui & Chengyu Wang
College of Resources and Environmental Sciences, Jilin Agricultural University, Changchun, PR China

ABSTRACT: Two crude oil-degrading bacteria strains was isolated from contaminated soil obtained from Jilin oilfield. Phenotypic features, physiological and chemotaxonomic characteristics, and phylogenetic analysis of 16s rDNA sequence revealed that the isolates belong to the genus of *Agrobacterium*, and named strain 2# and 3#, respectively. The influence of different conditions on crude oil degradation of the two strains was investigated by UV spectrophotometer and Gas chromatography. In this study, the optimal conditions for crude oil degradation were at initial pH 7.0, 8.0% NaCl concentration and 35°C, and the maximum degradation rate were up to 56.94% and 65.32%, respectively. In simulation experiment, the degradation rate of the 3# strain was up to 78.53%. These results indicated that the 3# strain could be better used for actual bioremediation of crude oil contaminated soil.

1 INTRODUCTION

Crude oil continues to be used as the principle source of energy. During the economic developing, petroleum produces, wastewater excrete. Crude oil is important environmental contaminants, and lead to carcinogenic mutagenesis and deformity through the food chain (Xiang Zhou, Zhi-Jun Xin et al., 2013, 23:386–393). Therefore, crude oil pollution is key discussed issues around the world. Due to the microbial treatment were without adding chemical reagents, low energy consumption and no secondary pollution, it has became the focus of study. Microbial bioremediation has received increasing attention for a number of advantages over other remediation technologies(Qinghua Qiu, Ha Nipa et al., 2013). Crude oil is complex mixture, include saturated hydrocarbon, aromatic hydrocarbon and colloid and asphalt. Due to their toxicity to microorganisms (L Y He, Z Dang, X Tang, et al., 2010, 30(6):1220–1227), and inhibit microbial absorption the pollutants. Therefore, it is especially important to isolate crude oil degradation bacteria with the capacity to produce biosurfactants (Xiangsheng Zhang, Dejun Xu, 2012, 209:138–146). Many crude oil-degrading bacterial strains from crude oil contaminated environments, and with efficient crude oil degradation capacity.

Two crude oil-degrading bacteria strains, isolated from Jilin oilfield. Using morphological, biochemical and physiological characterization and 16s rDNA sequencing, the two strains were identified as *Agrobacterium* (*Agrobacterium sp.*). At the same time, study on degradation characteristics of two strains.

2 MATERIALS AND METHODS

2.1 *Screening and identification of crude oil degrading bacteria*

Crude oil degrading bacterium, using crude oil as sole carbon source and energy source, was isolated from contaminated soil sample of Jilin oilfield. Phenotypic features, physiological and

Table 1. The simulation experiment groups.

Group number	Content of crude oil (g/kg)	Treatment method
A	30.6	Sterilized soil without experimental bacteria
B	30.6	Sterilized soil with experimental bacteria
C	30.6	Unsterilized soil without experimental bacteria
D	30.6	Unsterilized soil with experimental bacteria

chemotaxonomic characteristics, and phylogenetic analysis of 16s rDNA sequence were identified (Tao Wu, Wen-jun Xie, 2012).

2.2 *The influence of different environmental factors on the crude oil degradation*

The influence of initial pH value, salt concentration and temperature on the crude oil degradation was evaluated by UV spectrophotometer. The crude oil degradation rate of the optimal condition was evaluated by GC-FID (Oluwafemi S. Obayori, Sunday A. Adebusoye et al., 2009, 45:243–248).

2.3 *The determination of crude oil degradation capacity*

The residual oil in the flasks was recovered according to Zhang's methods (Xiangsheng Zhang, et al., 2012, 209:138–146).

Using UV spectrophotometer at 225nm, measured the absorbance values, n-hexane as reference. The degradation rate (η) was calculated as follows:

$$\eta = (m_1 - m_2)/m_1 \times 100\%$$

where m_1 = the initial crude oil weight and m_2 = the treatmented crude oil weight.

The samples' hydrocarbon content was analyzed by GC-FID (Hong Chang, et al., 2013,7(2)).

2.4 *Simulation experiment*

The grouping, crude oil content and treatment method of simulation experiment were listed in Table 1. The crude oil degradation rates were evaluated at 0, 5, 10, 15, 20, 25 and 30 d.

3 RESULTS AND DISCUSSION

3.1 *Screening and identification of the crude oil degradation bacteria*

The two isolated crude oil degradation strains were designated as 3# and 2#. The Phenotypic features, physiological and chemotaxonomic characteristics, and phylogenetic analysis of 16 s rDNA sequence revealed that the isolates belong to the genus of *Agrobacterium*.

3.2 *The influence of environmental factors on the crude oil degradation*

3.2.1 *Effect of Initial pH on the biodegradation rate*

The maximum degradation rate of two strains were 49.87% and 42.37% respectively at the pH 7.0, as shown in Figure 1. These data illustrate that the strains are adaptable to pH7-8 culture fluids. The acidic or alkaline condition could affect the microbial degradation enzyme secretion and growth, and hinder microbial activity and nutrient absorption (Jinyang Feng et al., 2010).

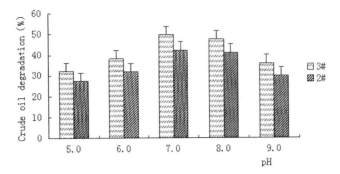

Figure 1. The influence of initial pH value on crude oil degradation.

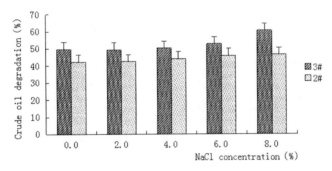

Figure 2. The influence of the concentration of NaCl on crude oil degradation.

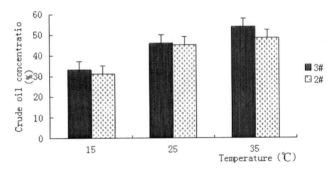

Figure 3. The influence of temperature on the crude oil degradation.

3.2.2 *Effect of salt concentration on the biodegradation rate*
As shown in Fig. 2, with the salt concentration increasing, the crude oil degradation rate of the two isolates were raised. The maximum degradation rate at the 8.0% salt concentration were up to 60.49% and 46.37% respectively. This may be due to the changes of cell osmotic pressure (Chen Li, 2009),

3.2.3 *Effect of temperature on the biodegradation rate*
With the cultured temperature increasing, the crude oil degradation rate of the two strains showed an upward trend, and the maximum degradation rate of the two isolates were up to 53.79% and 48.37% respectively at 35°C (As shown in Fig. 3). Previous study indicated lower temperature

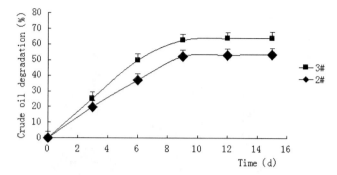

Figure 4. The crude oil degradation rate under the optimal conditions.

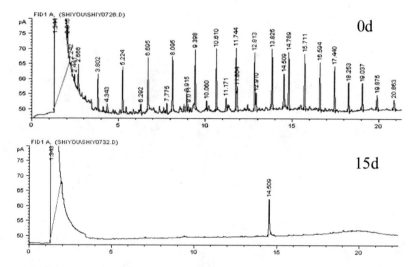

Figure 5. The GC-FID results of the crude oil degradation of the 3# strain at 0 d and 15 d.

inhibits the production of microbial enzymes, thus affects microbial metabolism, and decrease the crude oil degradation capacity (Rahman et al., 2002, 18: 257–261).

3.3 *The determination of crude oil degradation rate of the optimal conditions*

As shown in Fig. 4, on the conditions of the optimal pH, salt concentration, and temperature, the crude oil degradation rate of the two strains, 3# and 2#, were up to 65.32% and 56.94% after 15 d incubation, respectively.

Under the optimal conditions, the crude oil degradation capacity of the 3# strain at 0 and 15 d were evaluated by gas chromatogram (GC-FID). As shown in Fig. 5, all chromatographic peak almost disappeared at 15 d. The crude oil is consist of saturated hydrocarbon, aromatic hydrocarbon, resins and asphaltenes. After 15 d biodegradation, saturated hydrocarbon chromatographic peak entirely disappeared (Mu-tai Bao et al., 2012, 30(6):1220–1227).

As shown in Fig. 6, with the biodegradation time increasing, the crude oil degradation rate of the four groups were increased. The volatilization and content of crude oil in the Group A decreased. the crude oil degradation rate in the Group B was significantly increased, while the group C only rely on original indigenous bacteria, and the crude oil degradation rate was lower than the group B. In the Group D, the crude oil degradation bacteria and the original indigenous bacteria may have a symbiotic relationship, and the crude oil degradation rate was up to 78.53%.

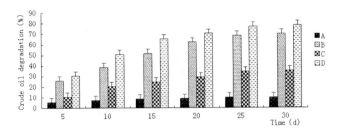

Figure 6. The crude oil degradation rate of divided groups.

4 CONCLUSION

(1) Two bacterial strains were screened and isolated from crude oil contaminated soil. The 3# and 2# strains were *Agrobacterium (Agrobacterium sp.)*.
(2) The optimal degradation conductions were at pH 7.0, 8.0% NaCl concentration and 35°C.
(3) The crude oil degradation rate of the 3# strain in the simulation experiments was up to 78.53%. Therefore, it could be better used for actual bioremediation of crude oil pollution

ACKNOWLEDGEMENT

This research was supported by Environmental Protection Department of jilin Province (Gant No. 2012025).

REFERENCES

Chen Li, 2009. Studies on the biodegradation characteristics of Oil-degrading bacteria and mixed culture.
Hong Chang, Maiqian Nie, 2013. Effects of rhamnolipid on oil degradation by strain NY3. Chinese Journal of Environmental Engineering, 7(2).
Jinyang Feng, Xiaode Zhou, 2010. Experimental study on factors influencing biodegradation of crude oil.Chinese Journal of Environmental Engineering.
L Y He, Z Dang, X Tang, et al., 2010. Biodegradati on characteristics of crude oil bymixed bacterial strains [J]. Acta Scientiae Circumstantiae, 30(6):1220–1227.
Mu-tai Bao, Li-na Wang et al., 2012. Biodegradation of crude oil using an efficient microlial consortium in a simulated marine environment. Marine Pollution Bulletin, 64(6):1177–1185.
Oluwafemi S. Obayori, Sunday A. Adebusoye et al., 2009. Differential degradation of crude oil (Bonny Light) by four Pseudomonas strains. Journal of Environmental Sciences. 45:243–248.
Qinghua Qiu, Ha Nipa et al., 2013. Study on the Screening and Degradation Condition of Oil Degradation Mixed Bacteria Agents, Chinese Agricultural Science Bulletin.
Rahman, J. Thahira-Rahman et al., 2002. Towards effcient crude oil degradation by a mixed bacterial consortium. Bioresource Technology. 18:257–261.
Xiang Zhou, Zhi-Jun Xin et al. 2013. High efficiency degradation crude oil by a novel mutant irradiated from Dietzia strain by $^{12}C^{6+}$ heavy ion using response surface methodology. BioresourceTechnology, 23:386–393.
Xiangsheng Zhang, Dejun Xu, 2012. Isolation and identification of biosurfactant producing and crude oil degrading Pseudomonas aeruginosa strains, Chemical Engineering Journal, 209:138–146.

The challenge of water resources and environmental impact of China mainland unconventional gas exploration and development

Ting Yue, Sherong Hu, Jichao Peng, Fuhe Chu & Yu Zhang
China University of Mining and Technology, Beijing, China

ABSTRACT: China mainland is abundant in unconventional gas resources and has large development potential. The exploration and development of unconventional gas are on track, but the resulting water resources shortage and environmental and ecological problems are worrying. In this article, the author firstly analyzed the water consumption of three types of unconventional gas resources (tight gas, shale gas and coalbed methane (or CBM for short)) in the process of their exploration and development in China mainland. Then, combined with the present distribution of water resources, the author found that water resources supply will be one of the bottlenecks which restrict the development of China mainland unconventional gas. Finally, the author suggested that we should take effective measures, like optimizing adjustment and effective deployment of water resource, popularizing technology of no or less water fracturing, cycle utilization of waste water after processing, improving laws and regulations and establishing the supervision mechanism, to strengthen the management of water resources and environmental protection, and to achieve sustainable and harmonious development with environment of unconventional gas exploration and exploitation in China mainland.

1 INTRODUCTION

China mainland is abundant in unconventional gas resources and has large development potential. The recoverable resources of tight gas are $8.8 \times 10^{12} \sim 12.1 \times 10^{12}$ m^3, shale gas $15 \times 10^{12} \sim 25 \times 10^{12}$ m^3, CBM 10.9×10^{12} m^3, tight oil $13 \times 10^8 \sim 14 \times 10^8$ t. The recoverable shale oil resources are 160×10^8 t. Oil sand also has a certain resource potential (Jia Chengzao et al., 2012). In January 2013, the National Energy Work Meeting advocated more emphasize on developing shale gas, CBM and other unconventional gas resources to stabilize overall energy balance between supply and demand for the whole year. It's expected that the unconventional gas resources exploration and development efforts will be increasing. However, this without doubt will bring greater pressure to the balance between supply and demand of water resources in China mainland. Unfortunately, some important unconventional gas development plannings, such as Shale Gas Development Planning (2011–2015) and so on, make no explicit mention of water resources challenge and environmental impact of China mainland unconventional gas exploration and development.

Our predecessors have made some discussions about water-intensive and pollution problems of shale gas exploitation (Xia Yuqiang, 2010; Chen Li & Ren Yu, 2012) or have metioned the environmental problems (Wang Lansheng et al., 2011). In fact, the environmental and ecological problems during the process of unconventional gas resources exploration and development are very complex (Yue Ting et al., 2013). Presently, there is lack of systematic support and specific analysis on the challenge of water resources and environment impact of unconventional gas exploration and development in China mainland. On the basis of predecessors' research results and a large number of data, this paper analyzed the balance between water supply and consumption in the process of tight gas, shale gas and CBM exploration and development in China mainland, and further proposed the corresponding suggestions and measures from different angles, which may contribute to the sustainable development of unconventional gas resources in China mainland.

2 THE CHALLENGE OF WATER RESOURCES

In recent years, the development conditions of unconventional gas resources in China mainland are getting mature, and the exploration and development work have been carried out (Wang Xiyun et al., 2013). In 2012, the production of the previously mentioned three types of unconventional gas was 445×10^8 m^3, accounting for 41.75% of the total gas production in China mainland, of which tight gas (also known as low permeability gas) production was 320×10^8 m^3, net increase was 64×10^8 m^3, constituting a 25% year-on-year growth; CBM production was 125×10^8 m^3, increasing by 10×10^8 m^3 and 8.8%; Shale gas production reached 0.5×10^8 m^3. Predictably, in the next one or two years, the production of these three kinds gas may account for half of China mainland natural gas production. Combined with the development assumption of gas production (Table 1), it's can be seen that, over the next 10 to 20 years, China mainland natural gas production will increase dramatically, which is of great significance to meet the shortage of conventional gas production in China mainland. However, in terms of the development of exploration means and mining technology, unconventional gas resources exploration and exploitation will inevitably bring China mainland a heavy burden of water resources supply.

Up to now, tight gas has become one of the main forces of China mainland natural gas production, and in a long period of time, tight gas is expected to act as a leader of unconventional gas exploration in China mainland. For instance, Sulige field in Ordos Basin is the most successful development tight sandstone gas field in China. In southern district, water consumption is mainly concentrated on the drilling construction stage, followed by the exploration stage and operation stage, and the retirement stage can be neglected. In the construction stage, the drilling process needs to consume a large amount of water, for example, each well needs 225 m^3 water for drilling fluid, 81064.2 m^3 for mud system, and about 900 m^3 for fracturing and well washing. We can estimate that the total water consumption of 990 wells on three items above mentioned will be 8.14×10^7 m^3 (Table 2). Due to the unique situation of China mainland, at present and in a rather long period afterwards, we may still give priority to densely vertical wells and waterflooding development in the process of tight sandstone gas reservoir exploitation, and the water consumption in unit area will still be very huge.

Shale gas production is a process of high intensive water. The single-well drilling and fracturing operation usually needs about 500×10^4 gallon of water, depending on the concrete basin structure and geological structure. Fracturing process needs a lot of water mixed with sand and chemicals injected in a well to extract the gas, so its water consumption is the largest. Nearly all the rest of water is used for drilling stage (water is the main component of drilling fluid), and the field dust removal and drilling equipment cleaning and flushing only use a small amount of water. Although more and more water was recycled and reused, the drilling operation still needs a lot of fresh water, because salt water may be likely to spoil facility and cause formation damage, and thus reduce the successful probability of drilling. Therefore, the industry demand for fresh water has become an increasingly serious problem, which intensifies the water supply pressure and competition for

Table 1. The development assumption of gas production in China mainland (Qiu Zhongjian et al., 2012).

Age	Natural Gas ($\times 10^8$ m^3)	Conventional gas ($\times 10^8$ m^3)	Unconventional gas ($\times 10^8$ m^3)	Proportion (%)	Tight gas ($\times 10^8$ m^3)	CBM ($\times 10^8$ m^3)	Shale gas ($\times 10^8$ m^3)
2020	2300	1100	1200	52.17	800	300	100
2030	3800	1500	2300	60.53	1000	700	600

water abstraction licensing, especially in the areas with serious water scarcity and large water requirement.

In the process of CBM mining, people take the water-discharging and pressure-dropping to reduce static pressure in the coal-bed. It will consume lots of water and last longer time (ZHANG Qiang et al., 2012). According to statistics (Cynthia A.R., 2003), in America CBM mining area, single-well water discharge is 3.98~66.62 m^3/d, total gas-water ratio is 5.75 m^3/10^3 m^3. In China "twelfth – five" plan, CBM surface well gas production was designed up to be 160×10^8 m^3, then the water discharge calculated by 5.73 m^3 per 1000 m^3 CBM would be as high as 0.92×10^8 m^3/year.

The water resources of China mainland are scarce and unevenly distributed. The world bank report of water resources shows that, in 2007, China mainland per capita renewable water resources were estimated to be 2156 m^3/year, only accounting for a quarter of the world average. On the whole, China mainland has been facing severe water crisis, and the regional and seasonal maldistribution of water resources have exacerbated the scarcity.

Based on the statistical analysis of China mainland annual precipitation data from 1961 to 2012 (Figure 1), it can be seen that the annual precipitation average basically maintains 580~620 mm, but the amplitude of interannual variability of precipitation has increased significantly since 2000. This may cause more severe droughts in water-deficient areas, while more frequent flood disasters

Table 2. The statistics of water consumption in different development stages (Zhang Bo, 2009).

Development phase	Uses	Water Consumption per Well	Water Consumption
1st stage (exploration)	Drilling fluid	225 m^3/well	900 m^3
	Mud system	81064.2 m^3/well	324256.8 m^3
	Fracturing and well washing	900 m^3/well	3600 m^3
	Total	82179.2 m^3/well	328716.8 m^3
2nd stager (construction)	Drilling fluid	225 m^3/well	222750 m^3
	Mud system	81064.2 m^3/well	8.03×10^7 m^3
	Fracturing and well washing	900 m^3/well	8.19×10^5 m^3
	Base life	972.5 m^3/cluster wells	1.07×10^5 m^3
	Total	740585.3 m^3/cluster wells	8.15×10^7 m^3
3rd stage (operation)	Base life and Gas gathering Station life	0.1 m^3/(d·person)	20.4 m^3/d
	Total	0.1 m^3/(d·person)	20.4 m^3/d
4th stage (retirement)	Base life and Gas gathering Station life	0.1 m^3/(d·person)	20.4 m^3/d
	Total	0.1 m^3/(d·person)	20.4 m^3/d

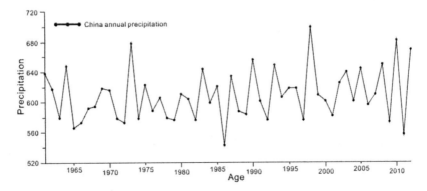

Figure 1. Statistical chart of China mainland annual precipitation data from 1961 to 2012.

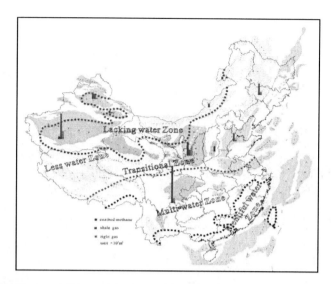

Figure 2. The distribution of China Mainland unconventional gas and water resources.

on both sides areas of the rivers in China mainland. Lack of water or less water areas account for large proportion in China mainland, the dramatically periodical decrease of annual precipitation will seriously affect the short-term balance between supply and demand of water resources, then affect the exploration and development process of unconventional gas resources. For example, in 2010, the five provinces in southwest China mainland (Sichuan, Chongqing, Guizhou, Yunnan and Guangxi), with 40% shale gas reserves, occurred seasonal water shortage, which led to the people suffering severe drought disaster for six months. It's a reminder that the drought provinces need good management of water resources during the process of large-scale shale gas development.

Although these data mainly reflect the characteristic and trend of China mainland precipitation, they also can clearly indicate that the pressure between supply and demand of China mainland water resources has increased in the last ten years.

Combined with Figure 2, we can see that the distribution of water resources in China mainland has the characteristic of ladder-like, which reduces gradually from south to north. The water resources in the southeast coasts are pretty rich. This has a direct relationship with too many water areas and abundant rainfall in the region. Middle and west inland on the south of the Yangtze River belongs to multi-water zone, second only to the southeast coasts, where rivers and lakes dot and water sources spread. The northwest region tends to be lack of water or to have less water, especially in the Tarim Basin and the areas adjacent to the border with Outer Mongolia where both desert and gobi are common. There is basically water shortage in North China, and from east to west, water shortage gets gradually serious. The water resources in northeast region are relatively abundant, but very unequally distributed, with the feature of "the north better than south, the east better than west, the edge better than interior".

Regions rich with unconventional gas reserves in China mainland, in general, are usually the areas with the serious problem of seasonal water shortage. The temporal and spatial distribution of water resource in China mainland doesn't match with the enrich areas of unconventional gas resources. For example, while North China and northeast region have 26% of shale gas resources, there is water shortage on the whole, and water per capita in North China is lower than the national average, only 700 m^3/year. The downstream of the Yangtze river and southeastern region is very rich in water, but has only 18% of the shale gas reserves in China mainland. There is no doubt that water problem will restrict the development of unconventional gas resources in China mainland. Specifically, except in Sichuan Basin, located in the south of the Yangtze river, water resources are abundant, in other basins, such as Tarim Basin, Ordos Basin, Junggar Basin, Tuha Basin, Bohai

Bay Basin, Erlian Basin, Qinshui Basin and Songliao Basin, especially in Ordos Basin, the main area with unconventional resources, the present situation of water resources is worrisome.

3 THE ANALYSIS OF ENVIRONMENTAL IMPACT

Another important issue of unconventional gas exploration and development is environmental pollution.

Although China mainland shale gas resources are rich, but the shale gas industry is still at the stage of studying and exploring, and there is no shale gas production now. According to the results of China Shale Gas Resources Potential Surveying Assessment and Core Area Evaluation and the instructions of the Strategic Action Framework of Ore-Search Breakthrough (2011–2020) issued by the Office of the State Council, [2011] 57th, China mainland is intensifying shale gas exploration. By the end of 2011, Chinese petroleum enterprises had fracturing tested 15 shale gas vertical wells, and nine of them produced shale gas. This not only shows that we have preliminary mastered the shale gas vertical well fracturing technology, but also confirm that China mainland has broad prospects of shale gas development. Therefore, through the analysis on environment impact of shale gas exploration and development, and the lessons from technologically-advanced countries, we can draw the conclusion that China mainland should follow the way that environment assessment is the antecedence to the development of shale gas. Based on the investigation about a series of environmental problems in the process of America shale gas exploration and development, the author thinks that the environmental impact of shale gas exploration and development mainly has these aspects: ① Shale gas exploitation needs to consume large amounts of water. On the basis of the existing industrial and agricultural water consumption, largely increasing water demand of shale gas exploitation may cause cumulative effects on the environment, such as rivers and other surface water and groundwater resources depletion, loss of aquifer storage capacity and water quality deterioration, as well as the sustainable utilization of regional water resources reduction. ② Environmental pollution generated by noise, waste water, waste gas, earthquake and mining accidents during mining. ③ Environmental pollution caused by chemical additives (including more than 250 kinds of chemical products, and toxic substances benzene and lead) of hydraulic fracturing fluid. According to statistics, flowback water accounted for 30% to 70% of the pressure fluid volume, in addition to containing the high-pressure liquid chemical additive components, but also contain hydrocarbons, heavy metals, and total dissolved solids (Total Dissolved Solids, Acronym TDS), of which TDS component comprises calcium, potassium, sodium, chlorides, carbonates and the natural radioactive substances from gas formation, such as uranium, thorium and their decay products, etc. These high-salinity flowback water, if not promptly treated, may leak to damage the groundwater system, water quality and the environment, which should not be underestimated (Xia Yuqiang, 2010; Office of Fossil Energy and National Energy Technology Laboratory, US Department of Energy, 2009).

CBM development in China mainland advances rapidly. In 2010, the amount of CBM downhole drainage had reached 75×10^8 m^3, and the utilization amount was 23×10^8 m^3, respectively increased by 226% and 283% than 2005; the amount of CBM ground development had reached 15×10^8 m^3, realizing commodity amount of 12×10^8 m^3 (Wang Xiyun et al., 2013). The process of CBM exploration and development is generally divided into three stages: exploration, trial production and exploitation stage. In addition to domestic sewage, each stage will also correspondingly produce a certain amount of waste water, for instance, the discharged water containing drilling fluid in the process of drilling in exploration stage, the composition complex fracturing fluid including 11 categories, more than 20 kinds of chemicals in trial production stage and a large amount of produced water in exploitation stage (Liang Xiongbing et al., 2006). The water tests in Jincheng district of Qinshui Basin show that, the produced water in the study area has obviously high salinity. A certain elements contents in produced water are not very stable, and some ion contents even vary greatly in the beginning and the mid to late period of the mining. This part of produced water has large volume and long duration. If the produced water flows into the underground, rivers or irrigated farmlands, it will affect local environment and seriously pollute the underground water layer (Wang Zhichao et al., 2009).

In recent years, China mainland has proven that the geological reserves of tight gas are 3000×10^8 m^3, and the production increases 50×10^8 m^3/year. Tight gas industry has formed a preliminary scale in development. In 2011, tight gas production reached 256×10^8 m^3, which accounting for a quarter of the national gas production. Tight gas has played an important role in natural gas exploration and development in China mainland (Qiu Zhongjian et al., 2012). At present, the development method of tight sandstone gas is different from shale gas, so the pollution of it is less than shale gas. For example, in the southern district of Sulige gas-field, the waste water produced in drilling stage is 19.8×10^4 m^3 and in fracturing is 32.67×10^4 m^3, and domestic sewage is about 9.6×10^4 m^3. In addition, other stages produce far less waste water than the construction stage. According to the results of shallow groundwater quality test in 2008, it's found that the shallow groundwater contamination in south Sulige block of Ordos Basin mainly come from domestic pollution sources. Combined with the application of waste water recycling and other measures, the water pollution in tight gas development can be basically under control.

4 ADVICES

China mainland unconventional gas resources exploration and exploitation have brought a lot of pressure to the balance between supply and demand of water resources, and have some degree impact on environment at the same time. In order to achieve sustainable development of unconventional gas resources exploration and development in China mainland, we can strengthen water resources management and environmental protection according to the following several aspects:

1) Optimizing adjustment and effective deployment of water resources

   We should take engineering measures and non-engineering measures, such as South-to-North Water Diversion, Water Transferring from West to East, Introducing Sea Water from Bohai Sea to West China, etc., and adjust industrial structure to improve the temporal and spatial distribution of water resources and create favorable conditions for unconventional gas development in China mainland.

2) Popularizing technology of no or less water fracturing

   One of the tendencies of fracturing technology development is the effective control on water consumption of horizontal well fracturing, and the present hydraulic fracturing technique will be replaced gradually by more environmentally friendly and more effective water-saving fracturing technology, such as the supercritical CO_2 system gas extraction technology proposed by Academician Shen Zhonghou at the Workshop on the Effective Exploitation Engineering and Technology of Unconventional Natural Gas in September 2011, and the liquefied petroleum gas (LPG) fracturing technology invented by Canada Gas Frac company. These will help to solve the problem of huge water consumption of unconventional gas fracturing to a large extent.

3) Cycle utilization of waste water after processing

   Cycle utilization of waste water after processing can effectively reduce the consumption of water resources, and thus become a very effective way to alleviate water shortage. At present, the water recirculation and reutilization technology in foreign countries is mature, with perfect related supporting facilities, lower costs and prominent effects. According to statistics, this technology can reduce the water consumption of shale gas to $1.512 \times 10^6 \sim 1.89 \times 10^6$ m^3/well and the recovery rate will reach 30~70%.

4) Improving laws and regulations and establishing the supervision mechanism

   China's current environmental laws and regulations are still focused on traditional gas industry, and have not yet clearly covered the unique water challenge of unconventional gas such as shale gas and so on. Therefore, China needs to develop the environmental regulations and laws on the exploration and exploitation of shale gas and other unconventional gas resources as soon as possible. In addition, combining with the characteristics of our country, we'd better set up supervision system, which covering the whole process of exploration and development, including drilling, production, wastewater treatment and gas well abandonment and seal, to effectively strengthen the environmental protection in the process of unconventional gas resources development. Then we should speed up the research on all-directional supervision system, which including groundwater, geology, soil and ecology and other aspects, and formulate the related laws and management regulations timely to ensure supervision is in place and development can be controlled.

5 CONCLUSIONS

According to the above, we can draw the following conclusions:

1. At present, the proven unconventional gas enrichment basins in China mainland, except Sichuan Basin, are commonly lack of water resources. However, each stage needs a large amount of water in the process of unconventional gas exploration and development. In addition, these stages always produce a series of environmental pollutions.
2. Putting forward some advices to strengthen the management of water resources and environmental protection, including improving laws and regulations, developing and governing at the same time, and monitoring environment problems in the whole process of China unconventional gas resources exploration and development.

REFERENCES

Chen Li, Ren Yu. 2012. Analyzing the Environmental Implications of Shale Gas Exploitation [J]. Environment and Sustainable Development, 9(3):52–55.
Cynthia A. Rice. 2003. Production waters associated with the Ferron coalbed methane fields, central Utah: chemical and isotopic composition and volumes [J]. International Journal of Coal Geology, 56 (issues 1–2):141–169.
Jia Chengzao, Zheng Min, Zhang Yongfeng. 2012. Unconventional hydrocarbon resources in China and the prospect of exploration and development [J]. Petroleum Exploration and Development, 39(2):129–136.
Liang Xiongbing, Cheng Shenggao, Song Lijun. 2006. Water pollution and control in CBM exploration and development [J]. Environmental science & Technology, 29(1):50–51+63.
Office of Fossil Energy and National Energy Technology Laboratory, US Department of Energy. 2009. Modern shale gas development in the United States: A primer [R]. Oklahoma: Gound Water Protection Council.
Qiu Zhongjian, Zhao Wenzhi, Deng Songtao. 2012. Roadmap of tight gas and shale gas development [J]. China Petrochem, (17):18–21.
Wang Lansheng, Liao Shimeng, Chen Gengsheng, et al. 2011. Bottlenecks and countermeasures in shale gas exploration and development of China [J]. Natural Gas Industry, 31(12):119–122.
Wang Xiyun, Dong Xiucheng, Pi Guangin, et al. 2013. Policy research on China unconventional oil and gas resources development [J]. Economic & Trade Update Sun, (1):68–69.
Wang Zhichao, Deng Chunmiao, Lu Wei, et al. 2009. Quality analysis on water from Coal Bed Methane borehole in Jincheng area [J]. Coal Science and Technology, 37(1):122–124.
Xia Yuqiang. 2010. The Challenges of Water Resource and the Environmental Impact of Marcellus Shale Gas Drilling [J]. Science & Technology Review, 28(18):103–110.
Yue Ting, Hu Sherong, Peng Jichao, et al. 2013. Environmental and ecological problems in the process of shale gas exploration and development [J]. China Mining Magazine, 22(3):12–15.
Zhang Bo. 2009. The influence evaluation of gas-field exploitation on groundwater environment in South Sulige [D]. Jilin University.
Zhang Qiang, Wang Zixiang, Li Mingliang, et al. 2012. Effects of discharged water in gas production phase of CBM Wells in Jincheng district [J]. Journal of Green Science and Technology, (9):154–156.

Advanced Engineering and Technology – Xie & Huang (Eds)
© 2014 Taylor & Francis Group, London, ISBN 978-1-138-02636-0

Optimization of 4-chlorophenol in the electrochemical system using a RSM

H. Wang
College of Environmental Science and Engineering, Beijing Forestry University, Beijing, China

Z.Y. Bian
College of Water Sciences, Beijing Normal University, Beijing, PR China

ABSTRACT: Pd-Fe/graphene catalysts used for the Pd-Fe/graphene gas-diffusion cathodes were prepared using a modified Hummer protocol and a formaldehyde reduction method. The electrochemical degradation of 4-chlorophenol (4-CP) was investigated using a diaphragm electrolysis system. The best electrolyze condition was investigated by changing the current density, electrolyte concentration, initial pH, and reaction time using by response surface methodology (RSM). The best condition was as follows: the initial pH was 7.00, electrolyte (Na_2SO_4) concentration was 0.02 mol/L, the current was 27 mA/cm^2, the reaction time was 100 min. The degradation efficiency of cathodic and anodic compartment was more than 98% and 95%, respectively.

1 INTRODUCTION

4-chlorophenol is resistant to physical, chemical or biological treatments and will cause hazardous influence on living organisms as well as human beings (Rivera-Utrilla et al. 2012). Due to its great hazard and toxicity, chlorophenol can hardly be removed by traditional treatment process such as biological methods. For its ease of control, amenability to automation, efficiency and environmental compatibility, the advanced eletrocatalytic oxidation process has attracted a great deal of attention recently (Zhao et al. 2012). Graphene as a new carbon material are one of the most exciting materials because of their unique electronic, chemical, and mechanical properties (Fu et al. 2013). Furthermore, the electrochemical performance of the graphene can be improved by loaded specifically chosen metal catalysts. In this paper, Pd-Fe/graphene catalyst was prepared, and the best condition of the 4-chlorophenol degradation was investigated by changing the current density, electrolyte concentration, initial pH, and reaction time using response surface methodology (RSM).

2 MATERIALS AND METHODS

2.1 Material and methods

Pd-Fe/graphene catalysts used for the Pd-Fe/graphene gas-diffusion cathodes were prepared using a modified Hummer protocol and a formaldehyde reduction method. The electrochemical degradation of 4-chlorophenol (4-CP) was investigated using a diaphragm electrolysis system sequentially fed with hydrogen gas and air over the Pd/graphene gas-diffusion cathodes. The 4-CP was identified in the electrolyzed solutions using high-performance liquid chromatography (HPLC, Shimadzu, Japan).

Table 1. Range of different factors investigated with central composite design.

Variables	Ranges and levels				
	−2	−1	0	1	2
Current density (A) X_1	0.30	0.45	0.60	0.75	0.90
Electrolyte concentration (mol/L) X_2	0.02	0.04	0.06	0.08	0.10
Initial pH X_3	3	5	7	9	11
Reaction time (min) X_4	40	60	80	100	120

2.2 Experimental design

RSM is a statistical method being useful for the optimization of chemical reactions and/or industrial processes and widely used for experimental design (Güven et al. 2008). As Table 1 shows RSM was employed to assess individual and interactive effects of the four main independent parameters (current density, electrolyte concentration, initial pH and reaction time) on the removal efficiency. A central composite design (CCD) was employed for the optimization of degradation of chlorinated organic compounds (Muhamad et al. 2013). A second-order empirical relationship between the response and independent variables was derived. Analysis of variance (ANOVA) showed a high coefficient of determination value ($R^2 = 0.823$).

3 RESULTS AND DISCUSSION

The composite experimental design, with five axial points and four replications at the center point leading to a total number of 30 experiments, was employed for response surface modeling. The experimental results and predicted values for color removal efficiencies are presented in Table 2. According to the above experimental results, the outcomes of model analysis of variance in the cathode and anodic compartments are shown in Table 3 and Table 4.

The behavior of this system and optimum values of selected variables were analyzed by RSM according to the composite experimental design. A semi-empirical expression in Eq. (1) and (2) in the cathodic and anodic compartments was obtained from the data analysis using the statistical graphics software expert design.

$$Y_1 = 93.18 - 0.74A - 0.22B + 2.01C + 2.04D - 0.38AB - 0.26AC - 0.78AD - 1.95BC$$
$$+ 0.000BD - .071CD - 0.75A^2 + 0.48B^2 - 2.68C^2 - 1.03D^2 \qquad (1)$$

$$Y_2 = 92.76 - 0.64A - 0.71B + 2.35C + 2.59D - 0.25AB - 1.03AC - 1.35AD - 0.54BC$$
$$+ 0.41BD + 0.48CD - 1.42A^2 - 0.078B^2 - 3.16C^2 - 1.26D^2 \qquad (2)$$

where Y is the predicted response; A, B, C, and D refer to current density, electrolyte concentration, initial pH, and reaction time, respectively.

The adequacy (statistical significance) of a quadratic model was tested through F- and p-values. It has long been known that a large F-value indicates that most of the variation can be explained by a regression equation whereas a low p-value (<0.05) indicates that the model is considered to be statistically significant (Silva et al. 2011). The software reads: the Model F-value of 5.0 implies the model is significant; there is only a 0.19% chance that a "Model F-Value" this large could occur due to noise.

Table 2. Response values of central composite design.

	Coded levels				Actual value				Removal efficiency (%)	
	X_1	X_2	X_3	X_4	X_1	X_2	X_3	X_4	Cathodic	Anodic
1	−1	−1	−1	−1	0.45	0.04	5	60	85.6	81.5
2	1	−1	−1	−1	0.75	0.04	5	60	85.0	83.5
3	−1	1	−1	−1	0.45	0.08	5	60	87.5	77.3
4	1	1	−1	−1	0.75	0.08	5	60	88.5	87.3
5	−1	−1	1	−1	0.45	0.04	9	60	91.1	88.7
6	1	−1	1	−1	0.75	0.04	9	60	91.2	90.1
7	−1	1	1	−1	0.45	0.08	9	60	87.6	86.7
8	1	1	1	−1	0.75	0.08	9	60	90.7	87.6
9	−1	−1	−1	1	0.45	0.04	5	100	85.8	84.1
10	1	−1	−1	1	0.75	0.04	5	100	91.4	90.2
11	−1	1	−1	1	0.45	0.08	5	100	93.7	91.9
12	1	1	−1	1	0.75	0.08	5	100	87.1	85.5
13	−1	−1	1	1	0.45	0.04	9	100	96.2	94.3
14	1	−1	1	1	0.75	0.04	9	100	91.4	90.2
15	−1	1	1	1	0.45	0.08	9	100	94.3	93.0
16	1	1	1	1	0.75	0.08	9	100	91.2	90.0
17	−2	0	0	0	0.15	0.06	7	80	91.9	91.1
18	2	0	0	0	0.9	0.06	7	80	85.7	80.0
19	0	−2	0	0	0.6	0.02	7	80	95.8	94.2
20	0	2	0	0	0.6	0.10	7	80	91.7	87.6
21	0	0	−2	0	0.6	0.06	3	80	76.3	74.3
22	0	0	2	0	0.6	0.06	11	80	85.9	82.9
23	0	0	0	−2	0.6	0.06	7	40	81.4	79.7
24	0	0	0	2	0.6	0.06	7	40	93.9	92.6
25	0	0	0	0	0.6	0.06	7	80	93.4	93.4
26	0	0	0	0	0.6	0.06	7	80	93.4	93.5
27	0	0	0	0	0.6	0.06	7	80	93.4	93.4
28	0	0	0	0	0.6	0.06	7	80	93.3	93.3
29	0	0	0	0	0.6	0.06	7	80	93.5	93.5
30	0	0	0	0	0.6	0.06	7	80	91.9	89.4

As Figure 1 shows, the good correlations between the actual and predicted values of 4-CP removal confirm the adequacy of the model to predict 4-CP removal. Current density, electrolyte concentration, initial pH, and reaction time were statistically significant factors. To sum up, the model can be used in the Pd-Fe/graphene gas-diffusion cathode system.

The best condition was as follows: the initial pH was 7.00, electrolyte (Na_2SO_4) concentration was 0.02 mol/L, the current was 27 mA/cm², the reaction time was 100 min. The degradation efficiency of cathodic and anodic compartment was more than 98% and 95%, respectively.

4 CONCLUSIONS

The composite experimental design, with a total number of 30 experiments, was employed for response surface modelling. Maximum removal efficiency was predicted and experimentally validated. The optimum current density, electrolyte concentration, initial pH and reaction time were found to be 27 mA/cm², 0.02 mol/L, 7.00 and 100 min, respectively. The degradation efficiency of cathodic and anodic compartment was more than 98% and 95%, respectively. This study clearly showed that RSM was one of the suitable methods to optimize the operating conditions.

Table 3. Analysis of variance (ANOVA) in the cathode compartment.

Source	Sum of squares	Degree of freedom	Mean square	F-value	P-value	Significant
Model	476.56	14	34.04	5.00	0.0019	Significant
A	13.10	1	13.10	1.92	0.1858	
B	1.13	1	1.13	0.17	0.6900	
C	96.80	1	96.80	14.21	0.0019	
D	100.31	1	100.31	14.72	0.0016	
AB	2.30	1	2.30	0.34	0.5698	
AC	1.07	1	1.07	0.16	0.6978	
AD	9.71	1	9.71	1.43	0.2510	
BC	14.31	1	14.31	2.10	0.1678	
BD	0.00		0.00	0.00	1.0000	
CD	0.08		0.08	0.01	0.9150	
A^2	15.29	1	15.29	2.24	0.1549	
B^2	6.39	1	6.39	0.94	0.3482	
C^2	196.37	1	196.37	28.82	<0.0001	
D^2	28.86	1	28.86	4.24	0.0574	
Residual	102.19	15	6.81			
Lack of Fit	100.31	10	10.03	26.59	0.0010	Significant
Pure Error	1.89	5	0.38			
Cor Total	578.75	29				

Table 4. Analysis of variance (ANOVA) for the quadratic model in the anode compartment.

Source	Sum of squares	Degree of freedom	Mean square	F-value	P-value	Significant
Model	697.09	14	49.79	4.46	0.0034	Significant
A	9.80	1	9.80	0.88	0.3639	
B	11.95	1	11.95	1.07	0.3175	
C	132.85	1	132.85	11.89	0.0036	
D	161.20	1	161.20	14.43	0.0017	
AB	0.97	1	0.97	0.09	0.7727	
AC	16.81	1	16.81	1.50	0.2389	
AD	29.34	1	29.34	2.63	0.1260	
BC	4.69	1	4.69	0.42	0.5267	
BD	2.72	1	2.72	0.24	0.6287	
CD	3.74	1	3.74	0.33	0.5716	
A_2	55.26	1	55.26	4.95	0.0419	
B_2	0.17	1	0.17	0.01	0.9046	
C_2	274.08	1	274.08	24.53	0.0002	
D_2	43.33	1	43.33	3.88	0.0677	
Residual	167.62	15	11.17			
Lack of Fit	154.08	10	15.41	5.69	0.0343	Significant
Pure Error	13.53	5	2.71		0.0034	
Cor Total	864.71	29			0.3639	

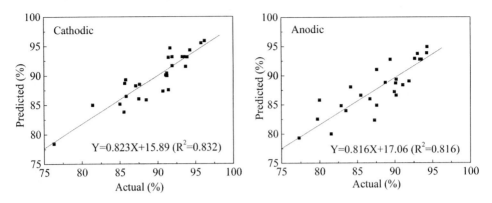

Figure 1. Actual and predicted removal of 4-CP in the cathodic and anodic compartments.

ACKNOWLEDGMENTS

This work was supported by the Beijing Higher Education Young Elite Teacher Project (No. YETP0773), the Fundamental Research Funds for the Central Universities (No. TD2013-2 and 2012LYB33), the National Natural Science Foundation of China (Grant 51278053 and 21373032), the Program for New Century Excellent Talents in University (No. NCET-10-0233), and the Beijing Natural Science Foundation (No. 8122031).

REFERENCES

Fu, X.G. Liu, Y.R. Cao, X.P. Jin, J.T. Liu, Q. & Zhang, J.Y. 2013. FeCo-N_x embedded graphene as high performance catalysts for oxygen reduction reaction. *Applied Catalysis B: Environmental* 130–131: 143–151.

Güven, G. Perendeci, A. & Tanyolac, A. 2008. Electrochemical treatment of deproteinated whey wastewater and optimization of treatment conditions with response surface methodology. *Journal of Hazardous Materials* 157: 69–78.

Muhamad, M.H. Abdullah, S.R.S. Mohamad, A.B. Rahman, R.A. & Kadhum, A.A.H. 2013. Application of response surface methodology (RSM) for optimisation of COD, NH3eN and 2,4-DCP removal from recycled paper wastewater in a pilot-scale granular activated carbon sequencing batch biofilm reactor (GAC-SBBR). *Journal of Environmental Management* 121: 179–190.

Rivera-Utrilla, J. Sánchez-Polo, M. Abdel daiem, M.M. & Ocampo-Pérez, R. 2012. Role of activated carbon in the photocatalytic degradation of 2,4-dichlorophenoxyacetic acid by the UV/TiO$_2$/activated carbon system. *Applied Catalysis B: Environmental* 126: 100–107.

Silva, G.F. Camargo, F.L. & Ferreira, A.L.O. 2011. Application of response surface methodology for optimization of biodiesel production by transesterification of soybean oil with ethanol. *Fuel Process.Technol.* 92: 407–413.

Zhao, H.Y. Wang, Y.J. Wang, Y.B. Cao, T.C. & Zhao, G.H. 2012. Electro-Fenton oxidation of pesticides with a novel Fe$_3$O$_4$@Fe$_2$O$_3$/activated carbon aerogel cathode: High activity, wide pH range and catalytic mechanism. *Applied Catalysis B: Environmental* 125: 120–127.

Integrated watershed management framework for abating nonpoint source pollution in the Lake Dianchi Basin, China

Chongyun Wang, Mingchun Peng, Rui Zhou & Shuhua Yang
Institute of Ecology and Geobotany, Yunnan University, Kunming, China

Zhaorong He & Changqun Duan
School of Life Science, Yunnan University, Kunming, China

Guoqiang Zhi, Zongxun Li & Landi He
Kunming Institute of Environmental Sciences, Kunming, China

ABSTRACT: Nonpoint source (NPS) pollution has produced persistent ecological problems for water resources around the world. Lake Dianchi is a typical eutrophic freshwater lake in China. As there has become better control of point site pollution in recent years, NPS pollution is now increasingly associated with worse water quality. Here, we develop an integrated watershed management framework to address water pollution in this region based on a recent NPS pollution survey, systematic analysis, and land use change trends in the basin. The watershed was divided into three zones in term of NPS pollution occurrences and features, as well as into three arenas that integrated land use management and practical instruments of NPS pollution control. In addition, prophylactic zonal eco-designs were outlined to reduce NPS pollution. In general, this holistic approach will help reduce NPS pollution at the watershed scale in highland lakes of similar characteristics.

1 INTRODUCTION

Eutrophication is the leading environmental problem for China's freshwater lakes (Jin, 2001; State Environmental Protection Administration of China, 2000; Liu and Qiu, 2007; Sun and Zhang, 2000), and Lake Dianchi (Yunnan Province) is one of the worst eutrophic freshwater lakes in China (State Environmental Protection Administration of China, 2000; Jin et al., 2008). In the 1950s, Lake Dianchi was classified as an oligo- to mesotrophic clearwater lake (Jin, 2003) but has become increasingly eutrophic since the 1960s given the increased industrialization, urbanization, land reclamation, population growth, wastewater input, and phosphorus flow imbalance, among other contributing factors. Large amounts of anthropogenic nutrient inputs began in the 1970s (Wang, et al., 2009), and since 1985, Lake Dianchi has experienced severe cyanobacterial blooms (dominated by *Microcystis* spp., Li, et al., 2007). Presently, this lake is one of the key state-controlled lakes with an inferior water grade quality (worse than Grade V; Jin et al., 2008). In the 1990s, research and efforts to control eutrophication began. In recent years, great efforts have been made to improve point-source depuration technologies and to reinforce control policies (Jin et al., 2008; Guo and Sun, 2002; Guo, 2007; Qian, et al., 2007; Lu and Wang, 2008; Lu, et al., 2009; Medilanski, et al., 2006; 2007). Therefore, nonpoint source (NPS), or diffuse, pollution has emerged as the more problematic source of pollution in this region. There is even an Administration of Lake Dianchi, which has been charged with focusing on pollution control in the lake. Meanwhile, many action plans, programs, and measurements related to NPS pollution prevention and remediation have been implemented by different governmental departments or institutions sharing responsibility and authority for urban drainage, agricultural practices, land development, and other NPS pollution contributing factors. However, improvement and maintenance of the integrity in NPS pollution control for watershed management is still lacking. Here, we outlined an integrated watershed management (IWSM, FAO, 1986) framework for NPS pollution control, based on a recent NPS pollution survey, systematic analyses, and land use change trends in the basin. Generally, we hope this study will help decision-makers and administrative departments effectively control and reduce NPS pollution at the watershed scale and implement more holistic land management practices.

2 STUDY AREA

Lake Dianchi (24°40′–25°02′N, 102°36′–102°47′E) is located in the central part of the Yungui Plateau, in southwestern China, in the central portion of the Lake Dianchi Basin. This basin is semi-closed and ranges in altitude from 1887.5 m (the highest lake water level) to 2980 m. Lake Dianchi is a fault lake extending north to south. The lake is shaped like a shoe, 40 km long from the north to the south, and 7.5 km wide from the east to the west (it is the sixth largest freshwater lake in China). Total water surface area is 298 km^2 with a capacity of 1.29×10^8 m^3. It is typical of a shallow freshwater lake in a late development stage. Its maximum depth reaches 10.3 m with an average of 4.4 m (State Environmental Protection Administration of China, 2000); such shallow lakes are more prone to eutrophication than deep lakes (Qin et al., 2006).

The entire watershed has an area of 2 980 km^2. The western bank of the lake is bordered by steep escarpments of the Xishan Mountains. The northern, southern, and eastern parts of the lake basin consist of flat delta, alluvial plain, mesa, hillocks, hills, and rugged low and middle mountains. Kunming City, the capital of the Chinese Province of Yunnan, lies on the northern shore of Lake Dianchi, and new urban areas are expanding to east (Chenggong). Overall, 29 main streams flow into the lake. These rivers run through towns, fields, and phosphorus mining areas, and as such, they bring a large amount of sediments and pollutants into the lake. The lake water flows out through the Haikou River to the Jinshajiang River (Upper Yangzi River).

The Lake Dianchi basin has a subtropical monsoon climate, with a distinct wet and dry season (alternatively controlled by southwest monsoon and subtropical continental air mass). The annual rainfall varies from 797 to 1007 mm year^{-1}, with annual evaporation of 1 870–2 120 mm year^{-1} (Jin, 2003). The dry season is from October until the end of May. Approximately 70–80% of the annual rainfall is deposited during the remaining part of the year. Kunming is called "Spring City" for its moderate temperature, with annual mean temperature of around 14.7°C, and does not have cold winters (monthly mean temperature is 7.4°C in January) or hot summers (19.6°C in July). The lake plays an important role in the regulation of local climate, acting as a heat "sink" in summer and heat "source" in winter.

3 METHODOLOGY

3.1 Data survey and collection

Intrinsic features of the basin, including land use/land cover, slope, and soil attributes, affect water quality by regulating sediment and chemical concentrations (Basnyat, 2000; Lewis and Grimm, 2007). We first made a data collection, including climatology, cartography, land use, forestry, hydrology, agriculture, environment protection, and planning. These data inquiries were distributed to different departments and in personal correspondence to collect existing data. A scale of 1:50 000 was chosen for the basic map frame. SPOT 5 imageries (5 m resolution of panchromatic band) of 2009 and 2010 were used to derive the ecosystem composition and update the land use/land cover change.

In total, sixteen sub-watersheds were divided based on the hydrological units allocated in the SWAT (Soil and Water Assessment Tool; Arnold et al., 1998). In addition, the zonation of NPS pollution was divided into three endocentric zones (the head water catchment zone, the transition zone, and the lakeside zone). Doing this made the administrative contacts and measurements to be carried out in different controlling zones and in specific sub-watersheds more manageable.

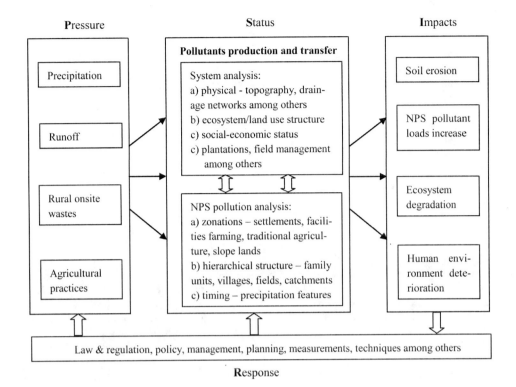

Figure 1. PSR model for NPS pollution analysis.

3.2 *PSR model for NPS pollution analysis*

The degradation of lake ecosystems is driven by processes that occur in the surrounding terrestrial ecosystem. NPSs deliver pollutants, mostly anthropogenic input nutrients such as nitrogen and phosphorus, to surface waters within watershed from diffuse origins (Carpenter et al., 1998). In general, NPS pollutants came from erosion, atmospheric deposition, runoff, and agricultural contaminants. Agricultural runoff is the dominant source of sediments, phosphorus, and nitrogen pollutants into aquatic ecosystems (USEPA, 2008; 2009). To understand the general information of NPS pollution in the Lake Dianchi Basin, we performed a systems analysis based on a Pressure-Status-Impacts-Response model (Figure 1).

There is a high degree of uncertainty about nonpoint emissions and their fate in NPS pollution (Shortle et al., 1998), including unpredictability and uncertainty driven by different weather patterns. We also noted that the groundwater table increased after precipitation, such that nutrients in the cultivated soil layer leached into to groundwater and were potentially exported to the lake water. Neglecting to collect sewerage in rural areas is common in China; therefore, on-site waste produced by local residents is problematic in NPS pollution control. In 2008, there were 780 490 farmer residents in the watershed. Total nitrogen (TN) produced by rural families was 796.9 t, total phosphorus (TP) was 257.2 t, and chemical oxygen demand (COD) was 8 809 t in the whole year. Most of these pollutants entered the environment without treatment. Additionally, NPSs of pollution were characteristic of inherent spatial heterogeneity that resulted from biogeochemical, topological, ecological and social-economic conditions (Shortle et al., 1998). Zonation and the hierarchical structure of NPS pollution needs to be understood in space and in time. That is, it is important to understand the critical source areas and times of NPS pollution in the watershed (MaClain, et al., 2003; Novotny, 2003).

3.3 Holistic control of NPS pollution in the Lake Dianchi Watershed

Because Lake Dianchi is a semi-closed plateau lake with endocentric drainage systems, three spheres were identified for NPS pollution control: the NPS pollution occurrence sphere, the planning and administrative sphere, and the technologies and measurements implementation sphere (Figure 2).

First, we divided the watershed into three territory zones. There are nine main reservoirs (four are drinking water sources) and other small ponds and pools in the watershed, which gathered the runoff from the upper-slope catchments. In this upper and headwater catchment zone, runoff and the main NPS pollutants were cut off and transferred to agricultural irrigation systems or depleted by drinking water supply systems. After water is used in urban areas, which are cumulatively populated by nearly three million people, it goes to wastewater treatment plants (eight main plants and another three are under construction). The middle transition zone, including urban and most of the agricultural systems, is where the most NPS pollutants are produced. Runoff and the irrigation water from the upper zone are enriched with agricultural pollutants and flows into the last, lakeside, zone. NPS pollutants in agricultural areas contribute much more than the urban ones to the lake; the latter being considerably reduced by urban grasslands, vegetated river banks, city rivers, and wastewater treatments. With the enforcement of lake water body protection, common housing and agriculture practices were made unlawful on the lakeshore. Furthermore, livestock and poultry were prohibited with 200 m of the riparian zones of the lake as well as streams and reservoirs. Wetland conversion and swamp forestation have been ongoing. Strict enforcement of ecological restoration in the riparian zone occurs, which can increase denitrification rates (Kaushal, et al., 2008). Thus, NPS pollutants, especially from agriculture contribute greater pollutant loads in the transition zone than in the other two zones.

Different planning, programs, and measurements have been conducted by different governmental departments in each zone. Management planning and administrative enforcement were directed and strategic, in contrast to technologies and measurements that were specific and practical for

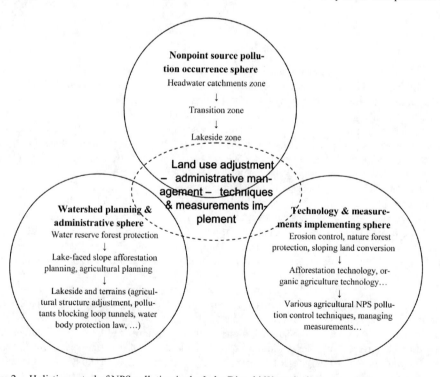

Figure 2. Holistic control of NPS pollution in the Lake Dianchi Watershed.

actual reductions and on-site controls. Three spheres could be integrated at different scales of space and time, linking land use planning and adjustment in the long-term and on a broad scale (Figure 2). Administrative management should be conducted in a practical way, and controlling techniques and measurements should be implemented at specific sites. These interfaces and links will be helpful in bridging the gap among scientific technologies, administration, and society.

4 RESULTS

4.1 *NPS pollution diagnosis and IWSM framework for NPS pollution control*

Root causes and endpoint effects of NPS pollution were determined by Sources-Flowpaths-Sinks model. For example, headwater streams exert control over nutrient export to lakes (Peterson et al., 2001). We recognized four critical aspects that were important in nutrient overload and NPS pollution control: ecosystem structures and services; land use/land cover; source-sink pattern of NPS pollutants at landscape level; and ecological barriers in terms of best management practices (BMPs, Ripa et al., 2006; Mitsch et al., 2001) (Figure 3).

4.1.1 *Ecosystem structure and service*

Ecosystem structure determines its functions. NPS pollutant loads increase with ecosystem degradation in ecosystem functioning and service. Ecosystem services do not always simply improve with increased forest coverage. For instance, many planted forests of exotic *Acacia dealbata* died during the drought in 2009. Cropland production increased with chemical fertilizer inputs, which results in scarifying soil sustainability.

4.1.2 *Land-use*

NPS pollution in watersheds strongly depends on land use (Basnyat et al., 2000; Gergel, 2005; Ripa et al., 2006). Patterns of agricultural land use were variable in the watershed, including rice paddies, facility farmlands of flowers and vegetables, non-irrigated slope croplands, terrain croplands, and orchards. Land use alternation happened rapidly over the last few decades. Lands are collective in China and farming families have the right to free land cultivation. However, rice paddies have been replaced gradually by facility croplands in the past 20 years. This increased local farmers' income, but increased the output of NPS pollutants, especially when farmers pursued high production in the short term. This unsustainable practice led to soil degradation and increased NPS pollution. In 2008, 3621 t of TN and 731 t of TP was produced in agricultural fields. 68% of TN and 75%

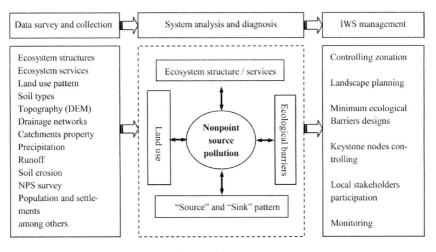

Figure 3. IWSM framework for NPS pollution control.

of TP were contributed by chemical fertilizers. Concurrently, with increasing urbanization, a new urban area is expanding on the lake's east bank, where most of the former facility farmlands were concentrated. Today, agricultural practices are centered at the south of the watershed.

With the prosperous growth of market economy in China, local farmers plant the crops that meet the market demands, trying to make more money from their croplands. Since 2009, concentrated livestock farms were forbidden in the watershed. Non-flower cultivation and limited vegetable cropland are part of the recent environment protection planning. Half of the cropland (400 000 mu, and 15 mu = 1 hectare) will be reduced soon. Heavily polluted agriculture will be gradually moved out of the watershed. Thus, governmental direction and policy will be considerably more important than before. Alternative plantations, agricultural ecotourism, organic agriculture, precise fertilization, and on the like would reduce NPS pollution loads. All of these policies may occur as long as there is no negative consequence for the income of local farmers.

4.1.3 Landscapes

The pattern of patchiness in land use/land cover determines the occurrence, movement, deposition, transformation and detention of NPS pollutants. Thus, it is necessary to understand the landscape-level sources and sinks of NPS pollutants' (Basnyat et al., 1999; Gergel, 2005; Chen et al., 2006; Zhang et al., 2009). Certain areas of a landscape have the capacity to function as "sources" (contributing) or "sinks" in the movement of nutrients, as determined by physical, ecological, and anthropogenic factors. In sink areas, biogeochemical processes transform inorganic nutrients, preventing movement of N or P into receiving waters (Gold, et al. 2001; MaClain, et al., 2003; Kellogg, 2010). Forests can act as sinks, and as the proportion of forest inside a contributing zone increases (or as agricultural land decreases), the levels of N downstream will decrease (Basnyat, 2000). Rural settlements, croplands, and orchards are strong contributors to the NPS pollutants.

Landscape planning, which includes a choice of afforestation sites, ecological corridor design, and rural environmental improvement, is important for controlling NPS pollution (Matthew et al., 2008). Viewshed analysis was conducted along the perimeter highway around the lake to optimize the landscape restoration design. The priority of afforestation sites depended on the vegetation status and ecological condition, as well as on visibility factors.

4.1.4 BMPs

Ecological barriers, including structured or non-structured BMPs, reduced the input of pollutants into lake water (Mitsch et al., 2001). Circle wetlands and vegetation belts are under construction along the lakeshore. A "Neo-villages" Project was established in 2009 in rural areas. A clean act plan for improving the rural human environment was initiated in the watershed by the government of Kunming City, enforced by a detailed environmental management program. Collection tunnels and wastewater treatment systems have been planned for each administrative village.

Minimum ecological barriers for NPS pollution control include wooded landscapes, vegetative filter strips, restored wetlands along the lakeside, or ecological ditches and eco-pools in croplands. Thus, it is urgent to apply BMPs in "hot spots" where impact reduction would not be enough for sustainable land use. In the south part of the Lake Dianchi Basin, there are phosphate mineral and industries where vegetation restoration on abandoned mining sites is crucial to control phosphorus loss.

NPS pollution is a diffusive process. A semi-closed lake system such as Lake Dianchi has certain critical nodes for transferring pollutants. Three loops of road systems around Lake Dianchi were under construction, along with the highway perimeter, an old circle road and recreation trails along the lakeshore. A closed tunnel system will be finished along the highway perimeter, which will collect and obstruct runoff sediments and pollutants.

4.2 Zonal eco-designs in NPS pollution control

4.2.1 Headwater protection zone

Headwater streams convey water and nutrients to larger streams and lakes. Thus, despite their relatively small dimensions, they play a disproportionately large role in N transformation on the landscape (Peterson et al., 2001). In this zone, the main objective was to control soil erosion and reserve water (Figure 4). Water in the reservoirs supports the social and economic development of the basin. Since 1998 and 1999, headwater forests have been under strict protection by the Natural Forest Protection Program (NFPP) and the Sloping Land Conversion Program (SLCP), respectively (Zhang et al., 2000; Ouyang et al., 2008). Furthermore, forest coverage has increased over the past few decades with the afforestation programs. However, most of these forests are monocultures of tree species such as *Eucalyptus globulus* (another common exotic species in the region), *Acacia dealbata*, *Pinus armadii*, and *Cupressus duclouxiana*. In the recent afforestation program, some native trees, such as *Cyclobalanopsis glaucoides*, *Photinia semiserrata*, *Cornus capitata*, and *Alnus nepalensis*, were planted in mixtures. However, ecological designs to facilitate vegetation regeneration are still necessary, which heretofore have been based on primitive community composite features and successional processes. Thus, forest ecosystem function and service will be improved.

4.2.2 *Transition zone*
Urban, as well as much of the concentrated rural, ecosystems are located in the transition zone (Figure 4). In the urban areas, rain water is gradually separated from wastewater sewage systems, and water saving and recycling programs are carried out in apartment complexes and governmental buildings. Urban river networks have been reformed, river silts were cleaned up, and vegetated corridors are designed along each river bank. In the rural areas, collective wastewater treatment systems have been built for each village, in addition to artificial wetlands and eco-tunnels. The recent nationwide "Neo-villages" project will improve the rural human environment and reduce diffused pollution in the countryside. Regarding croplands, multiple measurements and technologies were promoted, such as precise or balanced fertilization to reduce on-farm nitrogen and phosphorus use, intercropping, biological controls of pests, conservation tillage, farm waste recycling, impermeable agricultural drainage wells and sinkholes; and these actions will save irrigation systems and improve nutrition detention, among other benefits. We established a pilot area in the Chaihe River sub-basin in the south, and local families have been applying these control techniques.

4.2.3 *Lakeside zone*
The lakeside zone is the last source of NPS pollutants into the lake. In the 1950s, with the reclamation of agricultural lands on the riparian zone of the lake, a large area of wetlands was transformed into cultivated land. The lakeshore was solidified to control bank erosion to protect lakeside croplands and buildings. Under the strict water protection policy, agricultural activities were prohibited in the riparian zone and replaced with riparian buffers of natural and created wetlands and swamp forests. Restoration of wetlands was a main priority for the off-site control of agricultural NPS pollution. With more area shifted to restored wetlands and forested zones, pollutants will be more effectively absorbed and reduced, and the stability of lake banks will be improved.

Agricultural canals and ditches distributed among the rural transition zone and lakeside zone are important routes of preferential flow in agricultural systems (Haygarth et al., 1999; Figure 4). These networks are mitigating tools for NPS pollution loads and could be implemented for wetland restoration (Liu et al., 2010). Agricultural runoff and NPS pollutants transferred into the lake might be a sink and source for dissolved or particulate nutrients (Kröger, et al., 2008). Eco-ditches without solid bottoms would improve the reduction of nitrogen and phosphorus pollutants, and the resulting silts may be used as organic fertilizers. Eco-ponds or man-made wetlands may be constructed at important nodes of ditch networks to intercept the streams (Mitsch et al., 2001).

5 DISCUSSION AND RECOMMENDATIONS

Pollutants in runoff and seepage from agricultural land contribute significantly to water pollution in many areas around the world (Carpenter et al., 1998; Shortle et al., 1998; Ouyang et al., 2008; Bhuyan et al., 2003). With the prosperous economic growth and population pressures in China, the natural environment is facing a great deal of negative ecological impacts and challenges. The

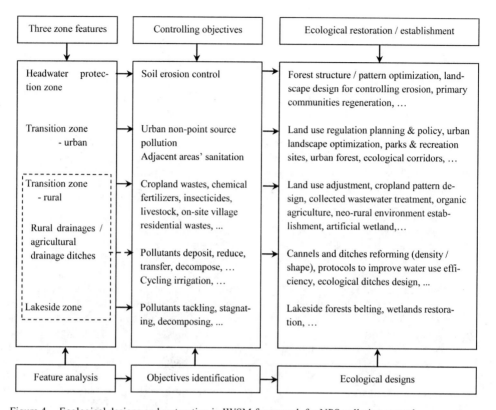

Figure 4. Ecological designs and restoration in IWSM framework for NPS pollution control.

"National Key Sciences and Technology Program for Water Solutions" was initiated in 2009 by the Ministry of Environment Protection (MEP) of the P.R.C. (former State Environmental Protection Administration of China, SEPA) to protect against the impairments of rivers and lakes. This three "five-year plan" program (2005–2020) will aim to tackle the water pollution situation in China. Lake Dianchi is one of the program sites, and our work is a part of that plan. However, there is no program in China similar to the USEPA Section 319 (established by the 1987 amendments to the Clean Water Act), to fund community-based efforts for fighting NPS pollution (USEPA, 2008; Hardy and Koontz, 2008). The health of highland lake ecosystems determines the downstream watershed eco-safety and water resources (FAO, 1986). A watershed protection plan is an important strategy for effectively protecting a watershed and thereby restoring aquatic ecosystems and protecting human health (Bhuyan et al., 2003; Hardy and Koontz, 2008). IWSM plans have been developed to provide a sound scientific solution for NPS pollution control (EPA, 2008; FAO, 2005). Scientifically sound options in ecosystem management practices help policy decision makers in balancing the relationship between development and sustainability. In addition, long-term monitoring of NPS and stakeholder participation in protecting against NPS pollution should be increased in the future.

Even though we considered the "integrated watershed" in this study, Lake Dianchi actually has extended "extra watersheds" through water division projects from the Zhangjiuhe and Niulanjiang Rivers to meet local water demand. Such sources outside of the catchment also contribute NPS pollutants into the basin. Southwest China includes mountainous regions with fragile eco-environments and intense human population pressures. Because of long-term over-burden, over-cultivation and disturbance, soils in this area have become heavily degraded (Ren and Peng, 2003). We hope that the framework developed here might be an example in creating NPS pollution control solutions for freshwater lake ecosystems with similar characteristics.

ACKNOWLEDGEMENTS

This work was supported by the "National Key Sciences and Technology Program for Water Solutions" (2008ZX07102-004).

REFERENCES

Arnold, J. G., R. Srinivasin, R. S. Muttiah, and J. R. Williams. 1998. Large area hydrologic modelling and assessment: Part I. Model development. *Journal of American Water Resources Association* 34(1):73–89.
Basnyat P., L. D. Teeter, B. G. Lockaby, and K. M. Flynn, 2000. The use of remote sensing and GIS in watershed level analyses of non-point source pollution problems. *Forest Ecology and Management*, 128(1–2):65–73.
Basnyat, P., L. D. Teeter, K. M. Flynn, and B. G. Lockaby. 1999. Relationships between landscape characteristics and nonpoint source pollution inputs to coastal estuaries. *Environmental Management* 23(4):539–549.
Bhuyan, S. J., J. K. Koelliker, L. J. Marzen, and J. A. Harrington. 2003. An integrated approach for water quality assessment of a Kansas watershed. *Environmental Modelling & Software* 18:473–484.
Carpenter, S.R., N. F. Caraco, D. L. Correll, R. W. Howarth, A. N. Sharpley, and V. H. Smith.1998. Nonpoint pollution of surface waters with phosphorus and nitrogen. *Ecological Applications* 8(3):559–568.
Chen L., B. Fu and W. Zhao. 2006. Source-sink landscape theory and its ecological significance. *Frontiers of Biology in China* 3(2):131–136.
FAO. 1986. Strategies, approaches and systems in integrated watershed management, from http://www.fao.org/docrep/006/ad085e/ad085e00.htm
FAO. 2005. Preparing the next generation of watershed management programmes projects: Asia, from http://www.fao.org/docrep/009/a0270e/a0270e00.htm
Gergel S. E., 2005. Spatial and non-spatial factors: When do they affect landscape indicators of watershed loading? *Landscape ecology* 20(2):177–189
Gold, A.J., P.M. Groffman, K. Addy, D.Q. Kellogg, M. Stolt, and A.E. Rosenblatt. 2001. Landscape attributes as controls on ground water nitrate removal capacity of riparian zones. *Journal of the American Water Resources Association* 37:1457–1464.
Guo H., and Y. Sun. 2002. Characteristic analysis and control strategies for eutrophicated problem of the Lake Dianchi. *Progress in Goegraphy* 21(5):500–506. (in Chinese)
Guo L. 2007. Doing battle with the green monster of Taihu Lake. *Science*, 2007, 317:1166
Hardy S. D. and T. M. Koontz. 2008. Reducing nonpoint source pollution through collaboration: policies and programs across the U.S. States. *Environmental Management* 41:301–310.
Haygarth, P.M., A.L. Heathwaite, S.C. Jarvis1, T.R. Harrod. 1999. Hydrological factors for phosphorus transfer from agricultural soils. *Advances in Agronomy* 69:153–178.
Heng H. H. and N. P. Nikolaidis. 1998. Modeling of nonpoint source pollution of nitrogen at the watershed scale. *Journal of the American Water Resources Association* 34(2):359–374.
Jin X, S Lu, X Hu, X Jiang, and F Wu. 2008. Control concept and countermeasures for shallow lakes' eutrophication in China. *Front. Environ. Sci. Engin. China* 2(3):257–266.
Jin X. 2003. Experience and lessons learned brief for Lake Dianchi, from http://www.worldlakes.org/uploads/Dianchi_12.26.03.pdf
Jin X., (Ed.) 2001. *Techniques for Control and Management of Lake Eutrophication*. Beijing: Chemical Industry Press. (in Chinese)
Kaushal, S. S., P.M. Groffman, P.M. Mayer, and E. Striz, and A.J. Gold. 2008. Effects of stream restoration on denitrification in an urbanizing watershed. *Ecological Applications* 18(3):789–804.
Kellogg, D. Q., A. J. Gold, S. Cox, K. Addy, and P.V. August. 2010. A geospatial approach for assessing denitrification sinks within lower-order catchments. *Ecological Engineering* 36:1596–1606.
Kröger R., M. M. Holland, M. T. Moore, and C. M. Cooper. 2008. Agricultural drainage ditches mitigate phosphorus loads as a function of hydrological variability. *Journal of Environmental Quality* 37:107–113
Lewis, D.B. and N.B. Grimm.2007. Hierarchical regulation of nitrogen export from urban catchments: interactions of storms and landscapes. *Ecological Applications* 17(8):2347–2364
Li H., G. Hou, D. Feng, B. Xiao, L. Song, and Y. Liu. 2007. Prediction and elucidation of the population dynamics of Microcystis spp. in Lake Dianchi (China) by means of artificial neural networks. *Ecological Informatics* 2:184–192.
Liu L., M. Peng, C. Wang, Y. Dong. 2010. Analysis on landscape characteristics of agricultural ditches in Lake Dianchi Basin. *Agricultural Science & Technology* 11(4):184–188.

Liu, W, and R Qiu. 2007. Water eutrophication in China and the combating strategies. *Journal of Chemical Technology and Biotechnology* 82(9):781–786

Lu S., P. Zhang, X. Jin, C. Xiang, M. Gui, J. Zhang, and F. Li. 2009. Nitrogen removal from agricultural runoff by full-scale constructed wetland in China. *Hydrobiologia*, 621:115–126.

Lu W., and H. Wang. 2008. Role of rural solid waste management in non-point source pollution control of Dianchi Lake catchments, China. *Front. Environ. Sci. Engin. China* 2(1):15–23.

Matthew W. D., J.T. Maxted, P.J. Nowak, and M. J. Vander Zanden. 2008. Landscape planning for agricultural nonpoint source pollution reduction I: a geographical allocation framework. *Environmental Management* 42:789–802.

McClain, M. E., E. W. Boyer, C. L. Dent, S. E. Gergel, N. B. Grimm, P. M. Groffman, S. C. Hart, J. W. Harvey, et al. 2003. Biogeochemical hot spots and hot moments at the interface of terrestrial and aquatic ecosystems. *Ecosystems*, 6:301–312.

Medilanski E., C. Liang. H.-J. Mosler, R. Schertenleib, and T. A. Larsen. 2006. Wastewater management in Kunming, China: a stakeholder perspective on measures at the source. *Environment & Urbanization* 18(2): 353–368.

Medilanski E., L. Chuan, H. -J. Mosler, R. Schertenleib, and T. A. Larsen. 2007. Identifying the institutional decision process to introduce decentralized sanitation in the city of Kunming (China). *Environ. Manage.* 39:648–662.

Mitsch W. J., J.W Day, J. W. Gilliam, P.M. Groffman, D. L. Hey, G. W. Randall, N. Wang. 2001. Reducing nitrogen loading to the Gulf of Mexico from the Mississippi River Basin: strategies to counter a persistent ecological problem. *BioScience* 51(5):373–388.

Novotny, V. 2003. The next step: Incorporating diffuse pollution abatement into watershed management. Pages 0-1–0-7 in Proceedings of the 7th International Specialized Conference on Diffuse Pollution and Basin Management, 18–22 August 2003, Dublin, from http://www.ucd.ie/dipcon/docs/keynote_paper.pdf

Ouyang, W.F. Hao,X.Wang,H. Cheng. 2008. Nonpoint Source Pollution Responses Simulation for Conversion Cropland to Forest in Mountains by SWAT in China. *Environmental Management* 41:79–89.

Peterson B. J., W. M. Wollheim, P. J. Mulholland, J. R. Webster, J. L. Meyer, J. L. Tank, E. Martí, W. B. Bowden, et al., 2001. Control of nitrogen export from watersheds by headwater streams, *Science*, 292:86–90.

Qian Y., X. Wen, and X. Huang. 2007. Development and application of some renovated technologies for municipal wastewater treatment in China. *Front. Environ. Sci. Engin. China* 1(1):1–12.

Qin B., L. Yang, F. Chen, G. Zhu, L. Zhang, and Y. Chen. 2006. Mechanism and control of lake eutrophication. *Chinese Science Bulletin* 51(19): 2401–2412.

Ren H. and S. Peng. 2003. The practice of ecological restoration in china: a brief history and conference report. *Ecological Restoration* 21(2):122–125.

Ripa M. N., A. Leone, M. Garnier, and A. Lo Porto. 2006. Agricultural Land Use and Best Management Practices to Control Nonpoint Water Pollution. *Environmental Management* 38(2):253–266.

Shortle J. S., D. G. Abler, and R. D. Horan. 1998. Research issues in nonpoint pollution control. *Environmental and Resource Economics* 11(3–4):571–585.

State Environmental Protection Administration of China. 2000. *Water Pollution Prevention and Control Plans and Programs for Huaihe, Haihe Rviver, Liaohe River, Taihu Lake, Chaohu Lake and Dianchi Lake in China*. Beijing: China Environmental Science Press. (in Chinese)

Sun, S., and C. Zhang. 2000. Nitrogen distribution in the lakes and lacustrine of China. *Nutrient Cycling in Agroecosystems* 57:23–31

U.S. Environmental Protection Agency. 2008. Handbook for Developing Watershed Plans to Restore and Protect Our Waters, from http://water.epa.gov/polwaste/nps/handbook_index.cfm.

U.S. Environmental Protection Agency. 2009. National water quality inventory: Report to Congress, 2004 cycle report. Office of Water. Washington, DC. 20460, from http://water.epa.gov/lawsregs/guidance/cwa/305b/2004report_index.cfm

Wang F., C. Liu, M. Wu, Y. Yu, F. Wu, S. Lü, Z. Wei, and G. Xu. 2009. Stable isotopes in sedimentary organic matter from Lake Dianchi and their indication of eutrophication history. *Water Air Soil Pollution* 199:159–170.

Zhang J., J. Jiang, Z. Zhang, Q. Shan, G. Chen, Y. Wang, Y. Xu, H. Wu, and A. Abarquezd. 2009. Discussion on role of forest to control agricultural non-point source pollution in Taihu Lake Basin – based on Source-Sink analysis. *J. Water Resource and Protection* 1:345–350.

Zhang P., G. Shao, G. Zhao, D. C. Le Master, G. R. Parker, J. Dunning, and Q. Li. 2000. China's Forest Policy for the 21st Century. *Science* 288(5474):2135–2136.

Study and analysis on spatio-temporal variation of water quality of the middle and lower reaches of Hanjiang River

Lei Song
School Law, Kunming University of Science and Technology, Kunming, China

Tao Li
Hubei Environmental Science Research Academy, Wuhan, China

ABSTRACT: Based on monitoring data of Hanjiang River water quality during 2001–2011 on different sections (8 monitoring sections), three great harmful pollution indexes, i.e. CODMn, NH3-N and TP, are picked, analyzed and compared using single factor index analysis method. The results shows that: (1) Temporal variation: Water quality of the middle and lower reaches of Hanjiang River in recent years sees a trend of deteriorate. Density of phosphorus has an apparent augment. (2) Spatial variation: from higher to lower reaches, density of pollutant in the main section of the Hanjiang has rising. Therefore, there are a lot to be done on the pollution control of Hanjiang River to ensure the coordinated development of economy and environment.

1 INTRODUCTION

The single pollution index method which using a simple and intuitive graphic, the regression analysis based on parametric tests and non-parametric tests research method, these three methods are the most commonly used methods on study of river water quality trends. With the monitoring data of Hanjiang River reaches the main control site water quality, single pollution index method is used to analyze water quality status by in recent years and CODMn, NH3-N and TP three indicators are studied.

With a length of 1570 km and the area of 159,000 square kilometers, Hanjiang River, which is originated in the Qinling Mountains, is the largest tributary of the Yangtze River. It flows through Shaanxi and Hubei provinces, and ended into the Yangtze River in Wuhan. Hanjiang River can be divided into higher reaches which is the part above Danjiangkou, middle reaches, Danjiangkou to Nianpanshan, and the rest from Nianpanshan to Wuhan Han River estuary the lower reaches.

Hanjiang River is the pearl of Hubei Province. It is an important base of grain and cotton production and freshwater aquaculture in China. The lower reaches of Hanjiang River is rich in mineral resources and flows through a densely populated area with booming economic development. However, with the rapid economic development of the Han River, accelerated urbanization and water resource utilization, the total wastewater discharged into the Han River increases yearly and most of which are not even decontaminated, resulting in obvious deterioration of Hanjiang River water quality in recent years. Therefore, the study of the Hanjiang River water quality changes in recent decades has a profound significance for both the Hubei sustained economic development in the future and Han River water quality protection and water resource management.

2 MATERIALS AND METHODS

There are 19 river water quality monitoring sites along where Hanjiang River flows through. (Figure 1).

With data collected in above sites during 2001–2011's three water period (moderate period, flood period, low period), nine monitoring factors (dissolved oxygen, permanganate index, BOD5, ammonia nitrogen, total phosphorus, total mercury, lead, volatile phenol, oil) are evaluated using GB3838-2002 Class III surface water quality standards. Results are as follows: (Table 1 and table 2).

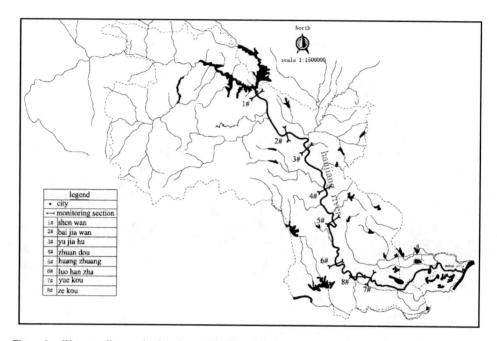

Figure 1. Water quality monitoring sites on Hanjiang River.

Table 1. 2001~2011 Hanjiang River Water Quality.

Class	I			II			III		
	F	L	M	F	L	M	F	L	M
Occurred Time	2	0	5	60	62	65	24	21	16
Percent (%)	0.8	0	1.9	22.8	23.6	24.7	9.1	8.0	6.1

Class	IV			V			>V		
	F	L	M	F	L	M	F	L	M
Occurred Time	1	3	0	0	0	0	0	2	2
Percent (%)	0.4	1.1	0	0	0	0	0	0.8	0.8

*F: Flood period; L: Low period; M: Moderate period.

Table 2. Superscale of Hanjiang River Water Quality during 2001–2011.

Period	Flood	Low	Moderate
Superscaled Count	19	18	12
Percent (%)	7.2	6.8	4.6

3 RESULT AND ANAYSIS

3.1 Overall water quality of Hanjiang River and Pollution Analysis

As can be seen from Table 1, middle and lower reaches of Han River are basically II, III water quality, at a percent of 94.3% (248 out of 263).

As from table 2, 49 in total exceeded yearly criteria, with a relation of flood period > low period > moderate period.

A spatial variation can be seen in Hanjiang River water quality. During 2001~2011, Zhongxiang and its above together with Tianmen and its below parts shows a better water quality, more than 93.9% are within II or III quality criteria. 42.4% of the part from Zhongxiang to Tianmen cannot reach the standard of II class: 33.3% time in Zhongxiang exceeded II class with ammonia and BOD5 causing most pollution and 45.5% in Tianmen with phosphorus as major pollution

3.2 Temporal variation of middle and lower reaches of Hanjiang River water quality

CODMn, NH3-N and TP are used as three indicator in the analyses of the water quality of two typical sections of Hanjiang River (Laohekoushenwan and Qianjiangzekou) during three water time (Figure 2~Figure 4).

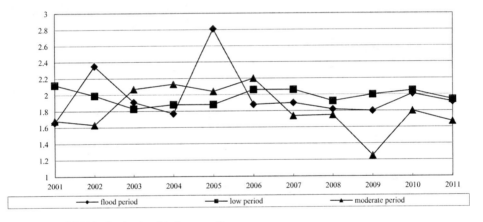

Figure 2. a) CODMn density (mg/L) change at Shenwan.

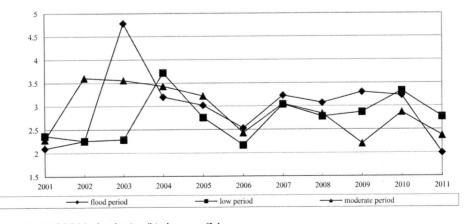

Figure 2. b) CODMn density (mg/L) change at Zekou.

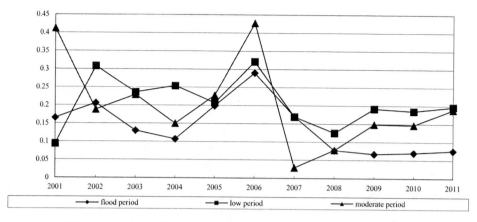

Figure 3. a) NH3 density (mg/L) change at Shenwan.

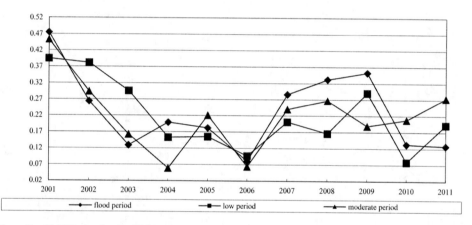

Figure 3. b) NH3 density (mg/L) change at Zekou.

CODMn: CODMn index basically stay the same in Laohekoushenwan, and only got a small fluctuation in Qianjiangzekou.

NH3: Changes Laohekoushenwan can be divided into three phases: from 2001 to 2006, NH3 index goes up and reached the maximum in 2006; 2006 to 2008, goes down; 2008~2011, stays essentially unchanged.

For Qianjiangzekou, 2001 to 2005 goes down, and reach the minimum level; 2005 to 2009 gradually increased; 2009 to 2011 decreased again. From water time aspect, the index during flood period is lower than that of low water period.

TP: Total phosphorus in Laohekoushenwan rises overtime. And in Qianjiangzekou, it went up at first then fall back and up again in recent years.

3.3 Spatial variation of Hanjiang River water quality

As it is shown in Figure 5, the average density of CODMn during 2001–2011 at different sites varies a lot. From the higher reach to lower reach, Shenwan-Yujiahu density goes up dramatically, Yujiahu-Zekou basically stays steady. Yukou has the highest density among all. And among the three period, low water period makes a better water quality than the others.

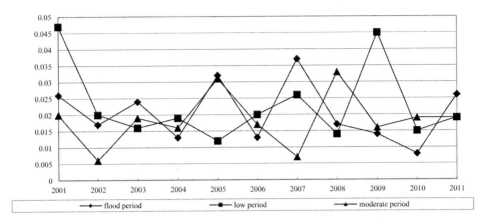

Figure 4. a) TP density (mg/L) change at Shenwan.

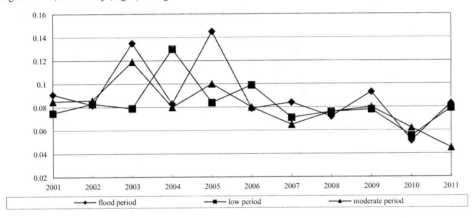

Figure 4. b) TP density (mg/L) change at Zekou.

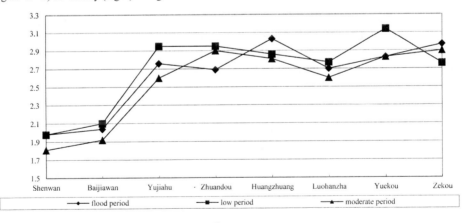

Figure 5. CODMn density (mg/L) change at Hanjiang.

Figure 6 focus on NH3 level. Shenwan-Luohan has an increase trend and Luohan made the highest density. Luohan to Zekou shows a slight decline. Water quality during flood period is better than low and moderate period.

Figure 7 shows the variation on TP. From Shenwan-Yujiahu, TP density goes up, and Yujiahu-Zekou, TP goes down. Flood period has better water quality.

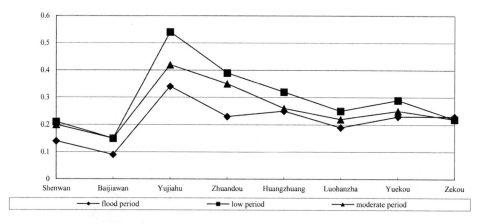

Figure 6. NH3 density (mg/L) change at Hanjiang.

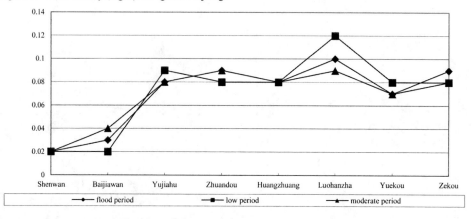

Figure 7. TP density (mg/L) change at Hanjiang.

Overall, density of different pollutant varies a lot from one another. From middle to lower reach, density of overall pollutant increased, thus lead to wore pollution lower reach.

4 REASONS FOR THE WATER QUALITY VARIATION OF MIDDLE AND LOWER REACHES OF HANJIANG RIVER

Zhongxiang to Tianmen city and Tianmenluohanzha part of Hanjiang River suffers an inferior water quality. Indicators shows a majority of pollutant is TP, secondly NH3. Pollution during flood period is worse. The main reasons are as follows:

1. Pollutant cumulative. Middle and lower reaches of Hanjiang River continuously get contaminated from daily residential wastewater from upper cites like Xiangyang and Tianmen as well as industry wastewater and eventually exceed its self-purification ability.
2. Agricultural source pollution. Hanjiang River flows through Jianhan Plain where agricultural activity is frequent and intense. Meanwhile, the plain has limited vegetation and rainfall has a great impact. Massive fertilizer usage and flat earth surface make it easy to make unit overload. It can be inferred from the fact that most severe pollution happen during low water period and that ammonia made the main pollution.
3. Rich phosphate resources. Phosphate mineral resource is rich in middle and lower reaches of Hanjiang River. Dissolution of phosphate and mining pollution is another important factor resulting in the superscale phosphate in water.

5 SOLUTION

1) Advertise to improve the awareness of environment protection of general public. Government should, through a variety of media, inform the public the current water quality conditions, trends, potential threats and hazards, so that people are fully aware of water pollution prevention and the direct relationship between society and the individual. Thus establish a correct concept of environmental protection, and consciously protect the environment by behaviors like water save, pollution fines, industrial structure adjustment, use of green products, etc. In addition to raising public awareness of environmental protection, it is also needed to make public aware of protecting their right such as environmental reporting, environmental monitoring, environmental claims, etc.
2) Speed up the construction of urban sewage treatment facilities. During the past 10 years, residential wastewater has become a major source of pollution affecting the water quality of the Han River. Therefore, it is urgent to strengthen the construction and operation of sewage treatment facilities around the city, raise the standard from GB18918-2002 class B to class A.
3) Strengthen the treatment of urban industrial pollution. It is necessary to continue conform "One control, Two Goals" actions, stick with the "Hubei Hanjiang River Water Pollution Control Ordinance" to strictly control new industrial pollution sources. Through the optimization of industrial structure, strengthening of environmental management, and the adoption of cleaner production technology and advanced wastewater treatment process and other measures, industrial wastewater emissions shall be controlled effectively.
4) Speed up the ban of phosphorus. Excessive of total phosphorus is an important factor responsible for the "water bloom" happened in the middle and lower reaches of Hanjiang River in 1990s. To prevent this from happening again, phosphate and its derivatives shall be strictly controlled. In 2000, the ban suggested by Hubei Province in "Hubei Hanjiang River Water Pollution Control Ordinance" is still necessary today.

    Agricultural non-point source load will change as seasonal agricultural production, the study of non-point sources of water pollution and its contribution rate during different water period will give the government a better understanding and guidance on the control of non-point pollutions. It is also necessary to strengthen the regulation on phosphate mineral and its production management.
5) Strengthen tributaries water pollution treatment. Improve the water quality of Tangbaihe. Manhe, Xiaoqinghe and other major tributaries so that less polluted water joins the main stream and thus reduce the pollution of Hanjiang River.

REFERENCES

Chen Dajun, 1996. *North Water Transfer Project on the Impact of Hanjiang River*: 51
Chenyong Shu & Wu Fucheng, 2004. *Xiangjiang River in the past 20 years changes in water quality analysis. Resources and Environment in the Yangtze River*: 13
Liu Peitong, 1977. *Introduction to Environmental Science*: 165–175 Higher Education Press
Yin Kuihao, 2001. *North Water Transfer Project on the Hanjiang River "bloom" effect*: 31–36
Zhang jiuhong, 2004. *Hanjiang River Water Pollution Situation and trend analysis. Water Resources Protection*

# Different ratio of rice straw and vermiculite mixing with sewage sludge during composting

M. Yu, Y.X. Xu, H. Xiao, W.H. An, H. Xi, Y.L. Wang & A.L. Shen
Environmental Resources and Soil Fertilizer Institute, Zhejiang Academy of Agricultural Sciences, Hangzhou, Zhejiang, China

J.C. Zhang & H.L. Huang
College of Resources and Environment, Hunan Agricultural University, Changsha, Hunan, China

ABSTRACT: To identify and prioritize the addition ratio of rice straw and vermiculite as raw materials during sewage sludge composting, the physico-chemical parameters and the changes of microbial communities analyzed by phospholipid fatty acids (PLFA) were investigated. Four runs containing sewage sludge were used: adding only rice straw (Run A; addition ratio (dry weight) rice straw:vermiculite = 1:1 (Run B); addition ratio (dry weight) rice straw:vermiculite = 3:1 (Run C) and adding only vermiculite (Run D). The results showed that the physico-chemical parameters except C/N ratio among four runs have not obvious difference. The totPLFA was abundance in Run C. While in Run D it reached the peak value on day 6, then decreased and was lower than that in other runs after day 12. Based on the PLFA markers of Gram-positive bacteria (G^+), Gram-positive bacteria (G^-) and fungi, the dominant speices were respectively G^- in Run A, fungi and G^+ in Run B, fungi and G^+ in Run D before day 18. In the later stage all speices began decreasing, exept for a slight increase of bacteria G^+ and G^- in Run A appeared. It suggested that the addition ratio in Run C was more proper.

1 INTRODUCTION

The amount of sewage sludge generated from sewage treatment plants is around 30000000 tons discharging moisture content 80% per year with the development of economy in China (Chen 2012). Moreover, rice is an important crop in China. Its residue is usually burnt in the field in autumn which causes environmental problem. Similar reports have issued in Japan (Arai 1998 and Torigoe 2000). An attractive alternative to recycling such waste is composting. Due to the characteristics of sewage sludge, high water content, organic matter content is not high, small porosity, high viscosity and so on, therefore using sewage sludge as raw materials composting, usually with compost amendment. The characteristics of rice straw and sewage sludge are complementary for composting; besides, the optimal conditions for blends to reach maximum microbial activity in the first steps of composting have also been reported (Roca-Pérez 2005). On the other hand, vermiculite is a good compost amendment as it is absorbent and breathability. During composting, it is necessary to know how the physical, chemical and biological parameters evolve over time to improve the process (S Amir 2009, Roca-Pérez 2009). According to the report of Iranzo et al. (2004), However, we found no works which evaluate the composting process and the compost quality using these raw materials exclusively. Therefore, it is necessary to study the evolution of the physical and chemical parameters during the process, and also the effect of microbial community succession on material degradation during composting (Lhadi 2004). However, only 10% or less microorganisms in an environmental sample can be successfully cultured on agar media, as plating method is selective and unable to give a representative picture of microbial flora (Kristin 2003). PLFA analysis is that the most microorganisms contain phospholipids in their membranes when they are alive, and different microbial communities differ in their fatty acid compositions. Although few PLFAs can be considered as absolute signature substances for a single species or even a specific group of organisms, it is possible to get indications of changes in major groups, like fungi, actinomycetes, gram-positive (G^+) and gram-negative (G^-) bacteria (Frostegård 1996). PLFA analysis can give more information of living microbes as PLFAs in their membranes are not stored but turned over relatively rapidly during metabolism (Kristin 2003, Yu 2009).

In this study, composting mixtures of sewage sludge with different proportions of rice straw and vermiculite were set up. The objective was to identify the optimal proportion of raw materials by analyzing the physico-chemical parameters and microbial community structures using PLFA analysis, and to elucidate the relationship among the physico-chemical parameters, composting materials and microbial community.

2 MATERIALS AND METHODS

2.1 *Composting experimental set up materials collection and processing*

Sewage sludge was collected from sewage treatment plant in Hangzhou. Rice straw was collected from suburban areas, which was dried and cut to 10–20 mm lengths. Vermiculite was bought from local flower market.

Four runs as an experimental composting system with a weight of about 40 kg were set up in this study. There was 25 kg sewage sludge in each run. The total weight of rice straw and vermiculite was 10 kg in each run. The ratios of rice straw and vermiculite were set as follows: adding only rice straw (Run A), rice straw:vermiculite = 1:1 (Run B), rice straw:vermiculite = 3:1 (Run C) and adding only vermiculite (Run D). Moisture was monitored and adjusted to about 55% during the first fermentation phase and about 45% during the second fermentation phase by the addition of sterile water. To provide some aeration, the mixture was turned twice a week in the first 2 weeks and then once a week afterwards.

2.2 *Analysis*

The temperature in the center of the composting materials was monitored. Samples were taken at various stages of the composting process (0, 3, 6, 9, 12, 15, 18, 24, 30 and 42 days) to measure the C/N ratio. The moisture content of samples was determined after drying at 105°C for 24 h. The dried samples were ground and analyzed for total organic carbon (TOC) by dry combustion. And the total nitrogen (TN) was measured by Kjeldahl method (Bremner 1982). The germination index was analyzed by the method of Tiquia 1996 using radish seed.

PLFA analysis was conducted as described by Yu 2009. Analyses were done using the results for individual PLFA with carbon chain lengths from 13 to 18.

3 RESULTS AND DISCUSSION

3.1 *Physico-chemical parameters*

The changes of temperature during the composting are shown in Fig 1(a). The composting process could be divided into two phases: (i) the first fermentation phase (day 0 to day 12) (ii) the second fermentation phase (day 13 to day 42). High temperature (exceeding 50°C) was maintained for at least 10 days in the compost heaps during the first fermentation phase in Run A–C.

The pile temperature in Run D increased rapidly in the first three days, and the maximum temperature exceeded 70°C. But the thermophilic stage did not keep a long time, and in the later stage the pile temperature in Run D was lower than that in other runs. It suggested that vermiculite did have strong absorbent and breathability which can make moisture content proper for pile fermentation at the beginning, but it can not provide carbon source for the growth of microorganism.

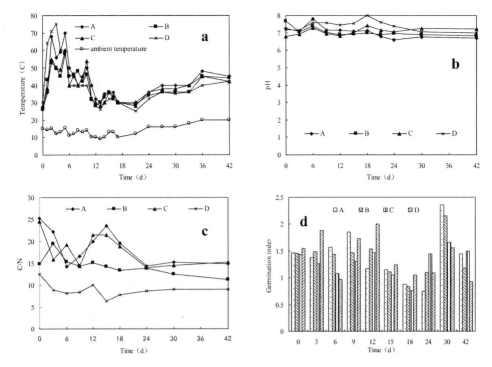

Figure 1. Changes of physico-chemical parameters during composting process.

Abdelmajid (2005) reported that more heat output probably was the result of biological activity in compost. So the result in this study indicated that the biological activity in Run D was lower.

The C/N ratio changed from 25 to 10 in Run A and C during the composting process (Fig. 1b). While the results in Run B and D which the adding ratio of vermiculite was little higher showed no obvious differences and were below 15. This observation suggests that the addition of vermiculite must be moderate, Too much will be not conducive to adjust the initial C/N of composting for fermentation reaction sustained.

The change of pH value was shown in Fig. 1c. No significant change and differences were found. According to the previous researchers, a germination index value of 80% has been used not only as an indicator of the disappearance of phytotoxicity in compost (Tiquia 1996). The germination index of radish was almost exceeded 80% during composting time in all four runs (Fig. 1d). There was little change of the index in Run B and C over time, while fluctutation in Run A and D. The results showed that rice straw and vermiculite need be added simultaneously for sewage sludge composting.

3.2 Changes in microbial community structure

In the experiment, 45 patterns of PLFA were detected during the composting, and 37 kinds were used for analyzed. The results were shown in Fig. 2.

Total PLFA (totPLFA) has been suggested as a measure of the total microbial biomass, and it has been shown to correlate well with the biomass measurements using fumigation extraction technique in municipal solid waste compost in Hellmann (1997)'s work. In this experiment, the totPLFA was used to suggest the changes in microbial biomass as illustrated in Fig. 2a.

The changes of totPLFA content were different among four runs. Thus, the proportion of the raw materials had a significant effect on the microbial biomass in the processes. In the first 3 days the

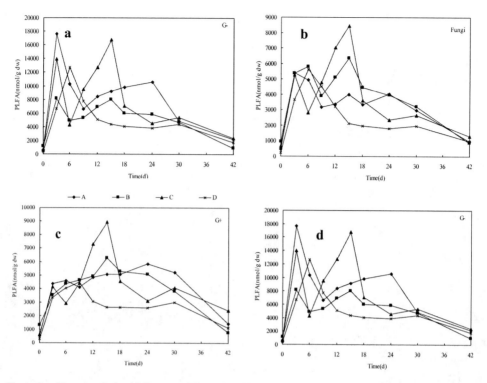

Figure 2. Changes of microbial communities analyzed by phospholipid fatty acids (PLFA) during composting process.

totPLFA content increased quickly among Run A-C. Run D, which was not added with rice straw, had a lower totPLFA content than the other runs, and then it peaked on day 6. It indicated that rice straw can stimulate the growth of other microbe and improve the biological activities during sewage sludge composting as reported by Roca-Pérez 2005.

The main changes in microbial community structure were determined by marker fatty acids. PLFA 18:1ω9c and 18:2ω6,9 were used as the marker for fungi; the gram-positive bacteria (G^+) was estimated by the sum of the following fatty acids: i15:0, a15:0, i16:0, i17:0,; the gram-negative bacteria (G^-) was represented by the sum of: 16:1ω5, 16:1ω7t, 16:1ω9c, cy17:0, 18:1ω9t, 18:1ω7, cy19:0 (Frostegård 1996).

As Fig. 2(b–d) shown, the dominant speices were respectively G^- in Run A, fungi and G^+ in Run B, fungi and G^+ in Run D before day 18. In the later stage all speices began decreasing, while a slight increase of bacteria G^+ and G^- appeared in Run A which was not add vermiculite. It suggested that the addition ratio in Run C was more proper.

4 CONCLUSIONS

The PLFA analysis method was successfully used to analyze the changes of microbial community structure and biomass during sewage sludge composting. It has been proved that the proportion of the raw materials had a significant effect on the microbial biomass in the processes. Rice straw can stimulate the growth of other microbe and improve the biological activities, while vermiculite can adjust the moisture content and ventilation during sewage sludge composting. Rice straw and vermiculite need be added simultaneously for sewage sludge composting, and the addition amount of rice straw should be higher than that of vermiculite.

ACKNOWLEDGEMENTS

The study was financially supported by the National Natural Science Foundation of China (51108423, 51108178) and the National Natural Science Foundation of Zhejiang Province (Y5100234).

REFERENCES

Abdelmajid J. 2005. Chemical and spectroscopic analysis of organic matter transformation during composting of sewage sludge and green plant waste. Int Biodeterior Biodegrad 56: 101–108.
Amir S. 2010. PLFAs of the microbial communities in the composting mixtures of agro-industry sludge with different proportions of household waste. International biodeterioration & biodegradation. 64: 614–621.
Arai, T. 1998. Bronchial asthma induced by rice. Int. Med. 37: 98–101.
Bremner J. M. 1982. Nitrogen total. In Page, A. L., Miller, R. H., Keeney, D. R. (Eds.), Methods of Soil Analysis, Part 2: 371–378. Am. Soc. Agron. Madison
Frostegård Å. 1996. The use of phosphlipid fatty acid analysis to estimate bacterial and fungal biomass in soil. Biol Fertil Soils 22: 59–65.
Hellmann B. 1997. Emission of climate-relevant trace gases and succession of microbial communities during open-window composting. Appl Environ Microbiol 63: 1011–1018.
Iranzo M. 2004. Characteristics of rice straw and sewage sludge as composting materials in Valencia (Spain). Bioresource Technol 95: 107–112.
Kristin S. 2003. Comparison of signature lipid methods to determine microbial community structure in compost. J Microbiol Methods 55: 371–382.
Lhadi E. K. 2004. Co-composting separate MSW and poultry manure in Morocco. Compost Sci. Util. 12: 137–144.
Roca-Pérez L. 2009. composting rice straw with sewage sludge and compost effects on the soil-plant system. Chemosphere 75: 781–787.
Tiquia SM. 1996. Effects of composting on phytotoxicity of spent pig-manure sawdust litter. Environ Pollut 93: 249–256.
Torigoe, K. 2000. Influence of emission from rice straw burning on bronchial asthma in children. Pediatr. Int 42: 143–150.
Yong C. 2012. Sewage sludge aerobic composting technology research progeress. AASRI Procedia 1: 339–343.
Yu M. 2009. Microbial communities responsible for the solid-state fermentation of rice straw estimated by phospholipid fatty acid (PLFA) analysis. Process Biochemistry 44: 17–22.

Heavy metal contamination in surface sediments of the Pearl River Estuary, Southern China

Meiying Song
Department of Chemistry, College of Life science and Technology, Jinan University, Guangzhou, China

Pinghe Yin
Department of Chemistry, College of Life science and Technology, Jinan University, Guangzhou, China
Research Center of Analysis and Test, Jinan University, Guangzhou, China

Ling Zhao
Department of Environmental Engineering, Jinan University, Guangzhou, China

Zepeng Jiao
Research Center of Analysis and Test, Jinan University, Guangzhou, China

Shunshan Duan
Research Center of Hydrobiology, College of Life science and Technology, Jinan University, China

ABSTRACT: Sediment samples collected at six freshwater discharge outlets of the Pearl River Estuary were analysed for metal contamination The result showed that the concentration ranged (in mg/kg dry weight): Fe, 30,140–51,920; Cd, 0.23–1.09; Cr, 48.7–109; Pb, 22.3–70.2; respectively. Cd and Pb showed a similar spatial distribution in the sediment which reflected similar anthropogenic origins. While the concentrations of Cr in the sediments of all the sites were similar. Further more, the enrichment factor (EF) and the geoaccumulation index (I_{geo}) were used to comprehensively evaluate the pollution degree of heavy metals. I_{geo} indicated the sediment of the PRE had to be considered as 'unpolluted' to 'moderately polluted', while the EF of the sediment would be ranked as 'minor enrichment'. This findings will be useful in proposing measures for strategic environmental control in estuary.

1 INTRODUCTION

The Pearl River Delta (PRD) is a highly developed and industrial area of China, Since 70's, the economy and society in the Pearl River Delta region have been undergoing a great development and billions tons of total wastewater were discharged every year. These wastes were mostly transported and deposited in the Pearl River Estuary (PRE) which is created by inflows of freshwater to the South China Sea. The PRE is made of eight freshwater discharge outlets (Humen, Jiaomen, Hongqimen, Hengmen, Jitimen, Modaomen, Hutiaomen and Yamen) whose sediments become one of the most important records to reveal environmental changes and anthropogenic impacts of the PRE. Estuarine areas are complex and dynamic aquatic environment (Morris et al. 1995), the accumulation of heavy metal contaminants in its freshwater discharge outlets sediments can pose serious environmental problems to the surrounding areas. Cd, Cr and Pb are an important toxic pollutant source in the PRD and are not biodegradable in general, it is therefore significant to comprehensively understand the distribution and contamination of Cd, Cr and Pb in the sediments of its freshwater discharge outlets.

Numerous studies have dealt with heavy metal concentration, distribution pattern sources, chemical, chemical forms and contamination (Ip et al. 2007, Yu et al. 2010, Chen et al. 2012) in the PRE. To our knowledge, few studies have comprehensively researched the distribution and risk assessment of Cd, Cr and Pb in sediments of the six freshwater discharge outlet (Humen, Jiaomen, Hongqimen, Jitimen, Yamen and Modaomen) of the PRE. on the basis of distributional characteristics of heavy metals in water of the Pearl River (Zhao et al. 2011), the present study was to further analysis the distribution and contamination of heavy metals (Cd, Cr and Pb) in the sediments of the PRE, which will be useful in proposing measures for strategic environmental control in the PRD.

2 MATERIALS AND METHODS

2.1 Sediment sampling and pre-treatment

Seven sediment samples were collected in the freshwater discharge outlets of PRE In January 2013, which were Jiaomen (S1) (22°36′82″N, 113°39′26″E), Hongqimen (S2) (22°35′02″N, 113°37′16″E), Humen (S3) (22°49′12″N, 113°35′36″E) Humen (S4) (22°48′87″N, 113°34′75″E), Modaomen (S5) (22°11′00″N, 113°24′34″E), Yamen (S6) (22°11′79″N, 113°05′45″E) and Jitimen (S7) (22°11′16″N, 113°18′86″E) The sediment samples were taken with a gravity corer, for physical and chemical analysis sediment samples were carefully stored in polyethylene bags. The samples were air-dried, then large debris and fragments of shells were removed before grounding. All of the sediment samples were then ground in an agate grinder until the sediments all sieved through a 60 meshes (0.3 mm) sieve, the particles with size less than 0.3 mm were obtained.

2.2 Analysis of heavy metal concentration

The sediment samples were digested in concentrated nitric acid, Hydrofluoric acid and Hydrogen peroxide of 30% to dissolve heavy metals in solution (Yang et al. 2012). The heavy metals (Pb, Cr, Fe) concentration of the solutions were measured by Inductively Coupled Plasma-Atomic Emission Spectrometry (ICP-AES, Perkin Elmer Optima, 2000DV), and Cd was measured by Inductively Coupled Plasma-Mass Spectrometry (ICP-MS, Thermo Fisher scientific X series 2). For the analytical quality control, reagent blanks, standard reference materials (GBW07333) from the State Oceanic Administration of China, and sample replicates were inserted in the analysis. The result showed that there was no sign of contamination in the analysis and all of the relative standard deviations of the replicate samples were <10%. The recovery rates for the heavy metals and major elements in GBW07333 were higher than 80%.

2.3 Environmental risk assessment

In order to assess the environmental risk comprehensively, two different levels, i.e. accumulation level (I_{geo}), and the enrichment factor (EF) were selected. Geo-accumulation index (I_{geo}) (Müller 1969) is a classical assessment model for indicating heavy metal accumulation in sediments, and calculated using the following model:

$$I_{geo} = \log_2[C_i/(1.5 \times B_i)] \qquad (1)$$

where C_i is the content of measured metal "i" in the samples, B_i is the crustal shale background content of the metal "i" (Taylor 1964), the constant of 1.5 is introduced to minimize the variation of background values due to lithogenetic origins, and I_{geo} is a quantitative index of metal enrichment. We referred geochemical background values (as mg/kg) as described by Lau et al. (1993) 0.4 for Cd, 35.0 for Cr and 25.0 for Pb. A seven level classification of I_{geo} is defined as: unpolluted (<0), unpolluted to moderately polluted (0–1), moderately polluted (12), moderately to strongly polluted (2–3), strongly polluted (>3), strongly to extremely polluted (3–4), and extremely polluted (>4) (Müller 1969).

The enrichment factor (EF) is a good tool to differentiate the metal source between anthropogenic and naturally occurring (Adamo et al. 2005). According to this technique metal concentrations were normalized to the textural characteristic of sediments. In this study, Fe was used as a reference

Table 1. The content of metals in sediments from the freshwater discharge outlets of PRE.

Site		Cd (mg/kg)	Cr (mg/kg)	Pb (mg/kg)	Fe (g/kg)
S1	Range	0.40–0.72	68.1–91.6	39.3–63.8	45.3–51.9
	Mean ± SD	0.59 ± 0.04	80.9 ± 2.89	46.2 ± 1.44	49.7 ± 1.50
S2	Range	0.29–0.70	57.6–84.3	22.3–84.3	41.8–45.1
	Mean ± SD	0.51 ± 0.16	71.3 ± 7.15	38.0 ± 11.5	43.5 ± 1.87
S3	Range	0.23–0.81	56.6–86.2	29.6–70.2	30.1–42.7
	Mean ± SD	0.42 ± 0.08	71.7 ± 5.61	49.1 ± 13.0	36.5 ± 1.86
S4	Range	0.40–0.60	63.9–87.4	37.1–70.2	39.3–43.7
	Mean ± SD	0.47 ± 0.01	76.0 ± 9.06	50.6 ± 10.9	41.4 ± 2.41
S5	Range	0.56–0.88	49.5–109	24.9–53.3	34.6–40.2
	Mean ± SD	0.70 ± 0.02	80.0 ± 21.3	38.1 ± 4.93	37.6 ± 3.50
S6	Range	0.45–0.54	60.7–80.3	34.0–40.2	41.1–48.8
	Mean ± SD	0.50 ± 0.04	71.7 ± 7.97	36.6 ± 2.29	44.9 ± 1.66
S7	Range	0.65–1.09	48.7–78.1	43.4–56.2	36.3–44.1
	Mean ± SD	0.87 ± 0.06	67.4 ± 11.4	49.9 ± 7.94	40.7 ± 0.64
Overall range		0.23–1.09	48.7–109	22.3–70.2	30.1–51.9
Overall mean ± SD		0.57 ± 0.16	74.2 ± 9.27	44.1 ± 8.70	42.0 ± 4.51

SD, standard deviation

element because its natural concentration is an order of magnitude higher than the concentration of the toxicologically relevant metals (Wedepohl, 1995). The EF is defined as:

$$EF = (X/Fe)_{sediment} / (X/Fe)_{crust} \qquad (2)$$

where X/Fe is the ratio of the heavy metal (X) to the Fe. EF < 1 indicates no enrichment, EF < 3 is minor enrichment, EF = 3–5 is moderate enrichment, EF = 5–10 is moderately severe enrichment, EF = 1025 is severe enrichment, EF = 25–50 is very severe enrichment, and EF > 50 is extremely severe enrichment.

3 RESULTS AND DISSCUSSION

3.1 *Metal concentrations in sediments of the Pearl River Estuary*

The mean, standard deviation, and range of the metal concentrations in sediments of the Pearl River Estuary are summarized in Table 1. As we all known, Fe is an abundant metal in the earth crust, as shown in Table 1, the concentrations of Fe in all the seven sites are similar, suggesting that the concentrations of Fe are strongly controlled by natural weathering process (Zhang et al. 2007). Overall, the concentrations ranged (in mg/kg dry weight): Fe, 30,100–51,900; Cd, 0.23–1.09; Cr, 48.7–109; Pb, 22.3–70.2; respectively. Like that of their average concentrations in the upper continental crust (Taylor & Mclennan, 1995), the mean concentration of Cd in this study was the lowest. But compared to other rivers or bays, the average concentration of Cd (0.57 mg/kg) in this study was higher than Bohai Bay (0.22 mg/kg) (Gao & Chen, 2012) but lower than Yangtze River (1.5 mg/kg) (Yang et al. 2009) and Danube (1.2 mg/kg) (Woitke et al. 2003). Compared with the reports about sediments from the domestic and the overseas rivers or bays, the average concentration of Cr (74.2 mg/kg) in this study was higher than this found in sediments of the overseas, such as the Danube (64.0 mg/kg) (Woitke et al. 2003), and Masan Bay (67.1 mg/kg) (Hyun et al. 2007), whereas lower than this found in sediments of the domestic, such as the Bohai Bay (101.4 mg/kg) (Gao & Chen, 2012) and Yangtze River (87.8 mg/kg) (Yang et al. 2009), this was contributed to the higher pollution of Cr in our country. While the average concentration of Pb (44.0 mg/kg) recorded in this study was higher when compared with Bohai Bay (34.7 mg/kg), but was comparable to

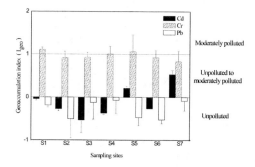

Figure 1. Geoaccumulation index (I_{geo}) of heavy metals in sediments of the PRE.

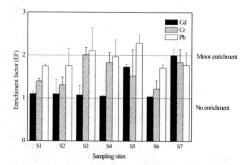

Figure 2. The enrichment factor (EF) of heavy metals in sediments of the PRE.

other rivers or bays, such as Yangtze River (45.2 mg/kg), Danube (46.3 mg/kg), and Masan Bay (44.0 mg/kg), this may because the surrounding areas of these rivers are heavily urbanized zones.

3.2 Distribution of heavy metals in the sediments of the PRE

Table 1 presents the mean and range concentrations of metals in sediments of all sites studied in freshwater discharge outlets of the PRE. The result showed that the mean sediment concentration of Cd was in the range of 0.23–1.09 mg/kg and the concentration of Cd in S7 was the highest one, its mean concentration was 0.87 mg/kg (Table 1). The concentration of Cd in S5 was also higher than other sites except S7. The results indicated that the pollution of Cd in S5 and S7 was more serious than other sites. Cd and Pb showed a similar spatial distribution in the seven studied areas of PRE. The concentrations of Pb in the sediments of S3, S4 and S7 were higher than other sites of this study. There are many factories and shipside in the surround of S3, S4 and S7, which will produce a lot of heavy metal pollution, this might be attributed to the serious pollution of Pb in these three sites. These sites of PRE also received high strength untreated industrial wastewater and domestic wastewater from upstream. As shown in Table 1, a 3-fold difference between the highest and lowest concentrations of Pb was recorded in these sediments, While the concentrations of Cr in the sediments of the seven sites of this study were similar, the concentration varied from 48.7 to 109 mg/kg.

3.3 Environmental risk assessment of heavy metals in the sediments of PRE

Numerous sediment quality guidelines have been developed to deal with environmental concerns, and two of them were chosen to assess the contamination extent of heavy metals in sediments of the freshwater discharge outlets (Figure 1 and Figure 2). Geoaccumulation index (I_{geo}) (Figure 1) showed that the freshwater discharge outlets of PRE were not polluted ('unpolluted') by Pb and Cd except S5 and S7 ('Unpolluted to moderately polluted'), while most sediments were 'Unpolluted to moderately polluted' by Cr with the exception of sediments from S1, S4 and S5 ('moderately polluted'). Overall, an order of relative risk significance of heavy metals in sediments was Cr > Cd > Pb based on the calculation of I_{geo} (Figure 1). However, the large standard deviations indicated that a great variation existed in each site and the heavy metal contamination in sediments might derive from point source pollution of anthropogenic activity (Landre et al. 2011).

Another commonly used criterion to evaluate the heavy metal pollution in sediments is the enrichment factor (EF), the mean EF values of the metals studied were obtained (Figure 2). Higher EF values of Pb and Cr were found in the sediments of S3 and S4, which received a huge amount of metallic discharges from nearby industrial plants and sewage discharges. Sediments in S7 also showed high degree of metal contamination with EF values being lightly less than that of S3 and S4. The EF values for Cr, Pb and Cd were all larger than one but less than three which indicates the minor degree of metals contamination. Cr had the highest EF values among the three metals studied, and overall the mean EF decreases in the order is Cr > Pb > Cd.

The geo-accumulation index (I_{geo}) and the enrichment factor (EF) provided coincident risk assessment results for the three selected heavy metals (Cd, Cr and Pb), as we hypothesized, in the sediments of the PRE (Figure 1 and Figure 2). Therefore, the two indices together provide a comprehensive risk assessment for sediments from the PRE at two different levels.

4 CONCLUSIONS

The present study showed clear anthropogenic inputs of heavy metal in the PRE. The results indicated the concentration of Cr was higher than Pb and Cd. Cd and Pb showed a similar spatial distribution in the seven study area of PRE, this reflected their similar anthropogenic origins in the estuarine areas. Assessment of sediment contamination with heavy metals in the PRE taking account of the geo-accumulation index (I_{geo}) and according to the enrichment factor (EF), suggested that the PRE has to be considered as 'moderately polluted', while the EF of the PRE would be ranked as 'minor enrichment'. However, this study found the combination of the two risk assessment indices did provide a comprehensive understanding of sediment quality in the PRE. And efforts should be needed to protect the freshwater discharge outlets of estuary from pollution which can reduce the environmental risk of estuary and its adjacent sea.

ACKNOWLEDGEMENTS

This study was financially supported by the unite fund project of National Natural Science Foundation of China-Guangdong: Research of ecological chemical behavior and toxicology effect of the environmental hormonal compounds in the Pearl River estuary (No. U1133003).

REFERENCES

Adamo, P. Arienzo, M. Naimo, D. Sranzione, D. 2005. Distribution and partition of heavy metals in surface and sub-surface sediments of Naples city port. *Chemosphere* 61: 800–809.

Chen, B.W. Liang, X.M. Xu, W.H. Huang, X.P. Li, X.D. 2012. The changes in trace metal contamination over the last decade in surface sediments of the Pearl River Estuary, South China. *Science of the Total Environment* 439: 141–149.

Gao, X.L. & Chen, C.T.A. 2012. Heavy metal pollution status in surface sediments of the coastal Bohai Bay. *Water research* 46: 1901–1911.

Hyun, S. Lee, C.H. Lee, T. Choi, J.W. 2007. Anthropogenic contributions to heavy metal distributions in the surface sediments of Masan Bay, Korea. *Marine Pollution Bulletin* 54: 1059–1068.

Ip, C.C.M. Li, X.D. Zhang, G. Wai, O.W.H. Li, Y.S. 2007. Trace metal distribution in sediments of the Pearl River Estuary and the surrounding coastal area, South China. *Environment Pollution* 147: 311–323.

Landre, A.L. Winter, J.G. Helm, P. Hiriart-Baer, V. Young, J. 2011. Metals in Lake Simcoe sediments and tributaries: do recent trends indicate changing sources? *Journal of Great Lake Research* 37 (suppl. 3): 124–131.

Lau, M.M. Rootham, R.C. Bradley, G.C. 1993. A strategy for the management of contaminated dredged sediment in Hong Kong. *Journal of Environmental Management* 38: 99–114.

Morris, A.W. Allen, J.L. Howland, R.J.M. Wood, R.G. 1995. The estuary plume zone: source or sink for land-derived nutrient discharges? *Estuarine, Coastal and Shelf Science* 40: 387–402.

Müller, G. 1969. Index of geoaccumulation in sediments of the Rhine river. *Geological Journal* 2: 109–118.

Taylor, S.R. & McLennan, S.M. 1995. The geochemical evolution of the continental crust. *Reviews of Geophysics* 33: 241–265.

Taylor, S. 1964. Abundance of chemical elements in the continental crust: a new table. *Geochimica et Cosmochimica Acta* 28: 1273–1285.

Wedepohl, K.H. 1995. The composition of the continental crust. *Geochimmica et Cosmochimica Acta* 59 (7): 1217–1232.

Woitke, P. Wellmitz, J. Helm, D. Kube, P. Lepom, P. Litheraty, P. 2003. Analysis and assessment of heavy metal pollution in suspended solids and sediments of the river Danube. *Chemosphere* 51: 633–642.

Yang, Y.Q. Chen, F.R. Zhang, L. Liu, J.S. Wu, S.J. Kang, M.L. 2012. Comprehensive assessment of heavy metal contamination in sediment of the Pearl River Estuary and adjacent shelf. *Marine Pollution Bulletin* 64: 1947–1955.

Yang, Z. F. Wang, Y. Shen, Z. Y. Niu, J. F. Tang, Z.W. 2009. Distribution and speciation of heavy metals in sediments from the mainstream, tributaries, and lakes of the Yangtze River catchment of Wuhan, China. *Journal of Hazardous Materials* 166: 1186–1194.

Zhang, L.P. Ye, X. Feng. H. Jing, Y.H. Ouyang, T. Yu, X.T. Liang, R.Y. Gao, C.T. Chen, W.Q. 2007. Heavy metal contamination in western Xiamen Bay sediments and its vicinity, China. *Marine Pollution Bulletin* 54:974–982.

Zhao, L. Quan, Z.Z. Yin, P.H. Liu, Y.F. Wu, X.R. 2011. Distributional characteristics of heavy metals in water of surface and subsurface microlayers from Guangzhou section of Pearl River. *Chinese Journal of Spectroscopy Laboratory* 3: 1182–1186 (in Chinese).

The design of water pollutant emission allowance control scheme in the city of Shenzhen, China

Xiao-dan Xiao & Xiao-ming Ma
School of Environment and Energy, Peking University Shenzhen Graduate School, Shenzhen, China

Pei-shan Song
Guangdong Provincial Academy of Environmental Science, Guangzhou, China

ABSTRACT: In order to achieve the reduction in water pollutant in the city of Shenzhen, here in this paper the Water Pollution Emission Allowance Control Scheme (WP EACS) is proposed. It works on the "emission cap control and balance" principle that it divides allowances into such four categories as tradable allowance of major industrial sources, confined allowance of minor industrial sources, confined allowance of publicly owned wastewater treatment works and reserved allowance of government, and keeps them break-even and cyclical balanced. Besides, the scheme constitutes three main parts: emission allowance initial allocation, emission allowance overall balance and emission allowance flow management. It has a profound influence on quantitative and dynamic management in water pollutant control.

1 INTRODUCTION

Coordination of socio-economy and water environment has become a global issue of concern as human populations grow, industrial and agricultural activities expand. Since 1972, the United States has enacted several federal programs to address water pollution problems, such as Clean Water Act, Total Maximum Daily Loads and Water Quality Trading (Leatherman et al. 2004, Hoornbeek et al. 2013). Japan firstly implemented the Total Pollutant Load Control (TPLC) policy in water pollution management of Seto Inland Sea, Tokyo Bay and Ise Bay in 1970s (Zhao et al. 2007), while the European Union (EU) enacted the Dangerous Substances Directive in 1976 and put forward the Pollution Reduction Programmes (PRPs) to realize the reduction of water pollutants (European Commission-DG Environment 2003). Moreover, the Chinese Environmental Protection Administration proposed the total amount control plan for main pollutant discharge in 1996, officially bringing total pollutant load control strategy into the pollution control system (Song 2000).

In conclusion, countries have a tendency to take pollutant discharge into control by rationing emissions for key emission sources. Among these projects, the US Acid Rain Program and the NO_x Budget are the most remarkable ones, having achieved substantial reductions in emissions of sulfur dioxide and nitrogen oxides from power plants in the United States (Lauraine 2005, Alex 1999, Guang 2011). In order to adapt to the development of pollutant allowance control and make good use of cap and trade policies to control emissions, we explore a new model of water pollutant allowance control for China, which will be discussed as below.

2 WATER POLLUTION EMISSION ALLOWANCE CONTROL SCHEME

Water Pollutant Emission Allowance Control Scheme (WP EACS) implements an innovative emission cap control and balance program to reduce water pollutant emissions. It is quite different from emissions cap and allowance trading program (cap and trade). While cap and trade program allows all of individual sources to trade their allowances or bank them for future (Swift 2000), WP EACS divides allowances into such four categories as tradable allowance of major industrial sources, confined allowance of minor industrial sources, confined allowance of publicly owned treatment works (POTWs) and reserved allowance of government, among which only the major industrial sources' allowances can be traded. Besides, all kinds of emission allowances are in line with two balance conditions: emission allowance break-even status and emission allowance cyclical balance, which are similar to balance of payments (John 2006) and cyclical balance fiscal policy (Burnside & Meshcheryakova 2004). And WP EACS constitutes three parts: emission allowances initial allocation, flow management and overall balance.

2.1 *Emission allowance initial allocation*

As only the major industrial sources' allowances can be traded in the WP EACS, building an initial allocation method to determine emission cap of the city, emission allowances of industrial sources and publicly owned treatment works is not so important as that in emission trading system. But it should also comply with the principles as follow: fair, efficiency, transparency and integrated information. Based on these rules, the emission cap of the city is calculated via the top-down method. It depends on country/province's total emission control target. The emission allowances of industrial sources and POTWs are calculated via the bottom-up method. According to the historic emissions parameter, potential emission reduction parameter and industry growth parameter, the allowances are allocated to industrial sources under a grandfathering scheme, which uses the parameters in 2010 as the baseline reference period (European Commission 2003, Martinez & Neuhoff 2005). And the POTWs' emission allowances are allocated according to their volume of wastewater treatment and emissions performance parameter.

2.2 *Emission allowance flow management*

As emission allowances have been classified into four categories, WP EACS creates three emission allowance accounts management platforms for flow management, including tradable allowance accounts management platform for major industrial sources, confined allowance accounts management platform for minor industrial sources and POTWs, and credit allowance accounts management platform. For one thing, government allocates tradable allowance accounts or confined allowance accounts for current pollutant sources. For another, it sets up reserved allowance accounts for newly affected enterprises' emission allowance allocation. Not until the reserved allowances were used up could government register the credit allowance accounts to borrow allowances from future additional reduction projects. All of the industrial sources and POTWs should register their accounts on their own platform, and make emission allowance flow statements to keep track of their emission allowances.

2.3 *Emission allowance overall balance*

Emission allowance overall balance, the core of WP WACS, focuses on the cyclical balance and components of the overall emission allowance. At the beginning of a regulation period, according to the initial allocation, government should establish the initial total balance sheet and emission allowance balance sheet, which show components of emission cap and emission allowances of the city. They are "snapshots of a region's water pollution control level", different from emission allowance flow statement, which dynamically reflects the situation and flow of current emission allowances. And at the end of the period, the final sheets should be established as well, based on pollutant sources' allowance accounts. They show the overall result of emission allowance cyclical balance in the regulation period.

Table 1. The environmental capacity of chemical oxygen demand (COD) and ammonia-nitrogen (NH$_3$-N).

River	COD environmental capacity (t/a)		COD load over-limit ratio***	NH$_3$-N environmental capacity (t/a)		NH$_3$-N load over-limit ratio***
	Natural water	Water diversion		Natural water	Water diversion	
Shenzhen River	1269.3	4071.9	1.1	1269.3	4071.9	12.8
Maozhou River	920.9	925.1	2.2	920.9	925.1	7.2
Guanlan River	996.5	1630.4	0.2	996.5	1630.4	6.6
Longgang River	1147.8	1117.6	0.4	1147.8	1117.6	2.5
Pingshan River	624.2	553.0	−0.1	624.2	553.0	0.7
Other rivers	1831.1*	3090.5**	0.4	52.5*	108.7**	13.5

*calculated below 90% guaranty rate of the rainfall runoff; **calculated according to the quantity of water diversion and river flow; ***calculated according to the pollution source census data of 2010.

3 CASE STUDY AND SCENARIOS

3.1 Water environment parameters

Due to its geographical condition, there are no great rivers or lakes in Shenzhen. According to the Shenzhen ecological environmental bearing capacity study (Shenzhen Urban Residential Environment Committee 2009), its urban pollution load goes far beyond the water environmental carrying capacity, as shown in Table 1. It can be concluded that Shenzhen's environmental capacity of chemical oxygen demand (COD) is 18,178.3 t/a, and environmental capacity of ammonia-nitrogen (NH$_3$-N) is 526.0 t/a. Moreover, NH$_3$-N emissions exceed the carrying capacity more seriously than COD emissions.

3.2 Pollution sources parameters

So far, three censuses on pollution sources within the scope of Shenzhen have been completed respectively in 2007, 2009 and 2010. According to the data analysis, COD and NH$_3$-N emissions mainly come from living and industrial sources rather than agricultural sources. Moreover, living sources have the highest reduction quantity in the pollutant sources, while industrial sources have the highest reduction rate.

Furthermore, both of COD and NH$_3$-N emissions have the characteristic of concentrated emission. According to the census in 2009, which included 6164 enterprises, there are only 1933 enterprises with more than 1 t average annual COD emissions, but their cumulative emissions account for 94.5% of the city's total COD emissions. And 1102 enterprises annually emitted more than 0.1 t NH$_3$-N, accounting for 97.2% of the total NH$_3$-N emissions.

Last but not the least, there are 32 POWTs (include centralized and decentralized treatment) in Shenzhen in 2010, with a total capacity of 3,228,000 t/d. The later POWTs build, the more advanced technology they use, and the higher disposal rate they have. Generally, centralized POWTs have higher efficiency and larger capacity than the decentralized ones.

4 RESULTS AND DISCUSSION

4.1 Initial allocation

As we have divided Shenzhen's point sources' water pollutant emissions into four categories according to the WP EACS, in this part, the initial allocation of their emission allowances will be discussed as follow.

4.1.1 Emission cap of the city

During the twelfth five-year period (2011–2015), Shenzhen's total COD emissions should fall down 23.1% compared to that of the year of 2010, including a 23.3% drop in living and industrial sources; while the total NH_3-N emission should decrease by 23.9%, with a 24.3% drop in living and industrial sources (Shenzhen Urban Residential Environment Committee 2012). Besides, we assume that total COD and NH_3-N emissions would reach the requirement of water environmental capacity in 2040. Therefore, the COD emissions' annual reducing rate should be 6% and the NH_3-N emissions' should be 12% during 2016–2040.

4.1.2 Emission allowances of industrial sources

Not all of the industrial sources are included in WP EACS. Because of their characteristic of concentrated emission, only the larger ones are the target objects of WP EACS. Besides, for the reason of promoting the industrial structure, enterprises in the industry that has a higher pollutant emission per unit of output are also included in the target objects. More specifically, the target objects of COD emission control contains enterprises that discharge more than 1.0 t COD emissions annually or that take part in the COD emission trade. On the other hand, enterprises with an average annual emission that's greater than 0.1 t NH_3-N, or those included in the industry with more than 0.04 kg NH_3-N per ten thousand dollars are on the list of target objects of the NH_3-N emission control.

In accordance with the principle of integrated information, the classification and calculation of emission allowances are both based on the pollution census database and the COD emission trading enterprise database. Computation of each district's industrial sources' annual emission allowances during 2010–2040 is shown clearly in Figure 1. It shows that the drop in the COD and NH_3-N emission allocation will be 26.4% and 26.6% in 2015 compared to the discharge in 2010, achieving a higher target than that stipulated in the "Twelfth Five-year Plan".

For newly affected industrial sources, enterprises that generate more than 30 m^3/d wastewater will be involved in the WP EACS. We make a prediction of newly affected industrial sources' emissions in each district in Shenzhen, based on its enterprises' pollutant emission coefficient and growth of gross value of industrial output. As a result, the forecast quantity of newly affected industrial sources' emissions are about 1677.8t COD and 211.3t NH_3-N per year during 2011–2015, making up 14.4% and 14.7% of total amount of confined industrial sources' emissions in 2010. By referring to reserve capacity proportion of new enterprise in current emission control systems at home and abroad, and taking Shenzhen's planning of economic development into consideration, we limit districts to 6% of their current industrial sources' emission allowances as reserved allowances for newly affected industrial sources. So the total forecast amount of newly affected industrial sources' emission allowances are about 2919.4t COD and 370.1t NH_3-N from 2011 to 2015, much lower than the prediction.

4.1.3 Emission allowances of POWTs

According to the POWTs' volume of wastewater treatment and emissions performance parameter, their total emission allowances are 28,607.6 t/a COD and 5181.7 t/a NH_3-N in 2010. And actually Shenzhen have set their allowance cap of 93,735.7 t/a COD and 13,958.1 t/a NH_3-N in 2010. So publicly owned treatment works' emission allowance allocation meets the principle to be below their control target.

4.2 Flow management and overall balance

Emission allowance initial allocation is the first step of emission allowance overall balance control, in which the city and districts' emission cap and emission allowances have been calculated, and their initial emission allowance balance sheets have been established as well. And then for different categories of emission allowances, their flow management has been set apart via the way of emission allowance flow statement on different account management platforms. Lastly, emission

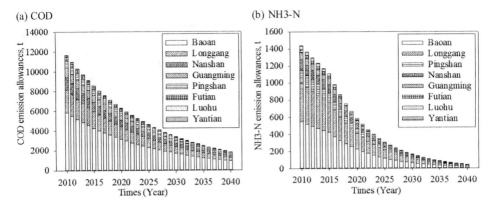

Figure 1. Each district's annual emission allowances of industrial sources during 2010–2040 (t).

allowance overall balance control requires to do the terminal balance accounting, collect the emission allowance flow statements, do emission adjust accounts, as well as establishing the terminal total balance sheet and emission allowance balance sheet.

5 CONCLUSION

The Water Pollutant Emission Allowance Control Scheme creatively brings the strategies of classified management and overall balance control of allowances into the water pollutant allowance control system. It is not only an innovation in management method for total pollutant load control, but also a breakout in quantitative and dynamic management in water pollutant. Reductions in water pollutant emissions committed by cities in the plan of total pollutant load control can be achieved by individual sources through different strategies in this scheme, like reducing emissions by investing in abatement technologies or for key industrial sources buying/selling emission allowances. According to the pilot project of Shenzhen's COD and NH_3-N emissions allowance control, WP EACS is appropriate to control the whole city's pollutant emissions, and is significant to realize systematic management of water pollutant emission allowances.

However, an in-depth study of the following aspects should be promote in the future as well: (1) internal relation and interaction among different types of emission allowances; (2) external effect of emission allowance initial allocation; and (3) guarantee measures for the implementation of this scheme.

REFERENCES

Alex, F., Robert, C. & Roger, R. 1999. The NO_x Budget: market-based control of tropospheric ozone in the northeastern United States. *Resource and Energy Economics.* 21:103–124.

Burnside, C. & Meshcheryakova, Y. 2004. Cyclical Adjustment of the Budget Surplus: Concepts and Measurement Issues. *Fiscal Sustainability in Theory and Practice: A Handbook, World Bank,* 113–32.

European Commission. 2003. Directive 2003/87/EC.

European Commission-DG Environment. 2003. Pollution Reduction Programmes in Europe.

Guang, Y. 2011. Tracking the US Acid Rain Program. *Environmental Protection.* 9:65–67.

Hoornbeek, J., Hansen, E., Ringquist, E., & Carlson, R. 2013. Implementing Water Pollution Policy in the United States: Total Maximum Daily Loads and Collaborative Watershed Management. *Society & Natural Resources,* 26(4), 420–436.

John S. 2006. *Economics.* Financial Times Prentice Hall.

Lauraine, G.C. & David, M.M. 2005. A Fresh Look at the Benefits and Costs of the US Acid Rain Program. *Journal of Environmental Management.* 77:252–266.

Leatherman, J., Smith, C., & Peterson, J. 2004. An Introduction to Water Quality Trading. *In Kansas State University, Department of Agricultural Economics, Risk & Profit Conference, Manhattan, KS, August 19e20.*

Martinez, K. K., & Neuhoff, K. 2005. Allocation of carbon emission certificates in the power sector: how generators profit from grandfathered rights. *Climate Policy,* 5(1), 61–78.

Song, G.J. 2000. China's total emission control and concentration control systems. *Environmental* Protection. (6):11–13.

Shenzhen Urban Residential Environment Committee. 2009. Shenzhen Ecological Environmental Bearing Capacity Study.

Shenzhen Urban Residential Environment Committee. 2012. Shenzhen Main Pollutant Total Load Control Plan during the Twelve Five-Year Period.

Swift, B. 2000. Allowance Trading and Potential Hot Spots-Good News from the Acid Rain Program. *Environment Reporter,* 31(19): 354–359.

Yan, G. & Wang, J.N. 2011. *China's Emission Trading Practice and Cases.* Beijing: China Environmental Science Press.

Zhao, H.L., Guo, Q.M.& Huang, X.Z. 2007. Japan's Total Water Environment Protection and Enlightenment of Total Control Technology and Policy: Inspection Report of Japanese Water Pollutant Load Control. *Environmental Protection.* 386:82–87.

Enhanced waste activated sludge digestion by thermophilic consortium

Y.L. Wang
Environmental Resources and Soil Fertilizer Institute, Zhejiang Academy of Agricultural Sciences, Hangzhou, Zhejiang, China

Z.W. Liang, S.Y. Yang & Y. Yang
College of Environmental and Resource Sciences, the Academy of Water Science and Environmental Engineering, Zhejiang University, Hangzhou, Zhejiang, China

ABSTRACT: Rapid increase in quality and quantity of municipal wastewater treatment recently resulted in a large amount of excess sludge. Thermophiles aerobic digestion is a promising technique, easy in operation and economical in maintenace. The thermphilic aerobic digestion sludge by thermophiles consortium isolated was effectively proved for solids dissolve. Inoculated with the consortium, at ultimate pH in thermophilic digestion kept on 7.5, the TSS and VSS removal rate were 33.22% and 44.66%, 14.53% and 24.73% respectively, higher than that in un-inoculated system. The content of SCOD, protein and ammonia all had a fast and large accumulating, and high content of ammonia helped remove pathogens and maintain good buffering. According to the sequence comparisons, it was derived from bacteria belonging to *Baclicus lincheniformis*, *Alcaligenes faecalis* and *Pseudomonas*, respectively.

1 INTRODUCTION

Wide employment of activated sludge process in the treatment of a variety of wastewater resulted in the production of a large amount of excess sludge. Thus, the excess sludge has been one of the most serious problems in activated sludge wastewater treatment. Treatment and disposal the sludge accounts for about even up to 60% of the total fee of wastewater treatment (Wei 2003). Recently, kinds of methods are used for sludge reduction, resolving the sludge problem fundamentally. However the common used techniques, such as ultrasonic, ozonation, mechanical destruction, alkali treatment, et al., are always high in cost and difficult in maintenance (Saby 2002). Compared to those physical and chemical methods, biological method has been paid much more attention. Among that, thermophiles aerobic digestion sludge is promising technique (Dumas 2010).

Thermophilic aerobic digestion was proposed to treat wastewater sludge firstly by Andrews and Kambhu (1973). The thermophiles used in the treatment could secrete extracellular-enzyme, such as protease, amylase, et al., which would depolymerize sludge flocs and then lyse cells in a short time. In that way, the macro-organics ran out of cells and into solution, being hydrolyzed into macro-molecule carboxylic acids, ammonia, CO_2 by thermophiles and the enzymes secreted (Appels 2008; Wei 2009). The predominant themphiles in the thermophilic aerobic digestion process is basically *Bacillus*, containing *Bacillus subtilis* (Sonnleitner and Fiechter, 1985), *Bacillus stearothermophilus* (Li 2009), *Brevibacillus Bacillus licheniformis* (Liu 2010), *Geobacillus* sp. (Hayes 2011) and so on. Hasegawa (2000) isolated a strain of thermophile *Bacillus stearothermophilus* SPT2-1 and inoculated it in a batch reaction. In that aerobic digestion, 40% of volatile suspended solids (VSS) were soluted by SPT2-1. Li (2009) isolated a new genus *Brevibacillus* named *Brevibacillus* sp. KH3 which could release protease. During the sterilized sludge digestion inoculated with KH3, maximum total suspended solids (TSS) removal rate was up to 32.8% after 120 h and during un-sterilized sludge digestion inoculated with KH3 TSS removal rate achieved 54.8%.

In this paper, strains of thermophiles were inoclulaed into un-sterilized sludge digestion process. And the differences between inoculated and un-inoculated digestion effect were compared and investigated. This study was to evaluate the biodegradation ability of aerobic digestion by using the inoclulaed thermophilic consortium.

2 MATERIALS AND METHODS

2.1 Thermophiles and cultivation

The thermophiles strains were isolated from sewage sludge at 55°C (Nosrati et al., 2007) and mixed cultured in Starch Liquid Medium with a shaker operated at 120 rpm at 55°C.

2.2 Bioreactor and operating conditions

The digestion experiment was carried out in a 2.2 L laboratory-scale ordered bioreactor. The working volume was designed as 2.0 L and temperature and impeller speed control and DO concentration monitoring were provided. 1.8 L sludge (obtained from Sibao Wastewater Treatment Plant, Hangzhou) was transferred into bioreactor with 0.2 L thermophilic consortium, while the system un-inoculated with 0.2 L deionized water. The operation temperature was 65°C and DO content was $2\,\mathrm{mg\,L^{-1}}$ kept for 120 h.

2.3 Analytical procedures

The samples were collected every 4 h in early stage and every 12 h after the first 12 h to determine TSS, VSS, soluble chemical oxygen demand (SCOD), NH_4^+-N, pH and protein concentration. TSS, VSS, SCOD and NH_4^+-N in the digestion system were measured according to standard methods (APHA, 1998). Samples were centrifuged for 10 min at 5000 rpm and the supernatant was filtered through a 0.45 μm mixed cellulose ester membranes. The value of pH was measured with a digital pH-meter. Protein concentration was determined by the modified Lowry method (Frolund et al. 1995).

3 RESULTS AND DISCUSSION

3.1 Changes of pH

In the un-inoculated system, the pH first increased by ammonia coming out and then tended to 8.0 in 60 h. While in the inoculated reaction, the pH increased rapidly in the early stage and tended to stabilize, keeping on 7.5. When the pH didn't change obviously, it found that the pH value in inoculated system was higher than that in un-inoculated. As many organisms like protein, fat, carbohydrate, et al. into solution because of lysis, the amount of macromolecule carboxylic acids hydrated accumulated a lot, which achieving balance with ammonia and lowering the pH value in thermophilic aerobic digestion process. Kim (2002) also found the pH always increased first and then decreased in conventional aerobic sludge digestion process, according to the conclusion in this paper. The ultimate pH keeping on 7.5 in thermophilic digestion was effective for the aerobic sludge digestion.

3.2 Removal efficiency of TSS and VSS

The main target of sludge digestion is to reduce the VSS content in sludge, while the VSS was composed mainly by bacteria, fungi, protozoon and metazoan. Fig. 1 and Fig. 2 shows the changes of TSS and VSS in two different systems. It's obvious that after inoculating thermohiles consortium, both TSS and VSS removal rate were improved. The TSS removal rates in inoculated and

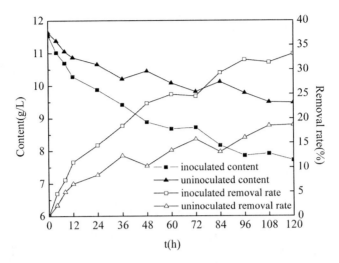

Figure 1. Thermophiles inoculation influence on TSS.

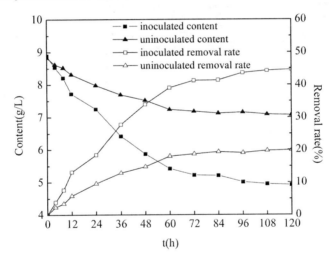

Figure 2. Thermophiles inoculation influence on VSS.

un-inoculated system were 33.22% and 18.69% respectively, 14.53% higher than without thermophiles consortium. In addition, the VSS removal rates in inoculated system achieve 44.66%, 24.73% higher than un-inoculated. Because of the extracellular enzyme, depolymerize sludge flocs and lyse cells could be realized quickly and the solids were dissolved as soon as the thermophlic aerobic digestion started, emerging efficient aerobic digestion. Zhao (2007) also used mixed thermophiles digesting sludge for 120 h and the TSS and VSS removal rate were increased by 12.25% and 19.22% respectively. While in this paper the TSS and VSS removal rate increased by 14.53% and 24.73%, illustrating the thermophiles consortium used in this experiment effectively enhanced the aerobic digestion sludge.

3.3 *Removal of SCOD*

After sludge depolymerization and cells lysis, abundant organic matters outflow into liquid phase, increasing the organics and SCOD in solution. The changes of SCOD (Fig. 3) shows the SCOD content inoculated with thermphiles consortium grew up rapidly in the first 12 h and reached

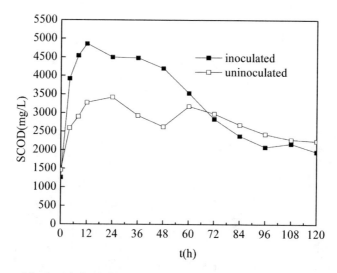

Figure 3. Thermophiles inoculation influence on SCOD.

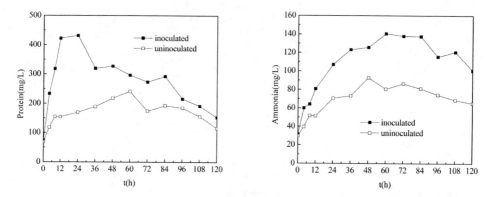

Figure 4. Thermophiles inoculation influence on protein and NH_4^+-N.

nearly 4900 mg/L. Then the SCOD content went down after kept on about 4500 mg/L and the speed lowered down as well. Although the SCOD without thermophiles consortium experienced the same course, the increasing rate and range were both smaller than inoculated system. Inoculating with thermophiles consortium brought much more predominant organisms in the system, so it needed more substances to grow and survive, which caused faster and larger ranged reduction in SCOD. That meant thermophiles consortium not only help depolymerizing sludge flocs and lysing cells but also reduced SCOD content in solution and organic loading in effluent.

3.4 *Removal of protein and NH_4^+-N*

The changes of protein content (Fig. 4) had the same trend with SCOD, Protein was one of the effluent organics, so its changes were in line with SCOD. This also improved the roles of themphiles consortium played in aerobic sludge digesion. With the transformation of protein, the ammonia content dramatically changed. Fig. 4 illustrated the changes in different systems. In inoculated system, ammonia content reached maximum value 140.58 mg/L in 60 h and maintained for nearly 24 h, then going down gradually. In the other system, ammonia content achieved 92.75 mg/L in 48 h and decreased right away. Protein hydrolyzation and amino acids deaminization would increase the ammonia content, while decreased protein content and NH_4^+-N transforming into NH_3 in high temperature environment made the ammonia content dropped down. Without consortium, low content of protein and hydrolyzation caused low content of ammonia. It was found high content of ammonia helped remove pathogens and maintain good buffering. It also proved thermophilic aerobic digestion sludge could promote effectively digestion of sludge.

4 CONCLUSIONS

This study proved the improvement of thermophilic aerobic digestion sludge by thermophiles consortium isolated. Inoculated with the consortium, the TSS and VSS removal rate were 33.22% and 44.66%, 14.53% and 24.73% higher than that in un-inoculated system respectively, better than some research achievements nowadays. Due to thermophiles consortium, the content of SCOD, protein and ammonia all had a fast and large accumulating, and high content of ammonia helped remove pathogens and maintain good buffering. Cells lysis was occurring during the course of the process, and the thermophiles consortium was belonging to *Baclicus lincheniformis*, *Alcaligenes faecalis* and *Pseudomonas*, respectively.

ACKNOWLEDGEMENTS

The study was financially supported by the National Natural Science Foundation of China (51108423) and the National Natural Science Foundation of Zhejiang Province (Y5100234).

REFERENCES

Andrews J.F., Kambhu K., 1973. Thermophilic aerobic digestion of organic solid wastes. U.S. EPA Report No. 620/2-73-061, Washington D.C.: PB-222–396.
APHA, 1998. Standard Methods for the Examination of Water and Wastewater, 20thed. United Book Press, USA.
Appels L., Baeyens J., Degreve J., Dewil R., 2008. Principles and potential of the anaerobic digestion of waste-activated sludge. Progress in Energy and Combustion Science 34(6), 755–781.
Dumas C., Perez S., Paul E., Lefebvre X., 2010. Combined thermophilic aerobic process and conventional anaerobic digestion: Effect on sludge biodegradation and methane production. Bioresource Technology 101(8) 2629–2636.
Frolund, B., Griebe, T., Nielsen, P.H., 1995. Enzymatic activity in the activated-sludge floc matrix. Applied Microbiology Biotechnology, 43(4), 755–761.
Hasegawa S., Shiota N., Katsura K., Akashi A., 2000. Solubilization of organic sludge by thermophilic aerobic bacteria as a pretreatment for anaerobic digestion. Water Science and Technology, 41(3), 163–169.
Hayes D., Izzard L., Seviour R., 2011. Microbial ecology of autothermal thermophilic aerobic digester (ATAD) systems for treating waste activated sludge. Systematic and Applied Microbiology, 34(2), 127–138.
Kim Y.K., Bae J.H., Oh B.K., Lee W.H., Choi J.E., 2002. Enhancement of proteolytic enzyme activity excreted from bacillus stearothemophilus for a thermophilic aerobic digestion process, Bioresource Technology, 82(3): 157–164.
Li X.S., Ma H.Z., Wang Q.H., Matsumoto S., Maeda T., Ogawa H.I., 2009. Isolation, identification of sludge-lysing strain and its utilization in thermophilic aerobic digestion for waste activated sludge. Bioresource Technology, 100(9), 2475–2481.
Liu S.G., Song F.Y., Zhu, N.W., Yuan H.P., Cheng J.H., 2010. Chemical and microbial changes during autothermal thermophilic aerobic digestion (ATAD) of sewage sludge. Bioresource Technology, 101(24), 9438–9444.
Nosrati M., Sreekrishnan T.R., Mukhopadhyay S.N., 2007. Energy Audit, Solids Reduction, and Pathogen Inactivation in Secondary Sludges during Batch Thermophilic Aerobic Digestion Process, Journal of Environmental Engineering, 133(5), 477–484.
Saby S., Djafer M., Chen G.H., 2002. Feasibility of using a chlorination step to reduce excess sludge in activated sludge process. Water Research, 36(3), 656–666.

Sonnleitner B., Fiechter A., 1985. Microbial flora studies in thermophilic aerobic sludge treatment. Conservation and Recycling, 8(1), 303–313.
Wei Y.S., Van Houten R.T., Borger A.R., Eikelboom D.H., Fan Y.B., 2003. Minimization of excess sludge production for biological wastewater treatment. Water Research, 37(18), 4453–4467.
Wei Y.S., Wang Y.W., Guo X.S., Liu J.X., 2009. Sludge reduction potential of the activated sludge process by integrating an oligochaete reactor. Journal of Hazardous Materials, 163(1), 87–91.
Zhao W.N., Li X.M., Yang Q., Zeng G.M., Zhang Y., Liao Q., Wang Z.S., 2007. Study on reduction behavior of excess sludge with solubilization by thermophilic enzyme. China Water&Wastewater, 23, 29–33.

The effects of reversible lane's application on vehicle emissions in Chaoyang Road in Beijing

Xu Yao & Li Lei
Beijing Jiaotong University, Beijing, China

Jiemeng Yang
Beihang University, Beijing, China

ABSTRACT: This paper first gives an overview of reversible lane and the relevant content of vehicles' pollutants' emission, and then makes use of the survey data to calculate various pollutants' emission systematically according to the actual condition of reversible lane in Chaoyang Road. Finally, an explanation about the application of reversible lanes effects on vehicle emissions will be given according to the contrast by the reversible lane before and after use.

1 INTRODUCTION

In order to relieve severe congestion of some sections in Chaoyang Road scientifically, Beijing municipal government make an important decision: the section between the 3rd Ring Road and the 4th Ring Road in Chaoyang Road (Jingguang Bridge to Ciyunsi Bridge) starts to carry out the reversible lane in daily evening peak (from 17 pm to 20 pm) since September 12, 2013 onwards. Thereafter, Beijing's first reversible lane gets into the trial operation stage.

The length of reversible lane in Chaoyang Road is 2.5 km, which sets 9 traffic lights totally. It has an isolation belt between the main road and auxiliary road in both directions. Vehicles can only select whether to get into or leave main road when passing the crossing, and it is not permitted to switch this in any other places. When the reversible lane is not enabled, it has 2 roads in main road in both directions, and one is bus lane for BRT2. It also required that you can only go straight when you are in the main road, left turn and U-tern is not permitted. While the auxiliary road in both directions set 3 lanes. Among them, some can execute the function of turning left, U-turn or turning right, others are only go straight. When the reversible lane is used, a lane in main road in the direction of east to west is changed.

2 AN OVERVIEW OF VEHICLES' POLLUTANTS' EMISSION

Urban vehicles' pollutants' emissions are various, including CO, HC, NO_x, SO_2, suspended particulate matter, aldehydes, etc. Among them, CO, HC and NO_x are the primary pollutants that account the most percentage. What the hazard they bring to us is also the most (Ying H. et al, 2001) (Jiming H. & Guangda M., 2002, p. 431–432).

2.1 *Carbon monoxide (CO)*

By the statistics of 2009, the vehicles' total CO emission in the whole nation is 31107 thousand tons, which has a 2.3% up than 2008. Among them, the total emission of small vehicles is 10745 thousand tons, and the sharing rate comes to 34.54% (Zhiliang Y., Qidong W. & Xianbao S., 2012, p. 40). In addition, the total emission of CO ranked first in various pollutants' emission.

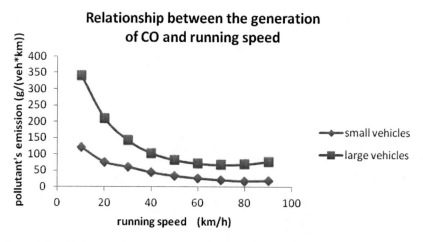

Figure 1. Relationship between the generation of CO and running speed.

The generation of CO has a close relationship with O_2's density. Only the air fuel ratio is smaller than 14.8 can the CO generates. However, the process of fuel's combustion is exothermic, therefore, the CO_2 that generates in the combustion can be reduced to CO in a small part though the fuel is completely combusted. So, the CO is always there in vehicles' exhaust (Jing Z., 2007).

In addition, the working condition of engine is another important factor which influences the generation of CO. In the idle condition, the air entered into the engine is not enough, so it can generate a large amount of CO. With the increasing of running speed and engine's temperature, CO's generation become smaller. However, the vacuum of vehicle's intake pipe will rise which make the air fuel ratio decline in decelerate process. Therefore, the generation of CO will increase. The relationship between the generation of CO and running speed is shown in Figure 1.

So, the total emission of CO in congestion is much bigger than other conditions if the other parameters are the same. That is to say, the application of reversible lane can decrease the emission in its direction.

2.2 Hydrocarbons (HC)

HC is another important pollutant in vehicles' exhaust. The greatest feature of it is that it constitutes intricately and the type is various. The total emission of HC in vehicle's exhaust comes out in front by the statistics of 2009. Among them, small vehicles are the mainly emission sources of HC which has a sharing rate of 27.69% (Zhiliang Y., Qidong W. & Xianbao S., 2012, p. 41).

HC, as a product of incomplete combustion, its emission is relatively more when vehicles are in idle condition. It is because of the density of mixed gas is concentrated that make the fuel combust incompletely, even if the air fuel ratio is more than 14.8. With the increasing of running speed, the fuel fully mixed, the generation of HC will decrease followed. The figure of relationship between the generation of HC and running speed is similar to figure 3.

So, the total emission of HC in congestion is much bigger than other conditions if the other parameters are the same. That is to say, the application of reversible lane can decrease the emission in its direction.

2.3 Nitrogen oxides (NO_x)

NO_x is a primary pollutant in vehicles' exhaust. By the statistics of 2009, the total emission of NO_x is 5298 thousand tons. Different from the pollutants above, heavy trucks are the mainly emission source that make a sharing rate of 39.55% (Zhiliang Y., Qidong W. & Xianbao S., 2012, p. 42).

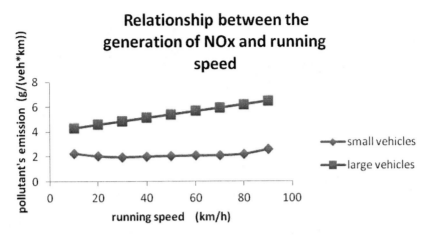

Figure 2. Relationship between the generation of NO_x and running speed.

The temperature and the rotate speed of engine play a decisive role in the generation of NO_x. When vehicles are in idle condition, the rotate speed of engine is also low, NO_x's generation gets higher with the speed up. When the engine's temperature gets into a high level, the generation of NO_x rapidly increases (Jincheng H. & Jie S., 2012, p. 21). The relationship between the generation of NO_x and running speed is shown in Figure 2.

In addition, the air fuel ratio and ignition time can also influence the generation of NO_x. Therefore, the frequent process of ignition and flameout in congestion must improve the total emission of NO_x. Thus, it will lead to the pollution of urban atmosphere. So, the application of reversible lane can decrease the emission in its direction.

3 THE CALCULATION OF VEHICLES' POLLUTANTS' EMISSION OF REVERSIBLE LANE IN CHAOYANG ROAD

3.1 *Survey method and data acquisition*

We take single factor comparative method to analyze the emission's change of some vehicles' pollutants between the reversible lane is not used and being used, and then explain the affect to improve the urban traffic environment when reversible road is used.

3.1.1 *The selection of survey time and sites*

People's travel behavior is determined by many factors, including age, sex, season, weather, etc. Therefore, the survey time's selection is essential to complex factors above. Since reversible lane in Chaoyang Road just opened a month ago, the season's affect should be ignored consequently. At the meanwhile, the road will be abstracted into a line source model. So, the volume of vehicles will play a decisive role in total emissions. Therefore, Friday evening peak should be selected as research subjects to acquire the peak that vehicle exhaust's affect to the road.

Regard to the survey site, Hongmiao intersection is the most concentrated areas of traffic flow in the range of reversible lane in Chaoyang Road according to author's site survey. Therefore, Hongmiao intersection should be selected as our survey site, the section between the intersection of Chaoyang Road and West Dawang Road and the 4th Ring Road, that is. Thus, the total emissions from vehicles of reversible lane in Chaoyang Road can be estimated.

3.1.2 *The selection of survey objects*

When survey time and sites are determined, we need to determine survey objects. In this paper, we choose all passing vehicles which comply with the provisions in section 3.1.1, including small passenger cars, large cars, trucks, city buses and so on.

3.1.3 *The selection of survey data*

In order to make the survey data more persuasive, survey averaging should be taken in order to minimize the error. And the following four should be investigated: cross traffic, the average stop delay, vehicle travel time and average running speed.

(1) Cross traffic: the most critical factor in the survey. In order to reduce the manual notation error, videos are used to record data. That is to say, recording interval of 10 minutes to record the flow of main road and auxiliary road respectively on Friday evening peak(17:00–18:00).

Then, the record can be collected to acquire cross traffic of the evening peak according to vehicle conversion factors. And take the average of multiple measurements as an analytical basis.

(2) Average stop delay: this parameter is mainly used for the idle condition of vehicle emissions. For the average stop delay investigation, the basic expression is used to calculate intersection delay (Chunfu S., 2004, p. 48):

Total delay = total number of parking * observation interval
Each suspended the average vehicle delay = total delay/suspended total number of vehicles.
Average vehicle stop delay form is shown in Table 2:

(3) Vehicle travel time: vehicle travel time related to the running speed of the vehicle, thereby affecting the amount of vehicle emissions. The floating car method can be used to measure east-west traffic.

(3) Average running speed: it related to the amount of pollutant emissions directly. It refers to the journey average speed during operation of vehicles on the road which removes its parking time. Calculated as follows:

$$\bar{v} = s / t_{tra} - t_{del} \qquad (1)$$

where, $\bar{v}$ – average running speed, km/h;
s – the length of the road measured, km;
t_{tra} – vehicle travel time, h;
t_{del} – average stop delay, h.

When the vehicle travel time and average delay are obtained, the average running speed can be easily calculated.

3.2 *Survey index*

It is most mainly to see the sharing rate of all kinds of pollutants when measuring the extent that the vehicle exhaust affect urban traffic environment. It refers to the emissions of various pollutants percentage in the total volume of emissions. For Beijing, petrol vehicles are our first choice for people's daily travel. Therefore, the emission of petrol vehicles in actual operation should be taken as reference standards to select analytic index. The emissions of various substances in petrol vehicles are shown in Table 1 (Zexing Z. et al., 2000):

Therefore, the primary pollutants CO and HC are selected as the evaluation index.

In addition, the buses and other large vehicles are also make a group cannot be ignored during the investigation, and they constitute a major source of NO_x pollutants. Therefore, NO_x's environmental influence should be considered. Thus, the final analytic index is determined as CO, NO_x and HC. This is also in line with nation's relevant regulations which are promulgated in "light vehicle exhaust emission standards".

Table 1. The volume fraction of various substances.

substances	emissions
N_2	83.6%
CO_2	12%
O_2	2.22%
CO	0.97%
H_2	0.23%
NO	$2.9*10^{-6}$
HC	$205*10^{-6}$
NO_2	$18*10^{-6}$

3.3 Data calculation

3.3.1 Vehicle emission factors

Vehicle Emission Factors refers to species vehicles emitted during its actual operation on the road. Its value is closely contacted with vehicle operating conditions and idle conditions. Urban vehicles' driving condition is very complex. Different vehicles' idling, acceleration, deceleration and constant speed traveling state has the differences in different cities or in lanes of different ranking (Xiugang L. et al., 2001). China has already provided a variety of experimental models of standard conditions in the experimental specification of exhaust pollutants. Among them, Guoxiang Li took the highway in Beijing as object, and conducted statistical work of vehicle operating conditions. The conclusion is that he considered that the national norm conditions can approximately reacts the actual vehicle operating conditions in Beijing (Guoxiang L. & Hua M., 1992).

In order to the simplify the calculating process, the emission factors determined by Zexing Zhou (2000, pp. 48–53) in the paper "The Study on Vehicles' Driving Condition and Pollutant Emission Factors in Beijing" are used to calculate. The forms are shown in Table 2 and Table 3.

3.3.2 Vehicle emission model

As we can see from above, vehicles' pollutants emission are significantly different when they are in idle condition and operating condition. Therefore, emissions should be calculated separately in two parts.

(1) The emission in idle condition:

$$PI_w = \frac{1}{3600} \sum_{j=1}^{n} EFI_{jw} * V_j * D \qquad (2)$$

where, PI_w=the emission of pollutant "w" in idle condition, g/h; EFI_{jw}=the idle emission factor of pollutant "w" of vehicle type "j", g/(h·veh); V_j=cross traffic of vehicle type "j", veh/h; and D=average stop delay, s.

(2) The emission in operating condition:

We take the formula of mobile linear source's intensity to calculate the emission in operating condition:

$$Q_w = \sum_{j=1}^{n} Q_{jw} = \sum_{j=1}^{n} Q_j * l * Ef_{jw} \qquad (3)$$

where, Q_w = the source intensity of pollutant "w" in the linear source road, g/h; Q_{jw} = he source intensity of pollutant "w" of vehicle type "j" in the linear source road, g/h; Q_j = cross traffic of

Table 2. Operating emission factors of pollutants.

Operating emission factors	Running speed	Small vehicles	Large vehicles
CO (g/(veh·km))	10	121.67	341.17
	20	76.18	210.95
	30	61.02	141.95
	40	44.27	103.49
	50	33.22	82.58
	60	25.84	71.72
	70	20.57	67.77
	80	17.59	69.70
	90	19.05	78.01
HC g/(veh·km)	10	13.53	25.46
	20	8.37	15.35
	30	6.64	9.87
	40	5.12	6.72
	50	4.15	4.92
	60	3.51	3.83
	70	3.05	3.18
	80	2.79	2.82
	90	2.88	2.66
NO_x g/(veh·km)	10	2.26	4.32
	20	2.03	4.59
	30	1.95	4.86
	40	2.01	5.13
	50	2.05	5.40
	60	2.08	5.67
	70	2.10	5.93
	80	2.21	6.20
	90	2.59	6.47

Table 3. Idle emission factors of pollutants.

Idle emission factor	Small vehicles	Large vehicles
CO g/(veh·h)	640.76	1183.06
HC g/(veh·h)	72.07	88.71
NO_x g/(veh·h)	7.34	10.4

vehicle type "j", veh/h; L = the length of the road, km; and Ef_{jw} = the operating emission factor of pollutant "w" of vehicle type "j", g/(h·veh);

(3) Total Emission

$$Q = Q_w + PI_w \qquad (4)$$

where, Q = the total emission of pollutant "w", g/h.

3.3.3 *Data calculation*

For purposes of calculation, the following two assumptions are used:

(1) The emission factors of all the small vehicles are the same, and so it is with the large vehicles.

Table 4. Cross traffic.

	Before (veh/h)				After (veh/h)			
	Small vehicles	Large Vehicles	BRT2	Routine buses	Small vehicles	Large Vehicles	BRT2	Routine buses
east → west	1488	31	22	103	1326	34	24	126
west → east	2460	42	28	139	2488	50	27	168

Table 5. Average running speed.

	Before (km/h)			After (km/h)		
	Routine buses	BRT2	Social vehicles	Routine buses	BRT2	Social vehicles
east → west	19.3	25.9	52.1	13.3	26.7	44.6
west → east	15.6	26.4	44.3	17.5	25.3	50.7

Table 6. Total operating emissions of vehicle pollutants (west → east).

Period	Sorts	CO	HC	NO_x
Before	Small vehicles	265464.75	30891.45	12484.5
	Large vehicles	9922.82	591.15	550.2
	BRT2	11675.3	825.3	333.2
	Routing buses	93215.49	6877.03	1546.38
Total		380278.35	39184.93	14914.28
After	Small vehicles	202025.6	25414.92	12751
	Large vehicles	10180.25	612.5	678.25
	BRT2	12050.1	851.18	318.94
	Routing buses	102272.1	7505.4	1900.5
Total		326528.05	34383.99	15648.69

(2) The vehicle emission factors in operating condition obey uniform distribution in the adjacent interval.

The average cross traffic and average running speed we measured are shown in Table 4 and Table 5.

Then we can calculate the total pollutants' emission (g/h) in Chaoyang Road according to the formulas in 3.3.2. The calculation results are shown in Table 6 to Table 10.

3.4 Pollutant emission reduction rate

Comparing the calculated results, we can draw the following conclusions:

(1) Reduction rate of operating emission:

On the whole, the effect that reversible lane's application to reduce the pollutants' operating emission is not very obvious. As it can be seen from the statistics in different direction, though reversible lane's application can improve the running speed to make pollutants' emission become smaller, it also increase the cross traffic that adds the pollutant source in the meanwhile.

(2) Reduction rate of idle emission:

As it can be seen from the table, the effect that reversible lane's application to reduce the pollutants' idle emission is very obvious. To the direction out of the city, congestion almost completely

Table 7. Total operating emissions of vehicle pollutants (east → west).

Period	Sorts	CO	HC	NO$_x$
Before	Small vehicles	117812.4	15151.56	7626
	Large vehicles	6223.25	363.48	423.15
	BRT2	9363.2	666.6	260.7
	Routing buses	56665.45	4135.45	1176.77
Total		190064.3	20317.09	9486.62
After	Small vehicles	129772.3	15481.05	6729.45
	Large vehicles	7978.95	500.65	446.25
	BRT2	9883.2	700.8	286.2
	Routing buses	93933	6967.8	1386
Total		241567.45	23650.3	8847.9

Table 8. Total idle emissions of vehicle pollutants (west → east).

Period	Sorts	CO	HC	NO$_x$
Before	Small vehicles	78813.48	8864.61	902.82
	Large vehicles	2484.42	186.29	21.84
	BRT2	1565.28	124.19	14.56
	Routing buses	8222.26	616.53	72.28
Total		91176.46	9768.62	1011.5
After	Small vehicles	31884.2	3586.20	365.24
	Large vehicles	1183.06	88.71	10.4
	BRT2	638.85	47.9	5.61
	Routing buses	3975.08	298.07	34.94
Total		37681.19	4020.88	416.2

Table 9. Total idle emissions of vehicle pollutants (east → west).

Period	Sorts	CO	HC	NO$_x$
Before	Small vehicles	28603.53	3217.2	327.66
	Large vehicles	1100.25	82.5	9.67
	BRT2	780.81	58.55	6.86
	Routing buses	3655.66	274.11	32.14
Total		34140.25	3632.37	376.63
After	Small vehicles	59475.34	6689.54	681.3
	Large vehicles	2815.68	211.13	24.75
	BRT2	1987.54	149.03	17.47
	Routing buses	10434.59	782.42	91.73
Total		74713.16	7832.12	815.25

Table 10. Total emissions of vehicle pollutants.

	CO	HC	NO$_x$	Total
Before	695659.36	72903.01	25789.03	797351.06
After	680489.85	69887.29	25728.04	776105.18

disappeared when the reversible lane is used. The idle emission reduces drastically followed. In the other direction, the congestion exacerbates. All the social vehicles are in the auxiliary road except BRT2, it makes vehicles' conflicts and average stop delay in the auxiliary road increase inevitably. Therefore, the idle emission rises drastically. However, the number of vehicles in the direction out

Table 11. Reduction rate of operating emission.

	Direction	CO	HC	NO$_x$
Reduction rate	west → east	−14.13%	−12.25%	+4.9%
	east → west	+27.09%	+16.41%	+6.73%
	total	−0.39%	−2.46%	−0.39%

Table 12. Reduction rate of idle emission.

	Direction	CO	HC	NO$_x$
Reduction rate	west → east	−58.67%	−58.84%	−58.85%
	east → west	+118.84%	+115.62%	+116.46%
	total	−10.31%	−11.55%	−11.29%

Table 13. Total reduction rate.

	Direction	CO	HC	NO$_x$
Reduction rate	west → east	−22.74%	−21.55%	+0.87%
	east → west	+41.07%	+31.45%	+1.99%
	total	−2.2%	−4.1%	−0.24%

of the city is far more than the other, so the synthesized effect is that the total emission is also reduces.

(3) Total reduction rate:

As it can be seen from the table, the reduction rate of various pollutants' emission is relatively lower. However, the reduction rate of pollutants in different directions is significantly different. In one hand, the application of reversible lane increase the traffic capacity of main direction, it makes the average running speed get higher by a large margin though the amount of vehicles is also increased. When vehicle's running speed at a relative high level, the total emission of CO and HC vary obviously in accordance with running speed. So, the superimposed effect makes the total emission reduce drastically while NO$_x$'s emission changed slightly. In the other hand, social vehicles mix up with routine buses in case of the reduction of traffic lanes, it not only decrease the average running speed, but also make the average stop delay longer. When vehicle's running speed at a relative low level, the total emission of CO and HC increase obviously. Therefore, the superimposed effect is the total emission increases though the amount of vehicles is decreased.

4 CONCLUSIONS

The application of reversible lane in Chaoyang Road has a positive effect on reducing vehicles' pollutants. However, the effect is limited because of the road condition. When the reversible lane is used, the pollutants' emission becomes smaller in the out-of-city direction while the opposite increases. If we can take some measures to transform the traffic mode of the opposite direction, it is sure that the reversible lane's application will reduce the vehicles' total emission by a large margin, and make a great step to improve the urban traffic environment.

REFERENCES

Chunfu S., 2004. *Traffic Planning*. Beijing:China Railway Publishing House.
Guoxiang L. & Hua M., 1992. A Survey and Tests of the Exhaust Emission Rates of In-use Vehicles in China. *Journal of Highway and Transportation Research and Development*, 9(1). pp. 54–58.
Jiming H. & Guangda M., 2002. *Air Pollution Control Engineering*. 2nd cd. Beijing: High Education Press.
Jincheng H. & Jie S., 2012. *Automotive Engine Emissions and Pollution Control*. Beijing: Science Press.
Jing Z., 2007. Influence of Traffic Signal System on Traffic Environment. M.E. Beijing: Beijing Jiaotong University.
Xiugang L. et al., 2001. Motor Vehicles' Exhaust Emission Factors for Urban Transportation Planning.*Journal of Traffic and Transportation Engineering,* December, 1(4). pp. 87–91.
Ying H. et al., 2001. The analysis on exhaust emission of motorcars on city road. *Shandong Internal Combustion Engine*, 2001(4). pp. 20–23.
Zexing Z. et al., 2000. The study of driving cycle and emission factor of vehicle in Beijing city. *ACTA Scientiae Circumstantiate*, 20(1). pp. 48–53.
Zhiliang Y., Qidong W. & Xianbao S., 2012. *Vehicle Energy Consumption and Pollutant Release and Control*. Beijing: Chemical Industry Press.

Research on system evaluation framework of green elderly community for Chinese population change in next decade

Zhi Qiu, Yan Ming & Jie Wang
College of Civil Engineering and Architecture, Zhejiang University, Hangzhou, Zhejiang, China

ABSTRACT: Aging population in China is getting severe in next coming decade. However, the current regulation of elderly community in China still essentially concentrates on accessibility, assurance of space and scale, and hardware construction of elderly community. Following the principle of construction safety, convenience, energy efficiency, and eco-friendly green elderly community, the author conducted a comprehensive analysis over evaluation principle for green living building and intelligent building focusing on outdoor/indoor comfort, intelligent living, especially, soft-supporting and management in order to develop and perfect the idea of green elderly community on a wider approach. Furthermore, this research intend to establish a systematic evaluation framework on macro level (society community), mezzo level (public space of community), and micro level (living unit), and also predict the development orientation and trend of green elderly community.

1 INTRODUCTION

1.1 Background

Viewing from the perspective of demographic trends, China is rapidly growing into an aging society. Because of China's long-term implementation of the family planning policy, the housing issue for the urban elderly population is predicted to be unprecedented severe in the next decade. Based on this knowledge, the society should pay more attention to the corresponding elements that greatly affect the living quality of the elderly, such as housing, medical, nursing, endowment insurance, and social services, and handle these issues properly.

In addition, the national policy on the urban elderly's housing issue is based on the idea "9073", that is, 90% of the elderly are to live at home and be taken care of by the family, 7% of the elderly are to live at home and be taken care of by the community, and 3% of the elderly live in institutes such as nursing homes. This policy has already begun to show the contradiction between itself and the changes of population structure. First of all, 97% of the elderly population depends on individuals and society for support, which will increase social pressure. Secondly, as a result of the long term implication of family planning policy, in reality more families are structured in a reverse triangle "421" structure, that is, one young or middle aged couple need to support four elders, and raise one child. This family structure leads to the decrease in proportion of the elderly able to live with their family and be taken care of by their children or grandchildren. Therefore, the housing issue of the elderly needs to be taken seriously, and relevant measures should be taken.

1.2 Literature review on the existing regulation for elderly community

Existing architectural design specifications regarding elderly residences mainly consist of "Code for design of residential building for the aged" of the national level, and local standards such as "Architectural Design Standards for the Facilities for the Elderly". Using "Specifications for the Architectural Design of Elderly Residences" as an example, this document sets a series of technical and economic indicators to be complied with when designing an elderly residence, from the base design and planning, interior design, building implements, to indoor environment. It emphasizes the technical measures needed to be paid special attention to in the interior design of a elderly residence, such as room layout and size, space requirements of entrances and exits, corridors, public staircases, elevators, apartment doors, halls, indoor corridors, toilets, kitchens, bedrooms, and balconies. However, from the analysis on each index, it is not difficult to see that the specifications on base design and planning, as well as interior design, almost all focus on barrier-free design and management of the scale of length and space to cater the elderly; the requirements on building implements and indoor environment begin to put comfort of indoor physical environment into consideration; however, popularization of intelligentization and operational management in the elderly communities are not discussed.

1.3 *Literature review on green elderly communities*

In 2011, Meisheng Ye published "Joint Evaluation System of China's Green Elderly communities", marking the beginning of a research boom on the "greening" process of China's elderly communities. The book takes a novel approach, combining the concepts of "the elderly" and "green". However, in the establishment stage of the evaluation system, the author paid more attention to the "Save Four Resources, Protect One Environment" aspect in the "greening" process than barrier free design and regulations of the scale. Moreover, the author of this paper believes that in the aspect of green elderly communities' operational management, service support such as "the university for the aged", "health and medicine", "geriatric nursing" and others is as important as intelligentization in the management of the communities, and should be given sufficient attention in the evaluation system.

2 EVALUATION INDEXES OF GREEN ELDERLY COMMUNITIES

The author attempts at establishing a more reasonable evaluation system for green elderly communities through comprehensive analysis on evaluation criteria for green residential buildings, interviews with personnel conducting construction or scientific researches on elderly residences in the production, academic, research and political fields, as well as the research results on Elderly Residence Planning. At the same time, the author establishes cooperation with Geentown Yile Limited and uses Greentown Wuya Elderly community as a platform, to implement and test the evaluation and analysis on elderly communities. The author extracts evaluation indexes overlooked by previous research respectively on the macro level – community, the mezzo level – apartment buildings, and the micro level – living space.

2.1 *On macro-level*

Evaluation on the macro level evaluates social environment such as policy, management, and facilities of green elderly communities. The evaluation needs to pay special attention to the following aspects.

2.1.1 *Service support*
From the perspective of Architecture, building green elderly communities is considered in general the building and application of landscape, space, materials and energy. However, care over the elderly is cultural, which requires additional services in order to design a new and comfortable life for the elderly, in addition to the commonly acknowledged infrastructure. Therefore, when building green elderly communities, we need to pay attention to the penetration and operation of relevant service institutes such as geriatric nursing centers, medical clinics, nursing homes, and the university for the aged.

In the university for the aged "Yile College" in the Wuya Community in Wuzhen, there are static and dynamic teaching areas, a chess room, an exhibition building, a meditation hall, a swimming

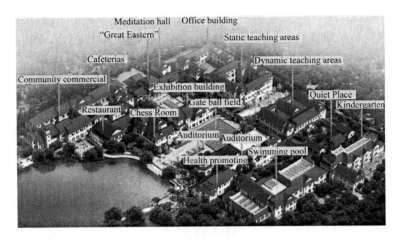

Figure 1. Layout Chart of the Greentown Wuya University for the Aged Operational. Provided by Greentown Yile Limited.

pool, a gate ball field, a medical building, as well as cafeterias and a kindergarten (Figure 1). The curriculum has also been well planned, there are in total five categories of courses: Health and Fitness (Exercise, Diet and Nutrition, Mental Health, Emotion Management), Humanities (Literature, History, Military, Politics, Language and Travel), Art (Calligraphy, Chinese Painting, Western Painting, Photography, Chinese opera, Singing, Instruments, Dance), Recreation and Sports (Chess, Ball Games, Swimming, Gymnastics, Martial Arts), and Life (Computer, Cooking, Fashion, Gardening, Crafts).

2.1.2 *Management*
The evaluation system for green buildings can be divided into the evaluation of two stages – the design stage and the operational stage. The author believes that the evaluation of green elderly communities should emphasize the operational stage. If we look into the development and operation of elderly communities in China and abroad, we can see that in a lot of cases, care for the elderly doesn't go beyond design. They neglect the elderly care services, medical support, nursing education and community intelligent management in the operational stage. In some cases the communities don't even actively organize activities for the elders. As a result, those elderly communities lose their original purpose of care for the elderly, and show no difference from a traditional residential community. In building elderly communities, supporting facilities such as the university for the aged, the geriatric nursing center, the medical clinic, and the modern "nursing home" should be carefully planned and implemented. Service facilities that can cater both the elderly and other age groups, such as the cultural recreation center, could also be considered. Moreover, the communities should actively develop and popularize the intelligent operational management system.

Take service facilities in the Greentown Wuya Community as an example, in this case, on top of the design and construction of the community and residential homes, the community pays a lot of attention on making the elderly community comprehensive. The Wuya Community is centered on the university for the aged (Yile College), and integrates it with the residential area, the business street, a resort, a medical park, and an international geriatric nursing center, as an effort in improving the quality of life of the elders from the service support and operational management level (Figure 2).

2.1.3 *Eco-friendly and energy saving*
Evaluation on this aspect should focus on the community's hygiene, space usage, safety, water environment, vegetation planning, as well as the energy-saving irrigation system and landscape

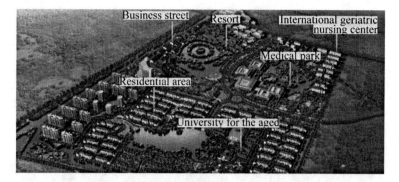

Figure 2. Aerial View of the Greentown Wuya Community. Provided by Greentown Yile Limited.

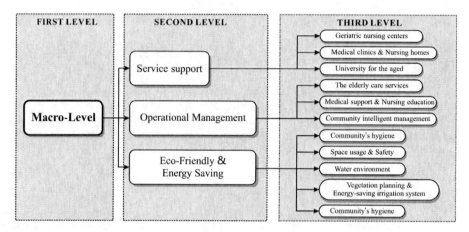

Figure 3. The evaluation framework of the macro-level.

design. It should use the evaluation standards for green residential buildings as reference, and improve the environment of green elderly communities on the technical level, such as saving energy, water, land and materials. Therefore, the evaluation framework of the macro-level could be summarized as the following program (Figure 3).

2.2 *On mezzo-level*

The mezzo-level targets at public areas such as the community's environment and the apartment buildings, including the "healing landscape", entrance area of the apartment buildings, hallways of the apartment building, elevators, and public corridors. The evaluation on this level should focus on safety, comfort, and the popularization of intelligent systems.

2.2.1 *Safety*

In terms of safety, we should focus on assessing the space's level mobility performance, vertical mobility performance, as well as space usage. In the aspect of the space's level mobility performance, we should pay attention to the difference in height in a flat space, such as the threshold or windowsill, and try to avoid elevation differences between two adjacent spaces. We also need to pay attention to the mobility safe-assistant equipment, such as railings and lighting, as well as the scale of movable space, that is, by managing the scale to avoid insufficient width. In the aspect of

Figure 4. All-Purpose Card. Provided by Greentown Yile Limited.

the space's vertical mobility performance, green elderly communities should popularize elevators, and properly plan barrier-free ramps, while paying attention to the width and angle. In terms of space usage, we should mainly assess whether there's enough space in the hallways and elevator waiting halls, whether the spatial stream lines are too long or crossing, and the decoration of space, such as the use of antiskid materials on the floor.

2.2.2 *Comfort*
In terms of comfort, firstly the green elderly communities should focus on building the "healing landscape". That is, in addition to the visual aspect, the residential community should make use of the characteristics of local ecological species, such as the scent of the plants, complement with diet, to achieve the effect of disease prevention, health building, and life lengthening through reasonable vegetation planning. Secondly, there should be relevant design and preliminary evaluation for the physical environment within a community, such as the outdoor solar environment, lighting and ventilation. In addition, in planning accessory facilities, the community should also pay attention to cater the elderly, such as installing a large amount of benches and pavilions, where the elderly can rest. When choosing materials for the benches, the community should use mild materials such as wood, and try to avoid cold concrete or stones.

In terms of comfort on the level of apartment buildings, we should pay attention to the lighting and ventilation of the corridors, and make sure that the apartment building is clean, with the help of the property management, in order to create a most comfortable public environment.

2.2.3 *Facility performance and intelligent systems*
Green elderly communities should improve facility performance and intelligent systems application to address the safety, comfort, and communication needs of the elderly, such as wireless network, integrated information platform, comprehensive service system, energy consumption and environmental monitoring system, monitoring and help system, vital signs monitoring system and other new technologies and products. From the macroscopic angle, the community needs to penetrate households with internet, manage service personnel and domestic helpers, monitor energy consumption, locate help seekers accurately, and utilize remote medical treatment, as well as develop new technologies such as multi-functional wheelchairs and intelligent light control (Figure 4) (Figure 5).

When the elderly encounters a sudden physical condition in the community activities, he could send distress signals through the monitoring and help system, which will enable him to be located through the vital signs monitoring system and the comprehensive service system, and receive preliminary medical treatment through the remote medical treatment system. In addition, the

Figure 5. Video Monitoring from Fixed Locations. Provided by Greentown Yile Limited.

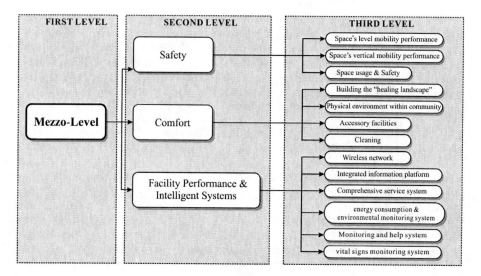

Figure 6. The evaluation framework of the mezzo-level.

monitoring system that covers the whole community is also very important. It can assist community staff in discovering the problem and providing necessary measures timely when the elderly fails to send distress signals. Therefore, the evaluation framework of the mezzo-level could be summarized as the program shows (Figure 6).

2.3 *On micro-level*

The micro-level refers to the interior of living space, that is, the interior of an apartment. We should pay special attention in handling this space since it is the most used by the elders. The evaluation indexes on this level mainly include the follows.

2.3.1 *Safety and convenience*
In terms of safety, we should focus on assessing the space's mobility performance, as well as space usage. From the aspect of the space's mobility performance, we should pay attention to

the differences in height in a flat space, such as the threshold or windowsill, and try to avoid elevation differences between two adjacent spaces. We also need to pay attention to the width of the space to avoid inconvenience when passing. In addition, we need to pay attention to the mobility safe-assistant equipment, such as railings and lighting, and evaluate whether the floor lighting is sufficient for the elderly, measured by the elderly's average eyesight. In terms of space usage, we should mainly assess whether the spatial streamlines are reasonable, including whether the spatial streamlines are too long, whether the streamlines conflict with each other, or whether the streamlines conflict with the facilities. For example, it would cause problems if the shower room/toilet is too far from the bedroom, or if there's no toilet in the bedroom suite.

2.3.2 *Comfort in the interior physical environment*

In terms of interior physical environment of the green elderly apartments, we should emphasize on the application and evaluation on daylighting, as well as the indoor sound, light, and thermal environment. The elderly have left their work, and shifted their life center from the office to their residences. "Home" is the place where the elderly spend the most time during their retired years. Therefore, we need to pay more attention at building comfortable indoor physical environment. Apart from performing evaluation in strict accordance with the "Evaluation Standard for Green Building" (GB/T 50378-2006), we should also take the characteristics of the elderly population into full consideration. For example, the elderly population's adaptability to temperature has been decreasing by age, so we should try to sustain a constant indoor temperature. At a certain temperature, from the energy saving point of view, we could consider bringing the "municipal central heating in the winter" to the south, or at least installing geothermal heating or solar heating when building new green elderly communities.

2.3.3 *Eco-friendly and energy saving*

From the perspective of energy saving, environmental protection and sustainable development, green elderly communities should also pay attention to energy saving and environmental protection. For example, the elderly communities should actively promote the application of renewable energy in the apartments. They should classify water used by households into drinking water and domestic water, and actively promote the use of non-traditional water resources such as reclaimed water and rainwater. Also, through the use of water-saving appliances and equipment, water saving rate can exceed 8%. In addition, as mentioned before of the indoor physical environment, although we need to create a comfortable environment with constant temperature for the elderly, we need to pay proper attention to the energy consumption of air conditioning or heating.

2.3.4 *Facility performance and intelligent systems*

In terms of facility performance and penetration of intelligent systems, we still need to think from the aspect of the elderly residences' safety and comfort, for example, the use of the gas alarm device, the emergency call device, the recording and monitoring device, the intercom device, and the pace of life abnormity sensor. Take the pace of life abnormity sensor as an example. Body movement sensors are installed in the bedrooms and bathrooms to monitor the use of certain household equipment. The device analyzes the monitoring data through software and sends alarm signals to relevant personnel when detecting abnormity. Therefore, the evaluation framework of the micro-level could be summarized as the following diagram (Figure 7).

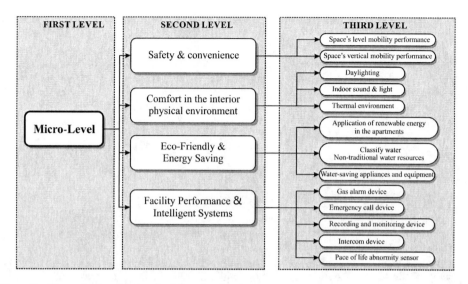

Figure 7. The evaluation framework of the micro-level.

3 CONCLUSION

In recent years, China is paying increasing attention to the housing situation of the elderly, and promotes this process through real estate development activities, such as the Beijing Yuecheng Group's development of specialized nursing institutes, as well as Hangzhou Greentown Group's development and exploration on elderly communities. However, no development in any field is achieved overnight, and is always an advancing process of twists and turns. Development should also continuously address the society's major development themes and demand. Creation of the concept "green elderly communities" is one of such development, attempting at integrating "the elderly" and "green". While delivering care for the elderly, the communities are also compatible with the sustainable development of energy saving and environmental protection.

Considering the life cycle of a building and a family, green elderly communities of the future is likely to apply to all ages and groups. That is, the population structure will be consistent with the social population structure. In this way, the communities are no longer specialized "elderly camps". On the premise of this concept, the author thinks that the essence of green elderly communities is to improve the quality of life of the elderly, and the environmental protection standards in current residential communities. For example, universities for the aged, medical clinics, nursing centers, nursing homes could be added to the existing facilities of kindergartens, bus stops, and business services. In addition, intelligent systems should also be given sufficient attention and promotion, while combined with the related concepts of green building, in order to make green elderly communities a new type of community that can cater both the old and the young, and is eco-friendly and sustainable.

REFERENCES

Code for Design of Residential Building for the Aged. ГB/T 50340-2003.
Architectural Design Standards for the Facilities for the Elderly. (2000) 0047. Shanghai.
Nie M.S., Yan Q.C., Gordon P.A. 2011. *United Evaluation Affirm System on Green Elderly Community in China.* Beijing: China Architecture & Building Press.
The Center for Housing Industrialization. 2012. *Construction Principles and Technology Guide Rules of Intelligence System in Elderly Community.* Beijing: China Architecture & Building Press.
Evaluation standard for green building. ГB/T 50378-2006.

The survey of EEWH–Daylight evaluation system by LEED-Daylight standard – Taking Beitou Library as an example

Y.M. Su & Y.S. Wu
Graduate Institute of Architecture and Urban Design, National Taipei University of Technology, Taipei, Taiwan

ABSTRACT: The developing of urban caused amount of energy consumption, many of the sustainable building rating systems have taken numerical simulation as one of the rating indicators to reduce damage to the environment during construction. However, the current green building rating system of Taiwan, called EEWH, has not included any numerical simulation items but assessed window area and daylight area of daylight. This research has employed indicators from LEED-Daylight to simulate a case of the Taipei Public Library Beitou Branch, which awarded Diamond Certification set of EEWH. This study trends to conclude with some recommendations on Taiwan EEWH Interior Daylight Simulation by comparing the requirements for daylight assessment between EEWH and LEED.

1 INTRODUCTION

The BREEAM (Building Research Establishment Environmental Assessment Method), established in 1990, is the first green building rating system in the world and had a great impact on the global development of green buildings. Accompanied with the rapid development of IT systems, the application of numerical simulation has made it possible for such buildings to simulate the environmental quality prior to construction completion. Therefore, many green building rating systems also employ numerical simulation as part of their ratings in order to optimize the efficiency and energy conservation during the design phase. However, due to the lack of numerical modeling for Taiwan EEWH, this research employed LEED–Daylight rating standard to simulate results for the Taipei Public Library Beitou Branch, which awarded Diamond Certification set of EEWH. This study trends to conclude with some recommendations on Taiwan EEWH Interior Daylight Simulation by comparing the requirements for daylight assessment between EEWH and LEED.

1.1 *EEWH*

Taiwan's green building rating system, established in 1999, is the first subtropical green building rating system. It is composed of 4 key categories such as Ecology, Energy Saving, Waste Reduction and Health (EEWH) and 9 indicators. Daylight related indicators are included in the Daily Energy Saving and Indoor Environment requirements. The energy saving requirements for EEWH is 20% higher than current technical building regulations in Taiwan, and the rating process for certification is simple and quick, which has made green building certification more popular and less costly.

1.2 *LEED*

Leadership in Energy and Environmental Design (LEED) is a Green Building Rating System which provide standard checking tool of ecology performance of building, and divided into different certified methods with different building type and lifecycle.[2] The Rating System of LEED contains: Sustainable-Sites, Water Efficiency, Energy & Atmosphere, Material & Resources, Indoor

Table 1. Comparison with LEED, EEWH and Building Technical Regulations of Taiwan.

Assessment Item	LEED	EEWH	Building technical regulations
Simulation	Yes	None	None
Measurement	Yes	None	None
Simulation Setting Date	September 21st at 9 am and 3 pm	None	None
Grid Size	10 feet grid	None	None
Illumination Standard	25 lx to 500 lx	None	None
Daylighting Distance	0.15 < WFR < 0.18	3 times	None
Daylighting Area	None	None	1/5–1/8
Not Included Window Area	30 inch (76.2 cm)	None	50 cm
Measurement After Construction	Yes	None	None

Environment Quality and Innovation & Design Process. LEED also developed different version around the world. While 7 nations use internal LEED version, 48 states in US use LEED rating system, even made LEED as legal standard to put into practice.

1.3 *Daylight assessment item of LEED, EEWH and building technical regulations of Taiwan*

By comparing LEED [3], EEWH, and the current Technical Building Regulations in Taiwan, it is found that the application of LEED can assure the efficiency of the buildings with simulations, regulations, and verification upon completion. However, application of EEWH can fill the gaps for current regulations in Taiwan and additionally regulate Daylight Depth and Skylight Coverage in order to avoid the problem of solar radiation resulting from designs with too many openings. While the current regulations in Taiwan have specifications about daylight areas and building gaps, the lighting quality of the building environment can be assured with the regulations based on domestic research by academic professionals.

2 CASE SELECTION AND MODELING

2.1 *Case selection*

This study selected Beitou Library in the northern Taipei which awarded Diamond Certification set by EEWH as the simulation case. It located in Hot Spring Park, northern Taipei (Coordinates: 25°N, 121°E, Figure 1(a) and (b)) and build with two storeys above the ground and one storey underground. (Figure 1(c))[4] There are many windows on the east and north sides of the Taipei Public Library Beitou Branch building (Figure 1(c)), with 2.8 meter high window opening for each floor, 1.8 meter wide outer corridor for sun-shielding [5], staircases down to the basement in the east side, and void design for the second floor slab and large-sized windows (up to the roof with the height of 9 meters). The south-west side is the main entrance with spaces for shorter visits and designed for faculty spaces with few window openings. This research took Taipei Public Library Beitou Branch as an example case and intended to discuss its efficiency for indoor daylighting quality, and evaluate its EEWH Checklist through LEED-Daylight numerical simulation.

2.2 *Modeling*

This research used Autodesk Series Software as main software to build 3D model, setting material and simulated. First, we edited 2D graphic in Autodesk AutoCAD, and build 3D model in Autodesk Revit. Finally, the 3D model was imported into Autodesk Ecotect Analysis to do daylight simulation (Figure 2). Autodesk Ecotect Analysis provide lighting, thermal, energy using and acoustic analysis

Figure 1. Location of Beitou Library (a) Location of Taipei (b) Relative Location of Site to Taipei (c) Northern-Side of Beitou Library (d) Eastern-Side of Beitou Library.

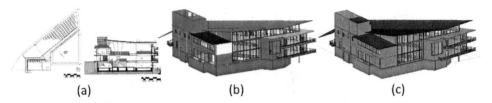

Figure 2. Modeling Process from 2D to Simulating Model (a) 2D Graphic (b) 3D Model in Revit (c) 3D Model in Ecotect.

which included natural and artificial light levels, shadowing, hourly thermal comfort, monthly space loads and environmental impact, acoustic reflections, and reverberation time as well [6,7]. One of the biggest merits of this tool was its compatibility with most three dimensional file types, include complex model and allows doing different analysis in the same model in the same Graphical User Interface (GUI).

3 SIMULATION SETTING AND RESULT

3.1 Simulation setting

This research employed LEED-Daylight simulation settings, in which the modeling was set for the indoor illuminance at 9 am and 3 pm on the day of September 21st (autumn equinox), with sun path shown in Figure 3(a) and (b), at the 25°N parallel for the Taipei area. The global solar radiance was based on domestic research by academic professionals [8,9] with global illuminance in clear sky condition between 60,000 to 100,000 lux. The simulated global illuminance was set to be 80,000 lux in clear sky conditions, and the natural daylight illuminance at the operational height of 1 m above ground on the first floor with 30 cm grid was simulated. However, the bathroom spaces on the southwest side and the storage spaces were excluded. See Figure 3(c) for grid setting.

3.2 Result

The conclusions of the simulation based on the above settings are as follows:

3.2.1 Result of 1st floor

The simulations of illuminance on the first floor included (see Figure 4): (a) illuminance at 9 a.m. and (b) illuminance at 3 p.m. The illuminance of South-East side, which is the entrance of building with partial opening, was over 500 lux. The diffuse light at northern side made better uniformity of illuminance, however the illuminance of area near the window was over 500 lux, whereas the illuminance within the area covering from 5.5 m to the windows almost equaled to zero. The illuminance area of west side with the opening of only three 120-centermeter-high windows was relatively small, and the illuminance level is mostly over 500 lux, leaving the rest of the area in a

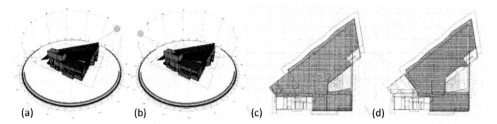

Figure 3. Simulation Setting (a) Sun Path at 9 a.m, Sep 21st (b) Sun Path at 3 p.m, Sep 21st (c) Grid Setting of 1st Floor (d) Grid Setting of 2nd Floor.

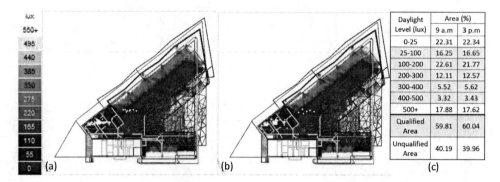

Daylight Level (lux)	Area (%)	
	9 a.m	3 p.m
0-25	22.31	22.34
25-100	16.25	16.65
100-200	22.61	21.77
200-300	12.11	12.57
300-400	5.52	5.62
400-500	3.32	3.43
500+	17.88	17.62
Qualified Area	59.81	60.04
Unqualified Area	40.19	39.96

Figure 4. Results of Simulation on 1st floor (a) Illuminance Level at 9 a.m, Sep 21st (b) Illuminance Level at 3 p.m, Sep 21st (c) Illuminance Level Interval.

short distance away with illuminance dropping dramatically down below 25 lux; as for the west side facilitated with staircases connecting the first floor to the second floor and the rooftop skylight, the illuminance had also exceeded the standard of 500 lux based on the simulation result.

According to Figure 4(c) which illustrated the illuminance distributed at each grid for 9 a.m. and 3 p.m. on the first floor, the areas with illuminance below 25 lux accounted for 22.3% at the highest rank, meaning that there were still spaces on the first floor unable to take in the natural daylight; whereas the areas with illuminance over 500 lux accounted for 17.7% or so. In summary, the areas with illuminance between 25–500 lux accounted for 60%, leaving the rest 40% of the areas with illuminance below 25 lux or over 500 lux.

3.2.2 *Result of 2nd floor*

The simulations of illiminance on the second floor included (see Figure 5): (a) illuminance at 9 a.m. and (b) illuminance at 3 p.m. The audio-visual spaces located on the south side, with the natural daylight covering 1/2 of the space, the illuminance for 1/5 of the space was over 500 lux. Some natural daylight might be taken in through the four small windows in the south side with relatively minor yet even illuminance. There was much better daylight through the relatively large window opening on the north side. Moreover, with the tilted roof skylight the daylight depth for the north side of the second floor even reached to 1/2 with an evener illuminance than the first floor. The staircases to the first floor also had better daylight taken in through the roof skylight However, due to solar direct radiation, the illuminance was over 500 lux.

According to Figure 5(c), it can be seen that the spaces with illuminance of 0–25 lux had been reduced down to 12.34 %, and with illuminance over 500 lux accounted for 16.78%. Spaces with illuminance between 25–500 lux accounted for 70.88%, while spaces with illuminance below 25 lux or over 500 lux accounted for 29.12% of the whole space. The illuminance on the second floor was evener than that on the first floor in general.

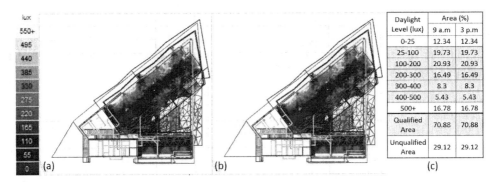

Figure 5. Results of Simulation on 2nd floor (a) Illuminance Level at 9 a.m, Sep 21st (b) Illuminance Level at 3 p.m., Sep 21st (c) Illuminance Level Interval.

4 CONCLUSION

After the simulation of daylight illiminance level of Taipei Public Library Beitou Branch is shown above, we could concluded in (a) the difference between standard of LEED-Daylight and EEWH-Daylight, and (b) the simulation setting suggestion of EEWH-Daylight by LEED-Daylight.

4.1 *The difference between standard of LEED-daylight and EEWH-daylight*

According to the results of the simulations, the insolation of Taipei Public Library Beitou Branch reached the standards of daylight requirement based on Building Technical Regulations in Taiwan and EEWH. However, according to LEED-Daylight Rating System, the room space with illuminance of 25–500 lux accounting for 65% of the entire space failed to reach the LEED-Daylight standards. The fact that the spaces with illuminance under 25 lux on the first floor accounted for 22% of the first floor area, while 12% for the second floor, as well as the large-sized window opening on the north side took in too much daylight with illuminance over 500 lux, made Taipei Public Library Beitou Branch unable to reach LEED-Daylight standards. Based on the results of this research, it is found that there were only lower limits set for daylight insolation yet no upper limits according to the current regulations in Taiwan and EEWH. On the other hand, LEED-Daylight aims to enhance the evenness of natural daylight, improve visual comfort, and assure occupational functionality through the indoor illuminance. Generally speaking, for places like Taiwan in tropical climates, the emphasis should be more on the degrees of daylight insolation and the setting of upper limits, as compared to countries in temperate climates, in order to avoid the concentrated daylight insolation from causing discomfort and energy consumption. Therefore, it is recommended to set specifications for the upper limits of daylight insolation and improve evenness regulations to avoid too much energy consumption.

4.2 *The simulation setting suggestion of EEWH-Daylight by LEED-Daylight*

In simulation settings, LEED-Daylight uses the autumn equinox day as the simulation basis, however, due to the dramatic illuminance differences between summer and winter in the Taiwan, it is recommended to furthermore take the summer equinox and winter equinox into consideration too, in order to have a more comprehensive understanding about the indoor environment in different seasons through numerical simulation. Designers may proceed with optimization at the early design stages to assure the quality of the buildings and to better achieve the goals of Ecology, Energy Saving, Waste Reduction and Health.

ACKNOWLEDGEMENT

This paper represents part of the results obtained under the support of the National Science Council, Taiwan, ROC (Contract No. NSC101-2627-E-027-001-MY3).

REFERENCES

Ministry of Interior of Taiwan 2012. ABRI, Green Building Evaluation Manual – Basic Version. Architecture and Building Research Institute
Information on: http://www.usgbc.org/
Information on http://greenexamguide.com/
Information on http://www.forgemind.net/xoops/modules/news/article.php?storyid=863
Information on Bioarch Architecture.
Nedhal AT. and Sharifah FSF. 2012. Energy-efficient envelope design for high-rise residential buildings in Malaysia. In *edited by Architectural Science Review* Vols. 55, No. 2, pp. 119–127.
Arbi GS. 2010. Comparison of PV System Design Software Packages for Urban applications. In *IEEE, and Pragasen Pillay*, Fellow IEEE.
Ting-Fa D, Zhi-Wen Z, Kun-Lin L. and Qian-Ni Y. 2012. The Application of BIM to Energy Efficient Design – A Case Study of Taipei MRT Wanda Line Station LG05. In *Technic of Rapid Transit System Half-year Journal*, Vols. 47, pp. 47–54.
Zhong-Zhi Z. and Zhao-Ren L. 2008. The Evaluation of Daylight Performance of Window Integrated with Daylighting Components, *5th Taiwan Creativity Forum – Creativity and Culture of Architecture*.

Slaughterhouse wastewater treatment by hydrolysis and Bardenpho process

A. Li, Y. Zhang, H. Guo & T. Pan
Beijing Municipal Research Institute of Environmental Protection, China
State Environmental Protection Engineering (Beijing) Center for Industrial Wastewater Pollution Control

ABSTRACT: Slaughter wastewater was a kind of typical organic wastewater that can be biodegraded, but with high concentration of organic matter and nitrogen materials, treatment of this kind of wastewater remained difficult. Basing on the characteristics of the slaughter wastewater, this study adopted hydrolysis + Bardenpho reactors to treat slaughterhouse wastewater and realized process optimization by selecting the appropriate process parameters. Impacts of different hydraulic retention times (HRT) on the removal efficiency of COD_{Cr}, NH_4^+-N and TN were studied so as to evaluate the feasibility of combined hydrolysis and Bardenpho process in treating slaughterhouse wastewater. The results showed that the concentrations of COD_{Cr}, NH_4^+-N and TN in effluent were 24~28 mg/L, 0.01~1.5 mg/L and 10.2~12.4 mg/L respectively with met with the B emission limits of "Sketch of water pollutant emission standards" (DB11/307-20××): COD ≤ 40 mg/L, NH_4^+-N ≤ 2.0 mg/L, TN ≤ 20 mg/L.

1 INTRODUCTION

Wastewater produced during the process of slaughtering and producing meat contained large amount of bloodiness, fur, ground meat, viscera, undigested food, and faeces, etc. Slaughterhouse wastewater with reddish brown and distinct stench was a kind of typical organic wastewater and was rich in protein,fat as well as salinity. In general, COD_{Cr}, NH_4^+-N and TN concentrations of this kind of wastewater were 1500~2000 mg/L, 150~250 mg/L and 150~250 mg/L respectively. Therefore, it was difficult to treat (Chen Lie, Zhou Xingqiu, Gao Feng, 2003; Yu Feng, Chen Hongbin, 2005). At present, the main treatment processes of slaughterhouse wastewater were activated sludge process, such as sequencing batch reactor (SBR), biological contact oxidation, anaerobic sequencing batch reactor (ASBR) (Zhang Wenyi, 2002), expanded granular sludge bed (EGSB) reactor (Munez L A, Martinez B, 1998) and upflow anaerobic sludge bed (UASB) reactor (Zhao Yingwu, Li Wenbin, Gong Min, 2007), etc. which all suffered from several problems, especially the low removal efficiency of TN (generally less than 85%). Along with the more and more strict emission requirement for slaughterhouse wastewater treatment station in Beijing, the above processes could hardly meet with the relative discharge criteria. In conclusion, with comprehensive evaluation of characteristics, research status and discharge standard of slaughterhouse wastewater, combined hydrolysis and Bardenpho process was adopted to carry out experimental research of treatment slaughter wastewater. The organic matters of slaughter waste water were mainly protein and fat which were difficult to be used by aerobic bacteria because they belonged to long chain great molecule organic substances. However, after broken down into small molecule organic substance such as amino acids, carbohydrates,etc by enzyme, they can be utilized easier. So it was necessary to set hydrolysis acidification process before Bardepho process.

Experimental results showed that this combined process not only had good removal effect for COD_{Cr} but also for TN and the slaughterhouse wastewater after treatment can achieve corresponding pollutant discharge standards of "Sketch of water pollutant emission standards"

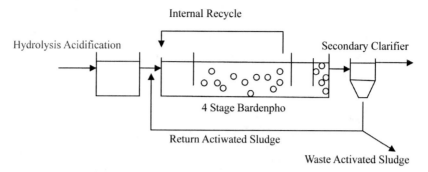

Figure 1. Schematic diagram of experimental set-up.

Table 1. The water quality of experiment.

Water quality index	Unit	Wastewater quality	Processing demands
COD_{Cr}	mg/L	1400~2500	≤30
NH_4^+-N	mg/L	150~200	≤1.5
TN (mg/L)	mg/L	200~250	≤15

(DB11/307-20××). It provided reliable basis for the same type of slaughter wastewater treatment project.

2 EXPERIMENTAL MATERIALS AND METHODS

2.1 Experimental apparatus

Experiment devices mainly composed of hydrolysis acidification reactor, Bardenpho reactors and settling basin as shown in figure 1. The effective volume of hydrolysis reactor was 4.8 L. The water was fed from bottom to top. Two rows water distributors were arranged at the bottom of hydrolysis reactor so as to which can realize water distribution uniform. Four-stage Bardenpho process was composed of four consecutive reactors which were primary anaerobic, primary aerobic zone, secondary anaerobic and secondary aerobic zone. Effective volumes of each Bardenpho reactor were 6.4 L, 19.2 L, 3.2 L, 1.6 L and the time ratio for hydraulic retention time of each reactor was 4:12:2:1. The design parameters of sedimentation tank was that effective height for sedimentation was 0.25 m and precipitation time was 3 h.

2.2 Wastewater quality and inoculum

The water used in the test were taken from a slaughterhouse in Beijing. The concentrations of various pollutants were constantly changing. COD, NH_4^+-N and TN of effluent should achieve B emission limits of "Sketch of water pollutant emission standards" (DB11/307-20××).

The water quality of experiment was shown in table 1.

Inoculated sludge in hydrolysis acidification reactor was anaerobic granular sludge (pH7.18, particle size of sludge is 1~2 mm). Inoculated sludge in Bardenpho reactor was aerobic sludge from reclaimed water treatment station of Beijing Municipal Reaearch Institute of Environmental Protection (pH8.28). The quantity of initial inoculated sludge was 2000 mgMLSS/L.

2.3 Analytical methods

The pH of each reactor were monitored by pH meter S-100 (SUNTEX, Taiwan). Water temperature and dissolved oxygen(DO) were monitored by portable dissolved oxygen meter (LDOTM, America). The water quality analysis refered to the literature: COD_{Cr} was monitored by standard potassium dichromate method; NH_4^+-N was monitored by Nessler's reagent spectrophotometric method; TN was monitored by alkaline potassium persulfate digestion ultraviolet spectrophotometric method; NO_2^- was monitored by N-1-naphthyl-ethylenediamine spectrophotometric method; NO_3^- was monitored by UV spectrophotometry; MLSS was monitored by gravimetric method.

3 EXPERIMENTAL PROCESS

3.1 Experimental start-up

Inoculated sludge of about one eighth the volume was put in the reactor and then filled the reator with slaughter wastewater. Stirred 8h continuously to make good contact of sludge and waste water, after that let it sit 1 h and drained overlying water. Finally, adopting the way of continuous water intake made into the sludge acclimatization stage. The start-up of hydrolysis was finish when removal rates of COD_{Cr} and NH_4^+-N in effluent reached steady state (Fluctuate was less than 5%).

The inoculated sludge was aerated for 3 days in slaughter wastewater. Let the reactor stand and then replaced 1/5 of water every day. The entire device entered into the phase of sludge acclimatization stage when water was inflowed consecutively. The experimental start-up was finished when COD_{Cr} of effluent was stable and the removal rate of COD_{Cr} was able to achieve at least 80%.

3.2 Experimental operation

After the success of the experimental start-up system entered into a stage of stable operation. The internal recycle (IR) at the end of the primary aerobic zone brought nitrate back to the primary anoxic zone (400% of daily flow). The returned activated sludge (RAS) was about 100% of the daily flow. Slaughterhouse wastewater was inflowed by peristaltic pump homogeneously and consecutively. Effluent overflowed from sedimentation tank. The two anaerobic pools achieved a state of completely mixed by stirring with stirrer. Stirring speed was low, 40 rad/min, making dissolved oxygen at 0.2~0.5 mg/L. Two stage aerobic pools achieved a state of completely mixed by aeration. Keep the pool dissolved oxygen in 2.0~4.0 mg/L by adjusting the aeration. The sludge concentration of primary aerobic zone was about 3500 mg/L and in secondary aerobic zone was about 3000 mg/L when the system was at the stage of experimental stable operation period. Water inflow for design were 0.4 L/h, 0.6 L/h, 0.8 L/h. The whole experiment process was divided into three phases that the inflow of the first stage was 0.4 L/h for 8 days, the inflow of the second stage was 0.6 L/h for 9~21 days, the inflow of the third stage was 0.8 L/h for 22~36 days. The operating condition of the experiment for treatment slaughter wastewater using hydrolysis + Bardenpho process was observed by the method of increasing water inflow and reducing hydraulic retention time.

The experimental parameters of Bardenpho device during the running were shown in table 2.

4 RESUTS AND DISCUSSION

4.1 COD removal efficiency of system

Under the different inflow, removal efficiencies of COD in each pool with hydrolysis + Bardenpho process were shown in figure 2.

Experiment showed that the removal efficiency of COD was excellent by the hydrolysis + Bardenpho process. COD_{Cr} concentration of inflow was between 1286 and 2054 mg/L, the average was 1760 mg/L. COD_{Cr} concentration of effluent was 24~28 mg/L that can achieve B

Table 2. The experimental parameters of Bardenpho during the running.

Parameter indexes	A1	O1	A2	O2
Water temperature (°C)	20.0~24.0			
pH	7.45~8.10	7.61~8.16	7.48~7.93	7.53~8.30
DO(mg/L)	0.2~0.5	Forepart 2.5~4.8 Midpiece 2.2~4.4 Back end 1.1~2.8	0.2~0.5	2.8~6.5

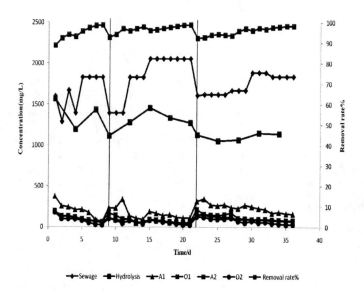

Figure 2. COD removal efficiency in the process of hydrolysis + Bardenpho.

emission limits of "Sketch of water pollutant emission standards" (DB11/307-20××). The total removal rate of COD_{Cr} was 97.66~98.68%.

Nonbiodegradable macromolecular organic matter could be degraded to small molecular organic matter that could improves biodegradability of wastewater. At the same time, COD_{Cr} was partial removed by hydrolysis and the removal rate was 38%. Removal effect of COD_{Cr} was obvious in primary A/O and removal rate could reach 80.8~95.6%, the average was 90.4%. Removal effect of COD_{Cr} was 9.8~60.5% in secondary A/O and the average was 30.2%.

The change of water inflow did not change the trend of COD_{Cr} removal, but it had a certain influence on the removal rate. The removal rate of COD_{Cr} was slightly lower with the water inflow gradually increase, but later the removal rate would increase slowly and tended to be stable. The removal rate of COD_{Cr} was 98.68% at the stage of low water inflow, 98.14% at the stage of mid water inflow and 97.66% at the stage of high water inflow when the system was at stability. In terms of hydraulic retention time, hydraulic retention time reduced with the water inflow increase that caused that microbial degradation was not complete and the removal rate of COD_{Cr} decreased. In terms of microbes, the sudden change of water inflow led to microbial inadaptation. But microbes gradually adapted to the environment with the passage of time and the biodegradability for COD_{Cr} gradually increased and tended to be stable.

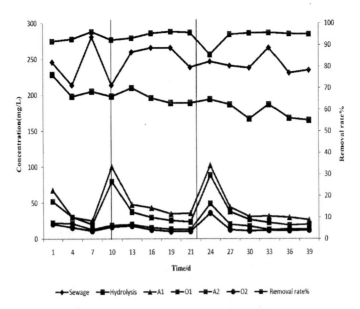

Figure 3. TN removal efficiency in the process of hydrolysis + Bardenpho.

4.2 TN removal efficiency of system

Experiment showed that the removal efficiency of TN was excellent by the hydrolysis + Bardenpho process. TN concentration of inflow was 214~281 mg/L, the average was 246 mg/L. When the system was at steady state TN concentration of effluent was 10.2~12.4 mg/L that could achieved B emission limits of "Sketch of water pollutant emission standards" (DB11/ 307–20××). The total removal rate of TN was 94.85~96.16%. The removal rate of TN was 38% by the hydrolysis. Removal effect of TN was obvious in primary A/O and removal rate could reach 79.9~90.2%. Removal effect of TN was 40.8~67.0% in secondary A/O. Water inflow changes had a certain impact on TN removal, but TN concentrations of effluent in each pool gradually decreased and tended to be stable in the end after microbes gradually adapted to the environment.

Protein and other nitrogenous organic compounds were translated into NH_4^+-N under the action of ammonification by ammonifier and microbes could consume part of NH_4^+-N to finish their metabolic activity in the hydrolysis zone. Therefore, concentration of TN reduced 22.4~29.8% by the combined effect of ammoniation and microbial consumption. Removal efficiency of TN was remarkable in primary A/O and most of the TN was removed in this stage. Nitrification was that nitrobacteria converted NH_4^+-N into nitrate nitrogen mainly happening in the aerobic zone. Denitrification reaction was that denitrifying bacteria translated nitrate nitrogen into N_2 when the internal recycle brought back lots of NO_3^- and NO_2^- to the primary anaerobic zone. But TN in effluent of primary A/O failed to meet the relevant provisions of effluent water quality standards. Subsequently, waste water flowed into the secondary A/O and nitrate nitrogen was removed through denitrification. It further strengthened the denitrification effect.

4.3 NH_4^+-N removal efficiency of system

It was observed that the average of NH_4^+-N concentration was 166 mg/L. NH_4^+-N concentration of effluent was 0.01~1.5 mg/L that could achieved B emission limits of "Sketch of water pollutant emission standards" (DB11/307-20 ××). The total removal rate of NH_4^+-N was 97.39~99.99%. Removal effect of NH_4^+-N was obvious in primary A/O and removal rate could reach 63.8~99.6% and the average was 87.8%. Removal effect of NH_4^+-N was 52.4~99.2% in secondary A/O and the average was 78.9%. Under the different water inflow, the removal efficiency of NH_4^+-N was similar

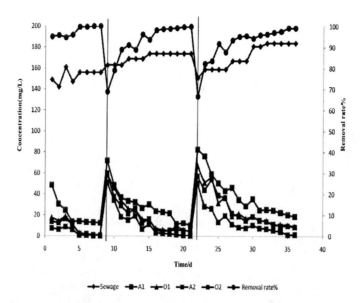

Figure 4. NH_4^+-N removal efficiency in the process of hydrolysis + Bardenpho.

to the removal efficiency of COD in system. It showed that NH_4^+-N concentration of effluent rose slightly and removal rate declined precipitously with the increase of water inflow. The removal rate of NH_4^+-N rose slowly and was tending towards stability over time.

4.4 The migration of nitrogen in the experimental system

4.4.1 The mechanism of strengthening denitrification in system

Macromolecular organic nitrogen in the wastewater was converted into small molecules by hydrolysis acidification and it created good conditions for subsequent denitrification process. In the primary A/O process, nitrate nitrogen in the return of mixing liquid carried on denitrification under the action of denitrifying bacteria using carbonaceous organic material as carbon source in the primary anaerobic pool. That nitrogen organic matter were ammoniated and nitrified in primary aerobic zone.As the same time, N_2 produced by denitrification released into the environment by aerated stripping in primary aerobic zone. In the secondary A/O process, when the mixture from primary aerobic pool got into the secondary anaerobic pool, denitrifying bacteria used endogenous metabolism matter of the mixture for further denitrification. N_2 produced by denitrification released into the environment by aerated stripping in secondary aerobic zone.That was not only improves sedimentation characteristic of sludge,but also the ammonia nitrogen from endogenous metabolism could be nitrified in the secondary aerobic pool.

4.4.2 The migration of nitrogen in the experimental system

Under the condition of water inflow for 0.8 L/h, the dynamic change situation of NH_4^+-N, NO_3^-, NO_2^- and TN of all stages was studied in steady state of experimental system. It was shown in figure 5.

As we could see from figure 5, the form of nitrogen were mainly ammonia nitrogen (163 mg/L, 69.4%) and organic nitrogen in the waste water. When the waste water with a lot of blood, grease, wool, meat scraps and skeletal debris, etc got into the reactor, organic nitrogen matter such as protein were transformed into ammonia nitrogen under the action of ammonification by various kinds of ammonifying bacteria. Ammonia nitrogen was first used by bacteria for growth and the remaining form dissolved inorganic ammonia nitrogen in water (Lin Xiaoli, 2007). TN and NH_4^+-N

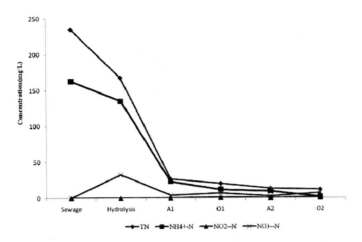

Figure 5. The migration of nitrogen in the experimental system.

of waste water were decreased. TN was decreased by 28.5% with the effect of ammoniation and denitrification. NH_4^+-N was decreased by 17.2% with the effect of ammoniation and consumption by microbe. Hydrolysis pool belonged to facultative anaerobic environment existing some nitrosobacteria and nitrobacteria. NH_4^+-N was degraded to nitrite nitrogen by nitrosobacteria and further to nitrate by nitrobacteria. So concentration of NO_3^- increased in this stage as shown in the figure 5.

The concentrations of various kinds of nitrogen decreased significantly when waste water got through primary anaerobic. TN decreased by 84.3% with dilution and denitrification. NH_4^+-N decreased by 83.1% with dilution. NO_2^- reduced from 0.07 mg/L to almost zero and NO_3^- decreased by 88.8%.

TN and NH_4^+-N in wastewater got a further degradation in secondary A/O reactor. The migration process of various forms of nitrogen was similar to the primary A/O. Removal rate of TN and NH_4^+-N were 41.3% and 81.2% respectively after passing secondary A/O reactor with the effluent water quality of primary A/O as a benchmark.

5 SUMMARY

Experimental results showed that hydrolysis + Bardenpho process had good removal effects on organic matter and nitrogen material in slaughterhouse wastewater. System for the removal rate of COD could achieve 97.66~98.68%. The concentration of COD in effluent could stabilize between 24 mg/L and 28 mg/L under the condition of the higher COD concentration. System had a good removal effect for TN.The removal rate of TN reached up to 96.2% and TN in effluent could be maintained at about 10.8 mg/L. System for the removal rate of NH_4^+-N could reach 97.39~99.99%, NH_4^+-N in effluent maintained at a relatively low level (0.01~1.5 mg/L). The concentrations of COD, NH_4^+-N and TN could achieve respectively B emission limits of "Sketch of water pollutant emission standards" (DB11/307-20××) under the condition of water inflow for design. This study has supplied some reliability data for treating slaughterhouse wastewater.

REFERENCES

Chen Lie, Zhou Xingqiu, Gao Feng, etc. The present situation and progress of treatment technology for slaughter wastewater. *Industrial water and Wastewater*, 2003, 34(6):9–13.

Lin Xiaoli. The cause analysis and improvement measures for rising of NH_4^+-N with SBR for treatment slaughterhouse wastewater. *Environmental Protection Science*, 2007, 33(2):29–30.

Munez L A, Martinez B. Anaerobic treatment of slaughterhouse wastewater in an expanded granular sludge bed(ECSB) reactor. *Water Science and Technology*, 1998, 40(8):99–101.

State environmental protection agency, *water and wastewater monitoring analysis method editorial committee. 3*. Beijing:China Environmental Science Press, 1989:195.

Yu Feng, Chen Hongbin. Treatment technology and the application progress of slaughter wastewater. *Environmental Science and Management*, 2005, 30(4):84–87.

Zhang Wenyi. Experimental study of treatment slaughter wastewater by anaerobic sequencing batch reactor. *Transactions of the Chinese Society of Agricultural Engineering*, 2002, 18(6):127–130.

Zhao Yingwu, Li Wenbin, Gong Min. Treatment slaughter processing wastewater by A/O contact oxidation process. *Water supply and drainage*, 2007, 33(11):68–70.

Effects of the construction of large offshore artificial island on hydrodynamic environment

Na Zhang, Chun Chen & Hua Yang
Tianjin Research Institute of Water Transport Engineering, Ministry of Communications, Tianjin, China

ABSTRACT: Based on construction of Dongjiang free trade port in Tianjin, a large offshore artificial island will be built on the east side of Dongjiang port covering an area of 40 km². In the paper, by using of the mathematical model, the impact of offshore artificial island's construction on the marine environment and artificial island's own environment problem are studied including of the change of the tidal current field before and after construction of the large offshore artificial island, the impacts of offshore artificial island's construction on pollutant dispersion, the water exchange ability for Tianjin harbor basin and the coastal resort before and after construction of the large offshore artificial island. Research revealed that the construction of offshore artificial island has had a limited effect on the marine environment and it can be provides basic evidence for the approval of the dongjiang large offshore artificial island.

1 INTRODUCTION

With the rapid development of China's economy, a long coastline was intensively and relentlessly exploited. In order to exploit ocean areas and resources, the construction of artificial island shows an increasing tendency. In order to construct Dongjiang free trade port in Tianjin, a large offshore artificial island will be built on the east side of Dongjiang covering an area of 40 km². The project is located in the east of Dongjiang, north of North Breakwater. The location of the project is shown in figure 1. The Tianjin coastline basically is industry coastline, from north to south including Center fishing port, Binhai tourist development area, Tianjin port, Lingang economic area, Nangang industry area and other large reclamation projects. The construction of the artificial island will have a certain impact on the surrounding marine environment. In the paper, mathematic model is used to study effects of the construction of large offshore artificial island on hydrodynamic environment.

2 METHODS

2.1 *Tide numerical model*

The tidal current field at project sea area are studied by the plane two-dimensional numerical model which developed by Danish Hydraulics Research institute (MIKE21).

$$\frac{\partial \zeta}{\partial t} + \frac{\partial uh}{\partial x} + \frac{\partial vh}{\partial y} = hS \tag{1}$$

$$\frac{\partial u}{\partial t} + \frac{\partial u^2}{\partial x} + \frac{\partial uv}{\partial y} = fv - g\frac{\partial \zeta}{\partial x} + \frac{\tau_{sx} - \tau_{bx}}{\rho h} + E_x\left(\frac{\partial^2 u}{\partial x^2} + \frac{\partial^2 u}{\partial y^2}\right) - \frac{1}{\rho}\left(\frac{\partial S_{xx}}{\partial x} + \frac{\partial S_{xy}}{\partial y}\right) + u_s S \tag{2}$$

$$\frac{\partial v}{\partial t} + \frac{\partial uv}{\partial x} + \frac{\partial v^2}{\partial y} = -fu - g\frac{\partial \zeta}{\partial y} + \frac{\tau_{sy} - \tau_{by}}{\rho h} + E_y\left(\frac{\partial^2 v}{\partial x^2} + \frac{\partial^2 v}{\partial y^2}\right) - \frac{1}{\rho}\left(\frac{\partial S_{yx}}{\partial x} + \frac{\partial S_{yy}}{\partial y}\right) + v_s S \tag{3}$$

Figure 1. The layout of the project.

where ζ is tide level, means the distance between water surface and datum plane; x and y are the Cartesian co-ordinates; h is water depth; g is gravitational acceleration; u, v are the velocity components in the x and y; t is time; f is Coriolis parameter; ρ is the density of water; S is the magnitude of the discharge due to point sources and (u_s, v_s) is the velocity by which the water is discharged into the ambient water; S_{xx}, S_{xy}, S_{yx} and S_{yy} are components of the radiation stress tensor; (τ_{sx}, τ_{sy}) and (τ_{bx}, τ_{by}) are the x and y components of the surface wind and bottom stresses; E_x and E_y are the horizontal eddy viscosity in the x and y.

2.2 Water exchange and pollutant dispersion numerical model

The conservation equation for a scalar quantity is given by

$$\frac{\partial(hc)}{\partial t} + \frac{\partial(uhc)}{\partial x} + \frac{\partial(vhc)}{\partial y} = \frac{\partial}{\partial x}(h \cdot D_x \cdot \frac{\partial c}{\partial x}) + \frac{\partial}{\partial y}(h \cdot D_y \cdot \frac{\partial c}{\partial y}) - F \cdot h \cdot c + S \quad (4)$$

where c is the concentration of the scalar quantity; u, v are the velocity components in the x and y (m/s); h is water depth (m); D_x, D_y are the horizontal diffusion coefficient in the x and y (m²/s); F is the linear decay rate of the scalar quantity (s^{-1}); S: $Q_s, (c_s - c)$; Q_s: is the magnitude of the discharge due to point sources (m³/s/m²); c_s is the concentration of the scalar quantity at the source.

3 MODEL CALIBRATION

In this paper, the two-dimensional tidal current mathematical model of MIKE 21 was used to study the flow field. In order to improve the accuracy of the simulation results, two level nested models are used in hydrodynamic model. The big model range is the whole of Bohai and the small range is Bohai coast. The open boundary of the large model lies between Dalian Tiger Beach and Yantai and is controlled by water level. The open boundary of the small model is provided by large model. The computational domain is divided in unstructured triangular grid. The tidal current mathematical model is verified based on the field data of the spring, medium and neap tide on Oct. 2012. The survey included 3 tidal stations and 11 flow station. The data are processed to provide continuous during the calibration period. Figure 2 and Figure 3 shows the time series simulated the tidal level, flow velocity and flow direction compared with the field measurements at two locations. It can

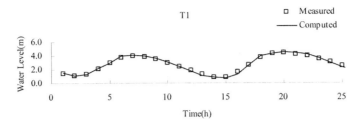

Figure 2. Verification of water level at T1 station.

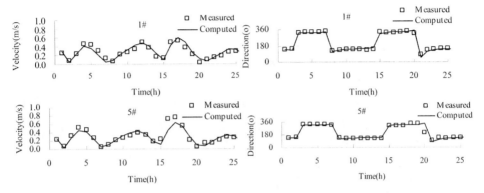

Figure 3. Verification of flow velocity and direction process during Spring on Oct. 2012 at 1# and 5# station.

be seen that the calculated tidal level is consistent with the measured data in both magnitude and phase, the deviation being within 0.1 m. The computed velocity and flow direction are also in good agreement with field data in both magnitude and phase.

4 RESULTS

4.1 Research on tidal current

Figure 4 show velocity vector before and after the project during flood tide. Figure 5 show the change magnitude of average speed during flood tide and ebb tide before and after the project ("+" denotes increase and "−" denotes reduction).

According to the analyses of current velocity and current direction before and after the project, the following conclusions are reached:

(1) The project has limited effect on the hydrodynamics of this sea area. It only influences the flow motion in a local area.
(2) The current speed decrease in the southeast side and the lake waters of second artificial island. The maximum decrease value of average speed outside of the breakwater of Tianjin Port is larger than 0.5 m/s. The current speed at the area between the main channel of the lake waters with the gate of the Tianjin port and the area near the northeast corner of second artificial island increase. The maximum increasing value of average speed is larger than 0.2 m/s.
(3) The change magnitude of average speed is less than 0.05 m/s at the entrance of the diamond island. The current speed is increased at the east gate and decreased at the south gate.
(4) The change magnitude of the average speed larger than 0.02 m/s' range is 18.6 km east–west direction and 16.8 km north-south direction.
(5) The maximum value of cross velocity is 0.19 m/s in the channel at the entrance before the project. The cross velocity is 0.01 m/s–0.10 m/s in internal channel and 0.11 m/s–0.26 m/s in

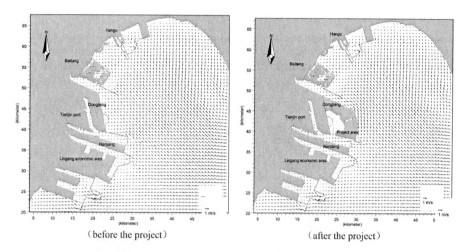

Figure 4. The velocity vector before and after the project during flood tide.

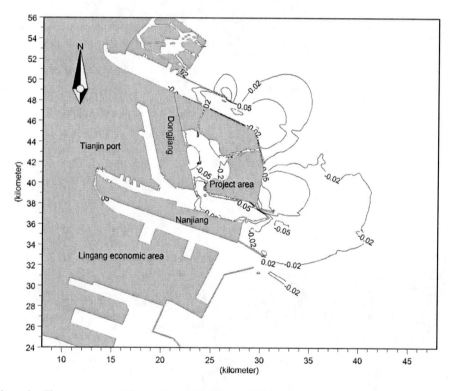

Figure 5. The average speed change during flood tide and ebb tide before and after the project.

external channel before the project. The cross velocity in internal channel increases from 10+0 to 16+0, while the cross velocity in external channel decrease after the project. The maximum increasing and decrease values of cross velocity are 0.12 m/s and 0.09 m/s, respectively.

(6) The cross velocity is 0.01 m/s–0.13 m/s in internal Dagusha channel of the channel turning section and 0.10 m/s–0.31 m/s external Dagusha channel of the shannel turning section after the project. Compared with cross velocity before the project, the change magnitude of cross velocity is less than 0.02 m/s.

The cross velocity is 0.03 m/s–0.36 m/s in internal channel of the channel turning section of Beitang port. Compared with cross velocity before the project, the change magnitude of cross velocity is less than 0.08 m/s.

The cross velocity in the channel of Central port is basically not affected by the project.

4.2 Research on water exchanges

The initial concentration of the water inside the hypothesis is inferior to the outside, no external load. It can appoint the external concentration of 0.5 units and the internal concentration of 1.0 unit. The open boundary is zero gradient.

Based on the analyses of the internal concentration before and after the project, the following conclusions are reached:

(1) Water exchange is more quickly on the mouth of the water channel than away from the water channel. The mean water exchange rate of the waters inside the second island is 38% after 5 days, 64% after 10 days and 82% after 30 days.
(2) The mean water exchange rate of the waters inside Tianjin port before the project is 26% after 5 days, 34% after 10 days and 37% after 30 days. The mean water exchange rate of the waters inside Tianjin port after the project is 29% after 5 days, 40% after 10 days and 43% after 30 days. The water exchange rate inside the Tianjin port increases.
(3) The mean water exchange rate of the waters is 72% after 5 days, 87% after 10 days and 92% after 30 days inside the diamond island before the project and is 71% after 5 days, 86% after 10 days, 91% after 30 days after the project. Visibly, water exchange in the diamond island dropped about 1% during the same period of time.

4.3 Research on pollutant dispersion

According to the of "Tianjin Quality Information Bulletin" in 2004, the total amount of COD into the sea from Yongding new river estuary is 53429 ton and the discharge into the sea in non – flood season is about 40 m^3/s which is average value of the discharge of normal flow year. In this simulation, simulation time is 60 day, before 30 days of continuous source and after 30 days of no COD emissions.

Figure 6 is COD diffusion maximum range before and after the project. It can be seen from the chart that the pollutant concentration all exceed 5 mg/L in the waters from the sewage outfall to open seas whose areas are 62.4 km^2 before the project and 43.9 km^2 after the project. In Beitang coastal tourist area, the pollutant concentration all exceed 4 mg/L in waters around the outlets whose areas are 2.2 km^2 before the project and 3.2 km^2 after the project.

Overall, after the project, COD diffusion concentration maximum range decreases, while range of larger than the 4 mg/L concentration is increased slightly in the coastal tourism. According to Regulations of the people's Republic of China National Standard "water quality standards", the coastal tourism sea area belong to the third class water quality, COD should less than 4 mg/L. Suggest setting a floodgate in the Yongding new river side exit of coastal tourism. Close the gate in the discharge period to ensure the water quality in the region.

5 CONCLUSIONS

A 2-D mathematic model of the tidal current, water exchange and pollutant dispersion has been built to simulate the change of the tidal current field before and after construction of the large offshore artificial island, the impacts of offshore artificial island's construction on pollutant dispersion and the water exchange ability.

- Modeled calculations demonstrate that modeled tidal elevation, current velocity, current direction are good agreement with the field measurements of time series.

Table 1. COD concentration contours envelope area before and after the project.

Tidal types	Case	The area of different concentrations/km²			
		≥2 mg/L	≥3 mg/L	≥4 mg/L	≥5 mg/L
Spring	Before the project	121.94	62.39	51.35	46.55
	After the project	90.44	43.86	29.18	25.06

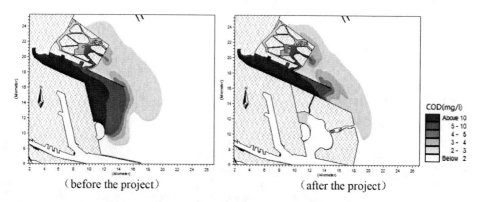

Figure 6. The maximum diffusion range of COD before and after the project.

- After the second artificial island is finished, the tidal current velocity between the north breakwater of Lingang economic area and Beitang port is changed within the offshore to Tianjin Port 12km outside the door. The current speed is increased at the east gate the diamond island and decreased at the south gate of the diamond island.
- After the second artificial island is finished, COD diffusion concentration maximum range decreases, while range of larger than the 4 mg/L concentration is increased slightly in the coastal tourism. Suggest setting a floodgate in the Yongding new river side exit of coastal tourism. Close the gate in the discharge period to ensure the water quality in the region.
- Because the tidal current velocity increased at the gate of Tianjin port after the project, the exchange rate between the main waterway of the second artificial island and the water gate of Tianjin port increased.

REFERENCES

Sun L.C. & Zhang N. & Chen C. 2011. Sediment research of the muddy coast Tianjin Port. China Ocean Press, Beijing.
MIKE21& MIKE 3 FLOW FM hydrodynamic and transport module Scientific Documentation, MIKE by DHI 2009.
Lu H.B. & Chen M.L. & Song J.Y. 2012. Influence of reclamation on seawater exchange based on numerical simulation. Port & Waterway Engineering, 9: 149–154.

The thinking and exploration on building a characteristic apartment system for the aged in city

Jinghua Dai, Zhu Wang & Zhi Qiu
College of Civil Engineering and Architecture, Zhejiang University, Hangzhou, Zhejiang, China

ABSTRACT: With the aging process increasing and silver industry advancing in China, the urbanized apartments for the aged are developing at a high speed. As an entry point, this paper pays equal attention to material life and spiritual life and also focuses on the differences between different life stages of the elderly. And then, taking account of the current situation of civil elderly apartments and their real needs, a concept of characteristic apartments for the aged is proposed, which helps set up a diversiform and optional service platform and supporting facilities as well as constitute an all-around and multilevel system so as to improve endowment environment and adjust the market supply and demand.

1 INTRODUCTION

By the year 2013, the aged population in China will become 202 million, breaking through the 200 million mark. And the degree of Chinese population ageing will rise up to 14.8% (Wu, 2013). And it is realized that Chinese aged population may reach a peak towards the middle of the century, which will account for 1/3 of the total population of Chinese people and exceed the sum of aged population in developed countries. In face of the great wave of white hair, the trend of aging cannot be avoided. At the same time, the traditional multi-generational families decrease rapidly while the empty-nest, disabled elders and some old people with hypophrenia increases largely. The traditional supporting model for the aged begins to transform into social provision for the old. The elderly apartment is a product of old people's needs to cope with the changes of age structure and adapt to the shifts of Chinese family types. Thus, it's extremely urgent to research on the current situations and future direction of apartments for the aged.

2 DEFINITION OF THE TOPIC

2.1 Concept

Apartment for the aged, as a kind of housing type with integrated management, provides services of foods, cleaning, culture and entertainment as well as the medical care. It is built specifically for the old people and is consistent with psychophysical characteristics of the elders. Different from traditional nursing institution for the aged, this type of apartment is the extension and complement of aged housings. It's equipped with specialized life service system and nursing system, built by social investment and run by the market, which provides different levels of housing and corresponding services for the aged with different self-care ability and income.

2.2 Scope of research

So far, the elderly security system and social resources support in town are far better than those in countryside in China, as a result of which there are sufficient basic conditions and promoting market in city. Therefore, the urban apartment is selected as topic of this paper, to maximize the optimize allocation of resources and to provide diversified services for the aged.

3 CHARACTERISTICS AND NEEDS OF THE AGED

The life stages of old people, defined as the period from aged to death, can be divided into four stages: 1) Independent period; 2) partial assist and nursing period; 3) assist and nursing period; 4) terminal period (Zhang & Chen, 2004) (Fig. 1). The trend of stage from the first to forth reflects differences of needs for social supports when they are in different stages. And different aging level corresponds to different self-care ability of daily living, and it is concluded as three types: the self-care elderly, the device-aided elderly and the nursing-cared elderly (JGJ122–99, 1999). As part of total population, the aged group's needs have the double characteristics of universality and specificity. Maslow's hierarchy of needs accurately summarizes various needs of human beings, that is mutual and the aged is no exception (Fig. 2). The elderly consists of every living and unique individual rather than a pronoun and it is also the society's epitome so we cannot take a part for the whole when we get to know and solve various related problems. Next we will try to learn the physiological functions, values and preferences of different life periods to analyze their internal needs.

3.1 *Independent period – characteristics and needs of the self-care elderly*

The old in independent period belongs to the self-care elderly, those who live on their own and don't need assistances from others (JGJ122–99, 1999). Most of them are in good health and retire not long ago, so the transition of social role and position depresses and abases them greatly. But in fact, these aged people are desperately eager to return to society and devote their remaining energy. Moreover, they hope to receive affirmation. Thus, in the special period, these elders should be recognized for their ability to work and contribution to society, and their individual personality traits and interests ought to be respected at first. Safe and comfortable environment, convenient for associating with others, should be developed to encourage them to take part in various social activities, such as cultural education, entertainment, skills training and so on.

3.2 *Partial assist and nursing period – characteristics and needs of the device-aided elderly*

As a typical representative of old people in this period, the device-aided elderly are those whose daily life need assist of handrails, walking sticks, wheelchairs or lifts (JGJ122–99, 1999). With

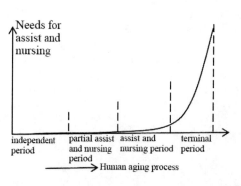

Figure 1. Variation of the elderly needs for assist and nursing in different life.

Figure 2. Maslow's hierarchy of needs.

age increasing, social activities of the elders decrease gradually, reserve capacity of every organ declines and their physiological function begins to be faced with obvious recession. The old in this stage has high requirements for rehabilitation, care, nursing, auxiliary equipment and instruments. And they also hope to participate in social activities and keep contact with others. For this type of old people, all-around facilities and services should be provided, as well as optional participation way and activity content, so as to improve their quality of life in spiritual and material view.

3.3 Assist and nursing period & terminal period – characteristics and needs of the nursing-cared elderly

The nursing-cared elderly is definitely those who live with the aid of nursing and instruments (JGJ122–99, 1999). Entering assist and nursing period, elders will experience the recession in both physiological and psychological mechanisms. Their ability to response to unfavorable conditions in external environment and resistance to disease tremendously declines, and then they may suffer from some chronic diseases, disability and dementia. Having lost autonomous ability and sense of independence, most elders in this period need to be taken care of thoroughly and carefully. For these people, assistance of different classes and types should be provided according to their type of illness and individual difference.

Terminal period is the final stage old people go through, when their disability or illness has no access to cure at the current technique level of medical treatment organization and there are only a few months left for them to live. However, terminal period doesn't mean waiting for death. It is a period of special life and their dignity ought to be respected and right be guaranteed. In general, demand of the elderly close to death can be divided into three types: preservation of life, relief from pain and death without pain or suffering. So when death is unavoidable, the warmest and most careful palliative care is necessary. Besides, spiritual consolation and care from relatives and medical workers also help the aged release their pain, eliminate psychological problems and go through the final moment without pain.

4 CURRENT SITUATION OF APARTMENT FOR THE AGED

4.1 Lack of facilities

By way of literature research and field survey, it is found that apartments for the aged run by government are in the majority and most of them are built far from urban areas with traffic inconvenient and facilities insufficient. This kind of apartment actually plays a role as 'hospice for the aged'. The selected remote site cuts off the elders' communication with friends and relatives, and in the meantime the backward and unsound supporting facility deprives their rights of participating in social activities and enjoying health care. If things continue this way, the lonely aged would become autistic and down in spirits. Meanwhile, body function of elders may decrease from bad to worse.

4.2 Single type and single structure

Although demands for apartments for the aged greatly exceed supply, there are plenty of vacant beds and apartments. According to a survey, most current elderly apartments are governed by collectivization and militarization, where the aged lives on an arranged schedule monotonously and they have no other choices. Traced backward the origin reasons, their planning is lack of periodic and hierarchical consideration. These institutions pay more attention to economic cost and number of beds, neglecting individualized care for elders in different life periods and improvement of quality of service. It leads to their poor initiative and physical & mental health condition. Furthermore, layout pattern of centralized nursing institutions may lead to a single age structure and has a subtle effect on elderly life and psychological state, which finally results in that apartment for the aged become a lifeless hospice. Undoubtedly, this type of supporting model for the aged would not be approved and chosen by elders, let alone becoming the mainstream development trend.

5 PONDERING OVER CHARACTERISTIC APARTMENT FOR THE AGED

Different from traditional social supporting model, which only provides fundamental services under modularization management, characteristic apartment for the aged is based on features and needs of elders in different life periods. It would improve elderly quality of living and provide diversiform places and supporting service to promote them to contact with and even reintegrate into society. Also, convenient and specialized medical care is available for elders.

5.1 *Integration of resources based on community*

From a view of environment behavioristic psychology, the mixed apartment for the aged is advocated. And it should be built relying on community, whose resources would be taken full advantage of, such as medical care, leisure and entertainment, the transportation network and geopolitical context. In this way, an all-around and multi-layered living service system is constructed, and the aged could live in a healthy and lively community synchronously developing with society.

5.2 *Strategies for different period of life*

5.2.1 *Independent period*

Considering elders in this period are in good health and they are eager to take part in social activities and achieve self-worth, they have the needs and ability to realize their potential and value. Hence, construction and processing of outdoor environment should be enhanced and the principle of comfort and safety must be observed when planning the elderly apartment. For example, large enough rest space and facilities are built in combination with field conditions, and cork and warm-toned wood is better to be chosen as material of seats. Handrails and backs ought to play a good supporting role. At the same time, seats and tables need to match well, which would be suitable for the aged to take part in some activities, such as chess and cards. The boundary of walking trails had better be covered by greening, which helps define the walking space, diverge traffic volume and increase security for elders to walk. Some elements, such as featured landscape and chairs, should be used to separate or recombine the field to build a space environment where elders will have the sense of belonging and could communicate with others easily (Fig. 3).

In addition, services on elders' recreational activities ought to get enriched and improved, and it's elaborated as the following three parts.

5.2.1.1 Accomplishment of the aged

Different planning scheme could be made for elders having different characteristics: old intellectuals continue to do some research or creation, the retired teacher may teach in local institute of

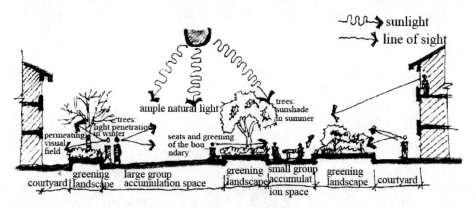

Figure 3. Schematic diagram of communication space Environmental Design.

technology, and the retired civil servants are likely to be engaged in some service work in their community. Other retired elders could also do some work that they are good at. Accordingly, the idea can be applied to design of apartment for the aged. For example, some multi-functional rooms are added for some elders to rent. In this way, not only the elderly self-worth could be achieved, but also it creates values for society.

5.2.1.2 Education for the aged

When planning an apartment, architects need to take it into consideration that learning field and related facilities should be built and fitted. Elderly university would be the main form, whose contents mainly include health care, painting & calligraphy, dance and music, photography and cooking, and so on. Besides, multifunction classroom, library reading room, and showrooms of different sizes and related facilities are necessary.

5.2.1.3 Entertainment for the aged

The aged also has high requirements for recreational facilities and entertainment venues. As an example, the entertainment center for aged people at high usage rate should include some entertainment places for talent shows, fitness and rest, teahouse and chess and card room, and provide chances for elders to communicate with others.

5.2.2 *Partial assist and nursing period*

It is the key of construction during this period to choose safe and convenient accessory equipment and create barrier-free environment for the aged. In terms of the external environment, the problem in treatment of ground is very important, such as altitude difference and material selection and paving method. By means of installation of ramp and handrail, pavement of anti-slip and smooth road, the risk can be minimized (Fig. 4). Secondly, sign system with strong guidance should be set up at the turn and end of walking path (Fig. 5), as well as combining appropriate lighting facility. Besides, elderly mobility scooter with high performance and stability is applied to take place of traditional manual wheelchairs, which will be the future trend. On the other hand, as for the internal environment, reasonable design of space environment and layout of the equipment are the focus during the construction of barrier-free bathroom, for instance, antiskid processing of the ground, installation of handrail and facilities including emergency call and mechanical ventilation, arrangement of bathing chairs and bathtub with extradition plates, etc.

Then, demands of the elderly for health care, nursing and rehabilitation need be taken into consideration to improve their physical quality, change their attitude towards the elderly support and slow down their hypofunction as far as possible by making room for geriatric disease policlinic, center for the elderly diseases control and prevention, medical health center, nursing center and mental health counselling room with well-equipped medical devices and specialized medical workers manned. In addition, the aged ought to be encouraged to participate in social activities properly according to their physical condition, in which way their physical pain and stress can be released so as to fulfill the ultimate goal of medical care for the aged and healthy endowment.

5.2.3 *Assist and nursing period & terminal period*

During assist and nursing period, the aged mainly depends on medical treatment and continuum of care, and the nursing grades should be delimitated according to the elderly illness and state of their lives. Medical and health technical personnel, mainly including doctors and nurses, and other professionally trained nursing personnel are able to provide integrated service, including treatment and nursing, for different types of the aged. Currently, some civil hospital and nursing institutions for the aged have begun the exploration and practice of the medical care mode, and most of them have set up nursing unit and specialized department for the elderly. So it is suggested that a suitable pattern of providing for the aged should be proposed after combining the advanced foreign nursing unit and current situation of China's development (Fig. 6, 7).

When entering terminal period, nursing mode for the elderly will be transformed into palliative care. Doctors can try to release the symptom by means of advanced technology and warm service and

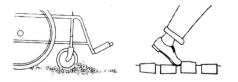

Figure 4. Pavement: inapplicable material and paving method.

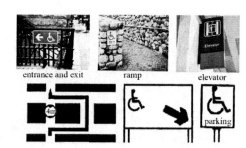

Figure 5. Sign system.

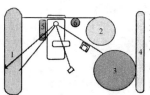

1 central working corridor
2 bathroom
3 activity room
4 external public traffic line
5 accessory equipment near the bed
6 storage space

Figure 6. Nursing unit near the bed.

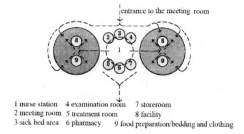

1 nurse station 4 examination room 7 storeroom
2 meeting room 5 treatment room 8 facility
3 sick bed area 6 pharmacy 9 food preparation/bedding and clothing

Figure 7. Organization diagram of the nursing unit.

they don't tend to give painful but little effective medical care to the aged. Meanwhile, comfortable daily nursing environment, where the elders could get intimacy and have the sense of belonging, should be created as far as possible. The palliative care should integrate all the related social resources. Doctors and nurses, family members, social workers, volunteers and psychological workers ought to take part in the process. Both physical and mental care from relatives and medical workers is suggested to given to the aged so that the elderly physical energy would be strengthened and they could gain confidence for the future.

6 CONCLUSIONS

The life stage of elders is a gradual process, during which they show different functional conditions and have corresponding characteristics and needs in different period. In addition, old people have differences in physical state and demands. Therefore, when apartments for the aged are planned, the elderly physical problems and spiritual needs should be paid more attention to. Personality and willingness of the aged also ought to be respected. By setting up an optional platform and taking full advantage of current social resources, diversified activities and service facilities would be available for the elderly, which is able to lead old people to adjust themselves, enrich their spiritual life, receive all-around and specialized medical care as well as improve their physical conditions. Only in these ways, could apartments for the aged be accepted by more elders and truly become welcome homes.

REFERENCES

JGJ122-99. 1999. Code for Design of Buildings for Elderly Persons. Beijing: China Architecture & Building Press.
Wu Yushao. 2013. China Report of the Development on Aging Cause (2013). Beijing: Social Sciences Academic Press.
Zhang Jin. & Chen Quan'an. 2004. Study on types of apartment for the aged. Shanxi: Shanxi Architecture. 30(19):28–29.

Effect of membrane characteristics of alginate-chitosan microcapsule on immobilized recombinant *Pichia pastoris* cells

Qi Li, Wei Deng, Qian Zhang, Yaling Zhao & Weiming Xue
School of Chemical Engineering, Northwest University, Xi'an, P.R. China

ABSTRACT: In order to overcome some shortage during traditional fermentation such as decrease of cell activity, low volumetric productivity and product inhibition, alginate-chitosan (AC) microcapsules encapsulated genetically engineered strain GS115-pPIC9k-ChiA4.0 were developed. Microcapsules with different chitosan molecular weight and different membrane formation time were prepared. Results showed that cells grew well in the AC microcapsules. The cultured GS115-pPIC9k-ChiA4.0 in AC microcapsules which modified with 100 kDa molecular weight chitosan and 10 min formation time showed the highest OD_{600} value (9.255) and the enzyme activities (37.5 U) respectively. Interestingly, when the molecular weight of chitosan decreased or the formation time increased, both the cell growth rate and enzyme activities of GS115-pPIC9k-ChiA4.0 in AC microcapsules decreased. Thus, the AC microcapsules we prepared were suitable for immobilizing GS115-pPIC9k-ChiA4.0 cells and will show a promising application in the field of cell immobilization.

1 INTRODUCTION

For decades, synthetic chemical insecticides have been used as a most powerful tool available for pest management. However, the abuse of them have being caused many serious problems such as environmental pollution, pesticide residue and pesticide tolerance, which have being threaten to environment and health of human being. Therefore, some novel and environmentally friendly methods for pest control are required. Chitinase is one of the potential proteins for pest control (Bhattacharya et al. 2007), which could degrade chitin in gut peritrophic membrane and exoskeleton of pest (Hirose et al. 2010). Based on this consideration, genetically engineered *Pichia pastoris* strain GS115-pPIC9k-ChiA4.0 was built which can express soluble chitinase with 63.6 kDa molecular weight (Xi et al. 2010). However, the traditional fermentation still has many problems, such as, difficulties inherent to high-density cultivation, low volumetric productivity, product inhibition, and microorganism contamination (Potvin et al. 2012). Thus, encapsulation method is a good way to solve these problems by building up a higher cell density in the bioreactor (Westman et al. 2012). Moreover, it's easy to separate cells from fermented liquid without contamination (Wang et al. 1997). Among the existing microcapsule systems, alginate-chitosan (AC) microcapsule has attracted particular attention because of their excellent properties such as biocompatibility, non-toxicity and biodegradability (Onishi & Machida. 1999; Holme et al. 2008; Lim et al. 2008). In this method, cells are enclosed in a semipermeable, spherical, and thin membrane which permits the required nutrients and cell products easily pass through it. Compared with other systems, AC microcapsules have several advantages, including easier microorganism separation, excellent mechanical strength, and better mass transfer performance.

The polyelectrolyte complex (PEC) membrane plays an important role in controlling the performance of AC microcapsules. In this paper, AC microcapsules with different chitosan molecular weight and different membrane formation time were prepared, and its effects on cell growth and metabolism were studied. Thus, this work will help us to know whether the AC microcapsules were suitable to encapsulate GS115-pPIC9k-ChiA4.0 cells and seek the optimum conditions for preparing AC microcapsules.

2 EXPERIMENTAL

2.1 *Materials*

Recombinant *Pichia pastoris* GS115-pPIC9k-ChiA4.0 strain which constructed and offered by professor Xi's group. Chitosan with different molecular weight (20 kDa, 50 kDa, 100 kDa) and degree of deacetylation (95%) was purchased from Golden-shell Pharmaceutical Corp. (Zhejiang, China). Sodium alginate with viscosity of 1.05 Pa·s was purchased from Chemical Reagent Corp. (Tianjin, China), Chitinase control (63 kDa, pI = 5.3) was obtained from Sigma Co.Ltd.. All other chemicals used were analytical grade.

2.2 *Preparation and culture of AC microencapsulated P. Pastoris cells*

15 g/L sodium alginate was sterilized by filtration through a 0.22 μm membrane filter. The sodium alginate solution was stored overnight at 4°C to facilitate deaeration before use. The recombinant *P. Pastoris* cells were suspended in sodium alginate solution at concentration of 1×10^3 cell/L. Then the suspension was extruded into 2 g/L calcium chloride solution using electrostatic droplet generator to form calcium alginate gel (CAG) beads. After been hardened for an hour, the CAG beads were put into 0.5% chitosan solution dissolved in 0.1 mol/L acetic acid solution at the volume ratio of 1:5 (beads:solution) to form membranes. After rinsed three times with distilled water, the AC microcapsules were obtained.

10 mL cell-loaded AC microcapsules were cultured within 50 mL MMGY medium in shake flasks in a shaking incubator at 30°C and a stirring speed of 200 rpm. When the OD_{600} value of the fermentation broth reached 2 to 6, AC microcapsules were transferred into 50 mL BMMY medium to produce extracellular chitinase.

1 L BMGY medium consisted of 10 g/L yeast extract, 20 g/L peptone, 0.1M phosphate buffer (pH 6.0), 4×10^{-5} g/L D-biotin (Japan), 100 mL glycerin. BMMY medium used 50 mL methanol instead of the 100 mL glycerin of BMGY medium and other components and their contents were identical with BMGY medium.

2.3 *Characterization methods*

The morphology of AC microencapsulated *P. Pastoris* was observed with XDS-1B Inverted Research Microscope (Chongqing optical instrument Corp., China). Cell concentration was determined by measuring absorbance at 600 nm using 722N spectrophotometer (Shanghai Precision & scientific Instrument Co.Ltd., China). Chitinase activity was determined by a dinitrosalicylic acid (DNS) method (Miller 1959). Measurement of chitinase in fermentation broth was conducted by SDS-PAGE gel electrophoresis. AC microcapsules were broken up using a chemical method as reference (Xue et al. 2004).

3 RESULTS AND DISCUSSION

3.1 *Morphology of cell-loaded AC microcapsules*

Figure 1 showed the optical images of cell-loaded AC microcapsules at different culture time. AC microcapsules were spherical and intact with smooth surface and they could maintain good spherical shape without breakage during 48-hour culture process. Black spherical spots in AC microcapsules (Fig. 1B) presented bacterial colonys. The number and the volume of spots were both increasing as culture time increased, which indicated that AC microcapsules provided a suitable microenvironment for cell growth.

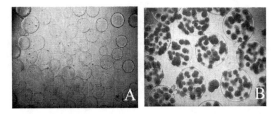

Figure 1. Optical images of cell-loaded AC microencapsules at different culture time. A: 0h (40×), B: 48 h (100×).

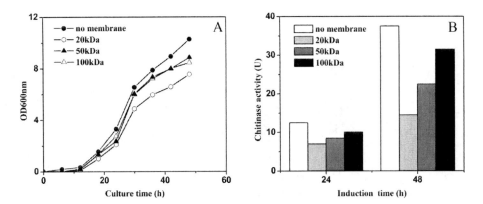

Figure 2. Growth kinetic curve of cells (A) and activities of chitinase secreted by cells (B) in AC microcapsules with different molecular weight chition.

3.2 *The effect of chitosan molecular on cell growth and metabolism*

Figure 2A was the growth profile of *P. Pastoris* cells entrapped into AC microcapsules with different chitosan molecular weight at the same membrane formation time. Due to lack of membrane resistance, it could be seen that the cell proliferation level of CAG beads sample was superior to that of AC microcapsules obviously. Cell proliferation level of AC microcapsules with 20 kDa chitosan was inferior to that of AC microcapsules with 50 kDa and 100 kDa chitosan. Cells entrapped into AC microcapsules with 50 kDa and 100 kDa chitosan had a pretty close growth profile. But most of the measuring points in the profile of cells entrapped into AC microcapsules with 50 kDa chitosan were slightly below than that with 100 kDa. Figure 2B was activities of chitinase secreted by GS115-pPIC9k-ChiA4.0 cells immobilized into microcapsules with different membranes. Generally, AC microcapsules with lower chitosan MW had a lower chitinase activity.

The formation of PEC membrane compressed the core volume of microcapsules, limiting the space for cells growth. Meanwhile, the membrane of microcapsules increased the diffusion resistance of oxygen and required nutrients (Zhao et al. 2013). These two aspects were both bad for cell proliferation. Chitosan with lower molecular weight had shorter pieces of molecular chain which is easier to diffuse deeper into calcium alginate beads and form electrostatic cross-linking membrane. Therefore, AC microcapsules with lower molecular weight chitosan could form thicker and denser membrane, but thinner and looser membrane with higher molecular weight chitosan.

3.3 *The effect of membrane reaction time on cell growth and metabolism*

Figure 3A was the growth profile of *P. Pastoris* cells entrapped into AC microcapsules with different membrane formation time. It could be seen that the cell proliferation level of calcium alginate beads was superior to that of AC microcapsules obviously. But calcium alginate beads were apt to break

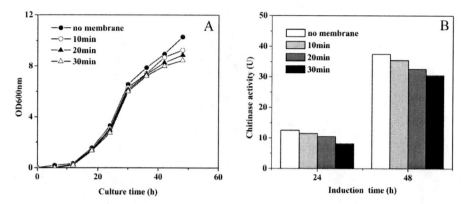

Figure 3. Growth kinetic curve of cells (A) and activities of chitinase secreted by cells (B) in AC microcapsules with different membrane reaction time.

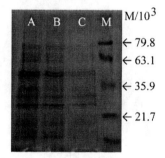

Figure 4. SDS-PAGE of crude enzyme extract with different ammonium sulfate saturation. (A: 70%; B: 50%; C: 30%; M: marker).

up in the MMGY and BMGY medium. Cell proliferation level of AC microcapsules with 10 min membrane formation time was superior to that of AC microcapsules with 20 min and 30 min. The activities of chitinase secreted by cells in AC microcapsules during culture process were shown in Figure 3B. The result showed that the activity of chitinase decreased as the membrane formation time increased.

The amount of chitosan participated in the membrane reaction increased with the membrane formation time increased. It led to a thicker membrane of microcapsules, which increased the mass transfer resistance and compressed the core volume of microcapsules. Thus, AC microcapsules with a longer membrane formation time had a lower cell growth rate and less chitinase amount transferring from inner space of microcapsules to outer medium.

3.4 *The SDS-PAGE analysis*

Treated with 30%, 50% and 70% (mass percent) ammonium sulfate, products of salting-out were obtained from cell culture supernatant respectively. Loading the salting-out samples and chitinase control onto polyacrylamide gel, SDS-PAGE was conducted and the result was as Figure 4. The result indicated that *P. Pastoris* cells immobilized into AC microcapsules could secrete extracellular target protein with 63.1 kDa molecular weight compared with chitinase control. Figure 4 also showed that there were other protein components in the fermentation broth which provided guiding significance for the separation of enzyme.

4 CONCLUSIONS

In this study, a kind of microcapsules loaded recombinant *P. pastoris* GS115-pPIC9k-ChiA4.0 cells with calcium alginate gel core and a layer of polyelectrolyte membrane was successfully prepared. The results showed that cells loaded within microcapsules could grow and metabolize well and activities of chitinase were maintained. Some factors during membrane preparation would give some important influence on mass transfer of microcapsules. In this paper, the influences of molecular weight of chitosan and membrane formation time on immobilized cell growth and chitinase transfer across membrane were discussed. When molecular weight of chitosan decreased or the membrane formation time increased, the thickness and the compact degree of membrane increased which led to smaller space of cells growth and larger mass transfer resistance. The growth rate of cells in AC microcapsules was also decreased.

REFERENCES

Bhattacharya, D., Nagpure, A. & Gupta, R.K. 2007. Bacterial chitinases: properties and potential. *Critical reviews in biotechnology* 27(1): 8–21.

Hirose, T., Sunazuka, T. & Omura, S. 2010. Recent development of two chitinase inhibitors, Argifin and Argadin, produced by soil microorganisms. *Proceedings of the Japan Academy Series B-Physical and Biological Sciences* 19(4): 453–466.

Holme, H.K., Davidsen, L. & Kristiansen, A. 2008. Kinetics and mechanisms of depolymerization of alginate and chitosan in aqueous solution. *Carbohydrate Polymers* 73(4): 656–664.

Lim, S.M., Song, D.K. & Oh, S.H. 2008. In vitro and in vivo degradation behavior of acetylated chitosan porous beads. *Journal of biomaterials science. Polymer edition* 19(4): 453–466.

Miller, G.L. 1959. Use of Dinitrosalicylic Acid Reagent for Determination of Reducing Sugar. *Anal Chem* 31: 426–428.

Onishi, H. & Machida, Y. 1999. Biodegradation and distribution of water-soluble chitosan in mice. *Biomaterials* 20(2): 175–182.

Potvin, G., Ahmad, A. & Zhang, Z.S. 2012. Bioprocess engineering aspects of heterologous protein production in Pichia pastoris: A review. *Biochemical Engineering Journal* 64: 91–105.

Westman, J.O., Ylitervo, P. & Taherzadeh. M.J. 2012. Effects of encapsulation of microorganisms on product formation during microbial fermentations. *Applied Microbiology and Biotechnology* 96(6): 1441–1454.

Wang, J.L., Liu, P. & Qian, Y. 1997. Biodegradation of phthalic acid esters by immobilized microbial cells. *Environment international* 23(6): 775–782.

Xi, Y.W., Zhao, J. & Wang Z.H. 2010. Expression of Chitinase Gene of HaSNPV, and Purification, Refolding of Its Recombinant Protein. *China Biotechnology* 30(2): 77–83.

Xue, W.M., Yu, W.T. & Liu, X.D. 2004. Chemical method of breaking the cell-loaded sodium alginate/chitosan microcapsules. *Chemical journal of Chinese* 25(7): 1342–1346.

Zhao, W., Zhang, Y. & Liu, Y. 2013. Oxygen diffusivity in alginate/chitosan microcapsules. *Journal of Chemical Technology & Biotechnology* 88(3): 449–455.

Working characteristics of natural circulating evaporator

Fengzhen Zhang, Yuesheng Zhong, Huaiming Du, Xingyong Liu, Linchun Yu & Hu Yang
*School of Material & Chemical Engineering, Sichuan University of Science & Engineering,
Zigong, Sichuan, China*

ABSTRACT: The working characteristics of Natural Circulating Evaporator (NCE) were analyzed and a mathematical model was established, revealing the internal law of transition from thermal energy to work. The results revealed that the mechanical energy needed by the circulation of fluid was less than 2% of input of external thermal energy, and the thermal energy inputted almost all contributed to the evaporation of liquid. Under the condition of low thermal energy input (Q < 500 W), the simulation values by the model were testified with experimental data. The errors between simulation values and experimental values of circulation flow rate and evaporation rate were within 15%, respectively.

1 INTRODUCTION

The driving force of the NCE is caused by density difference between the liquid flow in the downcomer and the vapor-liquid two-phase flow in the riser, such as thermosyphon, which is essentially the transition from thermal energy to work. As the flow inside the tube circulated without external force, it saves power consumption, improves the coefficient of heat transfer and reduces the deposition of fouling. Therefore, the NCE is widely used in the fields of chemical engineering, thermal power engineering etc.

The hydrodynamics and heat transfer characteristics of the vapor-liquid two phase flow in the NEC were studied by many researchers (Khodabandeh, 2005; Wang et al., 1996; Khodabandeh & Palm, 2002; Rahimi et al., 2010; Baars & Delgado, 2006) and some corresponding models of hydrodynamics and heat transfer were established to optimize the design and simplify the model establishment. It is confirmed that heat transfer in the riser is closely related to the circulation flow rate (Khodabandeh, 2005; Kandlikar, 1990). However, considering the complexity of the transition of thermal energy to work inside the NCE, the circulation flow rate was hard to predict. Currently, the circulation flow rate was calculated with heat balance, presuming the saturated evaporation (Chen et al., 1990); or the circulation flow rate can also be attained with momentum balance, using the pressure drop model of two-phase flow (Zhang et al., 2013).

This paper studied the working performance of the NCE on the essence of transition from thermal energy to work, predicted the circulation flow rate and evaporation rate, and revealed the allocation of energy.

2 THE HEAD OF NCE

Fig. 1 shows the driving force of natural circulation comes from the pressure head, which is caused by the different values between the vapor-liquid two-phase flow's average density in the riser ρ_m, and the liquid's density in the downcomer ρ_1. The pressure difference is determined by

$$\Delta P_T = (\rho_1 - \rho_m)gH \tag{1}$$

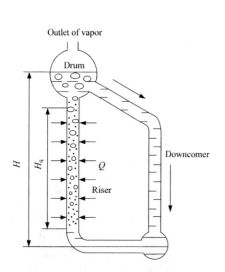
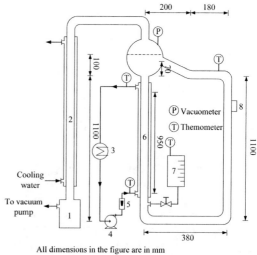

Figure 1. The schematic diagram of the circulation of the NCE.

1.Water collection tank, 2.Condenser, 3.Thermostat water bath, 4.Pump, 5.Flowmeter, 6.Riser of NCE, 7.Feeding system, 8.Ultrasonic flowmeter

Figure 2. Schematic diagram of the experimental setup.

where $\rho_l g H$ is the pressure head in the downcomer, $\rho_m g H$ is the pressure head in the riser, g is the gravity and H is the lifting height of the vapor-liquid two-phase flow in the riser. The ρ_m in the riser can be calculated as follows:

$$\rho_m = \varphi \rho_v + (1-\varphi)\rho_l \qquad (2)$$

where ρ_v is the density of vapor; φ is the section void fraction in the riser, which can be predicted with Armand-Treshchev equation:

$$\varphi = (0.833 + 0.167x)\beta \qquad (3)$$

where x and β are the mass void fraction and the volumetric void fraction of vapor-liquid two-phase flow, respectively. They are given as:

$$x = D/(D+G) \qquad (4)$$

$$\beta = \frac{D/\rho_v}{D/\rho_v + G/\rho_l} \qquad (5)$$

where D and G are the evaporation rate and the circulation mass flow rate of liquid, respectively.

3 THE PROCESS ANALYSIS OF NCE

Take the NCE as the control volume, the heat balance can be listed as follows:

$$I_{in,1} + I_{in,2} = I_{out,1} + I_{out,2} + W_e + Q_l \qquad (6)$$

where $I_{in,1}$ and $I_{out,1}$ are the enthalpies owned by the heating medium which flows into the jacket and out of the jacket, respectively. And their difference value Q is the thermal energy provided by the system, which can be calculated by:

$$Q = I_{in,1} - I_{out,1} = c_{p,1}\rho_l' V_s (t_{in} - t_{out}) \qquad (7)$$

where $c_{p,1}$, ρ'_l and V_s are the specific heat, density and volumetric flow rate of the heating medium, respectively.

$I_{in,2}$ is the enthalpy of saturated distilled water which flows into the evaporator from the feeding system. As the evaporating amount of the distilled water is equal to that supplied to the NCE, $I_{in,2}$ can be given as:

$$I_{in,2} = c_{p,2} DT_{sat} \tag{8}$$

where $c_{p,2}$ and T_{sat} are the specific heat of the distilled water and the saturation temperature corresponding to the operating pressure of the system.

$I_{in,2}$ is the enthalpy of vapor which is leaving the drum. W_e is the mechanical energy which makes the fluid flow in the riser. They can be calculated by:

$$I_{out,2} = c_{p,2} DT_{sat} + r_v D \tag{9}$$

$$W_e = \Delta P_T (G+D)/\rho_m \tag{10}$$

where r_v is the latent heat of vaporization.

Consequently, under the condition of ignoring the energy loss of the system Q_l, Eq. 6 can be transformed as:

$$Q = r_v D + \Delta P_T (G+D)/\rho_m \tag{11}$$

4 THE PRESSURE DROP OF NCE

Total pressure drop of the NEC ΔP_f consists of the pressure drop of the vapor-liquid two-phase flow in the riser ΔP_h and the other pressure drop in other parts ΔP_x:

$$\Delta P_f = \Delta P_h + \Delta P_x \tag{12}$$

where ΔP_h can be calculated with Lorkhart-Martinelli model (Lockhart & Martinelli, 1949). However, ΔP_x can be given as:

$$\Delta P_x = \sum \lambda_k \rho_l \frac{l_k}{d_k} \cdot \frac{u_k^2}{2} + \sum \zeta_k \rho_l \frac{u_k^2}{2} + \rho_m \zeta_b \frac{u_m^2}{2} \tag{13}$$

where λ_k and d_k are the resistance coefficient of the linear tubes apart from the riser and the inner diameter of the tube, respectively. ζ_k is the local resistance coefficient excluding the drum. u_k is the flow rate of the fluid in the tube. As the vapor and the fluid separated completely in the drum, the circulation mass flow rate in other tubes should be equal to that in the riser, which can be represented by G. u_k can be given as:

$$u_k = \frac{4G}{\pi d_k^2 \rho_l} \tag{14}$$

ζ_b is the local resistance coefficient of the drum. u_m is the flow rate of the vapor-liquid two-phase flow in the riser, which can be given as:

$$u_m = 4(D/\rho_v + G/\rho_l)/(\pi d_h^2) \tag{15}$$

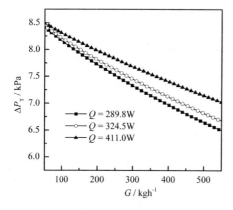

Figure 3. Effects of Q on the impact of ΔP_T.

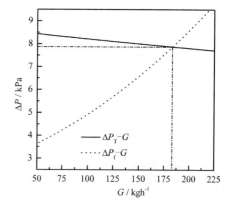

Figure 4. The confirmation of working point.

5 EXPERIMENTAL SETUP

The experimental setup is shown in Fig. 2. It's mainly consisted of the NCE, a pump, a thermostat water bath, a condenser, a vacuum pump and the feeding system. The NCE consisted of the riser with the jacket, the drum on top of the riser and the downcomer. The central position of the drum's outlet tube was linked to the top of the downcomer with an overflow pipe whose length was 70 mm. The riser was linked to the bottom of the downcomer with a glass tube. The length of the riser was 1100 mm which was the same as the downcomer's. The heating section of the riser was 900 mm. The glass tube at the bottom was 380 mm. The inner diameters of the riser, the downcomer and this glass tube were all 15 mm. And the inner diameter of the overflow pipe at the top was 20 mm.

The volume-measured distilled water was added into the evaporator. Thermostatic water was pumped to provide the heat through the jacket of the riser, with the measurement of a flowmeter. Meanwhile, the non-condensable gas was emitted out of the system by the vacuum pump. Abundant boiling bubbles were produced in the riser and then they entered the drum. The vapor separated from the liquid in the drum was condensed through the total condenser, collected in the water collection tank and weighed by the electric balance. The liquid in the drum overflow to the downcomer and reached the bottom of the riser through the glass tube. In order to keep the stability of the natural circulation, saturated distilled water was added into the evaporator through the feeding system to offset the evaporation loss of the water. The evaporating temperature and operating pressure within the system were measured. At the same time, the volumetric flow rate of hot water in the jacket V_s, the inlet and outlet temperature of the jacket (t_{in} and t_{out}) were also measured to attain the thermal energy provided by system. The liquid's circulation flow rate inside the downcomer was measured by ultrasonic flowmeter.

6 RESULTS AND DISCUSSION

6.1 The working characteristic curve of NCE

The flow and the heat transfer are coupled in the NCE. The thermal energy from heat source provided the power for the flow of the fluids in the riser. The relation of ΔP_T and G gained from Eq. 1–5 and Eq. 11 was the working characteristic curve, representing the ability to circulate the working fluids. Fig. 3 showed the functional relation under the condition that H was 1100 mm and with varified thermal provision. As can be seen, ΔP_T increased when G decreases, and the increasing of Q could also improve ΔP_T.

Table 1. Comparison of experimental data and simulated predictions.

$V_s/\text{L}\cdot\text{h}^{-1}$	$t_{in}/°C$	$t_{out}/°C$	Q/W	P_{op}/kPa	$G_{exp}/\text{kg}\cdot\text{h}^{-1}$	$G_{sim}/\text{kg}\cdot\text{h}^{-1}$	$\Delta/\%$	$D_{exp}/\text{kg}\cdot\text{h}^{-1}$	$D_{sim}/\text{kg}\cdot\text{h}^{-1}$	$\Delta/\%$
80	58.0	54.5	324.5	3.7	167	182	−9.0	0.469	0.495	−5.5
80	68.0	63.5	411.0	4.0	146	134	8.2	0.596	0.632	−6.0
80	55.2	51.8	315.3	3.5	166	187	−12.7	0.451	0.479	−6.2
100	69.0	66.5	289.8	8.2	186	201	−8.1	0.421	0.447	−6.2
100	77.5	74.0	396.3	7.9	147	144	2.2	0.578	0.616	−6.6

6.2 The pipe characteristic curve and the confirmation of the working point of NCE

The functional relation of ΔP_f and G gained from Eq. 12–15 was the pipe characteristic curve. Under circumstances of certain operating conditions and certain pipe lines, the curve reflected the relation between the flow rate and the head required by the working fluid in the evaporator. The flow head provided by the NCE overcame the flow resistance. As a result, the working performance parameters of the NCE were both determined by the power provided by the NCE and the pipe characteristcs. Actual cyclic flow G and the head ΔP were determined by the intersection point of the working characteristic curve ΔP_T-G and the pipe characteristic curve ΔP_f-G. Fig. 5 reflected the confirmation of the working point when the operating pressure was 3.7 kPa and the thermal energy input was 324.5 W. As seen in Fig. 4, the predicted mass flow and the head were 182 kg/h and 7850 Pa according to the model established.

6.3 The experimental results and the predicted results

Under certain operating conditions, working performance parameters like G, D, ΔP etc. could be predicted according to Eq. 1–15. Table 1 showed the comparison of the working performance parameters from the experiments and the model when the thermal energy input was less than 500 W. Experimental results of G and D were found with good agreement to those predicted by the model, with errors within 15%. From Table 1, when the average density ρ_m of the vapor-liquid two-phase flow was between 100 kg/m³ and 200 kg/m³, the required mechanic energy W_e could be calculated by Eq. 10, with its value less than 5W, which was less than 2% of the energy provided by the heat source. It reflected that the energy provided by the heat source almost fully contributes to the evaporation of the liquid.

7 CONCLUSIONS

(1) The working characteristics of NCE were analyzed and a mathematical model was established. The circulation flow rate and the evaporation rate could be predicted by the model, with their simulation results testified by the experimental data gained when the thermal energy was less than 500W.
(2) The internal law of transition of thermal energy to work was revealed. The thermal energy inputted almost all contributed to the evaporation of liquid and the mechanical energy needed by the circulation of fluid was less than 2% of input of external thermal energy.

ACKNOWLEDGEMENTS

The authors gratefully acknowledge the financial support of the Talents Import Funds of SUSE (NO. RC201207).

REFERENCES

Baars, A. & Delgado, A. 2006. Multiple modes of a natural circulation evaporator. *International Journal of Heat and Mass Transfer* 49: 2304–2314.

Chen, K., Chen, Y. & Tsai, S. 1990. An experimental study of the heat transfer performance of a rectangular two-phase natural circulation loop. *Experimental Heat Transfer – An International Journal* 3: 27–47.

Kandlikar, S.G. 1990. A general correlation for saturated two-phase flow boiling heat transfer inside horizontal and vertical tubes. *Journal of heat transfer* 112: 219–228.

Khodabandeh, R. & Palm, B. 2002. Influence of system pressure on the boiling heat transfer coefficient in a closed two-phase thermosyphon loop. *International Journal of Thermal Sciences* 41: 619–624.

Khodabandeh, R. 2005. Pressure drop in riser and evaporator in an advanced two-phase thermosyphon loop. *International Journal of Refrigeration* 28: 725–734.

Khodabandeh, R. 2005. Heat transfer in the evaporator of an advanced two-phase thermosyphon loop. *International Journal of Refrigeration* 28: 190–202.

Lockhart, R.W. & Martinelli, R.G. 1949. Proposed correlations for isothermal two-phase two-component flow in pipes. *Chemical Engineering Progress* 45: 39–48.

Rahimi, M., Asgary, K. & Jesri, S. 2010. Thermal characteristics of a resurfaced condenser and evaporator closed two-phase thermosyphon. *International Communications in Heat and Mass Transfer* 37: 703–710.

Wang, F., Ma, C., Bo, J. et al. 1996. Study on pressure drop characteristic of natural circulating two-phase flow system. *Journal of Tsinghua University (Sci & Tech)* 36: 23–26.

Zhang, P., Wang, B., Han, L. et al. 2013. Comparison of evaporating heat transfer models in two-phase thermosyphon loop. *CIESC Journal* 64: 2752–2759.

Advanced Engineering and Technology – Xie & Huang (Eds)
© 2014 Taylor & Francis Group, London, ISBN 978-1-138-02636-0

Guaranteed appropriate hydrolysis of meat by adding bromelain and zinc sulfate

Laping He
School of Liquor & Food Engineering, Guizhou University, Guiyang, China;
Key Lab of Agricultural and Animal Products Store & Processing of Guizhou Province, Guiyang, China

Lili Xue
School of Liquor & Food Engineering, Guizhou University, Guiyang, China

Qiujin Zhu
School of Liquor & Food Engineering, Guizhou University, Guiyang, China;
Key Lab of Agricultural and Animal Products Store & Processing of Guizhou Province, Guiyang, China;
Guizhou Pork products Research Center of Engineering Technology, Guiyang, China

Suansuan Song, Maohua Duan, Li Deng, Juan Chen & Xiaogang Zhang
School of Liquor & Food Engineering, Guizhou University, Guiyang, China

ABSTRACT: This paper reports the moderate tenderization of meat through the use of bromelain and its inhibitor at room temperature, which promotes the application of enzyme technology in the meat industry. The best enzymatic parameters for tenderizing pork are obtained by bromelain. Furthermore, zinc sulfate is found to be a suitable bromelain inhibitor at root temperature. The relative activity of bromelain is reduced to 3% of the original activity when 0.025 g/100 g of zinc sulfate is added to minced pork. The results show that the moderate tenderization of meat can be achieved by using bromelain under optimal conditions, followed by the addition of a safe inhibitor, zinc sulfate, to control bromelain activity to the desired degree. This study suggests that the enzyme technology resulting from the combination of bromelain and its inhibitor is useful in improving the quality of meat and in developing novel cured meat products.

1 INTRODUCTION

The palatability of meat is affected by various factors, and tenderness is cited as one of the most significant considerations. Tenderized meat can be got by protease including endogenous enzymes (Morgan et al., 1993, Sentandreu et al., 2002). However, fully suitable tenderization of meat couldn't gained by only inherent proteolytic enzymes (Goll et al., 2003) Therefore, studies on exogenous enzymes derived from plant, bacteria, and fungal sources have been conducted in the meat industry. The enzymes have been confirmed as successful in enhancing tenderness because of different reactions to the myofibrillar and connective tissue portions (Ashie et al., 2002, Chen et al., 2006, Foegeding and Larick, 1986).

Bromelain is an enzyme mixture that belongs to a cysteine protease from pineapple and can be used in tenderizing meat (Sullivan and Calkins, 2010). The safety of bromelain as a food additive has been confirmed by the United States Food and Drug Administration, the World Health Organization, and other countries and agencies. Thus, in this study, bromelain is selected as an agent in meat tenderization. However, one issue to be addressed in the application of bromelain in the meat industry is in controlling bromelain activity effectively to obtain the guaranteed tenderization of meat at room temperature. Otherwise, overtenderization will occur if bromelain activity is not inhibited.

Controlling bromelain activity with an inhibitor is an effective and simple process. Various bromelain inhibitors include zinc (Shukor et al., 2009) protein (Hatano et al., 1998), and others. However, to date, reports on using bromelain and its inhibitor zinc salts in meat tenderization are few. Thus, in this study, we attempt to find a safe and effective bromelain inhibitor from among several types of inexpensive zinc salts to control the degree of meat tenderization. This search for a guaranteed tenderizer not only adds value to the industry but also helps supply a more consistent and uniform product to consumers, thus increasing their overall satisfaction with the product quality.

2 MATERIALS AND METHODS

2.1 Meat sample preparation

Longissimus of refreshing raw pork was purchased from the local supermarket as the material for the sample. Before the detection procedure, the visible fat, tendon, and tendon knob was removed to obtain pure, lean meat. Then, longissimus paste was prepared by cutting and mixing the meat with a MT688 multifunction food processor (Shunde Aide Industrial Co. Ltd.). The paste was stored at −18°C to −21°C in a refrigerator before further treatment.

2.2 Medicine and reagents

This study used five types of enzymes: trypsin (0458, Amresco® LLC, OH., USA), pepsin (p 7000) and papain (p 3250) (Sigma-Aldrich Corp., St. Louis, Mo., USA), bromelain (1,200 μ/g, Guangxi Jiewo Biological Products Co., Ltd., China), and TG enzyme (1,200 μ/g, Jiangsu Taixing Biological Products Co. Ltd., China).

Iodine acetic acid and zinc sulfate were respectively bought from Aladdin Reagents (Shanghai) Co. Ltd. and Chengdu Jinshan Chemical Reagents Co. Ltd. Methanol and sodium hydroxide were provided by Shantou West Long Chemical Co. Ltd. and Tianjin Zhiyuan Chemical Reagents Co. Ltd., respectively. An electronic balance (FA2004N Shanghai Minqiao Precision Science Instruments Co. Ltd, China) was used during the entire study

2.3 Enzymatic tenderization of meat

A certain concentration of bromelain was added to the minced pork. Then, the combination was mixed for 10 min to distribute the protease solution uniformly in the meat and then achieve a maximum enzymatic reaction. A DK 98 II electrodermal constant temperature water-bath pot (Tianjin Taisite Instrument Co. Ltd., China) was used to obtain the optimal temperature condition.

2.4 Optimization of hydrolysis conditions

Bromelain concentration, hydrolysis temperature, and hydrolysis time were tested and identified via a one-variable-at-a-time approach to a single-factor experiment. The design is reported in Table 1.

A $L_9 (3^4)$ orthogonal design is employed to obtain the best combination of the critical parameters, namely, the levels of bromelain concentration, hydrolysis temperature, and hydrolysis time, which is listed in Table 4.

Table 1. Enzymatic pork conditions by bromelain.

Item	Invariant	Variation gradient enzymolysis conditions						
Concentration of bromelain (mg/g)	25°C, 30 min	0	0.05	0.1	0.15	0.2	0.3	0.5
Temperature (°C)	0.2 mg/g, 0.5 h	5	15	25	35	45	–	–
Time (h)	0.2 mg/g, 25°C	0.5	1	1.5	2	3	–	–

2.5 Enzymatic activity determination

Bromelain activity was measured via the Folin method (Michel and Francosi, 1986). One unit of protease activity (μ/g) was defined as the amount of enzymes required to liberate 1 μg tyrosine at 40°C and pH 7.5 per min.

2.6 Inhibition of bromelain activity by zinc sulfate

Before the determination of enzymatic activity, the bromelain solution was prepared with distilled water. Then, the inhibitory effect of zinc sulfate on bromelain activity was studied according to the following designed experiment.

2.6.1 Influence of zinc sulfate on bromelain activity in tenderizing minced pork

Minced pork weighing 100 g was incubated with 0.4 g bromelain (specific activity: 500,000 μ/g) within 5 min at 5°C, followed by incubation with 0.16 g zinc sulfate for 10 minutes at 5°C. The control sample was incubated without zinc sulfate. Then, the residual protease activity was measured according to the procedure described in Section 2.5

2.6.2 Influence of zinc sulfate on amino nitrogen in minced pork tenderized by bromelain

Tenderization of meat by bromelain (500,000 μ/g) was carried out according to 0.5 mg (enzyme)/g (meat) for a certain period under certain conditions. Subsequently, the meat containing bromelain was incubated with zinc sulfate according to the dosage of 0.025 g/100 g meat. The control sample was incubated without zinc sulfate (Levy, 1957). Then, the residual protease activity and the amino nitrogen were measured.

2.7 Other chemical and physical test methods

2.7.1 Water-holding capacity (WHC) and water loss rate (WLR)

After determining the MC of raw meat, the heating centrifugation method based on Wardlaw's method (Wardlaw et al., 1973) was employed to measure the WHC. First, a 10 g sample was added to the centrifugal tube, heated to 70°C for 20 min, and subsequently cooled to room temperature. Second, the sample was centrifuged at 3,000 r/min for 15 min. The WHC (g/100 g) was calculated after obtaining the sample weight with the removal of the supernatant. The formula is as follows:

$$WHC = 1 - \frac{M_1 - M_2}{M_1 C} 100\%,$$

where M_1 is the sample weight, M_2 the centrifuged sample weight, and C the MC of the sample.

WLR was measured with a homemade pressure gauge. A meat sample weighing 10.000 g was placed between a two-layer gauze, and then 12 layers of qualitative filter paper were added to the top and bottom of the gauze. Subsequently, the sample was placed on the modified steel ring of the compression apparatus to allow standardized meat expansion. Thereafter, pressure was increased to 35 kgf/cm^2 at unvaring speed, and was maintained for 5 min. After removal of the pressure, the sample was weighed immediately. The WLR (g/100 g) was calculated according to the following equation:

$$WLR = \frac{M_1 - M_2}{M_1} 100\%,$$

where M_1 is the weight of the initial sample, and M_2 is that of the pressurized sample.

2.7.2 Moisture content (MC), Nitrogen amino acids and Cooking loss (CL)

MC nitrogen amino acids and CL were measured by the Association of Official Analytical Chemists (AOAC 1997), the hygiene standards for soybean sauce (GB/T 5009.39-2003), and the method developed by Ockerman (Ockerman, 1985), repectively.

2.8 Statistical analysis

SPSS11.5 statistical software was used to analyze the data, which were generated from three trials for each experiment for comparison of the means.

3 RESULTS AND DISCUSSION

3.1 Influence of bromelain on pork quality and hydrolysis dynamics

3.1.1 Single-factor experiment on pork quality and hydrolysis dynamics

Tenderizing minced pork with bromelain was investigated to explore further the application of this material. The WHC increased with an increase in the enzyme concentration from 0 mg/g to 0.5 mg/g, which suggested an aggravated hydrolysis degree (Table 2). It shows that the WHC had a significant increase with the increase in bromelain concentration.

CL is an important index in meat processing quality evaluation. It affects not only the product yield but also the quality, color, flavor, tenderness, and moisture-holding capacity of the product. Table 2 indicates that the CL decreased with an increasing concentration of bromelain.

The lower bromelain solution (0 to 2 mg/g) was favorable in improving the WHC, and it reduced the CL (Table 2). Furthermore, the meat became too soft, its formability deteriorated, and its soup became turbid because a large amount of meat disintegrated when the enzyme concentration was over 0.2 mg/g. Thus, the suitable bromelain concentration was 0.2 mg/g.

The hydrolytic temperature of bromelain could affect the MC and WHC (Table 3). At 0.2 mg/g enzyme concentration and 0.5 h reaction time, the MC from reduced 72.86 g/100 g to 72.01 g/100 g with an increase in temperature. Thereafter, the WHC decreased from 82.36 g/100 g to 75.59 g/100 g due to the breakage of the meat's integrity caused by excessive hydrolysis by increasing temperature.

Table 2. Effect of bromelain concentration (mg/g) on MC (g/100 g), WHC (g/100 g) and CL (g/100 g) in meat.

Bromelain concentration	0	0.05	0.1	0.2	0.3	0.5
MC	75.49 ± 0.14^c	75.76 ± 0.18^b	75.82 ± 0.18^{ab}	75.85 ± 0.04^{ab}	75.91 ± 0.14^{ab}	76.03 ± 0.16^a
WHC	65.02 ± 2.61^d	68.92 ± 4.1^c	70.83 ± 1.69^{bc}	73.78 ± 1.2^{ab}	74.83 ± 0.81^a	74.52 ± 1.57^a
CL	32.69 ± 0.21^a	32.27 ± 0.54^a	30.56 ± 1.55^{ab}	26.83 ± 3.36^{cd}	24.27 ± 1.76^d	19.79 ± 0.81^e

Table 3. Effect of Temperature (°C) and Time (h) on MC (g/100 g) and WHC (g/100 g) in meat.

Temperature	MC	WHC	Time	MC	WHC
5	72.86 ± 0.16^a	82.36 ± 1.99^a	0.5	73.49	78.07 ± 3.567^a
15	72.93 ± 0.03^a	81.87 ± 1.27^a	1	74.54	76.59 ± 1.34^a
25	72.34 ± 0.10^b	80.19 ± 1.33^{ab}	1.5	74.7	75.56 ± 1.94^{ab}
35	72.1 ± 0.34^b	77.65 ± 1.96^{bc}	2	75.02	72.97 ± 2.36^a
45	72.01 ± 0.26^b	75.59 ± 0.77^c	3	75.06	68.61 ± 0.08^c

Note: There are significant differences between the different lowercase letters ($P < 0.05$) in the same column.

Additionally, the hydrolysis time of bromelain affected the MC and WHC (Table 3). The MC rose from 73.49 g/100 g to 75.06 g/100 g with an increase of the time under the condition of 0.2 mg/g enzyme and 25°C. Subsequently, the WHC decreased from 78.07 g/100 g to 68.61 g/100 g owing to the breakage of the meat's integrity caused by excessive hydrolysis with increasing time.

3.1.2 Composite factor experiment results on pork quality

Composite factor experiment was used to find the optimum factor-level combination for WLR and MC, respectively. Minced pork had the lowest WLR when bromelain concentration was 0.2 mg/g (Table 4). When the enzyme concentration reached 0.3 mg/g, the WLR was the highest. The R values indicated that the effects of three factors on the WLR were bromelain concentration (A) > enzyme hydrolysis temperature (B) > enzyme hydrolysis time (C). Because the experimental aim is to get the lowest WLR, the optimal factorlevel combination was A2B1C3, namely, the bromelain concentration (A) of 0.2 mg/g, the enzyme hydrolysis temperature (B) of 5°C, and the enzyme hydrolysis time (C) of 1.5 h.

The hydrolytic temperature of minced pork with bromelain had a significant impact on the MC ($p < 0.05$). When the hydrolytic temperature reached 5°C, the MC had the maximum. Hydrolysis at 25°C obtained the lowest MC. The R values indicated that the effects of three factors on the MC (see Table 4) were as follws: enzyme hydrolysis temperature (B) > enzyme hydrolysis time (C) > bromelain concentration (A). According to the K values, the highest MC in the minced pork could be got under the condition of the optimal factor levelof B1C2A1, namely, the bromelain concentration (A) 0.1 mg/g, the hydrolysis temperature (B) 5°C, and the hydrolysis time (C) 1 h.

3.2 Studies of bromelain inhibitor

The plant-derived cystein protease is characterized by non-specific action. It not only tenderizes meat but also degrades the meat's texture, often resulting in overtenderization (Weiss et al., 2010). Therefore, a nontoxic inhibitor of bromelain must be identified for the development of novel meat products at normal temperatures. A heavy-metal assay has been improved by using bromelain (Shukor et al., 2008), which provided information that was useful in discovering an inhibitor of exogenous bromelain.

In our previous study, a series of common food additives, including sodium chloride, potassium chloride, sodium nitrate, sodium nitrite, zinc acetate, zinc sulfate, and zinc chloride, were investigated for their inhibitory effects on bromelain. The results showed that zinc sulfate was the best bromelain inhibitor. Thus, merely the effect of zinc sulfate was examined in the following section.

Table 4. Orthogonal experimental results from the hydrolysis of minced pork.

Sample number	Concentration (A)	Temperature (B)	Time (C)	WLR (g/100 g)	MC (g/100 g)
1	1(0.1)	1(5)	1(0.5)	20.04 ± 0.25	75.47 ± 0.002
2	1(0.1)	2(15)	2(1)	19.30 ± 2.1	75.07 ± 0.003
3	1(0.1)	3(25)	3(1.5)	22.87 ± 1.18	73.97 ± 0.004
4	2(0.2)	1(5)	2(1)	19.89 ± 1.43	75.42 ± 0.002
5	2(0.2)	2(15)	3(1.5)	18.36 ± 1.61	74.74 ± 0.001
6	2(0.2)	3(25)	1(0.5)	19.90 ± 0.19	74.03 ± 0.003
7	3(0.3)	1(5)	3(1.5)	19.51 ± 0.14	75.19 ± 0.003
8	3(0.3)	2(15)	1(0.5)	24.62 ± 3.1	74.13 ± 0.001
9	3(0.3)	3(25)	2(1)	30.22 ± 3.1	73.56 ± 0.003
K1	20.737/74.837	19.813/75.360	21.520/74.543		
K2	19.383/74.730	20.76/74.647	23.137/74.683		
K3	24.783/74.293	24.330/73.853	20.247/74.633		
R	5.400/0.544	4.517/1.507	2.890/0.140		

Note: WLR stands for water loss rate and MC stands for moisture content (w/w).

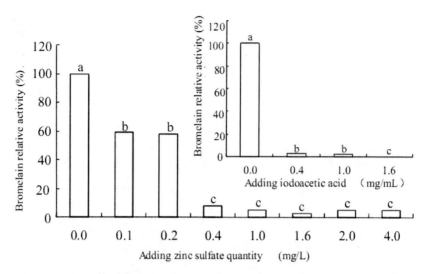

Figure 1. Effect of zinc sulfate and iodoacetic acid on the activity of bromelain. Note: There are significant differences between the different lowercase letters (P < 0.05) in per sub-graph.

3.2.1 *Effect of zinc sulfate on bromelain activity*

Iodine acetic acid has an inhibitory effect on bromelain (Gautam et al., 2010) but is unsuitable for food-industry application owing to its side effects. Zinc ion can inhibit the protease activity (Shukor et al., 2008). Thus, zinc sulfate was selected as a safe bromelain inhibitor from among diverse zinc compounds. The inhibitory effect of zinc sulfate on bromelain activity is shown in Figure 1. In the control group, the effect of iodine acetic acid on bromelain activity is also shown in the figure.

Figure 1 illustrates the effects of different concentrations of iodine acetic acid on bromelain activity. The inhibition of bromelain activity rose with the increase of iodine acetic acid concentration. When iodine acetic acid reached 0.08% (w/v), the bromelain activity was reduced to 0. Zinc sulfate showed a similar inhibition trend to bromelain. When the zinc sulfate concentration was 0.05 g/50 mL, the minimum relative activity was 5%.

3.2.2 *Effect of zinc sulfate on amino nitrogen content in the tenderization of minced pork*

The definite inhibition of zinc sulfate against bromelain activity in minced pork is indicated in Table 5. After the meat was tenderized for 1 h, zinc sulfate solution was added and maintained for 10 h at 25°C. The control sample was set without zinc sulfate based on the aforementioned procedures. The results showed that the amino nitrogen amount was 0.111 g/100 mL in the control sample and was only 0.02 g/100 mL in the sample with zinc sulfate. Thus, zinc sulfate had a significant inhibition against bromelain activity in minced pork.

3.2.3 *Dynamics studies on zinc sulfate against bromelain activity in minced pork*

Bromelain activity is influenced by pH, temperature, and storage time. Table 5 indicates that these factors did not significantly affect the inhibition of zinc sulfate of bromelain activity. Moreover, 0.4 g of zinc sulfate could reach the optimal effect and inhibit 1 g of bromelain activity. Zinc is involved in vital processes, such as protein synthesis as well as cellular respiration, and has been identified as an important factor in the repair of tissues (Pories et al., 1967). Moreover, zinc sulfate has been used as a food additive according to the GB 25579-2010 standard of China. Thus, it is a safe and efficient inhibitor of bromelain activity.

Dry-cured ham is a traditional product with a strong presence in the Mediterranean and East Asian markets. The potential function of ham in the context of nutrition has not been investigated

Table 5. Dynamics studies on zinc sulfate's inhibition of bromelain activity in minced pork.

Item	Amino nitrogen content (mg/g)	
	Pre-treatment	Post-treatment
Without ZnSO$_4$	11.50 ± 0.61^a	15.40 ± 0.60^b
With ZnSO$_4$	10.90 ± 1.11^a	12.80 ± 0.74^a

Item	Effect of ZnSO$_4$ on bromelain activity in meat at different temperatures (°C)		
Bromelain relative activity (%)	10	20	30
Without ZnSO$_4$	100 ± 6.56^a	100 ± 7.37^a	100 ± 7.23^a
With ZnSO$_4$	8 ± 3.06^b	5 ± 3.00^b	7 ± 2.08^b

Item	Effect of ZnSO$_4$ on bromelain activity in meat at different pH levels		
Bromelain relative activity (%)	5.8	6.2	6.6
Without ZnSO$_4$	100 ± 2.52^a	100 ± 13.12^a	100 ± 15.10^a
With ZnSO$_4$	50 ± 1.00^b	29 ± 5.13^b	61 ± 6.56^b

Item	Effect of ZnSO$_4$ on bromelain activity in meat at different times (min)		
Bromelain relative activity (%)	10	20	30
Without ZnSO$_4$	100 ± 2.65^a	80 ± 5.69^a	91 ± 9.24^a
With ZnSO$_4$	17 ± 11.00^b	24 ± 11.00^b	24 ± 6.00^b

Note: The data with different letters show significant differences ($P < 0.05$) in the same column for each item.

fully, and any innovative production methods can induce differences in composition. Fortunately, the addition of zinc ion was favorable in improving cure meat within an exact range, which could inhibit the formation of zinc-porphyrin, thus confirming the observations conducted on various cured meat products (Adamsen et al., 2006). Additionally, the traditional Italian dry-cured ham (Parma ham) with a stable red bright color was achieved by Zn–protoporphyrin IX without the use of nitrite and/or nitrate (Wakamatsu et al., 2004).

Therefore, zinc sulfate, as the inhibitor of bromelain, is potentially applicable on cured meat products at room temperature. Various issues have to be explored to promote the application of the bromelain co-inhibitor for innovative production. These issues include the function of nutrition, the establishment of quality standards, the differences in the composition at various processing stages, and so on.

4 CONCLUSION

Bromelain concentration, hydrolysis temperature and time influence the MC, WHC ($p < 0.05$) and CL of minced pork. The WHC of the control sample is lower than that of the enzyme hydrolysis of the meat, and the WHC rise with the increase of enzyme concentration. By contrast, the CL rate is reduced as the enzyme concentration increased.

The range values indicate that the impact of three factors on the WLR from highest to lowest are bromelain concentration, enzyme hydrolysis temperature, and enzyme hydrolysis time. The WLR reaches the lowest at the optimal factor level of A2B1C3. At the same time, the range values indicates that the effects of factors on the MC are hydrolysis temperature (B) > hydrolysis time (C) > bromelain concentration (A). In order to get the highest MC in the minced port, the optimal factor level combination is B1C2A1.

Zinc sulfate had a significant inhibitory effect on bromelain activity. The minimum activity is reduced to 5% under the condition of 0.05 g/50 mL of zinc sulfate concentration. The amino nitrogen amount is 0.111 g/100 mL in the control sample. On the contrary, it is 0.02 g/100 mL in the meat sample with zinc sulfate. The residual bromelain activity is inhibited. As a final conculsion, zinc sulfate can greatly reduce the excessive hydrolysis of the meat.

ACKNOWLEDGEMENTS

This study was partly financed by the Guizhou Agricultural Research Project during the 11th Five-year Plan period (QKH-NY-Z-2009-3029), the Special Fund of the Governor of Guizhou Province for Excellent Scientific, Technological, and Educational Talents (QKHRZ-(2010)07), and Guizhou Pork products Research Center of Engineering Technology (QKH-NG-Z-2013-4001).

REFERENCES

Adamsen, C., Møller, J., Laursen, K., Olsen, K. & Skibsted, L. 2006. Zn-porphyrin Formation in Cured Meat Products: Effect of Added Salt and Nitrite. *Meat Sci*, 72, 672–679.
Ashie, I., Sorensen, T. & Nielsen, P. 2002. Effects of Papain and a Microbial Enzyme on Meat Proteins and Beef Tenderness. *J Food Sci*, 67, 2138–2142.
Chen, Q., He, G., Jiao, Y. & Ni, H. 2006. Effects of Elastase from a Bacillus Strain on the Tenderization of Beef Meat. *Food Chem*, 98, 624–629.
Foegeding, E. & Larick, D. 1986. Tenderization of Beef with Bacterial Collagenase. *Meat Sci*, 18, 201–214.
Gautam, S., Mishra, S., Dash, V., Goyal, A. K. & Rath, G. 2010. Comparative Study of Extraction, Purification and Estimation of Bromelain from Stem and Fruit of Pineapple Plant. *Thai J Pharm Sci*, 34, 67–76.
Goll, D., Thompson, V., Li, H., Wei, W. & Cong, J. 2003. The Calpain System. *Physiol Rev*, 83, 731–801.
Hatano, K., Tanokura, M. & Takahashi, K. 1998. The Amino Acid Sequences of Isoforms of the Bromelain Inhibitor from Pineapple stem. *J Biochem*, 124, 457–461.
Levy, M. 1957. Titrimetric Procedures for Amino Acids: (Formol, Acetone, and Alcohol Titrations. *Meth in Enzym*, 3, 454–458.
Michel, L. & Francosi, L. 1986. Determination of Proteins and Sulfobetaine with the Folin-phenol Reagent. *Anal Biochem*, 157, 28–31.
Morgan, J., Wheeler, T., Koohmaraie, M., Savell, J. & Crouse, J. 1993. Meat Tenderness and the Calpain Proteolytic System in Longissimus Muscle of Young Bulls and Steers. *J Anim Sci*, 71, 1471–1476.
Ockerman, H. 1985. Quality Control of Post-mortem Muscles tissue. Dept. of Animal Science, Ohio State University.
Pories, W., Henzel, J., Rob, C. & Strain, W. 1967. Acceleration of Healing with Zinc Sulfate. *Ann Surg*, 165, 432–436.
Sentandreu, M., Coulis, G. & Ouali 2002. Role of Muscle Endopeptidases and Their Inhibitors in Meat Tenderness. *Trends Food Sci Tech*, 13, 400–421.
Shukor, M., Baharom, N., Masdor, N., Abdullah, M., Shamaan, N., Jamal, J. & Syed, M. 2009. The Development of an Inhibitive Determination Method for Zinc Using a Serine Protease. *J Environ Biol*, 30, 17–22.
Shukor, M., Masdor, N., Baharom, N., Jamal, J., Abdullah, M., Shamaan, N. & Syed, M. 2008. An Inhibitive Determination Method for Heavy Metals, Using Bromelain, a Cysteine Protease. *Appl Biochem Biotechnol*, 144, 283–291.
Sullivan, G. & Calkins, C. 2010. Application of Exogenous Enzymes to Beef muscle of High and Low-connective Tissue. *Meat Sci*, 85, 730–734.
Wakamatsu, J., Nishimura, T. & Hattori, A. 2004. A Zn–porphyrin Complex Contributes to Bright Red Color in Parma Ham. *Meat Sci*, 67, 95–100.
Wardlaw, F., Mccaskill, L. & Acton, J. 1973. Effect of Postmortem Muscle Changes on Poultry Meat Loaf Properties. *J Food Sci*, 38, 421–423.
Weiss, J., Gibis, M., Schuh, V. & Salminen, H. 2010. Advances in Ingredient and Processing Systems for Meat and Meat Products. *Meat Sci*, 86, 196–213.

Isolation and identification of a pectinase high-yielding fungus and optimization of its pectinase-producing condition

S.Y. Jiang, Y.Q. Guo, L.Z. Zhao, L.B. Yin & K. Xiao
Department of Biological and Chemical Engineering, Shaoyang University, Shaoyang, Hunan, China

ABSTRACT: A pectinase high-yield strain named BM was isolated from decayed navel orange. The strain was identified as *Fusarium oxysporum* according to its colony morphology and ITS-rDNA Sequence analysis. Then the phylogenetic tree constructed with other ITS sequences. Single factor and orthogonal test were used to investigate the enzyme production of the BM strain. The results of optimization showed that the maximum pectinase activity of 26.20 U/mL was obtained at 1% carbon source, 0.4% nitrogen source, a pH of 5, and temperature of 30°C.

1 INTRODUCTION

Microbial pectinases are heterogeneous group of enzyme that catalyses the hydrolysis of pectin which is responsible for the turbidity and undesirable cloudiness in fruits juices (Rehman et al. 2013). Pectinases are particularly attractive because they have a wide range of applications, such as in the field of food processing (Khan et al. 2013), textile industry (Bernava & Reihmane 2013), paper making (Jain et al. 2013), and in treatment of wastewaters (Zhang et al. 2013). Due to the low production and difficult extraction of natural products, pectinases are conducted using microbial fermentation method.

Peeling is the key technology of citrus processing. However, the traditional method of acid-base not only scattered rate is high, but also produces large amount of alkali wastewater to governance (Fava et al. 2013). Using enzymatic peeling becoming the hot research, and the microbial fermentation production of pectinase is very important. This study mainly aimed at isolating pectinase high-yield strains and optimized its production conditions of pectinase.

2 MATERIALS AND METHODS

2.1 *Microorganism and processing of navel orange powder*

The fungal isolates isolated from the decayed navel orange and soil. Navel orange powder was a by-product of citrus processing. For microbial cultivation and enzyme production, the navel orange peels were sundried, later oven dried and stored as navel orange powder.

2.2 *Media for isolation and enzyme production*

The enrichment medium (g/L): navel orange powder, 10; pectin, 1; peptone, 3; $(NH_4)_2SO_4$, 10; KH_2PO_4, 1; NaCl, 1; $MgSO_4 \cdot 7H_2O$, 0.6; $FeSO_4$, 0.01. The primary screening media (g/L): navel orange powder, 10; $(NH_4)_2SO_4$, 4; KH_2PO_4, 1; NaCl, 1; $MgSO_4 \cdot 7H_2O$, 0.6; $FeSO_4$, 0.01; Agar, 20. The secondary screening and enzyme production media (g/L): navel orange powder, 10; $(NH_4)_2SO_4$, 4; KH_2PO_4, 1; NaCl, 1; $MgSO_4 \cdot 7H_2O$, 0.6; $FeSO_4$, 0.01. All the media were sterilized at 121°C for 20 min.

2.3 Isolation and identification of strains

The decayed navel orange and soil sample were weighed into flasks with enrichment medium respectively, and then cultured at 160 rpm, 30°C for 48 h. From these, dilutions were subsequently made up to 10^{-6} and poured plating in primary screening media. After incubation, plates were observed for fungal growths and only fungal can grow using navel orange powder as the sole carbon source.

Colony morphology observed on the growth plates, and the cultured strains were examined by microscopy (Alix et al. 2012). Molecular identification was based on internal transcribed spacer (ITS). ITS of fungus ribosomal DNA was non-coding regions and showed great sequence polymorphism (Pinto et al. 2012). Genomic DNA was extracted with the DNA Purification Kit. The following primers were used to amplify eukaryotic ITS region: ITS_1 and ITS_4. The obtained sequences were compared with those from GenBank database. The aligned data set was used to create phylogenetic trees with MEGA Software Version 4.0. Neighbor joining algorithms were used to construct phylogenetic relationships (KESİCİ et al. 2013).

2.4 Enzyme extraction and enzyme activity assay

Stains which obtained through primary screening were inoculated into flasks with secondary medium and cultured at 30°C for 4 d. Fermented solutions were centrifuged and the supernatant was used as crude enzyme. Pectinase activity was determined by measuring the galacturonic acid with dinitrosalicylic acid (DNS) method (Almomani et al. 2013). One unit of pectinase activity was defined as the amount of enzyme that releases 1 μmol of galacturonic acid per minute under standard assay conditions, expressed in U/mL. All the experiments were conducted in triplicates and the average values were reported. The galacturonic acid standard curve was determined by the method recommended (Rehman et al. 2013).

3 RESULTS AND DISCUSSION

3.1 Screening for Pectinase Producing Strains

Five fungal isolates were obtained on primary screening agar plates. Isolates were designated as X4, X7, X10, X22 and BM. The five isolates were secondary screened, and BM strain showed the highest pectinase activity (Fig. 1). So the BM strain then selected for further experiments.

3.2 Identification of BM strain

As a result of observation, the BM strain showed white colonies on PDA plates. Training after 3 d, the BM strain gradually turned into the pale pink (Fig. 2). The colonies were 3–5 mm in

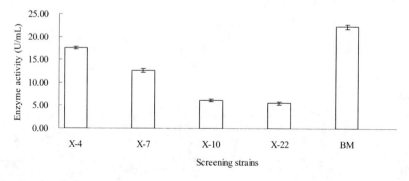

Figure 1. Results of primary screening.

height and small conidium were born on bottle (Fig. 3). The ITS region sequence of the BM displayed 99% similarity with the *F. oxysporum*. The phylogenetic tree constructed with other ITS sequences in Genbank (Fig. 4). Overall, a combination of morphology observation method and ITS sequence based molecular identification method were used to identify the BM strain. Both methods confirmed the BM strain was *F. oxysporum*.

3.3 *Optimization of fermentation process*

The medium set carbon source with 0.5%, 1%, 1.5% and 2% navel orange powder respectively, and determined the pectinase activity after training. It showed that the production of pectinase was directly related to the concentration of navel orange powder (Fig. 5). We added different concentration of ammonium persulfate and the pectinase production reached a maximum when 0.4% ammonium persulfate was used (Fig. 6). This may be due to carbon nitrogen ratio. Initial pH had significant effects on synthesis of enzyme. Enzyme activity reached the maximum when the pH between 4 and 5 (Fig. 7). Fermentation experiment was cultured respectively at different temperature. It showed that temperature had a great influence on the enzyme reaction speed, enzyme reaction of optimum temperature at 30°C (Fig. 8).

Figure 2. Colony morphology of the BM strain on PDA plate.

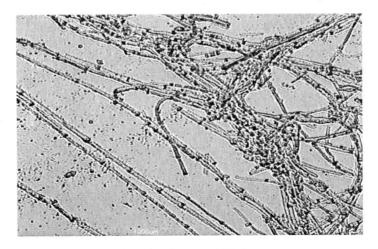

Figure 3. Morphology of the BM strain under microscope (400×).

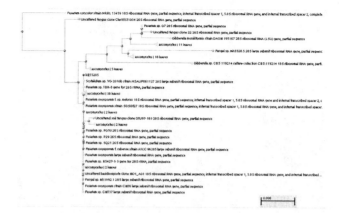

Figure 4. Phylogenetic tree of th BM strain.

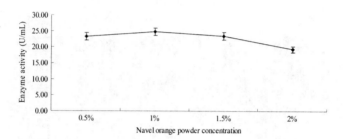

Figure 5. Effects of navel orange powder concentration on enzyme activity.

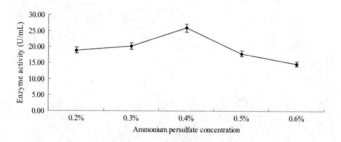

Figure 6. Effects of ammonium persulfate concentration on enzyme activity.

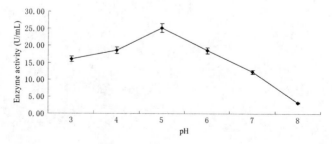

Figure 7. Effects of pH on enzyme activity.

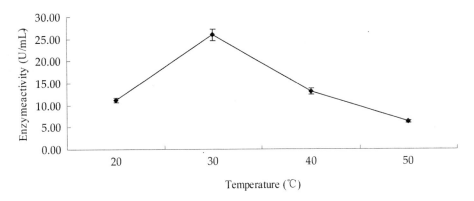

Figure 8. Effect of temperature on enzyme activity.

Table 1. The orthogonal experiment results.

Number	Carbon source	Nitrogen source	pH	Temperature	Enzyme activity (U/mL)
1	0.5%	0.2%	3	30	19.94
2	0.5%	0.4%	4	40	17.45
3	0.5%	0.6%	5	50	3.89
4	1.0%	0.2%	4	50	4.38
5	1.0%	0.4%	5	30	26.20
6	1.0%	0.6%	3	40	16.16
7	1.5%	0.2%	5	40	18.30
8	1.5%	0.4%	3	50	3.16
9	1.5%	0.6%	4	30	21.11
K_1	41.28	42.62	39.26	67.25	
K_2	46.74	46.82	42.94	51.91	
K_3	42.57	41.15	48.39	11.42	
k_1	13.76	14.21	13.09	22.42	
k_2	15.58	15.61	14.31	17.30	
k_3	14.19	13.72	16.13	3.81	
Rang R	1.82	1.89	3.04	18.61	

From the single factor experiment of carbon source, the BM strain's ability to use citrus navel orange powder is very strong. It may take into consideration in the large-scale fermentation production of pectinase that using cheap navel orange powder as the sole carbon source. According to the results of single factor experiment, do the orthogonal experiment for the culture conditions of carbon source, nitrogen source, pH and temperature. Results from the analysis showed that the culture temperature had a great influence on the pectinase producing of BM strain, followed is the influence of initial pH, carbon source concentration and nitrogen source concentration (Table 1). Above all, the BM strain expressed the maximum pectinase activity at the condition of 1% carbon source, 0.4% nitrogen source, the initial pH of 5 and 30°C. The pectinase activity reached the largest 26.20 U/mL.

4 CONCLUSION

The purpose of this research was to screen high-yield strains of pectinase from decayed navel orange and soil, and successfully obtained a high-yield strain named BM after initial and secondary screening. Through morphological observation and molecular identification, the BM strain confirmed as *F.oxysporum* and the phylogenetic tree was constructed.

Optimization assays of single factor experiments and orthogonal experiments were done to estimate the secreted pectinase by the BM strain and cultural conditions which expressed highest production of pectinase were studied. The BM strain produced pectinase maximally at the condition of 1% carbon source, 0.4% nitrogen source, the initial pH of 5 and 30°C in basal medium, the largest pectinase activity reached 26.20 U/mL. On the other hand, we not only can save the cost for mass production of pectinase, but also can realize the recycled clean processing. Above all, this study has economic significance and social significance in microbe large-scale fermentation production of pectinase, citrus processing.

REFERENCES

Alix, S., et al. 2012. Pectinase treatments on technical fibres of flax: Effects on water sorption and mechanical properties. *Carbohydrate Polymers* 87(1): 177–185.

Almomani, F. A., et al. 2013. Estimation of Jordanian isolated Pectobacterium cellulase, pectinase concentrations, RAPD polymorphism and their maceration activity. *Archives Of Phytopathology And Plant Protection* 23: 1–10.

Bernava, A. & S. Reihmane. 2013. Usage of Enzymatic Bioprocessing for Raw Linen Fabric Preparing. *Textiles and Light Industrial Science and Technology* 2(3): 48–51.

Fava, F., et al. 2013. New advances in the integrated management of food processing by-products in europe: sustainable exploitation of fruit and cereal processing by-products with the production of new food products (namaste eu). *New biotechnology* 30(6): 647–655.

Jain, R., et al. 2013. Enzymatic Retting: A Revolution in the Handmade Papermaking from Calotropis procera. *Biotechnology for Environmental Management and Resource Recovery* 33: 77–88.

KESİCİ, K., et al. 2013. Morphological and molecular identification of pennate diatoms isolated from Urla, İzmir, coast of the Aegean Sea. *Turkish Journal of Biology* 37: 530–537.

Khan, M., et al. 2013. Potential Application of Pectinase in Developing Functional Foods. *Annual review of food science and technology* 4(1): 21–34.

Pinto, F. C. J., et al. 2012. Morphological and molecular identification of filamentous fungi isolated from cosmetic powders. *Brazilian Archives of Biology and Technology* 55(6): 897–901.

Rehman, H. U., et al. 2013. Degradation of complex carbohydrate: Immobilization of pectinase from *Bacillus Licheniformis* KIBGE-IB 21 using calcium alginate as a support. *Food chemistry* 22(3): 8–12.

Zhang, J., et al. 2013. Enzymatic Treatment of Mulberry Bast to Pulping. *Applied Mechanics and Materials* 295: 351–353.

Study on flavonoids content and antioxidant activity of *muskmelon* with different pulp color

Xiyun Sun
Key Laboratory of Protected Horticulture, Ministry of Education/College of Horticulture, Shenyang Agricultural University, Shenyang, China
ShenYang Agricultural University, College of food, China

Tianlai Li
Key Laboratory of Protected Horticulture, Ministry of Education/College of Horticulture, Shenyang Agricultural University, Shenyang, China

Limei Jiang & Xinyue Luo
ShenYang Agricultural University, College of food, China

ABSTRACT: This paper mainly studied the differences of the total flavonoids contents between 11 Varieties muskmelons with different pulp color, and evaluate their antioxidant abilities in multiple antioxidant systems. The results showed that the contents of total flavonoids of Jingyu-4 and Jingyu-5 were higher than the other varieties. Meanwhile, they two have better antioxidation activity to develop health care beverage and related products.

1 INTRODUCTION

Muskmelon is a new variety of the genus of *cucumis* and a type of thick-skinned melon. It not only has high sugar and high quality, unique taste and flavor, but also has beautiful mesh texture, high commodity value and high economic benefits (Shen B., *et al.* 2012). Recently, many researches have shown that *muskmelon* contains abundant flavonoids substances (Deng B., *et al.* 2010). Therefore it is regarded as a top grade melon variety for its health function (Deng B., *et al.* 2011), moreover, its market prospect is broad.

The quality of *muskmelon* mainly consists of three aspects: commodity quality, flavor and nutritional quality (Guan X. 2006). Pulp color is one of the first visual indicators to evaluate commodity quality, furthermore, it determines how people like it. The color of the pulp is made of carotenoids and flavonoids which determine the nutrition and health to a certain extent. At present, pulp color has ivory, white, green and white, green, green and yellow, orange, orange red, etc. (Miao L., *et al.* 2009).There are different kinds and contents of flavonoids in different types of pulp.

In order to take fully advantage of these resources to breed functional varieties, we selected some some typical *muskmelons* to compare the content of flavonoids and antioxidant activity,as well as to select varieties with rich healthy function. I hope this paper can supply important experimental and theoretical basis for breeding programs and deep processing.

2 MATERIALS AND METHODS

2.1 *Experimental material*

Materials were *muskmelons* with different pulp color (as shown in the table 1). Samples were taken from horticulture experimental base of shenyang agricultural university.

2.2 Experimental reagents

Ethanol, $NaNO_2$, Al $(NO_3)_3$, NaOH, pH 6.6 phosphate buffer, potassium ferricyanide solution, trichloroacetic acid, ferric chloride, pH 8.2 Tris-HCl buffer, pyrogallol, orthophenanthroline solution, $FeSO_4$, H_2O_2, DPPH, alcohol.

3 EXPERIMENTAL METHODS

3.1 Preparation of raw material

Cut the peeled and seeded various muskmelons in pieces and then store them in the refrigerator.

3.2 Extraction of flavonoids

Chopped flesh (0.3*1) were wrapped into moist filter paper and then put into Soxhlet extractor extracted by 75% EtOH (80 mL) for 5 h. Determine the content of flavonoids and oxidation resistance as flesh cool.

3.3 Draw standard curve

Standard solution (100 μg/mL)—Weighed accurately 10 mg of rutin in a 100 mL volumetric flask. And added 30% ethanol solution to dissolve. Transfered 0.5, 1.0, 2.0, 3.0, 4.0, 5.0 mL of this standard solution respectively to 10 mL tubes, added 30% ethanol solution to 5 mL, added 0.3 mL $NaNO_2$ solution (5%), the mixture was allowed to stand for 6 min after shaking; then 0.3 mL $Al(NO_3)_3$ solution (10%), the mixture was allowed to stand for 6 min after us shaking, adding 4.0 mL NaOH solution (10%), then the mixture was allowed to stand for 6 min after us shaking; eventually, the final volume was adjusted to 10 mL with distilled water. The mixture was allowed to stand for 15 min and the absorption was measured at 510 nm, draw standard curve with concentration as the abscissa and the absorbance as the ordinate.

3.4 The determination of content

Transfer 2 mL of extract sample to a 10 mL tubes, added 30% ethanol solution to 5 mL, added 0.3 mL 5% $NaNO_2$ solution (5%), the mixture was allowed to stand for 6 min after us shaking; then 0.3 mL $Al(NO_3)_3$ solution (10%), the mixture was allowed to stand for 6 min after us shaking, adding 4.0 mL NaOH solution (10%), then the mixture was allowed to stand for 6 min after us shaking; eventually, the final volume was adjusted to 10 mL with distilled water. The mixture was allowed to stand for 15 min and the absorption was measured at 510 nm. at the same time, reagent blanks on the determination of absorbency. Carried out three parallel samples analysis.

3.5 Determination of flavonoids antioxidant activity from different kind of muskmelon

Reducing power: Prussian blue reaction method;
$O_2^{-\bullet}$ scavenging activity: the pyrogallol autoxidation method;
The determination of OH$^\bullet$ scavenging activity: adjacent dinitrogen fe method;
The determination of the ability to scavenge the DPPH$^\bullet$: spectrophotometric method.

4 RESULTS AND DISCUSSION

4.1 The comparison of flavonoids content from different varieties of muskmelon

4.1.1 The drawing of standard curve
Fig. 1 showed that, the relationship between the absorbance and the rutin concentration within 0–50 μg/mL was a linear with a high correlation ($R^2 = 0.9996$). Linear regression equation

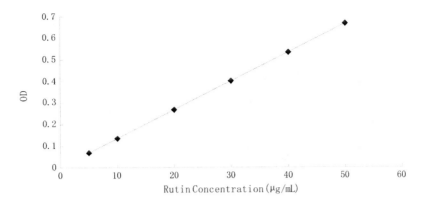

Figure 1. Rutin Standard Curve.

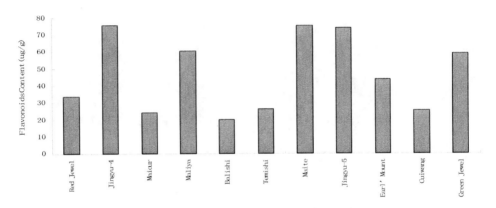

Figure 2. The comparison of flavonoids content from different varieties of *muskmelon*.

was: $y = 0.0133 + 0.0023x$. y was on behalf of the absorbance wavelength at 510 nm, x wais on behalf of the concentration of rutin.

4.1.2 *The comparison of flavonoids content from different varieties of* muskmelon

In this experiment, flavonoids content from different varieties of *muskmelon* were determined respectively (see Fig. 2). It showed that, flavonoids content of Jingyu-4, Maite, and Jingyu-5 were close and the highest among the eleven varieties, Balishi was the lowest. The three varieties which had the highest flavonoids content was 3.8 times of the variety which containts the lowest.

4.2 *The comparison of reducing power from different varieties of muskmelon*

In this experiment, reducing power from different varieties of *muskmelon* were determined respectively (see Fig. 3). Increasing absorbance of the reaction mixture indicates increasing reducing power. Maite had the highest reducing power, Second, was Red Jewel then Jingyu-4, Jingyu-5, Maiour, Green Jewel. Maliya, Balishi, Temishi, Earl' Mount, Cuiwang.

4.3 *The comparison of the ability to scavenge DPPH• from different varieties of muskmelon*

In this experiment, the ability to scavenge DPPH• from different varieties of *muskmelon* were determined respectively (see Fig. 4). It showed that, Earl'Mount had the highest Inhibitory rate of

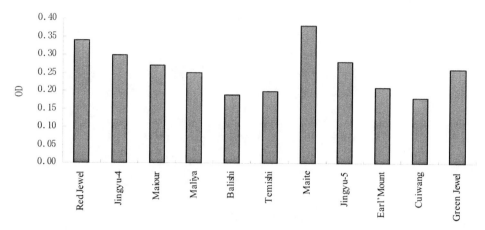

Figure 3. The comparison of reducing power from different varieties of *muskmelon*.

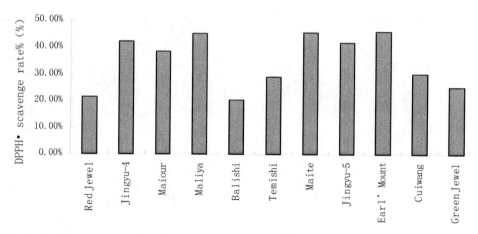

Figure 4. The comparison of the ability to scavenge DPPH· from different varieties of *muskmelon*.

DPPH• up to 45.83%, second was Maite, Maliya, Jingyu-4, Jingyu-5, Maiour, Cuiwang, Temishi, Green Jewel, Red Jewel, Balishi. The varieties which had the highest flavonoids content was 2.1 times of the variety which had the lowest.

4.4 *The comparison of OH• scavenging activity from different varieties of* muskmelon

In this experiment, OH• scavenging activity from different varieties of *muskmelon* were determined respectively (see Fig. 5). It showed that, Jingyu-5 had the highest Inhibitory rate of hydroxyl radical up to 40.00%, second is Jingyu-4, Maiour, Green Jewel, Temishi, Maliya, Red Jewel, Balishi, Cuiwang, Maite, Earl' Mount.

4.5 *The comparison of $O_2^{-\bullet}$ scavenging activity from different varieties of* muskmelon

In this experiment, $O_2^{-\bullet}$ scavenging activity from different varieties of muskmelon were determined respectively (see Fig. 6). It showed that, Jingyu-4 had the highest Inhibitory rate of Hydroxyl free radicals up to 40.00%, second was Green Jewel, Jingyu-5, Earl' Mount, Maliya, Maite, Red Jewel, Cuiwang, Maiour, Temishi, Balishi.

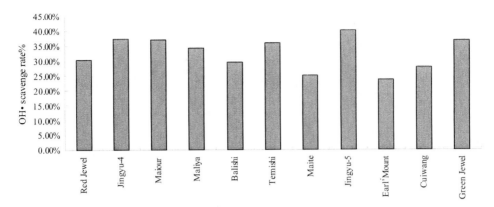

Figure 5. The comparison of OH• scavenging activity from different varieties of muskmelon.

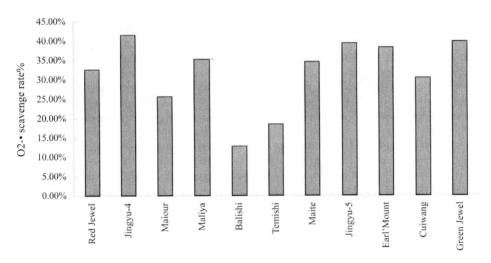

Figure 6. The comparison of O_2^{-}• scavenging activity from different varieties of muskmelon.

4.6 Correlational analyses between flesh color and flavonoids content

The color of flavonoids relates to the cross conjugation system and the type, number and substitutional positions of auxochrome ($-OH$, $-CH_3$). In general, flavone, flavonol and their glycosides vary from grayish yellow to yellow, chalcone varies from yellow to orange, and there is no such conjugate system or conjugate rarely in flavonone, flavanonol and isoflavone, so are not colored. The color anthocyanidins and their anthocyanins vary from different pH: red (pH < 7), purple (pH < 8.5), blue (pH > 8.5), and other colors.

The color of Maite, Jingyu-5, Earl' Mount, Cuiwang, Green J basically is green, there is a little of or no conjugated system, other varieties are orange red, orange (see Table 1). Cannot determine the general types of flavonoids.

5 CONCLUSION

The flavone contents of Jingyu-4 was the highest, the lowest was Balishi.

Table 1. Pulp color and flavonoids content.

No.	Varieties	Pulp Color	Flavonoids content (μg/g)
1	Red Jewel	jacinth	33.63
2	Jingyu-4	orange red	75.76
3	Maiour	reddish orange	24.17
4	Maliya	orange red	60.69
5	Balishi	orange red	20.10
6	Temishi	Yellow and green colors	26.37
7	Maite	viridis	75.21
8	Jingyu-5	Green	74.22
9	Earl' Mount	green	43.53
10	Cuiwang	Green and white	25.60
11	Green Jewel	Green and white	58.71

Maite's reducing power of ferric is the highest, the lowest was Cuiwang. Maliya had the hightest scavenging rate of DPPH free radical; Balishi had the lowest. Jingyu-4 had the highest $O_2^{-\bullet}$ scavenging activity, Balishi had the lowest. Jingyu-5 had the highest hydroxyl radical-scavenging activity, Earl' Mount had the lowest. The content of flavonoids from 11 varieties of muskmelon had a significant linear correlation with reducing power and $O_2^{-\bullet}$ scavenging activity, had no linear correlation with scavenge the 1,1-diphenyl-2-picryl-hydrazyl (DPPH) and hydroxyl radical-scavenging activity which indicated that content of flavonoids, had no linear correlation with antioxidant activity in vitro, muskmelon varieties which had the highest content of total flavonoid did not have the highest antioxidant activity. Therefore, screened the muskmelon varieties should be in consideration of both the content of total flavonoid and the antioxidant activity. The experiment choose the best varieties are Jingyu-4 and Jingyu-5 which had the highest content of total flavonoid as well as, higher antioxidant activity.

REFERENCES

Deng B., Li P., Nan W., et al. Effect of muskmelon on anti-oxidatiom function of liver in aged mice. *Acta Nutrimenta Sinica*, 2011, 33(6):594–596.
Deng B., Li P., Nan W., et al. Study on the total flavonoids content determination of muskmelon by the Spectrophotometry [J]. *Prev Med Chin PLA*. 2010. 28(5):344–346.
Guan X. Studies on Developmental Characteristics of Fruit Quality of Nettedmuskmelon (O'eumsimeot L.va. rertieualtus Naud [D]. 2006. Zhejiang University: hangzhou.
Miao L., Zhang Y., Jiang G., et al. Relationship between Melon Flesh Color and Carotenoid. Academic thesis abstract set of The 12th national watermelon melon research production coordination meeting [C]. 2009. 4.
Shen B., Xie X., Zhang Y., et al. Study on Agronomic and physiological characteristics of different pulp color muskmelon. *The north garden*. 2012(1):7–10.

Flavonoids intake estimation of urban and rural residents in Liaoning Province

Zhaoxia Wu
College of Food Science, Shenyang Agricultural University, Liaoning Province, China

Liu Yang
Foreign Language Teaching Department, Shenyang Agricultural University, Liaoning Province, China

Suijing Li
Department of Nutrition and Food Safety and School Health, Liaoning Provincial Center for Disease Control and Prevention, Liaoning Province, China

Wei Chen, Donghua Jiang, Yanqun Wang, Qi Zhang & Xuan Zhang
College of Food Science, Shenyang Agricultural University, Liaoning Province, China

ABSTRACT: Flavonoids, a class of vital phytochemicals, are commonly found in fruits, vegetables, soybean and soybean products. It's necessary to assessment the flavonoids intake when consideration of the relationship between flavonoids and health. This study is intended to estimate the overall flavonoids intake of urban and rural residents in Liaoning Province based on the Nutritional Survey in 2009 in Liaoning Province. The results showed that flavonoids could be detected in all foodstuffs with contents ranged from 1.37 ± 0.59 mg/100 g to 237 ± 1.29 mg/100 g. Among the vegetables detected, soybean and soybean products, and leaf vegetables had higher flavonoids contents. There is no difference between urban and rural residents in terms of vegetable consumption with the data of 365.6 g and 361.5 g respectively. However, compared to urban residents, rural residents consumed less fruits, with only 147.3 g, while the urban 171.4 g. The flavonoids intake by the urban residents reached 188.79 mg, and the rural 166.20 mg. The dominant reason for this difference attributed to the fact that urban residents consumed more soybean and soybean products than rural residents.

1 INTRODUCTION

Nowadays, the idea of using food to prevent and even to cure some kinds of diseases is not new. The nature provides us food comprised not only the traditional six nutrients but also abundant natural active compounds that could greatly promote our health. Among these active compounds, flavonoids are undoubtedly highlights for food scientists and consumers all around the world.

Flavonoids are a class of plant secondary metabolites with a 2-phenylchromen-4-one (2-phenyl-1,4-benzopyrone) structure. They are widely distributed in plants. Up to now over 5000 naturally occurring flavonoids have been characterized from various plants. As far as the functions of flavonoids are concerned, many good effects on human health have been confirmed and some achievements have been clinically applied and proved satisfactory results. These functions include scavenging free radicals, antioxidation (Aviram M, 2002; Piritta Lampila, 2009), vascular protection (Silvina B. Lotito, 2006; Mark F, 2008), estrogen-like effect (Rachel S, 2000; Ka-Chun Wong, 2013), anti-HIV action (De Clercq E, 2001; Andrae-Marobela K, 2013), antimicrobial activity (T.P. Tim Cushnie, 2005), reducing fluid retention in pre-menopausal women (S. Christie, 2004), anti-diabetic actions (Pon Velayutham Anandh Babu, 2013), and antitubercular activity (Akhilesh K. Yadav, 2013).

However, as far as total flavonoids amount intake in commonly eaten food, less information is available. For this reason, United States Department of Agriculture (USDA) establishes a laboratory to set up a database for flavonoids content of selected foods (Seema Bhagwat, 2013). Now the databases were expanded to comprise approximately 2900 foods from USDA's National Nutrient Database for Standard Reference (SR). In Europe, scientists developed a Combined Flavonoids Food Composition Database to assess intake of total flavonoids (Anna Vogiatzoglou, 2012). Nevertheless, fewer data of flavonoids content are available. In china, official database is limited except the isoflavone content of food in China Food Composition (2nd edition) published in 2009.

Plant food has a predominant position in Chinese diet. We assume that the flavonoids intake under this kind of dietary structure is much more than that in developed countries that more animal foods are consumed. However, it's difficult to measure flavonoids content for all the plant foodstuffs because the content is influenced by many factors such as geographical location, variety, growing environment, and planting techniques. So it's unreasonable to cited the data abroad directly. For this reason, we design our experiment based on the nutritional survey in 2009. Flavonoids contents for all the plant food consumed by people in Liaoning Province were measured and total flavonoids intake was estimated.

2 MATERIALS AND METHODS

2.1 Chemicals and reagents

Analytical standard rutin was provided by Sigma – Aldrich company. All reagents used in this study were of analytical grade. Rutin standard solution was prepared by dissolving 20 mg of rutin in 100 ml 70% ethanol. Serial dilutions were performed to acquire 5 rutin standard solution ranging between 0.1 mg/ml and 0.5 mg/ml. These standard solutions were used to obtain a calibration graph for calculating flavonoids content in fresh foodstuff.

2.2 Analytical procedure

Each 1.0 ml rutin standard solutions ranging between 0.1 mg/ml and 0.5 mg/ml were placed in glass tube. Adding 0.3 ml of 5% sodium nitrite solution (g/100 g) to each tube, shook up and stood for 6 min. Added 0.3 ml of 10% aluminum nitrate solution (g/100 g), shook up and stood for 6 min. At last added 4.0 ml of 4% sodium hydroxide solution (g/g) and 0.4 ml of 70% aqueous ethanol, shook up and stood for 12 min. Spectrophotometer was used to measure absorbance at 510 nm. The absorption value was proportional to the concentration of rutin standard solution. The linear regression equation was $y = 14.94x + 0.0057$, with correlation coefficient 0.9998. Value of X axis is the concentration of rutin solution, and value of Y is the absorption value at 510 nm.

2.3 Sample collection, flavonoids extraction, and quantification

According to the item of vegetables, fruits, soybean and soybean products in Liaoning Nutritional survey in 2009, we collected fresh foodstuffs from local markets or supermarkets for detection. All the items are listed in table 1.

10 g fresh sample was put into a mortar and ground for 15 min, and then removed to a beaker using 100 ml distilled water, following water bath heating for 30 min in constant temperature of 80°C and centrifugation at 6000 r/min for 30 min. The supernatant liquid was removed to another beaker. The precipitation was collected and mixed with 80% of aqueous ethanol at ratio of solid and liquid of 1:30 (g/ml). The mixture was water bathed at 80°C for 30 min, and centrifugated at 6000 r/min for 30 min. Repeated the procedure twice and collected all the supernatant. The supernatant was condensed under 0.1 MPa at 50°C. Then 1 ml condensed extraction solution was diluted to 5 ml with 70% of aqueous ethanol. According to above analytical procedure, 0.3 ml of 5% sodium nitrite solution, 0.3 ml of 10% aluminum nitrate solution, 4.0 ml of 4% sodium hydroxide

solution, 4.0 ml of 4% sodium hydroxide solution (g/g), and 0.4 ml of 70% aqueous ethanol was added to the glass tube one by one, and measured the absorbance at 510 nm. The flanonoid content could be calculated by use above regression equation.

2.4 Flavonoid intake estimation in Liaoning Province in 2009

Data of daily plant food consumption per capita in 2009 for Liaoning residents was provided by Liaoning Center for Disease Control and Prevention. The data was part of Nutritional and Healthy Survey for Liaoning Residents in 2009.

2.5 Statistical analysis

All data are expressed as mean ± S.E.M. For the same fruit or vegetable with different variety, the flavonoids content was measured for each variety and the average value is used to present the total flavonoids content.

3 RESULTS AND DISCUSSION

3.1 Favonoids contents in fruits consumed by Liaoning residents

The average consumption of plant foods, flavonoids contents, as well as the estimated flavonoids intakes are showed in Table 1.

As shown in Table 1, flavonoids could be detected in all the 18 fruits with content ranging from 2.35 mg/100 g to 62.37 mg/100 g. Among these fruits, pomegranate, hawthorn, bananas, orange, Lugan tangerine, longan, grape, apple, and kiwi fruit have higher flavonoids contents.

Flavonoids content in fruit is influenced by several factors such as climate, location, variety, maturity, individual difference, different part of fruit, as well as storage techniques. For example, compared to the mature kiwi fruit, immature one had less flavonoids content. The center of a watermelon had higher content than outer part.

3.2 Favonoid contents in vegetables consumed by Liaoning residents

Among all the vegetables, fresh leafy vegetables have higher flavonoids content ranging from 8.54 mg/100 g to 39.18 mg/100 g. Radish leaf has the highest content of 39.18 mg/100 g, while pickled Chinese cabbage has the lowest content of 2.75 mg/100 g. The result suggested that pickle or fermentation might reduce flavonoids content due to the proliferation and metabolism of microbes. The flavonoids content in fresh root vegetables ranged from 7.21 mg/100 g to 28.17 mg/100 g, and in fresh flower vegetable from 9.07 mg/100 g to 16.08 mg/100 g. For the fruit vegetables, flavonoids content ranged from 1.37 mg/100 g to 21.97 mg/100 g. The hot pepper has a much higher content.

Soybean and soybean products are famous for their higher flavonoids content. Our result confirmed this point. The content ranged from 32.78 mg/100 g to 54.36 mg/100 g that is much higher than other vegetables. The fresh green soybean is especially outstanding with the highest content of 54.36 mg/100 g. Accordingly, flavonoids content in soybean products is higher too. Soybean milk film occupied the first position with number of 237.30 ± 1.29 mg/100 g. Soybean milk is of the lowest content because a lot of water was added in the cooking procedure.

3.3 Flavonoid intake estimation

From the food consumed by Liaoning residents, urban residents ate much more soybean product like soybean milk film, soybean curd sheet, and soybean milk than rural residents. Vegetables consumption for urban and rural residents is 365.6 g and 361.5 g, respectively, and the result showed no difference. However, urban residents consumed much more fruit (171.4 g) than rural residents (147.3 g). As a result, urban residents ate more plant food. There are a lot of fresh fruits

Table 1. Flavonoids Contents in Vegetables and Fruits and Estimated Flavonoids Intakes in Liaoning Province.

		Food item	Totle flavonoid content (mg/100 g)	Average Food Consumption (g per capita)		Average intake of Flavonoids (mg per capita)	
				urban	rural	urban	rural
Fruits		pomegranate	62.37 ± 6.18	0.4	0	0.25	0.00
		hawthorn	41.58 ± 3.07	0	0.4	0.00	0.17
		banana	27.29 ± 7.62	11.4	6.7	3.11	1.83
		orange	20.97 ± 1.82	23.6	2.1	4.95	0.44
		Lugan tangerine	19.54 ± 5.35	2.3	0.4	0.45	0.08
		longan	18.70 ± 2.67	0.1	0	0.02	0.00
		grape	17.15 ± 3.98	10.3	17.3	1.77	2.97
		apple	16.97 ± 7.20	64.1	46.1	10.88	7.83
		kiwi fruit	15.90 ± 1.90	0.5	0.1	0.08	0.02
		xinjiang hami melon	9.07 ± 1.34	2.6	0.1	0.24	0.01
		peach	8.84 ± 3.61	7.8	22.1	0.69	1.95
		pear	6.41 ± 2.28	37.8	38.2	2.42	2.45
		fresh date	6.37 ± 0.70	2.5	5.4	0.16	0.34
		plum	5.34 ± 1.72	0.5	2.2	0.03	0.12
		watermelon	5.06 ± 1.55	3.5	5.2	0.18	0.26
		Chinese crabapple	4.76 ± 1.59	0	0.4	0.00	0.02
		persimmon	4.21 ± 1.27	3.8	0.6	0.16	0.03
		apricot	2.35 ± 0.45	0.2	0	0.00	0.00
Vegetables	Green leafy vegetables	chives	36.26 ± 0.28	4.9	1.78	2.36	1.78
		radish leaf	39.18 ± 1.33	0	0.00	0.43	0.00
		chrysanthemum	32.20 ± 0.55	0.3	0.10	0.13	0.10
		celery	32.20 ± 2.56	4.6	1.48	2.13	1.48
		spinach	23.57 ± 0.84	7.7	1.81	0.52	1.81
		mustard greens	26.76 ± 0.82	1.2	0.32	0.08	0.32
		bamboo shoot	23.69 ± 0.60	0	0.00	0.00	0.00
		garlic sprouts	22.40 ± 0.50	2.5	0.56	0.02	0.56
		Garlic Scape	18.82 ± 0.78	6.1	1.15	0.66	1.15
		coriander leaf	18.49 ± 0.82	0	0.00	0.13	0.00
		bok choi, green	20.31 ± 0.97	4.3	0.87	2.95	0.87
		rape	18.37 ± 1.16	2.5	0.46	0.68	0.46
		lettuce	12.00 ± 0.64	0.4	0.05	0.05	0.05
		bok choi/chinese cabbage	8.54 ± 0.84	78.9	6.74	6.31	6.74
		pickled Chinese cabbage	2.75 ± 0.31	7.7	0.21	0.47	0.21
	Flower vegetables	Broccoli	16.08 ± 1.16	0.9	0.14	0.32	0.14
		cabbage	8.46 ± 0.50	4	0.34	0.14	0.34
		cauliflower	9.07 ± 0.24	5.6	0.51	0.54	0.51
	Fruit vegetables	cucumber	21.97 ± 1.31	13.5	2.97	4.77	2.97
		eggplant	16.06 ± 1.36	30.2	4.85	5.75	4.85
		hot pepper	20.99 ± 0.76	5.6	1.18	1.45	1.18
		chayote	12.39 ± 1.41	0	0.00	0.01	0.00
		pumpkin	13.53 ± 0.76	7.1	0.96	0.78	0.96
		balsampear	18.96 ± 1.00	0.4	0.08	0.09	0.08
		chinese wax gourd	10.50 ± 0.46	4.5	0.47	0.16	0.47
		sweet pepper	10.50 ± 0.92	3.2	0.34	0.55	0.34
		sponge gourd	9.82 ± 0.27	0.1	0.01	0.13	0.01
		onion	7.93 ± 0.40	8.1	0.64	0.71	0.64
		zucchini, green	5.48 ± 0.31	4.0	0.22	0.07	0.22
		tomato	1.37 ± 0.59	4.0	0.24	0.19	0.24

(Continued)

Table 1. Continued.

	Food item	Totle flavonoid content (mg/100 g)	Average Food Consumption (g per capita)		Average intake of Flavonoids (mg per capita)	
			urban	rural	urban	rural
Root vegetables	carrot	28.17 ± 1.03	4.5	1.27	0.51	1.27
	radish (red)	8.32 ± 0.57	17.1	1.42	0.52	1.42
	radish (green)	8.09 ± 0.74	3.5	0.28	0.26	0.28
	white radish	7.52 ± 1.02	7.7	0.58	0.19	0.58
	radish (small red)	7.21 ± 1.06	0.4	0.03	0.00	0.03
Soybean and soybean products	soybean	97.24 ± 1.23	1	0.97	2.14	0.97
	mung bean	90.85 ± 1.55	0.1	0.09	0.09	0.09
	soybean milk film	237.30 ± 1.29	0	0.00	0.47	0.00
	bean sprout	222.78 ± 2.06	2.4	5.35	2.23	5.35
	dried bean curd	185.65 ± 1.95	1.4	2.60	3.16	2.60
	soybean curd	147.02 ± 1.26	59.5	87.48	68.22	87.48
	soybean curd sheet	136.99 ± 1.26	9.2	12.60	10.96	12.60
	soya-bean milk	85.17 ± 1.17	17.6	14.99	10.99	14.99
	fresh green soybean	54.36 ± 1.14	0	0.00	0.82	0.00
	kidney bean	55.41 ± 1.79	9.7	5.37	10.36	5.37
	haricot	45.64 ± 1.07	0.1	0.05	0.64	0.05
	cowpea	34.47 ± 0.72	3.5	1.21	3.27	1.21
	snow pea	32.78 ± 0.74	2	0.66	0.30	0.66
	pea	32.20 ± 2.56	0	0.00	0.03	0.00
Total			536.8	508.8	188.79	166.20

"0" means the average consumption is so tiny that could be ignored.

and vegetables available in local market with lower prices. That's a reason why local residents consumed many plant foods.

In term of the flavonoids intake, urban residents digested more flavonoids than rural residents. Although rural residents consumed almost equivalent vegetables, the flavonoids intake was less than that of urban residents. Urban residents digested 163.41 mg flavonoids per capita daily, and was 9.62% higher than 147.70 mg for rural residents. This is mainly because urban residents consumed more soybean and soybean products. Briefly, flavonoids coming from soybean and soybean products was 123.99 mg for urban residents, and only 98.98 mg for rural residents.

Our results also indicated that residents in Liaoning Province consumed more flavonoids than other adults in French (119.0 mg/day), Spanish (41.1 mg/day), American (18.6 mg/day), English (100.3 mg/day, excluding isoflavones), Belgian (34.8 mg/day, excluding isoflavones), Australian (24.13 mg/day, excluding isoflavones), and Finnish (53.3 mg/day, excluding flavones) populations that were studied previously (Guilan Li, 2013). However, we used spectro- photometry to detect total flavonoids content in fruits and vegetables and then calculated total flavonoid intake. Many investigations estimated flavonoids content based on USDA (Joanne M. Holden, 2005; RAUL ZAMORA-ROS, 2010; Chun OK, 2007; Janice E, 2011; Arabbi, P.R., 2004; Elizebete Wenzel de Menezes, 2011) in which five predominant subclasses of flavonoids – flavonols, flavones, flavanones, flavan-3-ols, and anthocyanidins were detected. So that was a reason why our data was somewhat higher than data of other countries. Moreover, we estimated the flavonoids from fruits and vegetables consumed by Liaoning residents. Tea might be another important resource for flavonoids. In other plant food such as grain, flavonoids could be detective. Although their flavonoids contents are not so high as in fruit and vegetables, Chinese people still consume many flavonoids from grains because grains are the predominant carbohydrates in

Chinese diet structure. So it's possible that Liaoning residents digest more flavonoids than our estimation above.

In our study, we detected flavonoids content in raw materials and then calculated flavonoids intake. However, the influence of cooking on flavonoids had not been considered. Our previous study indicated that the flavonoids contents of several vegetables increased from 26.57% to 135.19% after cooking, especially for stir-fry. In case of stew, some flavonoids reduced (Zhaoxia Wu, 2011). Another problem is that it's difficult to estimate the ways of cooking for most vegetables because there are so many cooking methods available in Chinese gastronomy. So it's necessary to give more details on cooking methods in later investigations.

Our results showed that urban presidents consumed more soybean and soybean products than rural residents. In area of Liaoning Province, the benefit of soybean is widely published due to the higher content of protein and calcium. For a country with limited stocks of meat and milk in the market, soybean is undoubtedly a good choice. This is not curious why soybean and soybean products often appear in Chinese dishes. Soybean milk is the most frequent breakfast food on the table for quite a lot family in Liaoning Province. In this point, the role of guidance of public opinion could not be substituted. For those preferring animal foods than plant foods, it's useful and necessary to communicate and encourage them to eat more fruits and vegetable to improve their flavonoids intake.

4 CONCLUSION

On the basis of Liaoning Nutritional Survey in 2009, we examed the flavonoids contents in all the fruits and vegetables consumed by Liaoning residents, and calculated the flavonoids intake. Our result showed that urban residents consume more fresh fruits and vegetables than rural resident. And daily total flavonoids intake per capita was 188.79 mg for urban residents, which was 10.36% higher than that of rural residents. The predominant reason was that urban residents consumed much more soybean and soybean products that contain plenty of flavonoids. Our estimation was almost equivalent to that in USA and the UK. Plant foods are predominant elements in Chinese traditional dishes. This dietary structure should be encouraged for it could provide enough flavonoids that are beneficial to the health of human being.

REFERENCES

Akhilesh K. Yadav, Jayprakash Thakur, Om Prakash, Feroz Khan, Dharmendra Saikia, Madan M. Gupta. 2013. Screening of flavonoids for antitubercular activity and their structure–activity relationships. *Medicinal Chemistry Research* 22(6):2706–2716.

Andrae-Marobela K, Ghislain FW, Okatch H, Majinda RR. 2013. Polyphenols: A Diverse Class of Multi-Target Anti-HIV-1 Agents. *Current Drug Metabolism* 14(4):392–413.

Anna Vogiatzoglou, Angela Mulligan, Robert Luben, Kay-Tee, Khaw, and Gunter G. C. Kuhnle. 2012. Development of a Combined Flavonoid Database for the Assessment of Flavonoid Intake in Europe using the EFSA Comprehensive European Food Consumption Database. *SFRBM 2012*, s89, doi:10.1016/j.freeradbiomed.2012.10.354.

Arabbi, P.R., Genovese, M.I., Lajolo, F.M., 2004. Flavonoids in vegetable foods commonly consumed in Brazil and estimated ingestion by the Brazilian population. *Journal of Agricultural and Food Chemistry* 52, 1124–1131.

Aviram M and Fuhrman B. 2002. Wine flavonoids protect against LDL oxidation and atherosclerosis. *Annals of the New York Academy of Sciences* 957:146–61.

Chun OK, Chung SJ, Song WO. 2007. Estimated dietary flavonoid intake and major food sources of US adults. *Journal of Nutrition* 137(5):1244–1252.

Dall'Agnol R, Ferraz A, Bernardi A.P, Albring D, C. Nör, Lamb L, Hass M. Von Posen G, Schapoval E.E. 2003. Antimicrobial activity of some Hypericum species. *Phytomedicine* 10:511–6.

De Clercq E. 2001. New developments in anti-HIV chemotherapy. *Farmaco* 56:3–12.

Elizabete Wenzel de Menezes, Nelaine Cardoso Santos, Eliana Bistriche Giuntini, Milana C.T. Dan, Maria Ines Genovese, Franco M. Lajolo. 2011. Brazilian flavonoid database: Application of quality evaluation system. *Journal of Food Composition and Analysis* 24:629–636.

Guilan Li; Yanna Zhu; Yuan Zhang; Jing Lang; Yuming Chen; Wenhua Ling. 2013. Estimated Daily Flavonoid and Stilbene Intake from Fruits, Vegetables, and Nuts and Associations with Lipid Profiles in Chinese Adults. *Journal of the Academy of Nutrition and Dietetics* 113:786–794.

Janice E. Maras, Sameera A. Talegawkar, Ning Qiao, Barbara Lyle, Luigi Ferrucci, Katherine L. 2011. Flavonoid intakes in the Baltimore Longitudinal Study of Aging. *Journal of Food Composition and Analysis* 24(8):1103–1109.

Joanne M. Holden, Seema A. Bhagwat, David B. Haytowitz, Susan E. Gebhardt, Johanna T. Dwyer, Julia Peterson, Gary R. Beecher, Alison L. Eldridge, Douglas Balentine. 2005. Development of a database of critically evaluated flavonoids data: application of USDA's data quality evaluation system. *Journal of Food Composition and Analysis* 18:829–844.

Ka-Chun Wong, Wai-Yin Pang, Xin-Lun Wang, Sao-Keng Mok, Wan-Ping Lai, Hung-Kay Chow, Ping-Chung Leung, Xin-Sheng Yao and Man-Sau Wong. 2013. Drynaria fortunei-derived total flavonoid fraction and isolated compounds exert oestrogen-like protective effects in bone. *British Journal of Nutrition* 10(3): 475–485.

Mark F. McCarty. 2008. Scavenging of peroxynitrite-derived radicals by flavonoids may support endothelial NO synthase activity, contributing to the vascular protection associated with high fruit and vegetable intakes. *Medical Hypotheses* 70:170–181.

Pon Velayutham Anandh Babu, Dongmin Liu, Elizabeth R. Gilbert. 2013. Recent advances in understanding the anti-diabetic actions of dietary flavonoids. *Journal of Nutritional Biochemistry xx (2013)* xxx–xxx, Available online at www.sciencedirect.com.

Piritta Lampila, Maartje van Lieshout, Bart Gremmen, Liisa Lahteenmaki. 2009. Consumer attitudes towards enhanced flavonoid content in fruit. *Food Research International* 42(1):122–129.

Rachel S. Rosenberg Zand, David J.A. Jenkins1, and Eleftherios P. Diamandis. 2000. Steroid hormone activity of flavonoids and related compounds. *Breast Cancer Research and Treatment* 62: 35–49.

Raul Zamora-Ros; Cristina Andres-Lacueva; Rosa M. Lamuela-Raventós; Toni Berenguer; Paula Jakszyn; Aurelio Barricarte; Eva Ardanaz; Pilar Amiano; Miren Dorronsoro; Nerea Larrañaga; Carmen Martínez; María J. Sánchez; Carmen Navarro; María D. Chirlaque; María J. Tormo; J. Ramón Quirós; Carlos A. González. 2010. Estimation of dietary sources and flavonoid intake in a Spanish adult population (EPIC-Spain). *Journal of the American Dietetic Association*:110(3):390–398.

Silvina B. Lotiton & Balz Frei. 2006. Consumption of flavonoid-rich foods and increased plasma antioxidant capacity in humans: Cause, consequence, or epiphenomenon? *Free Radical Biology & Medicine* 41:1727–1746.

S. Christie, A. F. Walker, S. M. Hicks, and S. Abeyasekera. 2004. Flavonoid supplement improves leg health and reduces fluid retention in pre-menopausal women in a double-blind, placebo-controlled study. *Phytomedicine* 11: 11–17.

Seema Bhagwat, David B. Haytowitz, Shirley I. Wasswa-Kintu, and Joanne M. Holden. 2013. USDA develops a database for flavonoids to assess dietary intakes. *Procedia Food Science* 2:81–86.

T.P. Tim Cushnie & Andrew J. 2005. Lamb. Antimicrobial activity of flavonoids. *International Journal of Antimicrobial Agents* 26:343–356.

Wu Zhaoxia, Chen Wei, Xu Yaping, Zhang Qi, Zhang Xuan. Level of flavonoids in common vegetables and the effect of cooking methods on content of flavonoids. *Science and Technology of Food Industry*. 2012(33). 9:372–374, 379.

Effect of 6-Benzyladenine, Gibberellin combined with fruit wax coating treatment on the lignification of postharvest Betel nut

Y.G. Pan
College of Food, Hainan University, Haikou, P. R. China

W.W. Zhang
College of Food, Hainan University, Haikou, P. R. China
Tangshan Dalu Agricultural Science and Technology Co. Ltd, Tanshan, Shangdong, P.R. China

ABSTRACT: The effects of 100 mg/L GA_3 + 100 mg/L 6-BA + 50% of fruit wax on the firmness and lignification of postharvest betel nut were examined during storage at 7°C. Coating treatment effectively delayed the increasing speed of fruit firmness and lignin content. And between them were positively correlated. At the same time, coating treatment also inhibited the activity of four kinds of key enzyme which could generate the lignin, PAL, 4-CL, CAD and POD. But coating treatment had a lesser inhibition effect on 4-CL and CAD activity compared to PAL and POD. Meanwhile, CAD activity showed a downward trend in the later storage; while in contrast, POD activity continued to rise.

1 INTRODUCTION

Betel nut (*Areca catechu* or 'Areca nut') is the fruit of the Areca catechu tree. Approximately 700 million individuals regularly chew betel nut worldwide (Yamada et al. 2013), most of them resident in Asia-Pacific regions. At the same time, more and more studies show that Betel nut chewing is related to the development of oral and esophageal cancer (Zhang & Reichart 2007, Wu et al. 2006, Gupta & Warnakulasuriya 2002). Betel nut chewing is easy to cause mechanical trauma to the oral mucosa, which may be the main reason for oral submucous fibrosis, the prophase of oral cancer. Betel nut chewing can easily cause oral cavity damage and it may be the result of tissue lignification. An increase of lignification in fruit tissue has been found in other fruits. For example, increased firmness of damaged mangosteen fruit was found to be related to enhanced lignins biosynthesis (Ketsa & Atantee 1998). The same phenomenon has also reported on bamboo shoot (Luo et al. 2008), loquat (Zheng et al. 2000).

At present, there has been little literature available in terms of the lignification of betel nut. Liu (2003) found lignin levels in green asparagus was inhibited by GA; consequently the green asparagus maintained excellent quality and high commodity rate. When immersed in 20 mg/g 6-benzylaminopurine (6-BA) and submitted to ultrasound wave (20 kHz) for 20 mins, the level of lignin in green asparagus was significantly reduced during the first 10 days (Wei & Ye, 2011). In our previous study, we found that 100 mg/L GA_3 + 100 mg/L 6-BA + 50% of fruit wax coating treatment could delay quality deterioration of betel nut fruit. At the same time, coating treatment could retard the rise of fruit firmness (Zhang et al. 2012). Therefore, the aims of this work were to further study the effect of GA_3, 6-BA combined with fruit wax coating treatment on the lignin generation and the activity of related enzymes in postharvest betel nut fruit, and to explore the effect mechanism of GA_3, 6-BA combined with fruit wax coating treatment on firmness of postharvest betel nut fruit.

2 MATERIAL AND MENTHOD

2.1 Plant materials

The sample betel nut fruits were obtained from a commercial orchard in Hainan Province in China. Fruits at commercial maturity were harvested carefully to minimize possible mechanical damage and transported to the laboratory within one day of harvest, and then they were selected for the uniformity of color, size and absence of defects.

2.2 Treatment

The fruit were rinsed with water, air dried, soaked in 0.02 % prochloraz for 5 minutes and left to dry. All fruit were distributed into two groups randomly. One group was dipped in the deionized water as the controls while the other was dipped in solution containing 100 mg/L GA_3 + 100 mg/L 6-BA + 50% of fruit wax (CFW fruit wax, Institute of agricultural products storage and processing, Gansu Academy of Agricultural Sciences). The dipping time of all samples was for 10 mins. After dried, all fruits were packed in a tray combining with polyethylene packaging film and stored at 7°C.

2.3 Methods

2.3.1 Firmness
The firmness of the fruit was measured with texture analyzer fitted with an 50-mm-diameter probe (Instron −5542) and at a traveling speed of $1.5 \text{ mm} \cdot \text{s}^{-1}$.

2.3.2 Lignin content
Lignin content was determined according to Morrison (1972) and Pan & Liu (2011) methods.

2.3.3 Enzymen activity
Phenylalanine hydroxylase activity was assayed following the method described by Koukol & Conn (1961); 4-CL activity was determined using a modified method of Bi et al. (1990); CAD activity was assayed following the method described by Goffner et al. (1992); POD was determined using a modified method of Li et al. (2000) and Chen & Wang (2002).

3 RESULT S AND DISCUSSIONS

3.1 Effect of GA_3 + 6-BA + fruit wax coating treatment on the firmness of betel nut

Tissue softening of fruits and vegetables is often the most apparent change that occurs after harvest. In betel nut, however, firmness tended to increase quickly. As shown in Figure 1, firmness increased both in fruits of the control fruits and of coated-treated fruits during storage. After 21 days of storage at 20°C, firmness of uncoated samples increased 2.1 times, whereas, coated fruits only increased 1.48 times. It was indicated that GA_3, 6-BA combined with fruit wax coating treatment may help to slow down the rise of Betel nut firmness.

3.2 Effect of GA_3 + 6-BA + fruit wax coating treatment on the lignin content in betel nut

Lignin is a complex polymer of phenylpropanoid residues mainly deposited in cell walls (Whetten & Sederoff 1995), which imparts rigidity to cell walls (Hu et al. 1999). The lignin content of Betel nut showed upward trend both in fruits of the control fruit and of coated-treated fruit during storage. At the end of storage, the lignin content of uncoated fruit increased 3 times, whereas, coated fruit only increased 2 times (Fig. 2). It indicates that coating can slow down the lignification of Betel nut.

In addition, a positive correlation can be seen significantly between lignin content and firmness of betel nut ($r = 0.98^{**}$). This suggests that fruit lignification is the main factor which leads to an increase of firmness in betel nut.

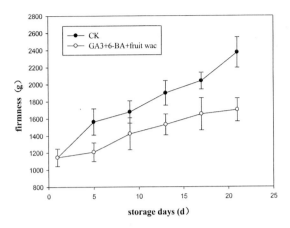

Figure 1. Effect of GA$_3$ + 6-BA + fruit wax coating treatment on the firmness of betel nut.

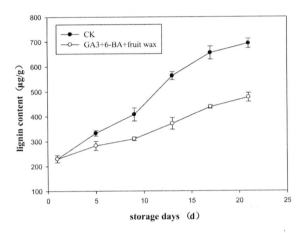

Figure 2. Effect of GA$_3$ + 6-BA + fruit wax coating treatment on the lignin content in betel nut.

3.3 Effect of GA3+6-BA+fruit wax coating treatment on PAL activity in betel nut

PAL, in catalysing deamination of phenylalanine to transcinnamate, is believed to be the critical enzyme controlling accumulation of lignin in plants (Boudet 2000). As shown in figure 3, the PAL activity of control fruit reached maximum on the fifth day. The PAL activity of tomato treated with GA$_3$ + 6-BA + fruit wax reached maximum on the seventeenth day. At the same time, except the first 3 days, the PAL activity of coated betel nuts was significantly lower than that of the control fruits throughout the storage. Therefore, GA$_3$ + 6-BA + fruit wax can inhibit the rising of PAL activity. Beyond doubt, inhibited PAL will help to reduce the generation of lignin.

3.4 Effect of GA3 + 6-BA + fruit wax on activity of 4-CL of betel nut

4-Coumarate:coenzyme A ligase (4CL) converts 4-coumaric acid and other substituted cinnamic acids nto corresponding CoA thio-lesters used for the biosynthesis of flavonoids, isoflavonoids, lignin, and so on (Ehlting et al. 1999, Beuerle et al. 2002). In the first five days, levels of 4-CL activity of the coated and control betel nuts only rose slowly and then increased sharply afterwards. It reached its maximum on the 9th day. Thereafter, 4-CL activity decreased quickly both in fruits of the control fruit and of coated-treated fruit. But 4-CL activity in coated fruits was lower than that in the control samples (Fig. 4). This suggests that coating treatment has inhibited effect on 4-CL activity to some extent.

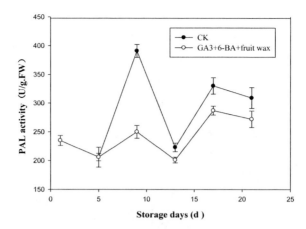

Figure 3. Effect of $GA_3 + 6\text{-}BA +$ fruit wax coating treatment on PAL activity in betel nut.

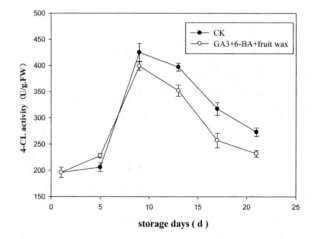

Figure 4. Effect of $GA_3 + 6\text{-}BA +$ fruit wax coating treatment on 4-CL activity in betel nut.

3.5 *Effect of GA3 + 6-BA + fruit wax on activity of CAD and POD of betel nut*

CAD catalyzes the last step of the monolignol pathway, while POD catalyzes the polymerization of monolignol to complete the process of lignification. (Imberty et al. 1985). CAD activity in coated fruit and control all offered the trend of rapidly rising and falling fast, and reached a peak in 9 and 13 days respectively (Fig. 5). Therefore, $GA_3 + 6\text{-}BA +$ fruit wax coating treatment can delay the rise of CAD activity. Unlike CAD activity change, POD activity in coated fruit and control all increased with the storage duration. POD activity of betel nut was suppressed by $GA_3 + 6\text{-}BA +$ fruit wax coating treatment (Fig. 6).

Research has shown that although the activity of CAD is strongly suppressed, fruits still can maintain normal level of lignin generation, which indicates that CAD may not the rate-limiting enzyme of the lignin synthesis (Wu et al. 1999). Our research also shows that activity of CAD decreased with the steady increase of lignin content in the later of storage. POD activity, in contrast, kept upward tendency in the whole process of storage. At the same time, There was a significant positive correlation between lignin content and CAD activity ($r = 0.91^{**}$). It is also true of postharvest bamboo shoots (Luo et al. 2007) and loquat (Cai et al. 2006). All these results show that the POD plays an important role in the process of lignification.

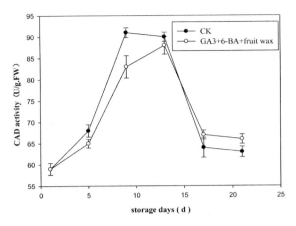

Figure 5. Effect of GA_3 + 6-BA + fruit wax coating treatment on CAD activity in betel nut.

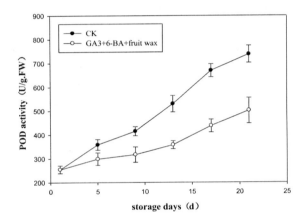

Figure 6. Effect of GA_3 + 6-BA + fruit wax coating treatment on POD activity in betel nut.

4 CONCLUSION

100 mg/L GA_3 + 100 mg/L 6-BA + 50% of fruit wax coating treatment effectively delayed the increasing speed of fruit firmness and lignin content in postharvest betel nut. And between them were positively correlated. At the same time, coating treatment also inhibited the activity of PAL, 4-CL, CAD and POD, which is thought to function in the synthesis of lignin. But coating treatment has a lesser inhibition effect on 4-CL and CAD activity compared to PAL and POD. Meanwhile, CAD activity shows a downward trend in the later storage; while in contrast, POD activity continued to rise. Accumulation of lignin in flesh tissue was also positively correlated with the activities of POD.

REFERENCES

Beuerle, T. & Pichersky, E. 2002. Enzymatic synthesis and purification of aromatic coenzyme a esters, *Anal Biochem*.302:305–312

Bi, Y.M. & Ou Yang, G.C. 1990. Properties of p-Coumarate CoA Ligase in Rice. *Plant Physiol Communication* (6):18–20.

Boudet, A. M. 2000. Lignins and lignification: selected issues. *Plant Physiol Biochem* 38:81–96

Cai, C., Xu C.J., Li X., Ferguson, I. & Chen, K.S. 2006. Accumulation of lignin in relation to change in activities of lignification enzymes in loquat fruit flesh after harves. *Postharvest Biol Technol* 40:163–169.

Chen, J.X. & Wang, X.F. 2002. *Experiment guidance of plant physiology*. Guang Zhou: South China university of technology press.

Ehlting, J., Buttner, D., Wang, Q., Douglas, C.J., Somssich, I.E. & Kombrink, E. 1999. Three 4-coumarate: coenzyme A ligases in Arabidopsis thaliana represent two evolutionarily divergent classes in angiosperms, *Plant J* 19:9–20.

Goffner, D., Joffroy, I., Grima-Pettenati, J., Halpin, C., Knight, M.E., Schuch, W. & Boudet, A.M. 1992. Purification and characterization of isoforms of cinnamyl alcohol dehydrogenase from Eucalyptus xylem. *Planta* 188:48–53.

Gupta, P.C. & Warnakulasuriya, S. 2002. Global epidemiology of areca nut usage. *Addict Biol* 7:77–83.

Hu, W.J., Harding, S.A., Lung, J., Popko, J.L., Ralph, J. & Stokke, D.D. 1999. Repression of lignin biosynthesis promotes cellulose accumulation and growth in transgenic trees. *Nature Biotechnol* 17: 808–812.

Imberty, A., Goldberg, R., & Catesson, A.M. 1985. Isolation and characterization of populus isoperoxidases involved in the last step of lignin formation. *Planta* 164:221–226.

Ketsa, S. & Atantee, S. 1998. Phenolics, lignin, peroxidase activity and increased firmness of damaged pericarp of mangosteen fruit after impact *Posth Biol Technol* 14:117–124.

Koukol, J. & Conn, E.E. 1961. The metabolism of aromatic compounds in higher plants. IV. Purification and properties of the phenylalanine deaminase of Hordeum vulgare. *J Biol Chem* 236: 2692–2698.

Knobloch, K.H. & Hahlbrock, K. 1977. 4-Coumarate: CoA ligase from cell suspension cultures of *Petroselinum hortense*. *Hoffm. Arch Biochem Biophys* 184:237–248.

Li, H.S., Sun, Q. & Zhao, S.J. 2000. *Plant physiological and biochemical principles and technology*. Bei-jin: Higher education press.

Liu, Z. 2003. *Studies on Physiological and biochemieal foundations and control technology of lignification in Green Asparagus*. Beijin: China Agrieultural Universit.

Luo, Z., Xu, X., Cai, Z. & Yan, B. 2007. Effects of ethylene and 1-methylcyclopropene (1-MCP) on lignification of postharvest bamboo shoot. *Food Chem* 105: 521–527.

Luo, Z., Xu, X. & Yan, B. 2008. Accumulation of lignin and involvement of enzymes in bamboo shoot during storage. *Eur Food Res Technol* 26:635–640.

Morrison, I.M.1972. A semi-micro method for the determination of lignin and its use in predicting the digestibility of forage crops. *J Sci Food Agric* 23(4):455–463.

Pan, Y.G. & Liu, X.H. 2011. Effect of benzo-thiadiazole-7-carbothioic acid S-methylester (BTH) treatment on the resistant substance in postharvest mango fruits of different varieties. *Afr J Biotechnol* 69(10):15521–15528.

Wei, X.Y. & YE, X.Q. 2011. Effect of 6-benzylaminopurine combined with ultrasound as pre-treatment on quality and enzyme activity of green asparagus. *J Food Processing and Preservation.* 35:587–595.

Whetten, R., & Sederoff, R. 1995. Lignin biosynthesis. *Plant Cell* 7:1001–1013.

Wu, I.C., Lu, C.Y., Kuo, C., Lee, K.W., Kuo, W.R., Cheng, Y.J., Kao, E.L., Yang, M.S.& Ko, Y.C. 2006. Interaction between cigarette, alcohol and betel nut use on esophageal cancer risk in Taiwan. *Eur J Clin Invest* 36:236–241.

Wu, R.L., Remington, D.L., Mackay, J.J., McKeand, S.E. & O'Malley, D.M. 1999. Average affect of a mutation in lignin synthesis in loblolly pine. *Theoretical and Applied Genetics* 99:705–710.

Yamada, T., Hara, K. & Kadowaki, T. 2013. Chewing Betel Quid and the Risk of Metabolic Disease, Cardiovascular Disease, and All-Cause Mortality: A Meta-Analysis. *Plos One* 8(8):1–10.

Zhang X., & Reichart P.A. 2007. A review of betel quid chewing, oral cancer and precancer in Mainland China. *Oral Oncol* 43:424–430.

Zhang, W.W., Pan, Y.G., Lu, C.S. Zhang Z.Y. & Zhu, P. 2012. The response surface methodology to optimize GA3, 6-BA and fruit wax optimal concentration ratio for betel-nut preservation. *Sci technol food industry* 33(15):345–348,352.

Zheng, Y.H., Li, S.Y. & Xi, Y.F. 2000. Changes of Cell Wall Substances In Relation to Flesh Woodiness in Cold stored Loquat Fruits. *Acta Phytophysiologica Sinica* 26(4):306–310.

Advanced Engineering and Technology – Xie & Huang (Eds)
© 2014 Taylor & Francis Group, London, ISBN 978-1-138-02636-0

A new approach for recycling potash from the mother liquor B in beet sugar industry

Si-Ming Zhu, Wei Long, Xiong Fu, Liang Zhu & Shu-Juan Yu
Institute of Light Industry and Chemical Engineering, South China University of Technology, Guangzhou, China

ABSTRACT: This work details on a novel means to extract and recycle potash from mother liquor B in beet sugar industry. The adsorption, elution and adsorption kinetics of K^+ on resin, the crystallization of potash in eluates were studied. Results showed that the adsorption capacity of K^+ in 50°C Bx mother liquor B on resin BK001 was 40.4 mg/ml wet resin at 60°C and a flow rate of 2BV/h. The elution efficiency of total K^+ on resin was higher than 98.6% using 450 ml of 0.166 mol/l Na_2SO_4 at 60°C and a flow rate of 2BV/h. The cause of eluting K^+ using Na_2SO_4 depends on the interaction between K^+ and SO_4^{2-} and the exchange between K^+ and Na^+. For recycling potash from the resin eluate, a fast cooling crystallization method is likely for industrial application for high yield and purity of potash. For the kinetics of adsorbing K^+, the ion exchange process is controlled by particle diffusion, the reaction order 1.389, the apparent activation energy 35.1 kJ/mol, and a kinetics model was established. Our findings suggested that the potash-recycling means might remit for potash deficiency in China, and showed potential for improving sugar yield without environmental pollution.

1 INTRODUCTION

Potassium is necessary for the function of all living cells and found in especially high concentrations within plant cells. The high concentration of potassium in plants, associated with comparatively low amounts of sodium there, resulted in potassium's being first isolated from potash, the ashes of plants, giving the element its name. For the same reason, heavy crop production rapidly depletes soils of potassium, and agricultural fertilizers consume 95% of global potassium chemical production (Greenwood, 1997). In sugar industry, an in-process product, mother liquor B contains about 2% (w/w) of potassium ions (K^+) and 1% (w/w) of sodium ions (Na^+), respectively. Because both ions are the melassigenic ions, lowering sugar yield in sugar industry, it is therefore important to remove these ions in order to increase the yield in sugar production (Elmidaoui, A., 2006).

There are several approaches available to remove these ions from mother liquor B, such as membrane filtration and ion exchange technique, e.g., Quentin and Gryllus method. Because of the high cost and the membrane pollution problem, the membrane filtration is not practical for industrial applications (Elmidaoui, A., 2006). Quentin method is a typical ion exchange technique for desalting mother liquor B, using resin (in Mg^{2+} form) as adsorbent and $MgCl_2$ as regenerant (Oldfield J F T., 1979). However, the method requires the magnesium resources that match to the needs for regenerating resin in the sugar mill. In Gryllus method, mother liquor B is used to regenerate the resin through desalting thin juice (rich in Ca^{2+}) in beet sugar mill (Gryllus, E., 1998; Gryllus, E., 1976). In our previous work, a combination method of the membrane and ion exchange resin method for desalting mother liquor B was applied (Zhu, S.M., 2011), i.e., an electrodialyzer (ED) was used to desalt thin juice and the ED brine rich in Ca^{2+} was used to regenerate resin in Ca^{2+} form. However, the combination method faced a bottleneck problem, i.e., the gross of Ca^{2+} in thin juice was not enough to exchange with K^+ and Na^+ in mother liquor B. Moreover the removal rate of alkaline metal ions in mother liquor B was only about 25%. We also tried the integration of extracting K^+ and the exchange of Na^+ in potassium-extracted mother liquor B and Ca^{2+} in thin juice on BK001 resin in beet sugar mill (Zhu, S.M., 2011). However, the method has some technological problems and is very difficult to be adopted in industrial practice due to the process is too complicated for industrialization. To overcome the problems mentioned above, a conceptual idea was raised for extracting potash from mother liquor B and meanwhile promote the sugar recovery rate, i.e., to stimulate the crystallization of sugar using Na^+ (having lower melassigenic coefficient) as a substitution of K^+. As a separative problem-solving scheme, the idea does not concern the exchange of Na^+ in potassium-extracted mother liquor B and Ca^{2+} in thin juice.

The objects of the study was to prove the hypothetical idea, including proposing the scheme of recycling potash from mother liquor B, studying the adsorption properties and kinetics of K^+ on resin, clarify the elution mechanism of K^+ on resin by high valent counter ions, and evaluate the crystallization of potash in eluate.

2 MATERIALS AND METHODS

2.1 *Chemical materials and main apparatus*

The mother liquor B (2-4% of K^+, 1-2% of Na^+, w/w) was from Xinjiang LVXIANG Sugar-Making Co., Ltd., Tacheng, China. Resin BK001 was kindly presented by Mrs. Sui-Hua Rong (Guangzhou Guanglianjin Chemical Engineering Co., Ltd), and pretreated according to Chinese standard method GB/T5476-1996 before use. All other chemicals were analytical grade and purchased from commercial sources.

The ion exchange column ($\Phi 22 \times 320$ mm) was made to our specifications by a glass apparatus company (Guangzhou Tianhe Precision Instruments Wholesale Division of the Department of Glass Co. Ltd).

2.2 *Determination*

The content of K^+ in eluate was determined using a flame photometer (FP640, Shanghai Precision Instruments Co., Ltd., Shanghai, China) (Zhang, P.J., 2010). The brix (Bx) was measured using an Abbe refractometer (RL3, Shanghai Precision Instruments Co., Ltd., Shanghai, China). The crystalline structure of potash salt was observed under a scanning electron microscope (TM3000, Hitachi Ltd, Japan). The eluate of K^+ was concentrated using a vacuum rotary evaporator (RE-52A, Shanghai YARONG Biochemistry Instrument Factory, Shanghai, China).

2.3 *Pre-treatment of resin BK001*

Resin BK001 of 100 ml was soaked in 200 ml of 8% NaCl (wt%) for 24h, then was rinsed using distilled water till the wash water was colorless transparent liquid. Then 400 ml of 4% HCl (wt %) was used to soak the resin for 4h, and distilled water was used to rinse the resin till the eluent was at pH7.0. Then the resin was soaked in 400 ml of 4% NaOH (wt %) for 3h, and rinsed using distilled water till the wash water was at pH7.0 (Zhu, S.M., 2010).

2.4 *Adsorption kinetics of K^+ using resin BK001*

5g resin BK001 was filled into a triangular flask of 250 ml, then a certain amount of mother liquor B was added into the flask, and the flask was placed in a thermostatic water bath with continuous shakes, every other 3 minutes samples were taken to analyze remaining K^+ content. The cation-exchange rate (F) of resin at a time was calculated using the following formula (1) (Zhang, P.J., 2010):

$$F = \frac{C_0 V_0 - (C_n V_n + \sum_{i=1}^{n-1} C_i V_i)}{MQ}, n \geq 2 \qquad (1)$$

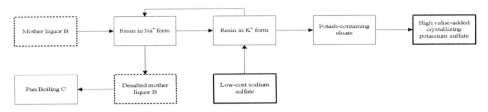

Figure 1. Schematic diagram of recycling potash from mother liquor B.

At present Moving Boundary Model was used to describe ion-exchange process, in addition the ion-exchange process was affected by 3 rate-controlling processes, liquid membrane diffusion, particle diffusion and chemical reaction. Among 3 processes, the slowest process was the main rate-controlling step (Xu, Z.G., 2010). The 3 rate-controlling processes in Moving Boundary Model were described using the following 3 formula (Zhang, P.J., 2010):

Liquid Film diffusion: $\ln(1-F) = -kt$ (2)

Particle diffusion: $1 - 3(1-F)^{2/3} + 2(1-F) = k_0 r^{-2} C_0^n \exp\{-\frac{E_a}{RT}\} \int_0^t \frac{C_i}{C_0} dt = kt$ (3)

Chemical reaction: $1 - (1-F)^{1/3} = \frac{K_0 C_0 t}{r} = kt$ (4)

To decide the rate-controlling step of ion-exchange reaction, experimental data were testified using afore-said modules.

3 RESULTS AND DISCUSSION

3.1 Scheme of recycling potash from mother liquor B

Fig. 1 shows the scheme of recycling K^+ from mother liquor B in beet sugar industry. K^+ in mother liquor B is concentrated on resin in Na^+ form, the resin is transformed into K^+ form, and the desalted mother liquor B flow to Pan Boiling C. Then K^+ on the resin is eluted using Na_2SO_4, the resin is transformed into Na^+ form again for the next cycle of adsorbing K^+ operations, and potassium sulfate is crystallized out of potash-containing eluent. In the process, the great advantages lie in 2 aspects. One is that high value-added potash is recycled using cheap Na_2SO_4. The other is the melassigenic coefficient of K^+ is greater than that of Na^+, the quality of mother liquor B in beet sugar is also improved by replacing K^+ with Na^+.

3.2 Adsorption of K^+ by resin BK001

Previous work discussed the optima of the dynamic adsorption of K^+ in mother liquor B by BK001 resin in Na^+ form (Zhu, S.M., 2011). The optimal conditions were as follows: the brixture of mother liquor B 50° Bx, reaction temperature 60°, flow rate 2 BV/h, the ratio of height to diameter (H/D) for resin BK001 in the column 8:1.

To test the optima and realize the extraction of K^+, an ion exchange column ($\Phi 24 \times 240$ mm) was filled with 90 ml of the BK001 resin in Na^+ form. The mother liquor B was introduced at the top of the column after being diluted with water to 50° Bx, heated to 60° with agitation (60 rpm), and the associated foam removed. The operating flow rate was 2 BV/h. The content of K^+ was determined in every 30 ml of eluate until the content remained constant (Zhu, S.M., 2011).

Under the optimal conditions, the kinetic adsorption curve is typical, as shown in Fig. 2. The breakthrough exchange capacity of resin BK001 is 40.4 mg/ml wet resin. Resin in Na^+ form was transformed into K^+ form while extracting K^+ from mother liquor B.

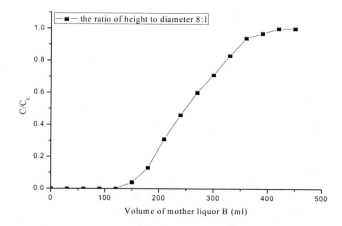

Figure 2. Effect of the kinetic adsorption curve of K^+.

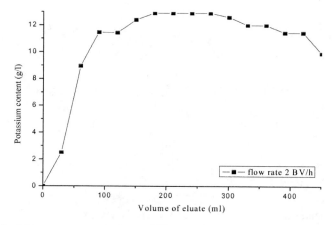

Figure 3. Effect of flow rate of eluate on the elution curve of K^+.

3.3 Elution of K^+

As an assumption, resin BK001 in K^+ form can be regenerated using high value counter ions like SO_4^{2-} (Bento L.S.M. 1997). To carefully verify the assumption, 10.6 g Na_2SO_4 was dissolved in 0.45 l water as regenerants. The bed volume (BV) of resin was 90 ml. The elution conditions are as follows: elution temperature 80°C, flow rate 2BV/h. The elution curve of K^+ is shown in Fig. 3, the K^+ in the eluate increased firstly and then decreased within 450 ml of eluate volume, like a typical parapola. The K^+ content in eluate reaches the maximum value 12.9 g/l when 180 ml of regenerant is used, and keeps at the content till 270 ml of regenerant is used. Collecting and mixing the 450 ml eluate, the content of K^+ is 5.03 g, equal to 98.6% of total potassium adsorption capacity 5.10 g by 90 ml of resin BK001.

So K^+ can be eluted by Na_2SO_4. Generally speaking, the regenerant of ion exchange resin is the mixture of 10% NaCl and 0.5% NaOH (Xu, Z.G., 2010). However, resin BK001 in Na^+ form can be regenerated by 2.36% of Na_2SO_4. The reason for it lies in the following: resin BK001 is a strong acid cation-exchange resin, a potassium adsorption layer is around the active group of the exchange agent because of the adsorption of K^+. During the desorption of K^+, the SO_4^{2-} in regenerating agent can attract K^+, and form a complex of $K-SO_4^-$ under the interference of water molecules, enabling Na^+ in liquid phase to approach the fixed-bed ion exchange active group on

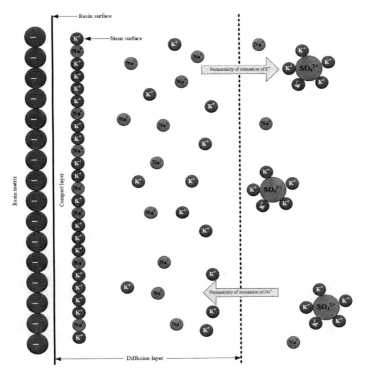

Figure 4. Double electrical layer model of regenerating resin using high valence state counter ions.

the skeleton frame of exchange agent to form the potential difference between two phases. Then the ion exchange between K^+ and Na^+ occurs and the adsorbed K^+ is replaced (see Fig. 4).

In short words, the integration of separation and coordination reaction of K^+ on BK001 resin ascribes to two facts: (1) Ion exchange—the content of Na^+ in the eluent is high enough to prompt the exchanging with K^+ on resin, i.e. to facilitate the exchange reaction balance K-R (resin matrix)+ $Na^+ \rightleftharpoons Na\text{-}R + K^+$ move rightward; (2) Coulomb attraction and salt-forming reaction—since K^+ is easier to bonded with SO_4^{2-} than Na^+, the coulomb attraction between the SO_4^{2-} in the eluent and K^+ on resin facilitates the elution of K^+ in exchange agent (Zhu, S.M., 2010; Bento L.S.M.,1997).

3.4 Crystallization of inorganic potash

The eluent containing 0.26~0.32 mol/l of K^+ was collected, then was concentrated to 2.08 mol/l of K^+ at 80°C. Then potash crystals were prepared through 3 crystallization methods: natural cooling crystallization, fast cooling crystallization and heat preservation crystallization. The crystallization time and mother liquor volume were 2h and 200 ml respectively in each process. Finally, the potash crystals were filtrated and dried by blowing under 50~53°C for 12h. Natural cooling crystallization meant that crystallization mother liquor at 80°C was cooled naturally to room temperature of 25°C and potash crystallized at 25°C for 2h. Fast cooling crystallization meant that crystallization mother liquor at 80°C was cooled fast using ice water and potash crystallized at 0°C for 2h. Heat preservation crystallization meant that crystallization mother liquor at 80°C was cooled to 40°C and kept at 40°C for 2 h of potash crystallization. To select a feasible crystallization method for industrialization amplification, the crystal form, crystal appearance, particle size distribution and crystal purity of crystallization K_2SO_4 from 3 crystallization methods were compared in Table 1 and Fig. 5.

According to Table 1, the sequence of crystallized raw K_2SO_4 crystals mass for 3 crystallization methods is fast cooling crystallization method > natural crystallization > heating preservation

Table 1. Effect of crystallizing methods on the yield, purity, feature and particle size distribution.

Crystallization method	natural cooling crystallization	Heat preservation crystallization	Fast cooling crystallization
K_2SO_4 (g)	11.96	8.96	16.25
K_2SO_4/Na_2SO_4 (%)	91.81	87.40	87.97
Percentage of greater than 20 mesh beads (%)	78.05	69.54	10.68
Percentage of 20 to 40 mesh beads (%)	15.94	26.51	43.20
Percentage of 40 to 60 mesh beads (%)	3.36	2.54	19.13
Percentage of less than 60 mesh beads (%)	2.54	1.51	26.81

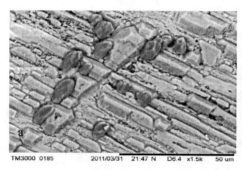
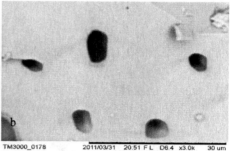

a Heat preseavation crystallization (×1500)   b Heat preseavation crystallization(×3000)

Figure 5. Microcosmic feature comparison of crystal plane of K_2SO_4 from different crystalling methods.

crystallization. The solubility of K_2SO_4 at 0°C, 20°C and 40°C is 7.4 g, 11.1 g and 14.8 g respectively, so the supersaturation degree of crystallization mother liquor varies following the variation of crystallization temperature.

The order from high to low of purity for K_2SO_4 crystals from different crystallization method is natural cooling crystallization, fast cooling crystallization and heat preservation crystallization. The highest purity of K_2SO_4 for the natural cooling crystallization method is 91.81%. The result in Table 1 shows that K_2SO_4 crystals contain a few Na_2SO_4 crystals, which may ascribe to remaining 4.2% (w/w) of Na_2SO_4 in crystallization mother liquor. Although the solubility of Na_2SO_4 at zero centigrade is 4.9 g, the Na_2SO_4 crystal may be embedded, adsorbed or adhered while K_2SO_4 crystallizing. Especially for heat preservation crystallization, many cavities lie in the surface of K_2SO_4 crystals, facilitating the embedding of impurities, as is shown in Fig. 5.

According to Table 1 and Fig. 5, particle size of K_2SO_4 crystals from different crystallization methods changes in the following order: natural cooling crystallization > heat preservation crystallization > fast cooling crystallization. Especially for natural cooling crystallization, 94.0% of K_2SO_4 crystals are greater than 40mesh, superior to commercial K_2SO_4 (Zhu, S.M., 2011; Zhang, P.J., 2010).

In addition, among 3 crystallization methods, K_2SO_4 crystals of natural crysallization methods have the largest particle size and highest purity, yet its crystals yield is in between. K_2SO_4 crystals of heat preservation crystallization have the smallest yield, lowest purity because of cavity embedding impurities (see Fig. 5), yet its particle size is in between; K_2SO_4 crystals of fast cooling crystallization have the smallest particle size and largest crystals yield, yet its purity is in between.

According to the yield, particle size, particle shape, purity and particle size distribution of K_2SO_4 crystals from different crystallization methods, among 3 crystallization methods the fast cooling crystallization method is the most likely for industrial application. Although the particle size of K_2SO_4 crystals is less for fast cooling crystallization, the yield is the most largest compare to other

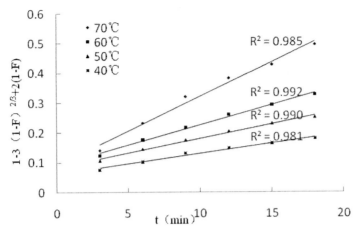

Figure 6. Fit between t with $[1 - 3(1 - F)^{2/3} + 2(1 - F)]$ under different temperature.

crystallization methods, and the purity of fast cooling crystallization is close to that of natural cooling crystallization.

3.5 Adsorption kinetics of K^+ by resin BK001

3.5.1 Effect of temperature on ion exchange process

The adsorption kinetics of K^+ by resin BK001 at 40°C, 50°C, 60°C and 70°C were studied. Formula (2), (2) and (3) were respectively used to fit test data. Results showed that, the formula $[1 - 3(1 - F)^{2/3} + 2(1 - F)]$ has better linear correlation to adsorption time. So the ion exchange process is controlled by particle diffusion. As shown in Fig. 6, the adsorbing reaction rate of K^+ increases as temperatures warm because of the increase in molecular motion.

The apparent rate constant k of potassium-adsorbing process at different temperature K^+ can be calculated according to the slope of each straight line in Fig.6. The calculations are substituted in the Arrihenius equation in logarithmic forms to calculate the slope "E_a/R" and intercept "A".

$$\ln k = \frac{E_a}{RT} + A \tag{7}$$

Calculated data of "ln k" are plotted versus "1/T" and results are shown in Fig. 7. According to Fig. 7. The linear fitting equation is $\ln k = 8.416 - 4.214 \times 10^3/T$ according to calculations of "ln k" and "1/T" in Fig. 7, the correlation coefficient is 0.958. Kinetic parameters have been calculated respectively from experimental data according to Equation (7): apparent activation energy (E_a) and apparent frequency factor (K_0) of the reactions. The calculated apparent activation energy E_a is 35.1 kJ/mol, indicating the potash-adsorbing process is easy. The calculated apparent fluency factor K_0 is 4.52×10^3 min^{-1}, showing the reaction rate constant under the interaction among other factors excepting the content of K^+.

3.5.2 Effect of K^+ concentration on ion exchange process

When the content of K^+ in mother liquor B were 0.047 mol/l, 0.071 mol/l and 0.156 mol/l respectively, the ion exchange process of K^+ on resin BK001 were studied. Formulae (3) was used to fit test data, and results are shown in Fig. 8. The correlation coefficients are all greater than 0.96 in Fig. 8, further testifying the ion exchange process is controlled by particle diffusion. As shown in Fig. 8, the ion-exchange rate increases with increasing content of K^+ in mother liquor B. The reason for it is that the reaction equilibrium of $K^+ + R\text{-}Na \rightleftharpoons R\text{-}K + Na^+$ shifts towards right according to related chemical reaction equilibrium theory (Eszterle M., 2009).

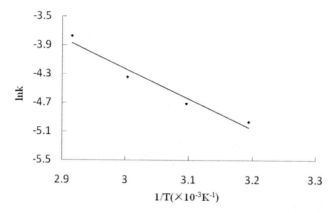

Figure 7. Diagram of Arrhenius.

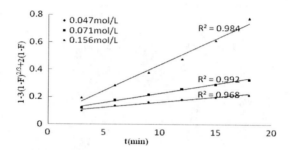

Figure 8. Fit between t with $[1 - 3(1 - F)^{2/3} + 2(1 - F)]$ under different K^+ Concentrations.

Table 2. Concentrations of Mother Liquor B and apparent rate constant.

$[K^+]$ (mol/l)	Lg$[K^+]$	K × 10^3	lg k
0.047	−1.328	7	−2.15
0.071	−1.149	13	−1.89
0.156	−0.807	37	−1.43

The apparent rate constant of ion-exchange reaction under different concentration of mother liquor B can be calculated according to different straight lines' slopes in Fig. 8. The relation among mother liquor B concentration and apparent rate constant is as shown in Table 2.

The apparent rate constant "k" is the slope of a straight line in Fig. 8. In addition, the apparent rate constant is proportional to a power function of the content of K^+ (see Equation (8)).

$$\lg k = b + n \lg C_0 \qquad (8)$$

In Equation (8), n is the reaction order of adsorption reaction, and C_0 is the content of K^+. The equation is $\lg k = 1.389 \lg [K^+] - 2.514$ after having fitted experimental data according to Table 2 and Equation (8). The reaction order of the adsorption process of K^+ is 1.389, and its linear correlation coefficient is 0.999, showing that the ion-exchange reaction rate increases with increasing K^+ concentration in mother liquor B.

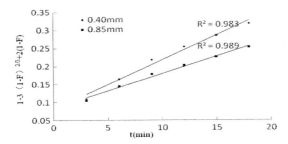

Figure 9. Fit between t with $[1 - 3(1 - F)^{2/3} + 2(1 - F)]$ under different resin particle size.

3.5.3 Effect of resin grain size on ion exchange process

To get the total kinetic equation, the adsorption process of K^+ on 0.4mm and 0.85mm particle size of resin BK001 was also studied respectively. Formulae (3) was used to fit test data, and results are shown in Fig. 9. The two correlation coefficients are all greater than 0.98 in Fig. 9, showing Equation (3) can fit the exchanging process well. The less diameter resin possesses larger contacting specific surface area, favoring the adsorption process of K^+ on resin BK001. So the ion-exchanging reaction rate increases with decreasing resin particle size.

3.5.4 Dynamics equation of ion exchange process

Given the above, the apparent rate constant "k" can be expressed using the following equation:

$$k = k_0 r_0^{-2} C_0^{1.389} \exp[-\frac{35100}{RT}], \quad k_0 = 8439 \tag{9}$$

So the total kinetic equation of the ion-exchanging reaction of K^+ in mother liquor B on resin BK001 is expressed as follows:

$$[1 - 3(1 - F)^{2/3} + 2(1 - F)] = 8439 \times r_0^{-2} [K^+]^{1.389} \exp[-\frac{35100}{RT}] \int_0^t \frac{C}{C_0} dt$$

4 CONCLUSIONS

A new idea was raised and tried to extract potash from mother liquor B and meanwhile promote the sugar recovery rate. The adsorption capacity of K^+ in 50° Bx mother liquor B on resin BK001 was 40.4 mg/ml wet resin under the optimal conditions of 60° of operating temperature and at a flow rate of 2BV/h. After enrichment and purification with ion-exchange resin, the elution efficiency of total potash was higher than 98.6% using 450ml of 0.166 mol/l Na_2SO_4 at 60° and a flow rate of 2BV/h. The existance of SO_4^{2-} promoted the elution of K^+ on resin BK001 compared to Cl^-, and the cause of eluting K^+ using Na_2SO_4 is the complex capacity between K^+ and SO_4^{2-} and the exchanging capacity K^+ and Na^+. Among 3 crystallization methods for recycling potash from the resin eluate, the crystallized raw potash mass is the largest for the fast cooling crystallization method, the potash purity is the highest for the natural cooling crystallization, and the fast cooling crystallization method is the most likely for industrial application for its highest yield and higher purity of potash. For the kinetics of adsorbing K^+, the ion exchange process is controlled by particle diffusion, the reaction order 1.389, the apparent energy of activation 35.1 kJ/mol, and the kinetics model was established as follows:

$$[1 - 3(1 - F)^{2/3} + 2(1 - F)] = 8439 \times r_0^{-2} [K^+]^{1.389} \exp[-\frac{35100}{RT}] \int_0^t \frac{C}{C_0} dt$$

So the idea of extracting and recycling K$^+$ was testified feasible without environmental pollution. The potash-recycling method raised the quality of the mother liquor B and show potential for improving the sugar yield and remitting for potash deficiency in China (Grbic J. 2008).

ACKOWLEDGEMENTS

This study was supported by "The National Natural Science Fundation" (NO. U1203183) and the Fundamental Research Funds for the Central Universities (NO. 2014ZZ0050).

REFERENCES

Bento L.S.M., 1997. Regeneration of decolorizing ion exchange resins: a new approach. Zucherind 122:304
Elmidaoui, A., Chay, L., Tahaikt, M., Menkouchi Sahli, M.A., Taky, M., Tiyal, F., Khalidi, A., Hafidi, My.R., 2006. Demineralisation for beet sugar solutions using an electrodialysis pilot plant to reduce melassigenic ions. Desalination 189:209.
Eszterle M., Gryllus E., 2009. Equilibrium state of the Ca/alkali (K, Na) ion exchange in model and factory solutions reveal the chemistry of the regeneration. Zuckerind 134:75.
Grbic, J., Jevtic-Mucibabic, R., Dokmanovic, N., 2008. The effect of nonsucrose compounds on sucrose crystal growth rates. Zuckerind 133:699.
Greenwood, N.N., Earnshaw, A., 1984. Chemistry of the Elements. Oxford etc.: Pergamon press
Gryllus, E., Delavier, H.J., 1975. BMA-Zsigmond-Gryllus-Verfahren, ein neues Verfahren zur Dunnsaftenkalkung. 2. halbindustrielle und Fabrik–Versuche. Zuckerind 25:554.
Gryllus, E., Eszterle, M., Anyos, E., 1998. The RDN regeneration method for decalcification of thin juice. Zuckerind 123:36
Oldfield, J.F.T., Shore, M., Gyte, D.W., 1979. Quentin process [J]. International sugar journal, 81:103.
Xu, Z.G., Wu, Y.K., Zhang, J.D., 2010. Equilibrium and kinetic data of adsorption and separation for zirconium and hafnium onto MIBK extraction resin. Trans Nonferrous Met. Soc. China 20:1527.
Zhang, P.J., Yu, S.J., Li, D., 2010. Kinetics of separating potassium ions from molasses vinasse by cation-exchange resin. Ion Exchange and Adsorption 26:439.
Zhu, S.M., Yu, S.J., 2011. 2011. International Conference on New Technology of Agricultural Engineering. Beijing: IEEE Press.
Zhu, S.M., Yu, S.J., Fu, X., Chai, X.S., Zhu, L., Yang, X.X., 2011. A modified Quentin method for desalting thin juice and mother liquor B in beet sugar industry. Desalination 268:214.
Zhu, S.M., Fu, X., Yu, S.J., Zhu, L., 2010. Coupled Regeneration of Decolourisation and Decalcification Resin in Sugar Industry and Reuse of Regenerated Waste Liquid. Journal of South China University of Technology (Natural Science Edition) 12:105.

Review of *Brettanomyces* of grape wine and its inhibition methods

Wenhua Lin, Xueyan You & Tao Feng
School of Perfume and Aroma Technology, Shanghai Institute of Technology, China

ABSTRACT: *Brettanomyces* produced a typically bad flavor containing 4-Ethylphenol (4-EP) and 4-Ethylguaiacol (4-EG) during winemaking process. The *Brettanomyces* flavor reduces the quality of grape wine, and causes serious economic losses in the grape wine industry. Prevention or inhibition for growth of *Brettanomyces* in wine involves aspects such as grapes quality, winery sanitation, control of sulfite, level of oxygen and humidity, use of uncontaminated barrels and so on. Metabolites of *Brettanomyces* also had an impact on acceptability and eating habit of people. This article reviewed *flavor characteristics of Brettanomyces and their formation, especially inhibition methods for undesirable flavor in wine.*

1 INTRODUCTION

Wine can easily produce some undesirable flavors during the process of aging, storage and transportation, thus reduce aroma and flavors of wine because of malic flavors being produced by microorganisms such as *Brettanomyces*. At the present, many researchers have studied some undesirable flavors in wine.

Brettanomyces yeast grew on the surface of grapes, however, it did not belong to the genera of yeasts found on the surface of grapes typically. Therefore, it was defined as *Brettanomyces* ssp. in the 1950s. *Brettanomyces* yeasts are responsible for special flavors including burnt plastic and bitter taste in wine or fermentation broth occurred and propagated after alcoholic fermentation during wine storage (Cynkar et al., 2007). *Brettanomyces* had large influence on wine across the world and reduced the flavor and sensory quality of wine, leading to serious economic loss for many wineries.

2 FLAVOR CHARACTERISTICS OF BRETTANOMYCES

Brettanomyces yeasts grew in the storage of wine in tank, barrel and bottle, contributing to the characteristic "bretty favor" which were described as smoke, creosote, plastic, burnt plastic, cow dung flavor and barnyard (Galafassi et al., 2011). A metallic bitterness can also be formed by them. Isovaleric acid, 4-ethyl phenol, 4-ethyl guaiacol and unidentified "burnt plastic" compound were responsible for bretty flavor in wine. 4-EP and 4-EG were responsible for poly phenols flavor or herbal and lilac taste or holiday spices flavor respectively. Concentration ratio between 4-EP and 4-EG had an important influence on wine, and the subtle diversity of 4-EP and 4-EG can be perceived by sensory evaluation teams and the consumers. And these flavors might completely cover varietal and regional flavor characteristics of wine so much that there is unpleasantly bitter in wine (Coulon et al., 2010). An inverse relation between fruity and bretty flavor perception can be showed by descriptive sensory evaluations. When presented at low concentration, bretty aromas that were often thought as a defect in some wine were considered as a positive attribute.

The bad flavor produced by *Brettanomyces* in wine have become hot issues across world. A Portuguese study of wine from different countries showed that concentration of 4-EP among over 25% of red wine has been more than threshold value 620 µg/ml. If concentration is higher than the threshold value, some customers will refuse to drink wine (Lesschaeve, 2007). The average concentration of 4-EP and 4-EG in wine influencing wine's flavor is 620 μg/ml and 140 μg/ml respectively, and the recognition threshold of 4-EP and 4-EG in wine is 770 μg/ml and 436 μg/ml respectively. The content and proportion of 4-EP and 4-EG in wine will entirely depend on the variety of grapes and wine types.

3 FORMATION OF FLAVOR CHARACTERISTICS OF BRETTANOMYCES

In wine, main purposes of *Brettanomyces* were to form volatile phenol compounds and "*Brettanomyces* characteristic flavor". *Brettanomyces* characteristic flavor was a special flavor formed by the reaction between volatile phenol and certain fatty acids. During the process of winemaking, the most critical period of volatile phenol production was aging, especially in barrels.

Volatile phenol compounds were the main indicators (p-coumaric acid, ferulic acid and caffeic acid) of *B. bruxeliensis* activity that should be responsible for off-flavors in red wine. The detailed volatile phenol formation way was as following: grape juice and grape fruit contained a lot of phenolic acids such as hydroxycinnamic acid, tartaric acid, anthocyanin, etc. and these phenolic acids were esterified into the corresponding esters, then generated the corresponding free acid under the action of cinnamyl ester enzymes. The decarboxylation of the free acid in hydroxy cinnamic acid (HCDC) formed the hydroxyl styrene substances (4-vinyl phenol, 4-vinyl guaiacol and 4-vinylcatechol) respectively, and then they were reduced to 4-ethylphenol (4-EP), 4-ethylguaiacol (4-EG) and 4-ethylcatechol by vinyl phenol reductase (Oelofse et al., 2008).

4 INHIBITION METHOD FOR BRETTANOMYCES BAD FLAVOR

Since *Brettanomyces* affected each wine producing region in the world and their quality, the research of bad flavor generated by *Brettanomyces* became a worldwide topic. And to suppress the bad flavor of *Brettanomyces* had a special significance for decreasing this kind of bad flavor worldwide and economic loss in wine industry.

Brettanomyces was highly susceptible to be contaminated by various elements from vineyard to wine filling, so many researchers and wineries spent much time to study the inhibition measures in wine industry. At present, inhibition methods reported were mainly as following: inhibition of production of *B. bruxellensis*; inhibition of *B. bruxellensis growth*; inhibition of the formation of flavor precursors. In the whole process of wine production, factors such as raw materials, equipment, alcohol content, sugar contents, acid contents, temperature, oxygen capacity and so on can be considered to suppress bad flavor of wine, and adding the inhibition reagents such as sulfur dioxide, caramel acid dimethyl (TM), ascorbic acid, benzoic acid etc. to the wine, which can't completely eliminate bad flavor of *Brettanomyces*, but can reduce concentration of bad flavor caused by *Brettanomyces*.

4.1 *Winery sanitation*

In the wine brewing process, if cleaning and sterilization of brewing equipment was not complete and contaminated oak barrel was used, *Brettanomyces* can exist in the stages of winemaking. So winemaking equipments must be demanded to clean and sterilize thoroughly, At the same time, the most important thing for winemakers was not to use barrels polluted by *Brettanomyces*. Researchers found that *B. bruxellensis* can penetrate barrel and reach to a depth of 8 mm in barrel where they also can grow and reproduce once barrels were polluted by *B. bruxellensis*, *Brettanomyces* yeast cannot be wiped out from barrels even after cleaning, sterilization, or even cutting barrels or other technical methods (Garde-Cerdán et al., 2008). For example, one treatment we used was to utilize ozone to treat polluted barrels, aiming to wipe out *Brettanomyces* but in fact just delay their growth.

4.2 Temperature

Some results showed (Brandam et al., 2007) the optimum temperature for the hydroxycinnamate decarboxylase or vinylphenol reductase activity of *B. bruxelluensis* D37 was 20 to 30°C. Reduced activity of enzymes was found at 15 to 20°C and inactive enzymes were seen at 30 to 40°C and 0 to 15°C. From the viewpoint of technology, flash-pasteurization at a temperature of 35 to 40°C treatments with wine can prevent wine from contaminating by *B. bruxelluensis* when aging in wooden barrels. The production rates of 4-ethylphenol of *B. bruxelluensis* was lower in temperatures from 16 to 30°C, so temperature affected production of 4-vinylphenol and 4-ethylphenol of *B. bruxellensis*. Researchers found that results of physical growth conditions of *B.bruxelluensis* strains isolated from wine was that more than one-third of the isolates could grow at 37°C, less than one-third grow at 10°C and the others grow at both 10°C and 37°C. According to these points, temperature of winemaking process should be controlled at 16 to 30°C, but fermentation temperature of *Saccharomyces* was another factor for wineries to consider, so fermentation temperature was generally between 20 and 30°C. Wineries kept stable fermentation temperature in this range, which can make the best wine. Therefore, wine industry controlled fermentation temperature at 20–30°C by using automatic refrigerating system, which not only made the best wine fermentation, but also effectively controlled emergency of bad flavor from *Brettanomyces*. Some researchers showed that wine aging shall be maintained during 10 to 12°C, and wine should be stored at 12–15°C, which can effectively inhibit growth and reproduction of *Brettanomyces*.

4.3 Alcohol content

Suarez et al. (Suárez et al., 2007) found a phenomenon that conversion rate of p-coumaric acid into 4-ethylphenol was higher when concentration of ethanol was low since *B. bruxellensis* had high tolerance to alcohol. According to statistics, tolerance ability to alcohol of *B. bruxellensis* reached 14–14.5%, which might be related to the capacity of *B. bruxellensis* to use ethanol as a carbon source. However, Benito et al. (Benito et al., 2009) described that *B. bruxellensis* seemed unable to convert p-coumaric acid into 4-EP when concentration of ethanol was more than 15% in the absence of all other limiting factors. It was suggested that alcohol with high concentration in wine could reduce production of 4-EP by *B. bruxellensis*. However, wineries considered similarly a high correlation between vitality of *Saccharomyces* and tolerance of alcohol (Vigentini et al., 2008). High concentration of alcohol in wine not only restrained the vitality of yeast, but also affected flavors of wine. Therefore, it was not reasonable only by increasing content of alcohol in wine to inhibit growth and reproduction of *Brettanomyces*, thus this method was rarely used in the process of wine production.

4.4 Acid content

Some studies showed that winemakers would select sour grapes as raw material or add some acid to wine to restrain growth and reproduction of *Brettanomyces*, and for example we knew that sorbic acid was widely used as an effective antifungal agent. Benito et al. (Benito et al., 2009) reported that *Brettanomyces* can be inhibited with a concentration of 900 ml/L of sorbic acid when pH was 3.6, but the fact was that content of sorbic acid just was 250 mg/L under the legal limiting condition. Similarly, some researchers indicated that some isolates grew at pH 2.5 and 94% of them grew at pH 2.0, which suggested that *B. bruxellensis* had a strong resistance to sorbic acid. At the same time, it was well-known that benzoic acid was widely used in soft drinks at concentrations between 100 and 200 mg/L, for inhibiting the growth of *Brettanomyces*, Benito et al. (Benito et al., 2009) reported that benzoic acid was an effective inhibitor of *Brettanomyces* under legal dose which was between 150 and 200 ml/L at a pH of 3.6 in musts, but the use of benzoic acid was not authorized and allowed in wine. Some reported that *Brettanomyces* was called acetate yeasts, which can grow in the whole process of alcoholic fermentation under the condition of allowable low sugar. Because they can overwhelm and replace main *Saccharomyces* to slow down the growth of *Saccharomyces* by producing acetic acid. However, there were also some studies that inhibiting growth of *B. bruxellensis* had nothing to do with the concentration of acid in wine body. Therefore, few wineries singly used these methods to inhibit bad flavors of wine.

4.5 Oxygen and moisture

According to the study (Röder et al., 2007), *B. bruxellensis* produced 7.2 g/L acetic acid under aerobic and grape juice culture conditions for 12 months. This result illustrated that there was a high correlation between content of oxygen and *B. bruxellensis*. Others investigated the effect of oxygen on growth of *B. bruxellensis*, acetic acid and production of ethanol and found that complete aerobiosis produced much acetic acid and little ethanol. Once over the oxygen transfer rate (OTR) of 105 mg O_2 $l^{-1} \cdot h^{-1}$, production of ethanol from *saccharomyces cerevisiae* inhibited glucose fermentation strongly, but acetic acid would be produced greatly, which suggested that the higher content of oxygen, the faster growth and production of acetic acid. And other studies showed that 4-Ethylphenol and 4-Ethylguaiacol of *Brettanomyces* metabolites in wine were positively correlated with the content of oxygen in liquor body, but negatively with humidity of cellar (Rayne and Eggers, 2008). Oxygen can promote growth of *B. bruxellensis*, and humidity can restrain growth of *Brettanomyces* due to lack of wine evaporation and certain contact between wine and oxygen. Studies found that manufacturing techniques such as crushing, juice separation and filtering could avoid exposing to the air directly, especially when nitrogen was added to clean up oxygen in wine aging process. Similarly, moisture of barrels should be set at 55% at the same time, which was beneficial to reduce the accessible area of wine and oxygen and inhibit growth of *B. bruxellensis*. Final humidity should be set and kept at 70% in the product storage environment. Technology such as barrels of bottled wine, eliminating oxygen with nitrogen etc. was much more popular.

4.6 Addition of sulfur dioxide

Some researchers found that plenty of sulfur dioxide was added to fermentation tank and maintained at a certain concentration in fermentation process to inhibit growth of harmful microbes such as *B. bruxellensis*. For example, there were increasingly many wineries in Australia using different levels of free sulfur dioxide in recent years (Godden and Gishen, 2005). and main preservative to inactivate *B. bruxellensis* is sulfur dioxide in wineries. The reason might be that a certain concentration of sulfur dioxide can bring *B. bruxellensis* into a state of survival but not breeding. The viable but non cultivable state (VBNC) was an adaptive strategy adopted by *B. bruxellensis* to resist stress existing in external environment. Then their cells may be able to recover once optimal conditions were offered (Serpaggi et al., 2012). If sulfur dioxide was used up in the aging process, harmful yeasts would still continue to grow in wine because oak barrels remained a nutritious place for slow-growing *B. bruxellensis* (Agnolucci et al., 2010). Therefore, wineries should add sulfur dioxide to fermentation tank and maintain a certain concentration in whole process of the winemaking. For example, many winery trials showed that blooms of *B. bruxellensis* were only prevented at the concentration of about 40 mg/L free sulfur dioxide in red wine, with 13.8% (v/v) ethanol and pH 3.42, matured in oak barrels (Barata et al., 2008). There is a phenomenon that *Brettanomyces* was most commonly identified in red wine but less frequently isolated in white wine (Dias et al., 2003), because white wine lacked precursor compounds which made them seem to have characters of *Brettanomyces* aroma and the loss of viability in white wine was largely due to the efficacy of sulfur dioxide at low PH. Generally speaking, SO_2 was one of the most effective inhibitors for *Brettanomyces*. And it was a suitable and mature method by adding sulfur dioxide to wine body during wine brewing process for its capacity to reduce the concentration of bad flavor of *Brettanomyces*.

4.7 Hydroxy cinnamic acid decarboxylase (HCDC)

Benito studied that hydroxy cinnamic acids (HCAs) decarboxylation with different activity from different *Brettanomyces* were used to ferment the original grape juice with full cumaric acid and enocianina. The purpose of them was to increase the production of styryl pyran cyanine, steady color and lustre of wine and reduce production of 4-EP. Morata et al. (Morata et al., 2013) studied the combined effect by using cinnamyl esterases (CE) and Hydroxy cinnamic acid decarboxylase + *saccharomyces cerevisiae* strains on the reduction of 4-EP production in red wine and found the result was that the lower production of 4-EP than not use. Using CE and HCDC + yeasts in wines can minimize content of ethylphenol precursors and reduce production of 4-EP due to CE can reduce the form of tartaric esters of HCAs in grapes to free acids and HCDC+ yeasts can transform hydroxycinnamic acids (p-coumaric acid) into stable vinylphenolic pyranoanthocyanins (VPAs) pigments during fermentation. It was a natural strategy to reduce the formation of 4-EPs in wine contaminated by *B. bruxellensis*. And Duncan et al (Harris et al., 2008) studied and found that optimal condition for hydroxycinnamate decarboxylase enzyme activity was 40°C and pH 6.0. Activity of enzymes could be enhanced by EDTA, Mg^{2+} and Cr^{3+}, while Fe^{3+}, Ag^+ and it can be inhibited by SDS completely. The strong activity of hydroxy cinnamic acid decarboxylase can make HCAs decarboxylization generate styrene, and styrene combined with anthocyanins can generate stable pyran anthocyanins. Styryl pyran cyanine can make color and lustre of wine more stable and the production of 4-EP and 4-EG less, and reduce concentrations of 4-EP and 4-EG (Agnolucci et al., 2009).

4.8 *Other methods*

It was essential for wineries to reduce production of off-flavors and inhibit the growth and reproduction of *B. bruxellensis* in the fermentation process. Some studies showed that adding clarified protein to the original grape juice can significantly reduce levels of *B. bruxellensis*. The amount of yeast and bacteria can be reduced largely after filtration and yeasts in a state of survival and not breeding were unable to get through the 0.45 μm filter membrane. Although filtration and clarification were generally considered to bring bad influence on quality of wine, the process can reduce level of *Brettanomyces* from the physical point of view (Conterno et al., 2006). Other method in wine storage was that wineries often added caramel dimethyl acid (Velcorin TM) to wine because Velcorin TM made *B. bruxellensis* inactive and other yeasts may start fermentation process again if wine contained some sugar or alcohol (Oelofse et al., 2008). Other studies show that *B. bruxellensis* lost the ability of reproduction at high concentration of glucose due to the fact that they cannot tolerate osmotic pressure (Röder et al., 2007). There were studies also reported that use the esterified cellulose can deplete 4-EP and 4-EG in wine (Larcher et al., 2012) and antimicrobial action of chitosan can restrain *Brettanomyces* in mixed culture fermentations.

At present, the most effective method would use a combination of these methods rather than use a single. For example, the growth of *Brettanomyces* can be reduced by cleaning and sterilizing equipment and raw materials, using barrels not polluted by *Brettanomyces* yet, adding proper amount of sulfur dioxide, unwinding barrel wine and utilizing seed culture medium etc.

5 CONCLUSIONS

Brett-associated metabolites in wine could be both negative and positive, which depended on their concentrations, personal preferences and of consumers' expectations. At low concentrations, compounds can contribute to complexity of wine aroma, but at high concentrations, these compounds can become an unpleasant odor for people. In a word, it was difficult to control these compounds in wine, as yeast can be found on grapes, in wineries, in barrels and bottled wines. Therefore more studies required to set up a systematic method to effectively prevent and control conditions in wine, and a series of quality and safety analysis methods in winemaking process, from the growing of grapes to final products.

REFERENCES

Agnolucci, M., Rea, F., Sbrana, C., Cristani, C., Fracassetti, D., Tirelli, A. & Nuti, M. 2010. Sulphur dioxide affects culturability and volatile phenol production by *Brettanomyces/Dekkera bruxellensis*. *International journal of food microbiology,* 143, 76–80.

Agnolucci, M., Vigentini, I., Capurso, G., Merico, A., Tirelli, A., Compagno, C., Foschino, R. & Nuti, M. 2009. Genetic diversity and physiological traits of *Brettanomyces bruxellensis* strains isolated from Tuscan Sangiovese wines. *International journal of food microbiology,* 130, 238–244.

Barata, A., Caldeira, J., Botelheiro, R., Pagliara, D., Malfeito-Ferreira, M. & Loureiro, V. 2008. Survival patterns of *Dekkera/bruxellensis* in wines and inhibitory effect of sulphur dioxide. *International journal of food microbiology,* 121, 201–207.

Benito, S., Palomero, F., Morata, A., Uthurry, C. & Su Rez-Lepe, J. 2009. Minimization of ethylphenol precursors in red wines via the formation of pyranoanthocyanins by selected yeasts. *International journal of food microbiology,* 132, 145–152.

Brandam, C., Castro-Martinez, C., Delia, M.-L., Ramon-Portugal, F. & Strehaiano, P. 2007. Effect of temperature on *Brettanomyces bruxellensis*: metabolic and kinetic aspects. *Canadian journal of microbiology,* 54, 11–18.

Conterno, L., Joseph, C. L., Arvik, T. J., Henick-Kling, T. & Bisson, L. F. 2006. Genetic and physiological characterization of *Brettanomyces bruxellensis* strains isolated from wines. *American Journal of Enology and Viticulture,* 57, 139–147.

Coulon, J., Perello, M., Lonvaud-Funel, A., De Revel, G. & Renouf, V. 2010. *Brettanomyces bruxellensis* evolution and volatile phenols production in red wines during storage in bottles. *Journal of applied microbiology,* 108, 1450–1458.

Cynkar, W., Cozzolino, D., Dambergs, B., Janik, L. & Gishen, M. 2007. Feasibility study on the use of a head space mass spectrometry electronic nose (MS e_nose) to monitor red wine spoilage induced by *Brettanomyces* yeast. *Sensors and Actuators B: Chemical,* 124, 167–171.

Dias, L., Pereira-da-silva, S., Tavares, M., Malfeito-Ferreira, M. & Loureiro, V. 2003. Factors affecting the production of 4-ethylphenol by the yeast *Dekkera bruxellensis* in enological conditions. *Food Microbiology,* 20, 377–384.

Galafassi, S., Merico, A., Pizza, F., Hellborg, L., Molinari, F., Piškur, J. & Compagno, C. 2011. *Dekkera/Brettanomyces* yeasts for ethanol production from renewable sources under oxygen-limited and low-pH conditions. *Journal of industrial microbiology & biotechnology,* 38, 1079–1088.

Garde-Cerd N, T., Lorenzo, C., Carot, J., Jabaloyes, J., Esteve, M. & Salinas, M. 2008. Statistical differentiation of wines of different geographic origin and aged in barrel according to some volatile components and ethylphenols. *Food Chemistry,* 111, 1025–1031.

Godden, P. & Gishen, M. 2005. the trends in the composition of Australian wine from vintages 1984 to 2004. *Australian and New Zealand Wine Industry Journal,* 20, 21.

Harris, V., Ford, C. M., Jiranek, V. & Grbin, P. R. 2008. *Dekkera* and *Brettanomyces* growth and utilisation of hydroxycinnamic acids in synthetic media. *Applied microbiology and biotechnology,* 78, 997–1006.

Larcher, R., Puecher, C., Rohregger, S., Malacarne, M. & Nicolini, G. 2012. 4-Ethylphenol and 4-ethylguaiacol depletion in wine using esterified cellulose. *Food chemistry,* 132, 2126–2130.

Lesschaeve, I. 2007. Sensory evaluation of wine and commercial realities: Review of current practices and perspectives. *American journal of enology and viticulture,* 58, 252–258.

Morata, A., Vejarano, R., Ridolfi, G., Benito, S., Palomero, F., Uthurry, C., Tesfaye, W., Gonz Lez, C. & Su Rez-Lepe, J. 2013. Reduction of 4-ethylphenol production in red wines using HCDC+ yeasts and cinnamyl esterases. *Enzyme and microbial technology.*

Oelofse, A., Pretorius, I. & Du Toit, M. 2008. Significance of *Brettanomyces* and *Dekkera* during winemaking: a synoptic review. *S. Afr. J. Enol. Vitic.*

Röder, C., Knig, H. & Frhlich, J. 2007. Species-specific identification of *Dekkera/Brettanomyces* yeasts by fluorescently labeled DNA probes targeting the 26S rRNA. *FEMS yeast research,* 7, 1013–1026.

Rayne, S. & Eggers, N. J. 2008. 4-Ethylphenol and 4-ethylguaiacol concentrations in barreled red wines from the Okanagan Valley appellation, British Columbia. *American Journal of Enology and Viticulture,* 59, 92–97.

Serpaggi, V., Remize, F., Recorbet, G., Gaudot-Dumas, E., Sequeira-Le Grand, A. & Alexandre, H. 2012. Characterization of the "viable but nonculturable"(VBNC) state in the wine spoilage yeast *Brettanomyces*. *Food microbiology,* 30, 438–447.

Suárez, R., Suárez-Lepe, J., Morata, A. & Caldern, F. 2007. The production of ethylphenols in wine by yeasts of the genera *Brettanomyces /Dekkera*: A review. *Food Chemistry,* 102, 10–21.

Vigentini, I., Romano, A., Compagno, C., Merico, A., Molinari, F., Tirelli, A., Foschino, R. & Volonterio, G. 2008. Physiological and oenological traits of different *Dekkera/Brettanomyces bruxellensis* strains under wine-model conditions. *FEMS yeast research,* 8, 1087–1096.

Extraction technology of crude polysaccharide from *Stropharia rugoso-annulata*

J.C. Chen, P.F. Lai, M.J. Weng & H.S. Shen
Agricultural Product Processing Research Centre, Fujian Academy of Agricultural Sciences, Fuzhou, Fujian, P.R. China

ABSTRACT: In this paper, aiming at optimization of the extraction of *Stropharia rugoso-annulata* crude polysaccharides, test factors and levels were firstly selected by single-factor tests. On the basis of previous results and according to the central composite test design principles, the method of response surface methodology with 3 factors and 3 levels was adopted. Response surface and contour are finally graphed with the extraction rate of *S. rugoso-annulata* crude polysaccharides as the response value. The results showed that the optimum condition described were as follows: extraction temperature was 75°C, extraction time was 20 min and solid-liquid ratio was 1:25 (g/mL). Under the optimized condition, the yield of crude polysaccharide from *S. rugoso-annulata* could reach 7.614%.

1 INTRODUCTION

Stropharia rugoso-annulata belongs to the Basidomycota, Hymenomycetes, Agaricales, Strophariaceae and Stropharia (Hawksworth 1995; Xu 2006). There are some studies showing that *S. rugoso-annulata*, one of the famous artificial cultivation of edible fungi in Europeans and Americans, has a high nutritive and developing value (Duan 2010; She et al, 2007; Wang et al, 2007). It was recommended to developing countries by FAO (Chen et al, 2010; Huang et al, 2012; Pang et al, 2007).

Fungal polysaccharide is an active polysaccharide, controlling division, differentiation, development and caducity in cell. It has a huge potential for development and the application prospect is extremely broad (Ye et al, 2007). Experimental investigation shows that polysaccharide from *S. rugoso-annulata* is mainly of pyranose, composed of 5 kinds of monosaccharides, including D-fructose, D-glucose and D-xylose, held together by α and β glucosidic bond. It has antitumor, antioxidation and alleviate fatigue effects and can effectively remove superoxide anion free radical and hydroxyl radical in vitro, while it also has some repairing effect on damaged organs of rat (Chen et al, 2011; Wang et al, 2007; Tao et al, 2007).

In this paper, we reported an extraction technology of crude polysaccharide (CP) from *S. rugoso-annulata* by Response Surface Methodology (RSM). The aim of this work is to provide the theory basis and the technical support for the further research of functional activity and development of products on CP.

2 MATERIALS AND METHODS

2.1 *Materials and chemicals*

S. rugoso-annulata was purchased from Sanming City, Fujian Province, China. Phenol, Sulfuric acid, Chloroform, n-butanol and Anhydrous ethanol are of analytical grade.

Table 1. Levels and code of variables chosen for Box-Benhnken design.

Factors	Levels		
	−1	0	1
x_1 Temperature/°C	60	70	80
x_2 Time/min	10	20	30
x_3 Ratio (g/mL)	1:20	1:30	1:40

2.2 Instrument and facility

High speed freezing centrifuge (Type: 5810R, Eppendorf AG), Uv-vis spectrophotometer (Type: TU-1901, Beijing general instrument co., LTD) were used.

2.3 Methods

Extraction: Fruit body was dried and grinded (100 mesh). The powder was put into a baffled flask with distilled water and then put the flask into the water bath mixer (600 rpm), the suspension was centrifuged (8000 rpm, 3 min), the sediment was extracted twice with the same method, supernatants of the 3 centrifuge were merged and concentrated at 50°C. Three volumes of ethanol were added for precipitation at 4°C overnight, the precipitation was obtained by centrifugation (8000 rpm, 20 min) and repeatedly washed with ethanol (80%) and acetone 3 times. The washed precipitation was redissolved with distilled water and removed free protein by Sevag method, and after ethanol precipitated and freezing dried, CP was obtained. The yield was calculated by weight.

Single factor design: Three main factors including ratio of solid to liquid, extracting time and temperature were studied by using single factor analysis method.

RSM experiment design: The optimal experiments were carried out on the factors including ratio of solid to liquid, time and temperature, with yield for response under 3 factors and 3 levels tests. The table of experiments factors and levels was listed below (Table 1).

3 RESULTS AND DISCUSSION

3.1 Effect of temperature on the yield of CP

(2.00 g) of mushroom powder was weighted, the parameters were set as temperature (40~90°C) ratio (1:50 (g/mL)) and time (30 min). Fig. 1 showed that the temperature and yield of CP was positive correlation.

3.2 Effect of time on the yield of CP

(2.00 g) of mushroom powder was weighted, the parameters were set as time (5~50 min) ratio (1:50 (g/mL)) and temperature (65°C). Fig. 2 showed that the yield was reached the maximum value, as the time was 30 min; There was not significant change over time.

3.3 Effect of solid-liquid ratio on the yield of CP

(2.00 g) of mushroom powder was weighted, the parameters were set as solid-liquid ratio (1:20~1:70 (g/mL)) time (30 min) and temperature (65°C). Table 2 indicated that the yield first increased then decreased with increasing the solid-liquid ratio, and the maximum value was on 1:40. Generally, the yield increases with the increase of solid-liquid ratio, but the costs of following process, concentration, would be risen rapidly. In view of the above, at the same time ensure the effect of extraction, the usage of water and the cost of concentration would be reduced.

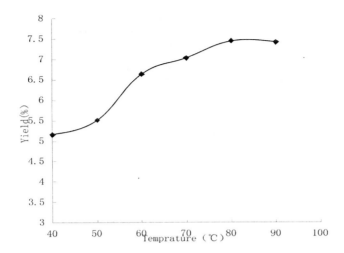

Figure 1. The effect of temperature on the extraction of CP.

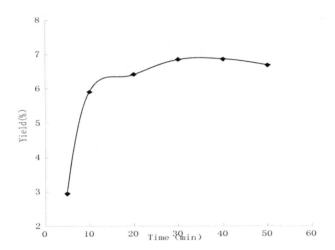

Figure 2. The effect of time on the extraction of CP.

Table 2. The effect of solid-liquid ratio on the extraction of CP.

Ratio (g/mL)	1:20	1:30	1:40	1:50	1:60	1:70
Yield/%	5.85	6.44	6.83	6.56	6.62	6.42

3.4 Results of RSM experiment

The results of RSM experiment were showed as Table 3.

3.5 Analysis of RSM

The quadratic regression equation of the yield of CP Y and temperature x_1, time x_2, ratio x_3 was expressed as follow: $Y = 7.37 + 0.61x_1 + 0.53x_2 + 0.52x_3 - 0.62x_1^2 - 0.60x_2^2 - 0.65x_3^2 - 0.03x_1x_2 - 0.12x_1x_3 - 0.10x_2x_3$. The model ($P < 0.0001$, lack of fit $P = 0.1754$) could be used to analysis and predict yield of CP. Fig. 3 showed that the extent of interaction was reflected by the

Table 3. Program and experimental results of RSM.

No.	x_1 Temp.	x_2 Time	x_3 Ratio	Yield/%
1	0	1	1	7.11
2	1	1	0	7.19
3	0	0	0	6.26
4	−1	0	1	6.12
5	−1	1	0	6.09
6	0	−1	−1	4.94
7	0	0	0	6.23
8	1	0	−1	6.34
9	0	1	−1	6.29
10	1	−1	0	6.29
11	−1	0	−1	4.83
12	0	0	0	6.19
13	0	0	0	6.18
14	0	−1	1	6.17
15	1	0	1	7.15
16	0	0	0	6.13
17	−1	−1	0	5.07

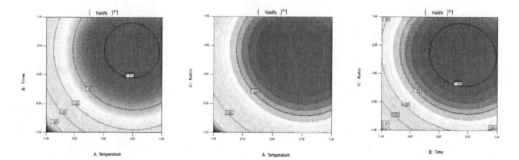

Figure 3. Response surface and contour plots of the interactions between two factors to the yield of CP.

shape of contour line, and oval was significant, while circle was opposite (Nilsang et al, 2005; Yi et al, 2009).

3.6 Verification of regression equation

The theoretical maximum yield of CP was 7.637% with the parameter (73.8°C, 22.7 min, ratio 1:26.2) calculated from the equation. The verifying parameter (75°C, 20 min, ratio 1:25) was adjusted according to the actual conditions. The verifying experiments were repeated 3 times. The results (yield = 7.614%) in keeping with predicted value indicated that the equation of regression had a high imitation degree.

4 SUMMARY

RSM was used to optimize the extraction of CP from *S. rugoso-annulata*. The model was tested in significant test by statistic method. The findings were modeled and predicted. The optimization of conditions were obtained. The results showed that the optimum condition described were as follows: temperature was 75°C, time was 20 min and solid-liquid ratio was 1:25 (g/mL). Under the optimized condition, the yield of CP from *S. rugoso-annulata* could reach 7.614%.

ACKNOWLEDGMENTS

The authors thank the Key Research Project Fund of Science and Technology Innovation Team from Fujian Provincial Department of Science and Technology (CXTD2011-16) for support.

REFERENCES

Chen, J.C. et al. 2010. "Extraction of flavonoids from *stropharia rugoso-annulata* farlow and the antibacterial activity", "Journal of Beijing Technology and Business University: Natural Science Edition", Vol. 28, No. 6, pp. 9–13. (in Chinese)

Chen J.C. et al. 2011. "Distribution of *stropharia rugoso-annulata* polysaccharide molecular weight and component sugar", "Scientia Agricultura Sinica, Vol. 44, No. 1, pp. 2109–2117.

Duan, Y.G. 2010. "Nutritional Quality Analyse, Extraction and Application of Flavonoids of Stropharia Rugoso-Annulata Farlow", "Thesis of Fujian Agriculture and Forestry University.

Hawksworth D.L. *et al.* 1995. "Ainsworth and bisby's dictionary of the fungi", "CAB International".

Huang, Q. *et al.* 2008. "Fungi Polysaccharides and it's Physiological Function", "Agricultural Product Resources" No. 5, pp. 36–40.

Huang C.Y. et al. 2012. "Study on optimal carbon sources for mycelia growt of *stropharia rugoso-annulata*", "Shandong Agricuhural Science", Vol. 44, No. 1, pp. 75–76.

Nilsang S. et al. 2005. "Optimization of enzymatic hydrolysis of fish soluble concentrate by commercial proteases", "Journal of Food Engineering", Vol. 70, No. 4, pp. 571–578.

Pang, Z.L. & Tian L., 2007. "Study on fermentation technical optimization of polysaccharide from *Stropharia rugoso-annulata*", "Jiangsu Agricultural Sciences", No. 1, pp. 157–158.

She, D.F. *et al*, 2007. "Analysis and comparison on the nutrients in Pleurotus nebrodensis and *Stropharia rugoso-annulata*", "Edible Fungi", No. 4, pp. 57–58.

Tao, M.X. et al. 2007. "Study on Antioxidant of Stropharia rugosoannulata Polysaccharide on Mouse Heart", "Food Science", Vol. 28, No. 0, pp. 529–531.

Wang, X.W. *et al.*, 2007. "Analysis on nutrient components and antioxidant components of *Stropharia rugoso-annulata*", "Edible Fungi", No. 6, pp. 62–63.

Wang X.W. 2007. "Nutrition Components Analyse, Extraction and Antioxidant Properties of Polysaccharide of *Stropharia Rugoso-annulata*", "Thesis of NanJing Normal University".

Xu, F. 2006. "Optimizing Liquid Cultural Conditions of *Stropharia Rugoso-annulata* Accumulated Selenium and Studies on Its Antioxidative Activities", "Thesis of Shandong Agricultural University ".

Ye, L.X. & Q.B. Liu, 2007. "Research Advancements of Fungus Polysaccharomyces", "Academic Periodical of Farm Products Processing". No. 5, pp. 26–28.

Yi, J.P. et al. 2009. "Optimization on Ultrasonic-assisted Extraction Technology of Oil from Paeonia suffruticosa Andr. Seeds with Response Surface Analysis", "Transactions of the Chinese Society of Agricultural Machinery", Vol. 40, No. 6, pp. 103–110.

Study on dietary fiber extracted from *bean curd residue* by enzymic method

Haitao Lai
School of Biotechnology, Jimei University, Xiamen, China

Guocheng Su & Wenjing Su
School of Biotechnology, Jimei University, Xiamen, China
Food Science and Technology Development Tast Center, Xiamen, China

Shaojun Zhang & Shunyon Shou
School of Biotechnology, Jimei University, Xiamen, China

ABSTRACT: Through a single factor and orthogonal design experiments of thermostable α-amylase on Bean curd residue, the optimal enzymolysis conditions with consistence, temperature and time was studied. The result showed that 1% thermostable α-amylase, temperature 100°C and time 2 h. Under the optimal enzymolysis conditions, satisfactory results have been obtained to dietary fiber extracted from bean curd residue and analysis on nutrient components in it.

1 INTRODUCTION

Dietary fiber, a kind of polymer that could not be digested by human digestive ferment, can be divided into two categories, soluble dietary fiber (SDF) and insoluble dietary fiber (IDF) (R.F. 1996) SDF is the dietary fiber that could not be digested by human digestive ferment, but could dissolve in hot water and then precipitate again when the solution is mixed into the ethanol with the volume four times as that of the solution, including pectin, alginic acid, agar, carrageenan, xanthan gum, synthetic sodium carboxymethylcellulose, and so on; while IDF is the dietary fiber that could neither be digested by digestive ferment nor dissolve in hot water, including cellulose, hemicelluloses, xylogen, protopectin, vegetable wax, chitin and chitosan of animality, and so on (H. & H. 1997; L. & L. & L. 2003; T. 2003 S. 2003.).

SDF and IDF play significant roles in maintaining health. For instance, IDF is effective on preventing obesity, astriction or intestinal cancer (M.O. & A.F. 2008); and SDF can lower lipid and cholesterol levels, and prevent diabetes with stimulating insulin secretion and slowing down the absorption of glucose in the small intestine (A.R. & V.J. (Eds.) 1990; J.G. & G.P. & K. & D. &, I.L. & G.R. 1993; G.V. & D. (Eds.) 1986); besides, SDF mostly hydrolyzed in the colon has great effectiveness on prevention of colon cancer (G. & Z. & W. & J. & C. 2013). Hence, dietary fiber is the seventh vital nutrient for human body, after protein, fat, sugar, vitamin, mineral substance and water (Z. & Y. 1995; L. & H. &, Y. & T. 2012; M. & C. & C. 2011).

Soy products contain high quantities of dietary fiber, but the residual is regarded as dross which no one would like to eat. Bean curd residue is the remainder of soybean, and the residue is always used to be cattle food or discarded. As a matter of fact, Bean curd residue is in rich of nutriment such as protein, fat, carbohydrate, vitamin, mineral substance and especially dietary fiber. Therefore, it makes sense to explore the way to extract dietary fiber, from bean curd residue.

2 MATERIALS AND METHODS

2.1 *Raw material*

Bean Curd residue was derived from Yin Xiang Group, Xiamen, China.

2.2 Instrument and reagent

DHG-9146A Series Heating and Drying Oven (Shanghai Jing Hong Laboratory Instrument Co. Ltd); Cary50 ultraviolet and visible spectrophotometer (South East Chemicals & Instrument Ltd.); pH211 Acidometer (HanaChina Co. Ltd).

Thermostable α-amylase (>4000 u/g); cellulase (>30 u/mg); glucose starch glycosidase enzymes (>100 u/mg); standard glucose solution (0.1%); dinitrosalicylic salicylic acid (DNS); 2% anthranone agent; starch substrate solution (mixture of 1 g amylum dried to constant weight with pH = 8.0 sodium citrate buffer solution, constant volume of 100 mL); ethanol (95%), All the reagents are analytical reagent (AR).

2.3 Experimental methods

2.3.1 Raw material preprocessing

The fresh bean curd residue was stored in the refrigerator overnight. And then the residue was unfreezed in the high temperature steam, homogenized, water scrubbed, centrifuged, dried, and reserved.

2.3.2 Enzyme activity test

DNS (W. & L. & W. & Z. & H. & G. 1998.) was used to test the enzyme activities of thermostable α-amylase.

2.3.3 Study of the optimum technological conditions to extract dietary fiber with the enzymolysis method of thermostable α-amylase

The extraction ratio of dietary fiber can be indicated by the total reducing sugar (TRS), and tested with the use of Anthrone-H_2SO_4 Colorimetry (L. 2006; G. 2008; S.S. & Y. 2002). The optimal enzymolysis conditions obtained. Through a single factor and orthogonal design experiments of thermostable α-amylase on Bean curd residue.

200 ul standard glucose solution (0.1%) was measured, put into a 10 mL pipet, and added with distilled water to 2.0 mL. 2.0 mL distilled water was put into another blank pipet. The two pipets of solution were handled in the same way as the following description. The researchers put accurate 8 mL cold anthracene ketone test solution into each pipet and shook up. The solution was heated by hot water for 6 minutes, water-cooled, and then kept within the room temperature for 10 minutes. The blank reagent was used as the benchmark to measure the optical density value, draw the absorption curve and calculate the λ_{max}. Using λ_{max} as incident light wavelength, the standard curve was draw.

3 g preprocessing bean curd residue and 150 mL pH = 8.0 sodium citrate buffer solution were put into a 500 mL beaker, and added by a little thermostable α-amylase. After a few minutes chemical reaction, the solution was mixed with 10 mL trichloroacetic acid (8%), heated by 60°C water for 30 minutes, and filtrated. Then the filter liquid was diluted to 62.5 times and its Absorbance was tested to estimate the production of SDF.

The filter residue was gathered into a 500mL beaker, added by 100 mL distilled water and 5 mL zinc sulfate solution. After 5 minutes chemical reaction, the mixture should be cooled promptly, added by 5 mL potassium ferrocyanide solution with the mixture was shaken, and then filtrated. The filter liquid was diluted to 12.5 times and its Absorbance was tested to estimate the production of IDF. The production of total reducing sugar would be the sum of those of SDF and IDF.

The linear regression model was applied:

$$y = bx + a \qquad (1)$$

with y being the optical density value (OD Value) and x being the concentration of standard glucose solution.

Based on this sample, the polysaccharide concentration C can be calculated with:

$$c = \frac{y-a}{b} \quad (2)$$

And then the amount of the polysaccharide can be calculated with:

$$\text{Polysaccharide content} = C \times D \times V \quad (3)$$

where D is the diluting times and V is the volume of the diluents.

2.3.4 Extraction of dietary fiber from bean curd residue

The abstraction of IDF and SDF from bean curd residue was implemented using AOAC method (S.S. & Y. 2002; AOAC International 1997). And the yield and total dietary fiber (TDF) amount were calculated.

15 g preprocessed samples of bean curd residue were put into each of two beakers, added by pH = 8.0 sodium citrate buffer solution and 0.15 g thermostable α-amylase and heated by 100°C water for 2 h, respectively. 20 mL trichloroacetic acid was put into one of the two beakers, and the solution was filtered after 30 min; the filter residue was rough IDF. The filtrate was mixed with 95% ethanol 4 times as its volume, heated by 60°C water for 30 min, and filtered. The filter residue washed with ethanol and acetone was rough SDF. The other beaker was then heated by 60°C water and added by 0.3 g trypsin. After 2 h, the pH of the solution was adjusted to 4.6, and 0.3 g cellulase was added to the solution. 2 h later, 0.3 g glucose starch nucleoside enzymes and another 2 h reaction were needed. The filter residue and filtrate were treated as the procedures described above, and dried. Then the pure SDF and IDF were obtained. The yield was calculated. Ultimately, several tests were implemented to the products, including Kjeldahl Method to test the protein, Soxhlet Extraction to test fat and Dry Ashing Method to test ash content (S.S. & Y. 2002).

3 RESULTS AND DISCUSSION

3.1 Enzyme vitality test

By testing the optical density value of 0.012 mol/L standard glucose solution with 490 nm by the use of Enzyme Activity Test Method, the researchers obtained the equation of linear regression $y = 0.0219x - 0.0379$, $r = 0.9995$. The enzyme vitality can be tested by examining the optical density values (OD values) of 0.03 g thermostable α-amylase, respectively, and putting the equation of linear regression to calculate. After the examination, the vitality of thermostable α-amylase is 4953 ± 123.4 u/mg (n = 3), and the amylase purchased reaches the standard of vitality higher than 4000 u/mg; Hence, which means the amylase is acceptable.

3.2 Optimum technological conditions to extract dietary fiber with the enzymolysis method of thermostable α-amylas

3.2.1 Absorption curve and standard working curve

The Figure 1 demonstrates that the maximum absorption wavelength reaches 580 nm, leading to the conclusion that the wavelength of the incident light should be 580 nm.

According to the Figure 2, the amounts of glucose, within the range from 0 to 8 mg/100 mL, have a linear relationship with the Absorbance, represented by the equation of linear regression $y = 0.0806x - 0.0024$, $r = 0.9996$, showing a favourable degree of fitness.

3.2.2 Single factor experiment of thermostable α-amylas

The single factor experiment of enzymolysis temperature, enzymolysis time and enzyme concentration was implemented. The experiment presents that: the best temperature is 90°C, based on

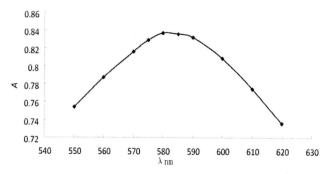

Figure 1. Absorption curve.

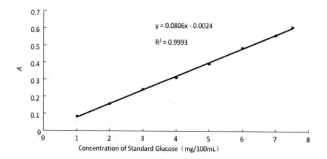

Figure 2. Standard curve.

Table 1. Levels orthogonal experiment factors.

	Enzyme concentration (%)	Temperature (°C)	Time (min)
1	1%	85	30
2	2%	90	60
3	3%	95	90
4	4%	100	120

which 85°C, 90°C, 95°C and 100°C were selected as the orthogonal experiment levels; the best time is 60 min, and as a result 30 min, 60 min, 90 min and 120 min were considered to be the orthogonal experiment levels; the optimum enzyme concentration is 3%, so that the researchers used 1%, 2%, 3% and 4% as the orthogonal experiment levels.

3.2.3 Orthogonal experiment of thermostable α-amylas

Based on the result of single factor experiment, the orthogonal experiment of three factors at four different levels was conducted. The result was showed in Table 1 and Table 2.

According to the range and variance analysis in Table 2, the ranking of the factors influencing the output of dietary fiber is: enzymolysis temperature > enzymolysis time > enzyme concentration. The optimal temperature, time and concentration are 100°C, 2 h, and 4%, respectively. Taking the producing cost into account, the researchers adopted 1% enzyme concentration in the experiment. And in that the optimal time reaches the maximum, another examination with longer time was implemented, of which the result was represented, the amount of TRS almost stayed constant after 2 h reaction. In sum, the optimal reaction condition of thermostable α-amylas is 100°C enzymolysis temperature, 2 h enzymolysis time and 1% enzyme concentration.

Table 2. Orthogonal experiment of Thermostable α-amylas.

No	Filtrate				Filter Residue				Total
	A	V (mL)	D (times)	Sugar (mg)	A	V (mL)	D (times)	Sugar (mg)	sugar (mg)
1	0.085	85	25	22.41	0	100	12.5	0.37	22.78
2	0.129	88	25	35.22	0	100	12.5	0.37	35.59
3	0.219	84	25	57.07	0.018	100	12.5	3.16	60.23
4	0.3	82	25	76.31	0.112	100	12.5	17.74	94.05
5	0.266	92	25	75.91	0.152	100	12.5	23.95	99.86
6	0.276	110	25	94.18	0.101	100	12.5	16.03	110.21
7	0.328	90	25	91.57	0.146	100	12.5	23.02	114.59
8	0.322	94	25	93.89	0.057	100	12.5	9.21	103.10
9	0.299	88	25	81.62	0.06	100	12.5	9.68	91.30
10	0.287	70	25	62.32	0.08	100	12.5	12.78	75.10
11	0.548	88	25	149.59	0.112	100	12.5	17.74	167.33
12	0.495	97	25	148.94	0.101	100	12.5	15.77	164.71
13	0.441	94	25	128.59	0.075	100	12.5	12.01	140.60
14	0.499	96	25	148.60	0.104	100	12.5	16.50	165.10
15	0.407	94	25	118.68	0.02	100	12.5	3.47	122.15
16	0.455	90	25	127.03	0.107	100	12.5	16.97	144.00
k_1	53.16	80.78	88.64						
k_2	106.94	91.37	96.5						
k_3	124.60	118.94	116.08						
k_4	142.96	131.58	126.47						
R	89.80	50.8	37.83						
Variance	1496.92	466.39	303.66						
F	3	3	3						
F-ratio	3.103	1.066	0.606						
F-critical	3.290	3.290	3.290						

Table 3. Product rate and component test.

	Rough-IDF (%)	Rough-SDF (%)	Rough-TDF (%)	Pure-IDF (%)	Pure-SDF (%)	Pure-TDF (%)
Productivity	30.91 ± 1.27	9.67 ± 0.26	40.58 ± 1.25	27.37 ± 0.80	9.26 ± 0.21	36.63 ± 0.80
Protein	7.63 ± 0.71	5.39 ± 0.32	6.51 ± 0.51	1.64 ± 0.08	undetected	0.82 ± 0.06
Fat	9.96 ± 0.57	undetected	4.97 ± 0.28	4.92 ± 0.47	undetected	2.46 ± 0.24
Ash content	2.03 ± 0.28	5.06 ± 0.076	3.55 ± 0.12	2.26 ± 0.28	7.50 ± 0.45	4.88 ± 0.37

3.3 Abstraction of dietary fiber from bean curd residue

Following the direction of AOAC method, the researchers abstracted IDF and SDF from the bean curd residue under the optimum technical conditions of thermostable α-amylase, and conducted protein, fat and ash content tests with 3 times parallel determination, of which the results are demonstrated in table 3.

Table 3 indicates that the amounts of protein and fat in rough dietary fiber are larger than those in pure dietary fiber, with those in pure SDF undetected, while the productivity of ash content in pure dietary is higher, with that in SDF being the highest. The result could be attributed to the fact that the abstraction of pure dietary fiber resorts to diversified enzyme, which leads to the thorough decomposition of starch, protein and fat. Therefore, the quantities of inorganic salt and oxide increase, and as a result the amount of ash content becomes larger.

The researchers also examined the yield of rough fiber and tested the protein, fat and ash content, of which the results were compared with those of the product (Table 4).

Table 4. Comparaison on raw material and Product.

	Rough TDF (%)	Protein (%)	Fat (%)	Ash content (%)
Raw material	49.17	26.14	13.02	4.89
Product	40.58	6.51	4.97	3.55

Table 4 presents that the abstraction of dietary fiber with enzyme decomposition method has a higher efficiency, with a almost complete abstraction of dietary fiber, apparent decreases of amounts of protein and fat, and a slight drop of ash content. This kind of dietary fiber is beneficial for human health, so that can be added into functional food and dietetic food.

4 CONCLUSIONS

This essay conducted the enzyme decomposition on bean curd residue, leftovers of bean curd, with thermostable α-amylase, and abstracted IDF and SDF with high yield and superior purity. The examinations of the products lead to a satisfactory result. The IDF and SDF abstracted can be used as food additives or in the food production. This application will not only turn the waste soy bean residue into something of value, which prevents the environmental pollution, but also become a profit-making source for a company.

REFERENCES

AOAC International. 1997. Official Methods of Analysis, 16th ed., 3rd revision. Method 991.43. AOAC International, Gaithersburg, MD.
Chen, G. & Jie, Z. & Mingfang, W. & Xuqian, J. & Ye, C. 2013. Researches on ExtractionCharacterization and Antioxidant of Soluble Dietary Fiber from Okara. Food Research and Development, 34 (10): 23–27.
Changhong, L. 2006. *Food analysis and experiment (the first version)*. Beijing: Chemical Industry Press.
Jinfeng, H. & Limin, H. 1997. Analysis of dietary fiber, *Food and Fermentation Industries*, 23 (5): 63–72.
Leeds, A.R. & Burley, V.J. (Eds.) 1990. Dietary Fibre Perspectives: Reviews and Bibliography 2. John Libby and Company Ltd., London.
Li, M. & Yongzhong, C. & Longsheng, C. 2011. Extraction and Functional Characteristics Analysis of Dietary Fibers. Academic Periodical of Farm Products Processing 244 (5): 15–18.
Lin, W. & Guosheng, L. & Linsong, W. & Zhihong, Z. & Jinhuai, H. & Huimin, G. 1998. The Optimal Conditions for Cellulase Activity Measurement with DNS Method, *Journal of Henan normal University (Natural Science)*, 26 (3): 66–69.
Muir, J. G. & Young, G.P. & O'Dea, K. & Cameron-Smith, D. & Brown, I.L. & Collier, G.R. 1993. Resistant starch-the neglected "dietary fiber"? Implications for health. Dietary Fiber, *Bibliography and Review* 1: 33–47.
Meiyan, L. & Qin, H. & Jiang, Y. & Dawei, T. 2012.The Utilizing of Okara in the Food Industry. Grain Science and Technology and Economy 37 (5): 44–46.
Nielsen, S.S. & Yianjun, Y. 2002. *Food analysis (the second version)*. Beijing: China Light Industry Press.
Owen, R.F. 1996. Food Chemistry (Third Edition) [ED], Marcel Dekker, Inc., New York: Third Edition.
Vahouny, G.V. & Kritchevsky, D. (Eds.) 1986. Dietary Fiber: Basic and Clinical Aspects. Plenum Press, New York.
Wei, L. & Chengmei, L. & Xiangyang, L. 2003. Present situation and prospect of studies on dietary fiber home and abroad, *Cereal and Food Industry*, 4: 25–27.
Weickert, M.O. & Pfeiffer, A.F. 2008. Metabolic effects of dietary fiber consumption and prevention of diabetes [J]. J. nutr, 138(3): 439–442.
Xuesen, T. 2003. Abstraction of Bean dregs fiber and experimental observation of clinical glucose lowering, *Henan Journal of Preventive Medicine*, 14(3): 136–137.
Xiangyang, G. 2008. *Food analyzing and testing (the first version)*. Beijing: China Metrology Press.
Yunxia, S. 2003. Study on the method of abstracting soluble dietary fiber from *Bean dregs*, Food Research and development, 24(3): 34–35.
Zhiliang, Z. & Liwei, Y. 1995. Producing Edible Fiber of High Quality with *Bean Dregs*, *Chinese Condimen.*, 17(2): 24–25.

Protective effects of *Portulaca oleracea* on alloxan-induced oxidative stress in HIT-T15 pancreatic β cells

J.L. Song
Department of Food Science and Nutrition, Pusan National University, Busan, South Korea

G. Yang
Department of Pharmacy, Northern Jiangsu People's Hospital Affiliated to Yangzhou University (Clinical Medical College of Yangzhou University), Yangzhou, Jiangsu Province, People's Republic China

ABSTRACT: This study was to investigate the protective effect of hot water extracts from *Portulaca oleracea* leaves (POWE) on alloxan-induced oxidative stress in hamster pancreatic HIT-T15 cells. The HIT-T15 cells were treated with alloxan (1 mM) for 1 h, and then co-incubated with the POWE for 24 h. POWE did not significantly exhibit cytotoxic effect and significantly reduced the alloxan-induced HIT-T15 cells damage in a concentration-dependent manner. The cellular levels of lipid peroxidation and endogenous antioxidant enzymes, including catalase (CAT), superoxide dismutase (SOD) and glutathione peroxidase (GSH-px) were also measured. POWE decreased the intracellular lipid peroxidation, and increased the activities of antioxidant enzymes. These results suggest that POWE exhibited cytoprotective activity against alloxan-induced oxidative stress in HIT-T15 cells through the inhibition of lipid peroxidation and stimulation of antioxidant enzymes activity.

1 INTRODUCTION

Diabetes mellitus (DM) is a chronic metabolic diseases, which was associated with the disorder of carbohydrate, protein and fat metabolism characterized by high levels of blood sugar resulting from defects in insulin secretion, insulin action, or both (Aring et al. 2005). DM has progressively become a serious public health problem worldwide. The global prevalence of DM is estimated to increase, from 4% in 1995 to 5.4% by the year 2025 (King et al. 1998). Reactive oxygen species (ROS) induced pancreatic β cells death has an important role in the pathogenesis of diabetes and also affects insulin secretion (Maiese et al. 2007). The ROS are particular responsible for oxidative stress include superoxide (O_2^-), hydroxyl radical ($^{\bullet}OH$), singlet oxygen ($^1O_2$), hydrogen peroxide (H_2O_2), nitric oxide (NO) and peroxynitrite ($ONOO^-$) (Halliwell 2006). Oxidative stress may induce the dysfunction of pancreatic β cells which decreased insulin secretion (Evans et al. 2003) and the development of diabetic complications, including retinopathy, nephropathy, neuropathy and vascular damage (Robertson et al. 2002, Rahimi et al. 2005). Currently, more than 400 traditional plant treatments for diabetes have been recorded (Bailey & Day 1989). It is possible that anti-diabetic components from those natural plants can be an ancillary medicine to the diabetes.

Portulaca oleracea is a warm-climate annual which is grown as an edible plant and distributed in many parts of the world. This plant was reported to exhibit muscle relaxant activity, neuropharmacological actions, wound healing, bronchodilatory, antioxidant, anti-inflammation effects, anti-diabetic and hepatoprotective activity (Lim & Quah 2007, Abas et al. 2006, Anusha et al. 2011). This study was to investigate the potential cytoprotective effects of POWE on alloxan-induced oxidative stress and also to elucidate the mechanisms underlying its protective effects in HIT-T15 cells.

2 MATERIALS AND METHODS

2.1 Chemical reagents

RPMI 1640 medium, fetal bovine serum (FBS), penicillin-streptomycin, 3- (4, 5-dimethylthiazol-2-yl)-2, 5-diphenyl tetrazolium bromid (MTT) and alloxan (ALX) were obtained from Sigma-Aldrich Co. (St. Louis, MO, USA). Dimethyl sulfoxide (DMSO), thiobarbituric acid, trichloroacetic acid and bicinchoninic acid were obtained from Tokyo Chemical Industry Co., Ltd (Shanghai, China). All other reagents were analytical grade.

2.2 Plant extract preparation

Fresh *Portulaca oleracea* leaves were purchased from a local market in Nanjing, China in October 2012. *Portulaca oleracea* hot water extracts (POWE) was prepared by boiling 100 g freeze-dried *Portulaca oleracea* leaves in 1 L distilled water for 4 h, followed by ultracentrifugation at 30,000 × g for 30 min, filtrating with 0.4-μm filter (Whatman International, Maidstone, Kent, UK), concentration by heat evaporation and freeze-drying. The POWE was redissolved in DMSO at a concentration of 50 mg/mL, and stored at 4°C until further study.

2.3 Cell culture

HIT-T15 Syria hamster insulin-secreting cells were obtained from the American Type Culture Col-lection (ATCC, Rockville, MD, USA). The cells were maintained in RPMI 1640 medium supple-mented 10% FBS and 1% penicillin-streptomycin in a humidified CO_2 incubator (Model 3154, Forma Scientific Inc, Marietta, OH, USA) with 5% CO_2 at 37°C.

2.4 Cell viability assay

Cell viability was using by MTT assay. Cells were seeded on 96 well plates at a density of 5×10^3 cells/well. After 24 h incubation, the cells were treated with ALX (1 mM) for 1 h, and then incu-bated with POWE (2.5–50 μg/mL) for 24 h. Following treatment, 100 μL MTT reagent (0.5 μg/mL) was added to each well and cells was incubated in a humidified incubator at 37°C to allow the MTT to be metabolized. After 4 h, the media was removed and cells were resuspended in formazan with 100 μL DMSO. The absorbance of the samples was measured at 540 nm using a microplate reader (model 680, Bio-Rad, Hercules, CA, USA).

2.5 Lipid peroxidation

Lipid peroxidation was evaluated by thiobarbituric acid (TBA)-reactive substances (TBARS) assay (Fraga et al. 1998). In brief, the treated cells were washed with cooled PBS, scarped into trichloroacetic acid (TCA, 2.8%, w/v) and sonicated. Total protein was determined by bicinchoninic acid (BCA) assay. The suspension was mixed with 1 mL TBA (0.67%, w/v) and 1 mL TCA (25%, w/v), heated (30 min at 95°C) and centrifuged (1,500 × g, 10 min at 4°C). TBA-reacts with the oxidative degradation products of lipids, to yield red complexes that absorb at 535 nm using a UV-2401PC spectrophotometer (Shimadzu, Kyoto, Japan).

2.6 Antioxidant enzymes activities

HIT-T15 cells were first treated with ALX (1 mM) for 1 h and then incubated with POWE (2.5–50 μg/mL) for 24 h. The treated cells were washed with PBS, detached by scraping and centrifuged, and the resulting cell pellet stored at −80°C. Cell pellets were thawed, resuspended in 300 μL cold lysis buffer (PBS, 1mM EDTA), homogenized and centrifuged (1,200 × g, 10 min, 4°C). The resulting supernatants were used for activity measurements. Cellular catalase (CAT), superoxide dismutase (SOD) and glutathione peroxidase (GSH-px) levels were determined using commercial

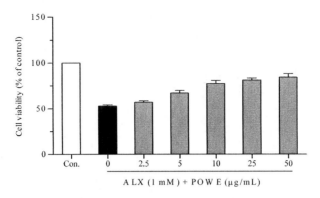

Figure 1. Effects of POWE on cell viability in 1 mM alloxan (ALX)-treated HIT-T15 cells.

assay kits (Beyotime Institute of Biotechnology, Jiangsu, China) according to the manufacturer's instructions. Enzyme activities were expressed as units (U) of enzyme activity per mg protein. Protein contents were determined by Bio-Rad protein assay kit according to the manufacturer's instructions (Bio-Rad, Hercules, CA, USA).

2.7 *Statistical analysis*

Data are presented as the mean ± SD. The SAS v9.1 software package (SAS Institute Inc., Cary, NC, USA) was used for the analysis.

3 RESULTS AND DISCUSSION

3.1 *Effects of POWE on ALX-induced oxidative damage in HIT-T15 cells*

Following treatment with various concentrations of POWE (2.5–50 μg/mL) for 24 h and the cell viability were determined by MTT assay. POWE did not exhibit any cytotoxicity and the cell viabilities were more than 90% (data not shown). Therefore, concentrations between 2.5 and 50 μg/mL were used for further studies. As shown in Fig. 1, POWE was able to reduce ALX (1 mM) induced HIT-T15 cell damage in a concentration-dependent manner.

3.2 *Effects of POWE on lipid peroxidation in ALX-induced HIT-T15 cells*

Oxidative stress participate in the toxic actions that increased the levels of malondialdehyde (MDA), which is a biomarker of insulin-producing cell membrane lipid peroxidation (Jain et al. 2006). As shown in Fig. 2, treatment with POWE was concentration-dependently reduced the ALX-induced lipid peroxidation in HIT-T15 cells, these results suggested that POWE was able to reduce lipid peroxidation may be due to its functions as a preventive antioxidants to savage initiating radicals.

3.3 *Effects of POWE on the antioxidant enzymes in ALX-induced HIT-T15 cells*

Pancreatic β cells have been reported that contains low levels of endogenous antioxidant enzymes, in particular GSH-px and CAT (Moriscot et al. 2000), indicating highly susceptible to oxidative stress induced β cell damage (Lenzen 2008). As shown in Table 1, ALX significantly decreased the activity of CAT, SOD and GSH-px in HIT-T15 cells. However, POWE treatment increased the activity of these antioxidant enzymes in ALX-treated HIT-T15 cells, indicating that POWE was able to reduce ALX-induced oxidative stress. Overproduced of free radicals are scavenged by endogenous antioxidant enzymes, including SOD, CAT and GSH-px. SOD catalyzes the conversion of O_2^- to H_2O_2 and H_2O_2 is further reduced H_2O by the activity of CAT or GSH-px. Certain studies

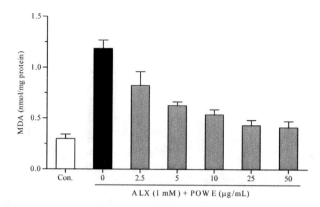

Figure 2. Effects of POWE on intracellular malonaldehyde (MDA) levels in 1 mM alloxan (ALX)-treated HIT-T15 cells.

Table 1. Effect of POWE on the activity of CAT, SOD and GSH-px in HIT-T15 cells exposed to ALX (1 mM).

Treatments	Concentrations (μg/mL)	CAT (U/mg protein)	SOD (U/mg protein)	GSH-px (U/mg protein)
Normal	–	2.32 ± 0.27	16.26 ± 0.44	5.87 ± 0.38
ALX + POWE	0	1.37 ± 0.17	7.98 ± 0.75	3.51 ± 0.26
	2.5	1.84 ± 0.27	8.54 ± 1.18	3.62 ± 0.17
	5.0	1.87 ± 0.15	9.74 ± 1.39	4.11 ± 0.36
	10.0	2.08 ± 0.25	11.40 ± 0.63	4.61 ± 0.18
	25.0	2.14 ± 0.07	11.72 ± 1.81	4.88 ± 0.20
	50.0	2.24 ± 0.15	12.76 ± 1.30	5.34 ± 0.20

have reported that the overexpression of Cu/Zn-SOD showed protective effect against nitric oxide-induced cytotoxic in human islets and INS-1 insulin-secreting cells (Kubisch et al. 1997) and ALX and streptozotocin-induced diabetes (Xu et al. 1999). In addition, administration of SOD and CAT were able to ameliorate the alloxan-induced pancreatic cell damage *in vitro* (Lenzen 2008). Furthermore, catalase also showed a protective effect to against H_2O_2- and streptozotocin-induced oxidative stress *in vivo* (Lortz & Tiedge 2003). Combinatorial overexpression of the CAT and GSH-px also revealed a protective effect against ROS-induced oxidative stress by increasing the activity of Cu/Zn SOD or MnSOD (Lepore et al. 2003, Mysore et al. 2005).

4 CONCLUSION

The results from the present study have demonstrated that *Portulaca oleracea* leaf hot water extract exhibit a protective activity against alloxan-induced cell damage in HIT-T15 hamster insulin-secreting cells. *Portulaca oleracea* leaf hot water extract was able to effectively reduce the alloxan-induced cell lipid peroxidation and prevent pancreatic β cell death by increasing the activities of intracellular antioxidant enzymes SOD, CAT and GSH-Px. These results also suggested that *Portulaca oleracea* can be used as a successful source of a natural anti-diabetic reagent and also as a possible food supplement in prevention and treatment of diabetes.

REFERENCES

Abas, F., Lajis, N. H., Israf, D. A., Khozirah, S., & Umi Kalsom, Y. 2006. Antioxidant and nitric oxide inhibition activities of selected Malay traditional vegetables. *Food Chemistry* 95(4): 566–573.

Anusha, M., Venkateswarlu, M., Prabhakaran, V., Taj, S. S., Kumari, B. P., & Ranganayakulu, D. 2011. Hepatoprotective activity of aqueous extract of Portulaca oleracea in combination with lycopene in rats. *Indian Journal of Pharmacology* 43(5): 563–567.

Aring, A.M., Jones, D.E. & Falko JM. 2005. Evaluation and prevention of diabetic neuropathy. *American Family Physician* 71(11): 2123–2128.

Bailey, C. J., & Day, C. 1989. Traditional plant medicines as treatments for diabetes. *Diabetes Care* 12(8): 553–564.

Evans, J.L., Goldfine, I.D., Maddux, B.A. & Grodsky, G.M. 2003. Are Oxidative Stress-Activated Signaling Pathways Mediators of Insulin Resistance and β-Cell Dysfunction? *Diabetes* 52(1): 1–8.

Fraga, C. G., Leibovitz, B. E., & Tappel, A. L. 1988. Lipid peroxidation measured as thiobarbituric acid-reactive substances in tissue slices: characterization and comparison with homogenates and microsomes. *Free Radical Biology and Medicine* 4(3): 155–161.

Halliwell, B. 2006. Reactive species and antioxidants. Redox biology is a fundamental theme of aerobic life. *Plant Physiology* 141(2): 312–322.

Jain, S. K., McVie, R., & Bocchini Jr, J. A. 2006. Hyperketonemia (ketosis), oxidative stress and type 1 diabetes. *Pathophysiology* 13(3): 163–170.

King, H., Aubert, R.E. & Herman, W.H. 1998. Global burden of diabetes, 1995–2025: prevalence, numerical estimates, and projections. *Diabetes Care* 21(9): 1414–1431.

Kubisch, H. M., Wang, J., Bray, T. M., & Phillips, J. P. 1997. Targeted overexpression of Cu/Zn superoxide dismutase protects pancreatic β-cells against oxidative stress. *Diabetes* 46(10): 1563–1566.

Lenzen, S. 2008. Oxidative stress: the vulnerable beta-cell. *Biochemical Society Transactions*, 36(Pt 3): 343–347.

Lenzen, S., Drinkgern, J., & Tiedge, M. 1996. Low antioxidant enzyme gene expression in pancreatic islets compared with various other mouse tissues. *Free Radical Biology and Medicine* 20(3): 463–466.

Lepore, D. A., Shinkel, T. A., Fisicaro, N., Mysore, T. B., Johnson, L. E., d'Apice, A. J., & Cowan, P. J. (2004). Enhanced expression of glutathione peroxidase protects islet β cells from hypoxia-reoxygenation. *Xenotransplantation* 11(1): 53–59.

Lim, Y. Y., & Quah, E. P. L. 2007. Antioxidant properties of different cultivars of Portulaca oleracea. *Food Chemistry* 103(3): 734–740.

Lortz, S., & Tiedge, M. 2003. Sequential inactivation of reactive oxygen species by combined overexpression of SOD isoforms and catalase in insulin-producing cells. *Free Radical Biology and Medicine* 34(6): 683–688.

Maiese, K., Morhan, S.D. & Chong, Z. Z. 2007. Oxidative stress biology and cell injury during type 1 and type 2 diabetes mellitus. *Current Neurovascular Research* 4(1): 63–71.

Moriscot, C., Pattou, F., Kerr-Conte, J., Richard, M. J., Lemarchand, P., & Benhamou, P. Y. 2000. Contribution of adenoviral-mediated superoxide dismutase gene transfer to the reduction in nitric oxide-induced cytotoxicity on human islets and INS-1 insulin-secreting cells. *Diabetologia* 43(5): 625–631.

Mysore, T.B., Shinkel, T.A., Collins, J., Salvaris, E.J., Fisicaro, N., Murray-Segal, L.J., Johnson, L.E., Lepore, D.A., Walters, S.N., Stokes, R., Chandra, A.P., O'Connell, P.J., d'Apice, A.J. & Cowan, P.J. 2005. Overexpression of glutathione peroxidase with two isoforms of superoxide dismutase protects mouse islets from oxidative injury and improves islet graft function. *Diabetes* 54(7): 2109–2116.

Rahimi, R., Nikfar, S., Larijani, B., & Abdollahi, M. 2005. A review on the role of antioxidants in the management of diabetes and its complications. *Biomedicine & Pharmacotherapy* 59(7): 365–373.

Robertson, R. P., Harmon, J., Tran, P. O., Tanaka, Y., & Takahashi, H. 2003. Glucose toxicity in β-cells: type 2 diabetes, good radicals gone bad, and the glutathione connection. *Diabetes* 52(3): 581–587.

Xu, B. O., Moritz, J. T., & Epstein, P. N. 1999. Overexpression of catalase provides partial protection to transgenic mouse beta cells. *Free Radical Biology and Medicine* 27(7): 830–837.

Study of gas-oil diffusion coefficient in porous media under high temperature and pressure

Ping Guo, Hanmin Tu, Anping Ye, Zhouhua Wang, Yanmei Xu & Zhipeng Ou
State Key Laboratory of Oil and Gas Reservoir Geology and Exploitation, Southwest Petroleum University, Chengdu, Sichuan, China

ABSTRACT: Effective gas-oil diffusion coefficients in porous media are studied under high temperature and pressure. Diffusion coefficients of gas – condensate oil system and gas – heavy oil system are tested in the full diameter core under 60°C and 20 MPa, respectively. The results show that the gas diffusion coefficient in PVT apparatus is obviously higher than that in the full diameter core. The gas diffusion coefficient in condensate oil is higher than that in heavy oil.

1 INTRODUCTION

Gas injection is an important way of enhanced oil recovery at present. The injected gas changes the properties of crude oil through interphase mass transfer to improved oil flow capacity. Gas-oil phase equilibrium experiments are basic work of gas injection design, which provide fluid PVT parameters for gas injection numerical simulation. The molecular diffusion coefficients are important parameters which describe the dynamic mass transfer speed between injected gas and formation crude oil. However, the effects of molecular diffusion on gas injection for low matrix permeability and high capillary pressure fractured reservoir have been rarely studied. Gravity drainage effect is restricted by gas oil density difference, and molecular diffusion would dominate. Therefore, reasonable determination of the molecular diffusion coefficients has great significance for the accurate evaluation of gas displacement efficiency.

Currently due to the calculation of molecular diffusion coefficient theory, there is no uniform method accurately determining the parameters only by experimental tests. There are two main experimental methods for determining molecular diffusion coefficient. One method is sampling fluid at different time and different diffusion distance, and then analyzing these samples to obtain gas concentration data, and deriving the diffusion coefficient with the corresponding mathematical model. The other method is deriving the diffusion coefficient with the corresponding mathematical model according to system pressure change and fluid density change caused by mass transfer. In recent years, with the development of modern test technology, indirect method has become the primary method to test the molecular diffusion coefficient.

Throughout these molecular diffusion coefficient research methods and results, they mainly has the following several inadequacy aspect: (1) Most of the molecular diffusion coefficient test methods do not consider the influence of porous medium. (2) Experimental test temperature and stress conditions are not up to the actual oil field gas injection requirements. (3) Most studies use sand filling tube instead of porous medium, without considering the actual reservoir core, which is only applicable to the single component of gas-liquid system.

This study carries out the real core diffusion experiment under 60°C and 20 MPa. Gas sample used in experiment is CO_2, oil samples are condensate oil and heavy oil. The purpose of this paper is to establish a new real core diffusion test method, and to test the multi-component injection gas diffusion coefficients.

Table 1. Composition of heavy crude oil and condensate oil.

Component	Heavy crude oil		Condensate oil	
	Volume percent %	Molar mass kg/kmol	Volume percent %	Molar mass kg/kmol
C_2	–	–	0.019	30.070
C_3	–	–	0.256	44.097
iC_4	0.057	58.124	0.239	58.124
nC_4	0.094	58.124	0.752	58.124
iC_5	0.405	72.151	1.153	72.151
nC_5	0.337	72.151	1.386	72.151
C_6	5.073	86.178	5.137	86.178
C_7	4.578	100.25	12.335	100.250
C_8	5.125	114.232	13.263	114.232
C_9	3.625	128.259	18.413	128.259
C_{10}	3.683	142.286	16.976	142.286
C_{11+}	77.02	256.313	30.077	198.640

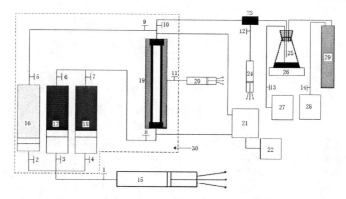

Figure 1. Procedure of diffusion experiment in porous media. 1–14—valve, 5—entrance pump, 16—intermediate container (gas), 17—intermediate container (oil), 18—intermediate container (formation water), 19—full diameter core holder, 20—confinement pressure pump, 21—numeric pressure sensor, 22—computer, 23—back pressure regulator, 24—back pressure pump, 25—extractor, 26—electronic balance, 27—oil chromatograph, 28—gas chromatograph, 29—aerometer, 30—incubator.

2 POROUS MEDIA DIFFUSION EXPERIMENTS

This paper uses the method that CO_2 diffusion with crude oil in full diameter core to study diffusion phenomena in porous media. Through experimental means to test the CO_2-oil system pressure changes in the porous medium, selecting outcropping full diameter core complete the diffusion experiment in porous medium. Experiment temperature was 60°C and pressure was 20 MPa. Diffusion gas sample is CO_2 commercial gas with composition of N_2 0.0796%, CO_2 98.181% and C_1 1.6939%. Diffusion oil sample is instead by gas free oil and gas free condensate oil of earth extractor, specific component are shown in Table 1. Full diameter sandstone core is used, with length of 9.941 cm, diameter of 7.554, porosity of 19.16%, and permeability of 20.3×10^{-3} um^2. The test procedure is presented in Figure 1.

Experiment procedures are as follows:

(a) Fluid and core heating and boosting: transfer the gas sample, oil sample and the formation water into the gas middle container, oil middle container and core picked up the suction at ordinary temperatures. Heating the middle container and core in incubator to required temperature. The entrance pump, confinement pressure pump and back pressure pump set to the required pressure.

(b) Oil sample transfer: open the valve1, valve3 and valve6 in order, read reading V1. Pump the pressure to be stable after entrance, read the reading V1 of entrance pump, then open the valve8, startup entrance pump, slowly inject a certain amount of oil sample, then stop entrance pump and read the reading V2 of entrance pump. The difference between V1 and V2 is roll-in oil sample volume.

(c) Gas sample transfer: close the valve8, valve6 and valve3 in order, open the valve2, valve5 and valve9 in order, startup entrance pump, slowly inject a certain amount of gas sample, then stop entrance pump and read the reading V3 of entrance pump. The difference between V3 and V2 is roll-in gas sample volume.

(d) Diffusion test: close the valve9, valve5, valve2, valve1 and entrance pump in order, start diffusion test, record time and pressure changes, when the pressure in rock does not change, the gas-oil has reached diffusion equilibrium, record equilibrium pressure and time. Confinement pressure in pump maintain constant in diffusion test.

(e) Gas composition test: open the valve1, valve4, valve7 and valve12 in order, start entrance pump adjusting pressure of intermediate container with formation water to step (d) equilibrium pressure and adjusting the back pressure pump pressure to equilibrium pressure; Set entrance pump for pump inlet mode, back pressure pump for retreat pump mode, both speed equal, speed as small as possible. At the same time open the two pumps, valve 8 and 10, roll out a certain amount gas sample from the core upper, test the gas oil ratio and composition, repeated measuring three times. Until the gas phase completely discharge, and then used the same way test underneath oil phase composition, test the gas oil ratio and composition, repeated test three times.

3 MATHEMATICAL MODELS

Porous medium containing non-equilibrium gas and liquid phase whose initial composition is known. During the whole experiment, temperature keeps constant. Gas and liquid maintain a balance in the interface. System pressure, volume and each phase composition will change over time, until eventually reached equilibrium state. In this paper pressure drop method is used to determinate multicomponent gas-oil molecular diffusion coefficient. Molecular diffusion coefficient calculation methods refer to the paper of Riazi (1996). According to Fick's second law, and combined with limited mass transfer model of Figure 2, the following equations are derived:

oil phase

$$\begin{cases} \dfrac{\partial C_{oi}}{\partial t} = \dfrac{\partial}{\partial z_o}\left[D_{oi}\dfrac{\partial C_{oi}}{\partial z_o}\right] \\ C_{oi}(z_o,0) = C_{oi}^1(z_o) \\ \dfrac{\partial C_{oi}(0,t)}{\partial z_o} = 0 \\ C_{oi}(L_o,t) = C_{obi} \end{cases} \quad (1)$$

gas phase

$$\begin{cases} \dfrac{\partial C_{gi}}{\partial t} = \dfrac{\partial}{\partial z_g}\left[D_{gi}\dfrac{\partial C_{gi}}{\partial z_g}\right] \\ C_{gi}(z_g,0) = C_{gi}^1(z_g) \\ C_{gi}(0,t) = C_{gbi} \\ \dfrac{\partial C_{gi}(L_g,t)}{\partial z_g} = 0 \end{cases} \quad (2)$$

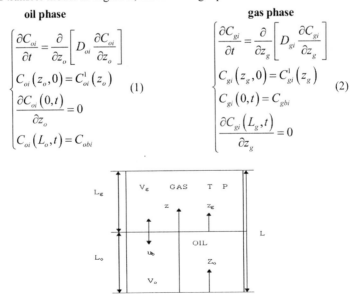

Figure 2. Limited boundary physical model.

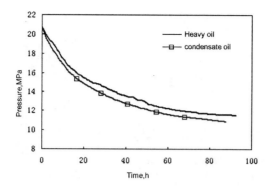

Figure 3. Relationship between diffusion pressure and time in porous medium.

C_{oi}^1, C_{gi}^1 are initial mole concentration of component i in oil phase, gas phase, respectively; C_{obi}, C_{gbi} are mole concentration of component i in oil phase, gas phase at the oil-gas interface, respectively; C_{oi} is mole concentration of component i in oil phase; D_{oi}, D_{gi} are diffusion coefficient of component i in oil phase, gas phase, respectively; t is time; L_o, L_g are length of oil phase, gas phase, respectively; z_o, z_g are positive distance of oil phase, gas phase, respectively.

Molecular diffusion coefficient of each component in the multicomponent gas-crude oil can be obtained by the equation (1) and equation (2). The specific equation solving steps refer to the paper of Guo et al. (2009).

4 RESULTS AND DISCUSSION

4.1 Equilibrium time

In full diameter core CO_2-condensate oil diffusion balance time is about 88.6 hours in 20 MPa, 60°C, while CO_2-heavy oil diffusion balance time is about 92.46 hours. It shows that due to the diffusion difference caused by the different properties of crude oil, the lower crude oil restructuring content, the bigger the diffusion coefficient and shorter system balance time. However in PVT tube CO_2-condensate oil diffusion balance time is about 16 hours in 20 MPa, 60°C, while CO_2-heavy oil diffusion balance time is about 27.6 hours (Guo et al. 2009). Thus it can be seen diffusion equilibrium time in full diameter core is much higher than in the PVT tube, indicating that the existence of the porous media greatly increases the diffusion equilibrium time, which is due to porous media random size distribution, connectivity, tortuosity in the porous medium. Molecular diffusion path in the porous medium is not the straight distance. Diffusion path is longer than in the PVT, so the oil and gas diffusion equilibrium time in porous media is more than in the PVT tube.

4.2 Pressure change

CO_2-condensate oil and CO_2-heavy oil in porous medium diffusion pressure changes with time relationship is shown in Figure 3. From the chart we can see that pressure drop of CO_2-heavy oil in porous medium diffusion in the initial period is more than the reduction of CO_2-condensate, but there is no significant pressure difference after balance is reached. It shows that crude oil component have certain effects on the diffusion. However, effects on the final diffusion extent are not big.

4.3 Calculation of diffusion coefficients

Take CO_2-condensate oil as an example, known from Figure 4, Figure 5, Figure 6, diffusion coefficients of each component in condensate oil gradually increased in the form of power function with the pressure drop, eventually the diffusion coefficient level off, the diffusion process reach equilibrium, and oil component diffusion coefficient in gas phase also have the same tendency, the

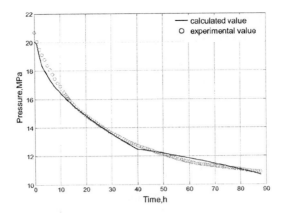

Figure 4. CO_2-condensate oil diffusion pressure fitting chart in porous media.

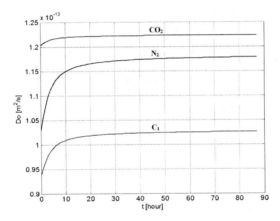

Figure 5. Condensate oil diffusion coefficient of each gas component in porous medium.

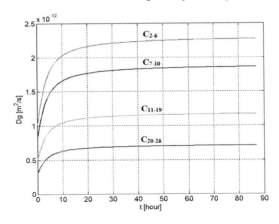

Figure 6. Gas diffusion coefficient of each component in condensate oil.

diffusion coefficient when equilibrium is reached is shown in Table 2. From the diffusion coefficient calculation results, we can see that at the same temperature and pressure condition, the gas phase components in condensate oil diffusion coefficient value is not the same, CO_2 is the maximum, N_2 is smaller, and C_1 is the minimum.

Table 2. Diffusion coefficient.

Component	Diffusion coefficient (final value)	Component	Diffusion coefficient (final value)
CO_2	1.223E-13	C_2–C_6	2.2689E-12
C_1	1.0265E-13	C_7–C_{10}	1.8594E-12
N_2	1.1783E-13	C_{11}–C_{19}	1.1683E-12
–	–	C_{20+}	7.0649E-13

5 CONCLUSIONS

(1) Test method of real reservoir core and multicomponent system of oil and gas molecular diffusion coefficient under high temperature and high pressure is established, the principle is simple and reliable. By gas phase component analysis, the accuracy of tests can be guaranteed.
(2) The CO_2 diffusion coefficient were tested in condensate oil and heavy crude in full diameter core under 60°C, 20 MPa. Compared with previous studies, the test use actual gas, oil system, and it is under high temperature and high pressure in the full diameter core, and more close to the actual gas injection process.
(3) Crude oil components have certain effects on the diffusion, but the effects on the final diffusion extent are limited.
(4) Gas diffusion coefficients in condensate oil are higher than in heavy crude.
(5) Diffusion coefficient of each component in condensate oil gradually increased in the form of power function and eventually level off, CO_2 is maximum, N_2 is smaller, and C_1 is minimum.

ACKNOWLEDGMENT

This paper is supported by the National Natural Science Foundation of China (No.: 51374179, "Theory and molecular dynamics study on CO_2-crude oil non-equilibrium diffusion considering capillary pressure and adsorption").

REFERENCES

Das, S. K.; Butler, R. M. 1996. Diffusion coefficients of propane and butane in Peace River bitumen. *Can. J. Chem. Eng.* 74, 985–992.
Etminan, S. R.; Maini, B. B.; Chen, Z. X; Hassanzadeh, H. 2010. Constant-Pressure Technique for Gas Diffusivity and Solubility Measurements in Heavy Oil and Bitumen. *Energy Fuels*. 24, 533–549.
Guo, P.; Wang, Z. H.; Shen, P. P.; Du, J. F. 2009. Molecular Diffusion Coefficients of the Multicomponent Gas-Crude Oil Systems under High Temperature and Pressure. *Ind. Eng. Chem. Res.* 48, 9023–9027.
Islas-Juarez, R.; Samanego V. F.; Luna, E.; Perez-Rosales, C.; Cruz, J. 2004. Experimental Study of Effective Diffusion in Porous Media. *SPE International Petroleum Conference in Mexico, Puebla Pue, Mexico, 2004*.
Oballa, V.; Butler, R. M.; 1989. An experimental-study of diffusion in the bitumen-toluene system. *J. Can. Pet. Technol.* 28, 63–90.
Riazi, M. R. 1996. A new method for experimental measurement of diffusion coefficients in reservoir fluids. *Journal of Petroleum Science and Engineering*. 14, 235–250.
Unatrakarn, D.; Asghari, K.; Condor, J. 2011. Experimental Studies of CO_2 and CH_4 Diffusion Coefficient in Bulk Oil and Porous Media. *Energy Procedia*. 4, 2170–2177.
Wen, Y.; Kantzas, A.; Wang, G. J. 2004. Estimation of diffusion coefficients in bitumen solvent mixtures using X-ray CAT scanning and low field NMR. *Canadian International Petroleum Conference, Alberta, Canada, 2004*.
Yang, C. D.; Gu, Y. G. 2006. A new method for measuring solvent diffusivity in heavy oil by dynamic pendant drop shape analysis. *SEPJ.* 11, 48–57.
Yang, D. Y.; Tontiwachwuthikul, P.; Gu, Y. G. 2006. Dynamic Interfacial Tension Method for Measuring Gas Diffusion Coefficient and Interface Mass Transfer Coefficient in a Liquid. *Ind. Eng. Chem. Res.* 45, 4999–5008.
Zhang, Y. P.; Hyndman, C.L.; Maini, B.B.; 2000. Measurement of gas diffusivity in heavy oils. *SPEJ.* 25, 37–475.

The significance of biomarkers in nipple discharge and serum in diagnosis of breast cancer

G.P. Wang, M.H. Qin, Y.A. Liang, H. Zhang & Z.F. Zhang
Department of Pathology, Rizhao People's Hospital, Affiliated Rizhao People's Hospital of Jining Medical College, Rizhao, Shandong Province, China

F.L. Xu
Department of Clinical Laboratory, Rizhao People's Hospital, Rizhao, Shandong Province, China

ABSTRACT: To explore the value of combined detection of tumor markers both in nipple discharge and serum in the diagnosis of breast cancer. The serum and nipple discharge from 179 patients including 43 cases of breast cancer and 136 cases of benign tumour was collected and tumor markers of CA153, TSGF, CA125 and CEA were detected. The nipple discharge and serum levels of the four biomarkers in bresat cancer patients were significantly higher than those in benign lesions ($P < 0.01$), and had a positive correlation with Ki-67, gread, lymph node metastas and tumor recurrence ($P < 0.05$), and negative correlation with ER and PR ($P < 0.05$). The sensitivity of the four markes in combination was 97.7% and specificity was 99.0%. The dynamic combined detection of the four biomarkers both in nipple discharge and serum are helpful to the stratification of preoperative patients and benefit to monitoring their recurrence, metastasis and clinical staging of tumors in clinic.

1 INTRODUCTION

Breast cancer is one of the commonly-encountered malignant tumors, and is the second leading cancer causing death in women. In recent years, its morbidity is on the rise year by year in the East and West, hence, how to treat breast cancer, effectively assess the therapeutic effect, correctively evaluate the prognosis and find the postoperative recurrence of patients with breast cancer in an early stage have been paid attention by more and more scholars at home and abroad (Deng et al. 2013, Sun et al. 2013, Zhang et al. 2013).

Serum has long been considered as a rich source for biomarkers and a large number of serum biomarkers such as cancer antigen 153 (CA153), malignant tumor specific growth factor (TSGF), cancer antigen 125 (CA125) and carcinoembryonic antigen (CEA) have been proposed. However, serum protein biomarkers are often not sensitive enough to be used for screening and early diagnosis because their levels refect tumor burden (Piura et al. 2010). Unlike the traditional tumor markers, nipple discharge autoantibodies to tumor antigens are detectable even when the tumor is very small, which makes them potential biomarkers for early cancer diagnosis. Moreover, the advantages of simple detection of nipple discharge using target antigen and secondary reagents, in contrast to tissue biopsy, traumatic examination, make it easy to construct a multiplex tumor-associated autoantibody assay. A panel of biomarkers may have better predictive performance than individual markers. Since breast cancer is a heterogeneous disease, it possibly requires a panel of several biomarkers to allow detection of the different subtypes. Hence, combined detection of tumor markers both in nipple discharge and serum is of great importance in the discovery and treatment of patients as well as the judgment on progression and prognosis of disease (Tarhan et al. 2013, Zhang et al. 2013).

In the study, the expression and signifcance of dynamically combined detecting indicators CA153, TSGF, CA125 and CEA both in serum and nipple discharge was explored, and their relationship with estrogen receptor (ER), progestrone receptor (PR), epidermal growth factor receptor-2 (HER2/neu), hypoxia inducible factor-1α (HIF-1α), Ki-67 proliferation index in judging the prognosis of breast cancer were explored too.

2 MATERIALS AND METHODS

2.1 General data

Forty-three patients diagnosed invasive breast cancer (mean age: 51.9 years; age range: 22–79) and 136 cases of benign lesions patients (mean age: 47.3 years; age range: 21–63) were collected during excision surgery at Rizhao people's Hospital from January 2002 to December 2012. Thirty cases of normal pregnant women were studied as a control group. Among patients diagnosed invasive breast cancer, there were 42 cases female and one case male, which had not been treated with Hormone endocrine therapy, anti-neoplastic chemotherapy or radiotherapy during the last six months. The diagnosis was verified by histological methods, and pathological categorization was determined according to the current World Health Organization classification system (Lakhani et al. 2012). All patients signed informed consents, and this study was approved by Local Ethical Committee.

2.2 Measure of biomarkers in patient blood serum and nipple discharge samples

We performed biomarkers of blood serum and nipple discharge analysis in 43 breast cancer patients. For CA153, CA125, TSGF and CEA analysis, 3ml heparinized blood and 0.5 ml of nipple discharge was drawn from each individual. The biomarkers were detected with electrochemiluminescence method in the clinical laboratory in Rizhao People's Hospital, and was compared with 30 cases of normal pregnant women. The cut off values of CA153, TSGF, CA125 and CEA in serum are 25.0 U/ml, 35.0 U/ml, 70.0 U/ml and 3.4 ng/ml, and those of nipple discharge are 35.0 U/ml, 40.0 U/ml, 95.0 U/ml and 9.8 ng/ml.

The relationship of the four biomarkers levels of blood serum and nipple discharge with mult biological parameters including the express of ER, PR, HER2/neu, HIF-1α, Ki-67 proliferation index and histological grade, region lymph node metastasis, distant metastasis and recurrence on files were assessed in order to study the clinical and pathological characteristics associated with breast cancer and improve the clinical diagnosis of breast cancer, especially in early stage of breast cancer.

2.3 Pathologic analysis

Breast tissue samples were fixed in 10% neutral buffered formalin and embedded in paraffin at 4°C for 24 hours. Tissue sections of 5 μm were deparaffinized and rehydrated using standard procedures. The specimens were examined under a binocular dissecting microscope. The pathological diagnosis was verified by histological methods independently by two pathologists, and pathological categorization was determined according to the current World Health Organization classification system (Lakhani et al. 2012). Immunhistochemical UltraSensitive™ S-P method was employed to detect the expression of ER, PR, HER2/neu, HIF-1α, VEGF and Ki-67. The pathologists were blinded to the subject's clinical history, and the results of the immunohistochemistry staining assay. The pathological reading was determined for each biopsy slide with an overall pathological diagnosis determined for each subject. The tumor grade was determined according to the modified Bloom-Richardson score. The grade is obtained by adding up the scores for tubule formation. Final grading scores were as follows: sum of point 3–5, final grade I; 6–7, II; and 8–9, III.

Immunoreactions were processed using the UltraSensitive™ S-P Kit (Maixin-Bio, China) according to the manufacturer's instructions, and signals were visualized using the DAB substrate, which stains the target protein yellow. The negative controls were used. The primary antibody was replaced with PBS, containing 0.1% bovine serum albumin at the same concentration as the primary antibody. The positive controls were tissues known to express the antigen being studied.

HIF-1α, ER & PR and Ki-67 immunoreactivity expression the percentage of cancer cells showed nuclear reactivity. The immunoreactive score (IRS) was obtained by multying the percentage of positive cells and the staining intensity. In brief, a proportion score was assigned that represents the estimated proportion of positive tumor cells on the entire slide. For each histological section, the percentage of positive cells was scored as 0 (<5%), 1 (6%–25%), 2 (26%–50%), 3 (51%–75%) and 4 (>75%), and the staining intensity was scored as 0 (negative), 1 (weak), 2 (moderate) and 3 (strong). Immunohistochemical results with an IRS of 0 were considered negative, 1–4 weak positive, 5–8 moderate positive and 9–12 strong positive. For ER, PR and Ki-67 expression the percentage of cancer cells showing a nuclear reactivity was recorded after inspection of all optical fields at 200 and the mean value was used to score each case. Tumors with expression of >1% of cancer cells were considered to be positive. In addition, cell membrane reactivity for HER2/neu the oncoprotein was evaluated following a similar approach and the mean value was used to score each case. Tumors expressing HER2/neu in >30% of the cancer cells were considered as positive.

2.4 Statistical analysis

SPSS ver 17.0 statistical software was used to detect the data, and the results were expressed with $(x \pm s)$. Measurement data between groups was compared with t test, while enumeration data with χ^2 test. The P value was considered to be significant if less than 0.05.

3 RESULTS

3.1 The levels of biomarkers in serum and nipple discharge

The nipple discharge and serum levels of CA153, TSGF, CA125 and CEA were shown in table 1. The nipple discharge and serum levels of the four biomarkers in the breast cancer patients were significantly higher than those in t benign lesions patients ($P < 0.01$) and healthy pregnant women ($P < 0.01$). The levels of the four biomarkers in nipple discharge were significantly higher than those in the serum ($P < 0.01$).

3.2 The relationship between tumor markers in nipple discharge and biological parameters within breast cancer group

The relationship of the four marker of CA153, TSGF, CA125 and CEA levels of discharge in breast cancer patients and the different clinical pathological factors were shown in table 2. Breast cancer tumor size and onset age had no signifcant association with nipple discharge level of the four biomarkers in nipple discharge among breast cancer group. The nipple discharge levels of the four biomarkers were markedly higher in those ER and PR negative patients compared to those positive

Table 1. The nipple discharge and serum levels of biomarkers.

Subgroup	n	CA153 (U/ml)	TSGF (U/ml)	CA125 (U/ml)	CEA (ng/ml)
Nipple discharge					
Breast cancer	43	128.2 ± 28.6	212.4 ± 45.7	115.7 ± 41.1	109.2 ± 30.9
Benign lesions	136	25.1 ± 6.1*	67.0 ± 15.0*	29.4 ± 7.2*	7.5 ± 1.8*
Healthy women	30	16.0 ± 3.2**	45.8 ± 10.4**	16.1 ± 4.8**	3.3 ± 0.9**
Serum					
Breast cancer	43	93.8 ± 21.8	145.5 ± 32.1	86.7 ± 21.4	43.3 ± 16.8
Benign lesions	136	18.6 ± 4.0*	56.0 ± 12.0*	20.3 ± 4.6*	2.4 ± 0.7*
Healthy women	30	14.2 ± 2.5**	39.8 ± 7.4**	15.8 ± 4.3**	2.3 ± 0.5**

*Compared with benign lesions, $P < 0.01$; **Compared with healthy pregnant women, $P < 0.01$.

Table 2. The impact of some personal parameters on nipple discharge levels of tumor markers within breast cancer group.

Subgroup	n	CA153 (U/ml)	TSGF (U/ml)	CA125 (U/ml)	CEA (ng/ml)
Age					
≥50 yr	29	129.0 ± 30.5	215.8 ± 38.9	116.4 ± 41.2	112.9 ± 31.1
<50 yr	14	127.7 ± 29.7	209.7 ± 35.5	76.5 ± 22.4	107.1 ± 28.4
Tumor size					
≤2 cm	27	126.5 ± 23.3	206.8 ± 36.7	114.0 ± 27.6	108.8 ± 25.2
>2 cm	16	129.0 ± 28.4	214.9 ± 40.0	116.9 ± 41.0	111.1 ± 32.9
ER					
positive	37	110.3 ± 26.7	189.2 ± 32.3	103.9 ± 26.1	99.7 ± 27.7
negative	6	156.6 ± 47.1	251.7 ± 53.9	156.5 ± 47.3	129.3 ± 40.9
PR					
positive	36	110.1 ± 20.9	193.6 ± 23.4	102.5 ± 29.8	89.3 ± 34.4
negative	7	153.6 ± 33.7	263.8 ± 53.2	154.5 ± 42.2	137.7 ± 33.4
HER2					
positive	14	151.4 ± 32.4	243.6 ± 54.8	120.7 ± 31.4	117.4 ± 29.7
negative	29	125.7 ± 26.1	187.3 ± 34.0	98.6 ± 20.8	85.6 ± 27.5
HIF-1α					
positive	33	272.4 ± 189.3	198.6 ± 54.7	258.6 ± 73.2	158.2 ± 55.2
negative	10	94.5 ± 27.4	87.6 ± 36.4	189.6 ± 27.3	99.0 ± 38.3
Ki-67					
>14%	10	165.2 ± 39.3	237.8 ± 46.5	139.2 ± 44.3	125.2 ± 33.4
≤14%	33	110.4 ± 25.8	201.8 ± 31.6	106.5 ± 25.2	103.6 ± 25.7
Grade					
III	17	312.6 ± 66.1	263.2 ± 68.4	165.3 ± 63.4	195.7 ± 56.3
I + II	26	68.2 ± 24.4	176.4 ± 23.7	96.4 ± 26.4	78.3 ± 14.7
Operative					
Preoperative	43	128.2 ± 28.6	212.4 ± 45.7	115.7 ± 41.1	109.3 ± 31.0
Postoperative	43	20.0 ± 11.0	46.9 ± 16.8	21.4 ± 7.8	5.5 ± 2.8
Lymph node metastasis					
positive	9	572.2 ± 105.3	228.2 ± 73.2	198.6 ± 47.3	158.3 ± 55.4
negative	34	90.6 ± 23.3	189.6 ± 26.3	143.6 ± 26.1	105.0 ± 28.3
Distal metastasis					
positive	9	572.2 ± 105.3	198.6 ± 47.2	158.2 ± 55.2	228.2 ± 73.1
negative	34	90.6 ± 23.3	143.6 ± 26.0	105.0 ± 28.3	189.6 ± 26.3
Recurrence					
positive	8	176.3 ± 105.3	236.7 ± 98.8	189.6 ± 98.8	171.8 ± 56.8
negative	35	107.4 ± 24.7	176.8 ± 32.3	101.4 ± 89.4	87.6 ± 23.2

patients. There were a significantly difference of the level between HER2/neu positive patients and HER2/neu negative patients, Ki-67 proliferation index ≤14% patients and >14% patients ($P < 0.05$). Furthermore, with increase in pathological grade, levels of CA153, TSGF, CA125 and CEA in nipple discharge gradually increased ($P < 0.01$). In addition, the levels of nipple discharge were significantly increased in those patients with lymph node metastasis and distal metastasis compared to those patients without ($P < 0.01$). Values with distal metastasis were notably higher than with region lymph node metastasis ($P < 0.01$). Last but not least the levels of the four marker in nipple discharge were significantly higher in those patients with HIF-1α positive patients compared to those negative patients ($P < 0.01$). The postoperative follow-up results revealed that positive HIF-1α and VEGF and levels of nipple discharge CA153, TSGF, CA125 and CEA in recurrence group

Table 3. The discriminative value of diagnosis in breast cancer (%).

Groups	Sensitivity	Specificity	Accuracy	Positive predictive value	Negative predictive values
Nipple discharge					
CA153	74.4	82.4	80.5	57.1	91.1
TSGF	69.8	83.1	79.9	56.6	86.7
CA125	72.1	83.8	81.0	58.5	90.5
CEA	69.8	86.0	82.2	61.2	90.0
Serum					
CA15-3	60.5	91.9	84.4	70.3	88.0
TSGF	62.8	91.2	84.4	69.3	88.6
CA125	55.8	90.4	82.2	64.9	88.6
CEA	53.5	89.0	80.5	60.5	85.8
Combination	97.7*	75.0	80.5	55.3	99.0*

*Compared with other items, $P < 0.01$.

were obviously higher than in non-recurrence group ($P < 0.01$). During the follow-up period from 6 months to 8 years, there were eight patients recurrent. The nipple discharge levels of CA153, TSGF, CA125 and CEA tumor markers in recurrence group were obviously higher than in non-recurrence group ($P < 0.01$).

3.3 *Discriminative value of combined detection of CA153, TSGF, CA125 and CEA both in nipple discharge and serum in diagnostic breast cancer*

The discriminative value of CA153, TSGF, CA125 and CEA in nipple discharge and serum and their combined detection in diagnostic breast cancer were shown in Table 3. The sensitivity of the four serum tumor markes in combination was only 69.8%, in contrast, the combined detection both in discharge and serum was 97.7%, and the negative predictive value was 99.0%. The sensitivity of combined detection both in nipple discharge and serum were significantly higher than other detection ($P < 0.05$).

4 DISCUSSION

Breast cancer is one of the most frequent malignancies in the world, and is the second leading cancer causing death in women (Siegel et al. 2012). In recent years, its morbidity is on the rise year by year in the world. Despite the widespread use of the screening methods, only 5% of the cases were diagnosed at the advanced stage. Tumor markers are among the prognostic factors and are of diagnostic value and can be used for follow-up.

Serum has long been considered as a rich source for biomarkers and a large number of serum cancer biomarkers such as CA153, TSGF, CA125 and CEA have been proposed. However, serum protein biomarkers show limited diagnostic sensitivity and specificity in stand-alone assays because their levels refect tumor burden (Piura et al. 2010). Unlike serum protein biomarkers, the traditional tumor markers, nipple discharge autoantibodies to tumor antigens are detectable even when the tumor is very small, which makes them potential biomarkers for early cancer diagnosis. The discharge from nipple is common, accounting for 5% of all breast related symptoms (Lang et al. 2007). Abnormal discharge is most commonly caused by benign conditions like intraductal papillomas, duct ectasia, papillomatosis, mastitis, fibrocystic changes (Zervoudis et al. 2010). The incidence of breast cancer in patients presenting with abnormal nipple discharge is between 6% to 21% (Dolan et al. 2010). A majority of patients with breast cancer who manifest with isolated discharge have an early stage disease associated with DCIS. Discharge owing to DCIS has been shown to be a marker for extensive disease, which often requires mastectomy to achieve adequate surgical margins (Parthasarathy et al. 2012, Chen et al. 2012). Discharge can be an early sign of breast cancer (Zervoudis et al. 2010, Chen et al. 2012). Diffusely spreading intraductal carcinomas which often have no clinically palpable breast mass can manifest as pathological discharge. Moreover, the advantages of simple detection of nipple discharge using target antigen and secondary reagents, in contrast to serum protein markers whose detection needs two different monoclonal antibodies, make it easy to construct a multiplex tumor-associated autoantibody assay. The few biomarkers that are tested for early detection of breast cancer, show limited diagnostic sensitivity and specificity in stand-alone assays (Kulasingam et al. 2009). Many biomarkers have been evaluated in breast diseases, which include CA153, TSGF, CA125 and CEA. Among them, CA153 and CEA are the most commonly used. CA153, a variant of mammary epithelial surface glycoprotein and an antigen related to breast cancer, in most cases can be considered as indicator to detect recurrence and metastasis of carcinoma. Studies have confrmed that as a sort of glycoprotein, CA125 can be indentifed by monoclonal antibody, and has a certain correlation with the genesis and development of breast cancer, whereas CEA as a sort of broad-spectrum tumor markers is applied to detect the postoperative recurrence rate of various epithelial tumors and judge the prognosis. If a cancerous tumor cell growing out of control is present in breast, the levels of CA153, TSGF, CA125 and CEA may increase as the number of cancer cells increase. In our study, the levels of the four biomarkers in nipple discharge were significantly higher than those in the serum ($P < 0.01$). In our study, the nipple discharge and serum levels of CA153, TSGF, CA125 and CEA in the breast carcinoma were significantly higher than those in breast benign lesions patients ($P < 0.01$). The nipple discharge high levels of the four biomarkers in breast carcinoma patients had a positive correlation with the poor differentiation, like high Ki-67, grade and clinical stage, lymphnode metastas, and tumor recurrence ($P < 0.05$), in contrast, negative correlation with the level of ER and PR. The clinical outcome in breast cance is affected by a number of factors, including HIF-1α status, ER and PR status, over-expression of *HER*2 and Ki-67, tumor grade, Lymph node metastasis and distant metastasis. Hypoxia, is a common feature of various cancers. Cells under hypoxic conditions develop numerous adaptive responses to hypoxic stress concurrently with altered expression of hundreds of genes that are regulated by hypoxia inducible factors (Semenza et al. 2010). HIF-1α plays an important role in the regulation of various genes associated with low oxygen consumption. Elevated expression of HIF-1α has been reported to be associated with tumor progression, invasion and metastasis in many cancers (Sueoka et al. 2013). Our results implicate the levels of CA153, TSGF, CA125 and CEA in HIF-1α positive patients with BREAST CANCER were significantly higher than those in HIF-1α negative patients.

The tumor-related indicators in all individuals cannot be detected systemically and comprehensively due to limited economics and techniques, thereby, combined detection of specifc tumor markers is of great importance. In this study, the sensitivity of the four serum tumor markes in combination was only 69.8%, in contrast, the dynamic combined detection both in discharge and serum was 97.7%, and the negative predictive value was 99.0%. The dynamic combined detection of the four tumor markers both in nipple discharge and serum are benefit to early diagnosis and interference and better prewarning markers for monitoring their recurrence and metastasis of breast cancer, but cannot increase the sensitivity of diagnosis of Precancerous lesions. A panel of biomarkers may have better predictive performance than individual markers. Since breast cancer is a heterogeneous disease, it possibly requires a panel of several biomarkers to allow detection of the different subtypes.

5 CONCLUSION

In this study, we analyze for the ?rst time the combined detecting CA153, TSGF, CA125 and CEA simultaneously in the same sample both in discharge and serum and the ER, PR, HER2, HIF-1a, Ki-67, and achieved a better application effect. Our results suggest that the levels of CEA, TSGF,

OPN and CA125 in nipple discharge were significantly higher than those in the serum, and the positive rate and the negative predictive value of the the dynamic combined detection of the four tumor markers both in nipple discharge and serum were significantly higher than other detection, which are benefit to early diagnosis and interference and better prewarning markers for monitoring their recurrence and metastasis of breast cancer.

REFERENCES

Chen, L. & Zhou, W. B. & Zhao, Y. et al. 2012. Bloody nipple discharge is a predictor of breast cancer risk: a meta-analysis. *Breast Cancer Res Treat* 132 (1): 9–14.

Deng, Q. Q. & Huang, X. E. & Ye, L. H. et al. 2013. Phase II trial of Loubo® (Lobaplatin) and pemetrexed for patients with metastatic breast cancer not responding to anthracycline or taxanes. *Asian Pac J Cancer Prev* 14: 413–417.

Dolan, R. T. & Butler, J. S. & Kell, M. R. et al. 2010. Nipple discharge and the efficacy of duct cytology in evaluating breast cancer risk. *Surgeon* 8 (5): 252–258.

Kulasingam, V. & Zheng, Y. & Soosaipillai, A. et al. 2009. Activated leukocyte cell adhesion molecule: A novel biomarker for breast cancer. *Int. J. Cancer* 125: 9–14.

Lakhani, S. R. & Ellis, I. O. & Schnitt, S. J. et al (eds), 2012. *WHO classification of tumours of the breast* [M]. Lyon, France: IARC.

Lang, J. E. & Kuerer, H. M. 2007. Breast ductal secretions: clinical features, potential uses, and possible applications. *Cancer control* 14: 350–359.

Parthasarathy, V. & Rathnam, U. 2012. Nipple discharge: an early warning sign of breast cancer. *Int J Prey Med* 3 (11): 810–814.

Piura, E. & Piura, B. 2010. Autoantibodies to tumor-associated antigens in breast carcinoma. *J Oncol* 264926, 2010.

Semenza, G. L. 2010. Defining the role of hypoxia-inducible factor 1 in cancer biology and therapeutics. *Oncogene* 29: 625–634.

Siegel, R. & Naishadham, D. & Jemal, A. 2012, Cancer statistics. *CA Cancer J Clin* 62: 10–29.

Sueoka, E. & Sueoka-Aragane, N. & Sato, A. et al. 2013. Development of lymphoproliferative diseases by hypoxia inducible factor-1alpha is associated with prolonged lymphocyte survival. *PLoS One* 8: e57833.

Sun, M. Q. & Meng, A. F. & Huang, X. E. et al. 2013. Comparison of psychological infuence on breast cancer patients between breast-conserving surgery and modifed radical mastectomy. *Asian Pac J Cancer Prev* 14: 149–152.

Tarhan, M. O. & Gonel, A. Kucukzeybek, Y. et al. 2013. Prognostic signifcance of circulating tumor cells and serum CA15-3 levels in metastatic breast cancer, single center experience, preliminary results. *Asian Pac J Cancer Prev* 14: 1725–1729.

Zervoudis, S. & Iatrakis, G. & Economides, P. et al. 2010. Nipple discharge screening. *Womens Health* 6: 135–151.

Zhang, S. J. & Hu, Y. & Qian, H. L. et al. 2013. Expression and Signifcance of ER, PR, VEGF, CA15-3, CA125 and CEA in Judging the Prognosis of Breast Cancer. *Asian Pacifc J Cancer Prev* 14 (6): 3937–3940.

Advanced Engineering and Technology – Xie & Huang (Eds)
© 2014 Taylor & Francis Group, London, ISBN 978-1-138-02636-0

Experimental study on minimum miscible pressure of rich gas flooding in light oil reservoir

Ping Guo, Zhipeng Ou, Anping Ye, Jianfen Du & Zhouhua Wang
*State Key Laboratory of Oil and Gas Reservoir Geology and Exploitation,
Southwest Petroleum University, Chengdu, Sichuan, China*

ABSTRACT: Slim tube test has been generally accepted as the standard method for minimum miscible pressure (MMP) determination. Gas injection volume is 1.2 pore volume (PV) in standard slim tube test procedure. In this paper, six slim tube tests are carried out with rich gas injection and lean gas injection. Gas-oil ratio (GOR) is measured in each test, and MMP is determined with different gas injection volume (1.2 PV and more than 1.2 PV) for the rich gas injection tests. The results show that, if the injected rich gas has high intermediate hydrocarbon content, there could be certain deviation in MMP determination with the standard procedure of 1.2 PV gas injection volume, and the oil recovery can be higher with the further increasing gas injection volume.

1 INTRODUCTION

Hydrocarbon gas injection has been successfully used as a popular enhanced oil recovery (EOR) process in many oil fields. Minimum miscible pressure is an important parameter for EOR of gas flooding oil reservoir. Miscible flooding is used when the reservoir pressure is greater than MMP. Conversely, immiscible flooding is conducted. Gas-oil interfacial tension becomes zero when the miscibility is reached. Mass transfer between gas and oil also reduces the interfacial tension even through miscibility is not achieved.

When the injected gas comes into contact with the crude oil, mass transfer of intermediate components occurs between gas and oil. According to the different kinds of the injected gas, there are three mass transfer mechanisms for gas flooding (Lee et al. 1988, Elsharkawy et al. 1992, Egermann et al. 2006). The first is the condensing gas drive. During this process, the intermediate hydrocarbons are transferred from the injected gas into the crude oil until miscibility is achieved. The second is the vaporizing gas drive. Vaporization of the intermediate components from the crude oil into the gas injected causes miscibility. The third is the vaporising/condensing gas drive. The light intermediate hydrocarbons ($C_2 \sim C_3$) of the injected gas condense into the crude oil in the leading edge of the gas front. While the heavy intermediate components ($C_4 \sim C_6$) of the crude oil vaporize into the gas after the front. Rich-gas flooding is the vaporising/condensing gas drive.

Several methods have been proposed for MMP determination. Empirical equations have been developed for MMP (Glaso 1985, Orr Jr et al. 1987, Yuan 2004). However, the empirical equations are normally used for the screening and feasibility study due to the low precision. Hence the analytical prediction methods are advanced (Wang & Orr Jr 1997, Yang et al. 2004). Slim tube test is generally accepted as the industrial standard method of MMP determination. Other experiments include rising-bubble experiment (Srivastava et al. 1994), interfacial tension method (Ayirala et al. 2003, Jessen & Orr Jr 2008) and core flooding experiment (Guo et al. 2001, Wang et al. 2010). Although the time of rising-bubble experiment and interfacial tension method is shorter than that of the slim tube experiment, it is hard to obtain the data of the gas breakthrough time, the gas-liquid components and the liquid density. Taking the high cost of core flooding experiment into consideration, thus the slim tube test is a better choice.

In this paper, the rich gas and lean gas are used to conduct the experimental slim tube test, respectively. The purposes of our work are as follows: (1) analysing GOR change for the injected

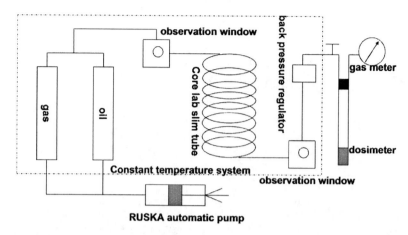

Figure 1. Flow diagram of slim tube experiment.

Table 1. Fluid properties.

Component	Sample Oil mol%	Rich gas mol%	Lean gas mol%
CO_2	0.009	0.09	1.42
N_2	1.823	5.05	0.39
C_1	27.09	63.69	88.29
C_2	4.643	10.61	5.02
C_3	3.466	6.46	2.23
iC_4	4.783	4.91	0.45
nC_4	2.805	2.28	0.92
iC_5	7.611	6.56	0.4
nC_5	1.716	0.16	0.32
C_6	3.039	0.11	0.33
C_7	3.978	0.01	0.19
C_8	6.013	0.07	0.04
C_9	3.271	–	–
C_{10}	2.521	–	–
C_{11}^+	27.233	–	–

rich gas and lean gas, respectively. (2) determining MMP using two methods (at 1.2 PV and more than 1.2 PV) for the rich gas injection.

2 EXPERIMENTAL APPARATUS AND PREPARATION

Six slim-tube experiments were performed to determine MMP. The experimental apparatus is pictured in Figure 1. Before each experiment, the slim tube was cleaned with the petroleum ether. The cleaning work did not finish until the inlet and outlet of the petroleum ether had the same color and composition. The cleaned slim tube was blown dry with N_2 or the compressed air, then it was baked under the required temperature for more than six hours. The porosity and permeability were determined, and pore volume of the baked slim tube was also calculated. The oil sample was saturated under the reservoir temperature and experimental pressure. The slim tube is 1800 cm long, with diameter of 0.466 cm, porosity of 35%, and permeability of 5 μm^2. Fluid properties are given in Table 1.

3 EXPERIMENTAL PROCEDURE

For each experiment, firstly the slim tube was saturated with oil under the required temperature and pressure. Secondly the intermediate container was filled with the injected gas, and balanced under the experimental temperature and pressure. Finally the back pressure was fixed to the pressure value which was required. The gas sample was pumped into the slim tube with a certain displacement speed by Ruska injection pump. The flooding experiment was finished after 1.2 PV of the injected gas. The oil production was measured by the automatic liquid collector at the regular intervals. The gas production was calculated by the automatic gas meter. The changes of the gas components were analysed by the gas chromatograph at regular intervals. In the experimental process, the experimental data were recorded on time, and we also observed irregularly the phase behavior and the fluid color.

4 EXPERIMENTAL CONDITIONS

The experiment temperature is 60.5°C, and the reservoir pressure is 13.05 MPa. Generally speaking, we should choose five points of the flooding pressure in order to determine the MMP, among which there must be 2 or 3 points where the oil recovery is more than 90% in the flooding process. There are six pressure points of the rich gas injection (11.53 MPa, 13.05 MPa, 14.60 MPa, 18.10 MPa, 20.50 MPa and 24.63 MPa) and of the lean gas injection (21.32 MPa, 32.46 MPa, 36.25 MPa, 40.39 MPa, 42.45 MPa and 45.21 MPa), respectively. The displacement rate was constant (0.125 ml/min). The injection pressure was the average pressure of the injection pump during the whole experiment. The back pressure, controlled by the automatic pump, was always the required displacement pressure, and its fluctuation was not more than 0.1 MPa. The injection volume was the actual volume measured by the injection pump under various pressures. The experiment was finished when the injected volume was 1.2 PV.

5 RESULTS AND DISCUSSION

Criteria for the minimum miscible pressure: MMP is determined by the cross point of straight lines of the pressure versus oil recovery at 1.2 PV injected for miscibility and immiscibility. We can see from Figure 2 and Figure 3, that the rich gas slim tube tests have been conducted at 11.53 MPa, 13.05 MPa, 14.60 MPa, 18.10 MPa, 0.50 MPa and 24.63 MPa. The pressures of miscible displacements are 18.10 MPa, 20.50 MPa and 24.63 MPa, with oil recoveries of more than 90%. Though the oil recovery is 89.106% at 18.10 MPa, brownish red oil produced becomes transparent light yellow up to white, so we think it is still miscible at this pressure. Immiscible displacements are indicated at other pressures. For lean gas injected, the pressures of miscible displacements are 40.39 MPa, 42.45 MPa and 45.21 MPa. The oil recovery gets higher with the increase of the injection pressure. The oil recovery is 91.263% at 21.32 MPa for the rich gas flooding when the experiment is finished, while 47.182% at 31.32 MPa for the lean gas flooding. Therefore, the recovery of the rich gas flooding is far greater than that of the lean gas flooding.

Even though rich gas injection does not form miscible, the oil recovery is also up to about 86% at the end of the experiment after a fixed volume of rich gas injected (more than the standard 1.2 PV) (see Figure 2). Shelton et al. (1975) (the recovery was 70% at 1.2 PV. The recovery was 98% at 4 PV) and Tiffin et al. (1991) (the recovery was 84.3% at 1.2 PV. The recovery was 96% at 1.5 PV) obtained the same conclusion. When the injected rich gas and crude oil come in contact, the intermediate components of the rich gas condense into the crude oil in the leading edge of the gas front. When the gas continues to move forward, the intermediate hydrocarbons are transferred from the crude oil into the gas. The oil is produced continually for this mechanism. Therefore, the oil recovery is high even though the rich gas and oil does not achieve miscibility.

Figure 4 and Figure 5 show the relationships between pore volume injected and GOR under various experimental pressures. The higher experimental pressure, the later the gas breakthrough

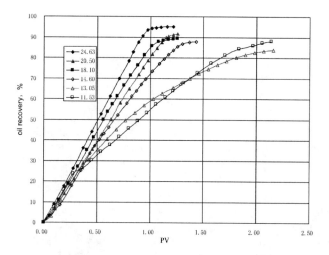

Figure 2. Relationship between recovery and rich-gas pore volume injected under the different pressures.

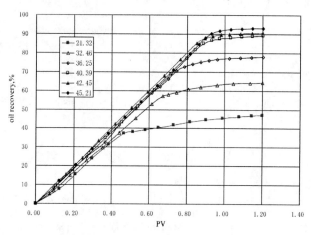

Figure 3. Relationship between recovery and lean-gas pore volume injected under the different pressures.

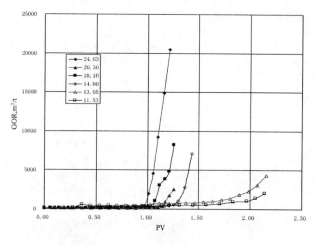

Figure 4. Relationship between GOR and rich-gas pore volume injected under the different pressures.

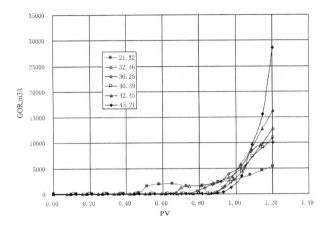

Figure 5. Relationship between GOR and lean-gas pore volume injected under the different pressures.

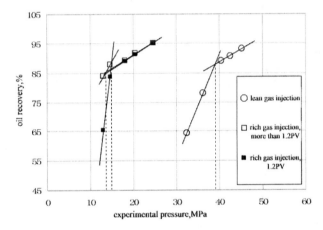

Figure 6. Comparison of the MMPs determined by the different methods and the different kinds of gas injected.

time. There are three stages of the curve for both immiscibility and miscibility: In the first and the second stages, GOR does not change significantly. While GOR level of the lean gas injected is higher, and its length of the smooth is shorter than that of the injected rich gas. GOR increases rapidly in the third stage. On the contrary, the rate of GOR rise for the immiscibility is slower than that for the miscibility.

Figure 6 shows the MMP determined by the pressure versus oil recovery at the standard 1.2 PV and at the end of the experiment, respectively. The MMP (14.80 MPa) of the rich gas injection at the standard 1.2 PV is far less than that (38.96 MPa) of the lean gas injection. That is because of the different components. The intermediate hydrocarbons (C_2–C_6) constitute 31.09% of the rich gas, much more than the 9.67% of the lean gas (Mihcakan & Poettmann 1994, Metcalfe 1982). The MMP (13.70 MPa) of the rich gas injection at the end of the experiment is lower than that (14.80 MPa) at the standard 1.2 PV. In the actual flooding well network, especially in the vicinity of the injection well, the gas volume injected is much higher than 1.2 PV. For the rich gas with high intermediate hydrocarbon content, the MMP determination by the recovery at 1.2 PV is higher than that at more than 1.2 PV. The oil recovery is still high with some more injected pore volume when the volume injected is more than 1.2 PV, even though miscibility is not achieved for rich gas flooding in light-oil reservoir.

6 CONCLUSION

(1) It is reasonable that the MMP has been determined by the relationship of the pressure and recovery at 1.2 PV for the lean gas injection, while the MMP at 1.2 PV has been higher than that at the end of the experiment (more than 1.2 PV) for the rich gas injection.
(2) For the lean gas injection, GOR has increased faster after the gas breakthrough, and no oil has been produced after 1.2 PV injected, while the curve of the GOR has maintained a long step which is relatively lower than for the lean gas injection. GOR has increased significantly after 1.75 PV injected for the rich gas injection.

ACKNOWLEDGMENT

This paper is supported by the National Natural Science Foundation of China (No.: 51374179, "Theory and molecular dynamics study on CO_2-crude oil non-equilibrium diffusion considering capillary pressure and adsorption").

REFERENCES

Ayirala S. C.; Rao D. N.; Casteel J. 2003. Comparison of Minimum Miscibility Pressures Determined from Gas-Oil Interfacial Tension Measurements with Equation of State Calculations. *SPE 84187*.

Egermann P.; Robin M.; Lombard J. M.; Modavi A.; Kalam M. Z.; 2006. Gas Process Displacement Efficiency Comparisons on a Carbonate Reservoir. *SPE81577*.

Elsharkawy A. M.; Suez Canal U.; Poettmann F. H.; Christiansen R. L. 1992. Measuring Minimum Miscibility Pressure: Slim-Tube or Rising-Bubble Method. *SPE 24114*.

Glaso. 1985. Generalized Minimum Miscibility Pressure Correlation. *SPE 12893*.

Guo P.; Liu J. Y.; Li S. L.; Xiong Y.; Zhou Z.; Zhao X. F.; Li Y. G.; Zhang X. L. 2001. Lab research of nitrogen and rich gas injection in XI-52 volatile reservoir. *Journal of Southwest Petroleum Institute*. 23(6): 44–47.

Jessen K.; Orr Jr. F. M. 2008. On Interfacial-Tension Measurements To Estimate Minimum Miscibility Pressures. *SPE 110725*.

Lee S. T.; Lo H.; Dharmawardhana B. T. 1988. Analysis of mass transfer mechanisms occurring in rich gas displacement process. *SPE 18062*.

Metcalfe R. S. 1982. Effects of Impurities on Minimum Miscibility Pressures and Enrichment Levels for CO_2 and Rich-Gas Displacements. *SPE 9230*.

Mihcakan M.; Poettmann F. H. 1994. Minimum Miscibility Pressure, Rising Bubble Apparatus and Phase Behavior. *SPE 27815*.

Orr Jr F. M.; Silva M. K. 1987. Effect of Oil Composition on Minimum Miscibility Pressure Part 2: Corretion. *SPE 14150*.

Shelton J. L.; Schneider F. H. 1975. The Effects of Water Injection on Miscible Flooding Methods Using Hydrocarbons and CarbonDioxide. *SPE 4580*.

Srivastava P. K.; Huang S. S.; Dyer S. B; Mourits F. M. 1994. A Comparision of Minimum Miscibility Pressure Determinations for Weyburn CO_2 Solvent Design. *PETSOC 94-50*.

Tiffin D. L.; Sebastian H. M.; Bergman D. F. 1991. Displacement Mechanism and Water Shielding Phenomena for a Rich-Gas/Crude-Oil System. *SPE 17374*.

Wang S. K.; Wei X. G.; Li Y. B. 2010. Physical simulation of long core by miscible flooding of enriched gas in zarzaitine oilfield. *Journal of Southwest Petroleum University* 32(4): 123–126.

Wang Y.; Orr Jr. F. M. 1997. Analytical calculation of minimum miscibility pressure. *Fluid Phase Equilibria* 139, 101–124.

Yang X. F.; Guo P.; Du Z. M. 2004. Calculation of minimum miscibility pressure with the line ananlysis. *Natural Gas Industry* 24(6): 98–102.

Yuan H.; John R. T.; Egwuenu A. M.; Dindoruk B. 2004. Improve MMP Correlations for CO_2 Floods Using Analytical Gas Flooding. *SPE 89359*.

The traumatic characteristics of inpatients injured in Lushan Earthquake

Y.M. Li, J.W. Gu, C. Zheng, B. Yang & N.Y. Sun
Chengdu Military General Hospital of People's Liberation Army, Chengdu, Sichuan, China

ABSTRACT: At 8:02 am on April 20, 2013, a strong earthquake of 7.0 magnitudes on the Richter scale occurred in Lushan County, Sichuan Province, China. The data of inpatients injured in Lushan Earthquake at Chengdu Military General Hospital of People's Liberation Army were collected using the "No. 1 Military Medical Project" hospital information system and the self-prepared questionnaire, to investigate the traumatic characteristics of these inpatients. In total, 65 inpatients were admitted to and treated at the hospital. There were 48 cases of closed trauma, accounting for 73.85%. The four most common injury sites were multiple trauma, lower limb, waist and abdomen, and chest and back. The four most common injury categories were skin and soft tissue injuries, fractures, combined injuries, and visceral injuries. The comprehensive and systematic understanding of the traumatic characteristics of inpatients injured in Earthquake provides important guidance for medical rescue following an earthquake.

1 INTRODUCTION

Earthquake disasters are unexpected, highly destructive, and commonly associated with secondary disasters. As a country with some of the worst earthquakes in the world, China faces a grim situation of disaster mitigation and relief (You et al., 2009). The comprehensive understanding of the injuries provides important guidance for medical rescue and relief following an earthquake (Chen, 2009, Li et al., 2011). Recently, Chinese scholars have analyzed the injury severity, sites, and other parameters of the victims of the Wenchuan (Xu et al., 2008, Yao et al., 2008, Zhou et al., 2008), and Yushu (Kang et al., 2012) earthquake. At 8:02 am on April 20, 2013, a strong earthquake of 7.0 magnitudes on the Richter scale occurred in Lushan County, Sichuan Province, China. After the earthquake, the military and local health workers actively participated in the medical relief. Chengdu Military General Hospital of People's Liberation Army (PLA) immediately sent two 20-person medical teams to the disaster area by airlift; these teams participated in the early treatment and psychological intervention of the earthquake victims, as well as performed related health and epidemic prevention work (Li et al., 2013). The hospital quickly activated its contingency plans and established emergency rooms to admit the earthquake wounded from earthquake area. Using the epidemiological method, this study investigated the general information, injury severity, sites, and categories of the inpatients following the Lushan earthquake.

2 MATERIALS AND METHODS

2.1 *Data sources*

The general information and clinical data of the Lushan earthquake victims admitted to Chengdu Military General Hospital, PLA were collected using the "No. 1 Military Medical Project" hospital information system (Sun and Xie, 1997). Referring to previous research (Xu et al., 2008, Yao et al., 2008, Zhou et al., 2008, Kang et al., 2012), the "Survey for inpatients attributable to the 4.20

earthquake in Lushan, Sichuan" was designed and used to collect information about the injury severity, site, and category for the inpatients. This questionnaire sought advice from 10 experts in our hospital in the departments of neurosurgery, orthopedics, general surgery, neurology, and others. The surveys were completed by face-to-face inquiry between the wounded patients and two medical staff.

During the medical rescue of this earthquake, our hospital had admitted 81 wounded patients. Excluding a new born baby and 15 wounded military personnel (traffic accident), 65 wounded patients, who were all local inhabitants, were included in the analysis.

2.2 Statistical methods

The data entry adopted the Epi-Data 3.02 and statistical analysis was performed using SPSS16.0. The measured variables were expressed as the mean ± standard deviation ($\bar{x} \pm sd$), while the counted variables were statistically described using frequency count and percentage. The difference of the measured variables between groups was compared using the t test, while the difference of the counted variables between groups was compared using the χ^2 test. Differences with $P < 0.05$ were considered statistically significant.

3 RESULTS

3.1 General information of the inpatients

The number of new patients admitted daily after the earthquake is shown in Figure 1. In total, 50 inpatients were admitted in the first three days, accounting for 76.92%. There were 30 male inpatients (46.15%). The average inpatients age was 45.74 ± 20.96 years old. The data of the age and marital status group are shown in Table 1.

3.2 Injury situations of the inpatients

Before the earthquake, 12 patients (18.46%) had preexisting diseases or poor mobility, including 5 patients with chronic obstructive pulmonary disease, 3 pregnant women, 3 patients with paraplegia, and one with heart disease. Of the inpatients, 63 (96.92%) were injured in the main shock. Indoor injuries occurred in 33 people (50.77%). Standing injuries occurred in 41 people (63.08%). Crushing injuries caused by collapsed buildings occurred in 28 people (43.08%). In terms of the rescue mode, 34 people (52.31%) received military rescue. The detail injury situations of the inpatients are shown in Table 2.

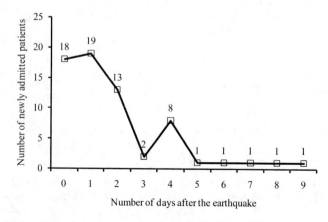

Figure 1. Distribution of the admission time of the Lushan earthquake inpatients.

Table 1. General information of the inpatients following the Lushan earthquake.

	All wounded	Male	Female	t/χ^2	P
Gender, n (%)	65	30 (46.15)	35 (53.85)		
Age, $\bar{x} \pm sd$	45.74 ± 20.96	43.57 ± 18.35	47.6 ± 23.07	0.771	0.440
Age by group, n (%)					
≤20	5 (7.69)	3 (10.00)	2 (5.71)	1.337	0.720
21–40	23 (35.38)	11 (36.67)	12 (34.29)		
41–60	20 (30.77)	10 (33.33)	10 (28.57)		
≥61	17 (26.15)	6 (20.00)	11 (31.43)		
Marital status, n (%)					
Unmarried	8 (12.31)	4 (13.33)	4 (11.43)	0.054	0.816
Married	57 (87.69)	26 (86.67)	31 (88.57)		

Table 2. The injury situations of the inpatients following the Lushan earthquake.

	n (%)
Before the earthquake	
Healthy	53 (81.54)
Unhealthy with poor mobility	12 (18.46)
Injury time	
Main shock	63 (96.92)
Other	2 (3.08)
Injury location	
Indoor	33 (50.77)
Outdoor	32 (49.23)
Injury position	
Lying	19 (29.23)
Sitting	5 (7.69)
Standing	41 (63.08)
Cause of injury	
Crushed by collapsed buildings	28 (43.08)
Fall down or bump when escaping	23 (35.38)
Accidental injury	14 (21.54)
Rescue mode[a]	
Self rescue	32 (49.23)
Mutual rescue	23 (35.38)
Local rescue	10 (15.38)
Military rescue	34 (52.31)

Note: [a]Some of the inpatients were rescued with a variety of modes.

3.3 Injury sites of the inpatients

The four most common injury sites were multiple traumas in 32 cases (49.23%) of all inpatients, lower limb injuries in 31 cases (47.69%), waist and abdominal injuries in 23 cases (35.38%), and chest and back injuries in 18 cases (27.69%). Among these wounded patients, 48 cases (73.85%) were closed injury. In terms of the injury severity, there were 8 mild cases (12.31%), 17 moderate cases (26.15%), 32 severe cases (49.23%), and 8 extremely severe cases (12.31%). The details of type and severity of the injury at each site are shown in Table 3.

Table 3. Distribution of injury sites of the inpatients following the Lushan earthquake (n (%)).

Injury site	n (%)	Injury type		Injury severity			
		Closed	Open	Mild	Moderate	Severe	Extremely severe
Multiple trauma	32 (49.23)	24 (75.00)	8 (25.00)	1 (3.13)	8 (25.00)	18 (56.25)	5 (15.63)
Lower limb	31 (47.69)	19 (61.29)	12 (38.71)	3 (9.68)	9 (29.03)	14 (45.16)	5 (16.13)
Waist and abdomen	23 (35.38)	18 (78.26)	5 (21.74)	2 (8.70)	7 (30.43)	11 (47.83)	3 (13.04)
Chest and back	18 (27.69)	14 (77.78)	4 (22.22)	2 (11.11)	3 (16.67)	10 (55.56)	3 (16.67)
Maxillofacial region	10 (15.38)	8 (80.00)	2 (20.00)		3 (30.00)	5 (50.00)	2 (20.00)
Hip and pelvis	10 (15.38)	7 (70.00)	3 (30.00)	1 (10.00)		6 (60.00)	3 (30.00)
Craniocerebral trauma	8 (12.31)	6 (75.00)	2 (25.00)	1 (12.50)	2 (25.00)	4 (50.00)	1 (12.50)
Upper limb	7 (10.77)	3 (42.86)	4 (57.14)		1 (14.29)	4 (57.14)	2 (28.57)
Spinal injury	6 (9.23)	4 (66.67)	2 (33.33)			4 (66.67)	2 (33.33)
Neck	3 (4.62)	2 (66.67)	1 (33.33)		1 (33.33)	2 (66.67)	
Total	65	48 (73.85)	17 (26.15)	8 (12.31)	17 (26.15)	32 (49.23)	8 (12.31)

Table 4. Distribution of the injury category of the inpatients following the Lushan earthquake (n (%)).

Injury category	n (%)	Injury type		Injury severity			
		Closed	Open	Mild	Moderate	Severe	Extremely severe
Skin and soft tissue injury	39 (60.00)	28 (71.79)	11 (28.21)	6 (15.38)	11 (28.21)	19 (48.72)	3 (7.69)
Fracture	38 (58.46)	30 (78.95)	8 (21.05)	4 (10.53)	7 (18.42)	23 (60.53)	4 (10.53)
Combined injury	28 (43.08)	21 (75.00)	7 (25.00)	4 (14.29)	3 (10.71)	18 (64.29)	3 (10.71)
Visceral injury	10 (15.38)	10 (100)		2 (20.00)	1 (10.00)	6 (60.00)	1 (10.00)
Crush injury	3 (4.62)	1 (33.33)	2 (66.67)		1 (33.33)	2 (66.67)	
Burn injury	3 (4.62)		3 (100)				3 (100)
Neural injury	2 (3.08)	2 (100)				2 (100)	
Total	65	48 (73.85)	17 (26.15)	8 (12.31)	17 (26.15)	32 (49.23)	8 (12.31)

3.4 *Injury categories of the inpatients*

The four most common injury categories were skin and soft tissue injuries in 39 cases (60.00%), fractures in 38 cases (58.46%), combined injuries in 28 cases (43.08%), and visceral injuries in 10 cases (15.38%). The details of type and severity of each injury category are shown in Table 4.

4 DISCUSSION

Chengdu Military General Hospital, PLA is a modern comprehensive Grade III Class A hospital in southwest China. Our hospital participated in the medical rescues during the "5.12" Wenchuan earthquake (Gu et al., 2009, Gu et al., 2010), "4.14" Yushu earthquake in Qinghai Province, "9.7" Yiliang earthquake in Yunnan Province, and "4.20" Lushan earthquake. During the relief operations for the Lushan earthquake, the military medical teams and hospital played a pivotal role, and have accumulated a large amount of clinical data in the process. The analysis of these clinical data is significant to the local and military hospitals in enhancing the ability to respond to major disasters and improving the treatment quality for emergency mass casualties.

After Lushan earthquake, the patients admitted to our hospital showed a male/female ratio of 1:1.17, which is lower than the 1:1.03 ratio of patients admitted to the West China Hospital after Wenchuan earthquake (Xu et al., 2008). Wounded patients between the ages of 21–60 years old accounted for 66.15% of the patients, and those older than 60 years of age accounted for 26.15%; these data demonstrate essentially the same age distribution seen in the Wenchuan earthquake victims (Liu et al., 2009, Yao et al., 2008). The male/female ratio and the age distribution of the patients provide a guiding significance for the configuration of the hospital beds and medical personnels required in the emergency ward following earthquakes.

Our investigation showed that 96.92% of patients were injured during the main shock, with 43.08% being crushed by objects such as collapsed buildings. On May 1, 2009, China promulgated and implemented the "Earthquake Disaster Mitigation Act", which specified that governments at all levels should organize publicity and education on earthquake disaster reduction to enhance the citizens' awareness and thus improve the social capacity for earthquake disaster reduction. Currently, earthquake disaster reduction education has been conducted gradually in schools, however, education is weak in urban communities and rural villages, which should arouse the attention of the relevant departments.

The injured person in earthquake could be rescued by various methods. Our investigation showed that military rescue saved 52.31% inpatients, while the proportions of patients saved by self rescue, mutual rescue and local rescue are 49.23%, 35.38%, and 15.38%, respectively. After the earthquake, people should immediately begin self rescue and mutual rescue. According to the international experience of disaster relief, external organizations can only rescue a small number of survivors; most survivors are saved by self rescue and locally organized mutual rescue (de Ville de Goyet, 2000). The medical rescue in an earthquake can be divided into four overlapping phases: (1) the first 2–3 days mainly address trauma; (2) the first two weeks mainly address the complications caused by trauma; (3) after two weeks, common medical problems such as obstetrics and gynecology, pediatrics, psychological problems, and so forth are addressed; and (4) the post-disaster period, which mainly addresses patient care and rehabilitation work (von Schreeb et al., 2008). Thus, if the medical institutions outside of the earthquake-stricken area can not immediately send rescue teams to the disaster area to begin rescue operations within two days after the earthquake, donations of money or urgently needed medical equipment and pharmaceuticals are recommended to support the disaster area (Stobbe, 2010).

Among the earthquake victims admitted to our hospital, 26.15% suffered open injuries, which is consistent with the findings in other study (Yao et al., 2008). Patients with severe and extremely severe injuries accounted for 61.54% of patients, which is higher than that of Wenchuan earthquake casualties (Yao et al., 2008, Zhou et al., 2008). This higher proportion in our data may result from the fact that after the Lushan earthquake, our hospital was designated as one of the hospitals to admit earthquake victims with severe injuries and our hospital is close to the airport and the highway; within 3 days after the earthquake, our hospital has admitted 28 casualties with severe injuries who were transported by air.

The four most common injury sites of the inpatients in Lushan earthquake were multiple traumas, the lower limbs, the waist and abdomen, and the chest and back. The four most common injury sites in Wenchuan County were the lower limbs, head, upper limbs, and spine (Zhou et al., 2008), while the four most common sites of the Wenchuan earthquake victims admitted to the West China Hospital and two hospitals in Deyang, Sichuan, were the lower limbs, upper limbs, head, and spine (Liu et al., 2009, Xu et al., 2008). Accordingly, the proportions of the inpatients with waist/abdomen and chest/back injuries admitted to the base hospital in Lushan earthquake are higher than those of the victims treated in the earthquake area and the base hospitals in Wenchuan earthquake.

The four most common injury categories of the inpatients injured in Lushan earthquake were skin and soft tissue injuries, fractures, combined injuries, and visceral injuries. The two most common injury categories of the Wenchuan earthquake victims admitted to a hospital in Deyang were fractures and skin and soft tissue injuries (Yao et al., 2008), while fractures were the common injury category among the Wenchuan earthquake victims admitted to the West China Hospital (Xu et al., 2008).

Using across-sectional survey, this study investigated the severity, site, and category of injuries of the Lushan earthquake victims admitted to our hospital. Through a comparative analysis, we found several earthquake-related regular patterns of hospitalized victims, which are expected to generate interest among relevant scholars and authorities.

ACKNOWLEDGMENTS

This study was supported by project of training research talents, China Postdoctoral Research Fund "2013M532123" and grant "2011YG-C12, 2013YG-B021" from Chengdu Military General Hospital PLA, Chengdu, Sichuan, China.

REFERENCES

Chen, Z. 2009. Review for Renewal-Lessons and Thoughts from the Wenchuan Earthquake Medical Rescue. *Chinese Journal of Evidence Based Medicine* 9(11): 1142.
De Ville De Goyet, C. 2000. Stop propagating disaster myths. *Lancet* 356(9231): 762–4.
Gu, J., Yang, W., Cheng, J., et al. 2010. Temporal and spatial characteristics and treatment strategies of traumatic brain injury in Wenchuan earthquake. *Emerg Med J* 27(3): 216–9.
Gu, J., Zhou, H., Yang, T., et al. 2009. Ultimate treatment for a patient with severe traumatic brain injury without intake for 192 hours after Wenchuan massive earthquake. *Chin Med J (Engl)* 122(1): 113–6.
Kang, P., Zhang, L., Liang, W., et al. 2012. Medical evacuation management and clinical characteristics of 3,255 inpatients after the 2010 Yushu earthquake in China. *J Trauma Acute Care Surg* 72(6): 1626–33.
Li, Q., Fu, B. & Liu, Y. 2013. Practice and reflection on medical rescue in Lushan Earthquake. *Hospital Administration Journal of Chinese People's Liberation Army* 20(7): 601–4.
Li, Y., Gu, J., Yang, B., et al. 2011. A number of medical problems and their enlightenments of the medical rescue after the Haiti earthquake. *Chinese Journal of Hospital Administration* 27(5): 353–4.
Liu, G., Wang, P. & Wang, S. 2009. Analysis of injury characteristics and treatment of hospitalized patients attributable to the 2008 Wenchuan earthquake in China: a report of 826 cases. *Chinese Journal of Trauma* 25(5): 446–50.
Stobbe, K. 2010. President's message. The earthquake in Haiti. *Can J Rural Med* 15(2): 50–1.
Sun, Z. & Xie, S. 1997. Establishment of the computer network in small hospitals with the No.1 Military Medical Project. *Hospital Administration Journal of Chinese People's Liberation Army* 4(4): 373–5.
Von Schreeb, J., Riddez, L., Samnegard, H., et al. 2008. Foreign field hospitals in the recent sudden-onset disasters in Iran, Haiti, Indonesia, and Pakistan. *Prehosp Disaster Med* 23(2): 144–51; discussion 152–3.
Xu, S., Cao, Y. & Wang, Y. 2008. Injury characteristics of the earthquake wounded admitted to the West China Hospital of Sichuan University. *Chinese Journal of Trauma* 24(10): 863–4.
Yao, Y., Zhang, L., Cheng, X., et al. 2008. Analysis of the remedy and injury characteristics of the wounded in China's Wenchuan earthquake. *Chinese Journal of Trauma* 24(10): 852–4.
You, C., Chen, X. & Yao, L. 2009. How China responded to the May 2008 earthquake during the emergency and rescue period. *J Public Health Policy* 30(4): 379–93; discussion 393–4.
Zhou, J., Wang, Z., Huang, X., et al. 2008. Injury characteristics of the earthquake-wounded and medical logistics management of Wenchuan County in the "5.12" earthquake. *Chinese Journal of Trauma* 24(7): 488–90.

Study on immunomodulatory activity of a polysaccharide from *blueberry*

Xiyun Sun & Xianjun Meng
College of food, ShenYang Agricultural University, China

ABSTRACT: The immunomodulatory effects of a polysaccharide (BBP3-1) extracted from *Blueberry* were evaluated by Delayed type hypersensitivity, lymphocytes proliferation and ability of peritoneal macrophage phagocytizing neutral red in vitro. The results showed that BBP3-1 high dose group (400 mg/kg of weight. D) indicated a extremely significant inhibition effect of on DTH of mice ($p < 0.01$), and could also promote T lymphocytes proliferation enhance and the ability of peritoneal macrophage phagocytizing neutral red ($p < 0.01$). It was concluded that the BBP3-1 could enhance the immunocompetence function of normal mice.

1 INTRODUCTION

Recently, plenty of studies have shown that many polysaccharides possessed the biological functions of antiviral, anti-aging, resistance to bone marrow suppression, anti-tumor and immunoregulation (Raveendran N, *et al.* 2004; Kim GY, Lee JO 2006; Wang Y. 2006), and as the drug had small side effects or adverse reactions to the body, therefore making, polysaccharide immunomodulatory effects research becoming a hot field of contemporary biological science (Wang J., *et al.* 2008.; Li J., *et al.* 2008; WU X., *et al.* 2010).

Blueberry, also known as *huckleberry* or *mustikka*, which belongs to the *Ericaceae Vaccinium* group of deciduous shrub plants, is the rage of the world due to its high health care value and is recommended by the FAO as one of the five major health fruits.

At present, the biological activity of *blueberry* extract is one of the popular issues studied by domestic and foreign scholars (Svarcova I, Heinrich J, Valentova K. 2007). There is however few report published about soluble polysaccharide physiological activity in *blueberry*. This paper demonstrates that *blueberry* polysaccharide (BBP3-1) possesses certain immunoregulatory activity through experiment of mice in vivo and cell in vitro.

2 MATERIAL AND EQUIPMENT

2.1 *Material*

BBP3-1 was prepared by this laboratory, preparation progress.

Beating after melting the 50 g frozen blueberry fruits, add water (material:water = 1:8) then agitating and digesting in 90°C constant temperature water bath for 1 h. Centrifuge the solution (5000 rpm, 10 min), after filtrate by gauze. After adding water, repeat the extraction procedure aforementioned. Combine the supernatants of twice extraction, rotary evaporated and concentrated to a minimum volume. Add ethanol (volume fraction: 80%) for 8 h's standing, centrifuge (5000 rpm, 10 min) again and collect the precipitation. Defatt after washing it three times with acetone, ether and ethanol, in that order, and collect the precipitation which lyophilized for 24 h.

Prepare the *blueberry* polysaccharide aqueous solution (10 mg/mL), load the 500 mL samples onto a polyamide chromatography column (wet packing column, 2.5 cm × 76.5 cm), equilibrium by distilled water (double than column volume) for decolorization and deproteinzation, after which, the column was eluted with 500 mL distilled water, flow rate was 1.7 mL/min, lyophilized the elutions.

After decoloration and de-protein, *blueberry* polysaccharide went thorough DEAE-52 cellulose colum chromatography and then washed out by distilled water, 0.1 mol/L NaCl, 0.3 mol/L NaCl, 0.5 mol/L NaCL in order, the four components of BBP1, BBP2, BBP3 and BBP4 were acquired at 11.77%, 14.08%, 34.65% and 10.35% yield respectively. BBP3 was further purified by going through Sephacryl S-300 chromatography and washing out by 0.1 mol/L NaCl to get BBP3-1 and BBP3-2 components. BBP3-1 was tested and verified as single polysaccharide through ultraviolet absorption spectrometry, Sephacryl S-300 chromatography and HPLC analysis.

The weight-average molecular weight (Mw) of BBP3-1 is 18643 Da. The constituents of the monosaccharide BBP3-1 are rhamnose, galactose and glucose at molar ratio of 2:3:4 and it is a β-polysaccharide of which the 1-6-glucopyranose chain as its main chain and it contains rhamnose, galactose and glucose.

SRBCs were aseptically collected from healthy adult sheep.

2.2 *Mainly reagent*

RPMI-1640 DMEM; DMEM/HIGH GLUCOSE (including ketonic acid sodium), HYCLONE; penicillin, streptomycin, USA; glutamine, AMRESCO; Hepes, MTT, ConA, Trypan Blue, Sigma Co.; fetal bovine serum, Hang Zhou Sijiqing Reagent Co.

2.3 *Mainly instrument and equipment*

CO_2 Incubator Thermo scientific 8000; Microplate Reader BIO-RAD 680; Inverted Microscope Olympus-CK2; Precision Electronic Balance TP-313 Sartorius Denver; Low Speed Centrifuge AnHui Zhongke ZhongjiaSC-2542; Clean Bench SuZhou AnTai SW-CJ-2F; Electric-heated Thermostatic Water Bath ShangHaiDK-S26; Vertical Pressure Steam Sterilizer ShangHai ShenAnLDZX-50KB; 96-well flat culture plate, 12-well flat culture plate, 96-well U-culture plate, (Costar, USA); culture flask, (Nunc, Denmark)

2.4 *Animals*

Kun Ming mice (female), aged 4–5 years, body weight around 20 g, purchased from Chinese Medial Sciences University.

3 METHOD

3.1 *Delayed type hypersensitivity (DTH)*

Divide 50 nos. of mice into 5 groups by random, each goup 10 nos., namely, control group, model control group, BBP3-1 high dose group (400 mg/kg of weight. D), BBP3-1 moderate dose group (200 mg/kg of weight. D) and BBP3-1 low dose group (100 mg/kg of weight. D). Intragastric administration of BBP3-1 were carried out for 15 days in a row at 0.2 mL for each mouse, while the same amount of normal saline were perfused for mice in control groups. In the 16th day, wash the fresh sterile sheep blood with normal saline 3 times and then inject 2% (V/V, prepared by normal saline) of hematocrit SRBC 20 μL into peritoneal cavity of each mouse to sensitise them. 4 days after sensitizaiton, measure the thickness of sole on both left and right back foot and get a mean value by measuring 3 times at the same position.

Inject hypodermically 20% (V/V, prepared by normal saline) of hematocrit SRBC 20 μL at the same meaured position and measure its thickness of sole after 24 hours and mark the DTH by the difference of sole thickness (sole swelling) between left and right back foot (Tang R., *et al.* 2009).

3.2 Spleen lymphocyte transformation: MTT

3.2.1 Preparation of cell suspension of spleen lymphocytes in mice:
(1) The mice were sacrificed by cervical dislocation, after which, disinfected in the ethanol for 5 min. Under the aseptic condition, the spleen was eviscerateed, Putted the germ-free spleen of the mice after weighing in RPMI-1640 medium without serum, then rinsed twice. Absorbed about 5 mL serum-free RPMI-1640 medium with a 5 mL syringe, and blowed cells out after inserting inside the spleen gently. Repeated this operation several times until the outer membrane of the spleen was transparent. Grinded the remaining spleen with a syringe and went through the 100 mesh sieve.
(2) After collecting all the cell suspension, centrifuged at 1000 rpm for 5 min, abandoned the supernatant. Added tris-NH_4Cl about 3 mL and blowed gently, then removal of the red blood cells, centrifuged at 1000 rpm for 5 min, at last abandoned the supernatant. Repeated this operation for 2 times; Resuspended the cellular precipitation with serum-free RPMI-1640 medium, blowed gently, and centrifuged at 1000 rpm for 5 min, then abandoned the supernatant.
(3) Resuspended the cellular precipitation with complete RPMI-1640 medium. Used the Trypan Blue staining to count living cells to more than 95%, and adjusted the cell density of 5×10^9/L, then made of the spleen lymphocytes suspension.

3.2.2 Experiment setting and testing
Added the groups of mice spleen lymphocyte suspension in 96-well cell flat culture plate (100 μL per well). Setting individual medication group (BBP3-1), combined medication group (BBP3-1 + ConA (5 ug/mL)), injected separately 100 μL into each well with RPMI-1640 complete culture liquid of varied BBP3-1 concentration or blended liquid of various concentrated RPMI-1640 complete culture liquid with ConA, making the final medication concentration at 3.125, 6.25, 12.5, 25, 50, 100 μg/mL respectively.

Meanwhile, positive control group (5 μg/mL ConA) and blank control group (RPMI-1640 complete culture fluid) was set up, each group with 4 wells.

Cultured the 96-well cell culture plate continuously 44 h in 5% CO_2 incubator at 37°C; Then removed the culture plate, added 20 μL MTT solution each well and continued culturing 4 h. Abandoned the supernatant after centrifuging at 1000 rpm for 10 min, and added 150 μL DMSO each well, then oscillated for 10 min until the purple crystal was completely dissolved; Measure the absorbance *OD* value at 570 nm, the mice spleen lymphocyte proliferation and the high-low of the conversion rate were reflected by OD_{570} values (Gao L., et al. 2008).

3.3 Peritoneal macrophages phagocytosis function: Neutral red phagocytosis experimental method

3.3.1 Preparation of cell suspension of peritoneal macrophages (PMΦ) in mice:
The mice were sacrificed by cervical dislocation, after which, disinfected in the ethanol for 5 min. Under the aseptic condition, the peritoneum was exposed, injected intraperitoneally with RPMI-1640 culture solution without serum, the abdomen irrigated adequately (to prevent the viscera from breaking). This was repeated 3∼4 times, the suspension of peritoneal macrophages (PMΦ) collected, centrifuged (1000 rpm of 5 min) and abandoned the supernatant, suspended the RPMI-1640 complete culture solution, counted the living cells more than 95% by TrtpanBlue dyeing, regulated PMΦ density 5×10^9/L.

The PMΦ cells suspension was cultured in 96-well flat cell culture plate (100 μL per well), in a humidified 5%-CO_2 atmosphere at 37°C for 2 h. The plate was removed and the supernatant abandoned and irrigated with the RPMI-1640 culture solution. 100 μL RPMI-1640 was added to the complete culture solution in each well, the cells in the plate well were purified mice PMΦ.

3.3.2 Experiment setting and testing
Took the purified PMΦ and set up the individual medication group (BBP3-1) and the combined medication group. Then injected separately 100 μL into each well with RPMI-1640 complete

culture liquid of varied BBP3-1 concentration or blended liquid of various concentrated RPMI-1640 complete culture liquid with ConA, making the final medication concentration at 12.5, 25, 50, 100 μg/mL respectively.

Set up positive control group (5 μg/mL ConA) and the blank control group (RPMI-1640 complete culture fluid) at the same time, there were 4 wells each group.

PMΦ were centrifuged at 1000 rpm for 10 min after culturing 24 h, than abandoned the supernatant. Add in the 96-well flat cell culture plate (100 μL per well). Added the 100 μL neutral red (1 g/L saline solution) each hole, then continued cultivating 30 min in 5% CO_2 incubator at 37°C; Centrifuged at 1000 rpm for 10 min and then putted away the supernatant. Joined 200 μL PBS each hole to wash plate and abandoned the supernatant, repeat this operation for 2 times; Added lysis liquid-acetic acid:anhydrous ethanol = 1:1(v:v) each hole by 100 μL, stewed overnight at room temperature to cell lysis; Measure the absorbance OD value at 540 nm wavelengths the next day, the strength of the PMΦ phagocytosis in mice was indicated by OD_{540} value (Gao L., et al. 2008).

4 RESULT AND ANALYSIS

Analysis variance and t-test were used. Data is reported as mean ± SD (standard deviation) and average. Statistical calculations were performed by SPSS software (version 12.0).

4.1 Effect of BBP3-1 on DTH of mice

DTH is the method used to detest the cell immune function. It is based on cellular immunity and is a cellular immunity response mediated by the specific sensitized T-cell.

It is seen in table 1 that there was no significant difference ($p > 0.05$) in BBP3-1 moderate dose group and BBP3-1 low dose group when a parallel was drawn with model control group. While relatively strong effect was appeared in BBP3-1 high dose group with extremely significant effect ($p < 0.01$), indicating a significant inhibition effect of BBP3-1 high dose group (400 mg/kg of weight. D) on SRBC-sensitized DTH of mice.

4.2 The effects of mice spleen lymphocyte transformation by BBP3-1 in vitro

It can be found from the table 2 that BBP3-1 can significantly boost the transformation capability of T lymphocyte when acting with normal mice splenic cells in vitro. Compared with control group, BBP3-1 individual medication group all reached extremely significant level ($p < 0.01$) and there exists some certain dose-effect relation between OD valve and BBP3-1 concentrations, that is to say that the lymphocyte transformation capability of normal mice gradually enhanced with the increasing of BBP3-1 concentrations.

When compared with ConA individual positive group, the BBP3-1 individual medication group was also showing a significant increasing trend with a significant difference level of $p < 0.05$ at

Table 1. Effect of BBP3-1 on delayed hypersensitivity (DTH) in mice ($\bar{x} \pm s$, n = 10).

Group	Dosage mg/kg weight.d	Toes Swelling mm
Normal control	0	0.025 ± 0.003
Model control	0	0.062 ± 0.005
HIGH Dosage	400	0.032 ± 0.002**
MID Dosage	200	0.056 ± 0.007
LOW Dosage	100	0.059 ± 0.004

Notes: Compared With Model control, *$p < 0.05$ and **$p < 0.01$.

12.5 μg/mL and an extremely significant difference level of $p < 0.01$, showing a better effect of BBP3-1 at a certain concentration than that of positive control group (ConA 5 μg/mL). There was a decreasing trend when BBP3-1 and ConA joint action group was compared with control group at concentrations lower than 25 μg/mL, demonstrating that a mutual inhibition effect was existed between BBP3-1 in vitro and ConA on the lymphocyte transformation capability of normal mice.

4.3 The effect of phagocytosis of peritoneal macrophage by BBP3-1 in vitro

It is showed in table 3 that a significant enhancement of the phagocytosis of macrophage when BBP3-1 acted directly with peritoneal macrophage of normal mice in vitro. Compared with control group, all reached extremely significant level ($p < 0.01$) and a certain dose-effect relation existed between OD valve and BBP3-1 concentrations. That is to say that the phagocytosis of macrophage of normal mice on neutral red is gradually increasing with the increasing of concentration of BBP3-1. However, the BBP3-1 individual medication group did not showed a better effect than positive control group. Compared with BBP3-1 individual medication group, the BBP3-1 and ConA joint action group also reached extremely difference level ($p < 0.01$), showing that a synergistic enhancement effect was existed in BBP3-1 in vitro and ConA joint action on the improvement of phagocytosis of peritoneal macrophage of normal mice.

Table 2. Effects of BBP3-1 on lymphocyte transformation in mice splenic cells in vitro.

Concentration μg/mL	Group	
	BBP3-1 OD_{570}	BBP3-1 + ConA OD_{570}
Control	0.224 ± 0.003	0.251 ± 0.004**
100	0.295 ± 0.005**△△	0.253 ± 0.006**↓↓
50	0.276 ± 0.005**△△	0.245 ± 0.004**↓↓
25	0.269 ± 0.006**△	0.222 ± 0.006↓↓
12.5	0.261 ± 0.003**△	0.215 ± 0.009↓↓
6.25	0.252 ± 0.007**	0.203 ± 0.006↓↓
3.125	0.236 ± 0.004**	0.195 ± 0.004↓↓

Notes: 1) Compared with control, *$p < 0.05$ and **$p < 0.01$; 2) Compared With Positive control, △$p < 0.05$ and △△$p < 0.01$; 3) Polysaccharides and ConA joint action group and Polysaccharide alone (same concentration) group compared with lower effect, ↓$p < 0.05$ and ↓↓$p < 0.01$.

Table 3. Effect of BBP3-1 on the function of PMΦ to swallow the Neutral red in vitro.

Concentration μg/mL	Group	
	BBP3-1 OD_{540}	BBP3-1 + ConA OD_{540}
Control	0.363 ± 0.005	0.868 ± 0.008**
100	0.468 ± 0.008**	0.859 ± 0.004**↑↑
50	0.455 ± 0.004**	0.753 ± 0.005**↑↑
25	0.437 ± 0.006**	0.661 ± 0.006**↑↑
12.5	0.387 ± 0.003**	0.592 ± 0.004**↑↑

Notes: 1) Compared with control, *$p < 0.05$ and **$p < 0.01$; 2) Polysaccharides and ConA joint action group and Polysaccharide alone (same concentration) group compared to promotion effect, ↑$p < 0.05$ and ↑↑$p < 0.01$.

5 CONCLUSION

BBP3-1 high-dose group (400 mg/kg of weight. D) had a significant inhibition effect on SRBC-sensitized DTH of mice; BBP3-1 in vitro had a significant boosting effect on the lymphocyte transformation capability of normal mice and this effect was much better than that in its positive control group under some certain concentration; BBP3-1 had a significant enhancement effect on phagocytosis of peritoneal macrophage of normal mice on neutral red. It could be preliminarily concluded that BBP3-1 is a good immunomodulator.

ACKNOWLEDGEMENTS

This work was supported by the National Natural Science Foundation of China (31301443) and the Dr. Start-up Foundation of Liaoning Province (20131105), China.

REFERENCES

Gao L., Wang J., Cui J., et al. 2008. Study on Immune Activity of Radix Cynanchi Bungei Polysaccharides. *Food Science*. 29(10):546–548.

Kim GY, Lee JO. 2006. Partial characterization and imminostimulatory effect of a novel polysaccharide-protein complex extracted from Phellinus linteus. *Biosci Biotechnol Biochem*, 70(5):1218–1226.

Li J., Huang Y., Liao R., et al. 2008. Effect of Sira itia Grosvenor polysaccharide on immunity of mice. *Chinese Pharm acological Bulletin*. 24(9):1237–1240.

Raveendran N, Sonia R, Reshma R, et al. 2004. Immune stimulating properties of a novel polysaccharide from the medicinal plant tinospom cordifolia. *Int Immunopharmacol*, 4(13):1645–1659.

Svarcova I, Heinrich J, Valentova K. 2007. Berry fruits as a source of biologically active compounds: the case of Lonicera caerulea [J]. *Biomed Pap Med Fac Univ Palacky Olomouc Czech Repub*, 151:163–174.

Tang R., Wu G., Li J., et al. 2009. Effect of Milk powder treasure on immunological function in Mice. *Chinese Journal of Veterinary Medicine*. 45(2):20–21.

Wang J., Pu Q., He K., Fu T., et al. 2008. In Vitro Immunomodulatory Effects of Polysaccharide from Cyathula officinalis Kuan. *Chinese Journal of Applied and Environmental Biology*. 14(4):481–483.

Wang Y. 2006. Research progress of traditional Chinese medicine polysaccharides. *Practical medical magazine*, 13(6):1021–1022.

Wu X., Zhang K., Zou M., et al. 2010. Effect of Corn Polysaccharide on Immunological Function in Mice. *Chinese Journal of Experimental Traditional Medical Formulae*. 16(17):187–188.

Effects of sodium additives on deacidification and weight gain of squid (Dosidicus gigas)

Y.N. Wang & S.W. Liu
Dalian Ocean University, Dalian, Liaoning Province, China

G.B. Wang
Dayaowan Entry-Exit Inspection and Quarantine Bureau, Dalian, Liaoning Province, China

Z.B. Li & Q.C. Zhao
Inspection and Manufacturing Technology Service Center, Dalian, Liaoning Province, China
Dalian Ocean University, Dalian, Liaoning Province, China

ABSTRACT: Four sodium additives (sodium carbonate, sodium bicarbonate, sodium citrate and sodium tripolyphosphate) on the deacidification and weight gain of giant squid fillets were carried out. The pH, sensory evaluation, weight gain, thawing loss and cooking loss of giant squid fillets were measured, and the orthogonal experiment was done. The results showed that sodium carbonate and sodium bicarbonate could increase pH value of squid fillets significantly. Sodium carbonate showed the highest weight gain, followed by sodium bicarbonate. Sodium citrate and sodium tripolyphosphate indicated higher thawing loss than sodium bicarbonate. Sodium bicarbonate and sodium carbonate showed lower cooking loss than sodium citrate and sodium tripolyphosphate. The orthogonal experiment demonstrated that optimal concentrations of sodium carbonate, sodium bicarbonate, sodium citrate and sodium tripolyphosphate were 0.35%, 0.4%, 0.5% and 0.5%, respectively.

1 INTRODUCTION

Squid is a high-protein, low-fat aquatic products, containing higher levels of abundant minerals such as calcium, iron and essential nutrients for human. At present, varieties of squid products are Peru squid, Pacific squid, New Zealand squid, Argentina squid, etc. In recent years, the yields of North Pacific squid and Argentina squid have fallen substantially. Abundant resources, stability yield and low price features of Peru squid have been gradually developed and used according to Li (2010). The squid processing only stay in simple stage. Development and utilization of giant squid fillets have become one of the most important branches in the field of the aquatic products processing. However, giant squid fillets were loose and acidulous. Deacidification and weight gain of giant squid fillets can improve the quality and stability of color. Wang (2002) proved that the effects of compound phosphate of minced products were better than single phosphate. In order to improve manufacturing technique and quality, a study of the sensory evaluation and consumers liking of the giant squid fillets products were also studied.

2 MATERIAL AND METHODS

2.1 Raw materials

Giantsquid fillets (*Dosidicus gigas*); production place: FRIO MAR S.A.C; catch area: Pacific Ocean FAOZone N°87. The giant squid fillets were stored at $-25°C$ and thawed in a cold room at

2–4°C for 2 h before filleting and skinning. Standardized samples 40 g were cut from the mantles of the skinless fillets.

2.2 pH

The pH was measured according to method of Benjakul (1997). Giant squid fillets were homogenized in 10 volumes water (w/v), and pH was measured using a pH meter (Leizi, Model PHS-3C Meter, China).

2.3 Sensory evaluation

The sensory of products after brining were evaluated by an expert panel of ten persons. The sensory was evaluated as acidity and texture.

2.4 Weight gain

The squid fillets and brine were mixed in 1:2 ratio (w:v) for 3 h at 0 to 10°C. The squid fillets were drained at room temperature (25°C) before the determination of weight gain.

$$X = \frac{M_2 - M_1}{M_1} \times 100\% \tag{1}$$

where M_1 = the weights of giant squid fillets before soaking, g; M_2 = the weights of giant squid fillets after soaking, g.

2.5 Thawing loss

The squid fillets were frozen ten days and placed at 4°C for thawing about 2 h.

$$X = \frac{M_1 - M_2}{M_1} \times 100\% \tag{2}$$

where M_1 = the weights of giant squid fillets before thawing, g; M_2 = the weights of giant squid fillets after thawing, g.

2.6 Cooking yield

The squid fillets were cooked by steaming for 3 min, and immediately drained at room temperature (25°C).

$$X = \frac{M_1 - M_2}{M_1} \times 100\% \tag{3}$$

where M_1 = the weights of giant squid fillets before cooking, g; M_2 = the weights of giant squid fillets after cooking, g.

2.7 Statistical analysis

Data were subjected to analysis of variance using SPSS17.0 procedures. Significant different were expressed at the $p < 0.05$ level. Statistics on a completely randomized design with three replicates were determined.

Table 1. pH changes on different concentrations of four sodium additives of squid fillets.

Additives	pH					
	0.25%	0.30%	0.35%	0.40%	0.45%	0.50%
Sodium carbonate	7.00±0.23	7.06±0.38	7.1±0.26	7.15±0.17	7.36±0.47	7.39±0.35
Sodium bicarbonate	6.72±0.12	6.84±0.21	6.94±0.28	6.88±0.08	7.28±0.20	7.20±0.29
Sodium citrate	6.345±0.01	6.35±0.01	6.41±0.11	6.42±0.16	6.43±0.23	6.43±0.11
Sodium tripolyphosphate	6.38±0.18	6.43±0.14	6.54±0.16	6.58±0.16	6.5±0.14	6.43±0.04

Table 2. The effects of concentrations of four sodium additives on sensory evaluation.

Additives	Sensory evaluation
Sodium carbonate	No sour and soft texture
Sodium bicarbonate	No sour, fresh and tender
Sodium citrate	No sour and soft texture
Sodium tripolyphosphate	Little sour and soft texture

3 RESULTS AND DISCUSSION

3.1 The pH and sensory evaluation

Four sodium additives are strong electrolytes and alkaline solution. The pH on different concentrations of four sodium additives of squid fillets was measured in table 1. With the same concentrations, the pH of sodium carbonate of squid fillets was the highest followed by sodium tripolyphosphate, sodium bicarbonate and sodium citrate. Sodium carbonate and sodium bicarbonate showed that pH of squid fillets increased significantly. Sodium citrate and sodium tripolyphosphate demonstrated that pH of squid fillets was still lower, but higher than the control group (pH 6.32). Considering weight gain, the concentrations of four sodium additives were determined to be around 0.5%. Li (2012) proved that the effect of sodium carbonate on deacidification was the most distinct, followed by sodium bicarbonate and sodium tripolyphosphate. The result of the experiment was the same as this experiment result. Sodium tripolyphosphate played an important role in the water retention. Li (2006) proved that sodium citrate can regulate pH value with buffer performance and chelate heavy metal. It was often used for food processing.

Different concentrations of additives on sensory evaluation were shown in table 2. The tastes of sodium tripolyphosphate were sourer than sodium carbonate, sodium bicarbonate and sodium citrate of squid fillets. Small concentration of sodium tripolyphosphate lead to low pH of squid fillets and the effect of deacidification was not obvious. The tastes of sodium bicarbonate of squid fillets were the best.

3.2 Weight gain

The effects of different concentrations of four sodium additives on weight gain were shown in figure 1. All the four sodium additives could increase the weight gain. With the increase of concentration of the sodium additive, weight gain was raised. Sodium carbonate showed the highest weight gain, followed by Sodium bicarbonate. Detienne and Wicker (1999) used the method of injection and also proved that sodium tripolyphosphate could make the pork weight increase.

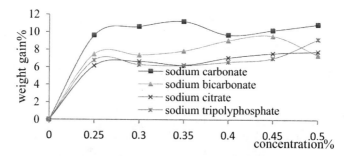

Figure 1. The effects of different concentrations of four sodium additives on weight gain.

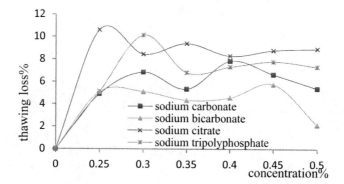

Figure 2. The effects of different concentrations of four sodium additives on thawing loss.

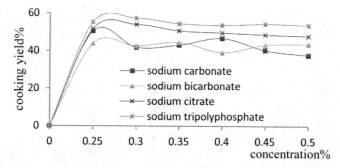

Figure 3. The effects of different concentrations of four sodium additives on cooking loss.

3.3 *Thawing loss*

The effects of different concentrations of four sodium additives on thawing loss were shown in figure 2. Sodium bicarbonate showed the lowest thawing loss and sodium citrate showed the highest thawing loss. Sodium citrate and sodium tripolyphosphate indicated higher thawing loss than sodium bicarbonate. Sheard (2004) proved that thawing loss, yield and cooking loss of compound phosphate had been improved obviously compared with single phosphate.

3.4 *Cooking loss*

The effects of different concentrations of four sodium additives on cooking loss were shown in figure 3. Sodium bicarbonate and sodium carbonate showed lower cooking loss. With

Table 3. Factors and levels of orthogonal experiment.

Level	Factors			
	A Sodium carbonate (%)	B Sodium bicarbonate (%)	C Sodium citrate (%)	D Sodium tripolyphosphate (%)
1	0.3	0.4	0.4	0.4
2	0.35	0.45	0.45	0.45
3	0.4	0.5	0.5	0.5

Table 4. Orthogonal experiment results.

Number	A	B	C	D	Weight gain (%)
1	1	1	1	1	9.98
2	1	2	2	2	10.44
3	1	3	3	3	11.01
4	2	1	2	3	11.82
5	2	2	3	1	10.98
6	2	3	1	2	11.42
7	3	1	3	2	9.97
8	3	2	1	3	10.66
9	3	3	2	1	9.26
K_1	10.48	10.51	10.62	10.07	
K_2	11.41	10.43	10.51	10.61	
K_3	9.96	10.49	10.65	11.16	
R	1.45	0.08	0.14	1.09	
the best level	A_2	B_1	C_3	D_3	

concentrations of four sodium additives increased, the cooking loss decreased but no significant difference In order to reduce the cooking loss effectively, compound additives could be used.

3.5 *Orthogonal experiment*

Taking weight gain as index and four sodium additives as factors, orthogonal optimization method was used According to the index of weight gain, the concentrations of sodium carbonate were from 0.3% to 0.4%, the concentrations of other three sodium additives were from 0.4% to 0.5% respectively Nine tests were carried out with the orthogonal experiment $L_9(3^4)$. The factors and levels of orthogonal experiment were shown in Table 3. The orthogonal experiment results were shown in table 4.

Orthogonal experiment results were shown in table 4. The best level of compound additive was $A_2B_1C_3D_3$. The orthogonal experiment demonstrated that optimal concentrations of sodium carbonate, sodium bicarbonate, sodium citrate and sodium tripolyphosphate were 0.35%, 0.4%, 0.5% and 0.5%, respectively. This optimal group was not in the orthogonal experiment, so the experiment should be verified. The results showed that the weight gain of the optimal group was 12.02%. It was higher than other nine groups. The pH of the optimal group was 7.49. The sensory evaluation results of nine groups were shown in table 5. The sensory evaluation of group 1, 4, 7 and optimal group were better. The weight gain of group 1, 4, 7 were 9.98%, 11.82% and 9.97%, respectively.

Table 5. The sensory evaluation results.

Index	Effectiveness									Optimal group
	1	2	3	4	5	6	7	8	9	
Acidity	+++	++	++	+++	+++	++	+++	++	+	+++
Colour	++	++	++	++	++	++	++	++	++	++
Flexible	+++	+++	+++	+++	++	+++	+++	+++	++	+++
Texture	+++	+++	+++	+++	+++	+++	+++	+++	+++	+++

*Acidity: +++ No sour; ++ Little sour; + Sour. Colour: +++ White; ++ More white; + Common white. Flexible: +++ Strong; ++ Weaker; +None. Texture: +++ Soft; ++ Common tough; + Tough.

4 CONCLUSIONS

The effects of compound additives on deacidification and weight gain were better than that of single additive. Compound additives could neutralize acid of squid fillets and increase the weight gain. The orthogonal experiment demonstrated that optimal concentrations of sodium carbonate, sodium bicarbonate, sodium citrate and sodium tripolyphosphate were 0.35%, 0.4%, 0.5% and 0.5%, respectively. At the same time, color, smell, taste and texture of the squid products were accepted by consumers and the results of this research can be applied to the production practice.

REFERENCES

Detienne, N.A. & Wicker, L. 1999. Effects of sodium chloride and tripolyphosphate on physical and quality characteristics of injected pork loins. *Journal of Food Science* 64(6):1042–1047.

Li, B.G. & Zhao, B.Q. 2006. The effects of phosphate, sodium citrate, sodium bicarbonate and heating on the color value of gardenia red pigment. *Modern Food Science and Technology* 22(4):99–103.

Li Y.H. 2010. Preliminary study on techniques of improving soft texture of Peru squid muscle. *Science and Technology of Food Industry* 31(8):251–254.

Li, S.F. 2012. Study on the deacidification and deodorization technique of tuna and product development. *Master's degree thesis of Zhejiang Ocean University*: 1–10.

Sheard P.R. & Tali A. 2004. Injection of salt, tripolyphosphate and bicarbonate marinade solutions to improve the yield and tenderness of cooked pork lion. *Meat Science* 68:305–311.

Soottawat, Benjakul et al. 1997. Physicochemical changes in Pacific whiting muscle proteins during iced storage. *Journal of Food Science* 62(4):729–733.

Wang, X.R. 2002. Study on water holding effects of mixed phosphates on surimi products. *Food Science and Technology* 9:50–51.

Advanced Engineering and Technology – Xie & Huang (Eds)
© 2014 Taylor & Francis Group, London, ISBN 978-1-138-02636-0

Preservation effects of active packaging membrane from allyl isothiocyanate molecularly imprinted polymers cooperating chitosan on chilled meat

Kuan Lu, Yunan Huang & Jian Wu
School of Liquor & Food Engineering, Guizhou University, Guiyang, China

Qiujin Zhu
School of Liquor & Food Engineering, Guizhou University, Guiyang, China;
Key Laboratory of Agricultural and Animal Products Store and Processing of Guizhou Province, Guiyang, China;
Branch Center of National Beef Processing Technology Research, Huishui, China

ABSTRACT: Active Packaging membrane with slow-release function was produced by allyl isothiocyanate molecularly imprinted polymer (AITC-MIPs) cooperating chitosan (AITC-MIPs-co-CS). The preservation effects of the membrane were investigated according to the physicochemical indexes on chilling meat stored at 4°C. At the same time, unpackaged group, pure chitosan coated membrane group and chitosan membrane including AITC inclusion complex group were set as the control samples. Preservation effects of AITC-MIPs-co-CS on chilled meat is superior to the other control groups, which can effectively inhibit the growth of bacteria and delay the rise of volatile base nitrogen (TVB-N), sulfur generation barbituric acid reagent (TBARS) and pH. After 11 d storage, pH is 6.48, TVB-N 15.12 mg/100 g, and total number of bacteria 4.52 lg CUF/g. The storage period is 3.67 times for the group 3 as long as the CK group. It can significantly prolong the shelf life of chilled meat by a packaging membrane of AITC-MIPs-co-CS.

1 INTRODUCTION

With people's living standards rising, the meat quality has been improved. Chilled meat is welcomed by consumers owing to its good flavor and nutrition (Gill, C.O. 1996). However, although the chilled meat stored at 0°C–4°C, it does not completely inhibit the growth of microorganisms. At present, there are meat preservation methods of high-pressure processing technology, frozen cryopreservation technology, add fresh food preservative and so on at home and abroad (Zhou, H. 2008; Economou, H. et al. 2009; Zhang, J.L. et al. 2010; Friedrich, L. et al., 2008). Among them, the natural preservative has gradually become the hot spot owing to its good safety and effect.

Allyl isothiocyanate (AITC) is a refinement of black mustard oil hydrolyzed by enzyme (Ko, J.A. & Kim, W.Y. 2012; Chin, H.W. et al. 1996). It can be used as an antimicrobial to kill common spoilage bacteria and pathogenesis (Olaimat, A.N. & Holley, R.A. 2013; Nadarajah, D. et al. 2005). In recent years, there are a lot of researches on AITC antimicrobial properties. However, its duration of action is short owing to its strong volatility and irritating (Ko, J.A. & Kim, W.Y. 2012). It can not only reduce the unpleasant smell of AITC, but also delay the release of the AITC by embedding AITC in β-cyclodextrin (β-CD) cavity [Lu, K. & Zhu, Q.J. 2012]. It is reported that the release time of AITC from AITC-β-CD complex is 120 h under different environmental relative humidity (Li, X.H. et al. 2007; Piercey, M.J. et al. 2012).

AITC molecularly imprinted polymers (AITC-MIPs) was prepared to further improve release time of AITC on the basis of the AITC inclusion complex. And then the combination of AITC-MIPs and chitosan (CS) with excellent film-forming properties were applied in packaging materials.

Further, the preservation effect of the packaging materials was studied on chilled fresh pork. It can provide experimental data for application of AITC inclusion compound used in food packaging.

2 MATERIALS AND METHODS

2.1 *Materials and equipment*

2.1.1 *Material*

AITC (99% purity), dimethyl sulfoxide (DMSO) and 2,4-toluene diisocyanate (TDI) were purchased from Xiya Chemical Technology Co., Ltd., Chengdu, China. β-CD (99% purity) was purchased from Sigma-Aldrich Company, USA. Chilled fresh pork was from the local supermarket (cold chain storage, Huaxi, Guiyang, Chian). CS was from Beijing Solarbio Science & Technology Co., Ltd., Beijing, China. Packing paper was from Yufeng Paper Products Co., Ltd., Xiong County, Hebei province, China. All the other reagents were of analytical grade.

2.1.2 *Equipment*

HJ-6A type magnetic stirrer was from Wenhua Instrument Co., Ltd., Jintan, Jiangsu Province, China. PHS-3C type pH meter was from Youke instruments Instrument Co., Ltd., Shanghai, China. TU-1810 UV spectrophotometer was from Purkinje General Instrument Co., Ltd., Beijing, China. TGL-16B centrifuge was from Anting Scientific Instrument Factory, Shanghai, China. HH-S6 constant temperature water bath was from Kewei Yongxing Instrument Co., Ltd., Beijing, China. SPX-150B-Z Incubator was from Boxun Industrial Co., Ltd., Shanghai, China. SHZ-type III circulating water pump was from Ya Rong Biochemical Instrument Factory, Shanghai, China. C-LM3 digital muscle tenderness instrument was from College of Engineering, Northeast Agricultural University, Haerbin, China.

2.2 *Method*

2.2.1 *Prepararion of AITC inclusion complex*

The reaction system for preparation of AITC inclusion complex was composed of 5 g β-CD dissolved in 150 ml distilled water and 2 ml mixture of AITC and ethanol (v/v, 1:1) was added. Then, it was kept on stirring at 45°C for 4 hours, followed by filtration to get AITC inclusion complex. The precipitate was washed repeatedly with 45°C water. Subsequently, it was put into an oven for 12 h. Next, it was ground and stored in a desiccator to reserve.

AITC content in AITC inclusion complex was measured (Padukka, I. et al. 2000), which was 83.86 μL/g.

2.2.2 *Preparation of AITC-MIPs*

The AITC-MIPs were prepared from DMSO, AITC, functional monomer β-CD materials, cross-linking monomer TDI by co-precipitation. The preparation process of a molecularly imprinted polymer was as follows: 5 g β-CD dissolved in 60 ml dry DMSO and 1 ml AITC were added to 150 ml Erlenmeyer flask under the protection of nitrogen. Then, β-CD and AITC were in full reaction at 45°C for 24 h. After that, the bubble in the reaction system was driven off by ultrasound, followed by the exhaust air by nitrogen. Next, 5 ml of TDI was added to the reaction system under the protection of nitrogen to conduct the reaction with vigorous stirring at 65°C for 24 h. The reaction solution was added to excessive acetone to give a milky white flocculent precipitate. The precipitate was obtained by vacuum filtration, washed 2–3 times with 50°C distilled water and acetone, respectively. Subsequently, it was dried in oven for 12 h. Next, it was ground and stored in a desiccator to reserve.

AITC content in AITC-MIPS was measured, which was 75.58 μL/g.

2.2.3 Preparation of packaging film

1 g CS was dissolved in 50 ml 1% (v/v) acetic acid to get 2% (w/v) coating solution. Air in the CS coating soultion was driven off by vacuum for 30 min. Then, it was evenly coated on 30 cm × 30 cm fresh paper, dried and used as the group 1. AITC inclusion complex 0.5 g was gradually added to the CS coating solution, mixed well and was out of air by vacuum for 30 min. It was evenly coated on 30 cm × 30 cm fresh paper, dried and used as the group 2. Similarly, the CS coating soultion containing AITC-MIPs was evenly coated on 30 cm × 30 cm fresh paper, dried and used as the group 3. The CK group was chilled meat without wrap.

2.2.4 Chilled fresh pork processing

Chilled meat 100 g was packaged into the groups 1 to 3; the CK group without plastic wrap. Each group was composed of 15 samples. They stored in a 4°C refrigerator. Indexes of chilled meat were measured from one sample of each group every day.

2.2.5 Measure of indicators during storage

2.2.5.1 Determination of pH

Addition of 10 g chilled meat sample grinding small to 100 ml distilled water was stirred for 5 min, and shocked for 20 min. After filtration, the pH of the filtrate was measured. Evaluation criteria were referred to some report (Warriss, P. D. 2010): ideal pH value of meat to start with is 5.8 to 6.3 for 48 hours after slaughter.

2.2.5.2 Determination of shear force

Refer to the method of Li (Li, C.B. et al. 2010). Remove the pieces on the subcutaneous fat and connective tissue, 80°C hot water bath until the center temperature of meat reach 70°C. The specimens were taken with a diameter of 1.27 cm sampler followed vertical muscle fiber direction. The fresh meat tenderness was measured by C – LM3 digital display type muscle tenderness meter.

2.2.5.3 Determination of thiobarbituric acid (TBA)

Addition of 10.0 g chilled meat grinding small to 50 ml 7.5% trichloroacetic acid was shock for 30 min, followed by filtration. Five ml filtrate was added to 5 ml 0.02 mol/L 2-thiobarbituric acid and remained in a boiling water bath for 40 min and then was taken out to cool for 1 h. Next, it was centrifuged 5 min at 1600 rpm. The supernatant was added to 5 ml chloroform, mixed well and remained stationary for a while. And then the supernatant was respectively measured at 532 nm and 600 nm by spectrophotometer.

2.2.5.4 Determination of volatile basic nitrogen (TVB-N)

TVB-N was measured according to GB / T 5099.44-2003, semi-micro diffusion method. Evaluation standard was as followed: TVB-N $\leq$ 15 mg/100 g for fresh meat (Fresh and frozen pork lean, cuts. GB/T9959.2-2008).

2.2.5.5 Determination of the total number of bacteria

The total number of bacteria was measured according to GB / T 4789.2-2010. Evaluation criteria were as followed: logarithm total number of colonies of (lg CUF) < 4 was for primary chilled meat, 4 < lg CUF < 6 for secondary fresh meat, and lg CUF > 6 for deterioration meat.

The end of the experiment was determined by primary criteria for each experimental indicator.

2.2.5.6 Chroma

Chilled meat was cut by approximate 50 mm × 50 mm. L (brightness value), a (red) and b (yellowness) was measured by WSD-III portable whiteness colorimeter, respectively.

Table 1. Standard of sensory evaluation of meat.

Group	Sensory score
Color (25′)	brown 0′, dull-red 1′–5′, dark red 6′–10′, red 11′–20′, light red 21′–25′
Flavor (25′)	stink 0′, smelly 1′–10′, ammoniacal odour 11′–15′, Acid smell 16′–20′, no abnormal odour 21′–25′
Elastane (25′)	poor 0′–5′, slightly elastane 6′–15′, good elastane 16′–20′, better elastane 21′–25′
Stickiness (25′)	dried-up 0′, dry 1′–5′, sticky 6′–15′, slightly sticky 16′–20′, normal 21′–25′

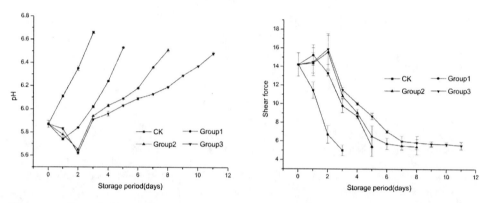

Figure 1. Effects of different packing conditions on pH and shera force of chilled meat.

2.2.5.7 *Sensory evaluation*

The standards of sensory evaluation for color, smell, elasticity, viscosity of chilled meat are shown in Table 1.

2.2.6 *Statistical analysis*

Each index was repeated three times, and the results were averaged. The experimental data were statistical analyzed by SPSS 11.5.

3 RESULTS AND DISCUSSION

3.1 *pH of different chilled meat groups*

After slaughter, it happened to produce lactic acid and other acidic by anaerobic glycolysis of glycogen in pig without supply of oxygen. Therefore, it affected the change of pH of pork, pH value can be used as important reference of evaluation of fresh meat quality (Warriss, P.D. 2010; Viljoen, H.F. et al. 2002).

However, when the pH drops to a maximum value, owing to protein decomposition in pigs to produce alkaline by the activation of the endogenous enzymes and microorganisms, it is caused that pH is gradually increased. As shown in Figure 1, initial pH of the chilled fresh pork is 5.87 for fresh meat. With the increase of storage time, pH generally decreased and then increased, which was in accordance with the report by Boakye (Boakye, K. & Mittal, G.S. 1993). As for the group 1, its pH was up to the critical point (5.74) on the first day. As for groups 2 and 3, on the second day, their pH reached the critical point, 5.65 and 5.62, respectively. Both of their critical pH were lower than that of the group 1. Thus, it suggested that chitosan had some extend of preservation to chilled meat (Piercey, M.J. et al. 2012), which was consistent with the report by Jeon (Jeon, Y.J. et al. 2002). It took more time to reach a pH critical point for the groups 2 and 3, because AITC inclusion complex and AITC-MIPS could make a release of AITC slow and make inhibition of the growth of microorganisms longer. The pH rise of group 3 was the slowest.

On the third day, the pH of the CK group was 6.66 close to the pH of the deterioration of meat (pH > 6.7). As for groups 1 to 3, their pH exceeded the fresh meat standard on the fifth day, eighth day and eleventh day, respectively. The increase of their pH was 0.66, 0.64 and 0.61, respectively. When the group 3 was stored for 11 d, its pH reached 6.48. The shelf life of the group 3 was 3.7 times as long as the control group. There were no significant for the shelf life between the group 1 and 2 ($P > 0.05$) while there were significant ($P < 0.05$) between the CK group and group 3.

3.2 Shear force of different chilled meat groups

The Figure 1 showed that the initial shear force of cold fresh pork was 14.21 N, CK group was on the decline, group 1, 2, 3 were generally increased and then decreased. The group 1 reached the peak on the 1 d, shear force was 15.53 N. The group 2 and 3 reached the peak on the 2 d, the shear force respectively were 17.54 N and 17.88 N.

On the third day, the shear force of the CK group decreased by 9.21 N were 2.08 times, 2.72 times and 3.41 times as group 1, 2 and 3, respectively. It proved that the packing processing of chilled meat had better preservation effect that can effectively extend the shelf life of chilled meat. The shear force of group 1 decreased faster than group 2 and group 3 during the storage, the shear force respectively were 5.39 N, 6.49 N and 8.68 N on the 5 d. The shear force of group 2 and group 3 decreased slowly. The shear force of group 2 was 5.38 N on the 8 d and group 3 was 5.53 N on the 11 d. There were significant ($P < 0.05$) between the CK group and group 1, 2 and 3.

3.3 Thiobarbituric acid (TBA) of different chilled meat groups

Thiobarbituric acid (TBA) content is an important indicator to determine whether the meat was fresh or not during its preservation process. Thiobarbituric acid reactive substances (TBARS) values of the samples increased with the increasing concentration of lactic acid and the days of storage (Sundar, S. & Zhang, M. 2006). As shown in Figure 2, the initial TBA content in chilled meat was 0.11 mg/kg. TBA content in each treatment group displayed an upward trend with the increasing storage time. There was significantly different ($P < 0.05$) for TBA content between the treatment groups, showed that there are significant differences in preservation effect of chilled meat of different treatment. On the third day, TBA content in the CK group increased 0.85 mg/kg, which was respectively 1.9 times, 3.4 times, 4.7 times as much as the group 1, 2 and 3. In addition, TBA content in the CK group was 0.97 mg/kg, close to metamorphic meat, while TBA content in the group 1, 2 and 3 did not reach 1 mg/kg.

On the fifth day, TBA content in group 1 was 0.97 mg/kg, while that in the group 2 and 3 was 0.53 mg/kg, 0.43 mg/kg, respectively. It showed that AITC can be delayed release in the group 2 and 3, which can effectively maintain the freshness of the chilled meat. On the eighth day, TBA content in group 2 was 0.94 mg/kg, while that in the group 3 only 0.63 mg/kg. It indicated that the release time of AITC was longest by formation of AITC-MIPs-co-CS, which could inhibit microbial growth to the maximum extent, maintain the freshness of the chilled meat and extend the shelf life. On the eleventh day, TBA content in group 3 was 0.92 mg/kg. The shelf life of the group 3 was 3.67 times as long as the CK group.

3.4 Volatile basic nitrogen (TVB-N) content in different chilled meat groups

Volatile basic nitrogen (TVB-N), alkaline nitrogen-containing substance, is produced by deamination and decarboxylation of protein by microbial. Its content and the total number of bacteria are indicators to judge the meat freshness. According to the national hygiene standards, TVB-N $\leq$ 15 mg/100 g reach to the index of fresh meat (Fresh and frozen pork lean, cuts. GB/T9959.2-2008). As shown in Figure 2, initial TVB-N content in chilled meat was 8.4 mg/100 g.

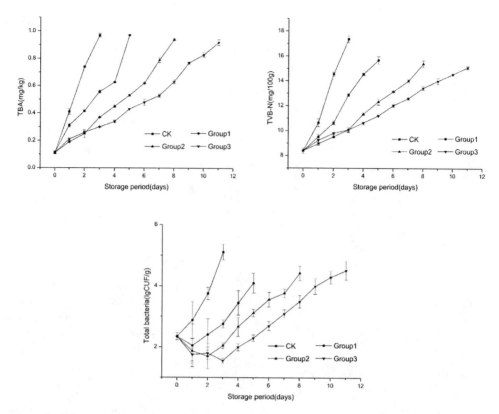

Figure 2. Effects of different packing conditions on the TBA, TVB-N and total bacteria of chilled meat.

TVB-N content in each treatment group displayed an upward trend with the increasing storage time. There was significantly different ($P < 0.05$) for TVB-N content between the treatment groups. TVB-N content of CK group rise was the fastest. On the third day, TVB-N content in the CK group was up to 17.36 mg/100 g, while TVB-N content in the group 1, 2 and 3 was respectively 12.88 mg/100 g, 10.08 mg/100 g and 10.08 mg/100 g, still belonging to fresh meat.

On the fifth day, TVB-N content in group 1 was 15.68 mg/100 g, while that in the group 2 and 3 was respectively 12.32 mg/100 g and 11.20 mg/100 g. It indicated that AITC could be delayed release in the group 2 and 3, and could effectively relieve the decomposition of protein in the meat to maintain the good shape and nutrients of the chilled meat. On the eighth day, TVB-N content in group 2 was 15.44 mg/100 g, that in the group 3 was 13.44 mg/100 g. It increased slowest in the group 3, and was slightly more than 15 mg/kg (15.12 mg/100 g) at storage time of 11 days. It showed that AITC could be better delayed to release by formation of AITC-MIPs-co-CS, suppress the rise of volatile basic nitrogen and maintain the protein content and nutrients of the chilled meat.

3.5 Total number of bacteria in different chilled meat groups

As can be seen in Figure 2, the initial total number of colonies in chilled meat was 2.35 lg CUF/g. Total number of colonies in the CK group displayed an upward trend with the increasing storage time. There was significantly different ($P < 0.05$) for total number of colonies between the treatment groups. The trend line of microbial growth is steep for the CK group. On the third day, the total number of colonies in the CK group was 5.11 lg CUF/g, while that in the group 1, 2 and 3 was respectively 2.76 lg CUF/g, 2.05 lg CUF/g and 1.55 lg CUF/g, all within the scope of fresh meat.

It showed that the total number of colonies in the group 1, 2 and 3 all initially decreased and then increased owing to the inhibition of microbial growth by chitosan and AITC.

The chilled meat in the group 1, 2 and 3 all occurred deterioration, when the storage time was respectively 5 d, 8 d and 11d. Correspondingly, the total number of colonies was respectively 4.1 lg CUF/g, 4.43, lg CUF/g and 4.52 lg CUF/g. On the whole, the microbial grown slowest in the group 3, its shelf life was 3.67 times as long as the CK group. In spite of the action of chitosan as food preservative would involve antioxidant and antimicrobial pathways (Serrano, R. & Bañón, S. 2012), the group 3 displayed that AITC could be better delayed to release by formation of AITC-MIPs-co-CS, and thus able to achieve a better preservation effect to extend the shelf life of chilled meat.

3.6 Chilled meat color value of different groups

3.6.1 L value of different chilled meat

The L value represents the brightness of the meat. As shown in Figure 3, initial L value of chilled fresh pork was 50.48 in the CK group and it displayed a downward trend. The L value of the groups 1 to 3 all initially increased and then decreased, which was consistent with the findings of Wang (Jeon, Y.J. et al. 2002). The L value of the group 1, 2 and 3 reached peak on the first day, on the second day and on the third day, the peak value of which was 51.91, 54.04 and 55.39. The reason was that pH changes will cause the loss of juices and the decease of water content, which lead to the initial increase of the brightness and then the decrease.

On the third day, the L value of CK group decreased 15.64, which was respectively 1.2 times, 2.6 times, and 2.5 times as much as the group 1, 2 and 3. During storage, the L value of the group 1 decreased faster, relative to group 2 and 3, indicating that AITC can delay the decline of L value of chilled fresh pork and maintain a high brightness. On the eight day, the L value of group 2 was 35.55, while on the eleventh day that of the group 3 was 34.77. It indicated that the brightness and good sensory shape of the chilled meat could be effectively maintained in the group 3 owing to the inhibition of loss of water in the meat by AITC-MIPs-co-CS, maintain good sensory shape. During storage, there was significantly different ($P < 0.05$) for the L value between the treatment groups. Moreover, on the fourth day later, the L value of group 3 was kept no significant difference ($P > 0.05$), which suggested that L value was delayed to decrease and the freshness of chilled fresh pork was maintained by AITC-MIPs-co-CS.

3.6.2 a value of different chilled meat

a value is the redness of meat, relative to the content of myoglobin in meat. As can be seen in Figure 3, initial a value of chilled meat was 11.94. During the storage, a value of the CK group, the group 1, 2 and 3 all initially increased and then decreased ($P < 0.05$).

On the third day, a value of the group 1 and 2 reached a peak, 17.66 and 17.40, respectively. On the fourth day, a value of the group 3 reached a peak of 17.30. After that, a value of each treatment group showed a downward trend, which was in accordance with the report(Farouk, M.M. 2007). That is because some of myoglobin is brought out owing the loss of juice and water early in the storage, which causes the increase of a value. When it is stored for a certain period, a low concentration of oxygen would prevent the formation of oxymyoglobin to the gradual formation of metmyoglobin, which causes the decline of a value. It could effectively reduce a value (Tasić, T. et al. 2013) and maintain good color by packaging process.

3.6.3 b value of different chilled meat groups

b value reflects the yellowness of the meat, which is a positive correlation to the degree of oxidation of meat. As shown in Figure 3, initial b of chilled meat was 7.73. During storage, the b value of each treatment group showed an overall upward trend ($P < 0.05$). On the third day, the b value of the CK group was 8.98, which was respectively 1.19 times, 1.95 times and 2.91 times as much as the group 1, 2 and 3. The reason is that it can effectively cut off the contact of chilled meat and oxygen and reduce the degree of fat oxidation of chilled meat by packaging.

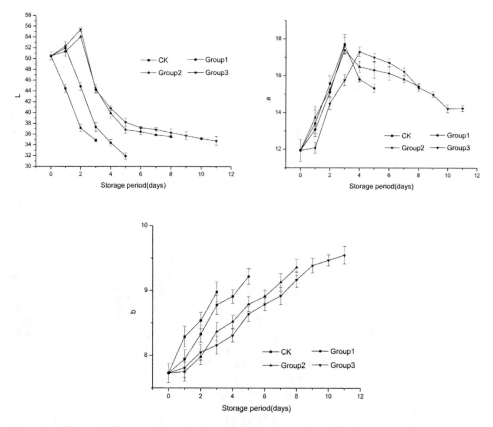

Figure 3. Effects of different packing conditions on the L, a and b value of chilled meat.

On the fifth day, b value of the group 1 increased 1.49, which was respectively 1.41 times and 1.64 times as much as the group 2 and 3. That is because AITC in the group 2 and 3 can be delayed to release and inhibit microbial growth and reproduction, which caused the decrease of fat oxidation and the preservation nutrients of chilled meat.

On the eighth day, b value of the group 2 increased 1.63, which was 1.13 times as much as the group 3. It suggested that AITC could be better delayed to release by formation of AITC-MIPs-co-CS, and thus caused better antibacterial effect and antioxidant capacity. It further delayed the oxidation of fat of chilled meat.

3.7 Sensory evaluation of chilled meat of different groups

As shown in Tables 2 and 3, the initial sensory score of chilled meat was 86 points with excellent quality. After 11d storage, sensory evaluation of the CK group was worst, its sensory score only for 3 points, while that of the group 3 was best, its sensory score 61 points. The results showed that it was group 3 > group 2 > group 1 > CK group for the sensory quality of each treatment group. It indicated that it was best for the delay of AITC release by AITC-MIPs-co-CS. Therefore, it could well maintain color, odor, elastic, viscous and the good sensory shape of chilled meat, extending the shelf life of the meat to improve the economic value by the packaging membrane of AITC-MIPs-co-CS.

It showed that it could extend the storage period of chilled meat and improve economic value by the packaging membrane of AITC-MIPs-co-CS.

Table 2. Sensory evaluation of chilled meat.

	Group	0 d	4th d	7th d	11st d
Color	CK	light red	dark red	dull-red	brown
	Group 1	light red	red	dark red	dull-red
	Group 2	light red	red	dark red	dull-red
	Group 3	light red	red	dark red	dull-red
Flavour	CK	no abnormal odor	acid smell	ammoniacal odor	stink
	Group 1	no abnormal odor	acid smell	acid smell	ammoniacal odour
	Group 2	no abnormal odor	light AITC flavour	light acid smell	ammoniacal odour
	Group 3	no abnormal odor	light AITC flavour	light AITC flavour	acid smell
Elastane	CK	better elastane	slightly elastane	poor	poor
	Group 1	better elastane	good elastane	slightly elastane	poor

Table 3. Sensory evaluation and score of chilled meat after 11 days.

Group	Sensory evaluation	Score
Initial sensory	light red, no abnormal odour, better elastane, normal	86
CK	brown, stink, poor, dried-up	3
Group 1	dull-red, ammoniacal odour, poor, dry	17
Group 2	dull-red, ammoniacal odour, slightly elastane, dry	28
Group 3	dark red, acid smell, good elastane, slightly sticky	61

4 CONCLUSION

These results indicated that the preservation effect of the group 3 was optimal. After storage 11d, the pH of the group 3 was 6.48, volatile basic nitrogen (TVB-N) 15.12 mg/100 g, the total number of colonies 4.52 and lg CUF/g. The storage time was 3.67 times for the group 3 as long as the CK group. The indicator values for the group 3 were just over fresh meat standards, indicating that the release of AITC could be better delayed by AITC-MIPs than by AITC inclusion complex. It can can effectively suppress the rise of the pH, decrease of TBARS and TVB-N, inhibit the propagation of microorganisms by the packaging membrane of AITC-MIPs-co-CS.

It is greatly affect for the edible value and the economic benefits of meat due to the short shelf life. To the date, there are few reports on the application of AITC-MIPS in the field of preservation. The results show that it can effectively keep chilled meat to meet primary fresh meat standard for a longer time, improve the quality stability of chilled meat and inhibit microbial growth and by packaging membrane of AITC-MIPs-co-CS. Therefore, it can better extend the preservation effect of chilled meat, effectively extend its shelf life and improve the value of processing of chilled meat by a packaging membrane of AITC-MIPs-co-CS, expanding the areas of application of the AITC-MIPS.

ACKNOWLEDGEMENTS

This work was partly supported by Governor of Guizhou Province for Excellent Scientific (QIANSHENGZHUANKEZI-2010-7) and National Natural Science Foundation of China for funding (31360373) and Key agricultural project of Guizhou province in 2011 (QIANKEHE-NY-Z-2011-3099).

REFERENCES

Boakye, K., Mittal, G.S. 1993. Changes in pH and water holding properties of Longissimus dorsimuscle during beef ageing, *Meat Science*. 34:335–349.

Chin, H. W., Zeng, Q.L., Lindsay, R. C., 1996. Occurrence and flavour properties of sinigrin hydrolysis products in fresh cabbage, *Food Science*. 61 101–104.

Economou, T., Pournis, N., Ntzimani, A., Savvaidis, I. N. 2009. Nisin-EDTA treatments and modified atmosphere packaging to increase fresh chicken meat shelf-life, *Food Chemistry*. 114:1470–1476.

Friedrich, L., Siró, I., Dalmadi, I., Horváth, K., Ágoston, R., Balla, C. 2008. Influence of various preservatives on the quality of minced beef under modified atmosphere at chilled storage, *Meat Science*. 79:332–343.

Farouk, M.M., M., Beggan, Hurst, S., Stuart, A., Dobbie, P. M., Bekhit, A. E.D. 2007. Meat Quality Attributes of Chilled Venison and Beef, *Food Quality*. 30: 1023–1039.

Fresh and frozen pork lean, cuts. GB/T9959.2-2008, The national standard of the People's Republic of China.

Gill, C.O. 1996. Extending the storage life of raw chilled meats, *Meat Science*. 43:99–109.

Zhou, H. 2008. Bioactive packaging technologies for extended shelf life of meat-based products, *Meat Science*. 78:90–103.

Jeon, Y.J., Kamil, J.Y.V.A., Shahidi, F. 2002. Chitosan as an Edible Invisible Film for Quality Preservation of Herring and Atlantic Cod, *Agri. and Food Chem.* 50:5167–5178.

Ko, J.A., Kim, W.Y., Park, H.J. 2012. Effects of microencapsulated Allyl isothiocyanate (AITC) on the extension of the shelf-life of Kimchi, Int. *Food Microbiol*. 153:92–98.

Lu, K., Zhu, Q.J. 2012. The recognition mechanism and evaluation method of molecular imprinting technique based on β-cyclodextrins, *Gongneng Cailiao/J. Functional Materials*. 43:10–15.

Li, X.H., Jin Z.Y., Wang, J. 2007. Complexation of allyl isothiocyanate by α- and β-cyclodextrin and its controlled release characteristics, *Food Chemistry*, 103:461–466.

Li, C. B., Chen, Y. J., Xu, X. L., Huang, M., Hu, T. J., Zhou, G. H. 2010. Effects of low-voltage electrical stimulation and rapid chilling on meat quality characteristics of Chinese yellow crossbred bulls, *Meat Science*. 72: 9–17.

Nadarajah, D., Han, J. H., Holly R. A. 2005. Inactivation of Esherichia coli O157:H7 in packaged ground beef by allyl isothiocyanate, *Food Microbiol*. 99: 269–279.

Olaimat, A.N., Holley, R. A. 2013. Effects of changes in pH and temperature on the inhibition of Salmonella and Listeria monocytogenes by Allyl isothiocyanate, *Food Control*, 34: 414–419.

Piercey, M.J., Mazzanti, G., Budge, S.M., Delaquis, P.J., Paulson, A.T. 2012. Truelstrup Hansen, L., Antimicrobial activity of cyclodextrin entrapped allyl isothiocyanate in a model system and packaged fresh-cut onions, *Food Microbiology*, 30:213–218.

Padukka, I., Bhandari, B., D'Arcy, B. 2000. Evaluation of various extraction methods of encapsulated oil from β-cyclodextrin-lemon oil complex powder, *Food Composition and Analysis*, 13:59–70.

Purnama Darmadji, Masatoshi Izumimoto. 1994. Effect of chitosan in meat preservation, *Meat Science* 38:243–254.

Sundar, S., Zhang, M. 2006. Effect of lactic acid pretreatment on the quality of fresh pork packed in modified atmosphere, *Food Engineering*, 72:254–260.

Serrano, R., Bañón, S. 2012. Reducing SO2 in fresh pork burgers by adding chitosan, *Meat Science*, 92:651–658

Tasić, T., Petrović, L., Ikonić, P., Lazić, V., Jokanović, M., Džinić, N., Tomović, V., Šarić, L., Effect of Storage in a Low Oxygen Gas Atmosphere on Colour and Sensory Properties of Pork Loins, *Packag. Technol. Sci.*, Article first published online: 7 FEB 2013, OI: 10.1002/pts.2011

Viljoen, H.F., de Kock, H.L., Webb, E.C. 2002. Consumer acceptability of dark, firm and dry (DFD) and normal pH beef steaks, *Meat Science*. 61: 181–185.

Warriss, P. D., 2010, An Introductory Text, CABI, ISBN: 9781845935931, *Meat Science -2nd ed*, Printed and bound in the UK by the Cambridge University Press, Cambridge.

Zhang, J.L., Liu, G.R., Li, P.L., Qu, Y. 2010. Pentocin 3-1, a novel meat-borne bacteriocin and its application as biopreservative in chill-stored tray-packaged pork meat, *Food Control*. 21:198–202.

Optimization of acid hydrolysis and ethanol precipitation assisted extraction of pectin from navel orange peel

L.B. Yin, K. Xiao, L.Z. Zhao, J.F. Li & Y.Q. Guo

Department of Biological and Chemical Engineering, Shaoyang University, Shaoyang, Hunan, China

ABSTRACT: In this study, acid hydrolysis and ethanol precipitation method was applied for pectin extraction from navel orange peel. The effects of solid-liquid ratio, extraction temperature, extraction time and extraction pH on the extraction yield of pectin were investigated. After single-factor experiment, four factor three level orthogonal experiment was carried out to optimize the extraction parameters and the optimal extraction conditions were determined (solid-liquid ratio 1:15, extraction temperature 80°C, extraction time 90 min, extraction pH 1.5) with the maximum pectin yield of 26.8%, which was confirmed through validation experiments.

1 INTRODUCTION

Pectin is polymers, widely found in higher plants cell wall and interstitial, usually white or light yellow powder, odorless, taste slightly sweet, slightly sour, soluble in water, slightly soluble in cold water (Dai & Shi, 2012, Kliemann et al., 2009, Kulkarni & Vijayan, 2010). Since pectin has excellent gel property and emulsion stability, it becomes an important food industry additive and had been widely used in candies, juice drinks and other foods (Iftikhar et al., 2013). In recent years, many studies have found pectin as a soluble dietary fiber with anti-diarrhea, diabetes and other effects, expanding the scope of its application (Bagherian et al., 2011, Infante et al., 2013). In addition, pectin is also indispensable auxiliary materials of some drugs, health products and antiproliferative activity (Lowther et al., 2013, Soultani et al., 2014, Concha et al., 2013). Pectin has been extracted from several resources, such as orange peel, apple pomace, cacao pod husk and sugar beet pulp (Ma et al., 2013, Yapo & Koffi, 2013). Orange peel and apple pomace are particularly abundant and contain high levels of pectic polysaccharides.

Navel orange planting acreage is large, with an annual output of tens of millions of tons in China. However navel orange peel containing 20% pectine is discarded, resulting in a waste of resources and environmental pollution. Therefore, extracting pectin from navel orange peel makes full use of resources, turns waste into treasure, and also protects the environment.

Currently, the technologies of extracting pectin mainly include alcohol precipitation method, ion exchange, salt precipitation and microbiological method (Methacanon et al., 2014). Alcohol precipitation method is one of the oldest industrial pectin production methods, and its basic principle is to convert non-water-soluble pectin in plant cells to water-soluble pectin in dilute acid. This method is more mature technology and its process conditions are relatively easy to control, and the pectin products are free of impurities. The main problem of salting method is that desalination is not complete which resulted from technological backwardness, resulting in decreasing viscosity and gel property of pectin. Capacity extracting pectin by ion exchange is small, and its cost is expensive (Guo et al., 2012). Requirements of process conditions of microbial method to extract pectin are strict, thus increasing its production costs and making it difficult to apply to actual production (Jiang et al., 2012, Masmoudi et al., 2012).

The main objective of this study is to develop an alcohol precipitation method extraction of pectin from orange peel and investigate the effect of process variables (solid-liquid ratio, temperature, time and pH).

2 MATERIALS AND METHODS

2.1 Raw materials and reagents

Fresh navel orange, cultivated in the south of Hunan province, was purchased from a local market (Shaoyang, China). The collected orange peel was first soaked in a water bath at 90°C for 5 min to inactivate its enzymes, then drying was carried out at 60°C in a vacuum drying oven until the water content was reduced to 15% (Kumar & Chauhan, 2010). It was then milled to a powder size 2~3 mm in an electric grinder. Finally, the vacuum-packed sample was stored for subsequent extraction.

All chemical reagents, including ethanol, hydrochloric acid, sodium hydroxide, etc., used in the experiments were analytical grade.

2.2 Extraction and purification of pectin

Take a clean and dry flask placed on the electronic balance, add a certain amount of navel orange peel powder and note the weight. Hydrochloric acid solution was added to pretreat the powder, and then it was boiled in thermostat water bath. After the same time interval, the flask was shaken to enhance the acid hydrolysis. Thereafter, the extracts from the peels rapidly cooled to 25°C to minimize heat degradation of the pectin and filtered through a whatman No. 1 filter paper. The filtered extract was centrifuged and the supernatant was precipitated with an equal volume of 95% (v/v) ethanol. The coagulated pectin mass were set into a clean petri dish to dry to constant weight at the temperature of 45°C.

2.3 Determination of pectin yield

The pectin yield (PY) was calculated from the following equation: $PY = (m_0/m) \times 100\%$.

Where m_0 (g) was the weight of dried pectin and m (g) was the weight of dried navel orange peel powder.

2.4 Single-factor experiments for determining the optimal conditions for pectin extraction

To determine the optimal solid-liquid ratio on the yield of pectin, different solid-liquid ratio at 1:10, 1:15, 1:20, 1:25, 1:30 (g/mL) was investigated. Then it was extracted for 90 min at the condition of pH 2.0, 80°C with stirring constantly. To confirm the extraction temperature on the yield of pectin, at the above optimal solid-liquid ratio, the extraction temperature was set from 60 to 100°C, and the extraction conducted at pH 2.0 for 90 min. To affirm the extraction time on the yield of pectin, the extraction procedure conducted at the different time (30, 60, 90, 120, 150 min). To verify the optimal extraction pH on the yield of pectin, different pH range from 1.0 to 3.0 was examined.

2.5 Orthogonal experiment for the optimal conditions for pectin extraction

According to the results of single factor experiments, a four factor three level orthogonal experiment was employed to investigate the individual and interactive effects of process variables on the pectin yield from navel orange peel.

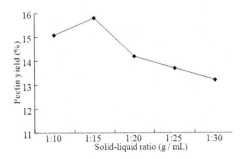

Figure 1. The effect of solid-liquid ratio on pectin yield.

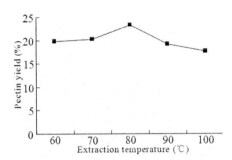

Figure 2. Effect of extraction temperature on pectin yield.

3 RESULTS AND DISCUSSION

3.1 *Effect of solid-liquid ratio on pectin yield*

The pectin yield increased with solid-liquid ratio increasing, and when solid-liquid ratio was 1:15, the maximum pectin yield reached 15.8% (Figure 1). With the further expansion of solid-liquid ratio, pectin yield began to decline. The use of large amounts of navel orange powder results resulted in a decrease of the viscosity of the solution; thus, the solvent circulates freely and made the contact with the navel orange peel power more efficient (Kumar & Chauhan, 2010). Excessive solvent resulted in hydrolysis of the pectin and excessively reducing of the viscosity of the extraction pectin solution made effect of sedimentation not ideal (Urias Orona et al., 2010).

3.2 *Effect of extraction temperature on pectin yield*

Extraction temperature is an important factor effecting pectin yield, five different temperatures (60, 70, 80, 90, 100°C) were tested in the single factor experiment. The result showed that the pectin yield rate were 19.8%, 20.3%, 23.3%, 19.2%, 17.6%, respectively. The optimal extraction temperature for pectin yield extraction was between 70 and 90°C (Figure 2). With the further increase of extraction temperature, pectin yield began to decline. With the temperature increasing above 80°C, pectin degraded and was not able precipitable with alcohol resulting in reduction of pectin yield.

3.3 *Effect of extraction time on pectin yield*

The pectin yield rates were found to be 15.2%, 16.8%, 23.4%, 19.4%, and 17.4% at the extraction time of 30, 60, 90, 120, 150 min, respectively. It showed that the optimal extraction time was 90 min (Figure 3). As the extraction proceeded, the concentration of the pectin in the solution increased

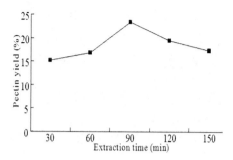

Figure 3. Effect of extraction time on pectin yield.

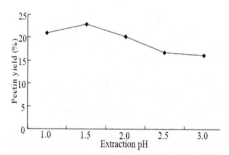

Figure 4. Effect of extraction pH on pectin yield.

and pectin yield progressively decreased for two reasons. First, the concentration gradient was reduced and, second, the solution became more viscous (Tang et al., 2011). A relatively long period of extraction time would cause a thermal degradation effect on pectin yield, thus causing a decrease in the amount precipitable by alcohol (Lerotholi et al., 2012).

3.4 Effect of extraction pH on pectin yield

The pH is also an important factor, which significantly effects the extraction of pectin yield. The pectin yield rates were found to be 20.9%, 22.9%, 20.2%, 16.8%, and 16.1% at the extraction pH of 1.0, 1.5, 2.0, 2.5 and 3.0, respectively. The higher extraction pH was not sufficient to hydrolyze the insoluble pectic constituents (Xu et al., 2011). In this study, the optimum pH value was at 1.5. Under pH of 1.5, saponification reaction and β-elimination reaction of ester pectin induced glycoside bond to fracture between galacturonic acid residues, to form the double bond between galacturonic acid C_4 and C_5, and to form conjugated double bond between carbonyl carbon and carbonyl oxygen, resulting in extraction yield declining (Zhang & Mu, 2011, Zhang et al., 2010).

3.5 Four factors and three levels orthogonal test results and analysis

Four factor three level orthogonal test L_9 (3^4), the use of orthogonal design assistant software (II V3.1, Professional Edition) and the results of the analysis were shown in Table 2.

By range value in the table above, which showed that the primary and secondary order of factors affecting pectin yield was B > C > D > A, namely pH > extraction time > extraction temperature > solid-liquid ratio. According to the K value, the optimal condition was $A_1B_2C_2D_1$, namely material liquid ratio of 1:15, pH of 1.5, extraction temperature of 80°C, the extraction time for 90 min. Which $A_1B_2C_2D_1$ optimal conditions were not in orthogonal table, the optimal extraction conditions were determined with the maximum pectin yield of 26.8%, which was confirmed through validation experiments.

Table 1. Four factor three level orthogonal experiment and its results analysis

Lever	A Solid-liquid ratio (g/mL)	B pH	C Temperature (°C)	D Time(min)	Pectin yield (%)
1	1 (1:15)	1 (1.0)	1 (70)	1 (60)	0.230
2	1	2 (1.5)	2 (80)	2 (90)	0.258
3	1	3 (2.0)	3 (90)	3 (120)	0.149
4	2 (1:20)	1	2	3	0.227
5	2	2	3	1	0.243
6	2	3	1	2	0.156
7	3 (1:25)	1	3	2	0.217
8	3	2	1	3	0.223
9	3	3	2	1	0.196
K1	0.637	0.674	0.609	0.669	
K2	0.626	0.724	0.681	0.631	
K3	0.636	0.501	0.609	0.599	
k1	0.212	0.225	0.203	0.223	
k2	0.209	0.241	0.227	0.210	
k3	0.212	0.167	0.203	0.200	
Range	0.004	0.074	0.024	0.023	
Order of priority	B > C > D > A				
Optimal decision	$A_1 B_2 C_2 D_1$				

4 CONCLUSION

Acid hydrolysis-ethanol precipitation method, as a traditional technology, was applied to extract pectin from navel orange peel and the extraction conditions were optimized using single factor experiment and four factor three level orthogonal experiment. Under the optimal conditions, the extraction yield of acid hydrolysis-ethanol precipitation method was 26.8%. In a word, acid hydrolysis-ethanol precipitation method is an efficient, timesaving, and eco-friendly method for extraction of pectin from navel orange peel, and is very promising to be applied to the industrial production of pectin.

REFERENCES

Bagherian, H., et al. 2011. Comparisons between conventional, microwave-and ultrasound-assisted methods for extraction of pectin from grapefruit. *Chemical Engineering and Processing: Process Intensification*, 50, 1237–1243.
Concha, J., Weinstein, C. & Ziga, M. E. 2013. Production of pectic extracts from sugar beet pulp with antiproliferative activity on a breast cancer cell line. *Frontiers of Chemical Science and Engineering*, 1–8.
Dai, Y. & Shi, H. 2011. Optimization on extraction pectin from orange peel. *China Food Additives*, 6, 9–14.
Dai, Y. & Shi, H. 2012. Optimization on extraction pectin by hydrochloric acid extraction and ethanol precipitation method from orange peel. *Northern Horticulture*, 113, 18–26.
Guo, X., et al. 2012. Extraction of pectin from navel orange peel assisted by ultra-high pressure, microwave or traditional heating: A comparison. *Carbohydrate Polymers*, 88, 441–448.
Iftikhar, T., Wagner, M. E. & Rizvi, S. S. 2013. Enhanced inactivation of pectin methyl esterase in orange juice using modified supercritical carbon dioxide treatment. *International Journal of Food Science & Technology*.
Infanter, B., Garcao, O. & Rivera, C. 2013. Characterization of dietary fiber and pectin of cassava bread obtained from different regions of Venezuela Caracterización del contenido defibra dietariay pectina decasabe obtenido dediferentes. *Rev. chil. nutr*, 40, 169–173.
Jiang, L., Shang, J., He, L. & Dan, J. 2012. Comparisons of microwave-assisted and conventional heating extraction of pectin from seed watermelon peel. *Advanced Materials Research*, 550, 1801–1806.
Kliemann, E., et al. 2009. Optimisation of pectin acid extraction from passion fruit peel (*Passiflora edulis flavicarpa*) using response surface methodology. *International Journal of Food Science & Technology*, 44, 476–483.

Kulkarni, S. & Vijayan, P. 2010. Effect of extraction conditions on the quality characteristics of pectin from passion fruit peel. *LWT-Food Science and Technology*, 43, 1026–1031.

Kumar, A. & Chauhan, G. S. 2010. Extraction and characterization of pectin from apple pomace and its evaluation as lipase (steapsin) inhibitor. *Carbohydrate Polymers*, 82, 454–459.

Lerotholi, L., Carsky, M. & IKHU OMOREGBE, D. 2012. A study of the extraction of pectin from dried lemon peels. *Advanced Materials Research*, 367, 311–318.

Lowther, S. A., et al. 2013. Process for preparation of over-the-counter gelatin or pectin-based drug delivery. US Patent 20,130,005,740.

Ma, S., et al. 2013. Extraction, characterization and spontaneous emulsifying properties of pectin from sugar beet pulp. *Carbohydrate polymers*, 98, 750–753.

Masmoudi, M., et al. 2012. Pectin extraction from lemon by-product with acidified date juice: effect of extraction conditions on chemical composition of pectins. *Food and Bioprocess Technology*, 5, 687–695.

Methacanon, P., Krongsin, J. & Gamonpilas, C. 2014. Pomelo (*Citrus maxima*) pectin: effects of extraction parameters and its properties. *Food Hydrocolloids*, 35, 383–391.

Soultani, G., et al. 2014. The effect of pectin and other constituents on the antioxidant activity of tea. *Food Hydrocolloids*, 35, 727–732.

Tang, P., Wong, C. & Woo, K. 2011. Optimization of pectin extraction from peel of dragon fruit (Hylocereus polyrhizus). *Asian Journal of Biological Sciences*, 4, 189–195.

Urias Orona, V., et al. 2010. A novel pectin material: extraction, characterization and gelling properties. *International journal of molecular sciences*, 11, 3686–3695.

Xu, R., Wu, W. & Nan, Y. 2011. Optimization of process parameters on pectin extraction from orange peel and preparation of pectin pilm *Journal of Tianjin University of Science & Technology*, 2, 70–76.

Yapo, B. M. & Koffi, K. L. 2013. Extraction and characterization of gelling and emulsifying pectin fractions from cacao pod husk. *Nature*, 1, 46–51.

Zhang, C. & Mu, T. 2011. Optimisation of pectin extraction from sweet potato (*Ipomoea batatas, Convolvulaceae*) residues with disodium phosphate solution by response surface method. *International Journal of Food Science & Technology*, 46, 2274–2280.

Zhang, K., et al. 2010. Optimization of Microwave-assisted extraction technology of pectin from lemon peel using response surface methodology. *Fine Chemicals*, 19, 111–115.

Optimization of ultrasonic-assisted enzymatic extraction of phenolics from broccoli inflorescences

H. Wu, J.X. Zhu, L. Yang, R. Wang & C.R. Wang
Qingdao Agricultural University, Qingdao, The P. R. China

ABSTRACT: An efficient ultrasonic-assisted enzymatic extraction (UAEE) technique was developed for the extraction of phenolics from broccoli inflorescences, which is a rich source of phenolic compound, without organic solvents. The enzyme combination was optimized by orthogonal test: cellulase 7.5 mg/g FW (fresh weight), pectinase 10 mg/g FW and papain 1.0 mg/g FW. The operating parameters in UAEE were optimized with response surface methodology using Box-Behnken design. The optimal extraction conditions were as follows: ultrasonic power, 440 W; liquid to material ratio, 7.0:1 mL/g; pH value of 6.0 at 54.5°C for 10 min. Under these conditions, the extraction yield of phenolics achieved 1.816 ± 0.0187 mg GAE/g FW. Compared with UAE and EAE, UAEE had expressed the better characteristic of the highest TPC value.

1 INTRODUCTION

Broccoli is a rich source of phenolic compounds such as flavonoids and hydroxycinnamic acid derivatives (Jahangir et al. 2009). It had been reported that broccoli was related to the reduction of the risk of chronic diseases including cardiovascular diseases and cancer (Matusheski et al. 2006). The main phenolic compounds content and composition of broccoli has been investigated, however, there were few studies related to the green and high-efficient extraction process of phenolic extracts from broccoli. Traditional methods such as conventional solvent extraction and acid extraction (Miean & Mohamed 2001) have been applied to isolate phenolics from plant materials. The major shortcoming of these techniques is the use of toxic organic solvent in the recovery of edible and essential flavonoids, which might lead to solvent contamination and strong corrosion due to the long-term exposure to organic solvents and acidic solutions. Other techniques, which contain ultrasonic-assisted extraction (UAE) (Wang et al. 2013), microwave-assisted extraction (MAE) (Jokić et al. 2012), enzyme-assisted extraction (EAE) (Gómez-García et al. 2012), and supercritical fluids extraction (SFE) (Santos et al. 2012) have also become ideal alternatives to these traditional techniques.

Compared with the traditional extraction techniques, UAE can reduce the solvent consumption, shorten the extraction time, improve the quality of extracts and have recently been widely employed in the plant phenolics extraction processes (Muñiz-Márquez et al. 2013). However, application of ultrasound in extraction of broccoli phenolics has not been reported. Cell wall degradation enzymes in EAE, such as cellulases, hemicellulase, pectinases, proteinase, or a combination of these, have been used for the pretreatment of plant materials to significantly improve the extraction efficiency of desirable phenolics (Pinelo et al. 2008). They can act on the plant cell wall that contains the bioactive constituents, catalyze the hydrolysis of them in turn, thus rendering the phenolic constituents more accessible to the extraction (Pinelo et al. 2006). However, using only a single extraction method to separate and purify the phenolic compounds could not lead to a conspicuous enhancement of the extraction yield. Hence, a green and effective ultrasonic-assisted enzymatic extraction (UAEE) technique, as a relatively novel combination method on the basis of UAE and EAE, may satisfy the high performance for the extraction of phenolic antioxidants from broccoli inflorescences. Recent researches provided another support for UAEE, which indicated that ultrasonic treatment of the cellulose fibers with a preceding or simultaneous enzyme incubation further improved the enzymatic reaction rate (Nguyen and Le 2013).

The objective of the present study was to evaluate the feasibility of UAEE for extraction of phenolic antioxidants from broccoli inflorescences and obtain the maximum yield. Meanwhile the effect of different extraction processes on extraction yield was studied. In this paper, the optimal enzyme combination was confirmed by a 3-level, 3-factorial orthogonal test and the multiple parameters of UAEE, such as liquid to material ratio, ultrasonic power, pH and temperature, were optimized using response surface methodology (RSM) by employing a 3-level, 4-variable Box-Behnken design (BBD).

2 MATERIALS AND METHODS

2.1 Materials and reagents

Fresh broccoli heads (Brassica oleracea L. var. italica) at maturity stage were purchased from a local supermarket in Qingdao City, China. Enzyme source: Commercial cellulase preparation (40000 U/g) was derived from Trichoderma reesei, the pectinase preparation (30000 U/g) was derived from Aspergillus niger and Papain preparation (200000 U/g) was derived from papaya. All the enzymes employed were obtained from Lihua Enzyme Preparation Technology Co., LTD. (Tianjin, China) and selected for this study after preliminary trial tests. All other chemicals and reagents were of analytical grade.

2.2 Sample preparation

Inedible parts (leaves and stems) of broccoli were removed and discarded. Edible parts (primary inflorescences) were washed thoroughly and cut into small pieces (∼3 cm diameter and ∼1 cm stalk; edible florets) with a sharp knife. All the pieces were well mixed, three replicates of broccoli samples were randomly selected for each treatment.

2.3 Determination of Total Phenolics Content (TPC)

The yield of TPC of the extracts was determined by the Folin-Ciocalteu method according to Singleton and Rossi (1965) with some modification. The reaction mixture was composed of 0.2 mL broccoli extractive solution after enzyme deactivation and centrifugation, 5.8 mL distilled water, 1 mL of Folin-Ciocalteu reagent and 3 mL of 20% sodium carbonate anhydrous solution. After incubation for 2 h at 30°C in darkness, the absorbance of samples was measured at 765 nm. The TPC was expressed in terms of gallic acid equivalents (GAE) based on the standard curve given in following equation ($Y =$ absorbance at 765 nm, $x =$ quality of gallic acid, linearity range 0–0.6 µg, $R^2 = 0.9992$):

$$Y = 0.7089x + 0.00361 \tag{1}$$

2.4 Selection and optimization of enzymes combination

UAEE was performed in an multipurpose thermostatic ultrasonic extraction processor (SY-1000E, Beijing hongxianglong biotechnology development Co., Ltd., Beijing, China), with one powerful probe-shaped ultrasonic transducer (Φ15 mm) and digital controlled sonotrode (20 kHz) which can change ultrasonic power (0–900 W) during extraction processing. The processor was also equipped with a heating/cooling system to control the ultrasonic temperature in the medium. In all of the following extraction processes, ultrasound launching duration was set as 2 s, then cooling 1 s.

Different enzymes or enzyme combinations were added to 100 g broccoli samples with a liquid to material ratio of 7:1 mL/g as follows: (A) cellulase 7.5 mg/g FW, (B) pectinase 10 mg/g FW, (C) papain 1.5 mg/g FW, (D) cellulase 7.5 mg/g FW & papain 1.5 mg/g FW, (E) pectinase 10 mg/g FW & papain 1.5 mg/g FW, (F) cellulase 7.5 mg/g FW & pectinase 10 mg/g FW, (G) cellulase 7.5 mg/g

FW, pectinase 10 mg/g FW & papain 1.5 mg/g FW. All UAEE processes were performed subjecting samples to ultrasonification with 500 W for 30 min at 50°C. Enzyme activity was terminated with heat treatment at 90°C for 30 s immediately following extraction treatments (Yang et al. 2010). The extractive solutions cooled to room temperature then were centrifuged 15 min at 302 g. The supernatants were collected for TPC analysis.

A 3^3 orthogonal test, which could economize trial-manufactured expenditure, time and work load, was used to design to optimize the enzyme combination. The three independent factors studied were the cellulase, A (7.5, 10, 12.5 mg/g FW), pectinase, B (7.5, 10, 12.5 mg/g FW) and papain, C (0.75, 1.0, 1.25 mg/g FW) which were coded at three levels as shown in Table 1.

2.5 *Optimization of UAEE process*

On the basis of the optimized enzyme combination, Box-Behnken design (BBD) design was used for optimization of UAEE process variables each at 3 levels with 29 runs including five replicates at the central point. The four independent factors studied were the liquid to material ratio, X_1 (4:1, 6:1, 8:1 mL/g), ultrasonic power, X_2 (300, 400, 500 W), pH, X_3 (4.5, 5.5, 6.5) and temperature, X_4 (50, 55, 60°C), which were coded at three levels as shown in Table 2. Experimental data were fitted to a second-order polynomial regression model containing the coefficient of linear, quadratic, and two factors interaction effects. Equation below is the general equation describing the second degree polynomial model:

$$Y = \beta_0 + \sum_{i=1}^{4} \beta_i X_i + \sum_{i=1}^{3} \sum_{j=i+1}^{4} \beta_{ij} X_i X_j + \sum_{i=1}^{4} \beta_{ii} X_i^2 \qquad (2)$$

where Y is the predicted response X_i, X_j are input variables which influence the response variable Y β_0 is the constant coefficient, β_i is the linear coefficient, β_{ij} is the interaction coefficient of two factors, and β_{ii} is the quadratic coefficient.

The effect of time variables (1–10, 15, 20 and 40 min) on TPC was studied for an attempt to describe different extraction phases and the optimal extraction time was selected in UAEE system under the condition optimized by response surface methodology.

2.6 *Comparison of different extraction methods*

The effect of different extraction methods on TPC was studied. Prepared broccoli samples of 100 g were added in the ultrasonic extraction container and the liquid to material ratio was set as 7.0:1 mL/g. Extractions were performed with the following conditions: (A) UAE: sample in water subjected to UAE with a power of 440 W for 10 min at 54.5°C; (B) EAE: sample in water, without ultrasound, subjected to a treatment of the optimized combination of enzymes incubated at 54.5°C for 2 h; (C) UAEE: sample in water subjected to a optimized UAEE process (Section 2.5). Enzyme activity was terminated with heat treatment at 90°C for 30 s immediately following extraction treatments. The extractive solutions cooled to room temperature then were centrifuged 15 min at 302 g. The supernatants were collected for TPC analysis.

2.7 *Statistical analysis*

All the experimental procedures were repeated thrice, and results obtained were expressed as mean values ± SD. Collected data were analyzed using analysis of variance (ANOVA) and Duncan's multiple comparisons of means in SPSS (version 17.0 for Windows 2000, SPSS Inc.). Statistical significance was established when $p < 0.05$. The quality of the response surface model was expressed by the R^2, lack of fit, Adj-R^2, Pre-R^2, coefficient of variation (C.V.) and its statistical significance was checked by F-test and P-value test (Ofori-Boateng and Lee 2013). The statistical analysis of the model was also performed in the form of analysis of variance (ANOVA) using Design-Expert 8.0.5.0 (Stat-Ease Inc., Minneapolis, USA). The statistical significance of the model and its variables was determined using a probability level of 5% ($p < 0.05$).

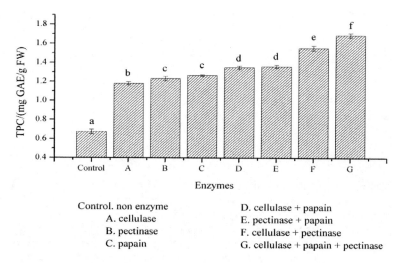

Figure 1. TPC from broccoli inflorescences subjected to different enzyme macerations. Values marked by the same letter are not significantly different (p < 0.05), similarly hereinafter.

3 RESULT AND DISCUSSION

3.1 *Enzyme combination and optimal ratio*

The yields of total phenols varied in response to the different enzyme combination in the experimental design. The results in Fig. 1 indicated that degrading enzymes could make the release of phenolics from the broccoli samples easier than the control, where ultrasonic treatment was used only, and a synergy formed by a combination of enzymes existed. Generally, phenolic compounds could be classified as non-cell-wall phenolics associated with vacuoles, the cell nucleus and cell-wall phenolics, which are bound to polysaccharides (Pinelo et al. 2006). Cross-linking of cell wall polysaccharides to phenolics is the main obstacle for the release of these cell-wall phenolics. Therefore, application of cell wall degrading enzyme preparations prior to extraction can obviously improve the extraction efficiency. Cellulose and Pectin were the two main components for cell wall polysaccharism of broccoli (Müller et al. 2003). Therefore, it was expected that cellulase and pectinase can degrade structural cellulose and pectin for releasing cell-wall phonetics. Protease (papain) was also used in this study and it played a significant role for releasing phenolics. The results suggested that protease might attacked specifically the protein network in the cell wall of plant materials including black currant juice press residues (Landbo and Meyer 2001). Thus, a combination of enzymes, cellulase, pectinase, and papain were chosen for further research.

The primary and secondary of the factors and the optimized combination of them are obtained by range analysis (Eq. 2, 3, 4) of orthogonal test. As shown in Table 1, which could reflect the effect of enzyme combination on TPC, the optimal ratio of the enzyme combination was cellulase of 7.5 mg/g FW, pectinase of 10 mg/g FW and papain of 1.0 mg/g FW, among which the effect of three factors were A > C > B, i.e., cellulase > papain > pectinase.

$$K_i^A = \sum \text{the amount of target compounds at A}_i \qquad (2)$$

$$k_i^A = K_i^A / 3 \qquad (3)$$

$$R_i^A = \max\{k_i^A\} - \min\{k_i^A\} \qquad (4)$$

Table 1. 3^3 Orthogonal test design and results.

Times	Factors			TPC/ (mg GAE/g FW)
	A Cellulase to material ration/ (mg/g)	B Pectinase to material ratio/ (mg/g)	C Papain to material ratio/ (mg/g)	
1	1 (7.5)	1 (7.5)	1 (0.75)	1.638
2	1 (7.5)	2 (10)	2 (1.0)	1.733
3	1 (7.5)	3 (12.5)	3 (1.25)	1.679
4	2 (10)	1 (7.5)	3 (1.25)	1.707
5	2 (10)	2 (10)	1 (0.75)	1.549
6	2 (10)	3 (12.5)	2 (1.0)	1.456
7	3 (12.5)	1 (7.5)	2 (1.0)	1.427
8	3 (12.5)	2 (10)	3 (1.25)	1.621
9	3 (12.5)	3 (12.5)	1 (0.75)	1.582
K_1	5.049	4.773	4.716	
K_2	4.713	4.902	5.022	
K_3	4.629	4.716	4.656	
k_1	1.683	1.591	1.572	
k_2	1.571	1.634	1.674	
k_3	1.543	1.572	1.552	
R	0.140	0.062	0.122	

3.2 Optimization of UAEE process

The enzyme combination with the optimized ratio, cellulase of 7.5 mg/g FW, pectinase of 10 mg/g FW and papain of 1.0 mg/g FW, was used for further optimization of UAEE process. The coded and uncoded levels of experimental factors and the results of extraction according to the factorial design were presented in Table 2.

Response surface regression models were fitted to the experimental data. The ANOVA data for response surface quadratic model has been summarized in Table 3. The model obtained for total phenolics in terms of GAE was highly significant with p value < 0.0001. The coefficient of determination (R^2) of the model was 0.9921, which indicated that more than 99.21% of the variability of responses was explained, asserting a good accuracy and ability of the established model within the limits of the range used (Li et al. 2013). The lack of fit for the model was insignificant relative to the pure error at 95% confidence level, which implies that the fitted model could be used to predict the TPC from UAEE broccoli extracts (Krishnaswamy et al. 2013). Additionally, Adj-R^2, Pre-R^2 and coefficient of variation ($C.V.$) were calculated to check the adequacy of the model. (Adj-R^2–Pre-R^2) < 0.2 indicated a high degree of correlation between observed and predicted data from the regression model (Erbay and Icier 2009). Furthermore, the coefficient of variation ($C.V.$) of less than 5% indicated that the model was reproducible. The final quadratic equation in terms of actual factors for TPC is given as follows:

$$Y = -15.67504 + 0.50296 X_1 + 0.013125 X_2 - 0.63691 X_3 + 0.54814 X_4 + 2.0625 \times 10^{-4} X_1 X_2 \\ + 0.062625 X_1 X_3 - 5.350 \times 10^{-3} X_1 X_4 + 4.650 \times 10^{-4} X_2 X_3 - 4.800 \times 10^{-5} X_2 X_4 + \\ 0.016500 X_3 X_4 - 0.049387 X_1^2 - 1.71425 \times 10^{-5} X_2^2 - 0.073675 X_3^2 - 5.422 \times 10^{-3} X_4^2 \quad (5)$$

where, X_1, X_2, X_3 and X_4 were liquid to material ratio, ultrasonic power, pH and temperature, respectively.

To visualize the combined effects of two operational parameters on the response, the response was generated as a function of two independent variables, while the other two variables were kept constant at the central level. The three-dimensional plots of multiple non-linear regression models

Table 2. 4^3 Box-Behnken design and experimental data for the responses.

Standard order	Run order	X_1	X_2	X_3	X_4	Y
1	11	4:1 (−1)	300 (−1)	5.5 (0)	55 (0)	1.388
2	18	8:1 (+1)	300 (−1)	5.5 (0)	55 (0)	1.474
3	1	4:1 (−1)	500 (+1)	5.5 (0)	55 (0)	1.390
4	2	8:1 (+1)	500 (+1)	5.5 (0)	55 (0)	1.641
5	28	6:1 (0)	400 (0)	4.5 (−1)	50 (−1)	1.738
6	5	6:1 (0)	400 (0)	6.5 (+1)	50 (−1)	1.627
7	24	6:1 (0)	400 (0)	4.5 (−1)	60 (+1)	1.474
8	12	6:1 (0)	400 (0)	6.5 (+1)	60 (+1)	1.693
9	13	4:1 (−1)	400 (0)	5.5 (0)	50 (−1)	1.411
10	7	8:1 (+1)	400 (0)	5.5 (0)	50 (−1)	1.700
11	27	4:1 (−1)	400 (0)	5.5 (0)	60 (+1)	1.435
12	22	8:1 (+1)	400 (0)	5.5 (0)	60 (+1)	1.510
13	26	6:1 (0)	300 (−1)	4.5 (−1)	55 (0)	1.585
14	29	6:1 (0)	500 (+1)	4.5 (−1)	55 (0)	1.601
15	19	6:1 (0)	300 (−1)	6.5 (+1)	55 (0)	1.510
16	25	6:1 (0)	500 (+1)	6.5 (+1)	55 (0)	1.712
17	8	4:1 (−1)	400 (0)	4.5 (−1)	55 (0)	1.574
18	10	8:1 (+1)	400 (0)	4.5 (−1)	55 (0)	1.489
19	17	4:1 (−1)	400 (0)	6.5 (+1)	55 (0)	1.383
20	3	8:1 (+1)	400 (0)	6.5 (+1)	55 (0)	1.799
21	14	6:1 (0)	300 (−1)	5.5 (0)	50 (−1)	1.470
22	15	6:1 (0)	500 (+1)	5.5 (0)	50 (−1)	1.664
23	4	6:1 (0)	300 (−1)	5.5 (0)	60 (+1)	1.435
24	9	6:1 (0)	500 (+1)	5.5 (0)	60 (+1)	1.533
25	20	6:1 (0)	400 (0)	5.5 (0)	55 (0)	1.869
26	23	6:1 (0)	400 (0)	5.5 (0)	55 (0)	1.846
27	6	6:1 (0)	400 (0)	5.5 (0)	55 (0)	1.822
28	21	6:1 (0)	400 (0)	5.5 (0)	55 (0)	1.839
29	16	6:1 (0)	400 (0)	5.5 (0)	55 (0)	1.827

were depicted in Fig. 2a–2f. Fig. 2a gave the interaction of ultrasonic power and liquid to material ratio at a pH of 5.5 and a temperature of 55°C. It was found that there was a rise first followed by a decline in TPC with an increase in ultrasonic power. By keeping power constant and increasing the liquid to material ratio, there still exists a rise first followed by a decline in the TPC. The maximum response value was obtained within a suited range of ultrasonic power and liquid to material ratio. Furthermore, the effect of independent and dependent variables on TPC in Fig. 2b–2f were the same as in Fig. 2a.

By carrying out parameter optimization on the basis of the built mathematical models, the maximum predicted yield of TPC was 1.8766 mg GAE/g FW obtained under the optimum extraction conditions: liquid to material ratio of 6.88:1 mL/g, ultrasonic power of 430.16 W, pH of 6.06 and temperature of 54.47°C. The values of the parameters were modified slightly in the verification experiments as following: liquid to material ratio of 7.0:1 mL/g, ultrasonic power of 440 W, pH of 6.0, and temperature of 54.5°C, because of the difficulty of practical operation under the predicted conditions. A actual value of 1.8656 ± 0.0071 mg GAE/g FW was obtained from the three parallel experiments, which were in close agreement with the predict values (1.8766 mg GAE/g FW). The results confirmed that the response model was adequate to reflect the expected optimization.

In the previous studies, a sufficient time of 30 min was selected for each experiment. However, the details of the mass transfer process during UAEE have not been researched, which may cause needless energy waste by prolonging the extraction time. Then, the extraction time should be optimized in the UAEE system. As shown in Fig. 3, obviously, there were three different phases in the whole UAEE process. The first phase is the rapid phase, this step occurred during the first

Table 3. ANOVA table for response surface quadratic model for yield of phenolics.

Source	Sum of Squares	df	Mean Square	F Value	p value
Model	0.69	14	0.049	125.20	<0.0001***
X_1	0.089	1	0.089	225.30	<0.0001***
X_2	0.038	1	0.038	97.53	<0.0001***
X_3	5.764E-003	1	5.764E-003	14.63	0.0019*
X_4	0.023	1	0.023	59.42	<0.0001***
$X_1 X_2$	6.806E-003	1	6.806E-003	17.28	0.0010*
$X_1 X_3$	0.063	1	0.063	159.29	<0.0001***
$X_1 X_4$	0.011	1	0.011	29.06	<0.0001***
$X_2 X_3$	8.649E-003	1	8.649E-003	21.96	0.0004**
$X_2 X_4$	2.304E-003	1	2.304E-003	5.85	0.0298*
$X_3 X_4$	0.027	1	0.027	69.11	<0.0001***
X_1^2	0.25	1	0.25	642.61	<0.0001***
X_2^2	0.19	1	0.19	483.89	<0.0001***
X_3^2	0.035	1	0.035	89.38	<0.0001***
X_4^2	0.12	1	0.12	302.55	<0.0001***
Residual	5.515E-003	14	3.939E-004		
Lack of Fit	4.146E-003	10	4.146E-004	1.21	0.4620
Pure Error	1.369E-003	4	3.423E-004		
Cor Total	0.70	28			
R^2	0.9921				
Adj R^2	0.9842				
Pred R^2	0.9626				
C.V. %	1.24				

***$p < 0.0001$, **$0.0001 \leq p < 0.001$, *$0.001 \leq p < 0.005$

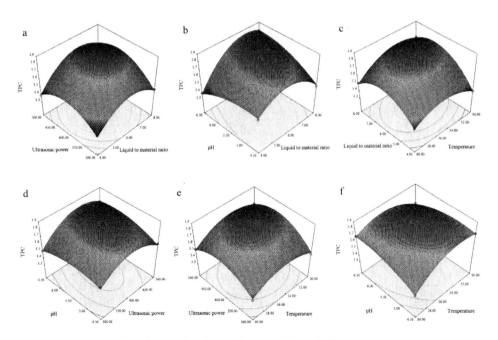

Figure 2. Response surface showing the effects of four variables on TPC.

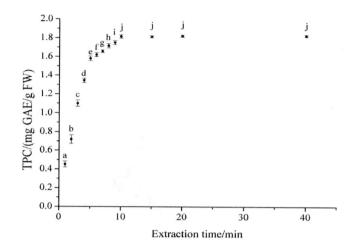

Figure 3. The effect of extraction time on the TPC.

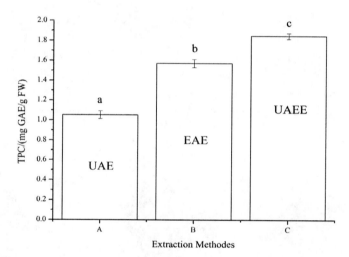

Figure 4. The effect of different extraction methodes on the TPC.

5 min of the process. During this stage, phenolics were rapidly transferred into the solvent with almost constant speed, most of the phenolics (86.6%) diffused into the solvent. This phase was similar to the rapid phase (Yang et al. 2010) described in previous studies. The second phase is the slower phase, which occurred during the 5 to 10 min period in the process. Relatively slower diffusion occurred in this phase which is likely ascribable to the exhaustion of phenolics contained from the matrix. The final phase is the steady phase. This phase occurred after 10 min of the UAEE procedure. In this phase, the extraction yield reached maximum and maintained steady. In order to obtain the maximum extraction yield of phenolics, 10 min was selected as the extraction time in UAEE procedure and maximum extraction yield of polyphenols in broccoli was 1.816 ± 0.0187 mg GAE/g FW under the optimized conditions of 7.0:1 mL/g, 440 W, pH 6.0, 54.5°C, 10 min.

3.3 *Comparison of different extraction methods*

Extraction is an important step in studies involving the discovery and isolation of active compounds of plant materials. As shown in Fig. 4, the TPC had significant difference among three extraction methods and the highest value of method (C) indicated that UAEE have a positive impact on phenolics extraction.

4 CONCLUSION

This is the first report on the feasibility of UAEE for extraction polyphenols from broccoli only with distilled water instead of any organic solvent. The enzyme combination was optimized by orthogonal test: cellulase 7.5 mg/g FW, pectinase 10 mg/g FW and papain 1.0 mg/g FW. The operating parameters in UAEE which were optimized with response surface methodology using Box-Behnken design were as follows: ultrasonic power, 440 W; liquid to material ratio, 7.0:1 mL/g; pH value of 6.0 at 54.5°C for 10 min. At the optimized parameters of UAEE in the present study, the extraction yield of polyphenols was 1.816 ± 0.0187 g GAE/kg FW and compared with UAE and EAE, UAEE had expressed the better characteristic of the highest TPC value.

REFERENCES

Erbay, Z., & Icier, F. (2009). Optimization of hot air drying of olive leaves using response surface methodology. Journal of Food Engineering, 91(4), 533–541.

Gómez-García, R., Martínez-Ávila, G. C. G., & Aguilar, C. N. (2012). Enzyme-assisted extraction of antioxidative phenolics from grape (Vitis vinifera L.) residues. 3 Biotech, 2(4), 297–300.

Jahangir, M., Kim, H. K., Choi, Y. H., & Verpoorte, R. (2009). Health-Affecting Compounds in Brassicaceae. Comprehensive Reviews in Food Science and Food Safety, 8(2), 31–43.

Jokić, S., Cvjetko, M., Božić, Đ., Fabek, S., Toth, N., Vorkapić-Furač, J., & Redovniković, I. R. (2012). Optimisation of microwave-assisted extraction of phenolic compounds from broccoli and its antioxidant activity. International Journal of Food Science and Technology, 47(12), 2613–2619.

Koh, E., Wimalasiri, K. M. S., Chassy, A. W., & Mitchell, A. E. (2009). Content of ascorbic acid, quercetin, kaempferol and total phenolics in commercial broccoli. Journal of Food Composition and Analysis. 22(7), 637–643.

Krishnaswamy, K., Orsat, V., & Gariépy Y. (2013). Optimization of Microwave-Assisted Extraction of Phenolic Antioxidants from Grape Seeds (Vitis vinifera). Food and Bioprocess Technology, 6(2), 441–455.

Landbo, A. K., & Meyer, A. S. (2001). Enzyme-assisted extraction of antioxidative phenols from black currant juice press residues (Ribes nigrum). Journal of Agricultural and food chemistry, 49(7): 3169–3177.

Li, Y., Fabiano-Tixier, A. S., Tomao, V., Cravotto, G., & Chemat, F. (2013). Green ultrasound-assisted extraction of carotenoids based on the bio-refinery concept using sunflower oil as an alternative solvent. Ultrasonics Sonochemistry, 20(1), 12–18.

Matusheski, N. V., Swarup, R., Juvik, J. A., Mithen, R., Bennett, M., & Jeffery, E. H. (2006). Epithiospecifier protein from broccoli (Brassica oleracea L. ssp italica) inhibits formation of the anticancer agent sulforaphane. Journal of Agricultural and Food Chemistry, 54(6), 2069–2076.

Miean, K. H., & Mohamed, S. (2001). Flavonoid (Myricetin, Quercetin, Kaempferol, Luteolin, and Apigenin) content of edible tropical plants. Journal of Agricultural and Food Chemistry, 49(6), 3106–3112.

Müller, S., Jardine, W. G., Evans, B. W., Viëtor, R. J., Snape, C. E., & Jarvis, M. C. (2003). Cell wall composition of vascular and parenchyma tissues in broccoli stems. Journal of the Science of Food and Agriculture, 83(13), 1289–1292.

Muñiz-Márquez, D. B., Martínez-Ávila, G. C., Wong-Paz, J. E., Belmares-Cerda, R., Rodríguez-Herrera, R., & Aguilar, C. N. (2013). Ultrasound-assisted extraction of phenolic compounds from Laurus nobilis L. and their antioxidant activity. Ultrasonics Sonochemistry, 20(5), 1149–1154.

Nguyen, T. T. T., & Le, V. V. M. (2013). Effects of ultrasound on cellulolytic activity of cellulase complex. International Food Research Journal, 20(2), 557–563.

Ofori-Boateng, C., & Lee, K. T. (2013). Response surface optimization of ultrasonic-assisted extraction of carotenoids from oil palm (Elaeis guineensis Jacq.) fronds. Food Science and Nutrition, 1(3), 209–221.

Duh, P.-D. (1998). Antioxidant activity of burdock (Arctium lappa Linne): Its scavenging effect on free-radical and active oxygen. Journal of the American Oil Chemists' Society, 75(4), 455–461.

Pinelo, M., Arnous, A., & Meyer, A. S. (2006). Upgrading of grape skins: Significance of plant cell-wall structural components and extraction techniques for phenol release. Trends in Food Science and Technology, 17(11), 579–590.

Pinelo, M., Zornoza, B., & Meyer, A. S. (2008). Selective release of phenols from apple skin: Mass transfer kinetics during solvent and enzyme-assisted extraction. Separation and Purification Technology, 63(3), 620–627.

Santos, S. A. O., Villaverde, J. J., Silva, C. M., Neto, C. P., & Silvestre, A. J. D. (2012). Supercritical fluid extraction of phenolic compounds from Eucalyptus globulus Labill bark. The Journal of Supercritical Fluids, 71, 71–79.

Singleton, V. L., & Rossi, J. A. (1965). Colorimetry of total phenolics with phosphomolybdic-phosphotungstic acid reagents. American Journal of Enology and Viticulture. 16(3), 144–158.

Wang, X. S., Wu, Y. F., Chen, G. Y., Yue, W., Liang, Q. L., & Wu, Q. N. (2013). Optimisation of ultrasound assisted extraction of phenolic compounds from Sparganii rhizoma with response surface methodology. Ultrasonics Sonochemistry, 20(3), 846–854.

Yang, J., Gadi, R., Paulino, R., & Thomson, T. (2010). Total phenolics, ascorbic acid, and antioxidant capacity of noni (Morinda citrifolia L.) juice and powder as affected by illumination during storage. Food Chemistry, 122(3), 627–632.

Yang, Y. C., Li, J., Zu, Y. G., Fu, Y. J., Luo, M., Wu, N., & Liu, X. L. (2010). Optimisation of microwave-assisted enzymatic extraction of corilagin and geraniin from Geranium sibiricum Linne and evaluation of antioxidant activity. Food Chemistry, 122(1), 373–380.

Author index

Ai, X.S. 183
Akwei, E. 165
An, W.H. 511
Anuar, S.A. 77

Bian, Z.Y. 487

Cao, J. 59, 437
Chen, C. 465, 567
Chen, H. 25, 49, 283
Chen, J. 153, 591
Chen, J.C. 641
Chen, K. 271
Chen, K.J. 323
Chen, L. 311
Chen, R. 95
Chen, W. 611
Chen, X. 25, 271, 283
Chen, Y. 445
Chen, Z. 89
Cheng, S. 143
Cheng, Su 143
Chi, W. 233
Cho, H.J. 43
Chu, F. 479
Cui, J. 459
Cui, K. 195
Cui, L. 437
Cui, S. 459, 473
Cui, W. 207
Cui, Z. 207

Dai, J. 425, 431, 573
Deng, L. 335, 591
Deng, S. 165
Deng, W. 579
Deng, Y. 19
Deng, Y.-S. 329
Ding, J. 95
Ding, S. 379
Ding, X. 201
Dong, L. 177
Dong, Q.J. 183
Dong, X. 353
Du, H. 585
Du, J. 673

Du, P. 213
Du, Q. 271
Duan, C. 493
Duan, M. 591
Duan, S. 517

Fan, S. 153
Fan, Y. 277
Fan, Y.B. 323
Fang, X. 65
Feng, L. 289
Feng, L.-N. 305
Feng, T. 635
Fu, X. 625
Fu, Y. 265

Gao, C.Q. 341, 347
Gao, G. 265, 271, 277, 323
Gao, J. 225
Gao, X. 207
Gao, Y. 239, 473
Gao, Z. 121
Gao, Z.-Y. 243
Gao, Z.Y. 183
Gong, X. 473
Gu, J.W. 679
Guo, H. 559
Guo, P. 659, 673
Guo, W.-K. 243
Guo, X. 465
Guo, Y. 195
Guo, Y.Q. 599, 707
Guo, Z. 425

Ham, S.W. 1
Hamid, N.H. 77, 115
Han, J. 383
Han, S.K. 37
He, H. 171
He, J. 171
He, L. 493, 591
He, Q. 305
He, X.-X. 289
He, Z. 493
Hou, D.Q. 133

Hu, C. 143
Hu, S. 479
Hu, X. 419
Huan, Q. 387
Huang, D. 299
Huang, H.L. 511
Huang, J.Y. 171
Huang, M. 159
Huang, Y. 697
Huang, Z. 453

Jang, J. 393
Jeong, J.W. 1, 7, 13, 31, 37, 43
Jiang, D. 611
Jiang, H.J. 109
Jiang, L. 445, 605
Jiang, S.-Y. 329
Jiang, S.Y. 599
Jiang, Z. 353
Jiao, Z. 517
Jin, S.Y. 359
Jin, Z. 271

Kang, Y. 201
Kao, C.-S. 393
Kay Dora, A.G. 115
Kim, M.H. 1, 7
Kou, C.-H. 393

Lai, H. 647
Lai, P.F. 641
Lee, E.J. 31
Lei, L. 535
Li, A. 559
Li, B. 437, 453
Li, C. 219
Li, D. 213
Li, H. 473
Li, J. 335
Li, J.F. 707
Li, Q. 425, 431, 579
Li, S. 611
Li, T. 503, 605
Li, T.-F. 393
Li, W. 271
Li, X. 159

Li, Y. 207, 387
Li, Y.H. 109
Li, Y.M. 679
Li, Z. 493
Li, Z.B. 691
Liang, B. 213
Liang, H. 251
Liang, J. 83
Liang, W. 159
Liang, X. 127, 147
Liang, Y.A. 665
Liang, Z.W. 529
Liao, C.G. 133
Lin, P.-Y. 393
Lin, W. 635
Lin, X.-D. 305
Liu, H. 59
Liu, J. 95
Liu, L. 341
Liu, Q. 445
Liu, S. 277
Liu, S.W. 691
Liu, S.X. 265
Liu, W.G. 373
Liu, X. 585
Liu, X.B. 347
Long, W. 625
Lou, N. 317
Lu, B.H. 165
Lu, G.H. 171, 177, 189
Lu, K. 697
Lu, X. 219, 289
Luo, F. 311
Luo, M. 445
Luo, X. 605
Luo, Y. 419
Lv, S. 153
Lv, Y. 251

Ma, F. 353
Ma, M. 473
Ma, N. 233
Ma, X. 523
Mao, Y. 189
Mei, R. 373
Meng, X. 685
Ming, Y. 545

Ou, Z. 659, 673
Ouyang, Y. 431

Pan, T. 559
Pan, X. 213

Pan, Y. 19
Pan, Y.G. 619
Pang, Q. 121
Park, J.Y. 13
Peng, J. 479
Peng, M. 493
Pu, Q. 59

Qin, L. 65
Qin, M.H. 665
Qiu, Z. 545, 573
Qu, M. 159
Quan, J.G. 365

Ren, F. 83
Ren, G. 207
Ren, S. 419

Sandoval-Solis, S. 183
Shen, A.L. 511
Shen, H.S. 641
Shen, M. 49
Shi, J. 251, 271
Shou, S. 647
Shu, X. 49
Song, J.L. 653
Song, L. 503
Song, M. 517
Song, P. 523
Song, S. 591
Su, G. 647
Su, W. 647
Su, Y.M. 553
Sun, B. 25, 283
Sun, C. 311
Sun, N.Y. 679
Sun, X. 605, 685
Sun, Y. 459

Tang, H. 335
Tang, J. 465
Tian, J. 65
Tong, J. 65
Tu, H. 659
Tu, J.J. 183

Wang, C. 425, 431, 459, 473, 493
Wang, C.R. 713
Wang, G.B. 691
Wang, G.P. 665
Wang, H. 487
Wang, J. 459, 473, 545

Wang, L. 305
Wang, L.C. 225
Wang, R. 713
Wang, S.-Y. 243
Wang, X. 399
Wang, Y. 611
Wang, Y.L. 511, 529
Wang, Y.N. 691
Wang, Z. 233, 437, 573, 659, 673
Wang, Z.Z. 133
Weng, M.J. 641
Wu, C. 83
Wu, H. 713
Wu, J. 697
Wu, J.Y. 383
Wu, R.W. 225
Wu, Y. 257
Wu, Y.S. 553
Wu, Z. 611
Wu, Z.Y. 171, 177, 189

Xi, H. 511
Xi, X. 399, 409
Xiao, H. 511
Xiao, K. 599, 707
Xiao, X. 523
Xie, L. 127, 147
Xie, W. 65
Xing, C. 437
Xing, J. 379
Xiong, M. 305
Xu, F. 359
Xu, F.L. 665
Xu, J. 379
Xu, J.-C. 329
Xu, J.X. 239
Xu, T. 233
Xu, Y. 659
Xu, Y.X. 511
Xue, L. 591
Xue, W. 579

Yan, B. 257
Yan, Y.J. 239
Yang, B. 679
Yang, G. 653
Yang, H. 121, 257, 567, 585
Yang, J. 251, 535
Yang, L. 611, 713
Yang, S. 271, 493
Yang, S.Y. 529

Yang, T. 49
Yang, Y. 171, 189, 409, 529
Yang, Z.Q. 109
Yao, X. 535
Ye, A. 659, 673
Ye, Y. 219
Ye, Z.Y. 365, 373
Yi, G. 317
Yi, X. 445
Yin, L.B. 599, 707
Yin, P. 517
Yin, Z. 233
You, X. 635
Yu, J. 207
Yu, L. 585
Yu, M. 511
Yu, S.-J. 625
Yuan, H. 383
Yuan, W.B. 71
Yuan, Y. 103
Yue, T. 479

Zeng, H.Y. 133
Zeng, S. 59
Zhai, J. 341, 347
Zhang, C. 71
Zhang, F. 585
Zhang, G. 89
Zhang, H. 665
Zhang, H.W. 165
Zhang, H.Y. 239
Zhang, J.C. 511
Zhang, J.H. 133, 189
Zhang, L. 25, 317
Zhang, N. 567
Zhang, Q. 257, 579, 611
Zhang, S. 647
Zhang, W.W. 341, 347, 365, 373, 619
Zhang, X. 353, 591, 611
Zhang, Y. 89, 399, 479, 559
Zhang, Y.H. 341
Zhang, Y.Q. 359

Zhang, Z.F. 665
Zhao, H. 453
Zhao, L. 517
Zhao, L.Z. 599, 707
Zhao, Q.C. 691
Zhao, Y. 219, 239, 579
Zhao, Y.H. 103
Zheng, C. 679
Zhi, G. 493
Zhong, X. 49
Zhong, Y. 283, 585
Zhou, H.-B. 289
Zhou, M.-S. 289
Zhou, R. 493
Zhou, X. 89
Zhou, Y. 311
Zhu, J.X. 713
Zhu, L. 625
Zhu, Q. 697
Zhu, Q.J. 591
Zhu, S.-M. 625
Zhu, Y. 127, 147